Fungal Diversity Research Series

Series editor
Kevin D. Hyde, Chiang Rai, Thailand

More information about this series at http://www.springer.com/series/10408

I.B. Prasher

Wood-rotting non-gilled Agaricomycetes of Himalayas

Springer

I.B. Prasher
Department of Botany
Panjab University
Chandigarh, India

ISSN 1608-6791 ISSN 2212-6236 (electronic)
Fungal Diversity Research Series
ISBN 978-94-017-9856-3 ISBN 978-94-017-9858-7 (eBook)
DOI 10.1007/978-94-017-9858-7

Library of Congress Control Number: 2015940571

Springer Dordrecht Heidelberg New York London

Printed on acid-free paper

Springer Science+Business Media B.V. Dordrecht is part of Springer Science+Business Media (www.springer.com)

This monograph is dedicated to my parents Late Prof.(Mrs) P. D. Ulfat and Mr. B. K. Ulfat in gratitude and for their affection and support

and

to my Ph.D. supervisor Late Prof. Gurdeep Singh Rawla, ex Chairperson, Department of Botany, Panjab University, Chandigarh, for introducing me to the fascinating world of fungi.

I.B. Prasher

Abstract

The wood-rotting (non-gilled) Agaricomycetes of Himalayas is a part of outcome of fungal forays conducted by the author; collecting and photographing different groups of fungi; along with the faculty of the Department of Botany, Panjab University, Chandigarh, for the last three decades. Some of the photographs are 30 years old. Extensive collections of wood-rotting Agaricomycetes made for the last 10 years along with the Ph.D. students have resulted in the compilation of this book. This work comprising of 167 genera and 488 species (including 6 varieties) is the first comprehensive compilation of these fungi from Himalayas and Assam hills. The characteristic features of all the families included in this work are given before dealing with the taxonomic treatment. The generic and species concept is based on morphological details and the keys to the genera belonging to different families are provided. All the species described are provided with standard description and line diagrams, whereas illustrations are provided for the species which have been photographed by the author. The species described by other workers have been provided with the original herbarium numbers and name of herbarium. The species not collected by the authors have been studied from the specimens deposited by the previous workers in the herbarium of the Botany Department, Panjab University, Chandigarh (PAN). All the specimens collected by the author are deposited in PAN. Besides taxonomic treatment, a chapter pertains to ecology, distribution and interrelationships between Eastern and North-Western Himalayan elements of these fungi. An introductory account of the forest vegetation of Himalayas from where these fungi have been collected is also provided. Out of the total 488 fungi described, 6 taxa (2 species and 4 varieties) are proposed as new to science, 40 species are recorded for the first time from Himalayas, 15 species are recorded for the first time from North-Western Himalayas and 6 species are recorded for the first time from Eastern Himalayas. In the present work the genera are differentiated on the basis of morphological and anatomical characteristics. Till the time when majority of the genera or representative of a group of genera (in case of complex) are phylogenetically analyzed, a classification based on molecular characterization is not feasible. Since the basic aim of the work is to provide a workable text – to be used for the floristic analysis and identification of the fungi belonging to this group – differentiation of genera has been made only on the basis of morphological and anatomical details. The genera have been placed in the families and orders in the text on the basis of their placement as given and followed in the mycobank, index fungorum and *Dictionary of the Fungi*. A list of the fungi included in the text is also provided at the end of the Chap. 1.

Chandigarh, India I.B. Prasher

Acknowledgements

The compilation of this work has been made possible by the generous support given by the chairpersons of Department of Botany for carrying out fungal forays for the last three decades to various destinations in Himalayas and to the Council of Scientific and Industrial Research (CSIR) India for grant-in-aid for 3 years for exploration of these fungi from N. W. Himalayas. I am especially thankful to Prof. A. S. Ahluwalia ex. Chairperson, Department of Botany, Panjab University, Chandigarh, for granting the permission to examine the specimens deposited in Herbarium of Department of Botany, Panjab University, Chandigarh, India (PAN). I am especially thankful to Ms. Lalita JRF in the CSIR-funded project for carrying out the collections and helping in the compilation of this manuscript. My Ph.D. students Rajnish Kumar Verma, Gargi Singh, Radha Chauhan, Deepali Ashok, Geetinder Kaur, Manju, Reena and Sushma have always helped me whenever I required. I am especially thankful to Senior Editor Prof. Zuzana Bernhart and Elisabete Machado, Editorial Assistant, for their kind support and co-operation throughout this project. I am also thankful to other staff at the production section for their help in this project. I am also extremely thankful to Prof. Kevin D. Hyde, Series Editor, for the publication of this monograph in the series.

Last but not least I am thankful to my wife Mrs. Amita Prasher and children Dr. (Ms) Suvidhi and Er. Ravi for their love and unconditional support.

I.B. Prasher

Contents

Chapter 1
Introduction

Abstract A brief introduction of the group is provided regarding taxonomy and their ecology along with the resources and present status concerning the deposition of these fungi and their live cultures. A list of fungi included, is also provided.

This monograph is intended to provide a workable manual for the identification of wood-rotting non-gilled Agaricomycetes from the Himalayas. The Himalayas in this text includes the mountains which are in the territorial jurisdiction of India as well as Royal Kingdom of Bhutan and also includes the Assam Hills. The necessity of this text was felt in the view of the fact that no comprehensive illustrated manual on wood-rotting non-gilled Agaricomycetes of Himalayas with updated nomenclature is available. The works of Bakshi (1971), Rattan (1977), Roy and De (1996) and Sharma (1995, 2012) pertains to either specific/general groups of fungi and or needs revision from nomenclature point of view as well as due to the fact that large number of species have been discovered and reported from different regions of the area. The other compilations are either checklist (Sharma 2000) or checklists with few descriptions pertaining to a specific region (Prasher 1998, 1999, 2008, 2012), Dhingra et al. (2011), Prasher and Ashok (2013), Prasher and Lalita (2013), Ashok and Prasher (2014), Lalita and Prasher (2014).

In light of the above mentioned fact this monograph has been compiled. This work is based on extensive exploration made by the author and his students for more than a decade in the Himalayas as well as critical study of the specimens deposited in the Herbarium of the Botany Department Panjab University Chandigarh (PAN) by the earlier workers and other herbaria where ever required. It is worthwhile to mention that major portion of Himalayan collections are deposited in PAN. The authors endeavour in the present compilation has been to present the detailed information regarding each genus/species covering their taxonomy, morphology and ecology. The present compilation actually represents a small percentage of these fungi; which may be occurring in the Himalayas but can be considered as a foundation for future study; since large expanse of the Himalayas remain unexplored till-to-date and requires a massive and systematic floristic exploration by the mycologists. The taxonomic data concerning these fungi pertains to the collections made and described till end of 2012 only.

I.B. Prasher, *Wood-rotting non-gilled Agaricomycetes of Himalayas*,
Fungal Diversity Research Series, DOI 10.1007/978-94-017-9858-7_1

1.1 About the Group

Wood-rotting non-gilled Agaricomycetes (referred to as "These fungi" now onwards in this volume) belong to class Agaricomycetes, sub-phyllum Agaricomycotina and phylum Basidiomycota. Agaricomycetes include fungi with hymenomycetous or gasteroid fructifications or basidiocarp and are characterized by perforate to imperforate parthenosomes and two to eight spored basidia (Hibbett et al. 2007; Kirk et al. 2008). The fungi with gasteroid type of fructifications have been placed in subclass Phallomycetidae (orders: Geastrales, Gomphales, Hysterangiales and Phallales) whereas those with hymenomycetous fructifications, belonging to orders Agaricales, Atheliales, Boletales have been assigned to subclass Agaricomycetidae. The fungi with hymenomycetous fructifications and belonging to orders Auriculariales, Cantharellales, Corticales, Gloeophyllales, Hymenochaetales, Polyporales, Russulales, Sebacinales, Thelephorales and Trechisporales have been placed in Agaricomycetes incertae sedis till the accumulation of sufficient data concerning molecular studies. Agaricomycetes as a group is equivalent to Homobasidiomycetes sensu Hibbett and Thorn (2001) plus Auriculariales and Sebacinales (Hibbett et al. 2007). These fungi have earlier been treated under Aphyllophorales by Patouillard (1900), Rea (1922), Bourdot and Galzin (1928), Donk (1964) and Talbot (1973) and other workers. Kirk et al. (2001) treated these fungi under subclass Agaricomycetidae. The classification of these fungi till the end of twentieth century by and large had been based on morphological and anatomical/histochemical parameters which have been followed in the major modern works on these fungi (Patouillard 1900; Rea 1922; Bourdot and Galzin 1928; Cunningham 1965; Donk 1960, 1962, 1964, 1974; Talbot 1973; Ryvarden and Johansen 1980; Gilbertson and Ryvarden 1987; Ryvarden 1990; Ryvarden and Gilbertson 1993, 1994; Núñez and Ryvarden 2001; Ryvarden et al. 2014). However, with the advent of the molecular fungal taxonomy started in 1990s with the analysis of PCR-amplified ribosomal RNA genes (White et al. 1990), a broad classification based on molecular fungal taxonomy has evolved with reassignment of the taxa (Hibbett and Thorn 2001). This has resulted in the intergeneric shifting of species, as well as reassignment of genera to different families, orders and classes. The fungi included in this work form fructifications (basidiocarps) on the wooden substratum, preceded by a vegetative phase in the form of mycelium inside the host as well as on the surface. These fructifications show varied forms and shapes. The simplest type is the radial growth of unlimited manner closely appressed to the host, on its under surface and bearing the hymenophore on the lower side. This type of fructification is termed as resupinate or effused e.g. *Poria*. The resupinate fructification may; after sometime in some cases; grow out of the substrate developing into characteristic forms like bract, shelf like or pileate (reflexed) fructifications or they may be of these types as such. The hymenium may line the walls of the cohering tubes or spread over smooth, teeth or corolloid, reticulate, meruloid and hydnoid surface. The presence of hymenium on the lower side towards the ground confers advantage with respect to spore dispersal as well as protection from environmental conditions.

Majority of these fungi grow on wood as saprophytes but some of these like *Aurificaria shoreae* and *Phellinus gilvus* (Schwein.) Pat., are also parasitic. These are capable of degrading both gymnospermous and angiospermous wood, the former termed as soft wood whereas the latter is termed as hard wood. The wood in the centre of the tree comprising of lignified xylem elements is called 'heart-wood' where as 'softwood' refers to the outer light-coloured region of the periphery of the wood. The heart-rot fungi continue to grow on the infected trees and do not kill the host whereas the soft-rot fungi growing on the sap-wood often kills the host caused by degradation of the sap-wood.

The lignin-degrading fungi cause white-rot which involves degradation of lignin and to a lesser extent the cellulose and hemicelluloses. This degradation reaction results in the formation of a bleached residue with greater percentage of cellulose. The brown-rot fungi utilize cellulose of the wood leaving a brown residue comprising of demethylated lignin (Jeffries 1987; Blanchette 1995) in the form of a powdery mass at later stages of degradation.

The degradation of lignocellulose of the tree as well as decayed wood by these fungi constitute one of the important events in the forest ecosystem, where by the carbon fixed by the green plants from the atmosphere is recycled back into the atmosphere. Most of the white as well as brown-rot fungi do not require lignin or cellulose for their growth in nature. Instead; for their survival and growth; they are dependent on simple plant polysaccharides which also provide them the energy (through oxidation) required for lignocellulolytic degradation. This physiological parameter of these fungi has been used beneficially in the biotechnological application like biopulping, employing fungi like *Phanerochaete chrysosporium & Ceriporiopsis subvermispora* (Akhtar et al. 1996; Wolfaardt et al. 1996), bio-bleaching, decolorization of industrial effluents, biodegradation of low molecular mass xenobiotic compounds, in the improvement of forage digestibility and in the production of many valuable chemicals like volatile organoleptic chemicals or their intermediate precursors and polynuclear aromatic hydrocarbon compounds- PAH's (Ralph and Catcheside 2002). The white-rot fungi have been found to be more effective than brown-rot fungi in all these applications (Bagley and Richter 2002). The major decay caused by brown-rot fungi is of coniferous trees in forest ecosystem, besides of wood in the buildings and other man-made infrastructure. The frequently occurring genera causing brown-rot of infrastructural wood are *Antrodia, Coniophora, Gloeophyllum, Neolentinus* and *Oligoporus*. Due to the degradation caused by brown-rot fungi, they have been used as test organisms world-wide by industries producing chemicals to check wood decay. Cultures of brown-rot fungi have been used world-wide for checking the efficacy of wood preservative chemicals before they are launched commercially in the market (Anonymous 1998). A lot of work has been done on bioconversion of biomass by these fungi (Kubicek et al. 2012). Extensive studies have been conducted on white-rot fungi than brown-rot fungi for their application in the industry and also the brown rot fungi are the little investigated ones as compared to white-rot fungi. Besides, due to their ability to degrade wood and their use in related biotechnological applications as discussed briefly above, they have been vigorously investigated for the production of bioactive metabolites for subsequent use in medicines.

1.2 Resources and Present Status

The major collections from the Himalayas have been deposited in the following herbaria in India by different workers:

(a) Herbarium Cryptogamae Indiae Orientalis (HCIO) IARI, New Delhi.
(b) Herbarium of Botany Department, Panjab University, Chandigarh, India (PAN).
(c) Herbarium of Botany Department, Punjabi University, Patiala (PUN).
(d) Mycological Herbarium of the Forest Research Institute, (FRI) Dehra Dun (D.D)
(e) Central National Herbarium, Botanical Survey of India, Calcutta (CAL).
(f) Herbarium of Botany Department, Calcutta University, Calcutta (CUH).
(g) Herbarium, Botanical Survey of India, Dehra Dun (BSD).
(h) Herbarium of Botany Department, Banaras Hindu University, Varanasi (BHUP).
(i) Herbarium National Botanical Research Institute, Lucknow (NBRI).
(j) Botany Group Plant Science Division, Agarkar Research Institute, Pune.
(k) Bose's Herbarium of Polyporaceae, Botany Department, R. G. Kar Medical College, Calcutta (CMS) now shifted to Presidency College Calcutta.

The collections in all these herbaria are well maintained. More than 80 % of the fungi (including type specimens) recorded from Himalayas and Assam hills are deposited in PAN. To the best of the knowledge of the author, major culture collections of these fungi is supported in the Indian Type Culture Collection (ITCC) at Indian Agricultural Research Institute (IARI), New Delhi; Agarkar Institute, Pune and in Microbial Type Culture Collection (MTCC) at Institute of Microbial Technology, Chandigarh (IMTech). The listing of culture at IMTech depicts 86 collections of wood-rotting Agaricomycetes. Majority of these resources are not obtained from the Himalayas but have been procured from the American Type Culture Collection (ATCC), Central bureau voor Schimmel cultures, Baarn (CBS), Deutsche Sammlung von Mikoorganismen and Zellkulturen Gmlott (DSMZ) & Imperial Mycological Institute Herbarium Culture Collection. The Indian collections deposited in MTCC are of A. Roy, D. S. Arora and A. Mitra, besides some cultures procured from Forest Research Institute (F.R.I) and ITCC. Around 150 collections of these fungi are being maintained in the Mycology and Plant Pathology Laboratory of the Botany Department, Panjab University, Chandigarh. However, this is a fraction of the vast resources available in Himalayas. This points out the state of affairs as far as maintenance/deposition of live cultures of wood-rotting Agaricomycetes in India is concerned. Not much attention has been paid towards culturing of these fungi from Himalayas and elsewhere in the country.

A total of 488 species of these fungi spreading over 167 genera are being recorded from the study area, where as 6 species and varieties are proposed as new to science. Forty species are recorded and described for the first time from Himalayas where as 15 and 6 species are recorded and described for the first time from North-Western and Eastern Himalayas respectively.

1.3 List of New Species & Varieties and New Records

1.3.1 New Species and Varieties

Leptosporomyces thindii Prasher & Lalita sp. nov.
Peniophorella microtsugae Prasher & Lalita sp. nov.
Leptosporomyces roseus Jülich, Willdenovia, Beih. 7: 208, 1972, var. *macrosporus* Prasher & Lalita var. nov.
Radulomyces confluens (Fr.) M.P. Christ., Dansk bot. Ark. 19(2): 230 (1960) (Fr.) M.P. Christ., var. *macrobasidiata* Prasher & Lalita var. nov.
Sidera lenis (P. Karst.) Miettinen, in Miettinen and Larsson, Mycol. Progr. 10(2): 136 (2011) var. *minutispora* Prasher & Lalita var nov.
Datronia stereoides (Fr.) Ryv., Blyttia 25: 168 (1967) var. *microspora* Prasher and Lalita var. nov.

1.3.2 New Records for Himalayas/India

Aleurodiscus lapponicus
Amyloathelia crassiucula
Anomoporia bombycina
Antrodia sinuosa
Boidinia lacticolor
Ceraceomyces cystidiatus
Coniophora arida
Coniophora fusispora
Coriolopsis caperata
Dendrothele seriata
Dichomitus leucoplacus
Dichostereum kenyense
Diplomitoporus crustulinus
Erythricium laetum
Fibulomyces mutabilis
Flaviporus hydrophilus
Fomes extensus
Fuscoporia ferrea
Galzinia incrustans
Ganoderma multiplicatum
Hyphoderma macedonicum
Hyphodontia abieticola
Hyphodontia alienate
Hyphodontia barbajovis
Inonotus albertinii
Inonotus tenuicarinus

Leptosporomyces galzinii
Microporellus obovatus
Microporellus obovatus
Paullicorticium delicatissimum
Phanerochaete velutina
Phlebia interjacenoides
Phlebiopsis flavidoalba
Postia ceriflua
Postia mappa
Postia undosa
Resnicium friabile
Scytinostroma alutum
Vararia minidichophysa
Vararia rugosispora

1.3.3 New Records for North-Western Himalayas

Crustoderma dryinum
Daedalea dochmia
Hyphoderma sikkimium
Hyphodontia caulicystidiata
Hyphodontia propinqua
Hypochnicium caucasicum
Lenzites eximia
Peniophorella rude
Phanerochaete galactites
Phanerochaete sordida
Phlebia interjacenoides
Phlebia rufa
Sarcodontia spumea
Tomentella clavigera
Vararia sphaericospora

1.3.4 New Records for Eastern Himalayas

Fuscoporia torulosa
Junghuhnia luteoalba
Polyporus grammocephalus
Postia guttulata
Postia mappa
Trametes incerta

1.4 Lists of the Orders, Families and Genera Included in This Text

Order- Agaricales

Family- Amylocorticiaceae

Genus- *Amyloathelia*

A. crassiucula

Genus- *Amylocorticium*

A. indicum

Genus- *Amyloxenasma*

A. allantosporum
A. grisellum

Genus- *Ceraceomyces*

C. bizonatus
C. borealis
C. cystidiatus
C. fibuliger
C. reidii
C. sublaevis
C. tessulatus

Family- Cyphellaceae

Genus- *Chondrostereum*

C. purpureum

Family- Physalacriaceae

Genus- *Cylindrobasidium*

C. evolvens

Family- Pterulaceae

Genus- *Aphanobasidium*

A. subnitens

Genus- *Radulomyces*

R. confluens var. *macrobasidiata*
R. molaris

Family- Stephanosporaceae

Genus- *Cristinia*

C. helvetica
C. mucida

Order- Atheliales

Family- Atheliaceae

Genus- *Amphinema*

A. byssoides

Genus- *Athelia*

A. decipiens
A. teutoburgensis

Genus- *Athelopsis*

A. parvispora
A. subinconspicua

Genus- *Fibulomyces*

F. cystoideus
F. mutabilis

Genus- *Hypochniciellum*

H. ovoideum

Genus- *Leptosporomyces*

L. adnatus
L. galzinii
L. globosus
L. roseus var. *macrosporus*
L. thindii

Order- Boletales

Family- Coniophoraceae

Genus- *Coniophora*

C. arida
C. cordensis
C. fusispora
C. olivacea
C. puteana

Family- Hygrophoropsidaceae

Genus- *Leucogyrophana*

L. mollusca
L. olivascens
L. thimphina

Family- Serpulaceae

Genus- *Serpula*

S. himantioides
S. lachrymans

Order- Cantharellales

Family- Botryobasidiaceae

Genus- *Botryobasidium*

B. asperulum
B. candicans
B. danicum
B. subbotryosum
B. subcoronatum
B. vagum

Genus- *Botryohypochnus*

B. isabellinus

Family- Ceratobasidiaceae

Genus- *Scotomyces*

S. subviolaceus

Family- Hydnaceae

Genus- *Paullicorticium*

P. delicatissimum
P. indicum

Genus- *Sistotrema*

S. angustispora
S. binucleosporum
S. brinkmannii
S. lachrymisporum
S. porulosum
S. sernanderi
S. subtrigonospermum

Order- Corticales

Family- Corticiaceae

Genus- *Corticium*

C. confine

Genus- *Dendrothele*

D. alliacea
D. incrustans
D. seriata

Genus- *Erythricium*

E. laetum

Genus- *Galzinia*

G. ellipsospora
G. incrustans

Genus- *Laeticorticium*

L. expallens

Genus- *Licrostroma*

L. subgiganteum

Order- Gloeophyllales

Family-Gloeophyllaceae

Genus- *Gloeophyllum*

G. abietinum
G. carbonarium
G. sepiarium
G. striatum
G. subferrugineum

Family- Hericiaceae

Genus- *Dentipellis*

D. leptodon

Genus- *Laxitextum*

L. bicolor

Order- Gomphales

Family- Lentariaceae

Genus- *Kavinia*

K. alboviridis

Family- Hymenochaetaceae

Genus- *Aurificaria*

A. indica

Genus- *Fomitiporia*

F. robusta

Genus- *Fuscoporia*

F. contigua
F. ferrea
F. ferruginosa
F.senex
F. torulosa

Genus- *Hymenochaete*

H. fuscobadia
H. leonina
H. luteobadia
H. mougeotii
H. rubiginosa
H. semistuposa

Genus- *Inonotus*

I. albertinii
I. cuticularis
I. dryadeus
I. tabacinus
I. tenuicarinis

Genus- *Onnia*

O. circinata
O. tomentosa

Genus- *Phellinus*

P. adamantinus
P. allardii
P. caryophylli
P. fastuosus
P. gilvus
P. grenadensis
P. igniarius
P. johnsonianus
P. linteus
P. merrillii

P. nilgheriensis
P. sanfordii
P. setulosus
P. xeranticus

Genus- *Phylloporia*

P. pectinata
P. ribis
P. spathulata

Genus- *Porodaedalea*

P. pini

Genus- *Tubulicrinis*

T. chaetophorus
T. glebulosus
T. hamatus
T. strangulates
T. subulatus

Family-Rickenellaceae

Genus- *Peniophorella*

P. microtsugae
P. pallida
P. praetermissa
P. pubera
P. rude

Genus- *Resinicium*

R. bicolor
R. friabile

Family- Repetobasidiaceae

Genus- *Sidera*

S. lenis var. *minutispora*
S. lunata

Family Schizoporaceae

Genus- *Alutaceodontia*

A. alutacea

Genus- *Basidioradulum*

B. tuberculatum

Genus- *Hyphodontia*

H. abieticola
H. alienata
H. altaica
H. alutaria
H. arguta
H. aspera
H. barbajovis
H. caulicystidiata
H. crustosa
H. efibulata
H. hastata
H. juniperi
H. nespori
H. pallidula
H. papilosa
H. propinqua
H. sambuci
H. spathulata
H. stipata

Genus- *Oxyporus*

O. cervinogilvus
O. corticola
O. populinus
O. ravidus

Genus- *Schizopora*

S. flavipora
S. paradoxa

Genus- *Xylodon*

X. pruni
X. rimosissima

O-Polyporales

Family- Cystostereaceae

Genus- *Cystostereum*

C. murrayi

Family- Fomitopsidaceae

Genus- *Antrodia*

A. albida

A. gossypium
A. serialis
A. sinuosa
A. xantha

Genus- *Anomoporia*

A. bombycina

Genus- *Daedalea*

D. dickinsii
D. dochmia
D. flavida
D. gollanii
D. imponens
D. quercina
D. sulcata

Genus- *Dacryobolus*

D. costratus
D. karstenii
D. sudans

Genus- *Fomitopsis*

F. palustris
F. pinicola
F. rosea
F. rubida
F. rufolaccata

Genus- *Laetiporus*

L. sulphureus

Genus- *Parmastomyces*

P. corticola

Genus- *Phaeolus*

P. schweinitzii

Genus- *Postia*

P. caesia
P. ceriflua
P. guttulata
P. leucomallella
P. mappa
P. sericeomollis

P. undosa

Family- Ganodermataceae

Genus- *Ganoderma*

G. applanatum
G. lucidum
G. multiplicatum
G. resinaceum
G. sessiliforme

Family- Hyphodermataceae

Genus- *Intextomyces*

I. contiguus

Family- Meripileaceae

Genus- *Physisporinus*

P. rivulosus

Genus- *Rigidoporus*

R. crocatus
R. lineatus
R. microporus
R. ulmarius
R. vinctus

Family- Meruliaceae

Genus- *Abortiporus*

A. biennis

Genus- *Bjerkandera*

B. adusta
B. fumosa

Genus- *Cabalodontia*

C. queletii
C. subcretacea

Genus- *Conohypha*

C. grandispora

Genus- *Crustoderma*

C. dryinum

Genus- *Flaviporus*

F. hydrophilus

Genus- *Flavodon*

F. flavus

Genus- *Gloeoporus*

G. dichorus
G. thelephoroides

Genus- *Gyrophanopsis*

G. polonensis

Genus- *Hyphoderma*

H. argillaceum
H. clarusproprietas
H. macedonicum
H. occidentale
H. parvisporum
H. setigerum
H. sibiricum
H. sikkimium
H. singularibasidium
H. sporulosum

Genus- *Hypochnicium*

H. caucasicum
H. cystidiatum
H. geogenium
H. longicystidiosum
H. lundellii
H. punctulatum
H. sphaerosporum

Genus- *Irpex*

I. consors
I. lacteus
I. vellereus

Genus- *Junghuhnia*

J. collabens
J. luteoalba
J. nitida
J. zonata

Genus- *Mycoacia*

M. fuscoatra
M. stenodon

Genus- *Phlebia*

P. crassisubiculata
P. interjacenoides
P. livida
P. microspora
P. radiata
P. rufa
P. segregata
P. singularisa
P. subserialis
P. subochracea
P. thindii

Genus- *Radulodon*

R. americanus
R. erikssonii

Genus- *Scopuloides*

S. hydnoides

Genus- *Sarcodontia*

S. delectans
S. pachyodon
S. spumea

Genus- *Steccherinum*

S. ciliolatum
S. fimbriatum
S. laeticolor
S. ochraceum

Family-Phanerochaetaceae

Genus- *Antrodiella*

A. semisupina
A. zonata

Genus- *Byssomerulius*

B. corium

Genus- *Candelabrochaete*

C. himalayana

Genus- *Ceriporia*

C. viridans

Genus- *Ceriporiopsis*

C. gilvescens

Genus- *Phanerochaete*

P. affinis
P. deflectens
P. galactites
P. laevis
P. sordida
P. tuberculata
P. velutina

Genus- *Phlebiopsis*

P. darjeelingensis
P. flavidoalba
P. gigantea
P. himalayensis
P. ravenelii

Genus- *Porostereum*

P. crassum
P. spadiceum

Genus- *Rhizochaete*

R. filamentosa

Family- Polyporaceae

Genus- *Cinereomyces*

C. lindbladii

Genus- *Coriolopsis*

C. caperata
C. occidentalis
C. telfairii

Genus- *Daedaleopsis*

D. confragosa

Genus- *Datronia*

D. mollis
D. scutellata
D. stereoides var. *microspora*

Genus- *Dichomitus*

D. leucoplacus

Genus- *Diplomitoporus*

D. crustulinus
D. rimosus

Genus- *Earliella*

E. scabrosa

Genus- *Epithelopsis*

E. fulva

Genus- *Favolus*

F. tenuiculus

Genus- *Fomes*

F. extensus
F. fomentarius
F. johnsonianus

Genus- *Grammothele*

G. fuligo

Genus- *Hapalopilus*

H. croceus

Genus- *Hexagonia*

H. badia
H. tenuis

Genus- *Lenzites*

L. betulina
L. eximia
L. stereoides

Genus- *Lignosus*

L. sacer

Genus- *Lopharia*

L. cinerascens
L. papyrina

Genus- *Loweporus*

L. lividus
L. tephroporus

Genus- *Macrohyporia*

M. inflata

Genus- *Microporellus*

M. obovatus
M. violaceo-cinerascens

Genus- *Microporus*

M. affinis
M. xanthopus

Genus- *Navisporus*

N. floccosus

Genus- *Nigroporus*

N. durus
N. vinosus

Genus- *Nigrofomes*

N. melanoporus

Genus- *Pachykytospora*

P. papyracea

Genus- *Perenniporia*

P. fulviseda
P. martia
P. medulla-panis
P. mesoleuca
P. ochroleuca

Genus- *Polyporus*

P. arcularis
P. badius
P. blanchetianus
P. brumalis

P. grammocephalus
P. melanopus
P. streiformis
P. squamosus
P. varius

Genus- *Poria*

P. auricoma
P. conferta

Genus- *Pycnoporus*

P. cinnabarinus
P. coccineus
P. sanguineus

Genus- *Rhodonia*

R. placenta

Genus-*Skeletocutis*

S. amorpha
S. nivea

Genus- *Trametes*

T. cingulata
T. cotonea
T. cubensis
T. gibbosa
T. hirsuta
T. incerta
T. lactinea
T. menziesii
T. ochracea
T. palisoti
T. pubescens
T. versicolor

Genus- *Trametopsis*

T. cervina

Genus- *Trichaptum*

T. abietinum
T. biforme
T. byssogenum
T. fusco-violaceum

Genus- *Tyromyces*

T. lacteus
T. tephrus

Family-Xenasmataceae

Genus- *Xenasma*

X. subclematidis
X. tulasnelloideum

Genus- *Xenasmatella*

X. subflavido-grisea
X. vaga

Order- Russulales

Family- Amylostereaceae

Genus- *Amylostereum*

A. chailletii

Family- Bondarzewiaceae

Genus- *Bondarzewia*

B. mesenterica

Genus- *Heterobasidion*

H. annosum var. *indicum*
H. insulare

Family- Echinodontiaceae

Genus- *Echinodontium*

E. japonicum

Genus- *Laurilia*

L. sulcata

Family- Lachnocladiaceae

Genus- *Asterostroma*

A. cervicolor
A. muscicola

Genus- *Dichostereum*

D. effuscatum
D. kenyense

D. pallescens
D. rhodosporum

Genus- *Scytinostroma*

S. albocinctum
S. alutum
S. crassum
S. cystidiatum
S. duriusculum
S. ochroleucum
S. portentosum
S. pulverulentum
S. rhizomorpharum

Genus- *Vararia*

V. brevispora
V. minidichophysa
V. rugosispora
V. sphaericospora
V. vassilieve

Family- Peniophoraceae

Genus- *Gloiothele*

G. citrina
G. lactescens

Genus- *Metulodontia*

M. indica
M. nivea

Genus- *Peniophora*

P. cinerea
P. incarnata
P. limitata
P. pini
P. pithya
P. quercina
P. rhodocarpa
P. rufomarginata
P. violaceolivida

Family- Russulaceae

Genus- *Boidinia*

B. furfuraceum

B. laticolor

Family- Stereaceae

Genus- *Acanthofungus*

A. ahmadii

Genus- *Aleurodiscus*

A. amorphus
A. lapponicus
A. oakesii
A. taxicola

Genus- *Amylosporomyces*

A. echinosporus

Genus- *Chaetoderma*

C. luna

Genus- *Conferticium*

C. ochraceum

Genus- *Gloeocystidiellum*

G. clavuligerum
G. donkii
G. luteocystidium var. *brevisporum*
G. sulcatum

Genus- *Scotoderma*

S. viride

Genus- *Scytinostromella*

S. heterogenea
S. olivaceoalba

Genus- *Stereum*

S. acanthophysatum
S. gauspatum
S. hirsutum
S. ostrea
S. peculiare
S. rugosum
S. sanguinolentum
S. subtomentosum

Genus- *Xylobolus*

X. frustulatus
X. subpileatus

Family-Wrightoporiaceae

Genus- *Wrightoporia*

W. lenta

Order-Thelephorales

Family- Thelephoraceae

Genus- *Amaurodon*

A. viridis

Genus- *Pseudotomentella*

P. mucidula
P. tristis

Genus- *Thelephora*

T. atra

Genus- *Tomentella*

T. albomarginate
T. badia
T. bicolor
T. botryoides
T. brevispina
T. bryophila
T. calcicola
T. cinerascens
T. clavigera
T. coerulea
T. crinalis
T. ellisii
T. ferruginea
T. galzinii
T. griseoumbrina
T. griseoviolacea
T. himalayana
T. indica
T. kalatopii
T. lapida
T. lateritia
T. muricata

T. olivascens
T. pilosa
T. puberula
T. punicea
T. pyrolae
T. radiosa
T. scobinella
T. stuposa
T. subclavigera
T. subcorticioides
T. terrestris
T. testaceogilva
T. umbrinospora
T. unicusa

Order- Trechisporales

Family- Hydnodontaceae

Genus- *Brevicellicium*

B. olivascens

Genus- *Fibriciellum*

F. silvae-ryae

Genus- *Fibrodontia*

F. gossypina

Genus- *Sistotremastrum*

S. niveocremeum

Genus- *Subulicystidium*

S. longisporum
S. meridense

Genus- *Trechispora*

T. alnicola
T. candidissima
T. farinacea
T. fastidiosa
T. microspora
T. mutabilis
T. mollusca
T. praefocata

Class- Agaricomycetes
Incertae sedis

Genus- *Dendrophlebia*

D. crassispora

Genus- *Hallenbergia*

H. singularis

Genus- *Radulomycetopsis*

R. cystidiata

References

Akhtar A, Faye G, Bentley DL (1996) Distinct activated and non-activated RNA polymerase II complexes in yeast. EMBO J 15(17):4654–4664

Anonymous (1998) E10-91, standard method of testing wood preservatives by laboratory soil-block cultures. American Wood Preservers Association, Woodstock, p 335

Ashok D, Prasher IB (2014) Wood rotting non-gilled Agaricomycetes new to India. J N Biol Rep 3(1):04–08

Bagley ST, Richter DL (2002) Biodegradation by brown rot fungi. In: Osiewacz HD (ed) The Mycota, a comprehensive treatise on fungi as experimental systems for basic and applied research Vol X- Industrial applications. Springer, Berlin/Heidelberg, pp 327–341

Bakshi BK (1971) Indian Polyporaceae (on trees and timber). ICAR, New Delhi

Blanchette RA (1995) Degradation of the lignocellulose complex in wood. Can J Bot 73(Suppl 1):S999–S1010

Bourdot H, Galzin A (1928) Hymenomycetes de France. Sceaux, Marcel Bry, Dessinateur Imprimeur

Cunningham GH (1965) Polyporaceae of New Zealand. Bull N Z Dep Scient Ind Res, Wellington

Dhingra GS, Priyanka, Kaur J (2011) A Checklist of resupinate, non-poroid agaricomycetous fungi from North-East India and Bhutan. Synopsis Fungorum 29:22–70

Donk MA (1960) The generic names proposed for polyporaceae. Persoonia 1:173–302

Donk MA (1962) Notes on resupinate Hymenomycetes-VI. Persoonia 2:217–238

Donk MA (1964) A conspectus of the families of Aphyllophorales. Persoonia 3:199–324

Donk MA (1974) Checklist of European polypores. London, Amsterdam

Gilbertson RL, Ryvarden L (1987) North American Polypores, vol 2. Fungiflora, Oslo

Hibbett DS, Thorn RG (2001) Basidiomycota: Homobasidiomycetes. In: Mclaughlin DJ, McLaughlin EG, Lemke PA (eds) The Mycota, Vol VII, Part B. Systematics and evolution. Springer, Berlin, pp 121–168

Hibbett DS, Binder M, Bischoff JF, Blackwell M, Cannon PF, Eriksson OE, Huhndorf S, James T, Kirk PM, Lucking R, Lumbsch HT, Lutzoni F, Matheny PB, Mclaughlin DJ, Powelli MJ, Redhead S, Schochk CL, Spatafora JW, Stalpers JA, Vilgalys R, Aime MC, Aptroot A, Bauer R, Begerow D, Benny GL, Castlebury LA, Crous PW, Dai Y-C, Gams W, Geisers DM, Griffith GW, Gueidan C, Hawksworth DL, Hestmark G, Hosaka K, Humber RA, Hyde KD, Ironside JE, Kõljalg U, Kurtzman CP, Larsson KH, Lichtwardt R, Longcore J, Miądlikowska J, Miller A, Moncalvo JM, Mozley-Standridge S, Oberwinkler F, Parmasto E, Reeb V, Rogers JD, Roux C, Ryvarden L, Sampaio JP, Schüβler A, Sugiyama J, Thorn RG, Tibell L, Untereiner WA, Walker C, Wang Z, Weir A, Weiss M, White MM, Winka K, Yao Y-J, Zhang N (2007) A higher-level phylogenetic classification of the fungi. Mycol Res 111:509–547

Jeffries TW (1987) Physical, chemical and biochemical considerations in the biological degradation of wood. In: Kennedy JF, Phillips GO, Williams PA (eds) Wood and cellulosics: industrial utilisation, biotechnology, structure and properties. Ellis Horwood Limited, Chichester, pp 213–230

Kirk PM, Cannon PF, David JC, Stalpers JA (2001) Dictionary of the fungi, 9th edn. Wallingford, Oxon

Kirk PM, Cannon PF, Minter DW, Stalpers JA (2008) Dictionary of the fungi, 10th edn. Wallingford, Oxon

Kubicek K, Cerna H, Holub P, Pasulka J, Hrossova D, Loehr F, Hofr C, Vanacova S, Stefl R (2012) Serine phosphorylation and proline isomerization in RNAP II CTD control recruitment of Nrd1. Genes Dev 26(17):1891–1896

Lalita, Prasher IB (2014) Wood rotting Agaricomycetes from Uttarakhand – new to India. J N Biol Rep 3(1):29–37

Núñez M, Ryvarden L (2001) East Asian polypore. 2. Synopsis Fungorum. Fungiflora, Oslo

Patouillard N (1900) Essai taxonomique sur les families et les genres des Hymenomycetes, Paris

Prasher IB (1998) Fungi of Chandigarh. I B D, Dehra Dun

Prasher IB (1999) Fungi of Bhutan. I B D, Dehra Dun

Prasher IB (2008) Macro-fungal Biodiversity of Nanda-Devi Biosphere reserve and its in vitro conservation. Final Technical Report, MoEF

Prasher IB (2012) Wood-rotting Basidiomycetes of North-Western Himalayas- a monographic study and screening for lignocellulolytic enzymes. Final Technical Report. CSIR, New Delhi

Prasher IB, Ashok D (2013) A Checklist of wood-rotting fungi (non-gilled Agaricomycotina) of Himachal Pradesh. J N Biol Rep 2(2):71–98

Prasher IB, Lalita (2013) A Checklist of wood-rotting fungi (non-gilled Agaricomycotina) of Uttarakhand. J N Biol Rep 2(2):108–123

Ralph JR, Catcheside DEA (2002) Biodegradation by white-rot fungi. In: Osiewacz HD (ed) The Mycota X – industrial applications. Springer, Berlin/Heidelberg, pp 303–326

Rattan (1977) The resupinate Aphyllophorales of the North Western Himalayas. Bibliotheca Mycologica 60:427

Rea C (1922) British Basidiomycetes, Cambridge

Roy A, De AB (1996) Polyporaceae of India. I B D, Dehra Dun

Ryvarden L (1990) Type studies in the Polyporaceae 22. Mycotaxon 38:83–102

Ryvarden L, Gilbertson RL (1993) European Polypores 1, vol 6, Synopsis fungorum. Fungiflora, Oslo

Ryvarden L, Gilbertson RL (1994) European Polypores 2, vol 7, Synopsis fungorum. Fungiflora, Oslo

Ryvarden L, Johansen I (1980) A preliminary Polypore flora of East Africa. Fungiflora, Oslo

Ryvarden L, Melo I, Niemelä T (2014) Poroid Fungi of Europe, vol 31, Synopsis Fungorum. Fungiflora, Oslo

Sharma JR (1995) Hymenochaetaceae of India. Botanical Survey of India, Dehra Dun

Sharma JR (2000) Genera of Indian Polypores. Botanical Survey of India, Dehra Dun

Sharma JR (2012) Aphyllophorales of Himalayas. Botanical Survey of India, Dehra Dun

Talbot PHB (1973) Aphyllophorales I: general characteristics; thelephroid and cupuloid families. In: Ainsworth GC, Sparrow FK, Sussman AS (eds) The Fungi IVB. Academic, New York/London, pp 327–349

White TJ, Bruns T, Lee S, Taylo J (1990) Amplification and direct sequencing of fungal ribosomal RNA genes for phylogenetics. In: Innis MA, Gelfand DH, Sninsky JS, White TJ (eds) PCR protocols: a guide to methods and applications. Academic, New York, pp 315–322

Wolfaardt JF, Bosman JL, Jacobs A, Male JR, Rabie CJ (1996) Biokraft pulping of softwood. In: Srebotnik E, Messner K (eds) Biotechnology in the pulping and paper industry: recent advances in applied and environmental research. Proceedings of the sixth international conference on biotechnology in the pulp and paper industry, Vienna, pp 211–216

Chapter 2
Review of Literature

Abstract A detailed review of literature at the international and national level has been provided which throws light on the progress made since 1729 till to date in the concerned subject. Brief introduction of the important works concerning the subject has been provided.

2.1 International

The wood-rotting non-gilled agaricomycetes must have drawn the attention of man ever since the need for fire-wood arose, but it was Micheli (1729), who introduced the generic name *Polyporus* to include 14 species with centrally stalked fruit-bodies having pores on the underside of the pileus. Hill (1751) introduced the generic name *Stereum*. He described three species of *Stereum* and made a mention of 14 others. The first species is believed to be composed of two species of *Stereum* viz *S. hirsutum* (Wild. ex. Fr.) SF Gray and *S. purpureum* (Pers. ex Fr.) Fr. The genus was suppressed by Linnaeus (1753) who merged it into *Boletus*. Linnaeus (1764) recognized ten genera of fungi i.e. *Agaricus, Boletus, Hydnum, Phallus, Clathrus, Halvella, Peziza, Clavaria, Lycoperdon* and *Mucor* under the group, out of which seven are now regarded up 'Cryptogamia Fungi', which were mainly composite. The genus *Boletus* contained 14 species, out of which 7 are now regarded as members of poroid Agaricomycetes. The entire list of 93 species of fungi included only 9 polyporoid species: *Agaricus (Daedalea) quercinus, A. (Lenzites) betulinus, Boletus suberosus (Polyporus betulinus), B. (Fomes) fomentarius, B. (Fomes) igniarius, B. (Coriolus) versicolor, B. (Tramates) suaveolens, B. (Pycnoporus) sanguineus* and *B. (Coltricia) perennis*. He did not divide fungi into orders or families but in 1780 he grouped the genera *Agaricus, Boletus, Hydnum,* and *Phallus* in a section "Pileati" and the remaining six in "Pileo destitute". Linnaeus (1780) characterized *Agaricus* as "subtus lamellosus", *Boletus* as 'subtus porosus', as *Hydnum* as subtus echinatus which laid foundation of the families. It may be mentioned here that family Polyporaceae had its inception here. Schaeffer (1762–1774) illustrated about 330 species of fungi in his classical work and his ideas of classification are summarized in an index occupying seven (unnumbered) pages of fourth volume. He recognized only those genera already used by Linneaus and each genus constituded a separate tribe. All poroid species of polyporaceae were included in the tribe 'Boleti'. Willdenow proposed the

I.B. Prasher, *Wood-rotting non-gilled Agaricomycetes of Himalayas*,
Fungal Diversity Research Series, DOI 10.1007/978-94-017-9858-7_2

generic name *Thelephora* (Thaelaephora). Persoon (1801), in his '*Synopsis Methodica Fungorum*' divided fungi into classes "Angiocarpi" and "Gymnocarpi". The latter was divided into three orders i.e. Lytothecii, Hymenothecii, Naematothecii. The order Hymenthecii was divided into Agaricoidei, Boletoidei. Hydnoidei, Gymnodermata, Clavaeformes and Halvelloidei. An excerpt of this is as follows:

Agaricoidei – Hymenium lamellar or veined
Amanita
Agaricus
Merulius – Lamellae of somewhat swollen veins (superficial).
Cantharellus – Pileus distinct, stipitate, or dimidiate without stalk.
Serpula – Resupinate-effused, veined, reticulate above.
Gomphus – Fungi fleshy, elongated, pileus and confluent.
Boletoidei – Hymenium in variously prominent tubes.
Daedalea – Pileus (dimidiate) coriaceous-cordens forming sinuous, oblong and reticulate pores.
Boletus – Pileus evident, tubes completely rounded.
Fistulina – Tubes free.
Polyporus- branched- lobed in all directions, poroid.
Poria – effused-resupinate, poroid above (Tissue tubular-poroid).

Link (1809a, b) rejected Persoon's *Corticium* on the basis that resupinate habit of the fructification is not a distinctive character and represents a transitional stage in the development of the fungal fructification. He recognized only three genera *viz*: *Thelepora, Stereum* and *Merisma*. Persoon (1825) separated the poroid fungi into a definite group, section "porodermei" and included under it those fungi in which the hymenium is poroid, with the basidia lining the interior surfaces of the tubes. Thus Persoon seems to have been the first to make a serious attempt towards the classification of the fungi.

The next major classification to appear was Fries' "*Systema Mycologicum*" (1821, 1823, 1832) Vol. I–III and Elenchus Fungorum (1828). Fries (1821) divided fungi into four classes, viz. "Coniomycetes", "Hyphomycetes", "Gasteromycetes", "Hymenomycetes". The "Hymenomycetes" was segregated on the basis of exposed hymenium "Hymenium nudum". The Hymenomycetes was divided into six orders *viz*: "Pileati", "Clavati", "Mitrati", "Cupulati", "Tremellinae", and "Sclerotiaceae". The order "Pileati" included the genera: *Agaricus,Cantharellus, Merulius, Schizophyllum, Daedalea, Polyporus. Boletus, Fistulina, Hydnum, Sistotrema, Phlebia* and *Thelephora* Fries' "*Systema Mycologium*" Vol. 1 (1821) was adopted as the starting point for nomenclature of "fungi caeteri" by the International Botanical Congress at Brussels in 1910. But now the starting point for all fungi has been changed to Ed. I. V. 1753 (Linnaeus, *Speciesplantarum* ed. 1) by XIII International Botanical Congress- Sydney 1981. However, name of Fungi caeteri (excluding Myxomycetes) sanctioned by Fries (*Systema Mycologicum*, vol. 1–3 and *Elenchus Fungorum* vol. 1–2) are not affected by, and take priority over homonymous and synonymous names published earlier. In the "Systema Mycologium",

Fries arranged the members of Polyporaceae under three genera: *Merulius*, *Daedalea* and *Polyporus*. He also accepted the genus *Favolus* of *P. beauvois* as one of the three subgenera of *Polyporus*; the other two being *Microporus* and *Polysticta*. In "*Elenchus Fungorum*" Fries (1828) raised *Favolus* to generic rank. Kunze (1830) established the genus *Cyclomyces* for sessile fruit bodies with the tubes breaking up to form concentrically elongated pores. Fries (1836), in his "Genera Hymenomycetum", established the genus *Trametes* for species of *Daedaela* with rounded pores and the context of the pileus descending unchanged between the pores. In this work, Fries (1836) also recognized the genera *Cyclomyces* and *Hexagonia*. In "Epicrisis Systematics Mycologici", Fries (1836–1838) did not recognize any additional segregate, but in "Novae symbolae Mycologici", Fries (1855) divided the genus *Polyporus* into three subgenera: *Eupolyporus*, to include annual, fleshy, tough species; *Fomes*, to contain perennial species with woody indurated pileus and stratified tube layers; and *Poria* for resupinate species. Fries (1855) also established the genus *Polystictus* for a group of species with stipitate or sessile coriaceous fruit bodies covered with fibrous cuticle, formerly placed under *Polyporus*. The genus *Polystictus* was different from his genus *Polysticta* (1821: 342) meant to contain certain abnormal forms. However the genus "*Polystictus*" did not appear in "Hymenomycetes Europaei" of Fries (1874), which clearly indicates that he himself abandoned it. In his classical and last work on Hymenomycetes, Fries divided the Hymenomyceres into six orders on the basis of hymenial configuration. The following is Fries' own summary of classification on the hymenomycetes (1874).

Hymenomycetum sequentes disting ordines:

A. Hymenio effiggurato

    I. lamellato = Agaricini. Gen. I–XX
    II. poroso = Polyporei. Gen XXI–XXX
    III. aculeato
        1. varie protuberante = Hydnei. Gen. XXXI–XLI
        2. Hymenio laevigato
    IV. Horizontali infero = Thelephorei. Gen. XLII–XLVII
    V. verticali amphigeneo = Clavariei. Gen. XLVIII–LIV
    VI. supero; gelatinosi = Tremellinei. Gen. LV–LXIII

He included *Lenzites* under "Agaricini", marked by radially lamellate hymenophore; Boletus, *Fistulina, Polyporus, Trametes, Daedalea, Hexagonia, Favolus, Merulius, Porothelium*, and *Solenia* under "Polyporei", with tubular hymenophore; and *Irpex* under "Hydnei", with hydnoid hymenophore. He divided the genus *Polyporus*, containing 227 species, into five sections: "Mesopus", for centrally or eccentrically stipitate polypores with concolorous stipe base; "Pleuropus", for laterally or eccentrically stipitate, lignicolous polypores with blackish stipe base; "Merisma", for polypores with numerous pilei, imbricate, caespitose-connate; "Apus", for sessile, dimidiate pilei with a margin; and "Resupinati", for pilei posteriorly adnate to the substratum. Each of these sections (subgroups) was further split into smaller sections. Similarly, he divided each of the genus, *Trametes* and *Daedalea into "Apodes and Resupinatae"*

He also included the genera *Craterellus, Thelephora, Stereum, Auricularia, Corticium* and *Cyphella* under the thelephorei. The genus *Corticium* included the sub genera *Hypochnus* and *Coniphora*. He further rejected Léveillé's Hymenochaete and placed the species under *Stereum.*

It is evident that in "Hymenomycetes Europaei" Fries (1874) felt the necessity of grouping these genera in a somewhat different arrangement. It may be mentioned that real author behind many of the Fries concept and classification is Persoon but these are called Friesian because Fries was responsible for their maintenance and improvement (Donk 1964). The Friesian concepts excercised an enormous influence and became the traditional ones and have been perpetuated through Saccardo's "Sylloge Fungorum". Léveillé (1846) on the basis of anatomical details of the fructifications introduced the genus *Hymenochaete* based on the presence of setae and cystidia. He subdivided Hymenomycetes into subclasses Basidiosporii (Hymenomycetes proper) and Thecosporii (Ascomycetes).

Berkeley (1839–1867) while emphasizing the importance of structure of hymenium, described number of species (Berkeley 1839, 1847, 1850, 1851a, b, c, d, 1852a, b, 1854a, b, c, d, 1856, 1867). Tulasne (1853, 1872) also pointed out the significance of nature of hymenium as well as structure of basidium in the classification of Basidiomycetes.

Kickx (1867) used the generic names *Fomes* which was validly published earlier by Fries himself in 1849 in the form of a foot-note (fide Donk 1960: 215). Subsequent workers either split some of Fries' genera into several sections or raised several of his sections to generic rank to improve upon the Friesian arrangement. French Mycologist Gillet (1876–1878) divided the large genus *Polyporus* into *Merisma* (for branched stipitate species), *Physisporus* Chev. (for resupinate species), and *Fomes* (for woody perennial species). Rabenhorst (1874) on the basis of peculiar structures of the hymenium, segregated the genus *Aleurodiscus* from *Corticium*. Cooke (1879, 1880) again recognized the Léveillé's *Hymenochaete* and divided 'Fries' *Corticium* into two genera *viz*: *Corticium* (without cystidia) and *Peniophora* (with cystidia). Cooke (1884–1886) raised Fries' subgenus *Porea* to generic rank and contributed towards amplifying the ideas of generic unit for resupinate polyporoid species (Cooke 1884, 1885, 1886a, b, c). Quelet (1886) was one of first to separate the Friesian divisions of *Polyporus* based on form, surface modifications and consistency of the pileus. These were taken together with the few minor changes and raised to generic rank. He divided the genus *Polyporus* into 11 genera: *Caloporus, Leucoporus, Pelloporus, Cerioporus, Cladomeris, Placodes, Phellinus, Inodermis, Coriolus, Leptoporus* and *Poria*. The genus Irpex was referred to the Polyporaceae, because of its similarity with this family by Quelet for the first time.

Saccardo (1888) in his "Sylloge Fungorum" Vol. VI, adopted mainly the Friesian system and in addition to Fries' genera, also included the *Fomes* and *Poria*. He treated *Lenzites* under family Agaricineae and *Irpex* under Hydneae. In "Sylloge Fungorum" Vol. XIV, Saccardo (1899) gave a 'Conspectus Cohorticum et

Familiarum' in which he divided the fungi into five cohortes (groups of orders) *i.e.* Basidiomycetae, Ascomycetae, Phycomycetae, Myxomycetae and Deuteromycetae. The cohort Basidimycetae was divided Eubasidiae and Hemibasidiae. The former was further divided into Holobasidiae and Protobasidiae. While Holobasidiae was divided into three groups i. e. Angiocarpi, Hemiangiocarpi and Gymnocarpi. Under Gymnocarpi, he placed four families-Polyporaceae, Hydnaceae, Clavariaceae and Thelophoraceae. The family Polyporaceae was divided into two subfamilies-Boleteae and Polyporeae, whereas Thelephoraceae was divided into five subfamilies- Thelephoreae, Hypochneae, Hemigastreae, Hymenolicheneae and Exobasidieae. Massee (1889, 1890) introduced *Asterostroma* (star shaped setae in the context) *Thwaitesiella* (= *Lopharia*) with encrusted cystidia while emphasising the importance of micromorphology of the fructification.

Karsten (1876) used the Friesian classification of the Polyporaceae. Later (1879–1889) he departed from this classification and divided the Friesian genera into as many as 25 genera (Karsten 1879, 1881, 1882, 1883, 1887, 1889). However, only five genera *viz*: *Vararia*, *Xylobolus*, *Phanerochaete*, *Amphinema* and *Sterellum* were retained later. He emphasized the importance of characteristics like consistency of the fructification and context; pigmentation of the fructification, context and spores; character of the upper surface; and the presence or absence of the stipe etc. for the separation of various genera. He was the first to indicate as early as in 1882, that the genus *Lenzites* be united with Polyporaceae, which was referred to Agaricaceae since the times of Fries. The overall arrangement of the genera by Karsten was also artificial and far from natural and his work was essentially an extention of that of Fries. Bresadola (1881–1903), Quelet (1886), Hennings (1897) contributed to the study of the Hymenomycetes. Bresadola was the first to consistently use microscopic techniques in the taxonomy of Hymenomycetes (Slysh 1960). Schroeter (1889) also introduced certain changes in the Friesian system, mainly in the genus *Polyporus*. He adopted this genus originally conceived by Fries (1821) and after uniting it with *Trametes* he followed Karsten, and divided it into *Polyporus*, *Ochroporus* and *Phaeoporus* according to the tissue and spore colouration. Applying the same principle, he divided each of these genera *Merulius*, *Daedalea* and *Lenzites* into two, and thus established only nine genera (inclusive of Merulinei). This system though easier and simpler to use, was also as artificial as Karsten. Lloyd (1898–1925), in his "Mycological Notes", described many species in only 12 genera and did not accept most of the generic segregates of the earlier workers or contemporaries. Hennings (1900), mainly adhering to the Friesian system, recognized the same genera as by Saccardo and included *Lenzites* in the Polyporaceae. Murrill, in America, largely followed Karsten in the separation of genera. In his "North American Flora" (Polyporaceae) 1907, he recognized 74 genera of true polypores. He divided the Polyporaceae into four subfamilies: "Poriae" for the resupinate species; "Polyporeae" for annual, sessile or stipitate, light or dark coloured species; "Fomitae" for perennial, sessile or stipitate, light or dark coloured species; and "Daedaleae" for annual or perennial, sessile or stipitate fruiting bodies with daedaloid or lamellate hymenophore.

Smith (1908), in his synopsis of the "British Basidiomycetes", divided Basidiomycetes into two orders, Hymenomycetes and gasteromycetes. He segregated the former into six families, Agaricaceae, Polyporaceae, Hydnaceae, Thelephoraceae, Clavariaceae and Tremellinaceae. In the family Polyporaceae he placed 11 genera (*Boletus*, *Strobilomyces*, *Polyporus*, *Fistulina*, *Fomes*, *Polystictus*, *Trametes*, *Daedalea*, *Merulius*, *Poria*, *Porothelium*) to include the 231 species whereas in the family Thelephoraceae were placed 11 genera (*Craterellus*, *Thelephora*, *Stereum*, *Cladoderris*, *Hymenochaete*, *Peniophora*, *Corticium*, *Coniophora*, *Cyphella*, *Solenia* and *Exobasidium*) comprising of 145 species. Burt (1914–1926) a series of papers "The Thelophoraceae of North America" described 30 genera spreading over 600 species which included more than 200 species reported as new to science (Burt 1914a, b, c, 1915a, b, 1916a, b, 1917a, b, 1918a, b, c, 1919a, b, 1920, 1921, 1922, 1924, 1925, 1926). He followed the Friesian system and placed *Lenzites* near *Schizophyllum* under the Agaricaceae and the genus *Irpex* under Hydnaceae. Ames (1913), after a series of anatomical studies, recognized 16 genera to arrange about 130 species of Polyporaceae from Eastern North America. Cunningham (1927), in New Zealand, listed the polypores as representing five genera in subfamily Polyporeae of the Polyporaceae. Killerman (1928), in the second edition of the well known work published by Engler and Prantl (1928) divided Hymenomycetineae into seven families. Clements and Shear (1931) compiled the list of all genera of fungi published up to that time.

Corner (1932a, b, c) in his classical work, demonstrated the existence of hyphae of different types in the fruit-bodies of polypores and established the concept of hyphal systems. His concept was extensively developed by Cunningham (1946: 243) and utilized later in series of papers (1947–1950) and a monograph on New Zealand Polyporaceae (1965). Cunningham (1965) divided the family Polyporaceae into two subfamilies: Polyporoideae and Fomitoideae, the former with species having contextual hyphaewith hyaline walls, not darkening in aqueous KOH solution while the latter contained species having contextual hyphae with walls coloured, some shade of brown, darkening in aqueous KOH solution. The Polyporoideae was further divided into four tribes: Polyporeae, Flaviporae, Trametoideae and Ganodermae; and Fomitoideae into three tribes: Coltriceae, Phellineae and Fomiteae. A total of 32 genera were recognized for the New Zealand polypores.

Overholts (1953), in his monograph "The Polyporaceae of the United States, Alaska and Canada" prepared for publication by Lowe, followed the conventional genera and keyed out only ten genera. *Merulius* and *Poria* are not treated in this work and under the remaining eight genera he included 235 species for the North America. Lowe (1957, 1966) in his two very important and exhaustive treatments on "The genus *Fomes*" and "The genus *Poria*" described 68 and 159 species, respectively. He divides each of these two genera into five sections. Patouillard (1854–1926) was one of the earliest workers who realized that Fries' system of classification was wholly unnatural and needed an overall revision. He emphasized the importance of the microscopic characters of Hymenomycetes instead of hymenial configuration. His "Essai Taxonomique sur les families et les generes des Hymenomycetes" published in 1900 is his most outstanding contribution, in which he discarded the configuration of the hymenophore as a basis of classification of Hymenomycetes. He divided Basidiomycetes into: 'Basidimomycétes hétérobasidiés' and 'Basidiomomycétes homobasidiés'. The latter was divided into four families i.e. Exobasidiaces (adopted to parasitic mode of life), Aphyllophoraćes (gymnocarpic, hymenium indefinite), Agaricacés (hemiangiocarpic, hymenium definite) and Gasteromycetes (angiocarpic). He divided the 'Aphyllophoraces' into two tribes: 'Tribu des Clavariés' and 'Tribu des Porohydnes'. The Patouillard's classification of Aphyllophoraces is as follows:

Tribu des clavaries

- Serie des Thelephores
- Serie des clavaries
- Serie des Physalacries

Tribu des Porohydnes

- Sous- tribu: Cyphelles
- Sous- tribu: Odonties
  - Serie des odonties
  - Serie des Corticies
  - Serie des Stereums
- Sous- tribu: Pores
  - Groupe: Les Polypores Vrais
    - Serie des Ploypores
    - Serie des Leucopores
    - Serie des Leptopores
  - Groupe: Les Fomes
    - Serie des Tremetes
    - Serie des Igniaires
    - Serie des Placodes
  - Groupe: Les Merules
  - Groupe: Les Fistulines
- Sous- tribu: Hydnes
  - Serie des Mucronelles
  - Serie des Hydnes
  - Serie des Echinodonties
  - Serie des Phylacteries
  - Serie des Asterostromes

Patouillard's classification remained ignored until it was brought to the forefront by Rea (1922), Bourdot and Galzin (1928) and Donk (1933). Rea (1922), in "British Basidiomycetae", followed the system of Patouillard and divided Basidiomycetae into two divisions: Homobasidiae and Heterobasidiae. The former was divided into two subdivisions, Exobasidiineae and Eu-homobasidiineae. The latter was divided into three orders, Gasteromycetales (hymenium enclosed within a peridium at maturity). Agaricales (hymenium at first covered by a volva, or a ring, becoming fully exposed at maturity, usually inferior, and spread over the surface of gills, fleshy pores, or a smooth surface), and Aphyllophorales (hymenium exposed from the first, amphigenous, or unilateral, and spread over pores, teeth, anastomosing gills, or a smooth surface). The order Aphyllophorales was divided into two, Porohydnineae (receptacle pileate, stipitate, sessile or resupinate; hymenium inferior) and Clavariineae (receptacle erect, dendroid, coralloid, simple or branched, never pileate; hymenium more or less amphigenous). The Porohydnineae contained seven families, Polyporaceae, Polystictaceae, Meruliaceae, Fistulinaceae, Hydnaceae, Thelephoraceae and Cyphellaceae. Bourdot and Galzin (1928), in their work "Hymenomycetes de France", also followed the concept of Patouillard and recognised 17 genera for 231 species of polypores. They not only adopted Patouillard's classification but also introduced in it some essential corrections based on the microscopical examinations of their own material such as shape of the basidium, amylodity of the spore wall, presence of modified structures in the fructification, hyphal arrangement of the context and the absence/presence of clamp connection.

Donk (1931, 1933) proposed a different system of classification of the Aphyllophoraceae. He introduced three new genera *viz. Botryobasidium*, *Botryohypochnus* and *Gloeocystidiellum* in addition to the recognization of 29 genera of polypores for the Netherlands and distributed these in five subfamilies and four tribes: Phylacterioideae (tribe Boletopseae), Polyporoideae (tribes Polyporeae, Tyromyceteae and Daedaleae), Ganodermoideae, Hymenochaetoideae and Fistulinoideae. In the classification of Polyporaceae, Donk (1933) followed the principles introduced by Karsten and Patouillard and thoroughly revised and supplemented it with detailed microscopic observations.

Bondarstev and Singer (1941) placed 60 polypore genera in six suborders and eight families: Phylacteriineae (fam. Boletopsidaceae), Clavariineae (fam. Scutigeraceae), Polyporineae (fam. Polyporaceae), Cyphellineae (fam. Fistulinaceae), Corticiineae (fam. Corticiaceae and Meruliaceae), and Boletinaceae (fam. Strobilomycetaceae and Boletaceae). The family Polyporaceae was divided into five subfamilies: "Porioideae", included most of the resupinate species under 8 genera; "Tyromycetoideae", with 7 genera; "Fomitoideae", divided into six tribes containing 15 genera; "Polyporoideae", with two genera and "Corioloideae", divided into four tribes with 21 genera. Many new names used by them did not have latin diagnoses. Therefore, Singer (1944) published a short paper validating 11 genera which required such action.

Pilat (1936–1942) recognised 22 genera in the Polyporaceae which was divided into three subfamilies, Ganodermoideae (2 genera), Polyporoideae (17 genera) and Hymenochaetoideae (3 genera). He placed *Fistulina* in the Fistulinaceae. He followed partly the concepts of Patouillard and Donk and based these genera on anatomical and morphological character. Imazeki (1943) published a monographic treatment of polyporoid species in Japan and adjacent countries. In this treatment, he recognised 38 genera in four tribes and seven subtribes: Polyporeae (Trametineae, 6 genera; Coriolinae, 5 genera; Tyromycetinae, 5 genera; Piptoporinae, 3 genera), Ganodermeae with 3 genera, Mucronoporeae with 8 genera; and Cryptoporeae with 1 genus.

Pinto-Lopes (1952), in his study of the biosystematics of the Polyporaceae, set up seven subfamilies to include 22 genera. These were: Cladodendronoideae of uncertain position, 1 genus; Leptoporoideae, 6 genera; Trametoideae, 4 genera; Phaeochroideae, 4 genera; Cladomeroideae, of uncertain position, 1 genus; Mensularioideae, 2 genera; and Xanthochroideae, 4 genera. Most of the genera used by him have been emended from the original descriptions either by himself or by earlier workers. He found the system proposed by Cunningham (1947–1950), based on Corner's (1932a, b, c) hyphal system concepts, of little value and proposed a system based on the type of thickening of the hyphae forming the fruit-body as they matured and aged, along with the presence or absence and the type of clamp connections (Cunningham 1947, 1948a, b, c, 1949a, b, 1950). This system was criticized by Corner (1953) when he described in detail the hyphal system of 4 polyporoid species stressing the importance of hyphal systems.

Bondartsev's exhaustive treatment "The Polyporaceae of the Europe U.S.S.R. and Caucasia" (in Russian) appeared in 1953 in which he divided the artificial group "Polyporineae" into five suborders and six families: Phylacteriineae f. Boletopsidaceae; Clavariineae f. Scutigeraceae; Corticiineae f. Corticiaceae and Meruliaceae; Cyphellineae f. Fistulinaceae; and Polyporineae f. Polyporaceae. He included 54 genera in the Polyporaceae which was further divided into five subfamilies and ten tribes as follows: Porioideae with 8 resupinate genera: Tyromycetoideae with 7 genera; Fomitoideae: Piptoporae 2 genera, Ischnodermateae 3 genera, Fomiteae 2 genera; Phaeoleae 1 genus, Inonoteae 7 genera and Ganodermateae 1 genus; Polyporoideae with 2 genera; Corioloideae: Corioleae 12 genera, Oxyporeae 4 genera, Hirschioporeae 1 genus; and Daedaleae 4 genera. In this work, approximately 300 species and 200 of their varieties and forms are described.

Eriksson (1950, 1958a, b, c) critically studied the resupinate fungi from Sweden. He erected seven new genera belonging to Thelephoroid fungi.

Cunningham (1953–1957) published series of papers on Thelephoraceae of New Zealand. He published "Thelephoraceae of Australia and New Zealand" in (1964) and presented his system of classification. Cunningham (1965) published Polyporaceae of New Zealand.

Kotlaba and Pouzar (1957–1958), adopted the hyphal system technique of studying polypores and based their system on Bondartsev's work extended to include narrower generic limits by strict generic definitions. They accepted 48 genera based on their own studies and on that of others. Since their treatment covered species occurring in Czechoslovakia only, the system they adopted remains incomplete. They did not include key to the genera.

Donk (1960) gave an excellent detailed annotated nomenclature enumeration of polyporoid genera in "The generic names proposed for Polyporaceae". In this account he considered 229 generic names as validly published, 37 not validly published, and six excluded. Donk (1964), in his monumental work "A Conspectus of families of Aphyllophorales", recognized 21 families under Aphyllophorales. The classification adopted in this paper was derived from Patouillard's (1900) work. The family Polyporaceae Corda is considered exclusive of Fistulinaceae, Boletaceae, Ganodermataceaae, Bondarzewiaceae (*Bondarzewia*), Hymenochaetaceae (Polyporoid genera), Corticiaceae (*Porogamme, Sistotrema* spp., *Echinotrema, Cristella* spp., *Merulius, Gloeoporus, Athelia* spp., and *Phlebia* spp.), Coniophoraceae (Serpula), and Tremellaceae (*Aporium*).

Cooke (1940, 1953, 1959) gave an account of the genera of poroid fungi. His paper of 1953 deals with the genera of the Homobasidiomycetes (exclusive of the Gasteromycetes). In 1959, Cooke mentioned, "Species of fungi with pores or pore-like structure in which the hymenium is produced have been assigned to about 300 genera". Approximately 100 of these can be considered valid or usable, and some of these can no longer be considered polypores but are more nearly related to such provisional groups as the "Cyphellaceae", Meruliaceae, "Coniophoraceae", and possibly other groups. Some of these valid genera cannot be used because they are synonymous with others.

Christiansen (1960) published "Danish Resupinate Fungi part II Homobasiomycetes" including these fungi.

In Poland, Domanski (1965) produced a book Grzyby (Fungi) II covering resupinate polypores viz Polyporaceae I (resupinate), Mucronoporaceae I (resupinatae) distributed among 29 genera. He established two genera viz: *Incrustoporia* Dom. *Cerioporiopsis* Dom. Published in 1963. In the English translation (1972) of Grzyby (Fungi) II, he included new genus *Diplomitoporus* Dom., published in 1970. This work (1965) is based mainly on Bondartsev's system (1953) and Domanski accepted a number of genera described or emended by various workers like Donk, Kotlaba and Pouzar, Parmasto and Reid. Domanski (1965) followed mainly the system of Kotlaba and Pouzar (1957a, b) and divided the species with tubular hymenophore into six families: Fistulinaceae, Boletopsidaceae, Bondarzewiaceae, Polyporaceae, Ganodermataceae, and Mucronoporocaea belonging to order Aphyllophorales. He recongnised 48 genera for these families: Polyporaceae pileati (40 genera), Mucuronoporaceae pileati (4 genera), Fistulinaceae (1 genus), Bondarzewiaceae (1 genus), Ganodermataceae (1 genus), Boletopsidaceae (1 genus). He divided the pileate fungi of the family polyporaceae into various genera on the basis of an upto date interpretation of differences in the microscopic structure of fructifications and less on macroscopic features. His system of classification is the net result of opinions and taxonomic systems of Kotlaba and Pouzar (1957–1958), Teixeira (1962), Donk (1964) and other authors.

Parmasto (1968), in his monograph "Conspectus Systematis Corticiacearum" discussed interrelationships in Corticiaceae and related families. He divided Corticiaceae into 11 subfamilies, which were further divided into tribes. He followed the modern generic concepts. Parmasto (1970) published another monograph "The Lachnocladiaceae of the Soviet Union", in which he splitted family

Lachnocladiaceae into two tribes: Varariodae and Lachnocladiodae. He placed *Asterodon* in family Hymenochaetaceae.

Jülich (1972), in his "Monograhie der Athelieae (Corticiaceae, Basidiomycetes)" recognised 11 genera including 52 species in tribe Athelieae of Parmasto (1968). Later in a series of papers Jülich (1973–1976), he introduced 26 genera, 12 species and made 85 new combinations.

Talbot (1973) accepted all the 21 families discussed by Donk (1964) and added the families Lacnocladiaceae and Podoscyphaceae for the order Aphyllophorales. He keyed out 23 families. Pegler (1973a) keyed out 8 poroid families under Aphyllophorales: Fistulinaceae, Coniophoraceae (Polyporoid genera), Corticiaceae (Polyporoid genera), Bondarzewiaceae, Thelephoraceae (Polyporoid genera), Ganodermataceae, Hymenochaetaceae (Polyporoid genera) and Polyporaceae. In a separate paper, Pegler (1973b) has given a key to 123 polyporoid genera.

Locquin (1974), in his "De Taxia Fungorum" Vol I. Syllabus, gave an outline of classification of fungi. He divides 'Mycota' (Fungi) into all 11 subdivisions on the basis of thallus organisation (reduced, plasmoidal or hyphal), septation and nature of septa. These subdivisions are: Myxomycotina, Premycotina, Actinomycotina, Eubacteria, Zoomycotina, Phoromycotina, Uteromycotina, Acromycotina, Epimycotina, Perimycotina and Cladomycotina. The class Basidiomycetes was divided into three subclasses, Basidiomycetidae, Meristemobasdiomycetidae and Lycoperdellomycetidae on the basis of spore character. The subclass Basidiomycetidae, having simple spores was divided into nine orders: Cantharellales with hymenium continuous, smooth or plicate and stichic basidia (fam. Hydnaceae, Cantharellaceae, Clavulinaceae, Gomphaceae, Clavariaceae, Pterulaceae, Sparassidaceae, Bondarzewiaceae, Hybogasteraceae, Hericeaceae and Auriscalpiaceae); Poriales with hymenium continuous, poroid basidia (fam. Bankeraceae, Thelephoraceae, Scutigeraceae, Laetiporaceae, Meruliaceae, Lachnocladiaceae, Poriaceae, Coniophoraceae, Stereaceae, Podoscyphaceae, Coriolaceae and Corticiaceae); Polyporales with hymenium discontinuous, structured into pollen chambers or channels with dimple, channels independent or united, tough, coriaceous or suberose (fam. Cyphellaceae, Fistulinaceae, Schizophyllaceae, Polyporaceae, Ganodermataceae and Hymenochaetaceae); Boletales with hymenium discontinuous, structured into pollen chambers or channels with dimple, united, fleshy (12 families); Gauteriales, 2 families; Agaricales, 39 families; Amanitales, 5 families; Russulales, 4 families; and Hymenogastrales, 4 families.

Donk (1974), in his "Check List of European Polypores", listed all the polypores recorded from Europe under 67 genera.

Larsen (1974) published a monograph on the genus *Tomentella* recognising 72 species.

Rattan (1977) gave a consolidated account of 198 taxa in "The Resupinate Aphyllophorales of the North Western Himalayas". He followed the classification of Donk (1964) and Parmasto (1968) with modifications.

Eriksson and Ryvarden (1973, 1975, 1976), Eriksson et al. (1978, 1981, 1984) in their monograph "The Corticiaceae of North-Europe" Vol II–VI have critically described 95 genera comprising of 345 species species. As many as 9 genera, 22 species have been proposed as new to science along with 57 new combinations.

Hjortstam et al. (1987, 1988) published "The Corticiaceae of North Europe" Vol I and VIII

Ryvarden (1976a, b, c, d, 1978a, b) published Polyporaceae of North Europe. Ryvarden and Johansen (1980) published "A preliminary polypore flora of East Africa" describing 4 new genera, 3 new species and 43 new combinations.

Ryvarden (1991) published "Genera of Polypores" discussing the Nomenclatural and Taxonomic evaluation of all the genera proposed for Polypores since 1753. It also includes two new genera, two nom. nov. besides ten new combinations.

Ryvarden and Gilbertson (1993, 1994) have contributed significantly to the European flora in the form of "European Polypores Vol. 1 and 2". It contains exhaustive information regarding the methodology, systematic and literature concerning the European fungi.

Hallenberg (1978–1998) in series of papers has contributed immensely to the study of wood inhabiting fungi (Hallenberg 1978, 1979, 1980, 1981a, b, c, d, 1983a, b, c, 1984a, b, 1985, 1986, 1988a, b, 1996, 1998a, b).

Breitenbach and Kränzlin (1986) published Fungi of Switzerland (Vol. 2) which included Heterobasidiomycetes, Aphyllophorales and gasteromycetes a beautifully illustrated work.

Nüñez and Ryvarden (2000, 2001) published in two volumes East Asian Polypores.

Ryvarden et al. (2003) published "A critical checklist of corticoid and poroid fungi of Norway".

Ryvarden (2004) published Neotropical polypores Part 1 which concerned Ganodermataceae and Hymenochaetaceae.

Bernicchia (2005) published Polyporaceae s. l. describing these fungi from Europe with particular reference to Italy.

Bernicchia and Gorjon (2010) published "Corticiaceae s.l" covering all the corticoid fungi from Europe.

Some of the other important works include those of Nüñez and Ryvarden (1995, 1999) and Larsen and Cobb-Poulle (1992).

The other notable works who contributed towards the study of these fungi include:

Hoehnel and Litschauer (1906–1908), Ames (1913),Yasuda (1917), Overholts (1914–1942), Bresadola (1916), Petch (1925), Baxter (1926–1945), Emmons (1927), Litschauer (1928, 1944), Pilat (1936–1942), Overholts (1929, 1939), Donk (1931–1974), Humphrey and Lens (1931–1932), Shope (1931), Banerjee (1935–1942), Biggs (1937–1938), Stevenson and Cash (1936), Imazeki (1939), Neuman (1914), Lowe (1942–1976), Rogers (1943–1951), Rogers and Jackson (1943), Steyaert (1946–1980), Corner (1948, 1968), Jackson (1948–1950a, b, c), Nobles (1948, 1958, 1965), Wakefield (1948–1966), Boidin (1950–1956; 1981), Reid et al. (1958), White (1951), Lentz (1954, 1960), Maxwell (1954), Welden (1954–1966), Gilbertson (1956–1981), Parmasto (1965–1979), Eriksson (1958a, b, c), Fidalgo (1958), Svrcek (1958, 1960), Pouzar (1959), Lowe and Gilbertson (1961), Liberta (1962–1980),

Reid (1962), Lemke (1964a, b, 1965), Lombard and Gibertson (1965), Oberwinkler (1965–1977), Aoshima (1967), Domanski (1969–1974), Eriksson and Strid (1969), Hjortstam (1969–1981), Budington and Gilbertson (1973), Canfield and Gilbertson (1971), Niemelä (1972–1978), Ryvarden (1972–1985), Boidin and Lanquetin (1973–1980), Lowe and Lombard (1973), Burdsall and Gilbertson (1974), Larsson and Hjortstam (1974), Mass Gaesteranus (1974), Hallenberg and Ryvarden (1975), Gilbertson and Burdsall (1975), Lindsey and Gilbertson (1978, 1983), Lombard et al. (1975), Niemelä and Ryvarden (1975), Ginns (1976–1984), Ginns and Kokko (1976), Rattan and Abdullah (1977), Martin and Gilbertson (1977–1980), Hallenberg (1978–1981), Hjortstam and Ryvarden (1979–1982), Rattan et al. (1978), Eriksson and Hallingback (1979), Eriksson and Wall (1979), Johansen and Ryvarden (1979), Parmasto and Parmasto (1979), Boidin and Gilles (1982), Lazzari (1980), Leger (1980), Rattan and El-Buni (1981), Carranza (1982), Hjortstam and Stalpers (1982), Nakasone et al. (1982), Rajchenberg (1982, 1983, 1987, 2002), Blackwell and Gilbertson (1985), Blumenfeld and Wright (1984), Gomez and Ryvarden (1985), Lombard and Larsen (1985), Larsen et al. (1985), Morse and Gilbertson (1986), Hallenberg and Larsson (1993), Hallenberg and Küffer (2001), Larsson and Larsson (2003), Binder et al. (2005), Dhingra (2006), Moncalvo et al. (2006), Nakasone (2006), Hallenberg et al. (2007–2010), Kauserud et al. (2007), Gorjan and Hallenberg (2008), Ghobad-Nejhad et al. (2009, 2012), Ghobad-Nejhad (2011), Ranadive et al. (2011), Baltazar et al. (2012) and Gorjan (2012).

2.2 National

The floristic studies on wood rotting non-gilled Agaricomycetes in India started much earlier than its initiation in the Himalayan region of the country. The earliest available record of these fungi is based on the collections made by Koenig- a Danish missionary in Tranquebar, Tanjore (Tamil Nadu) from 1765 to 1785. Fries examined some of these collections which contained *Daedalea sinulosa* Klotzsch, *Trametes sinensis* Fr. (=*Hexagonia sinensis* Fr.) and *Polyporus lanatus* Fr. The collections of DF Didrichsen from Nicobar Island (1845–1847) studied by Fries (1855) included *Polyporus campbelli* Berk. and *Hexagonia macrostrema* Jungh. (= *Lenzites alutaceae* Cke.). In the early part of the nineteenth century, Wight collected a large number of fungi in India. These were examined by Klotzsch (1832, 1833) and Berkeley (1839). These included *Polyporus wightii* Kl. (=*Hexagonia apiaria* Pers.), *Hexagonia sinensis* Fr., *Daedalea sanguinea* Kl., *Polyporus pectinatus* Kl.(= *Fomes pectinatus* (Kl) Gill), *Poria arenaria* Kl., *Polystictus leoninus* Kl., *Daedalea latissima* Fr. and *Polyporus tenuis* Hook. Bélanger made a few collections which were described by Montagne (1846) and these included *Fomes peguanus* Mont. and *Daedalea elegans* Spreng. Perrottet made collections mainly from the Nilgiri Hills (Tamil Nadu) which were described by Montagne (1842) and Léveillé (1845, 1846). Montagne (1842) recorded *Polyporus arcularius* Fr., *P. inamomeus* Mont., *P. flabellum* Mont., *P. nilgheriensis* Mont., *P. sanguineus* (L.) Mey., *P. versicolor* (L.) Fr. and

P. xanthopus Fr., while Léveillé (1845, 1846) described nine species including *Hexagonia glabra* Lév., *H. aculeate* Mont., *H. tenuis* Hooker, *Polyporus adustus* (Willd.) Fr., *P. cinerascenns* Lév., *P. coriaceous* Lév., *P. inquinatus* Lév., *P. pectunculus* Lév., and *P. pinsitus* Fr.

The collections made by Sir J. D. Hooker and sometimes by Dr. Thompson his co-worker perhaps represent the first records of these fungi from Himalayas. These collections were made from Eastern Himalayas and Khasi Hills. These collections along with sketches of some of these fungi were sent to M. J. Berkley, who published them between 1850 and 1854 in his series "Decades of fungi". A total of 106 species of Agaricomycetes including 51 new species were published by him. Berkeley (1856) again published few Indian species like *Corticium leave* Pers. ex Fr. and *Hymenochaete cacao* Berk.

Sulpiz Curz, the then Curator of the Royal Botanic Garden, Calcutta, made extensive collections of fungi from Burma and Bengal and these were reported by Currey in 1874. He recorded about 40 species including that from India such as *Daedalea pruinosa* Lév., *D. tennuis* Berk., *Fomes hypoplastus* Berk., *Ganoderma lucidum* (Curt. ex Fr) Karst., *Polyporus picipes* Fr., *Hexagonia kurzii* Currey etc. Cooke at Kew (1884, 1891a, b) received and described a large number of Indian fungi which included species like *Daedalea andamani* Berk. in Herb., *D. flabellum* Berk. in Herb. (= *D. pruinosa* Lév.) *Polyporus grammocephalus* Berk., *Polystictus aethiops* Cooke, *Polyporus hobsoni* (Berk.) Cooke, etc. Massee (1889) described *Coniophora indica* Mass. from India. Towards the end of nineteenth centuary, W. Gollan made extensive collections from Uttarakhand which were later described by Hennings (1900, 1901). Some of the notable species described by him included *Fomes rimosus* Berk., *Ganoderma australe* (Fr.) Pat., *Merulius pseudolacrymans* P. Henn., *Polyporus bambusicola* P. Henn., *P. gilvus* Schw., *P. hispidus* (Bull.) Fr., *P.Schweinitzii* Fr., *Polystictus gollanii* P. Henn., *P. tomentosus* Fr. and others.

Massee (1889–1912) began the series "Fungi exoticii" in the Kew Bulletin which was also carried on by Wakefield (1916–1922). They described many Indian species. These included *Daedalea gollanii* Massee (= *Irpex zonatus* Berk.), *D. suberosa* Massee, *Lenzites adusta* Massee, *Polyporus indicus* Massee, *Polystictus gleadowii* Massee, *P. villosus* Massee, *Fomes elegans* Wakef. and *Lenzites beckleri* Berk., based on collections sent to Kew Herbarium by Butler, Gollan, Hutchings, Gamble and Hole. Theissen (1911) recorded many fungi in the collections received from Father E. Blatter, collected mainly from Bombay, Poona (Maharastra), Dehradun (Uttrakhand) and Shimla (Himachal Pradesh), and this report included 25 species of Agaricomycetes. He included eight species to the list of Indian Agaricomycetes which were *Ganoderma resinaceum* Boud., *Irpex canescens* Fr., *I. vellereus* Berk. and Br., *Polyporus cubensis* Mont., *Polystictus floridanus* Berk., *P. russogramme* Berk., *P. zeylanicus* Berk., and *P. zonatus* (Koen.) Berk. Saccardo (1888) recorded *Favolus brasiliensis* Fr., *F. boucheanus* Kl., *Merulius corium* Fr., *Polyporus haematinus* Berk. In Herb. from India. Bresadola (1920a, b) recorded 19 species of poroid fungi from India. He recorded *Polyporus molliculus* Bres., *Polystictus cuminghii* Berk., *P. filbula* Fr., *P. menziesii* Berk., and *Trametes acupunctata* Berk. for the first from India in "Mycological notes", "Mycological Letters" and other publications from 1898 to 1925, where he made many references to Indian species like *Daedalea bosei* Lloyd, *Favolus bengala* Bose in Lloyd, *Fomes durissimus* Lloyd, *F. pinicola* Fr., *F. velutinosus* Lloyd, *Hexagonia burchelli* Berk. in Lloyd, *Polyporus acervatus* Lloyd, *P. hookeri* Lloyd, *P. ikenoi* Lloyd etc.

Bose (1918–1946) was the first Indian mycologist to collect and study the polypores on a comprehensive scale, mostly from Bengal (Bose 1918, 1920a, b, 1921a, b, c, 1922a, b, c, 1923a, b, 1924, 1925, 1928a, b, 1934, 1937, 1944, 1946). In a series of papers "Polyporaceae of Bengal" I–XI and other publications he described a total of 143 species including 9 new species. Sundararaman and Marudarajan (1925) reported 11 species of polypores from Madras.

Butler and Bisby (1931) published for the first time "The Fungi of India". They recorded about 300 species of polyporoid fungi spreading over 15 genera namely: *Cyclomyces*, *Daedalea*, *Elmerina*, *Favolus*, *Fistulina*, *Fomes*, *Ganoderma*, *Gloeoporus*, *Hexagonia*, *Lenzites*, *Merulius*, *Polyporus*, *Polystictus*, *Poria* and *Trametes*. Mitter and Tandon (1932, 1938) reported 30 species and 2 forms of polypores from Naini Tal (Uttarakhand). Banerjee (1935a, b) published an account of Indian Thelephoraceae, including four genera (*Stereum* Pers. ex. Gray, *Hymenochaete* Lev., *Creterellus* Pers. and *Asterostromella* Hoehn. and Litsch.) and 24 species from West Bengal. Banerjee (1946, 1947) listed 45 and 91 species of Polyporaceae from Sikkim Himalayas, and Calcutta and suburbs, respectively. Bagchee and Bakshi (1950) and Bakshi and Bagchee (1950) described the occurrence, pathology, sporophore and cultural characteristics of 14 species of polypores occurring on Indian trees and Oaks. Thind and Adlakha (1956) and Reid et al. (1958) gave an account of 12 species from Mussoorie hills (Uttrakhand). Vasudeva (1960) revised and brought up to date the Butler and Bisby's "The Fungi of India" (1931), including all records of Indian fungi published up to 1952. This includes about 400 species of these fungi belonging to 21 genera in the post-independence Indian Union.

Further contribution to the knowledge of Indian Polyporaceae were made by various workers at Forest Research Institute, Dehra Dun. These included contributions from: Bagchee (1951–1961), Bagchee and Bakshi (1951), Bagchee and Singh (1960), Bagchee et al. (1954), Bakshi (1955–1958), Bakshi and Boyce (1959), Thind and Chatrath (1960), Bakshi and Choudhury (1961), Rehill and Bakshi (1965, 1966), Bakshi and Singh (1970), Bakshi et al. (1955, 1966), Singh (1961, 1966a, b), Puri (1956), Sehgal et al. (1966), Singh and Bakshi (1961) and Singh et al. (1961a, b). These publications mainly pertained to pathological aspect of the fungus giving little emphasis on taxonomy. Bakshi (1971) in "Indian Polyporaceae", included 355 species placed in 15 genera. He placed 102 species under 'Inadequately known species', 32 under 'Species noted by authorities from collections in India' and 14 under 'Incompletely known species' with 'Type inadequate or unavailable'. He thus described only 207 species. Roy (1968–1971) studied the anatomical features of eight species of Polyporaceae.

Prof. K.S Thind started the detailed study of these fungi in Himalayas. He published series of papers with his students from 1960 to 1980. Thind and Rattan (1968–1976) in a series of paper, published 59 species from the North-Western

Himalayas (Thind and Rattan 1968, 1970, 1971a, b, c, d, 1972, 1973a, b, 1976). Thind and Rattan (1970–1971), Thind and Dhanda (1979a, b, 1980a, b) are the other significant contributions.

Rattan (1977) in his monograph 'The Resupinate Aphyllophorales of the North-Western Himalayas' gave an account of 198 resupinate taxa of Aphyllophorales from the North-Western Himalayas.

Prasher (1999) recorded 306 species of fungi from Bhutan which also, included 18 species of these fungi. Prasher and Chander (2006, 2007) and Prasher et al. (2011, 2012) published and described polyporoid fungi from Nanda Devi Biosphere reserve, Himachal Pradesh and Uttarakhand.

Dhingra and Priyanka (2011) published and described119 species of non-poroid Agaricomycetes from Eastern Himalayas and Bhutan in the form of checklist.

Prasher and Ashok (2013) and Prasher and Lalita (2013) published checklists of Wood rotting Agaricomycetes from Himachal Pradesh and Uttarakhand listing 355 and 200 species of these fungi respectively. These represents first ever published consolidated checklists of these fungi from these states.

References

Ames A (1913) A consideration of structure in relation to genera of Polyporaceae. Ann Mycol 11:211–253

Aoshima K (1967) Synopsis of the genus *Daedalea* Pers. Fr Trans Mycol Soc Jpn 8:1–4

Bagchee K, Bakshi BK (1950) Some fungi as wound parasites of Indian trees. Indian Forest 76(6):244–253

Bagchee KD (1951) The fungus diseases of Sal (*Shorea robusta* Gaertn.) – I. Indian Forest Rec 1:11–23

Bagchee KD (1957) The Fungal diseases of Sal (*Shorea robusta* Gaertn.) – III. The root disease of Sal due to Polyporus shoreae Wakef. (Partridge Wood rot). Indian Forest Rec 1(9):185–195

Bagchee KD (1958) The fungal diseases of Sal (*Shorea robusta* Gaertn.) – V. The heart rot of Sal caused by *Trametes incerta* (Currey) Cooke. Indian Forest Rec 2(4):61–69

Bagchee KD (1960) The fungal diseases of Sal (*Shorea robusta* Gaertn.) – II. Secondary parasites of 'Sal'. Indian Forest Rec 1(8):97–184

Bagchee KD, Bakshi BK (1951) *Poria monticola* Murr on Chir (*Pinus longifolia* Roxb) in India. Nature (Lond) 167:824

Bagchee KD (1961) The fungal diseases of Sal (*Shorea robusta* Gaertn.) – IV. *Fomes caryophylli* (Rec.) Bres. A destructive heart rot of 'Sal'. Indian Forest Rec 2(3):25–58

Bagchee KD, Puri YN, Bakshi BK (1954) Principal diseases and decays of oaks and other hardwoods in India-II. Indian Phytopath 7:18–42

Bagchee KD, Singh U (1960) List of common names of fungi attacking Indian forest trees, timber and the herbaceous and shrubby under growths and list of cultures of forest fungi. Indian Forest Rec 1(10):199–348

Bakshi BK (1955) Diseases and decays of conifers in the Himalayas. Indian Forester 81:779–797

Bakshi BK (1956) Occurrence of *Polyporus squamosus* (Huds.) Fr. in India. Indian Phytopath 9:191–194

Bakshi BK (1957) Diseases of Khair (*Acacia catechu* Willd.) and their prevention. Indian Forester 83(1):41–46

Bakshi BK (1958) New records of Hymenomycetes in India. Indian Phytopath 11:88

Bakshi BK (1971) Indian Polyporaceae. ICAR, New Delhi

Bakshi BK, Bagchee K (1950) Principal diseases and decays of oaks in India. Indian Phytopath 3:124–139
Bakshi BK, Boyce JS (1959) *Polyporus shoreae* (root rot) on 'sal'. Indian Forester 85(11):656–658
Bakshi BK, Choudhury TG (1961) Sexuality and interfertility studies in *Polyporus abietinus*, *P. pargamenus* and *P. versatilis*. Can J Bot 89:997–999
Bakshi BK, Puri YN, Singh B (1955) Two decay fungi on conifers in the Himalayas. Indian J Mycol Res 1:75–80
Bakshi BK, Singh S (1970) Heart rots in trees. Rev For Res 3:197–251
Bakshi BK, Singh S, Singh U (1966) A new root rot disease complex in 'teak'. Indian Forester 92(9):566–569
Baltazar JM, Drechsler-Santos ER, Ryvarden L, Cavalcanti MAQ, Gibertoni TB (2012) Contribution to the knowledge of Polypores (Agaricomycetes) from the Atlantic forest and Caatinga, with new records from Brazil. Mycosphere 3(3):267–280
Banerjee SN (1935a) Thelephoraceae of Bengal-I. J Indian Bot Soc 14:13–48
Banerjee SN (1935b) A historical account of the classification of Thelephoraceae. Sci Cult 1:314–317
Banerjee SN (1942) Importance of anatomical characters of sporophores in the taxonomical study of Thelephoraceae of Bengal. J Indian Bot Soc 21:33–39
Banerjee SN (1946) Some higher fungi from Sikkim Himalayas. Sci Cult 11:444–445
Banerjee SN (1947) Fungus Flora of Calcutta and suburbs-I. Bull Bot Soc Beng 1:37–54
Baxter DV (1926) Some Porias from the region of the Lake States. Pap Michigan Acad Sci 6:67–76
Baxter DV (1932) Some resupinate polypores from the region of the Great Lakes. III. Pap Michigan Acad Sci 15:191–228
Baxter DV (1933) Some resupinate polypores from the region of the Great Lakes. IV. Pap Michigan Acad Sci 17:421–439
Baxter DV (1934) Some resupinate polypores from the region of the Great Lakes. V. Pap Michigan Acad Sci 19:305–332
Baxter DV (1935) Some resupinate polypores from the region of the Great Lakes. VI. Pap Michigan Acad Sci 20:273–281
Baxter DV (1936) Some resupinate polypores from the region of the Great Lakes. VII. Pap Michigan Acad Sci 21:243–267
Baxter DV (1937) Some resupinate polypores from the region of the Great Lakes. VIII. Pap Michigan Acad Sci 22:275–295
Baxter DV (1938) Some resupinate polypores from the region of the Great Lakes. IX. Pap Michigan Acad Sci 23:285–305
Baxter DV (1939) Some resupinate polypores from the region of the Great Lakes. X. Pap Michigan Acad Sci 24:167–188
Baxter DV (1940) Some resupinate polypores from the region of the Great Lakes. XI. Pap Michigan Acad Sci 25:145–170
Baxter DV (1941) Some resupinate polypores from the region of the Great Lakes. XII. Pap Michigan Acad Sci 26:107–121
Baxter DV (1942) Some resupinate polypores from the region of the Great Lakes. XIII. Pap Michigan Acad Sci 27:139–161
Baxter DV (1943) Some resupinate polypores from the region of the Great Lakes. XIV. Pap Michigan Acad Sci 28:215–234
Baxter DV (1944) Some resupinate polypores from the region of the Great Lakes. XV. Pap Michigan Acad Sci 29:85–109
Baxter DV (1945) Some resupinate polypores from the region of the Great Lakes. XVI. Pap Michigan Acad Sci 30:175–191
Berkeley MJ (1839) Description of exotic fungi in the collection of Sir WJ Hooker, from memoirs and notes of JK Klotzsch, with additions and corrections. Ann Nat Hist 3:375–401

Berkeley MJ (1847) Decades XV-XIX. Ceylon fungi. London J Bot 6:479–514

Berkeley MJ (1850) Decades of fungi. Decades XXV to XXX. Sikkim Himalaya fungi, collected by Dr JD Hooker. Hookers J Bot Kew Gard Miscellany 2:76–88

Berkeley MJ (1851a) Decades XXXII-XXXIII. Sikkim-Himalayan Fungi, collected by Dr JD Hooker. Hookers J Bot Kew Gard Miscellany 3:39–49

Berkeley MJ (1851b) Decade XXXIV. Sikkim-Himalayan Fungi, collected by Dr Hooker. Hookers J Bot Kew Gard Miscellany 3:77–84

Berkeley MJ (1851c) Decade XXXV. Sikkim-Himalayan Fungi, collected by Dr Hooker. Hookers J Botany Kew Gard Miscellany 3:167–172

Berkeley MJ (1851d) Decade XXXVI. Sikkim-Himalayan Fungi, collected by Dr Hooker. Hookers J Bot Kew Gard Miscellany 3:200–206

Berkeley MJ (1852a) Decades XXXVII, XXXVIII. Sikkim and Khassya fungi. Hookers J Botany Kew Gard Miscellany 4:97–107

Berkeley MJ (1852b) Decades XXXIX, XL. Sikkim and Khassya fungi. Hookers J Bot Kew Gard Miscellany 4:130–142

Berkeley MJ (1854a) Decades XLI, XLIII. Indian fungi. Hookers J Bot Kew Gard Miscellany 6:129–143

Berkeley MJ (1854b) Decades XLIV, XLVI. Indian fungi. Hookers J Bot Kew Gard Miscellany 6:161–174

Berkeley MJ (1854c) Decades XLVII, XLVIII. Indian fungi. Hookers J Bot Kew Gard Miscellany 6:204–212

Berkeley MJ (1854d) Decades XLIX, L. Indian fungi. Hookers J Bot Kew Gard Miscellany 6:225–235

Berkeley MJ (1856) Decades of fungi. Decas 1-62, nos. 1-620. Hookers Lond J Bot 3–8:1844–1856

Berkeley MJ (1867) Fungi of the plains of India. Intellect Observer 12:18–21

Bernicchia A (2005) Polyporaceae s.l., vol 10, Fungi europaei. Candusso, Alassio

Bernicchia A, Gorjon SP (2010) Corticiaceae s.l, vol 12, Fungi europaei. Edizioni Candusso, Alassio, pp 1–1008

Biggs R (1937) The species concept in *Corticium coronilla*. Mycologia 29:686–706

Biggs R (1938) Cultural studies in Thelephoraceae and related fungi. Mycologia 30:64–78

Binder M, Hibbett DS, Larsson KH, Larsson E, Langr E, Langr G (2005) The phylogenetic distribution of resupinate forms cross the major clads of mushroom-forming fungi (Homobasidiomycetes). Syst Biodivers 3:113–157

Blackwell M, Gilbertson RL (1985) A new species of *Inonotus* (Aphyllophorales, Hymenochaetaceae) on oak in Louisiana. Mycotaxon 23:285–290

Blumenfeld SN, Wright JE (1984) A new pileate species of *Junghuhnia* (Polyporaceae). Mycotaxon 19:471–478

Boidin J (1950a) Les reactifs sulfo-aldehydiques; leur interet pour la determination et la classification des Thelephoracees. Bull Soc Nat Oyonnax 5:72–79

Boidin J (1950b) Sur l' existence de races intersteriles chez *Gloeocystidium tenue* (Pat.). Bull Soc Mycol Fr 66:205–221

Boidin J (1950c) Sur la comportement nucleaire du carpophore et du mycelium en culture chez les *Peniophora* Cooke. CR Acad Sci 230:2328–2330

Boidin J (1950d) Sur la cytologie des *Stereum* Fr., en culture pure. CR Acad Sci 230:1096–1098

Boidin J (1950e) Sur la reaction a l'iode de la paroi sporique chez lez Thelephoracees et quelques genres voisins. Bull Soc Linn Lyon 19:133–135

Boidin J (1950f) Sur la reaction a l'iode de la paroi sporique des *Gloeocystidium* Karst., et des *Stereum* Fr. CR Acad Sci 230:461–462

Boidin J (1951) Disposition hemi-chiastobasidiee chez quelques Thelephoracees. CR Acad Sci 233:1667–1669

Boidin J (1953) Comportement nucleaire des spores, germinations et myceliums de quelques *Corticium* et *Gloeocystidiellum*. CR Acad Sci 236:2333–2335

Boidin J (1956a) Polarite dite Sexuelle et systematique chez les Basidiomycetes Thelephoracees. Rev Mycol 21:121–131

Boidin J (1956b) *Stereum ahmadi* sp. nov. (Basidiomycetes, Thelephoraceae). Biologia 2:217–226

Boidin J (1981) Nouvelles especes de Lachnocladiaceae du Canada (Basidiomycetes). Naturaliste Can 108:199–203

Boidin J, Gilles G (1982) Basidiomycetes Botryohypochnoideae du Gabon. Mycotaxon 14:280–304

Boidin J, Lanquetin P (1973a) *Podoscypha involuta* (Klotzsch) Imaz. est une espece composite (Basidiomycetes, Podoscyphaceae). Persoonia 7:141–150

Boidin J, Lanquetin P (1973b) *Vararia (Dichostereum) ramulosa* Nouvelle Espece Africaine (Basidiomycetes, Lachnocladiaceae). Bull Soc Linn Lyon 42:164–166

Boidin J, Lanquetin P (1975) *Vararia* subgenus *Vararia* (Basidiomycetes: Lachnocladiaceae). Etude speciale des especies d'Afrique intertropicale. Bull Soc Mycol Fr 91:457–513

Boidin J, Lanquetin P (1976) *Scytinostroma albocinctum* (Berk et Br) et *Phaeosarcum* nov. sp. (Basidiomycetes, Lachnocladiaceae). Kew Bull 31:621–628

Boidin J, Lanquetin P (1977a) Les Genres *Dichostereum* et *Vararia* en Guadeloupe (Basidiomycetes: Lachnocladiaceae). Mycotaxon 6:277–336

Boidin J, Lanquetin P (1977b) *Peniophora* (subg. *Duportella) malenconii* nov. sp. (Basidiomycetes, Corticiaceae), espece mediterraneenne partiellement intersterile avec son vicariant californien. Rev Mycol 41:119–128

Boidin J, Lanquetin P (1980) Contribution A Letude du Genre *Dichostereum* Pilat (Basidiomycetes: Lachnocladiaceae). Bull Soc Mycol 96:381–406

Bondarstev A, Singer R (1941) Zur Systematik der Polyporaceen. Ann Mycol 39:43–65

Bondartsav (1953) The Polyporaceae of European part of USSR and Caucasus. Leningrad

Bose SR (1918) Descriptions of the fungi in Bengal: I. In: Proceedings of the Indian Association for the Cultivation of Science, Jadavpur, Kolkata, pp 109–114

Bose SR (1920a) Descriptions of the fungi in Bengal: II. In: Proceedings of the Indian Association for the Cultivation of Science, Jadavpur, Kolkata, pp 136–143

Bose SR (1920b) Fungi of Bengal. Polyporaceae of Bengal-III. Bull Carmichael Med Coll 1:1–5

Bose SR (1921a) Polyporaceae of Bengal- IV. Bull Carmichael Med Coll 2:1–5

Bose SR (1921b) One new species of Polyporaceae and some Polypores new to Bengal. Ann Mycol 19:129–131

Bose SR (1921c) Two new species of Polyporaceae. J Indian Bot 2:300–301

Bose SR (1922a) Polyporaceae of Bengal- V. Bull Carmichael Med Coll 3:20–25

Bose SR (1922b) Une Polyporaceae nouvelle du Bengala. Bull Soc Mycol Fr 38:173

Bose SR (1922) Polyporaceae of Bengal: Polyporaceae of Bengal VI. In: Proceedings of the Indian Association for the Cultivation of Science, Jadavpur, Kolkata, p 55–62

Bose SR (1923a) Polyporaceae of Bengal-VII. In: Proceedings of the Indian Association for the Cultivation of Science for the year 1920-21, Jadavpur, Kolkata, pp 27–36

Bose SR (1923b) Une Polyporaceae nouvelle d'Inde. Bull Soc Mycol Fr 39:226

Bose SR (1924) Les Polyporacees du Bengale. Rev Path Veg Ent Agric Fr 11:134–149

Bose SR (1925) A new species of Polyporaceae from Bengal. Ann Mycol 23:179–181

Bose SR (1928a) Polyporaceae of Bengal-VIII. J Dep Sci Calcutta Univ 9:27–34

Bose SR (1928b) Polyporaceae of Bengal-IX. J Dep Sci Calcutta Univ 9:35–44

Bose SR (1934) Polyporaceae of Bengal-X. J Dep Sci Calcutta Univ 11:1–18

Bose SR (1937) Polyporaceae of Lokra Hills (Assam). Ann Mycol 35(2):119–137

Bose SR (1944) Importance of anatomy in the systematics of Polyporaceae. J Indian Bot Soc 23:153–157

Bose SR (1946) Polyporaceae of Bengal-XI. J Dep Sci Calcutta Univ 2:53–87

Bourdot H, Galzin A (1928) Hymenomycetes de France. Sceaux, Paris

Breitenbach J, Kränzlin F (1986) Fungi of Switzerland, vol II, Non-gilled fungi. Verlag Mykologia, Lucerne

Bresadola G (1881–1900) Fungi Tridentini novi nondum delineati, descripti et iconibus illustrati 1:1–114; 2:1–118
Bresadola G (1903) Fungi Polonici a cl. Viro B. Eichler lecti. Ann Mycol 1:65–121
Bresadola G (1916) Synonymia et adnotanda mycologica. Ann Mycol 21:221–242
Bresadola G (1920a) Selecta mycologica I. Diagnoses specierum novarum. Ann Mycol 18:26–58
Bresadola G (1920b) Synonymia et adnotanda mycologica. Ann Mycol 18:58–70
Budington AB, Gilbertson RL (1973) Some south western lignicolous Hymenomycetes of special interest. Southwest Nat 17:409–422
Burdsall HH, Gilbertson RL (1974) A new species of *Platygloea* occurring on *Peniophora tamaricicola* in Arizona. Mycologia 66:702–706
Burt EA (1914a) The Thelephoraceae of North America I. *Thelephora*. Ann Mo Bot Gard 1:185–228
Burt EA (1914b) The Thelephoraceae of North America II. *Craterellus*. Ann Mo Bot Gard 1:327–350
Burt EA (1914c) The Thelephoraceae of North America III. *Craterellus borealis and Cyphella*. Ann Mo Bot Gard 1:357–382
Burt EA (1915a) The Thelephoraceae of North America IV. *Exobasidium*. Ann Mo Bot Gard 2:627–658
Burt EA (1915b) The Thelephoraceae of North America V. *Tremellodendron, Eichleriella* and *Sebacina*. Ann Mo Bot Gard 2:731–770
Burt EA (1916a) The Thelephoraceae of North America VI. *Hypochnus*. Ann Mo Bot Gard 3:203–421
Burt EA (1916b) The Thelephoraceae of North America VII. *Septobasidium*. Ann Mo Bot Gard 3:319–343
Burt EA (1917a) The Thelephoraceae of North America VIII. *Coniophora*. Ann Mo Bot Gard 4:237–269
Burt EA (1917b) *Merulius* in North America. Ann Mo Bot Gard 4:305–362
Burt EA (1918a) The Thelephoraceae of North America IX. *Aleurodiscus*. Ann Mo Bot Gard 5:177–203
Burt EA (1918b) The Thelephoraceae of North America X. *Hymenochaete*. Ann Mo Bot Gard 5:301–372
Burt EA (1918c) *Corticiums* causing Pellicularia disease of the coffee plant, Hypochnose of Pomaceous fruits, and *Rhizoctonia* disease. Ann Mo Bot Gard 5:119–132
Burt EA (1919a) *Merulius* in North America. Supplementary notes. Ann Mo Bot Gard 6:143–145
Burt EA (1919b) The Thelephoraceae of North America XI. *Tulasnella, Veluticeps, Mycobonia, Epithele* and *Lachnocladium*. Ann Mo Bot Gard 6:253–280
Burt EA (1920) The Thelephoraceae of North America XII. *Stereum*. Ann Mo Bot Gard 7:81–248
Burt EA (1921) Some North American *Tremellaceae*, *Dacryomycetaceae* and *Auriculariaceae*. Ann Mo Bot Gard 8:361–396
Burt EA (1922) The North American species of *Clavaria* with illustrations of the type specimens. Ann Mo Bot Gard 9:1–78
Burt EA (1924) The Thelephoraceae of North America. Ann Mo Bot Gard 11:1–35
Burt EA (1925) The Thelephoraceae of North America XIV. *Peniophora*. Ann Mo Bot Gard 12(3):213–357
Burt EA (1926) The Thelephoraceae of North America XV. *Corticium*. Ann Mo Bot Gard 13(3):173–354
Butler EJ, Bisby GR (1931) Fungi of India, Scientific monograph no I. ICAR, Delhi
Canfield ER, Gilbertson RL (1971) Notes on the genus *Albatrellus* in Arizona. Mycologia 63:964–971
Carranza J (1982) Polypores new to Costa Rica. Mycotaxon 15:405–408
Christiansen MP (1960) Danish resupinate fungi. II. Homobasidiomycetes. Dan Bot Ark 19:57–383

Clements FE, Shear CL (1931) The genera of fungi. Hafner Publishing Co, New York
Cooke MC (1879) On *Peniophora*. Grevillea 8:17–21
Cooke MC (1880) Fungi of India. Grevillea 8:93–96
Cooke MC (1884) Some exotic fungi. Grevillea 12:85
Cooke MC (1885) Some exotic fungi. Grevillea 14:11–14
Cooke MC (1886a) Some exotic fungi. Grevillea 14:89–90
Cooke MC (1886b) Some exotic fungi. Grevillea 14:129–130
Cooke MC (1886c) Some exotic fungi. Grevillea 15:16–18
Cooke MC (1891a) Addition to *Daedalea*. Grevillea 19:71–75
Cooke MC (1891b) *Trametes* and its allies. Grevillea 19:98–103
Cooke WB (1940) A nomenclatorial survey of the genera of the pore fungi. Llyodia 3:81
Cooke WB (1953) The genera of the Homobasidiomycetes (exclusive of Gasteromycetes). Spec Publ Division of Mycology and Disease Survey, U S Department of Agriculture, Beltsville
Cooke WB (1959) Genera of pore fungi. Llyodia 22:163–207
Corner EJH (1932a) The identification of the brown-root fungus. Garden Bull Straits Settl 5:317–350
Corner EJH (1932b) A *Fomes* with two systems of hyphae. Trans Br Mycol Soc 17:51–81
Corner EJH (1932c) The fruitbody of *Polystictus xanthopus* Fr. Ann Bot (Lond) 46:71–111
Corner EJH (1948) *Asterodon*, a clue to the morphology of the fungus fruit body, with notes on *Asterostroma* and *Asterostromella*. Trans Br Mycol Soc 31:234–245
Corner EJH (1953) The construction of polypores. Phytomorphology 3:152–167
Corner EJH (1968) A monograph of *Thelephora* (Basidiomycetes). Nova Hedwigia Beih, J Cramer, Gebr Borntraeger, Stuttgart
Cunningham GH (1927) The Polyporaceae of New Zealand. Trans N Z Inst 58:202–250
Cunningham GH (1946) Notes on classification of the Polyporaceae. N Z J Sci Technol 28(4):238–251
Cunningham GH (1947) New Zealand Polyporaceae. 1. The genus *Poria*. Bull N Z Dep Sci Ind Res 72:43
Cunningham GH (1948a) New Zealand Polyporaceae. 2. The genus *Fuscosporia*. Bull N Z Dep Sci Ind Res 73:14
Cunningham GH (1948b) New Zealand Polyporaceae. 4. The genus *Coriolus*. Bull N Z Dep Sci Ind Res 75:10
Cunningham GH (1948c) New Zealand Polyporaceae. 5. The genus *Fomitopsis*. Bull N Z Dep Sci Ind Res 76:8
Cunningham GH (1949a) New Zealand Polyporaceae. 10. Revision of New Zealand species and records. Bull N Z Dep Sci Ind Res 81:24
Cunningham GH (1949b) New Zealand Polyporaceae. 11. The genus Irpex. Bull N Z Dep Sci Ind Res 82(8)
Cunningham GH (1950) New Zealand Polyporaceae. 12. The genus Merulius. Bull N Z Dep Sci Ind Res 83:12
Cunningham GH (1953–1957) Thelephoraceae of New Zealand I-XV. Trans Roy Soc N Z 81:165–188, 321–328; 82:271–327, 973–985; 83:241–245, 247–293, 621–636; 84: 201–231, 231–237, 237–268, 479–489; 85: 1–51, 91–99
Cunningham GH (1965) Polyporaceae of New Zealand. N Z Dep Sci Ind Res Bull No 164:1–304
Dhingra GS (2006) *Trimitiella* gen. nov. (Basidiomycetes) from Eastern Himalayas, India. Mycotaxon 97:125–128
Dhingra GS, Priyanka (2011) *Hallenbergia* (Agaricomycetes), a new corticioid genus. Mycotaxon 118:289–292
Domanski S (1965) Flora Polska. Grzyby (Fungi) 2: Podstawczaki (Basidiomycetes), Bezblaszkowe (Aphyllophorales), Żagwiowate I (Polyporaceae I), Szczeciniakowate I (Mucronoporaceae I). PWN, Warszawa
Domanski S (1969a) Wood-inhabiting fungi in Bialowieza virgin forests in Poland. VIII. *Schizopora phellinoids* (Pilat) comb. nov. and its diagnosis. Acta Soc Bot Pol 38(2):255–269

Domanski S (1969b) Wood-inhabiting fungi in Bialowieza virgin forests in Poland. X. *Filuloporia subvermispora* (Pilat) comb. nov. and its diagnosis. Acta Soc Bot Pol 38(3):453–464
Domanski S (1969c) Wood-inhabiting fungi in Bialowieza virgin forests in Poland. XI. *Incrustoporia tschulymica* (Pilat) comb. nov. and its diagnosis. Acta Soc Bot Pol 38(3):465–473
Domanski S (1970a) Wood-inhabiting fungi in Bialowieza virgin forests in Poland. XIII. *Diplomitoporus* Dom gen. nov. Acta Soc Bot Pol 39:191–207
Domanski S (1970b) Wood-inhabiting fungi in Bialowieza virgin forests in Poland. XIV Coriolus hoehnelu (Bres. In Hohn.) Bourd. and Galz. Acta Soc Bot Pol 39(3):521–530
Domanski S (1970c) Wood-inhabiting fungi in Bialowieza virgin forests in Poland. XV. Acta Soc Bot Pol 38(3):531–538
Domanski S (1970d) Wood-inhabiting fungi in Bialowieza virgin forests in Poland. XVI Coriolus foliaceo-dentatus (Nikol) Dom comb. nov. Acta Soc Bot Pol 39(4):701–709
Domanski S (1973) *Poria elongatea* Overh in Poland. Persoonia 7(2):155–160
Domanski S (1974) Etude des specimens de Trametes serpens (Fr ex Fr) de l'herbier de Karsten. In: Travaux mycologiques dédiés à R. Kühner. Société linnéenne de Lyon, pp 147–152
Donk MA (1931) Revisie van de Nederlands *Heterobasidiomycetae* en *Homobasidiomycetae-Aphyllophoraceae* I. Meded Ned Mycol Ver 18–20:67–200
Donk MA (1933) Revisie van de Nederlandse *Heterobasidiomyceteae* (uitgez. *Uredinales* en *Ustilaginales*) en *Homobasidiomyceteae–Aphyllophraceae*: II. Mededelingen van het botanisch Museum en Herbarium van de Rijksuniversiteit Utrecht 9:1–278
Donk MA (1949) New and revised nomina generica conservation proposed for Basidiomycetes (Fungi). Bull Bot Gdns Buitenz 18(1):83–168
Donk MA (1954) Notes on resupinate Hymenomycetes. I. On *Pellicularia Cooke*. Reinwardtia 2:425–434
Donk MA (1956) Notes on resupinate Hymenomycetes-III. Fungus 26:3–24
Donk MA (1957a) Notes on resupinate Hymenomycetes-IV. Fungus 27:1–29
Donk MA (1957b) Typification and later starting-points. Taxon 6:245–256
Donk MA (1958) Notes on resupinate Hymenomycetes-V. Fungus 28:16–36
Donk MA (1960) The generic names proposed for polyporaceae. Persoonia 1:173–302
Donk MA (1962) Notes on resupinate Hymenomycetes-VI. Persoonia 2:217–238
Donk MA (1964) A conspectus of the families of Aphyllophorales. Fungus 3:199–324
Donk MA (1966) Notes on European polypores-I. Persoonia 4(3):337–343
Donk MA (1967) Notes on European polypores-II. Notes on *Poria*. Persoonia 5(1):47–130
Donk MA (1969a) Notes on European polypores-III. Notes on species with stalked fruit body. Persoonia 5(3):237–263
Donk MA (1969b) Notes on European polypores-IV. On some species of *Ganoderma*. Kon Neth Acad Wetensch Proc Ser C Biol Med Sci 72(3):273–282
Donk MA (1969c) Notes on European polypores-V. On the typification of *Hexagonia pollini* per Fr. Taxon 18(6):663–666
Donk MA (1971a) Notes on European polypores-VI A. Kon Neth Acad Wetensch Proc Ser C Biol Med Sci 74(1):1–14
Donk MA (1971b) Notes on European polypores-VI B. Kon Neth Acad Wetensch Proc Ser C Biol Med Sci 74(1):15–24
Donk MA (1971c) Notes on European polypores-VII. Kon Neth Acad Wetensch Proc Ser C Biol Med Sci 74(1):25–41
Donk MA (1971d) Notes on European polypores-VIII. Persoonia 6(2):201–218
Donk MA (1971e) Notes on European polypores-IX. Some species of Hymenochataceae. Kon Neth Acad Wetensch Proc Ser C Biol Med Sci 74(1):405–421
Donk MA (1972a) Notes on European polypores-X. Kon Neth Acad Wetensch Proc Ser C Biol Med Sci 75(3):165–178
Donk MA (1972b) Notes on European polypores-XI. Kon Neth Acad Wetensch Proc Ser C Biol Med Sci 75(4):287–304

Donk MA (1973) Notes on European polypores-X. Kon Neth Acad Wetensch Proc Ser C Biol Med Sci 76(3):217–230
Donk MA (1974) Checklist of European polypores. North- Holland Publishing Company, Amsterdam
Donk MA (1951–63) The generic names proposed for Hymenomycetes I-IX, XII, XIII. Reprint: Verlag von J. Cramer, Lehre
Emmons CW (1927) The Thelephoraceae of Iowa. Iowa Univ Stud Nat Hist 12:49–89
Engler A, Prantl K (1928) Die Naturlichen Pflanzenfamilien, vol II. Verlag Von Wilhelm Engelmann, Leipzig
Eriksson J (1950) *Peniophora* Cke sect Coloratae Bourd. and Galz: taxonomical study with special reference to the Swedish species. Symb Bot Ups 10:1–76
Eriksson J (1958a) Studies in Corticiaceae (*Botryohypochnus* Donk, *Botryobasidium* Donk and *Gloeocystidiellum* Donk). Svensk Bot Tidskr 52:1–17
Eriksson J (1958b) Studies in Heterobasidiomycetes and Homobasidiomycetes-Aphyllophorales of Muddus National Park in North Sweden. Symb Bot Ups 14:1–172
Eriksson J (1958c) Studies of the Swedish *Heterobasidiomycetes* and *Aphyllophorales* with special regard to the family *Corticiaceae*. Institute of Systematic Botany, University of Uppsala, Sweden
Eriksson J, Hallingback T (1979) A study of the sulfocystidia variation in *Peniophora nuda* (Fr). Bres Goteborg Svampklubb Arsskrift 63–66
Eriksson J, Hjortstam K, Ryvarden L (1978) The Corticiaceae of North Europe-V. Fungiflora, Oslo, pp 890–1046
Eriksson J, Hjortstam K, Ryvarden L (1981) The Corticiaceae of North Europe-VI. Fungiflora, Oslo, pp 1051–1276
Eriksson J, Hjortstam K, Ryvarden L (1984) The Corticiaceae of North Europe-VII. Fungiflora, Oslo, pp 1281–1449
Eriksson J, Ryvarden L (1973) The Corticiaceae of North Europe-II. Fungiflora, Oslo, pp 59–286
Eriksson J, Ryvarden L (1975) The Corticiaceae of North Europe-III. Fungiflora, Oslo, pp 287–546
Eriksson J, Ryvarden L (1976) The Corticiaceae of North Europe-IV. Fungiflora, Oslo, pp 549–886
Eriksson J, Strid A (1969) Studies in the Aphyllophorales (Basidiomycetes) of Northern Finland. Ann Univ Turku 40:112–158
Eriksson J, Wall S (1979) A remarkable gall-producing fungus on *Laurus*. Böteb Swampkl 1979:1–10
Fidalgo O (1958) The Nomenclatorial status of Genus *Daedalea* Pers. Fr. and related Genera. Taxon 7:133–140
Fries EM (1821) Systema mycologicum I. Lundae, pp 1–520
Fries EM (1823) Systema mycologicum II. Lundae, pp 1–621
Fries EM (1828) Elenchus fungorum–I. Ernestus Mauritius, Greifswald
Fries EM (1832) Systema mycologicum III. Greifswald, pp 1–524
Fries EM (1836) Floram scanicam. Palmblad, Sebell and C, Uppsala
Fries EM (1838) Epicrisis systematis Mycologici. E Typographia Academica, Uppsala
Fries EM (1855) Novae symbolae mycologicae K Ventenskaps soc. Upsala, Nova Acta Reg Soc Sci, Upsaliensis III 1:17–136
Fries EM (1874) *Hymenomycetes* Europaei. Siva epicriseous systematis mycologici. E. Berling, Upsaliae
Ghobad-Nejhad M (2011) Updated checklist of corticoid and poroid basidiomycetes of the Caucasus region. Mycotaxon 117:508
Ghobad-Nejhad M, Hallenberg N, Hyvonen J, Yurchenko E (2012) The Caucasian corticioid fungi: level of endemism, similarity, and possible contribution to European fungal diversity. Fungal Divers 52:35–48

Ghobad-Nejhad M, Hallenberg N, Parmasto E, Koitranta G (2009) A first annotated checklist of Corticioid and Polypore Basidiomycetes of the Caucasus region. Mycol Balc 6:123–168

Gilbertson RL (1956) The genus *Poria* in the central Rocky Mountains and Pacific Northwest. Lloydia 19:65–85

Gilbertson RL (1962) Resupinate hydnaceous fungi of North America-I. Type studies of species described by Peck. Mycologia 54:658–677

Gilbertson RL (1963) Resupinate hydnaceous fungi of North America-I. Type studies of species described by Bresadola, Overholts and Lloyd. Mich Acad Sci Arts Lett Pap 48:137–149

Gilbertson RL (1964) Resupinate hydnaceous fungi of North America-III. Additional type studies. Mich Acad Sci Arts Lett Pap 49:15–25

Gilbertson RL (1965a) Resupinate hydnaceous fungi of North America- V. Type studies of species described by Berkeley and Curtis. Mycologia 57:845–871

Gilbertson RL (1965b) Some species of *Vararia* from temperate North America. Mich Acad Sci Arts Lett Pap 50:161–184

Gilbertson RL (1970) A new *Vararia* from Western North America. Madrono 20:282–287

Gilbertson RL (1973) Notes on some corticioid lignicolous fungi associated with snowbanks in southern Arizona. Persoonia 7:171–182

Gilbertson RL (1976) The genus *Inonotus* (Aphyllophorales: Hymenochetaceae) in Arizona. Mem N Y Bot Gard 28:67–85

Gilbertson RL (1979) The genus *Phellinus* (Aphyllophorales: Hymenochetaceae) in western North America. Mycotaxon 9:51–89

Gilbertson RL (1981) North American woodrotting fungi that cause brown rot. Mycotaxon 12:372–417

Gilbertson RL, Burdsall HH Jr (1975) *Peniophora tamaricicola* in North America. Mycotaxon 2:143–150

Gillet CC (1876–1878) Les Hyménomycètes ou Description de tous les Champignons qui Croissent en France. Ch. Thomas, Alençon

Ginns J (1976) *Merulius* s. s. and s. l., taxonomic disposition and identification of species. Can J Bot 54:100–167

Ginns J (1978) *Leucogyrophana* (Aphyllophorales): identification of species. Can J Bot 56:1953–1973

Ginns J (1979a) *Henningsia* (Polyporaceae) and a description of type species. Mycologia 7:305–309

Ginns J (1979b) The genus *Ramaricium* (Gomphaceae). Bot Notiser 132:93–102

Ginns J (1980) The genus *Flaviporus* Murrill (Polyporaceae). Can J Bot 58:1578–1590

Ginns J (1984a) *Mollicarpus* gen. nov. (Polyporaceae) with notes on *Coriolopsis byrsina, Phellinus crocatus* and *Polystictus crocatus* var. *Sibiricus*. Mycotaxon 19:71–80

Ginns J (1984b) *Griseoporia* a new genus for *Hexagonia carbonaria* (Polyporaceae). Mycotaxon 20(2):557–565

Ginns J, Kokko E (1976) Basidiospore germ pore and wall structure in *Coniophora* (Basidiomycetes: Aphyllophorales). Can J Bot 54:399–401

Gomez LD, Ryvarden L (1985) *Inonotus fimbriatus* nov. sp. (Hymenochaetaceae, Basidiomycetes). Mycotaxon 23:291–292

Gorjan SP (2012) Some species of *Hypodontia* s.l. with encrusted cystidial elements. Mycosphere 3(4):464–474

Gorjan SP, Hallenberg N (2008) New records of *Sistotrema* species (Basidiomycota) from the Iberian Peninsula. Sydowia 60:205–212

Hallenberg N (1978) Wood-fungi (Corticiaceae, Coniophoraceae, Lachnocladiaceae, Thelephoraceae) in North Iran. Iran J Plant Pathol 14:38–87

Hallenberg N (1979) Wood fungi (Polyporaceae, Ganodermataceae, Hymenochaetaceae, Cyphellaceae, Clavariaceae, Auriculariaceae, Tremellaceae, Dacrymycetaceae) in N Iran II. Iran J Plant Pathol 15:11–31

Hallenberg N (1980) New taxa of Corticiaceae from North Iran (Basidiomycetes). Mycotaxon 11:447–475

Hallenberg N (1981a) Basidium repetition in Conferticium (Corticeaceae, Basidiomycetes). Mycotaxon 12:468–472
Hallenberg N (1981b) *Phlebia centrifuga* Karst. (Corticeaceae, Basidiomycetes) – compatibility between specimens from Sweden and Canada. Goteborgs Svampklubbs Arsskrift 33–37
Hallenberg N (1981c) Synopsis of wood inhabiting Aphyllophorales of Africa 5: *Cystostereum artocreas*, new to Africa. Mycotaxon 2:135–141
Hallenberg N (1981d) Synopsis of Wood-inhabiting Aphyllophorales (Basidiomycetes) and Heterobasidiomycetes from N. Iran. Mycotaxon 12(2):473–502
Hallenberg N (1983a) Cultural studies in *Hypochnicium* (Corticiaceae, Basidiomycetes). Mycotaxon 16:565–571
Hallenberg N (1983b) *Hericium coralloides* and *H. alpestre* (Basidiomycetes) in Europe. Mycotaxon 18:181–189
Hallenberg N (1983c) On the *Schizopora paradoxa* complex (Basidiomycetes). Mycotaxon 18:303–313
Hallenberg N (1984a) A taxonomic analysis of the *Sistotrema brinkmanni* complex (Corticiaceae, Basidiomycetes). Mycotaxon 21:389–411
Hallenberg N (1984b) Compatibility between species of Corticiaceae s.l. (Basidiomycetes) from Europe and North America. Mycotaxon 21:335–388
Hallenberg N (1985) On the *Hypochnicium eichleri* complex (Basidiomycetes). Mycotaxon 24:431–436
Hallenberg N (1986) Cultural studies in *Tubulicrinis* and *Xenasmatella* (Corticiaceae, Basidiomycetes). Mycotaxon 27:361–375
Hallenberg N (1988a) Pairing tests with species of Aphyllophorales (Basidiomycetes) from two Phytogeographically isolated areas. Mycotaxon 42:355–386
Hallenberg N (1988b) Species delimitation in Corticiaeae (Basidiomycetes). Mycotaxon 31:445–465
Hallenberg N (1996) Four new species of Corticioid Fungi (Basidiomycotina, Aphyllophorales) from Argentina. Mycotaxon 57:117–123
Hallenberg N (1998a) Evolutionary processes on species level in wood inhabiting Basidiomycetes. Folia Cryptog Estonica Fasc 33:35–40
Hallenberg N (1998b) Phylogenetic studies in species of Coticiaceae growing on branches. Mycologia 90:640–654
Hallenberg N, Küffer N (2001) Long-distance spore dispersal in wood-inhabiting Basidiomycetes. Nord J Bot 21:1–6
Hallenberg N, Larsson E (1993) On taxonomy of *Phlebia livida*. Mycol Res 97:351–354
Hallenberg N, Nilsson RH, Antonelli A, S-H WU, Maekawa N, Norden B (2007) The *Peniophorella praetermissa* species complex (Basidiomycota). Mycol Res 111(12):1366–1376
Hallenberg N, Ryberg M, Nilsson RH (2008) *Pseudolagarobasidium* (Basidiomycota): on the reinstatement of a genus of parasitic, saprophytic and endophytic resupinate fungi. Botany 86(11):1319–1325
Hallenberg N, Ryvarden L (1975) Studies in the Aphyllophorales of Africa 5. *Cystostereum artocreas*, new to Africa. Mycotaxon 2:135–141
Hallenberg N, Yurchenko E, Ghobad-Nejhad M (2010) *Peniophora pseudonuda* is a synonym to *P. laeta*. Mycotaxon 112:153–162
Hennings P (1897) *Hymenomycetineae*. In: Engler A, Prantl K (ed) Die Naturl Pflanzenfam, pp 105–276
Hennings P (1900) Fungi Indiae Orientalis I. Hedwigia 39:150–153
Hennings P (1901) Fungi Indiae Orientalis II. Hedwigia 40:323–342
Hill J (1751) A general natural history 2. The history of plants. Osborne, London
Hjortstam K (1969) Studies in the Swedish species of the genus *Tomentella* (Thelephoraceae) I. Sven Bot Tidskr 63:491–495
Hjortstam K (1970) Studies in the Swedish species of the genus *Tomentella* (Thelephoraceae) II. Sven Bot Tidskr 64:421–428

Hjortstam K (1971) The genus *Paullicorticium* in Sweden (Basidiomycetes). Goteb Svampklubb Arsskrift 1971:6–12

Hjortstam K (1973) Studies in the Corticeaceae (Basidiomycetes) and related fungi of Vastergotland in South-West Sweden I. Sven Bot Tidskr 67:97–126

Hjortstam K (1978a) Enumeration of Corticiaceae and related fungi of Västergötland in SW Sweden. Göteb Svampklubb Årsskrift 1977–1978:17–26

Hjortstam K (1978b) Vedbeboende svampar fran Raback pa Kinnekulle. Sven Bot Tidskr 72:321–326

Hjortstam K (1980) Corticiaceae (Basidiomycetes)-VII. A synopsis of the genus *Amylocorticium Pouz*. Mycotaxon 11:430–434

Hjortstam K (1981a) Notes on Corticiaceae (Basidiomycetes)-VIII. Two new species of *Tubulicrini*. Mycotaxon 13:120–123

Hjortstam K (1981b) Notes on Corticiaceae (Basidiomycetes) IX. Three new combinations in *Hypochniciellum*. Mycotaxon 13:124–126

Hjortstam K, Larsson KH, Ryvarden L (1987) The Corticiaceae of North Europe, vol 1, Phlebiella, Thanatephorus–Ypsilonidium. Fungiflora, Oslo, pp 1–59

Hjortstam K, Larsson KH, Ryvarden L (1988) The Corticiaceae of North Europe, vol 7, Phlebiella, Thanatephorus–Ypsilonidium. Fungiflora, Oslo

Hjortstam K, Ryvarden L (1979a) Notes on Corticiaceae (Basidiomycetes) IV. Mycotaxon 9:505–519

Hjortstam K, Ryvarden L (1979b) Notes on Corticiaceae (Basidiomycetes) V. Mycotaxon 10:201–209

Hjortstam K, Ryvarden L (1980a) Studies in tropical Corticiaceae (Basidiomycetes) -I. Mycotaxon 10:269–287

Hjortstam K, Ryvarden L (1980b) Studies in tropical Corticiaceae (Basidiomycetes) -II. Mycotaxon 12:168–184

Hjortstam K, Ryvarden L (1981) Studies in tropical Corticiaceae (Basidiomycetes) -III. Two new species of *Laxitextum*. Mycotaxon 13:35–40

Hjortstam K, Ryvarden L (1982) Aphllophorales from northern Thailand. Nord J Bot 2(3):273–281

Hjortstam K, Stalpers JA (1982) Notes on Corticiaceae (Basidiomycetes) -XI. *Boidinia*, a new genus segregated from *Gloeocystidiellum*. Mycotaxon 14:75–81

Hoehnel F, Litschauer V (1906) Beitrag zur kenntnis der Corticieen I. Sitzber Akad Wiss Wien Math Nat Klasse 115:1549–1620

Hoehnel F, Litschauer V (1907) Beitrag zur kenntnis der Corticieen II. Sitzber Akad Wiss Wien Math Nat Klasse 116:739–852

Hoehnel F, Litschauer V (1908) Beitrag zur kenntnis der Corticieen III. Sitzber Akad. Wiss Wien Math Nat Klasse 117:1081–1124

Humphrey CJ, Leus S (1931) A partial revision of *Ganoderma applantum* group, with particular reference to its oriental variants. Phillipp J Sci 45:483–589

Humphrey CJ, Leus S (1932) Studies and illustrations in the Polyphoraceae III. Supplementary notes on the *Ganoderma applantum* group. Phillipp J Sci 49:159–184

Imazeki R (1939) Observations on Japanese fungi-III. Some hard and perennial Stereums in Japan. J Jpn Bot 15:578–588

Imazeki R (1943) The genera Polyporaceae of Nippon. Bull Tokyo Sci Mus 6:1–111

Jackson HS (1948a) Studies of Canadian Thelephoraceae I. Some new species of *Peniophora*. Can J Res 26:128–139

Jackson HS (1948b) Studies of Canadian Thelephoraceae II. Some new species of *Corticium*. Can J Res 26:143–157

Jackson HS (1949) Studies of Canadian Thelephoraceae IV. *Corticium anceps* in North America. Can J Res 27:241–252

Jackson HS (1950a) Studies of Canadian Thelephoraceae V. Two new species of *Aleurodiscus* of conifers. Can J Res 28:63–77

Jackson HS (1950b) Studies of Canadian Thelephoraceae VI. The *Peniophora rimicola* group. Can J Res 28:525–534

Jackson HS (1950c) Studies of Canadian Thelephoraceae VII. Some new species of *Corticium* Section Athelia. Can J Res 28:716–725

Johansen I, Ryvarden L (1979) Studies in the Aphyllorales of Africa VII. Some new genera and species in the Polyporaceae. Trans Br Mycol Soc 92(2):189–199

Jülich W (1972) Monographie der Athelieae (Corticiaceae: Basidiomycetes). Willdenowia 7:1–283

Jülich W (1973a) Monographie der Athelieae II. Persoonia 7:381–388

Jülich W (1973b) Studies in resupinate Basidiomycetes I. Willdenowia 7:3–8

Jülich W (1974) The genera of the Hyphodermoideae (Corticiaceae). Persoonia 8:59–97

Jülich W (1976) Studies in resupinate Basidiomycetes IV. Persoonia 8:431–442

Karsten PA (1876) Mycologia Fennica Pars. Tertia (Basidiomycetes). Bidr Kann Nat Folk Helsingfors 25:1–377

Karsten PA (1879) Symbolae ad mycologiam Fennicam VI. Meddel Soc Fauna Fl Fenn 5:15–46

Karsten PA (1881) Hymenomycetes finnici enumerati. Acta Soc Fauna Fl Fenn 21:1–40

Karsten PA (1882) Enumeratio Fungorum et Myxomycetum in Lapponia orientaliaestate. Not Fauna Fl Fenn 8:193–224

Karsten PA (1883) Symbolae ad Mycologiam fennicam. Medd Soc Fauna Fl Fenn 9:57–67

Karsten PA (1887) Fungi novi vel minus bene cogniti Fenniae et Galliae. Rev Mycol Toulouse 9:9–11

Karsten PA (1889) Kritisk öfversigt af Finlands Basidsvampar (*Basidiomycetes*, *Gastero*–and *Hymenomycetes*). Bidrag till Kännedom of Finlands Natur Folk 48:1–470

Kauserud H, Svegarden IB, Seatre G-P, Knudsen H, Stensrud O, Schimdt O, Doi S, Sugiyama T, Hogberg N (2007) Asian origin and rapid global spread of the destructive dry rot fungus Serpula lacrymans. Mol Ecol 16:3350–3360

Kickx JJ (1867) Flora cryptogamique des Flandres, vol 1. Bailliere, Paris, pp 1–521

Killerman S (1928) Hymenomyceteae. In: Engler A, Prantl K (eds) Die Naturlischen Pflazenfamilien, 2nd edn, vol 6. pp 99–283

Klotzsch JF (1832) Mycologishe Berichtigungen. Linnaea 7:193–204

Klotzsch JF (1833) Fungi exotici e collectionibus britan-norum. Linnaea 8:478–490

Kotlaba F, Pouzar Z (1957a) Poznamky y trideni europskych chorosu (Notes on the classification of European pore fungi). Ceska Mykol 11:152–170

Kotlaba F, Pouzar Z (1957b) Nove nebo malo zname chorose pro Ceskoslovensko II. Ces Mykol 11(4):214–224

Kotlaba F, Pouzar Z (1958) Nove nebo malo zname chorose pro Ceskoslovensko III. Polypori novi vel minus cogniti Czechoslovakiae III. Ces Mykol 12:95–104

Kunze G (1830) in Fries E. Ecologae fungorum. Linnaea 5: 512

Larsen MJ (1974) A contribution to the taxonomy of the genus *Tomentella*. Mycologia Memoir 4:1–145

Larsen MJ, Cobb-Poulle LA (1992) *Phellinus* (Hymenochaetales) a survey of the world taxa. Synop Fung 3:1–206

Larsen MJ, Lombard FF, Hodges CS Jr (1985) Hawaiian forest fungi V. A new species of *Phellinus* (Hymenochaetaceae) causing decay of *Casuarina* and *Acacia*. Mycologia 77:345–352

Larsson E, Larsson KH (2003) Phylogenetic relationship of Russuloid basidiomycetes with emphasis on aphyllophoralean taxa. Mycologia 95:1037–1065

Larsson KH, Hjortstam K (1974) *Luellia*, a new genus in the Corticiaceae (Basidiomycetes). Sven Bot Tidskr 68:57–63

Lazzari G (1980) Glossario Micologico in cinque lingue. Gruppo micologico G Bresadola, Trento

Leger JC (1980) *Hymenochaete spathulata* nov sp (Basidiomycetes, aphyllophorales). Bull Soc Mycol Fr 96:407–411

Lemke PA (1964a) The genus *Aleurodiscus* (sensu stricto) in North America. Can J Bot 42:213–282

Lemke PA (1964b) The genus *Aleurodiscus* (sensu lato) in North America. Can J Bot 42(6):723–768

Lemke PA (1965) *Dendrothele* (1907) vs *Aleurodiscus* (1963). Persoonia 3:365–367

Lentz PL (1954) Modified hyphae of hymenomycetes. Bot Rev 20:135–199

Lentz PL (1960) Taxonomy of *Stereum* and allied genera. Sydowia 14:116–135

Léveillé JH (1845) Champignons exotiques. Ann Sci Nat III Ser 3:38–71

Léveillé JH (1846) Description des Champignons de I and herbier du Museum de Paris. Ann Sci Nat 3 Ser 5:111–167, 249–304

Liberta AE (1962) The genus *Paullicorticium* (Thelephoraceae). Brittonia 14:1219–1223

Liberta AE (1966) On *Trechispora*. Taxon 15:317–319

Liberta AE (1968) Description of the nomenclatural types of *Peniophora* described by Burt. Mycologia 60:827–857

Liberta AE (1969) Description of the nomenclatural types of *Corticium* described by Burt. Nova Hedwig 18:215–233

Liberta AE (1973) The genus *Trechispora* (Basidiomycetes: Corticiaceae). Can J Bot 51:1871–1892

Liberta AE (1980) Notes on the genus *Subulicystidium*. Mycotaxon 10:409–412

Lindsey JP, Gilbertson RL (1978) Basidiomycetes that decay aspen in North America. Bibl Mycol 63:1–406

Lindsey JP, Gilbertson RL (1983) Notes on Basidiomycetes that decay Bristlecone Pine. Mycotaxon 18:541–559

Link HF (1809a) Observations in ordines plantarum naturales. Mag Ges Nat Freunde 3:39–40

Link HF (1809b) Nova plantarum genera e classe *Lichenum*, *Algarum*, *Fungorum*. Neues J für die Botanik 3:1–19

Linnaeus C (1753) Species plantarum. Stockholm

Linnaeus C (1764) Species plantarum. Vindobonae

Linnaeus C (1780) Systema plantarum. ed. novissima, Curr. J. J. Reichard, part IV

Litschauer V (1928) Neue Corticieen aus Osterreich. Osterr Bot Zeitschr 77:121–134

Litschauer V (1944) Beitrag Zur kenntnis der Gattung *Aleurodiscus*. Ann Mycol 42:1–23

Lloyd CG (1898–1925) Mycological notes. Lloyd Library and Museum, Cincinnati

Locquin MV (1974) De taxia fungorum I. UAE Mondedition, Paris

Lombard FE, Gibertson RL (1965) Studies on some western porias with negative or weak oxidase reactions. Mycologia 57:43–76

Lombard FE, Larsen ML (1985) *Phellinus bicuspidatus (*Hymenochaetales, Hymenochataceae) a new species associated with white sap rot of oak in Louisiana. Mycologia 77(1):55–61

Lombard FF, Burdsall HH Jr, Gilbertson RL (1975) Taxonomy of *Corticium chrysocreas* and *Phlebia livida*. Mycologia 67:495–510

Lowe JL (1942) The Polyporaceae of New York State (except *Poria*), vol 60, Technical publication of the State University College of Forestry at Syracuse University. New York State College of Forestry, Syracuse, pp 1–128

Lowe JL (1946) The Polyporaceae of New York State. The Genus *Poria*. Tech Publ State Univ Coll For Syracuse Univ 65:1–91

Lowe JL (1956) Type studies of polypores described by Karsten. Mycologia 48:99–125

Lowe JL (1957) Polyporaceae of North America: the genus *Fomes*. Tech Publ State Univ Coll For Syracuse Univ 80:1–96

Lowe JL (1958) The genus *Poria* in north America. Lloydia 21(2):100–114

Lowe JL (1963) A synopsis of *Poria* and similar fungi from the tropical regions of the world. Mycologia 55:453–486

Lowe JL (1966) Polyporaceae of North America. The genus *Poria*. Tech Publ State Univ Coll For Syracuse Univ 90:1–183

Lowe JL (1975) Polyporaceae of North America. The genus *Tyromyces*. Mycotaxon 2:1–82

Lowe JL (1976) On *Polyporus sobrius*. Kew Bull 31(3):753–755

Lowe JL, Gilbertson RL (1961) Synopsis of the Polyporaceae of the Western United States of Canada. Mycologia 53(5):474–511

Lowe JL, Lombard FF (1973) On the identity of *Polyporus lacteus*. Mycologia 65(4):725–732

Martin KJ, Gilbertson RL (1977) Synopsis of wood-rotting fungi on spruce in North America I. Mycotaxon 6:43–77

Martin KJ, Gilbertson RL (1978) Synopsis of wood-rotting fungi on spruce in North America II. Mycotaxon 7:337–356

Martin KJ, Gilbertson RL (1980) Synopsis of wood-rotting fungi on spruce in North America III. Mycotaxon 10:479–501

Mass Gaesteranus RA (1974) Studies in the genera *Irpex* and *Steccherinum*. Persoonia 7(4):443–581

Massee G (1889) A monograph of the Thelephoraceae-I. J Linn Soc Bot 25:107–155

Massee G (1889–1912) Fungi exotici I. Kew Bull 1898:113–136, 1898; Fungi exotici 2. Kew Bull 1899; Fungi exotici 3. Kew Bull 1901:150–169, 1901; Fungi exotici 4. Kew Bull 1906:91–94, 1906; Fungi exotici 6. Kew Bull 1907:121–124, 1907; Fungi exotici 8. Kew Bull 1908:216–219, 1908; Fungi exotici 10. Kew Bull 1910:249–253, 1910; Fungi exotici 13. Kew Bull 1912:189–191, 1912; Fungi exotici 14. Kew Bull 1912:253–255, 1912

Massee G (1890) Fungi exotici I. J Ldnn Soc Bot 27:95–205

Maxwell MK (1954) Studies of Canadian Thelephoraceae XI. Conidium production in the Thelephoraceae. Can J Bot 32:259–280

Micheli PA (1729) Nova plantarum genera. Firenza B Paperninii, Florence

Mitter JH, Tandon RN (1932) Fungus flora of Nainital-I. J Indian Bot Soc 11:178–180

Mitter JH, Tandon RN (1938) Fungi of Nainital-II. J Indian Bot Soc 17:177–182

Moncalvo JM, Nilsson RH, Koster B, Dunham SM, Bernauer T, Matheny PB, Porter TM, Margaritescu S, Weiβ M, Garnica S, Danell E, Langer G, Langer E, Larsson E, Larsson KH, Vilgalys R (2006) Cantharelloid clade: dealing with incongruent gene trees and phylogenetic reconstruction methods. Mycologia 98(6):937–948

Montagne JFC (1842) Cryptogammae Nilgherenses. Ann Sci Nat II Ser 18:12–23

Montagne JFC (1846) Champignons, in Belanger, voyage aux Indes Orientale pendant les annees 1825-1829 2:146–159

Morse JC, Gilbertson RL (1986) Taxonomy of *Fomitopsis rosea* complex (Aphyllophorales, Polyporaceae). Mycotaxon 25(2):469–486

Nakasone KK (2006) Dendrothele griseocana (Corticiaceae) and related taxa with hyphal pegs. Nova Hedwig 83:99–108

Nakasone KK, Burdsall HH Jr, Noll LA (1982) Species of *Phlebia* section *Leptocystidiophlebia* (Aphyllophorales, Corticiaceae) in North America. Mycotaxon 14:3–12

Neuman JJ (1914) The Polyporaceae of Wisconsin. Bull Wis Geol Nat Hist Surv 33, Sci Ser 10:1–206

Niemelä T (1972) On Fennoscandian Polypores II. *Phellinus laevigatus* (Fr) Bourd and Galz and *P. lundellii* Niemelä, n. sp. Ann Bot Fennici 9(1):41–59

Niemelä T (1975) On Fennoscandian Polypores IV. *Phellinus igniarius*, *P. nigricans* and *P. populicola*, n. sp. Ann Bot Fennici 12(3):93–122

Niemelä T (1977) On fennoscandian Polypores V. *Phellinus pomaceus* n. sp. Karstenia 17:77–86

Niemelä T (1978) On fennoscandian Polypores VI. *Antrodia plicata* n. sp. Karstenia 18:43–48

Niemelä T, Ryvarden L (1975) Studies in the Aphyllophorales of Africa-IV: *Antrodia juniperina* new for East Africa. Trans Br Mycol Soc 65(2):427–432

Nobles MK (1948) Studies in Forest Pathology VI. Identification of cultures of wood-rotting fungi. Can J Res(c) 26:281–431

Nobles MK (1958) Cultural characters as a guide to the taxonomy and phylogeny of the Polyporaceae. Can J Bot 36:883–926

Nobles MK (1965) Identification of cultures of wood inhabiting Hymenomycetes. Can J Bot 43:1097–1139

Nüñez M, Ryvarden L (1995) Polypores new to Japan I species of *Polyporus* with a note on *P. Hatanni*. Mycoscience 36:61–65

Nüñez M, Ryvarden L (1999) New and interesting polypores from Japan. Fungal Divers 3:107–121

Nüñez M, Ryvarden L (2000) East Asian polypore, vol I, Synopsis fungorum. Fungiflora, Oslo
Nüñez M, Ryvarden L (2001) East Asian polypore, vol II, Synopsis fungorum. Fungiflora, Oslo
Oberwinkler F (1965) Primitive Basidiomyceten. Sydowia 19:1–72
Oberwinkler F (1972) The relationships between the Tremellales and the Aphyllophorales. Persoonia 7:1–16
Oberwinkler F (1977) Species and generic concepts in the Corticiaceae. In: Clemençon H (ed) The species concept in Hymenomycetes. Cramer, Vaduz, pp 331–348
Overholts LO (1914) The Polyporaceae of Ohio. Ann Mo Bot Gard 1:81–115
Overholts LO (1915) The Polyporaceae of the middle-western United States. Wash Univ Stud 3:1–98
Overholts LO (1929) Research methods in the taxonomy of Hymenomycetes. Proc Intl Conf Plant Sci 2:b1688–b1712
Overholts LO (1933) The Polyporaceae of Pennsylvania I. The genus *Polyoporus*. Bull Pa Agric Exp Stn 298:3–28
Overholts LO (1935) The Polyporaceae of Pennsylvania II. The genera *Cyclomyces, Daedalea, Favolus, Fomes, Lenzites* and *Trametes*. Bull Pa Agric Exp Stn 316:3–16
Overholts LO (1939) The genus *Stereum* in Pennsylvania. Bull Torrey Bot Club 66:515–537
Overholts LO (1941) New species of Polyporaceae. Mycologia 33:90–102
Overholts LO (1942) The Polyporaceae of Pennsylvania III. The genus *Poria*. Bull Pa Agric Exp Stn 418:3–64
Overholts LO (1953) Polyporaceae of the United States, Alaska and Canada. University Michigan Press, Ann Arbor
Parmasto E (1965) Corticiaceae U.R.S.S. I. Descriptiones taxorum novarum. Combinationes novae. Eesti NSV Tead Akad Toim 14(2):220–233
Parmasto E (1967) Corticiaceae U.R.S.S. IV. Descriptiones taxorum novarum. Combinationes novae. Eesti NSV Tead Akad Toim 16:377–394
Parmasto E (1968) Conspectus systematis corticiacearum. Institutum zoologicum et botanicum Academiae scientiarum R.P.S.S. Estonicae, Tartu
Parmasto E (1970) The lachnocladiaceae of the Soviet Union. Inst Zool and Bot Estonia SSR Acad Sci Publ, USSR
Parmasto E (1973a) Modern problems of the systematics of Aphyllophorales. In: Abstracts of the reports of the 5th meeting of the members of the All-Union Botanical society, Kiev, pp 325–326
Parmasto E (1973b) *Inonotopsis* Parm. gen. nova (Hymenochaetaceae). Folia Cryptog Estonica 2:11–13
Parmasto E (1974a) *Peniophora laeta* (Fr.) Donk the significance of hydnoid outgrowths of the hymenium. Folia Cryptog Estonica 5:33–35
Parmasto E (1974b) On *Polyporus chozeniae* (Vassilk.) Parm. comb. nov. and related species. Folia Cryptog Estonica 5:35–39
Parmasto E (1974c) *Helicogonium jacksonii* (Dipodascales) found in Caucasus. Folia Cryptog Estonica 6:41–42
Parmasto E (1974d) *Favolaschia sachalinensis* Parm. sp. nov. ("Cyphellaceae", Hymenomycetes). Folia Cryptog Estonica 6:43–44
Parmasto E (1976) Studies on Yakutian fungi II. Eesti NSV Tead Akad Toim Biol 25(4):316–321
Parmasto E (1977) Studies on Yakutian fungi III Polypores Poriaceae s. l. Eesti NSV Tead Akad Toim Biol 26(1):15–28
Parmasto E (1978) Summary: on the dispersal of aphyllophoraceous fungi by basidiospores. Eesti NSV Teaduste Akadeemia Toimetised. Bioloogia 27(2):141–149
Parmasto E (1979) *Perenniporia ochroleuca* (Berk.) Ryv. in the Soviet Union. Mikol Fitopatol 13(2):102–104
Parmasto E, Parmasto I (1979) The Xanthochroic reaction in the Aphyllophorales. Mycotaxon 8:210–232
Patouillard N (1900) Essai taxonomique sur les familles et les genres des Hyménomycètes. Lucien Declume, Lons-Le-Saunier

Pegler DN (1973a) Aphyllophorales IV: Poroid families. In: The fungi, an advanced treatise, vol 4B. Academic, London

Pegler DN (1973b) The polypores – with keys to world genera and British species. Bull Br Mycol Soc 7(1):3–43

Persoon CH (1801) Synopsis methodica fungorum. Gottingen

Persoon CH (1825) Mycologia Europea. Palm, Erlangen

Petch T (1925) Notes on Ceylon Thelephoraceae, etc. Ann R Bot Gard Peradeniya 9:259–298

Pilat A (1936–1942) Polyporaceae I-III. In: Kavina C, Pilat A (eds) Atlas des champignons del' Europe. Praha

Pinto-Lopes J (1952) Polyporaceae, Contribuicao para a sua Bio-taxonomia. Memories da Sociedade Broteriana. J Cramer, Lehre

Pouzar Z (1959) New genera of higher fungi III. Ceska Mykol 13:10–19

Prasher IB, Chander H (2007) A preliminary report on the macro fungi and lichens of the Nanda Devi Biosphere Reserve (Uttranchal). In: Prasher IB, Sharma MP (eds) Advances in mycology and plant pathology. Bishen Singh Mahendra Pal Singh, Dehra Dun, pp 129–166

Prasher IB, Ashok D (2013) A Checklist of wood-rotting fungi (non-gilled Agaricomycotina) of Himachal Pradesh. J New Biol Rep 2(2):71–98

Prasher IB, Chander H (2006) Polyporoid fungi of Nanda Devi Biosphere Reserve. Panjab Univ Res J (Sci) 56:103–123

Prasher IB, Lalita (2013) A Checklist of wood-rotting fungi (non-gilled Agaricomycotina) of Uttarakhand. J New Biol Rep 2(2):108–123

Prasher IB, Lalita, Ashok D (2011) Polyporoid fungi of District Bilaspur (Himachal Pradesh). J Ind Bot Soc 90:268–273

Prasher IB, Lalita, Ashok D (2012) Polyporoid fungi of district mandi (Himachal Pradesh). J Ind Bot Soc 91: 204–212

Prasher IB (1999) Fungi of Chandigarh. IBD, Dehradun

Puri YN (1956) Studies on Indian *Poria*. J Ind Bot Soc 35(3):277–283

Quelet L (1886) Enchiridion fungorum. Lutetiae

Rabenhorst (1874) Funji Eur Exs No 1824. Hedwigia 13:184

Rajchenberg M (1982) El género *Coriolus* (Polyporaceae) en la Republica Argentina. Bol Soc Argent Bot 21:17–57

Rajchenberg M (1983) New South America resupinate polypores. Mycotaxon 16:500–506

Rajchenberg M (1987) New South America polypores. Mycotaxon 28(1):111–118

Rajchenberg M (2002) *Cryptosporiopsis lomati* sp. nov. on *Lomatia hirsuta* from southern Argentina. Sydowia 54:1–7

Ranadive KR, Vaidya JG, Jite PK, Ranande VD, Bhosale SR, Rabba AS, Hakimi M, Deshpande GS, Rathod MM, Forutan A, Kaur M, Naik-Vaidya CD, Bapat GS, Lamrood P (2011) Checklist of Aphyllophorales from the Western Ghats of Maharastra State, India. Mycosphere 2:91–114

Rattan SS (1977) The resupinate Aphyllophorales of the North Western Himalayas. Bibliotheca Mycologica 60, J Cramer, Germany

Rattan SS, Abdullah SK (1977) Studies on wood-rot fungi of Iraq. Indian Phytopathol 29(3):296–302

Rattan SS, Abdullah ZK, Ismail ALS (1978) Studies on fungi causing diseases and decays of trees in Iraq. Nova Hedwig 29:765–779

Rattan SS, El-Buni AM (1981) Some new records of wood decaying fungi (Aphyllophorales) from Libya. Nova Hedwig 34:153–161

Rea C (1922) British basidiomycetes. A handbook to the larger British fungi. Cambridge University Press, Cambridge

Rehill PS, Bakshi BK (1965) Studies on Indian Thelephoraceae II. Indian Species of *Peniophora* and *Corticium*. Indian Forest Bull Dehra Dun 242:1–31

Rehill PS, Bakshi BK (1966) Studies on Indian Thelephoraceae III. The genus *Stereum*. Indian For Bull Dehra Dun 250:1–20

Reid DA (1962) Notes on fungi which have been referred to the Thelephoraceae sensu lato. Persoonia 2:109–170

Reid DA, Thind KS, Adlakha KL (1958) The Thelephoraceae of Mussoorie Hills II. Trans Br Mycol Soc 41:129–134
Rogers DP (1943) The genus *Pellicularia* (Thelephoraceae). Farlowia 1:95–118
Rogers DP (1944) The genera *Trechispora* and *Galzinia* (Thelephoraceae). Mycologia 36:70–103
Rogers DP (1951) *Trechispora* and *Pellicularia*. Mycologia 43:111
Rogers DP, Jackson HS (1943) Notes on the synonomy of some North American Thelephoraceae and other resupinates. Farlowia 1:263–328
Roy A (1968a) Anatomy of Indian Polyporaceae I. *Trametes cingulata* and *Trametes persooni* Fr. Bull Bot Soc Bengal 22(1):45–54
Roy A (1968b) Anatomy of Indian Polyporaceae II. *Daedalea flavida* Lev. and D. *microzona* Lev. Bull Bot Soc Bengal 22(2):131–134
Roy A (1969) Anatomy of Indian Polyporaceae III. *Polyporus adustus* Willd. ex Fr. and *Polyporus ostreiformis* Berk. Bull Bot Soc Bengal 23(2):205–211
Roy A (1970) Anatomy of Indian Polyporaceae IV. *Polyporus calcuttensis* Bose. Bull Bot Soc Bengal 24:151–156
Roy A (1971) IAnatomy of Indian Polyporaceae V. *Polyporus anthelminticus* Berk. Visva-Bharati Ann (Sc) Part II 14:20–29
Ryvarden L (1972a) A note on the genus *Junghuhnia*. Persoonia 7:17–21
Ryvarden L (1972b) Studies on the Aphyllophorales of the Canary Islands with a note on the genus *Periniporia*. Nor J Bot 19:139–144
Ryvarden L (1972c) A critical checklist of the Polyporaceae in Tropical East Africa. Nor J Bot 19(3–4):229–238
Ryvarden L (1973a) New genera in the Polyporaceae. Nor J Bot 20:1–5
Ryvarden L (1973b) Some genera of resupinate Polypores with a Note on *Aleurodiscus norvegicus* nov. sp. Nor J Bot 20(1):7–11
Ryvarden L (1973c) Type studies in the Polyporaceae I. Tropical species described by C H Persoon. Persoonia 7(2):305–312
Ryvarden L (1974a) Type studies in the Polyporaceae II. Species described by M Billi. Bull Jard Bot Nat Plantentuim Belg 44:65–76
Ryvarden L (1974b) Type studies in the Polyporaceae III. Species described by Lars Romell. Sven Bot Tidskr 68:273–284
Ryvarden L (1974c) Studies in the Aphyllophorales of the Canary Island II. Some species new to the islands. Cuad Bot Canar 20:3–8
Ryvarden L (1975a) Studies in the Aphyllophorales of Africa II. Some new species from East Africa. Nor J Bot 22:25–34
Ryvarden L (1975b) Studies in the Aphyllophorales of Africa-3. Three new polypores from Azire. Bull Jard Bot Nat Belg Bull Nat Plantentium Belg 45:197–203
Ryvarden L (1976a) The Polyporaceae of North Europe, vol I, Albatrellus-incrustoporia. Fungiflora, Oslo
Ryvarden L (1976b) Type studies in the Polyporaceae 4. Species described by JF Klotzsch. Mem N Y Bot Gard 28:199–207
Ryvarden L (1976c) Type studies in the Polyporaceae 7. Species described by JM Berkeley from 1836-1843. Kew Bull 31:81–103
Ryvarden L (1976d) Type studies in the Polyporaceae 8. Species described by E Rostrup. Bot Tidskr 71:100–102
Ryvarden L (1977) Type studies in the Polyporaceae 9. Species described by EM Wakefield. Mycotaxon 5(1):331–336
Ryvarden L (1978a) Studies in the Aphyllophorales of Africa 6. Some species from eastern Central Africa. Bull Jard Bot Nat Belg 48:79–117
Ryvarden L (1978b) The Polyporaceae of North Europe, vol II, Innotus-Tyromyces. Fungiflora, Oslo
Ryvarden L (1979) *Porograme* and related genera. Trans Br Mycol Soc 73(1):9–19
Ryvarden L (1981a) Type studies in the Polyporaceae -12. Species described by FW Junghun. Persoonia 11(3):369–372

Ryvarden L (1981b) Type studies in the Polyporaceae -13. Species described by JH Leveille. Mycotaxon 13(1):175–186

Ryvarden L (1982a) Synopsis of genus *Wrightoporia*. Nord J Bot 2:145–149

Ryvarden L (1982b) The genus *Hydnochaete* (Hymenochaetaceae). Mycotaxon 15:425–447

Ryvarden L (1982c) *Fuscocerrena,* a new genus in the Polyporaceae. Trans Br Mycol Soc 79(2):279–291

Ryvarden L (1982d) Type studies in the Polyporaceae -11. Species described by JFC Montagne, either alone or with other authors. Nord J Bot 2:75–84

Ryvarden L (1983a) The genus *Navisporus* (Polyporaceae). Nord J Bot 3:411–413

Ryvarden L (1983b) Two *Perennioporia* species from Pakistan. Mycotaxon 27:517–520

Ryvarden L (1983c) Type studies in the Polyporaceae – 14. Species described by N Patouillard, either alone or with other Mycologists. Farlow Herbarium. Harvard University, Cambridge

Ryvarden L (1984) Type studies in the Polyporaceae – 16. Species described by JM Berkeley either alone or with oher mycologists from 1856 to 1886. Mycotaxon 20(2):329–363

Ryvarden L (1985a) Type studies in the Polyporaceae – 17. Species described by WA Murrill. Mycotaxon 23:169–198

Ryvarden L (1985b) A note on *Poria* and *Hexagonia* (Polyporaceae, Basidiomycetes). Mycotaxon 23:293–296

Ryvarden L (1991) Genera of polypores. Nomenclature and taxonomy. Synop fungorum 5:1–363

Ryvarden L (2004) Neotropical polypores, part 1 introduction, Hymenochaetaceae and Ganodermataceae. Synop fungorum 19:1–228

Ryvarden L, Gilbertson RL (1993) European polypores, vol 1, Synopsis fungorum 6. Fungiflora, Oslo, pp 1–387

Ryvarden L, Gilbertson RL (1994) European polypores, vol 2, Synopsis fungorum 7. Fungiflora, Oslo, pp 388–743

Ryvarden L, Johansen I (1980) A preliminary Polypore flora of East Africa. Fungiflora, Oslo

Ryvarden L, Stokland L, Larsson KH (2003) A critical checklist of corticoid and Poroid Fungi of Norway. Synop fungorum 17:1–109

Saccardo PA (1888) Sylloge fungorum omnium hususque. Comniterum VI. Patavii

Saccardo PA (1899) Sylloge fungorum omnium hususque. Comniterum XIV. Patavii

Schaeffer JC (1762–1774) Fungorum qui in Bavaria et Palatinatn circa Ratisbonam nascuntur icones, vol I–IV. Typis Zunkelianis, Ratisbonae

Schroeter J (1889) Die Pilze Schlesiens. In: Cohn F (ed) Kryptogamen-Flora von Schlesien. 3. Band, 1. Hälfte. J U Kern's Verlag, Breslau

Sehgal HS, Sen M, Bakshi BK (1966) Temperature relations of Indian Polypores. Indian Forest Rec. (n.s.). Forest Pathol 2(7):131–137

Shope PF (1931) The Polyporaceae of Colorado. Ann Mo Bot Gard 18:287–456

Singer R (1944) Notes on taxonomy and nomenclature of the polypores. Mycologia 36:65–69

Singh B (1961) Occurrence of *Trametes ravida* (Fr.) Pilat in India. Indian Forester 87(7):429–430

Singh B (1966a) Timber decay due to five species of Fomes, as new records in India. Indian Forester 92:653–655

Singh B (1966b) Studies on Indian Poria, II. Diagnosis of five new records in India. Indian Forester 92:680–683

Singh B, Bakshi BK (1961) New records of Fomes from India. Indian Forester 87(5):302–303

Singh S, Singh B, Bakshi BK (1961a) Fungus flora of South Andamans. Indian Forester 87(4):248–250

Singh S, Singh B, Bakshi BK (1961b) Merulius tremellosus (Schrad.) Fr. and its Ptyogaster stage. Can J Bot 39:1845–1847

Slysh AR (1960) The genus *Peniophora* in New York state and adjacent regions. State Univ NY Coll Tech Publ 83:95

Smith WG (1908) Synopsis of the British Basidiomycetes. British Museum (Natural History), London

Stevenson JA, Cash EK (1936) The new fungus names proposed by CG Lloyd. Bull Lioyd Libr Museum 35:1–209

Steyaert RL (1946) Une technique rapide permettant le montage an baume de material botanique. Parasitica 2:137–138

Steyaert RL (1947) A technique for obtaining quickly permanent mounts of nonembedded botanical material. Science 105:47–48

Steyaert RL (1961) Note on the nomenclature of fungi and incidentally, of *Ganoderma lucidum*. Taxon 10(8):251–252

Steyaert RL (1972) Species of *Ganoderma* and related genera mainly of the Bogor and Leiden Herbaria. Persoonia 7:55–118

Steyaert RL (1975) The concept and circumscription of *Ganoderma tornatum*. Trans Br Mycol Soc 65:451–467

Steyaert RL (1980) Study of some *Ganoderma* species. Bull Jard Bot Nat Belg 50:135–186

Sundararaman S, Marudarajan D (1925) Some polyporaceae of the Madras Presidency. Madras Agric Dept Year Book

Svrcek M (1958) Contribution to the taxanomy of resupinate fungi. Cesk Mykol 12:66–77

Svrcek M (1960) Tomentelloideae Cechoslovakiae–Genera resupinate familiae Thelephoraceae s. str. Sydowia 14:170–245

Talbot PHB (1973) Aphyllophorales I: general characteristics; thelephroid and cupuloid families. In: Ainsworth GC, Sparrow FK, Susman AS (eds) The fungi IVB. Academic, New York/ London, pp 327–349

Teixeira AR (1962) The taxonomy of Polyporaceae. Biol Rev 37:51–81

Theissen F (1911) Fungi aliquot Bombayenses a Rev. Ed. Blatter Collecti. Ann Mycol 9:153–159

Thind KS, Adlakha KL (1956) Thelephoraceae of Mussoorie hills- I. Indian J Mycol Res 2:57–64

Thind KS, Chatrath MS (1960) Polyporaceae of Mussoorie hills-I. Indian Phytopath 13:76–89

Thind KS, Dhanda RS (1979a) The Polyporaceae of India-IX. Kavaka 7:51–58

Thind KS, Dhanda RS (1979b) The Polyporaceae of India XII. The genus *Albetrellus*. Indian Phytopath 32(1):55–60

Thind KS, Dhanda RS (1980a) The Polyporaceae of India X. Kavaka 8:59–67

Thind KS, Dhanda RS (1980b) The Polyporaceae of India XIII. Indian Phytopath 33(3):380–387

Thind KS, Rattan SS (1968) The Thelephoraceae of North Western Himalayas. Indian Phytopath Soc Bull 4:15–24

Thind KS, Rattan SS (1970) The Thelephoraceae of India- III. The genus *Tubulicrinis* and *Hyphoderma*. Proc Indian Acad Sci 71:118–131

Thind KS, Rattan SS (1971a) Type studies in the Polyporaceae IV. The genus *Tomentella*. Indian Phytopath 24:32–42

Thind KS, Rattan SS (1971b) The Polyporaceae of India-V. Indian Phytopath 24(1):50–57

Thind KS, Rattan SS (1971c) The Polyporaceae of India-VII. Indian Phytopath 24(2):290–294

Thind KS, Rattan SS (1971d) The Polyporaceae of India-VIII. Res Bull Panjab Univ 22(1):29–36

Thind KS, Rattan SS (1972) Type studies in the Polyporaceae VIII. Res Bull Panjab Univ 23(3–4):125–136

Thind KS, Rattan SS (1973a) Type studies in the Polyporaceae X. Mycologia 65:1250–1258

Thind KS, Rattan SS (1973b) Type studies in the Polyporaceae VI, VII. Indian Phytopath 26:485–494; 528–536

Thind KS, Rattan SS (1976) Type studies in the Polyporaceae XI. Res Bull Panjab Univ 27:29–39

Tulasne LR (1853) Observations sur l'organisation des tremellinees. Ann Sci Nat Bot 19:193–231

Tulasne LR (1872) Nouvelles notes sur les fungi tremellini et lour allies. Ann Sci Nat Bot 15:215–235

Vasudeva RS (1960) The fungi of India (revised edition of Butler and Bisby's work). ICAR, New Delhi

Wakefield EM (1916–1922) Fungi exoticii-XX. Kew Bull 1916:71–77, 1916; Fungi exoticii-XXIV. Kew Bull 1918:207–210, 1918; Fungi exoticii-XXVI. Kew Bull 1922:161–165, 1922

Wakefield EM (1948) Taxonomic problems in Hymenomycetes. Trans Br Mycol Soc 30:152–160

Wakefield EM (1952) New and rare British Hymenomycetes (Aphyllophorales). Trans Br Mycol Soc 35:34–65

Wakefield EM (1960) Some species of *Tomentella* from North America. Mycologia 52:919–933

Wakefield EM (1966) Some extra-European species of *Tomentella*. Trans Br Mycol Soc 49:357–362

Welden AL (1954) Some tropical American Stipitate Stereums. Bull Torrey Bot Club 81:422–439

Welden AL (1958) A contribution towards a monograph of *Cotylidia* (Thelephoraceae). Lloydia 21:38–44

Welden AL (1960) The genus *Cymatoderma* (Thelephoraceae) in the Americas. Mycologia 52:856–876

Welden AL (1965) West Indian species of *Vararia* with notes on extralimital species. Mycologia 57:502–520

Welden AL (1966) West Indian species of *Asterostroma* with notes on extralimital species. Am Midl Nat 76:222–229

White LT (1951) Studies of Canadian Thelephoraceae VIII *Corticium galactinum* (Fr.) Burt. Can J Bot 29:279–296

Yasuda R (1917) Thelephoraceae, Hydnaceae and Polyporaceae von Japan (Vorlaufige Mitterlung). Bot Mag Tokyo 31:42–63

Chapter 3
Materials and Methods

Abstract The chapter is divided into five sections which include: the need to collect these fungi, their habitat, procedures to collect them, their processing for further action. The last two sections deal with the procedures to isolate them *in vitro* and their preservation and the examination of the specimens for identification. The prerequisites for the collection and Do's and Do not's concerning field study are discussed in brief. The various procedures for *in vitro* isolation and preservation are also given.

3.1 Why to Collect

Wood-rotting non-gilled Agaricomycetes have drawn the attention of systematic mycologists as well as biotechnologists worldwide due to their ability to cause wood decay as well as for their application as tools in biotechnological processes. In the context of the Indian scenario; particularly the Himalayas; not much attention has been paid on the taxonomy and systematic of these fungi except for the work done only at the selected research laboratories located at Departments of Botany Panjab University, Punjabi University, Calcutta University and at the regional stations of Botanical Survey of India and at Mycology Herbarium at Forest Research Institute, Dehra Dun. No major initiative has been taken for their in vitro conservation. Since, The Himalayas are one of the mega biodiversity hotspots, attention must be paid for extensive exploration of these fungi for floristic analysis, for their ability to degrade wood and cause diseases of forest trees and their sustainable use in biotechnological processes.

3.2 Where to Collect

These fungi have a wide spread distribution in the forest ecosystem. They are found growing luxuriantly on fallen logs and twigs in the forests as well as elsewhere, where there is sufficient moisture available for their growth. A large number of these

I.B. Prasher, *Wood-rotting non-gilled Agaricomycetes of Himalayas*,
Fungal Diversity Research Series, DOI 10.1007/978-94-017-9858-7_3

are found growing, producing their fructification on the living trees causing the heart-rot. Brown rot fungi are mostly confined to coniferous forests, where as the others are found growing on angiospermous trees or wood, causing decay. Fructifications of the larger fungi; which are generally pileate; are more prominent and easy to find on the tree trunk and branches of the trees, exposed surface of the fallen logs and on other exposed woody substrates than the resupinate fungi, which normally grow on the underside of the fallen logs, twigs or on the shady side of the parts of the trees. The underside of the decayed and rotten twigs on the forest floor are one of the richest sources of these fungi. Fallen twigs near the bank of the aquatic bodies including streams and rivulets also abound with the effused fructifications of these fungi. There is a marked difference between the flora of these fungi occurring on the trees and at their base from these found on decaying wood near the aquatic ecosystem. Some of these, like species of *Ganoderma* are particularly confined to the lower part of the tree trunk near the ground level, often forming a collar like growth as in case of *G. applanatum* (Fig. 3.1).

Some times during the rainy season especially during the monsoons, only the anamorphic stages of these fungi are found exclusively growing on specific trees. *Ptychogaster aurantiacus* Pat. (Fig. 3.2a), the anamorphic stage of *Laetiporus sulphureus* (Fr) Murr. (Fig. 3.2b) grows extensively on *Mangifera indica* in the foothills of N. W. Himalayas.

Critical observations are required to arrive at correct identification in such cases. Brown rot fungi have been found growing on man-made wooden infrastructure as doors, cabinets and other wood fixtures and cause enormous damage.

Fig. 3.1 *Ganoderma applanatum*

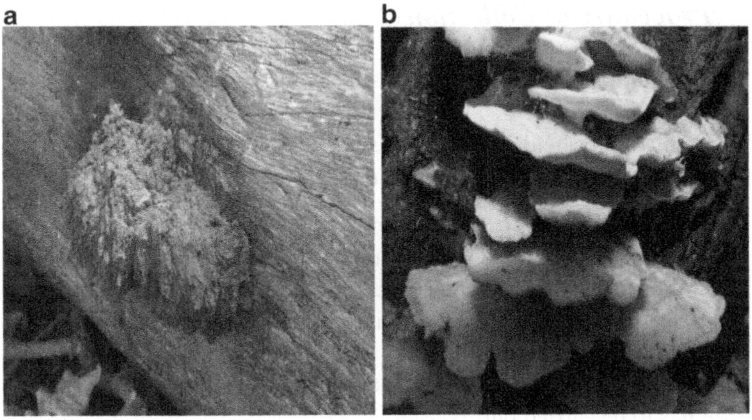

Fig. 3.2 (**a**) *Ptycogastor aurantiacus* (**b**) *Laetiporus sulphureus*

3.3 When and How to Collect

3.3.1 Collection Period

The most favourable season for making fresh collections of these fungi is in the monsoons, from the months of July to end of September. In some of the regions where the temperature does not fall much the collections of these fungi can be extended up to the mid of October. Although these fungi are found growing on wooden substrata throughout the year, these months are suitable for the collection of fertile and developing fructifications as well as collecting the mature ones with freshly developed hymenophore. The specimens collected are devoid of spores and hymenium during the summer months when the temperature is high with low moisture content and in the winters from mid October to end of February when the temperature is below normal for the growth of these fungi. Due to this reason most of the fungal forays particularly concerning these fungi are organized in the monsoon months and up to mid October. However, these fungi especially the resupinate ones are found growing on their substrate throughout the year in the microclimatic niches like shady places with sufficient moisture in the forest ecosystem in the himalayas, which favour their growth to some extent in the unfavourable months. There is a great variation in flora of these fungi during the pre-monsoon and in the winter months. Due to this reason, one or two extensive floristic surveys have to be conducted during these otherwise unfavourable periods of the year.

3.3.2 *Procedure of Collection*

3.3.2.1 Pre-requisites for Collection

Before moving out for field collection, the collector should check the following pre-requisite for the field collection. These include:

(i) Photo identity card from the employer (if one is working in a research/ Academic Institute along with an authorization to collect these fungi). This is essential since the regulation concerning conservation and protection of biodiversity has been issued by Ministry of Environment & Forests (MoEF), Government of India.
(ii) Permission sought from Conservator of Forests of the region where exploration has to be carried out.
(iii) Field diary.
(iv) Collection bags/baskets.
(v) Paper bags (preferably made up of brown paper or news paper bags depending upon the availability).
(vi) Hand lens.
(vii) Cutting tools (Knife, small axe, hammer and chisel). The tools should be properly covered.
(viii) Pencil/pen.
(ix) Stapler/paper pins, waxed paper.
(x) Compass for navigational purpose.
(xi) Portable light source which is required in some shady forest localities where the fungus is sometimes required to be photographed. Some ultra light portable lights are available which can be carried easily.
(xii) Ethanol/Acetone in a small bottle (to be carried by the collector in the field). It is the personal experience of the author that in the dense forests, one is prone to be mosquito/wasp/honey bee bites. An application of these reagents on the site of attack act as an antidote and gets one relieved of the pain as well as the inflammation.
(xiii) Equipment required for setting temporary laboratory, if moving for few days far off from the laboratory. These basically include: Storage boxes, portable dryer, insecticide, emergency light, portable microscope to examine temporary slides and a good digital camera with macro-photographic facility.

3.3.2.2 Collection of the Specimens in the Field

While collecting the specimens in the specialized niches, care must be taken to leave some of the specimens in the collection locality. This helps in further dispersal and in the long run helps conserve the diversity. One should collect the specimens without damage and care should be taken that fructifications should not be torn apart. They should be collected intact along with a part of the substratum. Each specimen should be placed in the paper bag along with the collection number, name of the host/substratum, locality on the label along with date of collection (pre attached to the paper bag with a stapler or gum). After putting the specimen in the paper bag, it should be folded on the top and sealed partially with a paper clip, so that the specimens do not fall in the basket or large bag during transportation to the temporary laboratory. Storage boxes with multiple chambers can be used to carry the delicate specimens. However in this case the collection number of the specimens along with the host/substratum should be noted on the labels to be put inside these chambers. The field data can be noted down in a diary. Plant diversity at the place of collection has to be noted down which is of great help while determining the inter-relationships if any at a later stage.

Do's and Do not's during the collection period:

1. Ask/employ locals in the vicinity of the forests if they know about these fungi. They can guide you in lesser time to specialised niches.
2. Devote more time in exploring a particular locality so as to explore it extensively than moving around in large areas. The resupinate fungi are most abundant on the underside of the fallen twigs/logs and broken branches still hanging from the living trees.

Do not's:

1. Do not use plastic bags for storage of the specimens in the field as well as later on because they encourage growth of moulds like *Penicillium*, *Aspergillus* etc. due to accumulation of moisture.
2. Do not put two taxa in one paper bag.
3. Do not put large number of fructifications in one open bag.
4. Do not risk yourself unnecessarily while collecting specimens from difficult niches.
5. Do not roll the large logs towards yourself while exploring fungi from the underside.
6. Do not put your uncovered hands in the cut/deep grooves of the tree trunks. You may be attacked by insects or snake.

3.4 Processing

3.4.1 Spore Print

Before drying the specimens, two specimens of the collection (if the specimens are small) or two pieces of collection (sufficiently moist) are selected and put on the black chart paper as well as glass slides (used for microscopy) with hymenial surface facing downwards. If the specimens are not moist, they are dipped in water for 30–60 min and then used for taking the spore print. The spore print is taken on both (black paper and slide) with the collection number of the specimen. The spore print

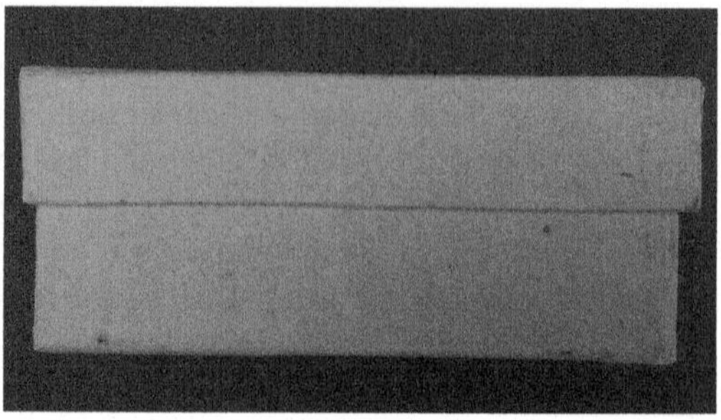

Fig. 3.3 Cardboard box

on the glass slide can also be used in culturing these fungi later on as well as in Scanning Electron Microscopic (SEM) studies of the spores. Both of these are covered with cellophane paper to avoid dust and contaminating spores of anamorphic fungi. The spore prints are checked after 10–12 h and packed. The slides with spore prints are covered with cellophane paper tied on both the ends with rubber bands whereas the paper with spore prints is folded four times and closed with adhesive tape. The paper used is of 4×4 inch which when folded makes to 1 inch square. These can be packed in a cardboard box Fig. 3.3.

3.4.2 Field Notes/Characters

A field performa can be used while collecting these fungi. The following points are important and are to be kept in mind while collecting in the field:

Most of the morphological features of the specimens are noted down while fresh and before drying.

The size of the specimen is noted as soon as possible because it changes to a larger extent in case of specimen with high moisture content upon drying. However, the fungi with perennial fructification change little. Similarly colour is compared with a standard colour chart. In our laboratory as also in the field laboratory, the colour of the specimens is compared with colour chart by Kornerup and Wanscher (1978). The other important morphological features to be noted down are the shape of the fructification, the surface of the pileus (if present) and the pore surface. Since most of the good digital cameras with fixed zoom lens have a macro-mode, one can photograph the pileus as well as pore-surface from close distance of the fresh specimen. The colour photographs of the specimens are taken before drying.

The specimens are described for macroscopic characters for identification as soon as possible, since in most of these fungi, particularly with effused fructifications, the growth of vegetative hyphae starts from the fructification after few hours of storage in the temporary laboratory which obstruct the hymenial layer. They should be examined as early as possible.

3.4.3 Culturing of the Fungi in the Field

It is better to culture the fungi *in vitro* on the PDA (Potato dextrose Agar) or MEA (Malt extract Agar) slants or petri plates in the temporary laboratory set at the main station within 24 h of the collection. The fungi are cultured from the germinating spores or from the tissue of the context. The cultures from the basidiospores are obtained by obtaining a spore print on the sterilized slide and then transferring these to the sterilized medium aseptically. A part of the inner tissue of the context is transferred with the help of a sterile needle to the sterile basal medium in petri plates or test tubes to obtain cultures. It is the personal experience of the author that these fungi are relatively easy to isolate *in vitro* when fresh than after few days of the collection. The fungi which could not be cultured in this way can be isolated from the spore prints taken on the slides. These can be isolated through single spore isolation techniques. The details regarding the various media and the isolation techniques are discussed in the section on *in vitro* isolation and preservation.

3.4.4 Preservation of the Specimens

The specimens are preserved by drying in the hot air oven. These are dried in an improvised wooden folding box like structure with three chambers. The base of the lower most chamber is fitted with hot air blower. The upper chambers are separated from the lower chamber by a horizontal wire gauge trays through which the hot air can pass. The upper most chamber is fitted with sliding cover divided into four sections. These can be opened or closed depending upon the specimens to be dried. The hot air after drying the specimens passes through the outlets in the upper chamber. The temperature in the upper chamber is maintained between 35 and 45 °C by adjusting the thermostat of the blower from position indicated on it. The blower is also replaced in some conditions by one or two electric bulbs of 60 W/100 W. The specimens retain their original colour in this method. The specimen put in the paper packet bags with labels can be placed as such in the dryer. This also rules out the mixing of specimens. After drying, these specimens are shifted to paper envelops of the size of 15 × 12 cm made of white computer stationary. Large number of genera of these fungi like species of *Ganoderma*, *Trametes* and other are infested with insects which make a powdery mass of fructification as a whole or a part of it. Commercially available insecticide like 1–4-Dichlorobenzene or Naphthalene balls are placed in each packet. These packets are than arranged in card board boxes of the size of 40 × 17 × 13 cm in a vertical position. In order to protect the collections from insect attack, insecticidal balls are also put in these boxes. Large specimens and those belonging to genera like *Trametes* which are most susceptible to insects are further placed in zip lock polyethylene bags. Before placement of these specimen packets, it is ensured that the fungi does not contain traces of moisture. In order

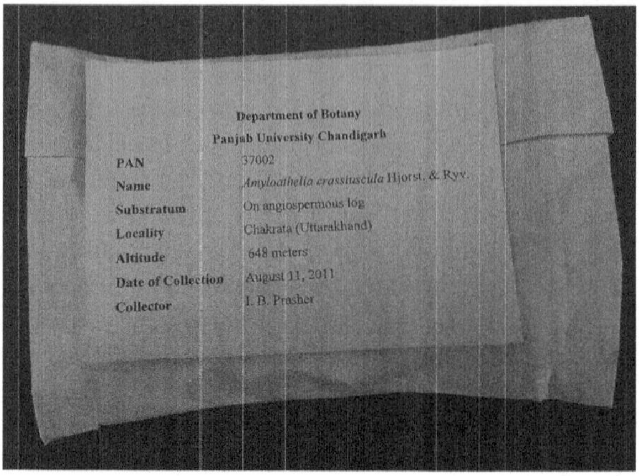

Fig. 3.4 Paper packet with label

to ensure this, the packets are placed for 1–2 h in the electric hot air oven at 45–50 °C. Each specimen paper packet is labelled with relevant data depicted in Fig. 3.4 which include: Name of the specimen, locality, substratum, Name of the determinator and Herbarium number.

3.4.5 Deposition of the Specimen

The specimens are deposited and arranged, in the herbarium after identification. One of the best ways to arrange the specimens is to arrange them alphabetically according to family, genera and species. The collections pertaining to species are arranged alphabetically with in genera. These are placed in card board boxes labelled with the name of the genera. The boxes pertaining to individual genera are arranged in the same fashion in a steel cupboard/almirah according to the family. A large family may be assigned to a single almirah/cupboard. The left out space may be utilized for the later on deposition of collections. The placement data is uploaded/stored in a computer in the herbarium which makes the retrieval easy and comfortable. The type specimens are normally stored separately for restricted access by the users only with the approval of the Curator of the herbarium.

3.5 *In Vitro* Isolation and Preservation

In order to study the decay characteristics and to study the physiological and biochemical parameters of these fungi; to be employed later on for biotechnological applications; pure and auxenic cultures are to be obtained. These are obtained by either single spore isolation technique or single hyphal tip isolation method. A brief procedure adopted for the single spore isolation (based on Choi et al. 1999) is briefly described below. A piece of fructification with hymenophore is cut from the fructification and is placed on the inside of the Petri dish using Vaseline. Alternatively it can be fixed with the help of double adhesive tape. The Petri plate is then sealed with parafilm and left horizontally over night at 25 °C. In case of fertile specimens, the basidiospores fall on the agar surface.

3.5.1 Single Spore Isolation

In this method either a part of the spore print (taken in the field) or the hymenial layer of the specimens is used for making a spore suspension in sterilized distilled water. This suspension is then poured in pre-sterilized petri-plates containing sterilized 2 % agar-agar medium (2 g agar-agar in 100 ml of water) with the help of a pipette. The number of spores/ml of the water is pre-determined with the help of Haemocytometer. The petri-plates with spore suspension are placed in upright position for 4–12 h depending upon the ability of the spores to germinate which can be determined before pouring the suspension. The petri-plates are inverted to decant the water after the pre-determined period required for germination of spores. The germinating spores are marked in black with fine marker pen on the inverted side of the petri-plates under low power (4 × objective) of the microscope. These are then lifted along with the agar with the help of a biscuit cutter needle and transferred to petri-plates or test tubes containing semi-synthetic sterilized media and incubated at optimum temperature.

An alternative method of obtaining the single spore cultures is to allow the spores to germinate and form a colony on the inoculated agar plates. When the individual colonies are about 2–5 mm in diameter, they are transferred to petri-plates or tubes containing the medium. The optimum temperature for most of these fungi has been found to be between 24 and 28 °C. It has been observed that suspension made from spore prints obtained in the field give better results (if they are properly covered) than spores obtained from the specimens collected few days before isolation.

3.5.2 Single Hyphal Tip Isolation

The cultures obtained from the hymenial tissue in the field contribute significantly to this procedure than the isolations from the specimens which have been collected few days earlier from their culturing. The success rate of the culturing the dried specimens is often not more than 40 %. The single hyphal tips can be marked on the inverted side of the petri-plates supporting cultures under low magnification of the microscope. These are than cut with biscuit cutter needle and transferred to petri-plates or test tubes containing synthetic/semi-synthetic media.

3.5.3 Composition of Different Media

The most commonly used media for isolation and maintenance of culture of these fungi are as follows along with their composition.

3.5.3.1 Potato Dextrose Agar Medium (PDA)

Composition	g/l
Potato (scrubbed and diced)	200.0 g
Agar	20.0 g
Dextrose	20.0 g
Distilled water	to make 1.0 L

Boil diced potatoes in 500 ml of distilled water, filter through cheese cloth, add distilled water to the filtrate to make 1.0 L. Add agar to filtrate and dissolve, then add dextrose.

Alternatively Potato Dextrose Agar medium concentrate manufactured by HIMEDIA, can be used. In this 39.0 g of the concentrate is to be mixed in 1.0 L of distilled water to get desired composition.

3.5.3.2 Yeast Extract Peptone Agar Medium (YEPA Medium)

Composition	g/l
Agar	15.0
Peptone	5.0
Yeast extract	3.0

Alternatively Dissolve 23.0 g powder concentrate to distilled water to make 1.0 L. The above medium is manufactured by Sisco Research Laboratories.

3.5.3.3 Malt Extract Agar Medium (MEA Medium)

Composition	g/l
Agar	18.0 g
Malt extract	20.0 g
Distilled water	to make 1.0 L

All the isolation/basal media mentioned above are sterilized at 16 lbs psi steam pressure at 121 °C for 30 min in an autoclave.

3.5.4 In Vitro *Preservation*

The cultures of these fungi are preserved in various ways. Some of the method employed include:

3.5.4.1 Storage of the Stock Culture

Storage of the stock culture in screw capped 1 inch diameter test tubes at ±4 °C in a frost free refrigerator. These cultures can be revived after 4 months.

3.5.4.2 Preservation of the Culture in Mineral Oil

Liquid paraffin, sterilized at 15 lbs psi for 2 h and cooled is added to the agar slants on which the fungal cultures are growing. Whole of the agar and fungal culture should be submerged in the oil. The tubes are kept in an upright position at a temperature of 15–25 °C. The level of the oil in the test tubes is checked at regular intervals. To revive the culture, a part of the culture submerged in mineral oil is removed and placed on the culture medium after draining the oil.

3.5.4.3 Immersion in Distilled Water

This is a low-maintenance and cost-effective method to preserve these fungi. The screw capped test tubes with cultures on the agar slants are completely submerged in sterilized distilled water in an upright position. Alternatively, the procedure involves cutting of the agar plugs from the edges of actively growing cultures and placing them in sterilized distilled water in screw capped tubes. It has been found that the fungal culture of these fungi survive for more than 2 years at 5 °C, in our laboratory which is in conformity with the findings of Marx and Daniel (1976) and Richter and Bruhn (1989). In our laboratory we subculture these isolates after 12–18 months. In order to save space, these culture plugs can also be stored in small, sterile, screw-cap cryovials and submerged in distilled water.

3.5.4.4 Freezing (After Ito 1991)

Discs are cut from the actively growing culture of these fungi on the petri-plates. These are aseptically placed in sterile cryotubes containing 10 % glycerol in distilled water. The cultures are preserved for up to 60 months at −80 °C.

3.5.4.5 Lyophilization (After Croan 2000)

This method is mostly applied to fungi which form large number of small propagules. The fungi cultured in our laboratory have not been preserved by this method.

3.6 Examination

The examination of fresh as well as dried specimens is accomplished by using the following mountants and stains to observe sterile and fertile elements microscopically. The microscopic details are combined with macromorphological features to describe a specimen for establishing an identity. For determining the identity of the specimen in the laboratory first of all macromorphological features are taken into consideration. These include: the size and the form of fructification; the pileus, its form, surface characteristics, shape and size of the pores; consistency/thickness of the context and tubes. The microscopic/micromorphological features are noted after sectioning the specimen along with the tubes, if present. These include: shape, size and dimensions of the hyphae, basidia, spores and other sterile elements. Sections from various zones of the fructification are observed for accurate and reliable conclusion. All the measurements concerning the elements of the fructification represent a mean of 30 reading concerning that element. Spores and hyphae are termed as amyloid if they turn blue or grey and dextrinoid if reddish brown. The following mounts and stains are used:

(a) 3 % KOH solution in water to record the colour and characters of the various structures *viz.*: hyphae, cystidia, basidia, basidiospores and setae etc.
(b) Cotton blue in lactic acid: it rapidly stains the cytoplasm of fungal cells. It is prepared by dissolving 0.01 g of cotton blue in 100 ml of 85 % of lactic acid while heating gently in a glass beaker. The cool solution is filtered and stored in a dark coloured bottle. For staining the hyphae, thin sections of the tissues of the context are revived in 3 % KOH solution, washed with excess of water and placed in cotton blue in lactic acid. The tissue is slightly warmed over a spirit lamp. This stain determines the cyanophyllous/acyanophyllous nature of the spores and other sterile structures depending upon whether the wall is stained blue or not.

Melzer's reagent: It is used to note the amyloidity or dextrinoidy of the various regions. Melzer's reagent is prepared as follows after Singer (1962): Iodine 0.5 g, Potassium iodide 1.5 g, Chloral hydrate 22.0 g and distilled water 20.0 ml.

Eosine and Phloxin: 1 g Eosine or Phloxin in 100 ml of water.

The sulfovanillin [distilled water 2.0 ml, Sulphuric acid (pure) 4.0 ml, Vanillin 0.5 g] is used for staining gloeocystidia. A fragment of dried/fresh specimen is put in the stain for 2–3 min and after crushing, it is observed. The structures stain black in the positive reaction.

All the measurements are taken in 3 % KOH solution. The diagrams have been made from a single collection. The structural details concerning hyphae, cystidia, basidia, basidiospores, basidiole and other sterile accessory structures have been drawn with the help of Olympus drawing attachment. The diagram of hyphae, basidia, cystidia and basidiospores have been drawn at a magnification of 2,000(×2,000) where as larger structures at ×400 or ×800 magnification. All the line diagrams have been provided scale. All the scales are equal to 10 μm.

The photographs have been provided only for specimens which have been photographed in the field or those whose morphology colour has not changed much after drying. No photographs have been given for the specimens studied from the herbarium. The following information is being provided for each genus in the order stated:

1. Name
2. Brief description
3. Reference to the legitimate report
4. Number of species reported in the world (after Kirk et al. 2008, Mycobank and Index fungorum)
5. Literature
6. Type species
7. Number of species being reported from the Himalayas.

The following information is being provided for individual species in the order stated:

1. Name of the species with reference to the published report. The synonyms/basionyms are given only for those species which have been recorded earlier under that name from Himalayas/India.
2. Description: Both macro and microscopic details. The description of each species is supplemented by the line drawings where ever possible coloured photograph of the specimen studied has been provided. The data concerning the frequency of occurrence (rare, common, abundant) along with the life form i.e. saprophytic/parasitic is also given.
3. Distribution: It includes the abbreviated name of the States (HP: Himachal Pradesh, UK: Uttarakhand, J & K: Jammu & Kashmir, N: Nepal, B: Bhutan, M: Manipur, Mi: Mizoram, WB: West Bengal, AP: Arunachal Pradesh, As: Assam, T: Tripura, Me: Meghalaya) along with stations.
4. Collection examined with herbarium number for the specimen studied along with abbreviated name of the collector (G. S. Rawla=RW/GSR, S. S. Rattan=R/SSR, R. S. Dhanda=D, M. P. Sharma=MPS, H. S. Khara=K/HSK, G. S. Dhingra=GSD, S. S. Virdi=SSV, I. B. Prasher=IBP and Lalita=L) Some of the collections have been deposited without the name of the collector. They are cited as such in the text. The collections examined (mentioned in this work) are those which have been worked out for making morphological, anatomical details. The other collections deposited in the herbarium are not mentioned.
5. Substratum.
6. Remarks

The photographs and line drawings provided in the monograph are the property/copy righted materials of the different agencies of the Government of India (U.G.C, CSIR (HRDG), MoEF) which is the outcome of the research projects undertaken by the author during the last 10 years. The author holds the right to publish this data; under the guidelines provided by different agencies; and is required to submit the published data to the agencies concerned.

References

Choi YW, Hyde KD, Ho WH (1999) Single spore isolation of fungi. Fungal Divers 3:29–38

Croan SC (2000) Lyophilization of hypha-forming tropical wood-inhabiting Basidiomycotina. Mycologia 92(4):810–817

Ito T (1991) Descriptive catalogue of IFO fungus collection XII. 91. *Trichobotrys effusa* (Berkeley & Broome) Petch. Research Communications Institute for Fermentation, Osaka 15:139–140, 144

Kirk PM, Cannon PF, Minter DW, Stalpers JA (2008) Dictionary of the fungi, 10th edn. Wallingford, Oxon

Kornerup A, Wanscher JH (1978) Methuen handbook of colour, 3rd edn. Eyre Methuen, London

Marx DH, Daniel WJ (1976) Maintaining cultures of ectomycorrhizal and plant pathogenic fungi in sterile water cold storage. Can J Microbiol 22:338–341

Ritcher DL, Bruhn JN (1989) Revival of saprotrophic and mycorrhizal basidiomycete cultures from cold storage in sterile water. Can J Microbiol 35:1055–1060

Singer R (1962) The Agaricales in modern taxonomy. Hafner, New York

http://www.indexfungorum.org/names/Names.asp

http://www.mycobank.org/Biolomics.aspx?Table=Mycobank_Advanced&Page=200&ViewMode=Basic

Chapter 4
Morphology

Abstract The general morphological details concerning the wood rotting non gilled agaricomycetes are discussed under the sections macromorphology and micromorphology. In the section concerning macromorphology, the vegetative and asexual stages, the general nature and type of the fructification, architecture of the fructification, is discussed. The micromorphology deals with the hyphae, covering various aspects like introduction of the hyphal system, type of hyphae and their spatial arrangement along with their role in the taxonomy of agaricomycetes. Besides the type of basidia, basidiospores and sterile structures (cystidia and their types) are discussed along with setae.

Morphological details of these fungi are studied with a naked eye or with a hand lens (macromorphology) which is primarily done before the study of different features with the help of microscope (micromorphology). The macromorphological studies help in the identification to a certain level only, whereas study of micromorphology is the only way to identify the genera and species correctly and exactly.

4.1 Macromorphology

4.1.1 Vegetative and Asexual Stage

The basidiospore germinates to produce the vegetative mycelium in the host after penetration. In some of the cases the asexual or anamorphic stages have been reported (Ryvarden 1991). The vegetative stage from which the basidiocarp/fructification develops on the host, is in the form of a network of branched hyphae inside the wood which can be living or dead. The aerial mycelium develops on the host surface from this endophytic mycelium. The aerial mycelium is in the form of mycelial mat or compact network of hyphae. The asexual and sexual stages may be distinctly separated as in the case of *Ptychogaster aurantiacus* (anamorph) and *Laetiporus sulphurous* (perfect stage).

I.B. Prasher, *Wood-rotting non-gilled Agaricomycetes of Himalayas*,
Fungal Diversity Research Series, DOI 10.1007/978-94-017-9858-7_4

4.1.2 *Fructification*

4.1.2.1 Type and Attachment

The mature fructification of these fungi have a characteristic type which is constant for a species. However, in some cases there may be transition from one type to the other with transition between resupinate & effused reflexed or between effused reflexed and pileate. The fructification may be semi-circular broadly attached, elongated broadly attached, dimidiate, cupulate-discoid with raised margins, flabelliform or spathulate (Figs. 4.1 and 4.2).

It may be resupinate or effused with upper surface attached to the substratum, sessile/bracket shaped laterally attached to the substratum and stipitate with pileus attached to the substratum by stipe. Depending up on the position of attachment of the stipe to the pileus the fructifications are termed as centrally stipitate, eccentrically stipitate and laterally stipitate.

The simplest type of fructifications are more or less effused with even, meruloid, poroid, warted or denticulate surface. The resupinate fructification is distinguished by the absence of sterile parts except margin. The effused reflexed species may sometimes develop pileus. The size of the fructifications vary from very minute to large. These are variously coloured.

The consistency; which is related to the structure as well as density of the tissue; also varies. The annual fructification with monomitic hyphal system are soft and usually shrink on drying, whereas the perennial fructifications, as in species of

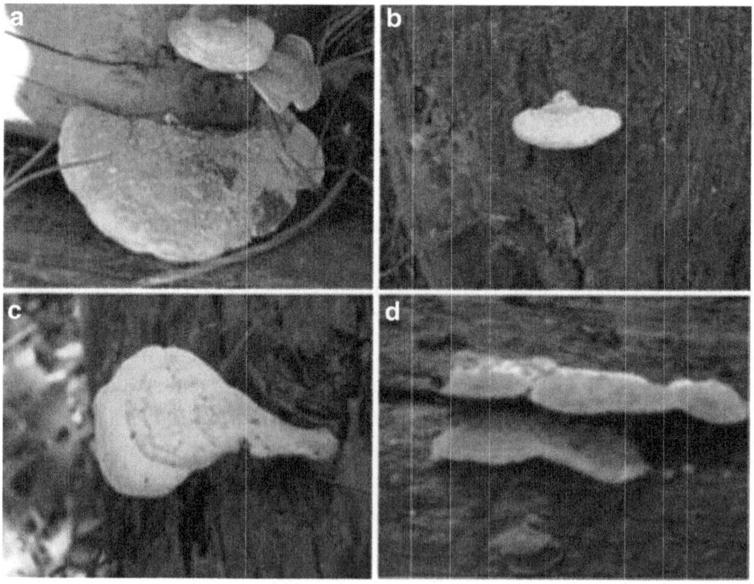

Fig. 4.1 (**a**) Broadly attached, (**b**) Dimidiate, (**c**) Flabelliform, (**c**) broadly attached

Fig. 4.2 (**a**) Triquetrous, (**b**) Applanate, (**c**) Convex, (**d**) Imbricate, (**e**) Resupinate, (**f**) Effused reflexed, (**g**) Ungulate, (**h**) Discoid, (**i**) Fan shaped-spathulate, (**j**) Stipitate

Phellinus (with abundance of skeletal hyphae) are woody in nature. The fructifications with trimitic hyphal system are tough. The resupinate fructifications may be loose to almost horny hard. They may be cotton like or floccose (Byssoid), mealy, minutely granulose with more or less discontinuous surface (Farinaceous), in the form of thin, usually pliable membrane over a loose subiculum (Athelioid/ Pellicular), membranous – more thicker than athelioid, closely adnate with waxy appearance (ceraceous) and dense and hard with warty appearance when fresh (phlebioid).

4.1.2.2 Architecture of the Fructification

4.1.2.2.1 Pileus

The pileus is protective in function. However, the resupinate type of fructifications are usually devoid of pileus. There is a wide variation in colour of pileus of different species. The color of the pileus changes with age in perennial genera/species (e.g., *Phellinus*) which are light colored where as those which are brightly or dark colored remain unchanged as with *Pycnoporous cinnabarinus* (Fig. 4.3).

The surface of the pileus may be glaborous to velvety, hirsuate tomemtose to villosa depending up on the type and density of the surface hairs. In the glaborous type, the surface may be smooth, shiny or dull and sometimes developing a black to reddish black cuticle as in *Ganoderma lucidum* (Fig. 4.4).

Fig. 4.3 *Pycnoporous cinnabarinus*

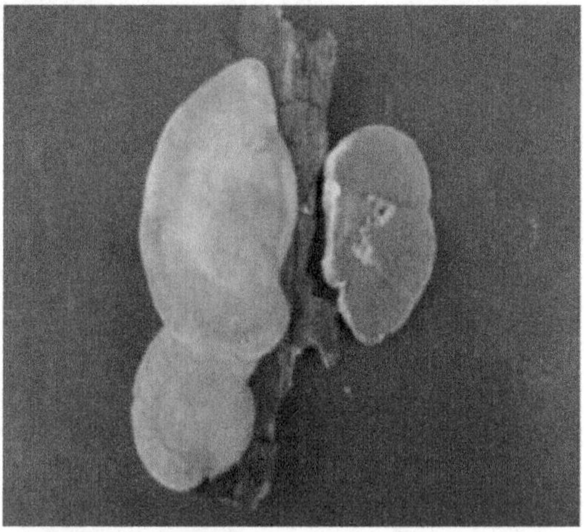

Fig. 4.4 *Ganoderma lucidum*

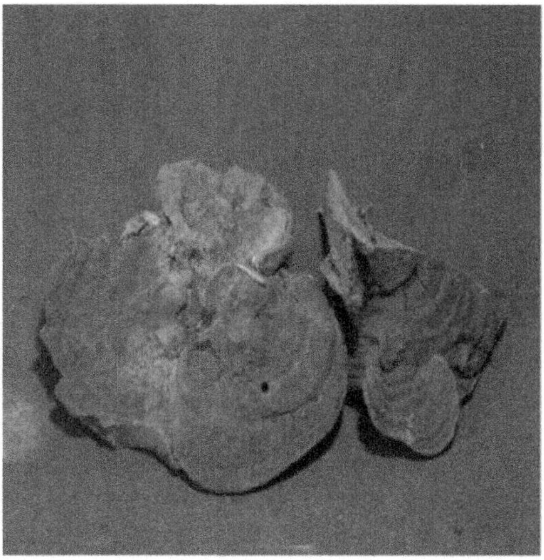

Fig. 4.5 *Phellinus fastuous*

The cuticle in this case consists of palisade like layer of hyphal endings. In the species of *Phellinus*, the surface is covered with more or less distinct hard and horny crust (Fig. 4.5).

The shape and type of hairs vary to large extent with the species and the change in atmospheric condition.

The glaborous pileal surface commonly supports sulcate zonation, indicative of different stages of growth, along with radial ridges which become more pronounced on drying. In *Phellinus rimosus* the surface of the fructification develops irregular cracks with age. Such a surface is termed as 'rimose'. The warts or tufts of hyphae on the surface of the pileus is due to the agglutinated hairs as in *Gloeophyllum sepiarium.*

In species like *Trametes versicolor* there is a distinct zonation on the pileus surface which is due to the falling of hairs of the pileus with age in a zone making the cuticle exposed. This dark colored cuticle makes a strong contrast with light colored tomentum (Fig. 4.6).

The color of pileus is an important character. It remains largely unchanged as in members of Hymenochaetaceae. The resistance to change in color in this case has been attributed to the presence of polyphenols which are resistant to degradation by natural causes. The increased pigmentation in basidiocarps with age results in increase in the intensity of the color.

4.1.2.2.2 Margin

In resupinate basidiocarps there are no sterile elements except margin. In some of these the margin become raised to give rise to discoid or cupulate type of fructifications. The texture of the margin varies with the species. It is continuous, soft and finally tomentose due to uniform growth of hyphae in majority of the fructifications. However, the rhizomorphs like strands of hyphae may develop in some species.

Fig. 4.6 *Trametes versicolor*

4.1.2.2.3 Stipe

The stipe may be central, lateral or eccentric. The stipe is concolorous with the pileus as in or it may be different.

4.1.2.2.4 Hymenophore/Pore Surface

The fructification is basically composed of context and hymenophore. The later bears the hymenium whereas, the context supports the hymenophore. The color of the hymenophore is peculiar and characteristic of the species. The shape of the pores may be circular, angular, oval, elongated, lamellate or hydnoid, irpicoid with irregular or flattened teeth, reticulate with network like ridges meruloid (network like folds), grandinoid (with small granules), odontoid with teeth or small spines, hydnoid (with prominent spines, tuberculate, sparse and irregular warts). The smooth type of hymenophore is fertile over the entire exposed surface except for the margins as in. The shape of the pore in many species do not remain stable and may change with age as is the case with species like *Daedalea quercina* and *Flavodon flavus*, where the pores may be circular and regular in young fructifications but become angular, oval, lamellate or hydnoid with age. The number of pores per mm or centimetre is recorded as an indicative of the size of the pores. In the case of elongated pores the size is taken tangentially to the margin.

4.1.2.2.5 Pore Tube

In majority of the species, the pore tubes are concolorous with the pore surface. The similarity or dissimilarity in the colour and consistency of the tubes and context varies with the species. In the poroid species the tube layer is organized as a layer of vertically placed tubes inseparably joined to the lower surface of the context in most of the cases. In perennial species of these fungi with every growth increment new tube layer is added over the surface of the old layer. The tubes open through the opening called pore surface.

4.1.2.2.6 Context/Subiculum

The sterile part of the fructification between the tubes and pileus surface is called context in poroid species whereas the sterile tissue between the hymenophore and the substratum is called subiculum in resupinate and non-poroid/hydnoid-species by some authors. In resupinate fungi, there is a layer of generative tissue/hyphae below the basidia usually growing vertically and branched. It lies between the basidia and subiculum. It is called subhymenium. It is very compact in some species and may be lacking in some resupinate taxa.

The context may be homogenous or heterogeneous with regard to consistency and colour. The homogenous context is radially structured whereas the heterogenous type is divided into upper soft and fibrous zone intermixing with pileal tomentum and lower hard zone without structural direction. The heterogeneous type of context is also termed as duplex context. The twin zones in duplex context in *Trametes versicolor* and in some species belonging to other genera are separated by a black zone. A mycelial core in the form of a spherical body close to the substrate, with a granular structure and impregnated with streaks or light coloured spots is present in the context of species like *Fomes fomentarius*. The subiculum is composed of loosely arranged hyphae which are more wider than the rest of the fructification and mostly grow parallel to the substratum.

4.2 Micromorphology

Microscopic examination of the specimens of these fungi reveal the micromorphology which is used to delimit genera and species. The different elements of the fructification which are included in the micromorphology are: Hyphae, Basidia, Basidiospores, Cystidia and sterile structures and asexual spores.

4.2.1 Hyphae

It was Ames (1913) who pointed the importance of hyphae in polyporoid fungi. However, Corner 1932 and 1953 introduced the currently used concept of hyphal system for the delimitation of taxa and emphasised the importance of hyphal system in the classification of these fungi. Donk (1964) introduced the term sclerified generative hyphae for clamped hyphae or apical/intercalary hyphae segments with thickened walls. Pouzer (1966) suggested the term 'vegetative hyphae' for binding and skeletal hyphae which often cannot be separated morphologically. The generative hyphae are septate. The nature of the septation; whether it is a simple septum or arises due to the formation of a clamp; is an important diagnostic character for the delimitation of taxa in Agaricomycetes. The clamps can be either simple or multiple clamps on one septa. The multiple clamps are of rare occurrence in corticoid fungi. The type of septation in a species; as in the case with majority of taxa; is either simple or clamped. But in some species of *Polyporous* Michel. ex. Adams; clamp connections and simple septa appear in the same hyphae. Similarly species like *Ceriporia excelsa* and *C. viridans* support multiple clamps and simple septa on the generative hyphae. In genera like *Athelia* and *Phanerochaete*, the hyphae in the subiculum are sparsely clamped whereas the hyphae in other parts of the fructifications are simple septate. The diameter, thickness of wall, type of septation, nature of branching and contents of the generative hyphae varies from species to species. They may be thin to thick walled to solid. Thick walled generative hyphae are called sclerified generative hyphae (Donk 1964) and have been reported in *Rigidoporous* and *Microporellus*. These often show broken clamps at the free ends as in *Hexagonia badia*. Irregular thickening of the walls of the hyphae have been reported in the basidiocarp of *Daedalea* spp. and also in hyphae produced in the culture and is considered a character of taxonomic value (Roy and De 1996). In the mature basidiocarps, the generative hyphae are best observed in the growing margin and the dissepiment edges. Teixiera (1956) used the term arboriform skeletal hyphae and acicular skeletal hyphae for the skeletal hyphae in Ganodermataceae. In the present text hyphae are broadly classified as: generative hyphae and vegetative hyphae. The later is further classified as skeletal and binding hyphae. A species is termed as monomitic; if only one type of hyphae are present, whereas dimitic and trimitic if two or three type of hyphae are present. In other words monomitic fructification are made up of only generative hyphae, dimitic refers to the presence of generative and binding or skeletal hyphae in the fructification. Whereas the trimitic refers to the presence of generative and vegetative (binding and skeletal hyphae) hyphae in the fructification. Fructification consistency in relation to hyphal system: fructification consistency has been found to be directly related to its hyphal construction. A monomitic fructification is usually soft, fleshy and shrivels on drying. A fructification with dimitic hyphal system comprised of generative and skeletal hyphae are normally ligneous whereas fructification with generative and binding hyphae are fibrous in consistency. Trimitic fructifications are normally suberose and coriaceous in consistency.

4.2.1.1 Generative Hyphae

All the fructifications contain the generative hyphae since these are the only hyphae present in the initial stages of the fructification development and in some cases entirely make the fructifications till maturity. The hymenium is made up of entirely by the generative hyphae. The vegetative hyphae (comprises of skeletal and binding hyphae) arise from the generative hyphae. In some species like *Daedalea*

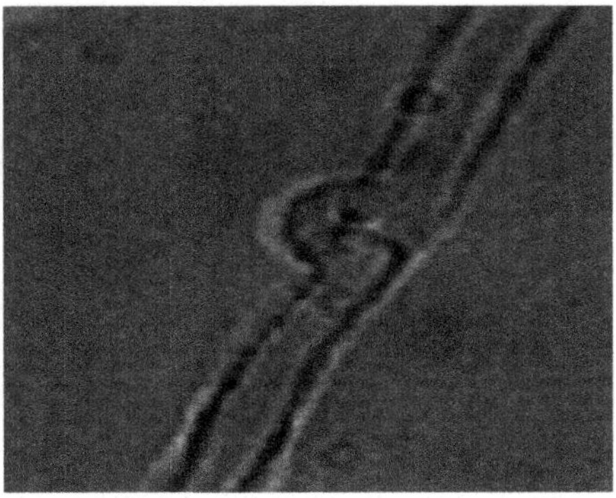

Fig. 4.7 Generative hyphae with clamp

astreiformis, *Fomitopsis officinalis*, *Rigidoporous vinctus* and *Amylosporus campbelli*, the generative hyphae are wide, thin walled and rarely showing septation. These are termed as Gloeoplerous hyphae. Their content stain deeply in Melzer's reagent. Hyphal system in a genus or a species is a stable character and is of great taxonomic importance (Fig. 4.7).

4.2.1.2 Vegetative Hyphae

These include the non-septate hyphae which originate from the septum or clamp. These are divided into binding and skeletal hyphae. The binding hyphae are normally of restricted growth whereas skeletal hyphae are of unlimited growth.

4.2.1.3 Skeletal Hyphae

These are thick walled, usually aseptate and unbranched. But in some species they are branched towards the distal ends as has been observed in *Trametes versicolor*. These have been termed as arboriform type of skeletal hyphae by Teixiera (1962). The skeletal hyphae normally arise from septum but may also arise from binding hyphae as in the case of *Microporous xanthopus* (Corner 1932). The thick walled skeletal hyphae become solid later on due to the obliteration of the lumen. The skeletal hyphae may develop septa towards the apex (Fig. 4.8).

4.2.1.4 Binding Hyphae

These are branched, solid to thick walled, and are of limited growth. The branches are long tapering. Binding hyphae normally arise from the septum e.g., in *Laetiporous* and *Perenniporia* but in some cases they also arise from the skeletal hyphae as side branches. Binding hyphae are very rare in corticoid fungi.

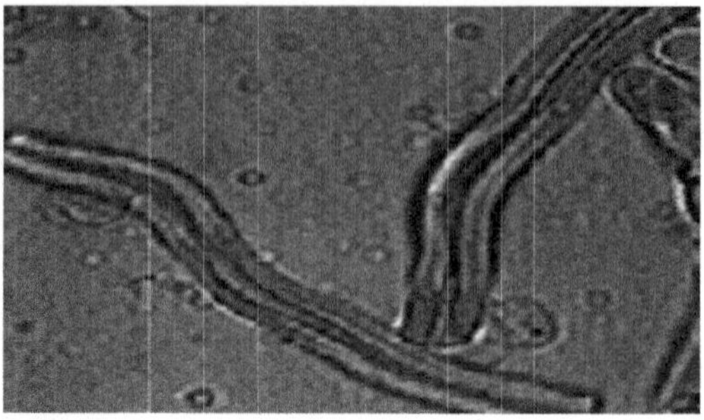

Fig. 4.8 Skeletal hyphae

Pileus cover in relation to hyphal construction: In monomitic species the tomentum is made up of generative hyphae. These hyphae are with considerably thickened walls and are much wider than hyphae forming the trama.

Hyphal structure in relation to context, trama and dissepiments: Hyphal structure between context and trama may vary in species. The context of a fructification varies with age with respect to hyphal construction. The young developing basidiocarp is generally entirely composed of generative hyphae whereas the developed fructification may contain only the vegetative hyphae.

The hyphae in trama are normally ornamented and parallel to each other. Central tramal region in some species of *Phellinus* is composed of vegetative hyphae and these are surrounded by generative hyphae forming the subhymenium which give rise to basidia. Skeletal hyphae may bend in to the hymenium to form cystidia. In addition to these, gloeoplerous hyphae and skeletal hyphae may also occur.

The dissepiments in most of the species are made up of sterile hyphal ends of either generative hyphae or generative hyphae mixed with skeletal hyphae.

Some of the thick walled hyphae in species of genera like *Daedalopsis* and *Skeletocutis* are encrusted. Similarly the hyphae present in the trama sometimes become modified only in the terminal part. Then these are called hyphidia. Dendrohyphidia (irregularly branched), Dichohyphidia (dichotomously branched), Asterohyphidia (stellate branching), Acanthohyphidia (bottle-brush like). Similarly when the hyphae with similar characters are present in trama then the word hyphae with corresponding prefix is used (Hjortstam et al. 1987) (Fig. 4.9).

4.2.2 Basidia

The cells of the hymenium where karyogamy followed by meiosis leading to the development of basidiospores on sterigmata occurs are called basidia. The basidia in most of these fungi are four spored and are clavate and hyaline. However in corticoid fungi there is great variation in size and shape of the basidium as well as in the number of stigmata which varies from 2 to 8. The occurrence of more than four sterigmata is a diagnostic character of all the species of genera like *Paullicorticium, Botryobasidium, Sistotremastrum* and *Sistotremella* and in some species of

Fig. 4.9 Binding hyphae

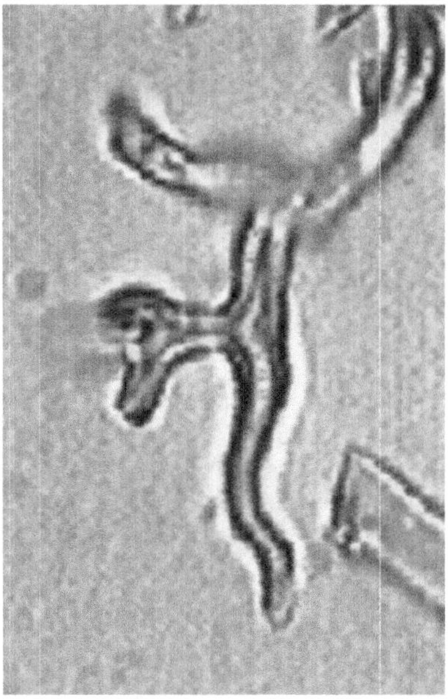

Sistotrema. The development of basidium is normally terminal but in some genera of Corticiaceae they may develop laterally. These are then called pleurobasidia (Fig. 4.10).

Internal proliferation of basidium is a diagnostic character of the genera like *Repepatobasidium, Repetobasidiellum, Galzinia* and species of *Gloeocystidiellum*. The basidia support a basal clamp if the generative hyphae in the context and trama are clamped.

4.2.2.1 Basidiole

Refers to immature and sterile basidium.

4.2.3 Basidiospores

Spore size, shape, ornamentation and colors are important characters of the genera and families. Shape of the spore also varies considerably. They may be globose, subglobose or ellipsoid etc. Cylindrical and allantoids spores are not common. These are also important in the delimitation of species. Basidiospores in large number of the genera are smooth but in some of the genera they are ornamented as in case of *Ganoderma*. The ornamentation in this case is in the form of prominent warts or protruberances or they are minutely warted as in case of *Heterobasidion* and *Amylosporus*. Besides the ornamentation, the thickness of the wall is also a

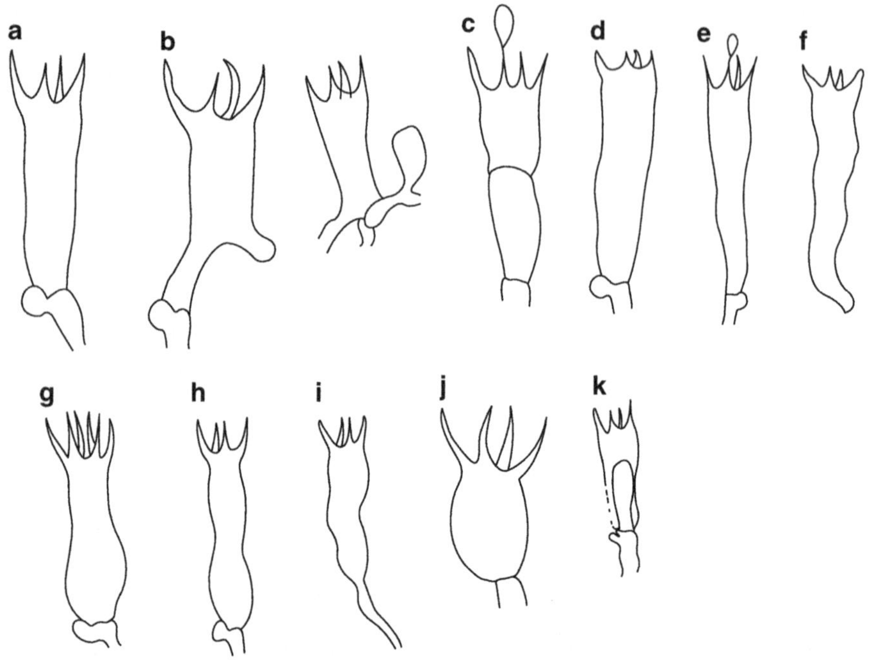

Fig. 4.10 Type of basidia (**a**) Terminal, (**b**) Pleurobasidium, (**c**) Secondarily septate, (**d**) Cylindrical, (**e**) Clavate, (**f**) Constricted, (**g**) urniform, (**h**) Utriform, (**i**) Stalked, (**j**) Subglobose, (**k**) Repetetive

distinctive feature as in the case with *Hypochniciellum* and some other corticoid genera. In genera like *Ceratobasidiom* the spore germinates to produce a sterigmata like structure which again bears a secondary spore. This process is called spore-repetition. Spore size also varies considerably (Fig. 4.11).

4.2.4 Cystidia and Other Sterile Elements

Cystidia occur in large number of genera and are an important characteristic for delimitation of genera. Based on their origin they are termed as Hymenial cystidia (originate in hymenial layer)- which are thin to thick walled, smooth or encrusted at the apex and are clavate to ventricose. Tramal/pseudocystidia arise in trama or subiculum. They are normally terminal ends of the skeletal hyphae which project into or beyond the hymenium. These are also called skeletocystidia and are sometimes found to occur in pores e.g., in *Hexagonia*. In addition to the hymenial and tramal cystidia various cystidial types and sterile organs (Fig. 4.12) reported to be associated are as follows:

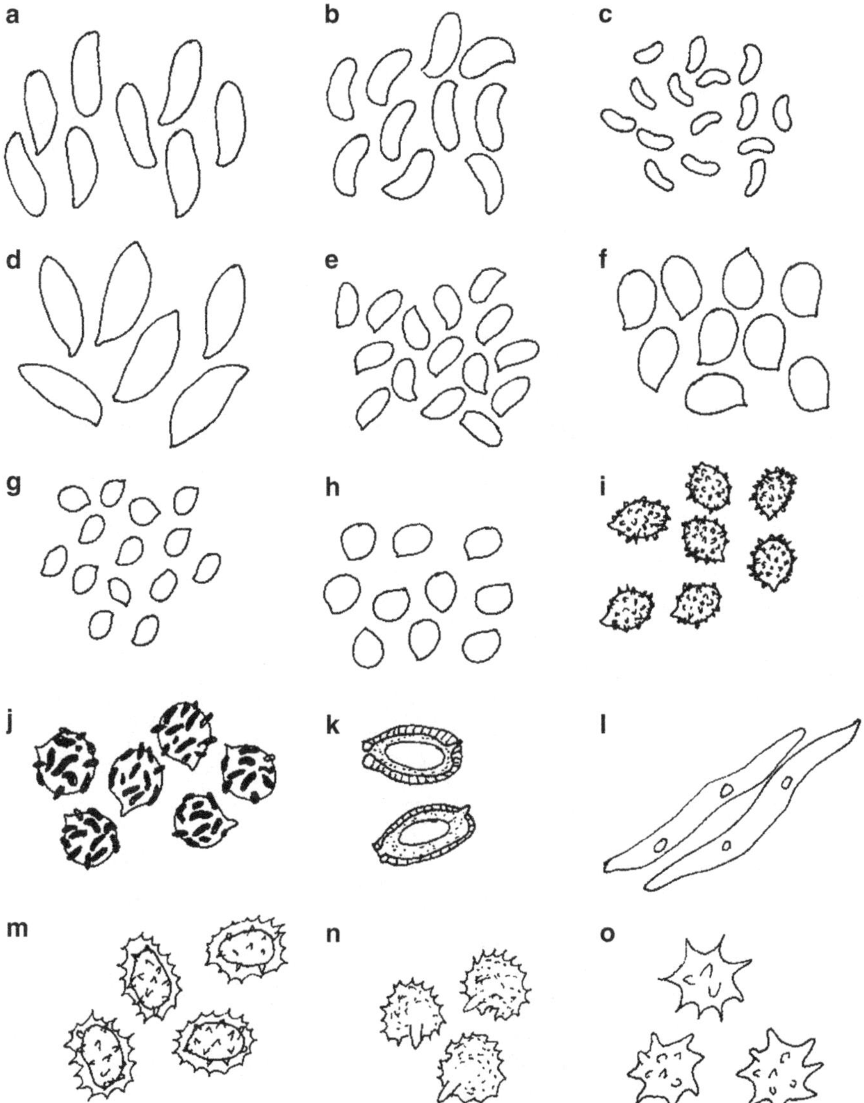

Fig. 4.11 Type of the spores: (**a**) Cylindric, (**b**) Allantoid, (**c**) Lunate, (**d**) Naviculare, (**e**) Oblong ellipsoid, (**f**) Broadly ellipsoid, (**g**) Subglobose, (**h**) Globose, (**i**) Warted, (**j**) Crested, (**k**) Ganodermatoid, (**l**) Sigmoid (Fusoid), (**m**) Spiny, (**n**) Irregular, (**o**) Tuberculate

4.2.4.1 Leptocystidia

These are thin to slightly thick walled, encrusted, cylindrical to fusoid or conical and mostly confined to hymenium. They may be sometimes constricted and are then called moniliform cystidia.

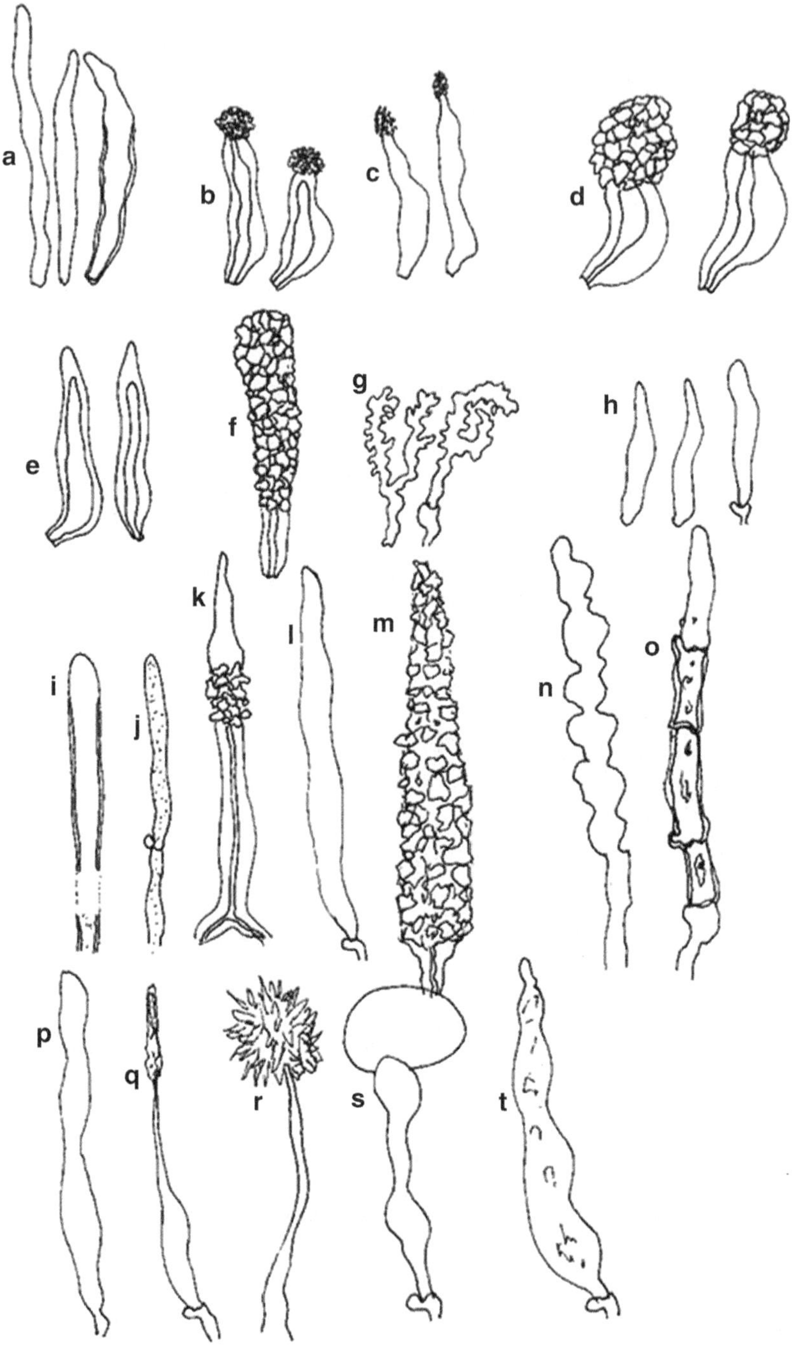

Fig. 4.12 Type of cystidia: (**a**) Smooth tubular, (**b**) Thick walled apically encrusted, (**c**) Thin walled apically encrusted, (**d**) Apically coarsely encrusted, (**e**) Ventricose, (**f**) Coarsely encrusted, (**g**) Dendrohyphidia, (**h**) Cystidioles, (**i**) Pseudocystidia, (**j**) Hyphoidcystidia, (**k**) Lycocystidium, (**l**) Leptocystidia, (**m**) Lamprocystidium, (**n**) Moniliform, (**o**) Septocystidia, (**p**) Subulate, (**q**) Lageniform, (**r**) Asterocystidium, (**s**) Halocystidium, (**t**) Schizopapillate

4.2.4.2 Metuloids

These are sometimes called Lamprocystidia. These are made up of two parts: basal thin walled with wide lumen and devoid of any encrustations and upper thick walled, conical part encrusted with crystals and with very narrow lumen. Sometimes the basal part is colored as in *Peniophora.*

4.2.4.3 Lycocystidia

In genera like *Tubulicrinis*, *Litschcuerella* and *Tubulicium* the cystidia have a thick refractive wall with capillary like lumen which widens near apex which is thin walled. These are 2-multirooted.

4.2.4.4 Gloeocystidia

These are hymenial cystidia and are hyaline, smooth, tubular structures with refractive contents e.g., *Heteroporus biennis.* They stain blue with sulphovanilline. Regularly constricted gloeocystidia are termed moniliform cystidia (Hjortstam et al. 1987).

4.2.4.5 Lagenocystidia

These are having a broader basal part and a needle like apical part which is encrusted e.g., species of *Hyphodontia.*

4.2.4.6 Cystidioles

These are hyaline, thin walled, smooth walled structures of the size of basidia but tapering at the apex beyond basidia.

4.2.4.7 Hyphidia

The modified terminal hyphae in the hymenium are termed as hyphidia.

4.2.5 Setae and Setal Hyphae

These are sterile, dark brown coloured prominent elements found in the hymenium and context of some species. They are of importance in the delimitation of species in some of the genera as in case of Hymenochaetaceae. They become prominent and dark coloured in KOH-Phloxine solution. Depending upon their position they are hymenial setae and tramal setae. Hymenial setae originate from the terminal ends of the generative hyphae at the level of basidia and normally project beyond the hymenial layer. The terminal part of these may be hooked or bent or straight. The characteristic features of these setae are constant for a species. The tramal setae originate in the trama where they are embedded and usually run parallel to the wall of the tubes. These are normally conical with the widest middle part and tapering towards the apex. These are star like as in *Asterostroma* with a central part having 3–7 radiating rays. These are conspicuous throughout the context and hymenium. Setal hyphae are transformed hyphal elements with setae like terminal parts which are embedded in the walls of the tubes and the context. These run straight and parallel to the wall of the tubes or sometimes their apex is bent into the hymenium. They can be distinguished from the tramal setae in not being swollen in the central part and being much longer. These are conspicuous in some species of *Inonotus* and in species belonging to *Phellinus*. They may or may not be present along with the hymenial setae.

References

Ames A (1913) A consideration of structure in relation to genera of Polyporaceae. Ann Mycol 11:211–253

Corner EJH (1932) The identification of the brown-root fungus. Gard Bull Straits Settlem 5:317–350

Corner EJH (1953) The constructon of polypores. Phytomorphology 3:152–167

Donk MA (1964) A conspectus of the families of Aphyllophorales. Fungus 3:199–324

Hjortstam K, Larsson KH, Ryvarden L (1987) The Corticiaceae of North Europe 1. Introduction and keys. Fungiflora, Oslo

Pouzer Z (1966) Studies in the taxonomy of the Polypores II. Folia Geobot Phytotax 1:356–375

Roy A, De AB (1996) Polyporaceae of India. International Book Distributors, Dehradun

Ryvarden L (1991) Genera of polypores. Nomenclature and taxonomy. Synopsis Fungorum Oslo Fungiflora 5:1–363

Teixiera AR (1956) Metodopara studio das hifas do carpoforo de fungos Poliporaceos. Publ Inst Bot S, Paulo, pp 1–22

Teixiera AR (1962) Microstructures do basidiocarpo e sistematica do genero *Fomes* (Fr.) Kickx. Rickia 1:15–93

Chapter 5
Forest Types in Himalayas

Abstract A detailed account of the forest types occurring in the different regions of the Himalayas *viz*: Eastern Himalayas and North-Western Himalayas is provided to give an account of the dominant tree vegetation along with the undercover flora covering the forest floor. A brief introduction regarding the classification of forest types provided by different authors is also provided. In addition to this the name of the localities from where the specimens have been collected along with dominant vegetational elements is given.

India is a tropical country with the tropic of cancer passing right through its median and its southern end. The climate of the Indian subcontinent is markedly influenced by the Indian Ocean in the south, the Himalayas in the north, extending from east to west. According to Puri (1960), the well-known phenomenon of latitudinal differences in vegetation witnessed in Europe and America is not visible in India. The natural vegetation of the country, in response to marked climatic variation and edaphic diversities, is rich and variegated (Mehra et al. 1971). Accordingly there is a great variation in nature and composition of forests. Champion (1936) published the first comprehensive classification of Indian vegetation, followed by Puri (1960) almost in the same pattern. Champion and Seth (1968) have classified the Indian forests into 16 types.

India has been divided into various floristic regions by different workers. Clarke (1898), Hooker (1907), Chatterjee (1939), Champion and Seth have divided India into 11, 9, 10 and 9 botanical regions respectively. The Western Himalayas (North-Western considered in this work) and Eastern Himalayas are part of these classified floristic regions.

The Eastern Himalayas receives more rainfall then the N.W. Himalayas but less snowfall. The N.W. Himalayas differ from the Eastern Himalayas in greater representation of conifers. In the N.W. Himalayas European elements are conspicuous where as the Malayan, Chinese and Burmese elements are more pronounced in the Eastern Himalayan vegetation.

Since the type of forest ecosystem plays an important role in the growth and mycogeographic distribution of fungi, it will be pertinent to discuss the vegetation of these regions from where collection of these fungi have been made.

I.B. Prasher, *Wood-rotting non-gilled Agaricomycetes of Himalayas*,
Fungal Diversity Research Series, DOI 10.1007/978-94-017-9858-7_5

5.1 The Eastern Himalayas

It is approximately 720 km long and comprises the portion between the Namcha Barba peak (7,750 m), east of which the river Brahmputra curves southwards, and the river Teesta in the west. In this region, there is very little of the sub Himalayan tract so that the Himalaya rises rather abruptly from the plain. The snowline in the region is at an elevation of 4,875 m, but glaciers come down by about 1,000 m below these limits. The rocks are of Pre-cambrian age. Conveniently, the surveyed areas can be classified into the following two parts:

5.1.1 The Eastern Himalayas

The Eastern Himalayas proper, which traverse Darjeeling (West Bengal), Sikkim and Arunachal Pradesh of the Indian territory and the Royal Kingdom of Bhutan.

5.1.2 The Assam Hills

The Assam hills i.e., the hills of undivided Assam state (excluding Arunachal Pradesh).

Following is the vegetational account of the areas from where the collections have been made in the Eastern Himalayas:-

5.1.3 The Eastern Himalayas

5.1.3.1 Darjeeling

Darjeeling region, during the monsoon season, receives average 2,600 mm of the total 3,165 mm of the average rainfall. Mean maximum yearly temperature, is about 16 ° C and the minimum temperature is 10 ° C. Humidity is almost near the saturation point (95 %) during the monsoon season.

Mehra and Bir (1964) have classified the vegetation of this region into five categories:

5.1.3.1.1 Low Hill Forests (Tropical Type, up to 900 m)

From the base town of Siliguri up to 900 m, the submountanous tracts are inhabitated by dense broad-leaved semi-evergreen forests. *Shorea robusta* is the dominant tree. Other prominent trees are: *Schima wallichii*, *Bauhinia purpurea*, *Cedrella toona*, *Salmalia malabarica*, *Lagerstroemia parviflora* and *Terminalia myriocarpa*.

Other associated trees are: *Garuga pinnata*, *Cedrela microcarpa*, *Amoora wallichii*, *Evodia mellaefolia* and *Eugenia formosa*.

5.1.3.1.2 Middle Hill Forests (Subtropical Type, 750–1,500 m)

These are characterised by evergreen species with the dense undergrowth is. *Castanopsis indica*, *C. tribuloides*, *Phoe bahainesiana*, *Schima wallichii*, are commonest species between 750 and 1,200 m. Other associated species include: *Drimycarpus racamusus*, *Juglans regia*, *Spondias lutea*, *Ficus cunia*, *Betula alnoides* etc. Large trees of *Quercus glauca, Q. spicata, Q. serrata* and *Q. griffithii* grow in dense formation between 1,200 and 1,600 m along with *Bambusa* spp. which are prominent near habitations.

These forests now occur in few patches because most of the mountains forests between 600 and 1,200 m, have been cut for tea plantation. *Cryptomeria japonica* forms dense growth from 1,200 to 2,400 m.

5.1.3.1.3 Upper Hill Forests (Warm or Wet Temperate Type, 1,500–2,700 m)

These are evergreen forests but with scattered deciduous species also occurring here and there. Oaks and laurels constitute the dominant vegetation. The forests are thick, extremely humid and have dense under growth. Between 1,800 and 2,100 m, *Quercus lamellose* is very common; between 2,100 and 2,400 m, *Q. pachyphylla* is common; between 2,400 and 2,700 m, common trees are *Michelia cathcartii*, *Magnolia campbellii*, *Quercus fenestrata* and *Castanopsis tribuloides. Acerc campbellii* and *Michelia excelsa* constitute the dominant vegetation. Oaks constitute bulk of the forest in upper areas. At certain places, *Rhododendron griffithianum* grows luxuriantly.

5.1.3.1.4 Rhododendron-Conifer Zone (Cold Temperate to Sub-alpine, 2,700–3,600 m)

In the lower region, often oaks and laurels are common but gradually upward *Rhododendron* replace these broad-leaved trees. The common species of *Rhododendron* are: *R. arboreum*, *R. campanulatum* and *R. grande*. Associated conifers are: *Taxus baccata*, *Tsuga brunoniana*, and *Abies densa. Tsuga brunoniana* is the dominant tree at an elevation of 2,700–3,000 m. *Abies densa* occurs in pure formation at certain places.

5.1.3.1.5 Alpine Scrubs (3,600–4,300 m)

At 3,600 m where the tree lines ends, and the exposed hill sides are covered by *Juniperus pseudo-sabina* and *J. recurva*.

5.1.3.2 Bhutan

The mountain ranges in Bhutan stretch mostly north-south. The Bhutan vegetation is a transition between the floras of Sikkim and West China. The forests of Bhutan can be classified into the following types:

5.1.3.2.1 Tropical and Subtropical Forests

The outer spurs receive heavy rainfall and are as a consequence densely forested with tropical and subtropical vegetation. The density of vegetation in the valleys, however, varies with the topography and the amount of rainfall. The interior of valleys, which are disposed in north-south direction receives the moisture-laden winds and bear a luxuriant vegetation. The laterally disposed valleys, which receive comparatively less rainfall, have bare slopes and support moist deciduous forests. *Shorea* and *Pinus* species are present here and there. *Shorea* when present, dominates the vegetation and occurs along with *Bauhenia* spp., *Careya arborea*, *Dillenia* spp., *Melastoma*, *Syzygium* species along with other.

In the tropical evergreen forests along the river banks, the dominant genera are *Dillenia*, *Duabanga*, *Hydnocarpus* and *Talauma*, with numerous climbers, epiphytes and orchids. In the *Quercus-Rhododendron-Schima* forests, *Quercus griffithii* is the dominant tree, associated with *Castanopsis* and *Engelhardtia*. *Pinus roxburghii* reaches a lower limit at about 600 m. On the dry slopes of the lower valleys *Euphorbia royleana* is the commonly met occurring species.

5.1.3.2.2 Temperate Forests

The *Rhododendron*, *Acer* and *Betula* species are the prominent trees. In some ranges, the forest is dominated by magnolias, oaks, *Strobilanthes* and others. In the inner valleys are also found the characteristic temperate conifer forests, consisting of *Abies spectabilis* (*A. densa*), *Larix griffithiana*, *Picea spinulosa*, *Pinus* spp. and *Tsuga dumosa* in the interior valleys. *Larix griffithiana* may occur in association with of *Rhododendrons*.

5.1.3.2.3 Alpine Forests

In the alpine zone, the vegetation varies between the moist alpine type on the south and the dry steppe type on the northern aspects. *Primula sikkimensis* is common in moist alpine zone.

5.1.3.3 Kameng (Arunachal Pradesh)

The vegetation of Kameng can be described under the following heads:

5.1.3.3.1 Tropical Forests

Species of *Ficus*, *Stercularia*, *Syzygium*, *Terminalia*, along with *Dubanga grandiflora*, constitute the main vegetation in the semi-evergreen forests at about 900 m. *Caryota* sp. and the screw-pine *Pandanus* also found occur in the forests of this zone.

5.1.3.3.2 Subtropical Forests

Mixed forests of *Ficus-Castanopsis-Callicarpa* occur on the lower ridges and *Schima-Castanopsis-Engelhardtia-Saurauja* association on the higher ridges. *Rhododendron-Lyonia* forests occur on the drier aspects of hills, and in the deep river valleys *Albizia*, *Morus* and bamboos are wide spread.

5.1.3.3.3 Temperate Forests

The hill tops and valleys in the temperate region support, mixed forests of *Acer*, *Betula*, *Juglans*, *Magnolia*, *Michelia*, *Quercus*, *Rhododendron* and others along with *Bambusa* spp. The temperate forests at higher altitude have a different composition. These have a dominant vegetation of *Rhododendron*, *Pyrus* and *Tsuga* among the trees. The temperate conifer forests are chiefly comprise of *Pinus wallichiana*, associated with *Rhododendron*, *Quercus* spp. and *Lyonia*. Epiphytes, particularly orchids are very conspicuous in this zone.

5.1.3.3.4 Subalpine to Alpine Forests

The subalpine vegetation includes the *Abies spectabilis*, with shrubby and bushy Rhododendrons, Junipers, *Berberis*, *Cotoneaster*, *Salix* and others. The alpine zone is devoid of any trees.

5.1.4 The Assam Hills

The much broken Assam hills comprise of into serrated ridges and deep valleys so that most of the region south of the Eastern Himalayas proper appears to be an undulating land. The area receives 2,500–2,600 mm average annual rainfall, of which as much as 65.80 % (1632.20 mm) is received during the monsoon months of June to September.

The mean daily minimum and maximum temperature during monsoon months is between 24 and 31 °C.

The Garo, Khasi and Jaintia hills from the western part of this range of hills (sometimes referred to as trans range). The southern and the central parts of the area constitute the Shillong Plateau, about 1,200–1,930 m above mean sea level. Shillong peak (1,930 m). The Shillong Plateau is also known as Meghalaya (the abode of clouds). The Cachar, Tripura, Manipur and Naga are other main hills under this division.

The vegetation of all these hills can be broadly classified in the following types:

5.1.4.1 The Tropical Vegetation

It typically extends up to elevations of 900 m and comprises of evergreen and semi-evergreen forests, dry and moist deciduos forests and grasslands. Tropical evergreen forests are found in Assam valley, in the lower pats of Naga, Manipur and Tripura hills. The tallest trees which constitute the upper canopy are: *Canarium resiniferum, Artocarpus chaplasha, Dipterocarpus turbinatus, Ailanthus grandis, Kavea assamica, Euphorea longana, Phoebe goalparensis, Terminalia chebula, Toona ciliate, Dysoxylum binactriferum, Dubanga grandiflora* and *Dillenia indica*. The lower canopy is constituted by trees as: *Amoora wallichii, Ficus rumphii, Lagerstromoemia parviflora* and *Terminalia myrocarpa*. The forests floor is covered with profuse growth of herbs and ferns, the most conspicuous element being the tree fern *Cyathea* sp.

The deciduous forests include the sal forests, with *Shorea robusta* as dominating tree. Such forests were located in the lower elevations of Khasi and Garo hills of Shillong Plateau and in some parts of North-Cachar hills. The other associated trees are *Careya arborea, Kydia* calycina, *Sterculia villosa, Bombax ceiba, Terminalia* sp., *Bauhinia* sp., *Acacia* sp., *Albizia* sp., *Adina cordifolia* and *Gamelina arborea*. The herbaceous vegetation and ferns are less profuse.

5.1.4.2 The Subtropical Vegetation

This vegetation is found in the areas extending up to 1,500 m above mean sea level and include associations of *Castanopsis, Schima, Engelhardtia, Terminalia, Ficus, Michelia* and *Albizia*.

5.1.4.3 The Temperate Vegetation

It occurs at elevations from about 1,300 to 2,500 m. In several localities there is a mixed temperate vegetation with subtropical vegetation. *Albizia, Acer, Juglans, Quercus, Magnolia* and *Michelia* occur with *Rhododendron rubus* and sprinkling of *Arundinaria*.

The temperate vegetation in the Khasi and Jaintia hills of Shillong Plateau, especially originally studied by Hooker (1854) and then subsequently by Bor (1942) and Rao (1969). These represent relicts, amdist a much disturbed and altered vegetation, which is due to the devastating practice of JHUMING, a kind of primitive agricultural practices, involving a large scale cutting down and burning of trees before cultivation. The small pockets of 'Sacred Forests' left untouched due to religious beliefs, present a glimpse of the original forest that must have once extended through. They are present in saucer-shaped depressions, amidst rolling grasslands, and often have little mountain stream manding through them. Fagaceae with *Quercus* and *Castanopsis* spp., Rosaceae with *Rosa, Photinia, Eriobotrya, Pyrus, Prunus, Sorbus* and several other shrubby and herbaceous species, *Carylopsis, Albizia, Manglieta, Acer* occur in them. The forest floor has dense carpet of herbaceous vegetation. A comparable kind of temperate vegetation occurs at compatible altitudes of the Naga Hills.

An important result of human influence from pre-historic times in these hills, is the intrusion and spread of *Pinus insularis*. In the Khasi and Jiantia hills of the Shillong Plateau this pine makes its appearance at about 900 m and forms extensive pure grooves at higher altitudes, giving the landscape a parkland appearance with the inter spread rolling grasslands. These pine-woods may contains sometimes sprinkling of Symplocos, *Schiima wallichii* and *Schiima khasiana*. The floor underneath is thick with a carpet of pine needles, and as may be expected, is devoid of any vegetation, but rich in fungal growth.

5.1.4.4 The Sub Alpine Vegetation

It occurs at altitudes of 3,000–4,500 in the Naga and Manipur hills. The dominant trees are species of *Abies*, *Tsuga* and *Picea* with the dense bushy zone of *Rhododendron, Juniperus, Recurva, Berberis, Salix* and *Rubus.*

The Localities from Where Collection Have Been Made

The species described in this monograph have been collected from the forest around the following places in the Eastern Himalayas and adjoining hills by different workers. The main station is the one which is written first, the left hand side, with its altitude. The localities around the main station and the nearby smaller and less known places, along with the range of their altitude, are put within parentheses on the right hand side. The type of forest is also mentioned against each.

A. INDIA

I. West Bengal

1. Siliguri	60 m	(Rongtong, Sukhna, Sevoke, 60–260 m). Tropical angiospermic forests.
2. Kalimpong	1,209 m	(Aligarh, Labha, 1,200–1,500 m) subtropical angiospermic forests with patches of *Cryptomeria japonica*
3. Darjeeling	2,265 m	(Lloyd Botanic gardens, Zoological park, Lebong, GhoomTista road, Takdah cantt., Senchal lake, Tiger Hill, Tung, Sonada, Sukiapokhri, Simana, Chetaidhura, Merrybong, Menebhanjang, Tonglu, Batasi, Dhotrey, Palamajuha, 1,500–2,700 m). Temperate angiospermic forests. Pure formations of *Cryptomeria japonica*, mixed with Angiospermic vegetation at number of places.

II. Arunachal Pradesh (A.P.)

1. Tipi	300 m	Tropical angiospermous forests
2. Jamiri	1,200 m	(On Jamiri-Buragaon road, 1,200–1,500 m). Subtropical Angiospermous forests.

3. Rupa	1,500 m	(Rupa to Shergaon road, 2,100 m). Temperate mixed Forests.
4. Dirang	1,650 m	(Dirang-Bomdila road, Dirang, Gompha, 1,500–1,700 m). Temperate angiospermous forests; mixed with pine.
5. Shergaon	2,100 m	(Shergaon Basti side, Shergaon-Rupa road, 2,000–2,500 m). Temperate angiospermous forests, mixed at some places.
6. Bomdila	2,700 m	(Ne Bomdila, Wang Basti, Bombila-Munna road, 2,500–2,700 m). Temperate angiospermous forests; mixed with Pines.
III. Manipur		
1. Imphal	792 m	(Kangchup, 800–1,000 m) Subtropical forests with patches of *Pinus* spp.
2. Ukhrul	1,860 m	(Imphal road, Sirohi peak, 1,800–2,000 m) Temperate Forests, patches of coniferous and angiosperms forest at Sirohi.
IV. Tripura State:		
1. Agartala	60 m	(Sepaijula, Baramurah) Tropical Angiospermous forests.
2. Ambassa	100 m	Tropical Angiospermous forests.
V. Meghalaya State:		
1. Nongpoh	800 m	(Umling, Lailad, 450 m). Tropical Angiospermous forests.
2. Nayabungalow	1,000 m	(Barapani, Umiam lake, 1,200 m). Subtropical Forests of *Pinus insularis* and of mixed vegetation.
3. Cherrapunji	1,142 m	(Mowsmai Cave), Subtropical angiospermous Forests.
4. Tura (Garo Hills)	1,200 m	(Tura park, 1,200–1,400 m). Subtropical and angiospermous forests.
5. Jowai (Jiantia Hills)	1,400 m	(10 km on Garampani road, Ummulong, 1,400–1,500 m). Subtropical to temperate forest, predominantly of *Pinus insularis* sometimes with mixed vegetation.
6. Shillong (Khasi Hills)	500 m	(Risa colony, Upper Shillong, Elephant falls, Shillong Peak, Happy Valley, Sweet falls, Mawphlang, Malwai, 1,300–1,930 m). Temperate forests with pure formations of *Pinus insularis* or angiosperms or mixed vegetation.
VI. Assam state:		
1. Silchar	60 m	Tropical angiospermous forests.
2. Garampani	700 m	Mixed and pure angiospermous tropical forests.
3. Haflong	900 m	(Jatinga forest, Bagetar 300–900 m). Tropical angiospermous forests.
VII. Sikkim		
1. Mangan	1,200 m	(Singtik, Naga). Mixed subtropical forests.
2. Gangtok	2,000 m	(Tashi View point). Mixed Temperate forests.
B. BHUTAN		
1. Wangdi	1,250 m	(Punakha, 25 km from Wangdi towards TongsaDzong, 7 km from Wangdi towards Dochula (Mindley-gang), 1,250–2,250 m). Patches of Pine at 7th km; temperate angiospermous forests towards Tongsa Dzong.
2. Chimakothi	2,037 m	(Tsimalakha, Chukha, Bunakha, 2,000–2,500 m). Temperate forests. Pure angiospermous forest at Tsimalakha and Chukha, mixed at Bunakha.
3. Paro	2,095 m	(D'Dzong, Chailela, 2,050–2,700 m). Temperate forests. Patches of pure conifers, but predominantly mixed forests.
4. Thimphu	2,275 m	Newephu, Namesling, Begana, Chankaphug, Hongsu, Ustselpang, Dochula.
5. Ha Dzong	2,690 m	(10 km from ha on Thimphu road, Kana, 2,600–3,000 m). Temperate, predominantly mixed forests.

5.2 North Western Himalayas

This floristic region is about 800 km long and about 200–400 km wide and covers the states of Jammu and Kashmir (J. & K.), Himachal Pradesh (H. P.), Union Territory of Chandigarh (U. T.), Uttarakhand (U. K.), Sub mountainous parts of Punjab and Haryana states. Geographically this region lies between 29° to 35° N latitude and 74–86°E longitude.

The type of vegetation in a particular region depends largely on the climate, the soil and the past treatment. Rainfall and temperature, in addition to wind and topography, are the most obvious factors of the climate. The rainfall in India is largely monsoonic and the total annual rainfall is an important factor in determining the nature of vegetation. The average annual rainfall in the N. W. Himalayas is 1,020–2,050 mm. The interior region recieves less monsoon rainfall but heavy snowfall. Thus, there is a considerable variation between the climate of the outer and inner ranges. In general the rainfall is maximum in the eastern parts of N. W. Himalayas (Nainital), which receives the bulk of monsoonic winds, but it goes on decreasing as these winds move westwards. In addition to the direction of the monsoon, some other factors like altitude, location and the direction of mountain ranges play an important part in the distribution of rain fall. Thus Dharamshala (H. P.), which is on the western side, recieves the maximum rainfall, up to 3,200 mm annually, because of the direction and location of the mountains, the distribution of the rainfall is quite uneven and varies from 600 to 3,200 mm annually, of the total annual rainfall, 75 % falls during the rainy months of July to September. The vegetational zonation is governed by monsoon and follows other South-East Asian Countries like South-East China, Indonasia and Japan, although factors like soil, physiography, topography and biotic factors have a role in governing the flora of any specific area.

Snow falls during the winter months above 1,500 m. and may remain for few days to months depending up on altitude and location of area. At elevation of 3,000 m. the average snow fall is about 300 cm and lasts from December to March while places above 4,500 m. remain perpetually under snow.

Temperature is the most obvious factor of climate. The mean annual temperature exceeds 24 °C over the whole country with the exception of hill areas and the extreme north-west. Champion and Seth (1968) divided India into four zones on the primary basis of temperature as follows:

Zone		Mean annual temperature	Mean January temperature	Winter
I.	Tropical	Over 24 °C	Over 18 °C	None: no forest
II.	Subtropical	17–24 °C	10–18 °C	Definite but not severe; Frost rare
III.	Temperate	7–17 °C	−1 to 10 °C	Pronounced, with frost and some snow
IV.	Alpine	Under 7 °C	Under −1 °C	Severe, with much snow

These zones are not very strict and are greatly influenced by altitude, rainfall, edaphic and other ecological factors. The variation in altitude play a significant role in the determination of the temperature and distribution of rain. Miller (1950), while discussing the effect of altitude on climate, states "Height above sea level has a pronounced influence on climate, in many respects imitating the effects of increased latitudes" and this is in confermity with earlier generalization of Humboldt (1817) who found a correlation between the altitude and latitude and stated the successive altitudinal zones of the vegetation correspond to the latitudinal zones from the equater to the poles and that an elevation of above 100 m. on a tropical mountain would make a difference of 1° latitude. Puri (1960) pointed out that the rate of decrease of the temperature with elevation at the time of maximum day temperature in the month of January averages 3 °F per 100 ft. (304 m), while the rate of decrease of temperature with elevation at night or edge time of minimum temperature averages 1 °F per 1,000 ft. (304 m) in the Western Himalayas. The wild range of altitude, temperature and rain fall result in a diversified forest flora in this region. Mehra et al. (1971) gave a floristic account of some forest type of this region. The divided N. W. Himalayas into four climatic zones as given below:

1. Tropical zone
   In the foothills, the Siwaliks and up to 1,000 m. in the outer HimalayanRanges.
2. Subtropical zone
   1,000–1,800 m.
3. Temperate zone
   1,800–3,600 m.
4. Alpine zone
   Above 3,600 m.

The forests do not extend beyond 3,600 m. The author has adopted these climatic zones to discuss the vegetation of this region.

5.2.1 Tropical Zone

This zone comprises the planes and the adjoining foot hills which may assend up to 1,000 m above mean sea level (M. S. L.) in the Siwaliks and the outer range of N. W. Himalayas. It includes the areas around Jammu, Pathankot, Dunera, Chandigarh, Kalka, Dehradun and Haldwani. There are mainly two types of tropical forests that occur in this zone, *viz.* Tropical moist deciduous forest and the tropical dry deciduous forests. Tropical moist deciduous forests occurs in the areas where annual rain fall is between 1,300 and 1,500 mm, the maximum temperature is 44 °C and the minimum 1.8 °C. In this type, two subtypes can be distinguished: moist Sal-bearing forest (*Shorea robusta*) and moist mixed deciduous forests. In the former 'Sal' (*Shorea robusta*) is dominant species and occur in almost pure formation in Dehradun (Lachhiwala forest range), Nanital (Ranibag). On its upper limit and on the Northern aspects *Pinus roxburghii* often occur. Some associated angiosperms are *Adina cordifolia, Eugenia jambolana, Cordia dichotoma, Cedrella toona, Albizzia procera, Mallotos philippinensis, Cassia fistula.*

The shrubby elements are *Adhotoda vasica, Clerodendron viscosum* and *Woodfordia fruiticulosa.* In the other type, moist mixed deciduous forests, the forests are always mixed and no single species forms pure crop. Some common trees are *Adina cordifolia, Acacia catachu, Albizzia procera, Bombax ceiba, Lagerstroemia parbiflora, Delbergia sissoo, Zizyphus jujuba, Z. Xylopyrus* etc. The commonly associated shrubs are *Adhotoda vasica, Clreodendron viscosum* and *Helictera sisora.*

The areas with less rain fall and less humidity develop tropical dry deciduous forests. The areas around Chandigarh, Kalka, Pathankot, Dunera and Jammu possesses such type of forests. The common trees are *Bombax ceiba, Acasia catachu, A. arebica, A. Leucophloca, A. Modesta, Pyrus pashia, Cassia fistula, Zizyphus jujuba, Delbergia sissoo, Mangifera indica* etc., the common shrubs are *Woodfordia fruiticulosa, Adhotoda vasica, Euphorbia royleana, Lantana camara, Capparis deciduas, Zizyphus numularia* and *Carrisa opaca.*

5.2.2 Subtropical Zone

This is an intermediate zone between the tropical and montane-temperate zones and extends from 1,000 to 1,800 m above m. s. l. The average annual rain fall varies from 900 to 2,000 mm. Summer months are generally hot with maximum temperature up to 3,800 °C while the winter is moderately cold. Snow falls only for few days in mid-winter and that too in upper limits. This zone comprises "Montane subtropical forests". The vegetation is a mixture of tropical and temperate species.

Typically *Pinus roxburghii* forms pure crops which are never dense. On moist places scattered broad-leaved evergreen trees occur. At the upper limit *Quercus incana* is the most typical associated along with *Rhododendron arboratum* and *lynonia ovalifolia.* The othe scattered trees are *Eugenia jambolana, Pyrus pashia, Terminalia tomentosa, Ficus palmata, Rhussemia lata* and *Acer oblongum.* The shrubby component is of *Viburnum coriaceum, Rubus elllipticus, Inula cappa* etc.

5.2.3 Temperate Zone

The montane temperate zone extends from 1,800 to 3,600 m above m. s. l. In this region. The annual rain fall considerably and ranges from 1,000 to 300 mm. The climate is cold in winters with a maximum temperature reaching up to −10 °C and it is moderately hot in summer with maximum temperature up to 32 °C. The snowfall is moderate and occurs during the month of December to March. This zone comprises "Monsoon montane temperate forests" and these are not comparable with the true temperate forests of the Northern Hemisphere because the climate is not typically humid as is found in the true temperate forests. The montane temperate zone can be further divided into lower, middle and upper zones.

The lower temperate zone forests overlap the subtropical zone forests. *Quercus incana* is a dominant species and occupied the lowest zone of the temperate belt. The typical associates are *Rhododendron arboretum* and *Lyonia ovalifolia.* Associated conifers are *Cedrus deodara* and *Pinus roxburghii.*

The middle temperate zone has mixed Oak- Acer forests. *Quercus dilatata* is a dominant species. *Q. incana* is also fairly common and two may form a gregarious mixture with *Acer, Aesculus* and *Litsaea* etc. in some localities *Q. incana* is dominant while in others *Cedrus deodara* acquires dominance. Puri (1960) has called such mixed temperate forests as "Glory of the Himalayas". The other associated trees are *Aesculus indica, Acer caesium, Taxus baccata, Pinus excels, Abies pindrow, Picea smithiana.*

In the upper temperate zone, the mixed coniferous forest comprising *Abies pindrow* and *Picea smithiana* with scattered *Cedrus deodara* and *Taxus baccata* are predominant. At some places *Quercus semicarpifolia, Juglans regia, Prunus cornuta* and *Aesculus indica* form mixed crop with the above mentioned conifers. However, *Q. semicarpifolia* is observed to form pure formation in pockets in undisturbed area like Hattoo- Mountain, Baghi and Khadrala (H. P.). Common shrubs are: *Viburnum cotinifolium, V. foetans, Berberis* ssp., *Indigofera gerardiana, Rubus niveus* etc. At places *Cedrus deodara* forms almost pure crop between 2,100 and 3,200 m, often with little mixture of *Pinus wallichiana* and *Picea smithiana.* The associated broad- leaved components are oaks and rhododendrons. *Cupressus torulosa* also forms open type forests with scattered trees between 2,000 and 3,200 m, e.g., Naini Tal. These occur on rocks where the slope is steep. In the inner Himalayas, where the rainfall is less, *Pinus gerardiana* is predominant and forms pure crop.

5.2.4 Alpine Zone

The zone extend above 3,600 m above m.s.l. The snowfall is very heavy, the winter is severe and summer is short. The rainy period is only during the summer months, July to September. The vegetation in this zone is 'Alpine stony deserts', 'Alpine scrubs', 'Alpine meadows' and 'Alpine forests'. These extend from 3,000 m to the snow line. The vegetation is bushy in habit with short and much branched stems. Typically, alpine planets are spreading and appear like a carpet on the ground.

'Alpine stony deserts' are found just below the snow line. Rocks are generally covered with lichens. *Sedum crassipes* and *Primula minutissima* have been recorded from this stony desert. The 'Alpine pastures' show a luxuriant ground flora vegetation. Some of the species present in this zone are: *Anemone rivularis, Chenopodium album, Draba tibetica, Corydalis longipes, C. ramose, Thalictrum alpinum, Ranunculus hyperboreus, R. affinis, Aconitum nepellus, A. violaceum, Saxifraga sibirica, S. pallida, Sedum tibeticum* etc. Just below the 'Alpine meadows' are present 'Alpine scrubs' comprising dwarf willows, *Juniperus* sp. and *Rhododendron* sp. etc. At places, there are present 'Alpine forests', just below the 'Alpine scrubs', consisting of trees like *Abeis webbiana, Betula utilis, Quercus semicarpifolia, Rhododendron companulatum* etc.

It is pertinent to mention here that the Kashmir valley has somewhat different vegetation than the rest of the N. W. Himalayas. The difference in the vegetation of Kashmir valley is due to the fact that this area usually receives very little rain during summer months as PirPanjal ranges (about 4,600 m) on its southern side serve as a formidable barrier to the monsoon winds. The only rain which this region receives in summer months is brought about by the winds that escape into the valley through the Jhelum gorge, near Baramula. This region also receives a local rainfall during the months of January and February. Puri (1960) mentions "In the present vegetation of the Kashmir valley there are no oaks, laurels and several other species" and divides the vegetation of the valley and the neighbouring mountains into four zones:-

5.2.4.1 The Kashmir Valley Zone (1,525–2,135 m)

This zone supports a mixed vegetation of the broad- leaved trees and conifer species: *Juglans regia, Populus ciliate, P. nigra* var *fastigata, P. alba, Salix wallichiana, Platanus orientalis, Morus serrata, M. alba, Pinus excelsa and Cedrus deodara.*

5.2.4.2 The Coniferous Forest Zone (2,135–3,200 m)

The dominant tree in this zone is *Abies pindrow.* Other common trees and shrubs are *Picea smithiana, Taxus baccata, Aesculus indica, Acer caesium, Juglans regia, Prunus cornuta, Viburnum nervosum, Rosa macrophylla, Ribes* sp., *Berberis* sp. and *Lonicera* sp.

5.2.4.3 The White Birch Zone (3,200–3,600 m)

This zone is just above the coniferous zone and *Betula utilis* is a dominant feature in this zone. Sometimes *Rhododendron companulatum, Pyrus foliolosum, Salix denticulate* and *Syringa emodi* etc. are also associated with it in this zone.

5.2.4.4 The Alpine Zone (3,660–4,120 m)

The climate is extremely cold in this zone and it supports a thick vegetation of stunted trees of *Juniperus communis* and *J. souamata.* Other woody plants that survive here are *Rhododendron anthopogon* and *Lonicera* sp.

The following places in the N. W. Himalayas were visited during the course of present study. Various localities with more or less same type of vegetation and having geographical contiguity have been grouped together under the one heading. The main station is the one which is written first and the smaller and less known places around the main station are put within parentheses. The altitude, type of forest and climatic zone is also noted against each.

I. Jammu and Kashmir State (J. & K.)		
1. Jammu	(366 m)	Angiospermic, tropical dry deciduous forests
2. Batote	(1,500 m)	(Patnitop, Sanasar, Bhadarwah, Seoj, Kishtwar. 1,500–2,100 m) mixed coniferous forests.
3. Bhaderwah	(1,613 m)	Temperate zone
4. Srinagar	(1,587 m)	(Harvan, Wular lake, Nishant Garden, Shankracharya Hill, Verinag, Yusmarg. 1,587–1,829). Predominantly angiospermic, mixed at Shankracharya Hill and Yusmarg. Temperate zone
5. Sonamarg	(2,667 m)	(Glacier point). Mixed coniferous, at places there are pure formation of *Betula utilis*.
6. Gulmarg	(2,655 m)	Temperate zone (Tangmarg, Ningal Nallah, Khilanmarg, Alpatharlake. 2,000–4,350 m). Predominantly mixed coniferous, alpine scrub and alpine meadows (above Khilanmarg) Temperate zone
7. Pahalgam	(2,100 m)	(Baisaran, Shikargah, Batakote, Kokarnag, Chandanwari, Pissooghati. 2,100–3,200 m). Mixed coniferous and mixed forests.
8. Patnitop	(2,024 m)	Temerate zone
II. Punjab State		
1. Pathankot	(350 m)	(Dunera. 350–900 m) Angiospermic, tropical dry deciduous forests
III. Union Territory		
1. Chandigarh	(280 m)	(Shahpur, Khuda, Mullanpur, High Court, Engineering College, University campus). Angiospermic, tropical dry deciduous forests
IV. Haryana State		
1. Kalka	(450 m)	(Panchkula, Surajpur, Pinjore. 350–450 m). Angiospermic, tropical dry deciduous forests
V. Himachal Pradesh		
1. Dalhousie	(2,042 m)	(Banikhet, Baloon, Lover's walk, Bakrota hill, Panjpulla, Jandrighat. 1,500–2,100 m) Predominantly angiospermic, at places mixed forests. Subtropical to temperate
2. Khjjiar	(1,800 m)	(Lakarmandi, Kalatop, Daikund. 1,800 2,400 m) Predominantly coniferous forests. Temperate zone
3. Chamba	(1,400 m)	(Sara, Thali, Sahu, Kilorgala. 1,400–2,000 m). Lower reaches with predominantl, angiospermic, higher reaches with mixed forests. Subtropical zone
4. Dharamshala	(1,400 m)	(McleodGanj, ForsynthGanj, Triund, Yol. 1,200–2,700) Predominantly angiospermic (Oak) sometimes mixed with *Pinus roxburghii* or *Cedrus deodara*. Subtropical to temperatezone

5. Kulu	(1,500 m)	(Jari, Kasole, Manikaran, Pulga, Parvati valley. 1,500–2,600). Angiospermic at Lower reaches, mixed at higher altitude. Subtropical to temperate zone
6. Manali	(1,800 m)	(Naggar, JagatSukh, Gojra, Kothi, Gulaba, Rohtang Pass. 1,500–3,978 m). Coniferous and mixed forests. Temperate to alpine zone
7. Keylong	(2,440 m)	(Baralachha Pass. 5,000 m). Predominently angiospermic, *Salix* mixed with *Pinus geradiana*. At places Alpine meadows at Baralachha Pass.
8. Kasauli	(1,800 m)	(Dharampur, Sanawar, Solan. 1,530–1,800 m). Mixed forests, subtropical to temperate
9. Shimla	(2,202 m)	(Taradevi, Summer hill, Glen, Chadwick falls, Jakhoo hill, Mashobra, Chharabra, Kufri, the Retreat. 1,500–2,500 m). Mixed
10. Bashahr	(1,350 m)	forests, mostly coniferous at higher altitudes. Subtropical to temperate zone
11. Chail	(2,148 m)	Mixed forests, predominantly coniferous. Temperate zone
12. Narkanda	(2,700 m)	(Shillaroo, Hattoo-mountain, Baghi, Tatapani, Khadrala. 2,700–3,300 m). Mixed forests. At places pure oak formation (Hattoo-mountain) or pure deodar formation (Shillaroo) Temperate forests to alpine meadows
13. Sarahan	(2,100 m)	(Nichar, Techlesh, Jeori, Luri. 2,000–3,000 m). Mixed temperate forests
14. Kalpa	(3,000 m)	(Pooh. 3,000–3,050 m). Coniferous forests, pure formation of *Cedrus deodara* and *Pinus gerardiana*
VI. Uttarakhand (U.K.)		
1. Dehra Dun	(682 m)	(Lachhiwala forests range, Rishikesh side, Asarori, Yamuna River bank. 500–682 m). Angiospermic, predominantly 'Sal' forests. Tropical moist deciduous forests
2. Mussoorie	(2,100 m)	(The Park, Municipal Garden, Jabber Khet, Lal Tibba, Balarookhud, Camel's back road, Chakrata toll, Kempty falls, Dhanulti. 1,700–2,400 m). Mixed forests, temperate zone.
3. Nainital	(1,981 m)	(Land's end, Tiffon's top, Naina peak, Laria Kanta, snow view, Tanki, Kilbury, Sat Tal, Bhim Tal. 1,400–2,400 m). Predominantly angiospermic with oak as dominant tree, at places pure formation of *Cedrus deodara* (Naina Peak and Kilbury). Subtropical to temperate zone
4. Ranikhet	(1,850 m)	(Kalika, 1,850–1,900 m). Mixed forests, at places pure *Cedrus* forests. Temperate zone
5. Kausani	(1,798 m)	Mixed forests, Subtropical zone
6. Almora	(1,615 m)	Mixed forests, Subtropical zone
7. Hemkunt	(4,389 m)	Mixed forests, Subtropical zone

References

Bor NL (1942) Some remarks on the geology and the flora of the Naga and Khasi hills. 150th Aniv Vol Roy Bot Gard. Calcutta, pp 129–135

Champion HG (1936) A preliminary survey of the forest types of India and Burma. Indian Forest Rec (n s), Silva 1: 1–279

Champion HG, Seth SK (1968) A revised survey of the forest types of India. Government of India Publ, Delhi

Chatterjee D (1939) Studies on the endemic flora of India and Burma. J Asiat Soc Bengal Sci 5:19–67

Clarke CB (1898) Sub-subareas of British India. J Linn Soc Bot 34:1–146

Hooker JD (1854) Himalayan Journals, vol II. Murray, London

Hooker JD (1907) A sketch of the flora of British India. In: Imperial gazetteer of India, vol 3(1,4). Clarendon Press, Oxford, pp 157–212

Mehra PN, Bir SS (1964) Pteridophytic flora of Darjeeling and Sikkim Himalayas. Res Bull Panjab Univ Sci (n s) 15:69–182

Mehra PN, Gill BS, Sareen TS (1971) Floristic account of some forest types of the Western Himalayas. Res Bull Panjab Univ Sci (n s) 22:487–503

Miller AA (1950) Climatology. Methuen & Co Ltd., London

Puri GS (1960) Indian forest ecology Vol I & II. Oxford Book & Stationery Co., New Delhi

Rao AS (1969) The vegetation of Khasi and Jaintia hills. Proc Pre-congress Symposium -21st International Geographical Congress, Gauhati

von Humboldt A (1817) De distributione geographica plantarum secundum coeli temperiem et altitudinem montium. Prolegomena, Paris

Chapter 6
Taxonomy

Abstract The data concerning the taxonomic treatment is preceded by the synopsis concerning the characteristics features of the families which have been included in the text. A key/synopsis to the genera concerning each family is provided along with the key/synopsis of species treated under each genera for easy identification. Each description is provided with detailed morphological note along with data concerning distribution, collection number of the specimen (of which line drawing/illustration is provided) and remarks along with camera Lucida line drawings and about half of the species are provided with illustration.

6.1 Characteristic Features of Families Included

Family- **Amylocorticiaceae** Jülich, *Biblthca Mycol.* **85**: 354, 1981.
Fructification resupinate, loosely adnate, effused. Hyphal system monomitic. Generative hyphae branched, septate, clamped. Cystidia present or absent. Basidia clavate to subclavate, 4-sterigmate, with basal clamp. Basidiospores ellipsoid, amyloid or inamyloid, acyanophilous.

Family- **Cyphellaceae** Lotsy, *Vortr. bot. Stammesgesch.* **1**: 695, 1907.
Fructification resupinate, loosely adnate, effused. Hymenial surface smooth. Hyphal system monomitic. Generative hyphae septate, clamped. Cystidia present. Basidia narrowly clavate, 4-sterigmate, with basal clamp. Basidiospores subcylindrical to suballantoid to obliquely ellipsoid to lacrymoid, inamyloid, acyanophilous.

Family- **Physalacriaceae** Corner, *Beih. Nova Hedwigia* **33**: 10, 1970.
Fructification resupinate; hymenial surface smooth and tuberculate; margin fimbricate. Hyphal system monomitic. Generative hyphae distinct, clamped, branching from clamp cells. Cystidia fusiform, thin-walled, slightly projecting. Basidia clavate, 4-spored with basal clamp. Basidiospores obliquely ellipsoid, non-amyloid.

Family- **Pterulaceae** Corner, *Beih. Nova Hedwigia* **33**: 10, 1970.
Fructification resupinate, closely adnate, effused. Hymenial surface smooth to tuberculate. Hyphal system monomitic. Generative hyphae septate, clamped. Cystidia absent. Basidia clavate, more or less sinuous, 4-sterigmate, with basal clamp. Basidiospores globose to ellipsoid, smooth, acyanophilous, inamyloid.

I.B. Prasher, *Wood-rotting non-gilled Agaricomycetes of Himalayas*,
Fungal Diversity Research Series, DOI 10.1007/978-94-017-9858-7_6

Family- **Atheliaceae** Jülich, *Biblthca Mycol.* **85**: 355, 1981.
Fructification resupinate, loosely adnate, effused. Hymenial surface smooth. Hyphal system monomitic. Generative hyphae branched, septate, clamped. Cystidia present or absent. Basidia clavate to subclavate, 4-sterigmate, with basal clamp. Basidiospores ellipsoid, cyanophilous, inamyloid.

Family- **Stephanosporaceae** Oberw. & E. Horak, *Pl. Syst. Evol.* **131**: 162, 1979.
Fructification resupinate, membranous to subceraceous, effused to widely effused. Hymenial surface white to pale coloured, smooth to raduloid, waxy in appearance. Context composed of somewhat loosely woven hyphae in the basal part. Hyphal system monomitic. Generative hyphae branched, septate, clamps present, thin-walled, subhyaline. Cystidia or gloeocystidia absent. Basidia clavate-cylindrical to subutriform, 4-spored. Basidiospores ellipsoid to ovoid, thick-walled, subhyaline, smooth, cyanophilous, nonamyloid.

Family- **Coniophoraceae** Ulbr., *Krypt.-Fl. Anfäng.* **1**(3): 120, 1928.
Fructification resupinate, adnate, effused. Hymenial surface smooth to tuberculate. Hyphal system monomitic. Generative hyphae septate, without clamps. Cystidia present or absent. Basidia clavate to subclavate to subcylindrical, 4-sterigmate, without basal clamp. Basidiospores broadly ellipsoid to ovoid, smooth, cyanophilous, inamyloid.

Family- **Hygrophoropsidaceae** Kühner, *Bull. mens. Soc. linn. Lyon* **49**: 900, 1980.
Fructification resupinate, loosely adnate, effused. Hyphal system monomitic. Generative hyphae septate, clamped. Cystidia absent. Basidia clavate to subclavate to subcylindrical, 4-sterigmate, with basal clamp. Basidiospores ellipsoidal to ovoid, smooth, cyanophilous, inamyloid.

Family- **Serpulaceae** Jarosch & Bresinsky, *Biblthca Mycol.* **191**: 90, 2001.
Fructification resupinate, loosely adnate, effused. Hymenial surface smooth to meruloid. Hyphal system monomitic. Generative hyphae septate, clamped. Cystidia none. Basidia clavate to subcylindrical, 4-sterigmate, with basal clamp. Basidiospores ellipsoid to ovoid, smooth, cyanophilous, inamyloid.

Family- **Botryobasidiaceae** Jülich, *Biblthca Mycol.* **85**: 357, 1981.
Fructification resupinate, loosely adnate, effused. Hymenial surface hypochnoid. Hyphal system monomitic. Generative hyphae septate, with or without clamps. Cystidia present or absent. Basidia ovoid to clavate to subcylindrical, 4-sterigmate, with or without basal clamp. Basidiospores navicular to ellipsoid, smooth or echinulate, acyanophilous, inamyloid.

Family- **Ceratobasidiaceae** G.W. Martin, *Lloydia* **11**(2): 114, 1948.
Fructification resupinate, loosely adnate, effused. Hymenial surface hypochnoid. Hyphal system monomitic. Generative hyphae septate, with or without clamps. Cystidia absent. Basidia short subcylindrical to suboblong, 4-sterigmate, with or without basal clamp. Basidiospores globose, smooth, acyanophilous, inamyloid.

Family- **Hydnaceae** Chevall., *Fl. gén. env. Paris* **1**: 270, 1826.

Fructification resupinate, effused, loosely adnate. Hymenial surface smooth to tuberculate. Hyphal system monomitic. Generative hyphae septate, clamped. Cystidia none. Basidia obovate to pyriform to urniform, 6–8 sterigmate, with basal clamp. Basidiospores ellipsoid to subcylindrical to tetrahedral, smooth, acyanophilous, inamyloid generally uniguttulate.

Family- **Corticiaceae** Herter, *Krypt.-Fl. Brandenburg* (Leipzig) **6**(1): 70, 1910.

Fructification resupinate, adnate, effused. Hymenial surface smooth to somewhat tuberculate, cracked on drying. Hyphal system monomitic. Generative hyphae septate, with or without clamps. Dendrohyphidia present or absent. Cystidia present or absent. Basidia clavate to subclavate to suburniform, 4-sterigmate, with basal clamp. Basidiospores allantoids to ellipsoid to subglobose to globose, smooth, acyanophilous, inamyloid.

Family- **Gloeophyllaceae** Jülich, *Biblthca Mycol.* **85**: 368, 1981.

Fructification annual to perennial, resupinate to pileate, coriaceous to woody; abhymenial surface brown orange to dark brown, smooth to velutinate to hirsute, azonate to zonate. Hymenial surface poroid with straight to sinuous lamellae, reddish brown to dark brown. Hyphal system dimitic/trimitic. Generative hyphae clamps; skeletal hyphae dominant. Cystidia present or absent, smooth to encrusted. Basidia cylindrical to clavate, thin to thick walled 4-sterigmate, with basal clamp. Basidiospores ellipsoid to cylindrical to allantoid, subhyaline, thin-walled, smooth, inamyloid, acyanophilous.

Family- **Hericiaceae** Donk, *Persoonia* **3**(2): 269, 1964.

Fructification resupinate, reflexed to clavate or shortly stipitate and strongly branched. Hymenial surface smooth to tuberculate to toothed. Hyphal system monomitic. Generative hyphae clamped, branched. Gloeocystidia present or absent. Basidia clavate to subclavate, 4-sterigmate. Basidiospores globose to subglobose to ellipsoid, usually ornamented, amyloid.

Family- **Lentariaceae** Jülich, *Biblthca Mycol.* **85**: 375, 1981.

Fructification resupinate, soft-fragile, widely effused. Hymenial surface hydnoid, cream to brown; margin thinning, white fibrillose. Context subhyaline. Hyphal system monomitic, hyphae septate, clamped, teeth subulate. Cystidia and Gloeocystidia absent. Basidia clavate-cylindrical, 4-spored. Basidiospores ellipsoid, the walls minutely echinulate, cyanophilous, non-amyloid.

Family- **Hymenochaetaceae** Donk, *Bull. bot. Gdns Buitenz.* **17**(4): 474, 1948

Fructification resupinate to pileate, annual to perennial, sessile to stipitate, staining black with KOH solution; stipe lateral to eccentric to centric. Abymenial surface tomentose to hispid, usually brown. Hymenial surface smooth to poroid; context usually fibrous in texture. Hyphal system monomitic/dimitic/trimitic. Generative hyphae with simple septa. Cystidia absent. Basidia clavate to cylindrical or broadly ellipsoid, 4-spored. Setae present or absent. Setal hyphae present or absent. Basidiospores subhyaline to yellowish to brownish, usually smooth, rarely cyanophilous or dextrinoid.

Family- **Rickenellaceae** Vizzini, *Micol. Veg. Medit.* **25**(2): 144, 2010

Fructification resupinate, adnate, effused. Hymenial surface smooth to odontoid. Hyphal system monomitic. Generative hyphae septate, clamped, thin- walled. Cystidia present. Basidia clavate, 4-sterigmate, with basal clamp. Basidiospores ellipsoid to subcylindrical, apiculate, thin- walled, acyanophilous, inamyloid.

Family- **Repetobasidiaceae** Jülich, *Biblthca Mycol.* **85**: 388, 1981.

Fructification annual to perennial, resupinate, widely effused, separable. Margin white to creamish, thinning to narrow fimbriate. Pore surface white to creamish when fresh, darker and cracked on drying; pores small, round to angular. Context white, soft, thin, soft, homogenous. Hyphal system dimitic; generative hyphae hyaline, thin-walled, branched, septate, clamped, cyanophilous. skeletal hyphae subhyaline, thick-walled, branched, acyanophilous. Cystidia absent, cystidioles as branched hyphae with crystal incrustations at their tips project into the pore cavity. Basidia thin-walled, hyaline, clavate, 4-spored. Basidiospores hyaline, thin-walled, smooth, cylindrical to allantoids, non-amyloid.

Family- **Schizoporaceae** Jülich, *Biblthca Mycol.* **85**: 389, 1981.

Fructification corticoid, resupinate, adnate, effuse, annual or perennial, white, yellowish or brown. Hymenium poroid with round or angular, often irregular pores, or with sparse to dense conical, or flat and irregular, spines. Hyphal system monomitic or dimitic, skeletal hyphae when present often poorly developed, generative hyphae with thin or thick walls, always with clamp connections. Cystidia usually present, varied in form, usually capitate with a globose encrusted or resinous tip. Basidia cylindrical to ± urniform, with four well-developed sterigmata. Basidiospores ellipsoidal, hyaline, thin-walled, smooth, not staining in iodine.

Family- **Cystostereaceae** Jülich, *Biblthca Mycol.* **85**: 362, 1981.

Fructification effused-reflexed to totally resupinate, perennial, often stratose, membranous ceraceous to somewhat corky or woody; hymenial surface pale coloured, smooth to densely and prominently tuberculate, often cracking deeply and irregularly on drying. Hyphal system dimitic; generative hyphae clamped; skeletal hyphae present. Cystidia absent. Gloeocystidia present. Basidia 4-spores. Basidiospores nonamyloid, acyanophilous.

Family- **Fomitopsidaceae** Jülich, *Biblthca Mycol.* **85**: 367, 1981.

Fructification annual to perennial, resupinate to effused-reflexed to pileate; pilei sessile, solitary to imbricate, leathery to corky; abhymenial surface smooth to velutinate to rugose to crustose. Hyphal system monomitic/dimitic/trimitic ,clamps usually present. Cystidia present or absent. Basidia usually clavate, 4-sterigmate. Basidiospores ellipsoid to cylindrical to allantoid, smooth, inamyloid/amyloid/dextrinoid.

Family- **Ganodermataceae** Donk, *Bull. bot. Gdns Buitenz.*, **17**(4): 474, 1948.

Fructification annual to perennial, pileate, sessile to stipitate, woody to coriaceous; abhymenial surface brownish, azonate to zonate, with or without crust. Hymenial surface poroid; pores round to angular. Hyphal system trimitic. Generative hyphae with clamp connections, subhyaline. Skeletal and binding hyphae subhyaline to brown. Basidia clavate to pyriform. 4-sterigmate. Basidiospores brown, ovoid to ellipsoid, apex truncate, rarely reticulate or longitudinally crested, with a two-layered wall, the exospore relatively thin and subhyaline, smooth, the endospore thick, strongly pigmented and often ornamented.

Family- **Hyphodermataceae** Jülich, *Biblthca Mycol.* **85**: 373, 1981.
Fructification resupinate or effuse, thin or thick, membranous, gelatinous or waxy, smooth, pilose or warted, usually pale, yellowish or brownish, Hyphal system monomitic clamp connection present or lacking, cystidia usually present often thick walled and rarely encrusted. Hymenium smooth to tuberculate or spinose. Basidia usually large, cylindrical, clavate or urniform, sometimes becoming septate, usually 4-spored. Basidiospores globose, allantoids, cylindrical or lacrimiform, smooth or rarely verrucose not staining in iodine.

Family- **Meripilaceae** Jülich, *Biblthca Mycol.* **85**: 378, 1981.
Fructification annual to perennial, resupinate to pileate; pilei produced from a common base, fleshy to coriaceous; abhymenial surface smooth to velutinate to squamulose. Hymenial surface poroid, whitish to greyish. Pores round to angular, hyphal system mono or dimitic. Generative hyphae thin- to thick-walled, septate, with or without clamps, variable in width and wall thickness, in some strongly sclerified. Skeletal hyphae thick-walled to solid, without septa. Cystidia absent or present. Basidia clavate to cylindrical, 4-sterigmate. Basidiospores ellipsoid to ovoid to globose, thin-to- somewhat thick-walled, inamyloid, acyanophilous.

Family- **Meruliaceae** Rea, *Brit. basidiomyc.*: 620, 1922
Fructification resupinate to effuse; pileate or not, waxy or soft-textured, whitish to brown, smooth or tomentose, rarely reduced to a subiculum. Hyphal system monomitic or rarely dimitic, hyphae gelatinized in some species, clamp connections often present, thick-walled cystidia usually present. Hymenium smooth, radially folded, spinose or ± poroid. Basidia narrow, cylindrical or clavate, 2- to 4- spored. Basidiospores ellipsoidal, cylindrical or allantoid, hyaline, smooth, not staining in iodine. Anamorphs unknown.

Family- **Phanerochaetaceae** Jülich, *Biblthca Mycol.* **85**: 384, 1981.
Fructification resupinate to effused-reflexed to pileate. Hymenial surface smooth to poroid to irpicoid, hyphal system mono or dimitic. Generative hyphae thin- to somewhat thick-walled, clamp connections may or may not be present. Cystidia usually present, smooth to encrusted. Basidia clavate to cylindrical, subhyaline, thin-walled. Basidiospores ellipsoid to allantoid to cylindrical to subglobose, subhyaline, smooth, thin-walled, inamyloid, acyanophilous.

Family- **Polyporaceae** Corda, *Icon. fung.* **3**: 49, 1839.
Fructification polyporoid with a central, eccentric or lateral stalk, or resupinate; annual or perennial; fleshy, leathery, or strongly woody, the upper surface smooth, tomentose or scaly, sometimes zonate. Hyphal system monomitic,

dimitic or trimitic with skeleton-ligative hyphae, clamp connection frequent. Cystidia absent. Hymenium tubular or lamellate without setae. Basidia small, clavate, usually with 4-sterigmata. Basidiospores cylindrical to allantoids, thin-walled, hyaline, usually smooth, not staining in iodine.

Family- **Xenasmataceae** Oberw., *Sydowia* **19**(1–6): 25, 1966.
Fructification resupinate, waxy or gelatinous, effuse and often inconspicuous. Hyphal system monomitic, hyphae often gelatinized. Cystidia present or absent. Hymenium smooth. Basidia widely cylindrical to ± urniform, attached laterally at the base to the fertile hypha, clamp connections lacking. Basidiospores hyaline, often staining in iodine.

Family- **Amylostereaceae** Boidin, Mugnier & Canales, *Mycotaxon* **66**: 487, 1998.
Fructifications resupinate to effused-reflexed or occasionally pileate, coriaceous to cartilaginous; upper surface of the pileus smooth to strigose. Hymenial surface smooth. Context with or without a cuticle on the abhymenial side, light brown composed of compactly arranged hyphae. Hyphal system dimitic or monomitic. Generative hyphae with clamps. Cystidia thick-walled, brown and often covered with subhyaline crystals. Gloeocystidia absent. Basidia clavate, 4-spored. Basidiospores ellipsoid, thin-walled, smooth, subhyaline, amyloid.

Family- **Bondarzewiaceae** Kolt. & Pouzar, *Ceská Mykol.* **11**(3): 163: 1957.
Fructification annual to perennial; resupinate to effused-reflexed to pileate; sessile or rarely stipitate and club shaped, soft to tough to woody; abhymenial surface tomentose to crustose, often zonate. Hyphal system usually dimitic. Skeletal hyphae not or sparingly branched, clamp connections present or absent. Hymenial surface smooth to spiny to toothed to poroid. Gloeocystidia absent or present. Basidia subhyaline, thin-walled, clavate to urniform, 2–4 spored. Basidiospores globose to subglobose to ellipsoid, usually ornamented, thin-to- thick-walled, cyanophilous.

Family- **Echinodontiaceae** Donk, *Persoonia* **1**(4): 405, 1961.
Fructification perennial, resupinate, effuse or sessile and pileate, ± glabrous or tomentose, often zonate, woody or corky in texture, the inner tissue often reddish or brown. Hyphal system usually interpreted as dimitic or trimitic, though skeletal hyphae are sometimes not well differentiated, clamp connections present and thick-walled encrusted cystidia often also present. Hymenium smooth to warted or spinose. Basidia clavate, with 4 sterigmata. Basidiospores hyaline to pale yellow, ± ellipsoidal, smooth or echinulate, usually thin-walled, staining blue in iodine.

Family- **Lachnocladiaceae** D.A. Reid, Beih. *Nova Hedwigia* **18**: 45, 1965.
Fructification resupinate, effused, reflexed, pileate; annual to perennial, soft-textured or leathery, smooth or tomentose, fleshy typically yellowish or brown, hymenial surface smooth, warted or ridged. Hyphal system monomitic or dimitic; generative hyphae possessing or lacking clamp connections, dendrohyphidia or dichohyphidia or asterosetae present. Basidia cylindrical to clavate, 2–4 sterigmate. Basidiospores variously shaped, hyaline, usually thin- walled, smooth or ornamented, amyloid or inamyloid.

Family – **Peniophoraceae** Lotsy, *Votr. bot. Stammesgesch.* **1**: 687, 1907.
Fructification resupinate, effused, sometimes reflexed, waxy to leathery. Hyphal system monomitic. Generative hyphae usually with clamps. Cystidia present. Hymenial surface almost always smooth. Basidia clavate, 4-sterigmate. Basidiospores ellipsoid to allantoids, smooth, acyanophilus, inamyloid.

Family- **Russulaceae** Lotsy, *Vortr. bot. Stammesgesch.*: 708, 1907. Lotsy, *Truffe*, **ed. 2**: 708, 1907.
Fructification pileate and stipitate, the stipe central or rarely lateral and usually lacking a veil, or gasteroid (hypogeous or epigeous); usually fleshy and often brightly coloured. Flesh complex in organization, with islands of globose cells separated by filamentous (sometimes lactiferous) hyphae, not staining in iodine, usually with large hymenial cystidia arising from inner tissues. Hymenium usually lamellar, the gills free or decurrent. Basidia clavate, usually 4-spored. Basidiospores usually ± globose, hyaline to dark yellowish or rarely brown, with an outer ornamented layer that stains blue in iodine

Family- **Stereaceae** Pilát, *Hedwigia*, Beibl. **70**: 34, 1930.
Fructification resupinate, effused-reflexed or discoid, rarely stalked, the outer surface often zoned, usually leathery or waxy. Hymenial smooth to tuberculate or rugose. Hyphal system monomitic or dimitic, hyphae with or without clamps. Cystidia common. Acanthophyses present or absent. Basidia small and narrow, 4- spored, sometimes accompanied by ornamental paraphysis-like hyphae. Basidiospores broadly ellipsoid to ovoid to subcylindrical, hyaline, smooth, amyloid or inamyloid.

Family- **Wrightoporiaceae** Jülich, *Biblthca Mycol.* **85**: 393, 1981.
Fructification annual, resupinate, effused, soft to tough. Pore surface cream, ochraceous to fulvous, pores round to angular or irregular. Margin myceloid. Context thin. Hyphal system dimitic. Generative hyphae clamped, hyaline and thin-walled. Skeletal hyphae dominating, thick-walled and dextrinoid. Cystidia none. Basidiospores broadly elliptical, small, smooth or minutely warted, amyloid, thin to thick-walled.

Family- **Thelephoraceae** Chevall., *Fl. Gén. env. Paris* **1**: 84, 1826.
Fructification resupinate, effused-reflexed, rarely stalked, hymenial surface smooth to toothed. Hyphal system monomitic, clamps present or absent. Cystidial elements present or absent. Basidia clavate to subclavate to suburniform to subcylindrical, generally 4-sterigmate. Basidiospores globose to subglobose to angular, ornamented, inamyloid, acyanophilous.

Family- **Hydnodontaceae** Jülich, *Biblthca Mycol.* 85: 372, 1981.
Fructification resupinate, adnate, effused. Hymenial surface smooth to tuberculate. Hyphal system monomitic or dimitic. Generative hyphae branched, septate, with or without clamps. Cystidial elements present or absent. Basidia clavate to subclavate to subcylindrical, 4–6 sterigmate. Basidiospores globose to subglobose to broadly ellipsoid to suballantoid, smooth or verrucose, thin-walled, apiculate, inamyloid, cyanophilous or acyanophilous.

6.2 Descriptions

O- Agaricales
Family-Amylocorticiaceae

Key to genera

1. Basidiospores amyloid 2
1. Basidiospores inamyloid 3
2. Hymenial surface smooth to meruloid, fructification distinctly pelliculose when young but becomes membranous– ceraceous with age .... *Ceraceomyces*
2. Hymenial surface smooth, fructification thin, waxy gelatinous .... *Amyloxenasma*
3. Basidiospores broadly ellipsoid .... *Amyloathelia*
3. Basidiospores narrowly ellipsoid to cylindrical .... *Amylocorticium*

Amyloathelia Hjortstam & Ryvarden,
Mycotaxon 10(1): 201, 1979.

Fructification resupinate, loosely adnate, effused; hymenial surface smooth. Hyphal system monomitic; generative hyphae septate, clamped, thin-walled; hyphal cordons present; Cystidia absent. Basidia clavate to subclavate, thin-walled, 4-sterigmate, with basal clamp. Basidiospores broadly ellipsoid, thick-walled, smooth, apiculate, amyloid, acyanophilous.

Three species, wide spread

Lit.: Hjortstam & Ryvarden (*Mycotaxon* 10: 201, 1979).

Type species: *Corticium amylacium* Bourdot & Galzin, 1911

Habitat: Wood

Himalayas: One

Amyloathelia crassiuscula Hjortstam & Ryvarden, Mycotaxon 10(1): 204, 1979. Fig. 6.1a–d

Fructification resupinate, loosely adnate, effused, up to 240 µm thick in section; hymenial surface smooth, yellowish white to pale yellow when fresh, greyish yellow on drying; margin thinning, fibrillose, paler concolorous to indeterminate.

Hyphal system monomitic; generative hyphae up to 5.3 µm wide, septate, clamped, thin-walled; basal hyphae parallel to the substrate, less branched; subhymenial hyphae vertical, profusely branched. Hyphal cordons up to 15.0 µm, individual hyphae up to 4.5 µm wide. Cystidia absent. Basidia 12.0–15.6 × 6.2–6.8 µm, clavate to subclavate, 4-sterigmate, with basal clamp; sterigmata up to 4.0 µm long. Basidiospores 6.4–7.3 × 4.0–5.0 µm, broadly ellipsoid, apiculate, thick-walled, smooth, amyloid, acyanophilous.

Distribution: U.K.: Chakrata.

Collection examined: IBP 37002.

Substratum: On angiospermic log.

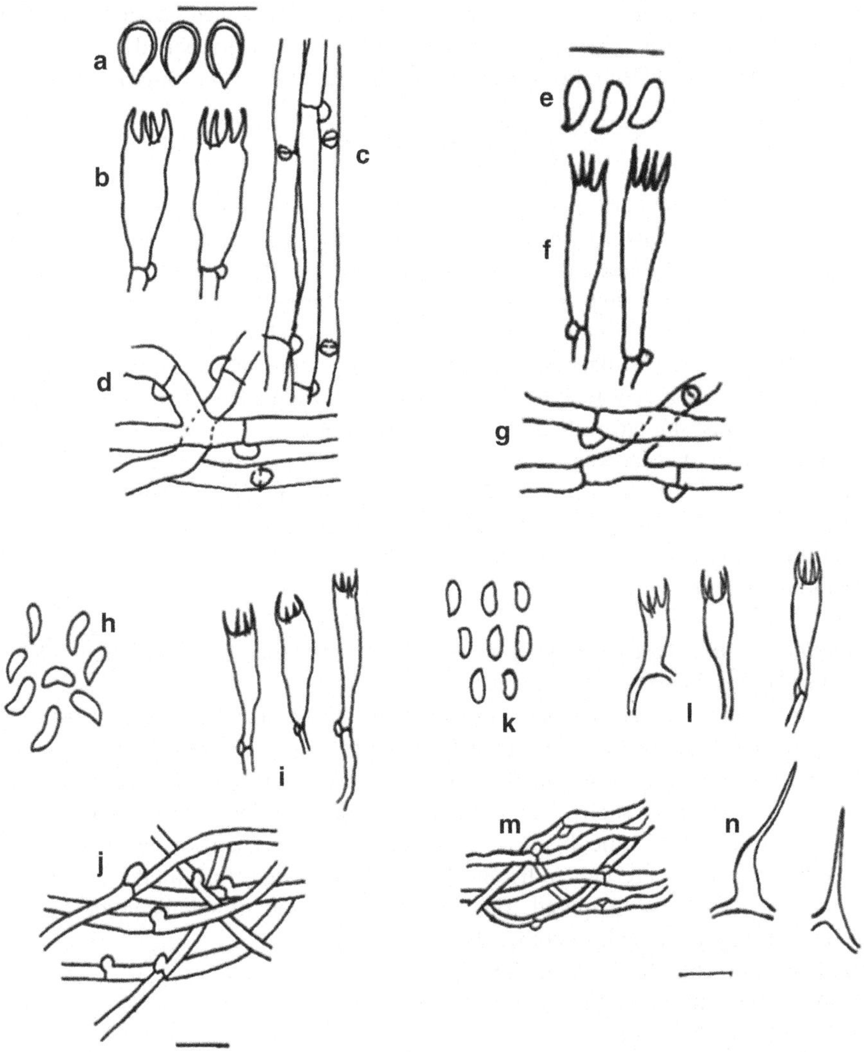

Fig. 6.1 (**a–d**) *Amyloathelia crassiucula* (**a**) Basidiospores, (**b**) Basidia, (**c**) Generative hyphae, (**d**) Hyphal cordons; (**e–g**) *Amylocorticium indicum* (**e**) Basidiospores, (**f**) Basidia, (**g**) Generative hyphae; (**h–j**) *Amyloxenasma allantosporum* (**h**) Basidiospores, (**i**) Basidia, (**j**) Generative hyphae; (**k–n**) *Amyloxenasma grisellum* (**k**) Basidiospores, (**l**) Basidia, (**m**) Generative hyphae, (**n**) Hyphoid structure

Remarks: This species is characterized by, amyloid thick-walled basidiospores and clavate to subclavate, 4-sterigmate basidia. Earlier, it has been reported by Hjortstam and Ryvarden (1979) from Norway. It is a new record for India/Himalayas.

Amylocorticium Pouz.,
Česká Mykol. 13:11 1959.

Fructification resupinate, pelliculose to submembranous, more or less atheloid when young, adnate; hymenial surface smooth, lighter coloured, continuous. Subiculum subhyaline in section, composed of loosely woven hyphae. Hyphal system monomitic, hyphae branched, septate, clamped, the walls thin but firm, subhyaline. Hyphal tissue turning purple in 3 % KOH solution. Cystidia, gloeocystidia or other ancillary structures absent, some thin-walled cystidioles may be present. Basidia clavate-cylindrical, 4-spored. Basidiospores narrowly ellipsoid to allantoids, smooth-walled, subhyaline, amyloid.

Eleven species, wide spread

Lit.: Gilbertson & Hemmes (*Mem. N. Y. Bot. Gdn*. 89: 81, 2004; Hawaii)

Type species: *Amylocorticium subsulphureum* (Karst.) Pouz., 1881

Habitat: Dead wood

Himalayas: One

Amylocorticium indicum K.S Thind & S.S. Rattan, Trans. Br. mycol. Soc. 59(1): 125 (1975). Plate 6.1a, Fig. 6.1e–g

Fructification resupinate, annual but revive in the next year, adnate, widely effused; hymenial surface yellow when fresh but fades to cream yellow on drying, smooth, continuous, not creviced; margin thinning, paler concolorous to concolorous, adnate. Subiculum subhyaline in section, composed of loosely woven hyphae.

Hyphal system monomitic; hyphae 2.2–3.5 µm wide, branched at wide angles, septate, clamps are present, walls are thin to thick-walled. Basidia clavate, 17.4–24.5×3.6–4.2 µm, 4-spored. Basidiospores ellipsoid to suballantoid, shortly apiculate, thin-walled, smooth, subhyaline, amyloid.

Distribution: H.P.: Kullu; J&K: Batote.

Collection examined: GSR 5043, SSR 5207.

Substratum: On log of *Cedrus deodara*.

Remarks: This species is characterized by yellow fructification, absence of cystioles and small ellipsoid to allantoid basidiospores. It is widely distributed in the coniferous forests of North Western Himalayas.

Amyloxenasma (Oberw.) Hjortstam & Ryvarden,
Syn. Fung. (Oslo) 20: 34, (2005)

Fructification resupinate, thin, waxy-gelatinous; hymenial surface smooth, light coloured; margin thinning to indistinct. Hyphal system monomitic; generative hyphae branched, septate, clamped. Cystidia or Gloeocystidia absent. Basidia clavate-cylindrical, 4–5 sterigmate. Basidiospores globose, ellipsoid or allantoids, acyanophilous.

Five Species, wide spread

Plate 6.1 (**a**) *Amylocorticium indicum*. (**b**) *Amyloxenasma allantosporum*. (**c**) *Amyloxenasma grisellum*. (**d**) *Ceraceomyces borealis*. (**e**) *Ceraceomyces cystidiatus*. (**f**) *Ceraceomyces fibuliger*

Lit.: Hjortstam & Ryvarden (*Syn. Fung.* (Oslo) 20: 34, 2005)
Type Species: *Amyloxenasma grisellum* (Bourdot) Hjortstam & Ryvarden, 2005
Habitat: Dead wood
Himalayas: Two
Key to species

1. Basidiospores slightly allantoid, 4.6–6 × 1.6–2.2 μm, smooth .... *A. allantosporum*
1. Basidiospores 3.5–4.7 × 1.8–2.6 μm, ellipsoid to suballantoid, smooth .... *A. grisellum*

Amyloxenasma allantosporum (Oberw.) Hjortstam & Ryvarden, Syn. Fung. (Oslo) 20: 34, 2005. Plate 6.1b, Fig. 6.1h–j

Fruitbody resupinate, closely adnate, effused thin, waxy-gelatinous; hymenial surface smooth, yellowish- grey to pastel grey; margin not differentiated. Hyphal system monomitic; generative hyphae indistinct, almost gelatinized, very narrow, 1–2 μm wide, branched, septate, clamped. Cystidia none. Basidia 7–20 × 4–6 μm, short cylindrical to clavate, 4-sterigmate; sterigmata up to 3.5 μm long. Basidiospores 4.5–7 × 1.5–2.5 μm, allantoids, smooth, thin walled amyloid, acyanophilous with oily contents.

Distribution: Bhutan: Thimphu, Begana.

Collection examined: GSD 19631.

Substratum: on trunks of *Picea abies*, *Pinus nigra*, *P. pinea* and *Abies alba*.

Remarks: Oberwinkler (1965) erected this species with small pleurobasidia and amyloid, allantoids basidiospores. It resembles Xenasma *griesella* in most of the characters except that it differs from the latter in having allantoids (4.5–7 × 1.5–2 μm) basidiospores in comparison to ellipsoid to subellipsoid (3.5–5 × 1.8–2.5 μm) basidiospores. This collection resembles the description of *X. allantospora* as given by Oberwinkler (1965) in most of the characters, except for the size of basidia.

Amyloxenasma grisellum (Bourdot) Hjortstam & Ryvarden, Syn. Fung. (Oslo) **20**: 34 (2005) = *Xenasmatella grisella* (Bourd.) Oberw., Syd. Ann. Mycol. 19(1–6): 35, 1965 = *Corticium grisellum* Bourd., Add. aux Cort. Rev. Sci. Bourd. Centr. Fr. 35(1): 17, 1922. Plate 6.1c, Fig. 6.1k–n

Fructification resupinate, closely adnate, effused, thin, membranous- ceraceous to ceraceous, waxy-gelatinous; hymenial surface smooth, yellowish- white when fresh, greyish orange to brownish orange on drying; margin undifferentiated.

Hyphal system monomitic; generative hyphae branched, septate clamped, narrow, 1–2.5 μm wide, indistinct to gelatinized. Cystidia absent, hyphoid structures with broadened base present. Basidia 6–10 (−16) × 4–5 μm, uniform to short cylindrical, pleurobasidiate, 4-sterigmate; sterigmata up to 5 μm long. Basidiospores 3.5–4.7 × 1.8–2.6 μm, ellipsoid to suballantoid, smooth, thin-walled, amyloid, acyanophilous.

Distribution: West Bengal: Darjeeling. Bhutan, H.P.: Kullu, U.K.: Karanprayag-Chakrata

Collection examined: GSD 19263, 19264, IBP 37006, L 37005.

Substratum: Decaying angiospermic stump.

Remarks: This species was first described by Bourdot (1922) as *Corticium grisellum*. Oberwinkler (1965) considered it in the genus *Xenasmetella*. He also considered *Corticium subilascens* Litsch., *C. pruina* Bourd. & Galz. and *C. subacinaeforme* Bourd. & Galz., as synonyms of this species. The species is characterized by small- sized basidia and ellipsoid to suballantoid, thin-walled, amyloid basidiospores. The present collections resemble closely the description of the species as given by Oberwinkler (1965). It is a new record for Himalayas.

Ceraceomyces Jülich,
Willdenowia. Beih. 7:146. 1972.

Fructification resupinate, distinctly pelliculose when young but becomes membranous-ceraceous with age, loosely adnate to separable, hymenial surface smooth to meruloid when fresh but wrinkles often disappear on drying, somewhat slimy to touch; margin thinning to sometimes rhizomorphic. Subiculum subhyaline in section, composed of loosely woven hyphae. Hyphae remain distinct and loose. In the basal part. Hyphal system monomitic, hyphae branched, septate, clamped, thin to moderately thick-walled, subhyaline. Gloeocystidia absent. Cystidia absent. Basidia clavate, 4-spored. Basidiospores thin-walled, smooth, non-amyloid.

Sixteen species, wide spread

Lit.: Legon (*Mycologist* **19**:167, 2005)

Type Species: *Ceraceomyces tessulatus* (Cooke) Jülich., 1878

Habitat: Wood

Himalayas: Seven

Key to species:

1. Cystidia present.. 2
1. Cystidia absent.. 4
2. Cystidia cylindrical to sub–cylindrical 3
2. Cystidia hyphoid...*C. sublaevis*
3. Cystidia encrusted with crystalline matter..................................... *C. cystidiatus*
3. Cystidia not encrusted.. *C. fibuligera*
4. Basidiospores ellipsoid/cylindrical, curved at the base 5
4. Basidiospores subglobose to ellipsoid .... *C. bizonatus*
5. Basidiospores up to 15.0 µm in length..*C. reidii*
5. Basidiospores up to 8.0 µm in length.. 6
6. Basidiospores up to 6.0 µm in length..*C. tessulatus*
6. Basidiospores up to 8.0 µm in length...*C. borealis*

Ceraceomyces bizonatus Dhingra & Avn. P. Singh Mycotaxon **106**: 399–401 2009. Fig. 6.2a–c

Fructification resupinate, loosely adnate, effused; cracks appear on drying; surface smooth to tuberculate, yellowish brown; margins indeterminately thinning.

Hyphal system monomitic; generative hyphae up to 2.6 µm wide, branched septate, clamped, thin-walled; basal hyphae running somewhat parallel to the substra-

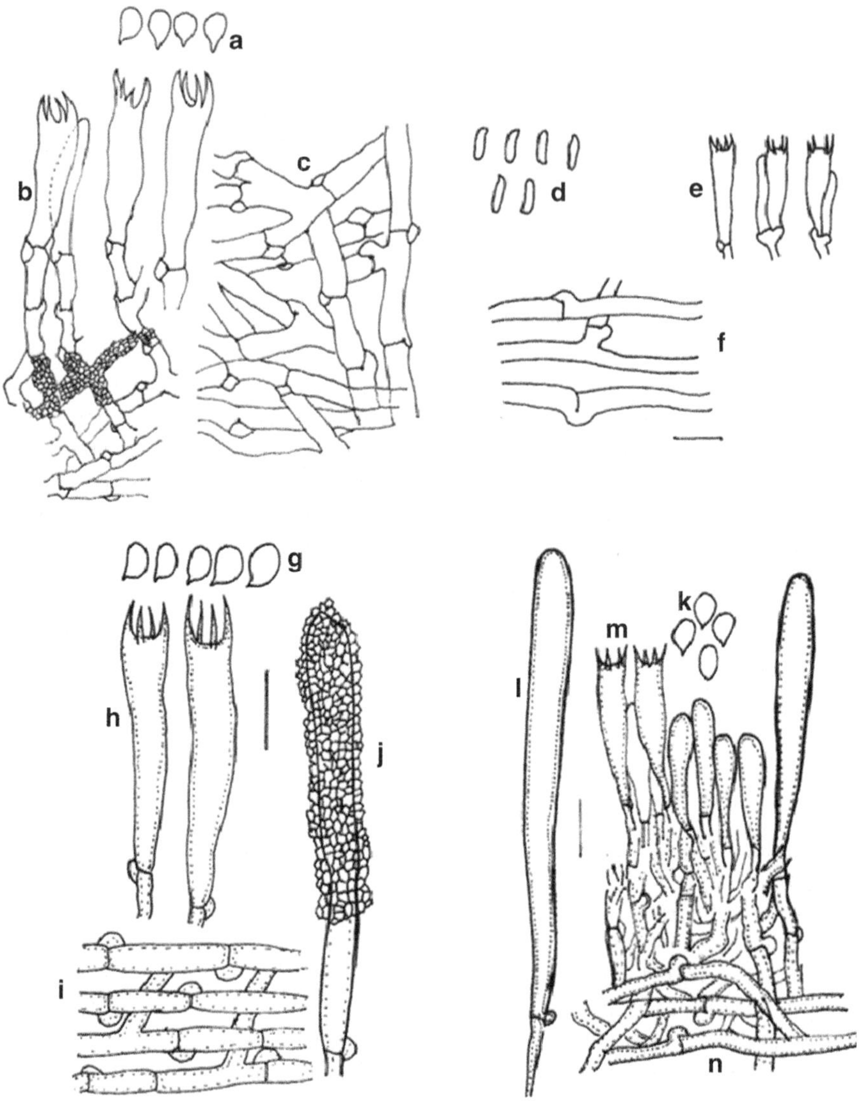

Fig. 6.2 (**a–c**) *Ceraceomyces bizonatus* (**a**) Basidiospores, (**b**) Basidia, (**c**) Generative hyphae; (**d–f**) *Ceraceomyces borealis* (**d**) Basidiospores, (**e**) Basidia, (**f**) Generative hyphae; (**g–j**) *Ceraceomyces cystidiatus* (**g**) Basidiospores, (**h**) Basidia, (**i**) Generative hyphae, (**j**) Encrusted cystidia; (**k–n**) *Ceraceomyces fibuliger* (**k**) Basidiospores, (**l**) Cystidia, (**m**) Basidia, (**n**) Generative hyphae

tum; subhymenium characteristic in having two zones- lower zone of vertical hyphae without encrustation followed by a peculiar zone of short celled almost isodiametric, encrusted hyphae. Basidia 16.3–23.8 × 3.1–5.5 μm, clavate, 4 sterigmate, with a basal clamp; sterigmata up to 3.7 μm long. Basidiospores 2.5–3.3 × 1.8–2.2 μm, ellipsoid to somewhat subglobose, smooth, thin-walled, inamyloid, acyanophilous.

Distribution: H.P.: Kullu- Manali- Kothi, Dalhousie.

Substratum: Decaying log of *Cedrus deodara.*

Collection examined: IBP 37006

Remarks: The species is characterized by two zoned subhymenium, hymenium with a lower zone of vertical hyphae and an upper zone of short-celled isodiametric encrusted hyphae.

Ceraceomyces borealis (Romell) J. Erikss. & Ryvarden, Cortic. N. Europe (Oslo) **2**: 205 (1973) = *Merulius borealis* Romell, Arkv. f. botanic 11(no.3): 27, 1911. Plate 6.1d, Fig. 6.2d–f

Fructification resupinate, loosely adnate, pellicular, thin, whitish when young, becoming ceraceous when mature, turning yellowish or pinkish-white, pale ochraceous or pale orange on drying, smooth initially but soon becomes folded and wrinkled; hymenial surface smooth to merulioid.

Hyphal system monomitic; generative hyphae branched, septate, clamped, clamps large and usually open; basal hyphae up to 5.8 μm wide, loosely interwoven. Cystidia none. Basidia 13.0–19.7 × 3.5–5.0 μm, clavate, 4-sterigmate, with a basal clamp. Basidiospores 4.3–8.0 × 1.6–2.25 μm, cylindrical, smooth, thin-walled, non-amyloid, acyanophilous, with oil drops.

Distribution: Arunachal Pradesh: West Kameng, Bomdila, Shergaon; West Bengal: Darjeeling, Batasi; Bhutan: Thimphu, Bunakha, Dochula; H.P.: Kinnaur; U. K.: Rudarprayag-Chamoli

Collection examined: GSD 19828, 19149, IBP 37007, L 37007

Substratum: Decaying gymnospermic log, decaying *Cryptomeria* log.

Remarks: This species is characterized by wrinkled, loosely attached fructification, generative hyphae with large and usually open clamps; clavate, 4-spored basidia and subcylindrical, thin-walled basidiospores. It was first described by Romell (1911) as *Merulis borealis* from Lappland. Eriksson and Ryvarden (1973) transferred it to *Ceraceomyces*. These collections resemble the description given by Eriksson and Ryvarden (1973). The species is new record for North Western Himalayas.

Ceraceomyces cystidiatus (J. Erikss. & Hjortst.) Hkortst., Sv. Bot. Tidskr. 67: 105, (1973) = *Corticium violascens* (Fr.) var. *cystidiata* J. Erikss. & Hjortst. In Jülich, Willd. Beih. 7:146, 1972. Plate 6.1e, Fig. 6.2g–j

Fructifications resupinate, effused, loosely adnate; hymenial surface smooth, whitish to greyish yellow when fresh, greyish orange on drying; margin thinning, fibrillose, paler concolorous to indeterminate.

Hyphal system monomitic; generative hyphae up to 4.1 µm wide, septate, clamped; basal hyphae parallel to substrate, less branched; subhymenial hyphae vertical, more branched. Cystidia 55.8 × 156.9 × 3.5–6.8 µm, cylindrical to subcylindrical, thin-walled, with basal clamp, encrusted; projecting up to 35.0 µm out of the hymenium. Basidia 25.8–35.0 × 5.6–5.9 µm, somewhat constricted, 4-sterigmate, with basal clamp; sterigmata up to 6.5 µm long. Basidiospores 3.4–5.0 × 2.8–3.8 µm, apiculate, ellipsoid, smooth, thin-walled, inamyloid, acyanophilous.

Distribution: U.K.: NDBR.

Collection examined: IBP 42292.

Substratum: Decaying gymnospermic log, Decaying *Cryptomeria* log.

Remarks: This species is characterized by ellipsoid, small sized basidiospores and cylindrical to subcylindrical, encrusted cystidia. It is a new record for Uttarakhand as well as Himalayas.

Ceraceomyces fibuliger (Thind & Rattan) Rattan, (as '*fibuligera*') Biblthca Mycol. 60, 248 (1977). Plate 6.1f, Fig. 6.2k–n

Fructifications resupinate, membranous- ceraceous to wholly ceraceous but become hard and brittle on drying, adnate, arising as small colonies, which may become effused later due to growth or confluence; hymenial surface cream to cream yellow or orchraceous, smooth to finely tuberculate, continuous, not creviced; margin thinning to fibrillose, white to paler concolorous, adnate.

Hyphal system monomitic; generative hyphae (1.3) 2.4–5.5 µm wide, branched, septate, clamped, clamps prominent and almost at all septa, thin to thick-walled, wall up to 1.5 µm thick. Hyphae in the lower part of the context are broad, comparatively thick-walled and tinted towards the hymenium. Gloeocystidia absent. Cystidia 50–90 × (3.5) 5–6 µm, cylindrical with obtuse apices, subhyaline, thin to slightly thick-walled (0.6 µm), often arising from the upper part of the context, just emerging of projecting up to 50 µm out of the hymenium, unincrusted or smooth.

Basidia 25–30 (41) × 4.5–5.0 µm, clavate, 4-spored, sterigmata up to 4 µm long. Basidiospores 4–5.5 × 2–3 µm, ellipsoid to broadly ellipsoid, minutely apiculate, thin-walled, sub hyaline, smooth, non-amyloid.

Distribution: H.P.: Khadrala- Mahasu, Kullu- Jagat Sukh, Shimla: Tara Devi.

Collection examined: SSR 5341; IBP 37008.

Substratum: On bark of *Taxus baccata*, cut stump *Cedrus deodara* and *Picea smithiana* wood

Remarks: This species is marked by the colour of the hymenial surface, presence of cystidia and the shape and size of basidiospores. It is closely related to *C. cystidiatus* (Erikss. & Hjortst.) Hjortst. and perhaps both are conspecific but in *C. cystidiatus* the subiculum is light brown and turns red violet in KOH. Moreover, the hymenial surface also shows deep cracking in to the polygons.

Ceraceomyces reidii (K.S. Thind & S.S. Rattan) S.S. Rattan, Biblthca Mycol. **60**: 252 (1977) = *Corticium redii* Thind & Rattan, Trans. Brit. mycol. Soc. **59**(1): 125 (1972). Fig. 6.3a–c

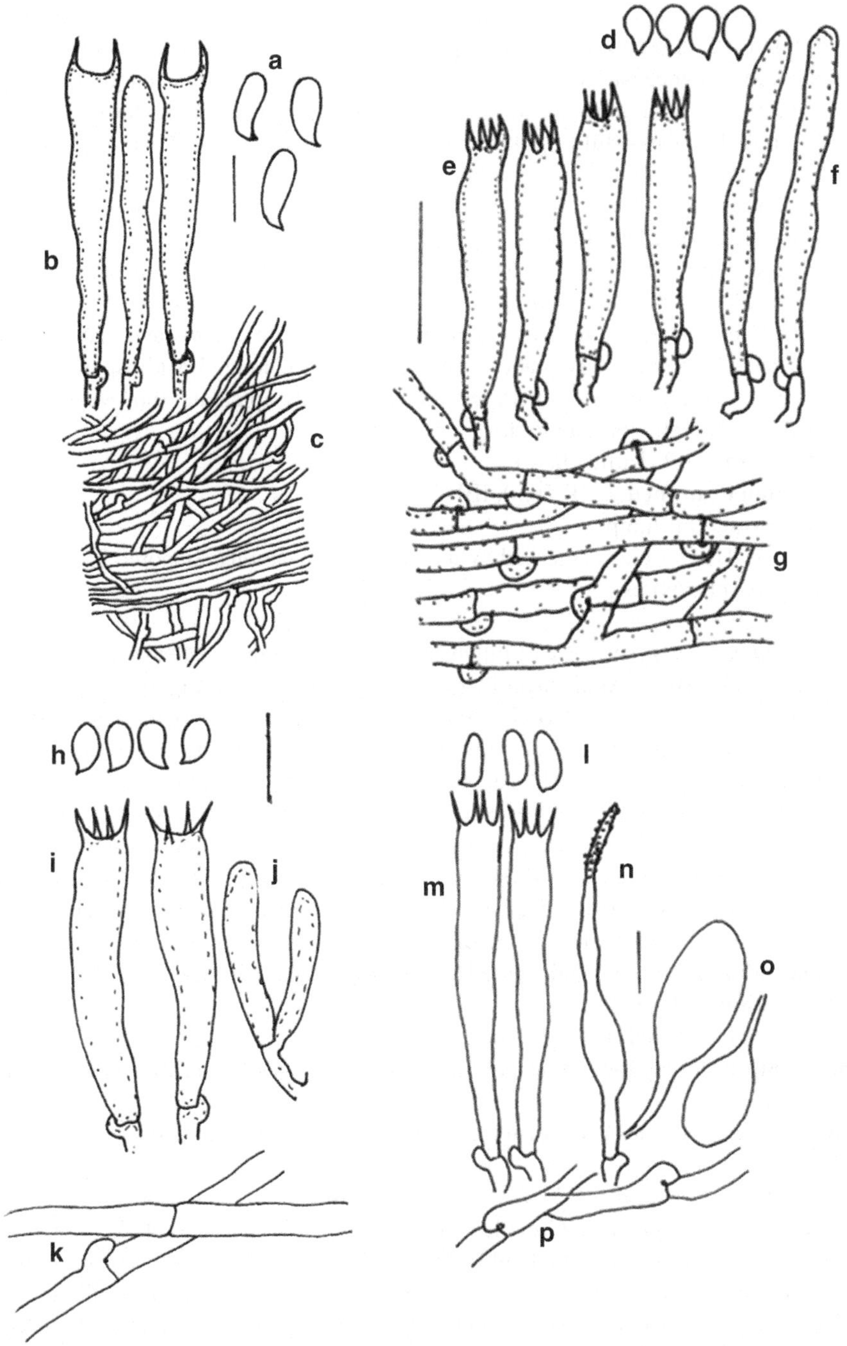

Fig. 6.3 (**a–c**) *Ceraceomyces reidii* (**a**) Basidiospores, (**b**) Basidia, (**c**) Generative hyphae; (**d–g**) *Ceraceomyces sublaevis* (**d**) Basidiospores, (**e**) Basidia, (**f**) Cystidia, (**g**) Generative hyphae; (**h–k**) *Ceraceomyces tessulatus* (**h**) Basidiospores, (**i**) Basidia, (**j**) Basidioles, (**k**) Generative hyphae; (**l–p**) *Chondrostereum purpureum* (**l**) Basidiospores, (**m**) Basidia, (**n**) Cystidia, (**o**) Hyphal vesicles, (**p**) Generative hyphae

Fructifications resupinate, membranous-ceraceous, separable, widely effused, up to 1 mm thick in section; hymenial surface deep cream yellow to light ochraceous, smooth to somewhat uneven, waxy and ceraceous, continuous; margin thinning, fibrillose to rhizomorphic, loosely adnate, white to paler concolorous. Rhizomorphs abundant, white, branched, usually arise from the basal part of the subiculum and penetrating the substratum. Subiculum subhyaline in section. Basal hyphae are somewhat distinct and loosely woven but some of them often become organised in to rhizomorphs.

Hyphal system monomitic; generative hyphae 2–3.5 µm wide, branched, septate, clamped, thin-walled, subhyaline. Cystidia absent. Basidia 44.5–55 × 6–7 µm, clavate to clavate-cylindrical, often proliferating from clamps and occurring in clusters, 2-spored, sterigmata slender up to 7 µm long. Basidiospores 11.5–14.5 × 4.5–6 µm, ellipsoid with curved base, shortly apiculate, thin-walled, subhyaline, smooth, non-amyloid.

Distribution: H.P.: Kullu- Jalori

Collection examined: GSR 5050, IBP 37009

Substratum: On stump

Remarks: This species is marked by the thick fructifications, 2-spored basidia and large basidiospores.

Ceraceomyces sublaevis (Bres.) Jülich, Willdenowia, Beih. 7: 147 (1972) = *Corticiumsublaeve* Bres., Annls mycol. 1(1): 95, 1903. Plate 6.2a, Fig. 6.3d–g

Fructification resupinate, loosely adnate, effused; hymenial surface smooth to tuberculate, orange to greyish orange when fresh, pale orange on drying; margin thinning, arachnoid to fibrillose, paler concolorous to indeterminate.

Hyphal system monomitic; generative hyphae up to 3.5 µm wide, septate, clamped, thin-walled; basal hyphae parallel to substrate, less branched; subhymenial hyphae vertical, branched. Cystidia 22.0–28.0 × 3.0 µm, somewhat sinous, thin-walled, smooth with basal clamp. Basidia 12.0–19.0 × 3.0–3.5 µm, clavate to subclavate, 4-sterigmate with basal clamp. Basidiospores 3.5–4.0 × 2 µm, apiculate, ellipsoid, thin-walled, smooth, inamyloid, acyanophilous.

Distribution: H.P.: Kullu.

Collection examined: IBP 37010.

Substratum: Decaying gymnospermic wood, bark of *Cedrus deodara*, *Cryptomeria* log.

Remarks: The species is characterized by small ellipsoid basidiospores.

Ceraceomyces tessulatus (Cooke) Jülich., Willdenovia, Beih. **7**: 154. 1972 = *Corticium tessulatum* Cooke, Grevillea 6 (no. 40): 132. 1878. Fig. 6.3h–k

Fructification resupinate, pelliculose to distinctly atheloid initially, becoming membranous-ceraceous with age, loosely adnate, often arising as small circular colonies which may coalesce and become widely effused; hymenial surface cream yellow to yellow when fresh fading to cream on drying, smooth, somewhat waxy and shiny, continuous; margin thinning, somewhat fibrillose, loosely adnate, white to paler concolorous.

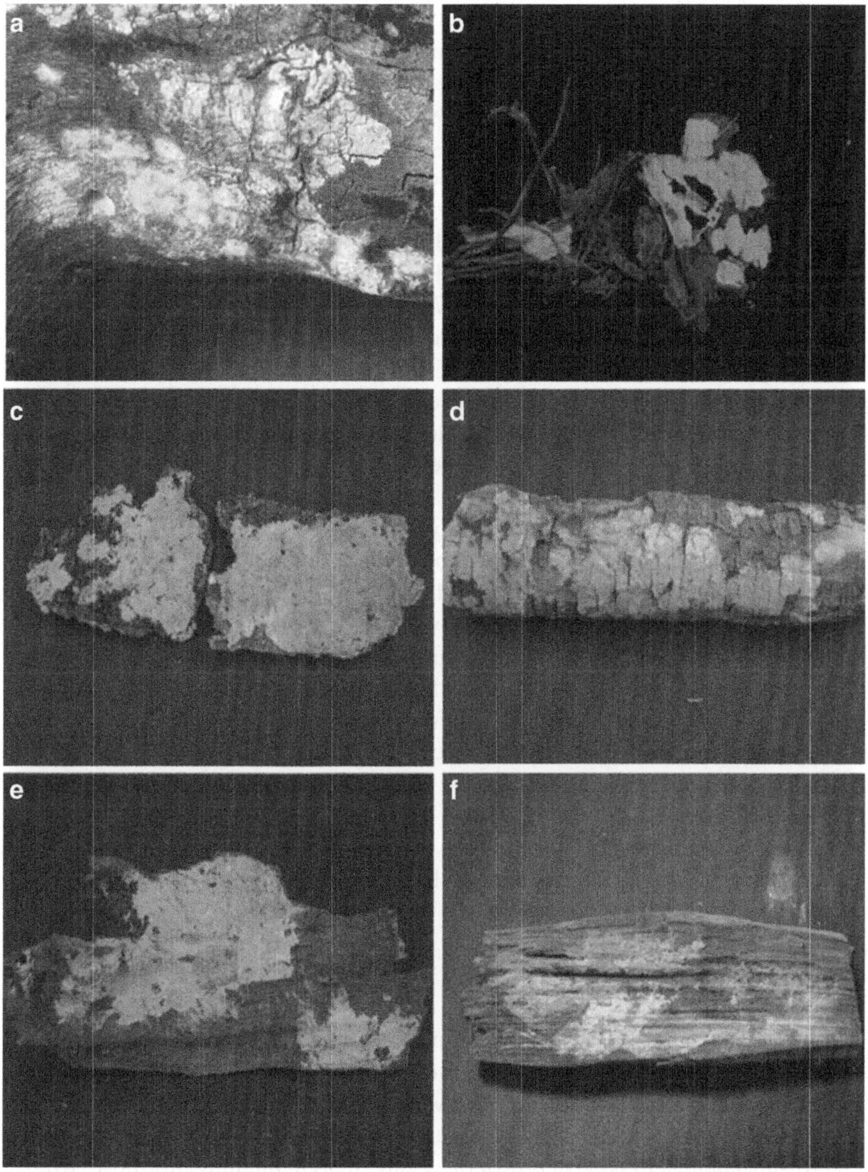

Plate 6.2 (**a**) *Ceraceomyces sublaevis*. (**b**) *Chondrostereum purpureum*. (**c**) *Cylindrobasidium evolvens*. (**d**) *Radulomyces confluens*. (**e**) *Cristinia helvetica*. (**f**) *Amphinema byssoides*

Hyphal system monomitic, hyphae loosely woven in the basal part but become more compact and ceraceous in the upper part, hyphae 3–5 (6.5) μm wide, branched at wide angles, septate, clamped, clamps single, but sometimes opposite and often proliferating in to new branches, the walls thin to moderately thick, subhyaline. Cystidia absent. Basidia 25–35 × 5.5–6.5 μm, clavate, often proliferating from clamps and occurring in clusters, 4-spored, sterigmata up to 3.5 μm long. Basidiospores 5.5–6 × 2.8–3.2 μm, ellipsoid with a slightly drawn out and curved base, shortly apiculate, thin-walled, subhyaline, non-amyloid.

Distribution: H.P.: Rohtang, Kullu, Keylong.

Collection examined: SSR 5747, IBP 37012.

Substratum: On log of *Abies pindrow*, rotten wood, twigs.

Remarks: The distinguishing features of this species are smooth and cream yellow to yellow hymenial surface, presence of rhizomorphs and shape and size of basidiospores. The Himalayan collection is quite typical of the species but differs in having slightly smaller basidiospores.

Family- Cyphellaceae

Chondrostereum Pouzar,
Česká 13(1): 17, 1959.

Fructification resupinate, effused, loosely adnate, reflexed, gelatinous when fresh, become horny on drying; hymenial surface smooth to tuberculate, margin thinning, fibrillose, paler concolorous to indeterminate; hyphal system monomitic; generative hyphae septate, thin to thick-walled, branched; cystidia fusiform to subfusiform, thin-walled with basal clamp; basidia narrowly clavate, 4-sterigmate; basidiospores subcylindrical to suballantoid, smooth, inamyloid, acyanophilous.

Five species, widespread

Lit.: Pouzar, Z. 1959. New genera of higher fungi III. 13(1): 10–19.

Type Species: *Stereum purpureum* Pers. 1794.

Habitat: Wood

Himalayas: One

Chondrostereum purpureum (Pers.) Pouzar, Česká Mykol. 13(1): 17 (1959) = *Stereum purpureum* Pers., Neues Mag. Bot. 1: 110 (1794). Plate 6.2b, Fig. 6.31–p

Fructification resupinate, effused, somewhat reflexed, gelatinous when fresh horny on drying; hymenial surface smooth to tuberculate, reddish brown when fresh, brownish orange to greyish brown on drying; margin thinning, fibrillose, whitish to paler concolorous to indeterminate.

Hyphal system monomitic; generative hyphae up to 4.3 μm wide, septate, clamped; basal hyphae thin to thick-walled, branched; subhymenial hyphae vertical, richly branched. Cystidia 65.0–78.0 × 4.0–7.0 μm, fusiform to subfusiform, thin-walled, smooth with basal clamp. Basidia 16.5–34.0 × 4.1–6.0 μm, narrowly clavate, 4-sterigmate with basal clamp. Basidiospores 5.5–7.5 × 2.4–4.5 μm, ellipsoid, thin-walled, smooth, inamyloid, acyanophilous.

Distribution: H.P.: Chamba, Shimla, Kullu.

Collection examined: SSR: 5123, 5582, 5600, 5731, IBP 370013.

Substratum: Dead and decaying angiospermic and gymnospermic wood and twigs.

Remarks: The species is recognised by colour of basidiospores, vesicular hyphae, fusiform to subfusiform cystidial elements.

Family- Physalacriaceae

Cylindrobasidium Jülich, Persoonia **8**(1): 72, 1974.

Fructification resupinate; hymenial surface smooth and tuberculate; margin fimbricate, subhymenium thickening; hyphal system monomitic; generative hyphae distinct, clamped, branching from clamp cells; cystidia fusiform, thin-walled, slightly projecting. Basidia clavate, 4-spored, basal clamp present, new basidia arise from the clamp cells. Basidiospores obliquely ellipsoid, thin-walled, non-amyloid.

Six Species, world-wide

Lit.: Jülich (*Persoonia* **8**: 72, 1974).

Type species: *Cylindrobasidium evolvens* (Fr.) Jülich, 1974.

Habitat: Wood

Himalayas: One

Cylindrobasidium evolvens (Fr.) Jülich, Persoonia 8(1): 72 1974 = *Basidioradulum evolvens* (Fr.) Parm. Consp. Syst. Cort. p 112. 1968. Plate 6.2c, Fig. 6.4a–d

Fructifications resupinate, membranous, adnate, widely effused; hymenial surface cream brown to light brown, smooth, continuous but often cracking areolately exposing the white context; margin thinning, adnate, paler concolorous. Subiculum subhyaline in section, composed of a basal zone of compactly arranged repent hyphae and an upper zone of semi-erect hyphae.

Hyphal system monomitic; generative hyphae 2.5–4 (5.0) µm wide, branched, septate, clamped, clamps prominent and almost at all septa, thin-walled, firm, subhyaline. Cystidia and gloeocystidia absent. Basidia 40–50 × 5–7.5 µm, clavate-cylindrical, 4-spored, sterigmata long. Basidiospores 7.5–9 × 4.5–6 µm, pip-shaped or ovoid, shortly apiculate, often occurring in groups of fours, thin-walled, subhyaline, smooth, non-amyloid, acyanophilous.

Distribution: H.P.: Kullu; Arunachal Pradesh: West Kameng, Bomdila; Bhutan: Thimphu, Nawephu, Dochula, Chakaphug, Paro, D'Dzong, Ha.

Collection examined: SSR 5680, GSR 5042, IBP 37673.

Substratum: On fallen twigs.

Remarks: The diagnostic features of this species are membranous fructifications with smooth hymenial surface, lack of cystidia or gloeocystidia, large basidia and pip-shaped basidiospores occurring in groups of fours.

Family- Pterulaceae

Key to genera

1. Fruiting body often waxy–gelatinous, Basidia clavate–cylindrical to suburniform. Basidiospores globose to ellipsoid to suballantoid, smooth or finely echinulate .... *Aphanobasidium*
1. Fructification not waxy–gelatinous, Basidia clavate, basidiospores ellipsoid to broadly ellipsoid, smooth .... *Radulomyces*

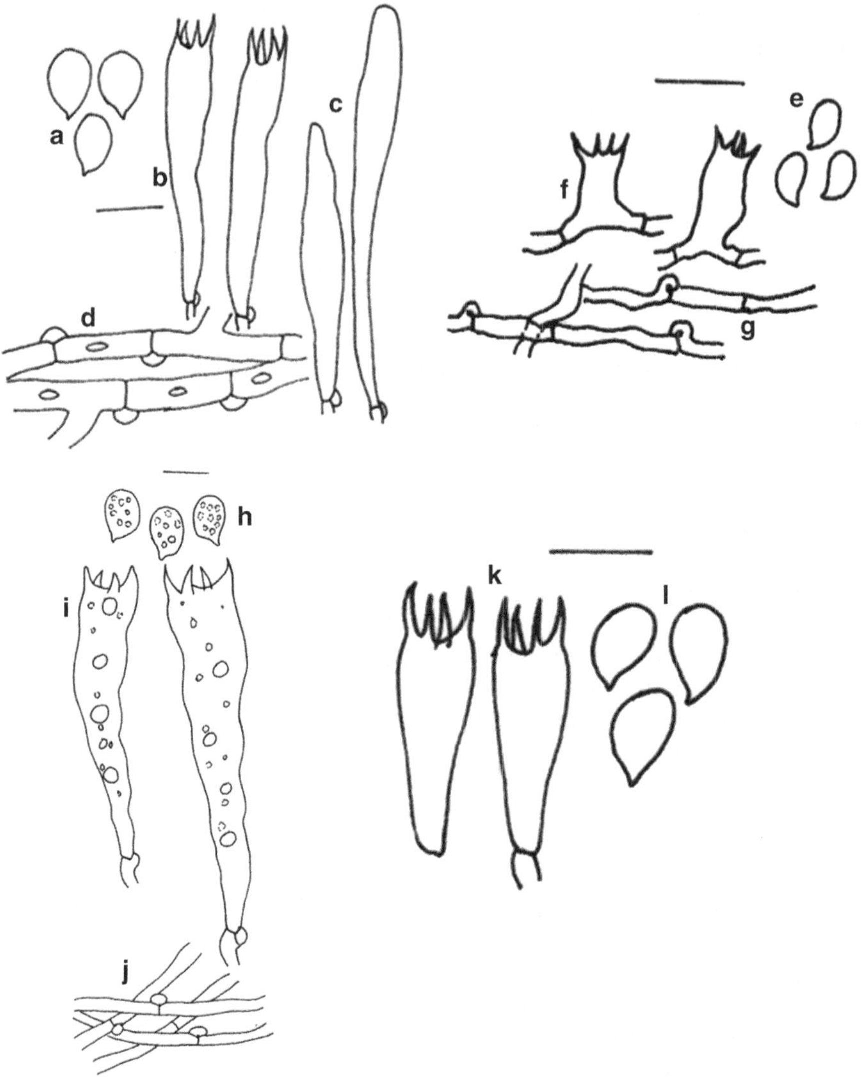

Fig. 6.4 (**a–d**) *Cylindrobasidium evolvens* (**a**) Basidiospores, (**b**) Basidia, (**c**) Cystidia, (**d**) Generative hyphae; (**e–g**) *Aphanobasidium subnitens* (**e**) Basidiospores, (**f**) Basidia, (**g**) Generative hyphae; (**h–j**) *Radulomyces confluens* (**h**) Basidiospores, (**i**) Basidia, (**j**) Generative hyphae; (**k**, **l**) *Radulomyces molaris* (**k**) Basidia, (**l**) Basidiospores

Aphanobasidium Jülich,
Persoonia 10(3): 326, 1979.

Fructification resupinate, usually occurring as a fine thin film of hyphae which is difficult to discern with unaided eye, often waxy-gelatinous; hymenial surface, smooth, lighter coloured; margin thinning to indistinct. Subiculum subhyaline or pale coloured, wholly agglutinated or ceraceous. Hyphal system monomitic, hyphae often collapsed, with or without clamps. Cystidia or gloeocystidia present or absent. Basidia clavate-cylindrical to suburniform, arising directly from the repent hyphae, pleurobasidiate, sometimes pedicellate 4-spored, Basidiospores globose to ellipsoid to suballantoid, smooth or finely echinulate, nonamyloid, sometimes amyloid.

Fifteen species, widespread

Lit.: Boidin et. al. (*BSMF* **119**: 333, 2003; subgen. *Aphanobasidium*)

Type Species: *Corticium subnitens* Bourdot & Galzin, 1928.

Habitat: Wood

Himalayas: One

Aphanobasidium subnitens (Bourdot & Galzin) Jülich, Persoonia **10**(3): 326 (1979) = *Xenasma subnitens* (Bourd. & Galz.) Liberta, Mycologia 52: 910. 1960. Fig. 6.4e–g

Fructification resupinate, membranous-ceraceous to ceraceous, closely adnate, arising as small, thin film-like colonies hardly visible to the unaided eye; hymenial surface white, smooth but appears farinose under the lens, not creviced; margin determinate, concolorous. Context composed of completely agglutinated few repent hyphae which are difficult to discern.

Hyphal system monomitic; generative hyphae septate, clamped, thin-walled. Cystidia absent. Basidia 10–12 × 5–6 μm, short cylindrical, without a basal septum and appearing bifurcate at the base (pleurobasidiate), 4-spored, sterigmata up to 3 μm long. Basidiospores 5.5–7 × 2.5–3 μm, ellipsoid to ellipsoid-ovoid, shortly apiculate, thin-walled, subhyaline, smooth, non-amyloid.

Distribution: H.P.: Mahasu- Narkanda.

Collection examined: SSR: 5599, IBP 37672.

Substratum: Bark of *Rhododendron arboretum.*

Remarks: This species is marked by very thin, film-like fructification, absence of cystidia or gloeocystidia, short cylindrical basidia and ellipsoid-ovoid, smooth basidiospores. The present collection is quite typical of the species but differs in having slightly narrow basidiospores as compared with those of the species (5–7 × 3.5–4.5 μm).

Radulomyces M.P. Christ.,
Dansk. bot. Arkiv. **19**(2): 230. 1960.

Fructification resupinate to slightly effused-reflexed, membranous to subceraceous, discern to widely effused; hymenial surface smooth to raduloid or rarely toothed, usually pale coloured. Subiculum subhyaline in section, composed of compactly arranged more or less agglutinated hyphae. Hyphal system monomitic, hyphae branched, septate, clamped, thin-walled, subhyaline walls thin often collapsing. Basidioles beaded or flexuous. Cystidia and Gloeocystidia absent. Basidia clavate, 4-spored. Basidiospores ellipsoid to broadly ellipsoid, thin-walled, subhyaline, smooth, non-amyloid, acyanophilous.

Ten species, widespread

Lit.: Gilbertson & Nakason (*Mycol.* **95**:467, 2003)

Type Species: *Radulomyces confluens* (Fr.) Christ., 1960.

Habitat: Dead wood
Himalayas: Two
Key to species

1. Basidia 50–87 × 11–15 μm, basidiospores 8–11 × 7–9 μm, broadly ellipsoid to subglobose, thin to somewhat thick walled .... *R. confluens*
1. Basidia 25–30 × 5.5 μm, basidiospores 8.5–10 × 5.5–6.5 μm, broadly ellipsoid to ovoid, thin-walled .... *R. molaris*

Radulomyces confluens (Fr.) M.P. Christ., Dansk bot. Ark. 19(2): 230 (1960) (Fr.) M.P. Christ., var. *macrobasidiata* Prasher & Lalita var. nov. Plate 6.2d, Fig. 6.4h–j

Mycobank MB812326

Fructification resupinate, closely adnate, effused, up to 350 μm thick in section, when wet watery-ceraceous and hygrophanous, somewhat tuberculate, greyish-yellow with rosaceous tint and smoothing on drying; margin normally fimbricate, thinning, whitish.

Hyphal system monomitic; generative hyphae clamped, thin-walled, up to 2.8 μm wide; basal hyphae mainly parallel to the substrate, sparsely branched, subhymenial hyphae richly branched and vertical. Basidia 50.0–87 × 11.0–15.0 μm, clavate, sinuous, with numerous oil drops, 4-sterigmate, with basal clamp. Basidiospores 8.0–11.0 × 7.0–9 μm, broadly ellipsoid to subglobose, smooth, thin to somewhat thick-walled, non-amyloid, acyanophilous, with oil drops or granular oily contents.

Etymology: Refers to the larger size of the basidia.

Distribution: U.K.: Nainital.

Collection examined: IBP 37675- PAN (Holotype), L 37677.

Substratum: On decaying angiospermic twigs, Bark of angiospermic twigs.

Remarks: The species is characterized by its hygrophanus fructification, clavate, sinuous basidia with oil droplets and ellipsoid-subglobose, thin to thick-walled basidiospores. This species was first described by Fries (1821) as *Thelephora confluens*. Christiansen (1960) transferred it to *Radulomyces*. The above collection resemble the description given by Eriksson et al. (1981) in most of the characters except the larger size of basidia, which are 50–87.5 × 11–15.5 μm as compared to 35–55 × 6.5–9 μm in the latter. On this basis it is treated as a new variety.

Radulomyces molaris (Chaill. ex Fr.) M.P. Christ. Dansk., bot. Ark. 19(no. 2): 232. (1960) = *Radulum molare* Chaill. ex. Fr., Elench fung. (Greifswald) 1: 151 (1828). Fig. 6.4k, l

Fructification resupinate, membranous to subceraceous, adnate, arising as small, colonies; hymenial surface cream yellow to yellowish brown, smooth to distinctly warted, warts large, hemispherical to dome-shaped, continuous, often cracking irregularly on drying exposing light coloured context.

Hyphal system monomitic, hyphae 1.5–3 µm wide, sparsely branched, septate, clamped, thin-walled, subhyaline, often collapsing, agglutinating and difficult to discern. Basidia 25–30 × 5.5 µm, clavate, 4-spored, sterigmata stout and up to 5.5 µm long. Basidiospores 8.5–10 × 5.5–6.5 µm, broadly ellipsoid to ovoid, shortly apiculate, the walls thin, subhyaline, smooth, nonamyloid, acyanophilous.

Distribution: H.P.: Kullu, Manali, Palampur; U.K.: Chamoli-NDBR, Gobindghai, Nainital.

Collection examined: SSR 5701, IBP 37014, L 37015.

Substratum: On stump of *Pinus excelsa*, bark of *Cedrus deodara*, stumps of *Pinus roxburghii*

Remarks: This species is marked by the raduloid hymenial surface, ceraceous and strongly cretaceous context, presence of clamps on the hyphae and large, ovoid basidiospores.

Family-Stephanosporaceae

Cristinia Parmasto,

Consp. System. Corticiac (Tartu): 48, 1968.

Fructification resupinate, membranous to subceraceous, widely effused; hymenial surface white to pale coloured, smooth to raduloid, waxy in appearance. Subiculum subhyaline in section, composed of somewhat loosely woven hyphae in the basal part. Hyphal system monomitic; generative hyphae branched, septate, clamped, thin-walled, subhyaline. Cystidia or gloeocystidia absent. Basidia clavate-cylindrical to subutriform, 4-spored. Basidiospores ellipsoid to ovoid, thick-walled, subhyaline, smooth, weakly cyanophilous, non-amyloid.

Seven Species, widespread

Lit.: Hjortstam & Grosse- Brauckmann (*Mycotaxon* 47:405, 1993; key)

Type species: *Hydnum helveticum* Pers., 1825.

Habitat: Decaying wood

Himalayas: Two

Key to species

1. Basidiospores less than 5.0 µm long, broadly ellipsoid to ovoid; basidia 15–19 × 5.6 µm.. *C. helvetica*
1. Basidiospores more than 5.0 µm long, ovoid to subglobose; basidia 30–45 × 6.6–8.2 µm... *C. mucida*

Cristinia helvetica (Pers.) Parmasto, Consp. Syst. Cort. 48. 1968 = *Hydnum helveticum* Pers., Mycol. Eur. 2: 184. 1825. Plate 6.2e, Fig. 6.5a–c

Fructification resupinate, membranous to membranous-ceraceous, loosely adnate, widely effused; hymenial surface cream to cream yellow or cream brown, granular to colliculose with hemispherical warts, continuous; margin thinning, loosely adnate, white to paler concolorous. Hyphal cordons abundantly present on the abhymenial side and often transferring the substratum, up to 75 µm thick, unbranched or occasionally branched, subhyaline to pale brown, composed of

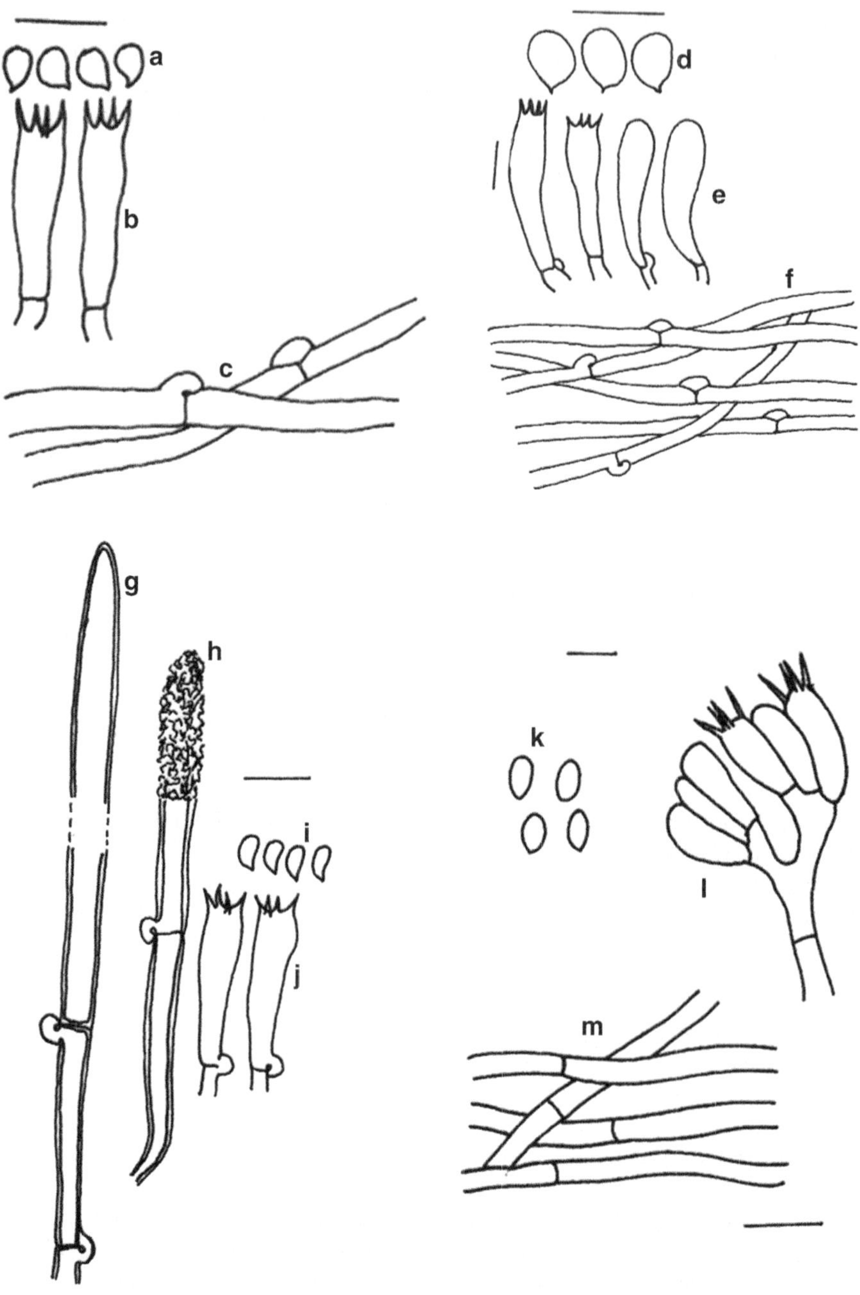

Fig. 6.5 (**a–c**) *Cristinia helvetica* (**a**) Basidiospores, (**b**) Basidia, (**c**) Generative hyphae; (**d–f**) *Cristinia mucida* (**d**) Basidiospores, (**e**) Basidia, (**f**) Generative hyphae; (**g–j**) *Amphinema byssoides* (**g**) Cystidia, (**h**) Encrusted cystidia, (**i**) Basidiosppores, (**j**) Basidia; (**k–m**) *Athelia decipiens* (**k**) Basidiospores, (**l**) Basidia, (**m**) Generative hyphae

compactly arranged agglutinated somewhat agglutinated hyphae. Subiculum subhyaline in section, composed of loosely arranged hyphae in the basal part but becomes compactly arranged towards the hymenium becoming wholly ceraceous with age.

Hyphal system monomitic; generative hyphae branched, septate, clamped, thin-walled, subhyaline, with few or none ampulliform swellings near the septa. Basidia subclavate to cylindrical, with basal clamp, 15.0–19.0×5.0–6.0 µm. Basidiospores small, broadly ellipsoid to ovoid, minutely apiculate, the walls thin to slightly thick-walled, subhyaline, smooth, weakly dextrinoid, often staining with cotton blue but not truly cyanophilous, 3.5–4×3.0–3.5 µm.

Distribution: H.P.: Kullu, Pulga, Narkanda, Mahasu; U.K.: NDBR.

Collection examined: SSR 5614.

Substratum: On log of *Abies pindrow*, wood under conifers.

Remarks: This species is characterized by loosely attached, membranous to subceraceous fructification, closely septate, clamped generative hyphae; short, sub cylindrical, 4-spored Basidia (15–19×5–6 µm) and small, subglobose to globose, cyanophilous Basidiospores (3.5–4.5×3–4 µm). Rattan (1977) was the first to report this species from India from the N.W. Himalayas. The above collection resembles the description of the species as given by Eriksson and Ryvarden (1975).

Cristinia mucida (Pers.) J. Erikss. & Ryvarden, Cortic. N. Eur., **3** *Coronicium-Hyphoderma* (Oslo) **3**: 311 (1975)=*Radulum mucidum* (Pers.) Bourdot & Galzin, *Bull. Soc. mycol. Fr.* **30**(2): 247 (1914). Fig. 6.5d–f

Fructifications resupinate, sub membranous to membranous, easily separable, often arising in small colonies; hymenial surface yellow when fresh fading to pale ochre in drying, smooth to finely tuberculate, continuous but occasionally cracking irregularly on drying; margin fibrillose to rhizomorphic, loosely adnate, white to paler concolorous. Subiculum subhyaline, composed of somewhat compactly arranged hyphae.

Hyphal system monomitic, hyphae 3.3–5.2 (7.8) µm wide, branched, septate, clamped, clamps prominent and almost at all septa, the walls thin to moderately thick (0.8 µm thick), subhyaline, weakly dextrinoid. Basidia 30–45×6.6–8.2 µm, clavate-cylindrical, 4-spored, sterigmata. Basidiospore 5.3–5.8×4.4–5 µm, ovoid to subglobose, shortly apiculate, the walls subhyaline to faintly tinted yellow-brown, slightly thick, non-amyloid, subcyanophilous to more or less cyanophilous.

Distribution: J&K: Gulmarg; U.K.: Chamoli.

Collection examined: SSR 5285.

Substratum: On wood under conifers.

Remarks: This species has basidiospores which have slightly thick and cyanophilous walls and on the basis of this feature, Eriksson and Ryvarden (1975) have transferred this species to *Cristinia*. Another closely related species is *Leucogyrophana mollusca* (Fr.) Pouz. but the wall of the basidiospores is more strongly tinted and it belongs to Coniophoraceae.

O- Atheliales
Family- Atheliaceae

Key to genera

1. Cystidial element always absent .... *Fibulomyces*
1. Cystidial element present or absent 2
2. Hyphoid septate, clamped cystidia present .... *Amphinema*
2. Cystidioles or gloeocystidia present or absent 3
3. Basidia basally stalked, basidiospores ellipsoid to allantoid .... *Athelopsis*
3. Basidia not basally stalked, basidiospores ellipsoid 4
4. Hymenial surface smooth .... *Athelia*
4. Hymenial surface smooth to grandinioid to tuberculate 5
5. Basidiospores thin wall, smooth, acyanophilous .... *Leptosporomyces*
5. Basidiospores thin to thick walled, distinctly cyanophilous .... *Hypochniciellum*

Amphinema P. Karst.,
Bidr. Kann. Finl. Nat. Folk 51: 228. 1892.

Fructification resupinate, pelliculose to typically atheloid; hymenial surface smooth, continuous, usually cream yellow or yellow; margin fibrillose to rhizomorphic. Subiculum pale yellow in section, composed of loosely woven hyphae. Hyphal system monomitic, hyphae usually branched at wide angles, septate, clamped, thin to thick-walled. Hymenium non thickening, sub-ceraceous. Cystidia cylindrical to hypoid, septate, clamped, thin to thick-walled. Gloeocystidia absent. Basidia clavate, 4-spored. Basidiospores ellipsoid to ovoid, thin to thick-walled, tinted yellow, smooth, non-amyloid.

Six species, widespread

Lit.: Sutton & Crouse (*Mycol. Res.* 101: 215, 1997)

Type Species: *Diplonema sordescens* P. Karst. 1889.

Habitat: Dead Wood

Himalayas: One

Amphinema byssoides (Pers.) J. Erikss., Symb. Bot. Upsal. 16: 112. 1958. Plate 6.2f, Fig. 6.5g–j

Fructification resupinate, loosely adnate, widely effused; hymenial surface yellow, smooth; margin fibrillose, loosely adnate, paler concolorous. Subiculum light yellow composed of loosely interwoven hyphae.

Hyphal system monomitic, generative hyphae branched at wide angles, thin to thick-walled, septate, clamps prominent at all septa, incrustation sub-hyaline. Hymenium non-thickening, subceraceous. Cystidia 75–140 × 4.5–6 μm, hyphoid to cylindrical, septate, clamped, arising in the subhymenial zone and projecting out of the hymenium walls, thin, sometimes impregnated with fine crystals. Basidia clavate, 4-spored, 18–25 × 5–5.8 μm. Basidiospores broadly ellipsoid, minutely apiculate, smooth, the walls tinted yellow, thin to thick-walled, 3.9–4.5 × 2.5–3.0 μm, non-amyloid.

Distribution: H.P.: Kullu, Dalhousie, Shimla; J&K: Gulmarg; Bhutan: Paro- D' Dzong, Ha, Thimphu.

Collection examined: GSR 5086, SSR 5287, IBP 37015.

Substratum: On angiospermous logs, on slash of *Pinus excelsa*, *Cedrus deodara*, *Abies pindrow*, *Picea smithiana*.

Remarks: This species is very common in North Western Himalayas on conifers. It is recognised by yellow atheloid fructification, typical hypnoid cystidia. The basidiospores are smooth with thick, tinted walls.

Athelia Pers.,

Traité sur les Champignons Comestibles: 57, 1818

Fructification resupinate, pellicular, white to whitish; hymenial surface smooth when dry, often slightly wrinkled when fresh. Hyphal system monomitic, with or out clamps, basal hyphae wide with thickened walls. Basidia short, club like with 2–4 sterigmata. Basidiospores smooth, rounded, ellipsoid or cylindrical, non-amyloid, acyanophilous.

Twenty eight species, widespread

Lit.: Okabe & Matsumoto (*Mycol. Res*. 107: 164, 2003), Jülich (Persoonia 10, 149, 1978, Key Lichenized spp.).

Type Species: *Athelia epiphylla* Pers., 1818.

Habitat: Wood

Himalayas: Two

Key to species

1. Hyphae with clamps, basidiospores 7.5–9 × 3.5–4 µm ....*A. teutoburgensis*
1. Hyphae without clamps, basidiospores 5.4–6 × 3–3.4 µm ....*A. decipiens*

Athelia decipiens (Höhn & Litsch.) J. Erikss., Sym. bot. Upsal. 16(1): 86, 1958 = *Corticium decipens* Höhn. & Litsch., Sber. Akad. Wiss. Wien, Math.-naturw. Kl., Abt.1 117(1): 1116, 1908. Plate 6.3a, Fig. 6.5k–m

Fructification resupinate, thin, white and pellicular; hymenial surface smooth; greyish white when fresh, becoming pale yellow after drying, margin thinning, paler concolorous to indeterminate.

Hyphal system monomitic and all hyphae septate without clamps, 3–5 µm in diameter, often with a grainy incrustation. Basidia small, about 12–15 × 5 µm, the sub-basidial cells often somewhat widened or swollen because of side branches developing from them. Basidiospores ellipsoid to ovate, 4.5–5.4 (6) × 3–3.4 µm, often glued together in groups of two or four.

Distribution: H.P.: Shimla; U. K.: Chakrata.

Collection examined: SSR 5750, IBP 37016, L 37017.

Substratum: On *Cedrus deodara*.

Remarks: The characteristic features of the species are loose adnate, fructification, clavate, 4-spored basidia and ellipsoid basidiospores which are smaller than other known species and all hyphae without clamps. The species is being recorded for the first time from Uttarakhand.

Plate 6.3 (**a**) *Athelia decipiens.* (**b**) *Athelopsis parvispora.* (**c**) *Fibulomyces cystoideus.* (**d**) *Fibulomyces mutabilis.* (**e**) *Hypochniciellum ovoideum.* (**f**) *Leptosporomyces thindii*

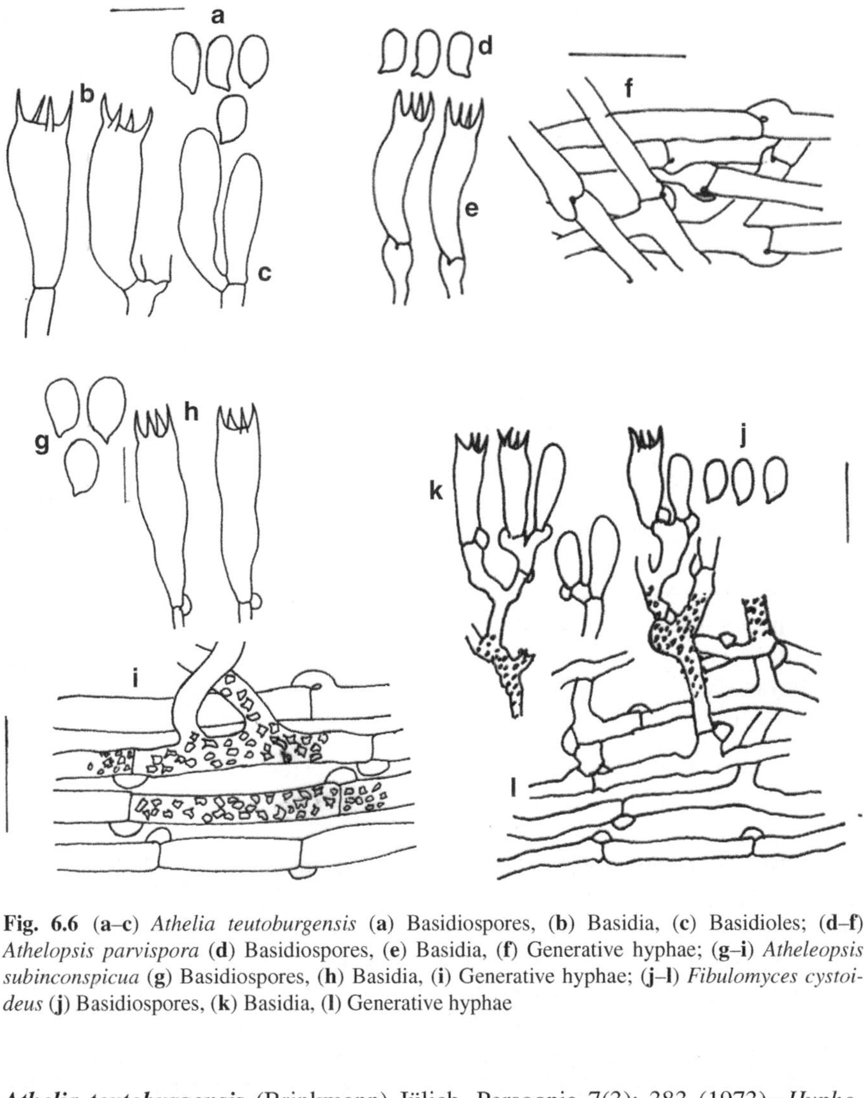

Fig. 6.6 (**a–c**) *Athelia teutoburgensis* (**a**) Basidiospores, (**b**) Basidia, (**c**) Basidioles; (**d–f**) *Athelopsis parvispora* (**d**) Basidiospores, (**e**) Basidia, (**f**) Generative hyphae; (**g–i**) *Atheleopsis subinconspicua* (**g**) Basidiospores, (**h**) Basidia, (**i**) Generative hyphae; (**j–l**) *Fibulomyces cystoideus* (**j**) Basidiospores, (**k**) Basidia, (**l**) Generative hyphae

Athelia teutoburgensis (Brinkmann) Jülich, Persoonia 7(3): 383 (1973)=*Hypoderma teutoburgense* (Brinkm.) Erikss., Symb. bot. upsal. (16): 100. 1958. Fig. 6.6a–c

Fructification resupinate, membranous to membranous ceraceous, adnate, widely effused, up to 75 μm thick in section; hymenial surface white to cream, smooth, continuous, creviced; margin thinning, adnate, white to paler concolorous. Subiculum subhyaline in section, very thin, composed of compactly arranged to more or less agglutinated hyphae.

Hyphal system monomitic, hyphae 2–4 μm wide, branched, septate, clamped, thin-walled, subhyaline, often collapsing and agglutinating and difficult to discern. Cystidia and gloeocysytidia absent. Basidia 20–30×6.5–8.5 μm, subutriform,

4-spored, sterigmata up to 6 μm long. Basidiospores allantoids 7.5–9 (10) × 3.5–4 μm, sub minutely apiculate, with obtuse end, sometimes curved, often with one or more guttules, thin-walled, smooth, subhyaline, non-amyloid.

Distribution: H.P.: Kullu, Rohtang; U.K.: Nainital-Kilbri, China peak.

Collection examined: SSR 5750, IBP 37017.

Substratum: On log of *Abies pindrow*.

Remarks: This species is marked by smooth and white hymenial surface and lack of cystidia and gloeocystidia. *H. cremeoalbum* (Höhn & Litsch.) Jülich is very similar but possesses definitely larger basidiospores (10–14 × 5–6.5 μm). It was recorded from Kullu by Rattan (1977).

Athelopsis Oberw. ex Parmasto,
Consp. Syst. Cort.: 41, 1968

Fructification resupinate, pellicular, smooth, pale yellowish, texture very loose. Hyphal system monomitic; hyphae thin-walled, clamps at all septa. Basidia stalked, clavate, 4-sterigmate. Basidiospores hyaline, ellipsoid to cylindrical, non-amyloid, acyanophilous.

Ten Species, widespread

Lit.: Kotiranta & Saarenoksa (*Ann. Bot. fenn.* **42**: 335, 2005; Finland)

Type Species: *Corticium glaucinum* Bourdot & Galzin 1928.

Habitat: Wood

Himalayas: Two

Key to species

1. Basidiospores narrowly ellipsoid, 4.3–4.7 × 1.5–1.9 μm .... *A. parvispora*
1. Basidiospores ellipsoid 4.7–6.4 × 2.6–3.8 μm .... *A. subinconspicua*

Athelopsis parvispora Singh A.P., Dhingra & J. Kaur, *Mycotaxon* **113**, 327–329. 2010. Plate 3b, Fig. 6.6d–f

Fructification resupinate, adnate, effused, thin, pellicular, almost athelioid; hymenial surface smooth, pale yellowish; margins indeterminately thinning.

Hyphal system monomitic; generative hyphae up to 3 μm wide, branched, septate, clamped; basal hyphae somewhat thick-walled, encrusted; subhymenial hyphae thin-walled, without encrustation. Basidia 11.0–16.0 × 3.0–4 μm, clavate, basally stalked, 4-sterigmate, with a basal clamp; sterigmata up to 4.3 μm long. Basidiospores 4.3–4.7 × 1.5–1.9 μm, narrowly ellipsoid, thin-walled, smooth, inamyloid, acyanophilous.

Distribution: H.P.: Kullu, Manali.

Collection examined: IBP 39010.

Substratum: On stump of *Quercus incana*.

Remarks: This species is characterized by pellicular fructification with smooth hymenial surface and clavate basidia which are stalked at the base. It is a closely related species to *Athelopsis subinconspicua* from which it differs in the smaller size of the spores.

Atheleopsis subinconspicua (Litsch.) Jülich, Persoonia 8(3):292 (1975) = *Corticium subinconspicuum* Litsch., Glasn. Skopsk. Naunn. Drustva 18: 178 (1938). Fig. 6.6g–i

Fructification resupinate, effused, loosely adnate; hymenial surface smooth, light grey to pale orange when fresh, greyish orange on drying; margin thinning, paler concolorous to indeterminate.

Hyphal system monomitic; generative hyphae up to 4.0 µm wide, clamped, septate, encrusted; basal hyphae thin to thick-walled; subhymenial hyphae thin-walled, highly branched. Cystidia absent. Basidia 7.6–15.0 × 3.0–4.0 µm, clavate, 4-sterigmate with basal clamp. Basidiospores 4.7–6.4 × 2.6–3.8 µm, apiculate, smooth, thin-walled, inamyloid, acyanophilous.

Distribution: H.P.: Solan, Shimla; U.K.: NDBR.

Collection examined: L 37017, IBP 37018.

Substratum: On the bark of *Cedrus deodara*, decaying gymnospermic wood.

Remarks: This species is characterized by stalked clavate basidia and ellipsoid basidiospores.

Fibulomyces Jülich,
Willd. Beih. 7: 178, 1972.

Fructification resupinate, pellicular to atheloid, hymenium smooth or meruloid, white or light cream or rose-coloured, rhizomorphs thin, sometimes present; hyphal system monomitic, hyphae thin-walled or basal hyphae somewhat thick-walled, with clamps at all septa with frequent branches. Basidia small narrowly clavate, with 4-sterigmata. Basidiospores ellipsoid to cylindrical or fusiform, thin-walled, smooth, non-amyloid.

Four species, widespread

Lit.: Hjortstam & Larsson (Windahlia 21, 1994).

Type Species: *Corticium mutabile* Bres. 1898

Habitat: Wood

Himalayas: Two

Key to species

1. Hymenial surface tuberculate, greyish–white to pale yellow; cystidia present .... *F. cystoideus*
1. Hymenial surface tuberculate, yellowish–white; cystidia absent ...... *F. mutabilis*

Fibulomyces cystoideus Dhingra, In Plant Diversity in India: 480, 2004. Plate 6.3c, Fig. 6.6j–l

Fructification resupinate, hypochnoid to subpellicular, effused; hymenial surface tuberculate, greyish-white to pale yellow; margins not differentiated.

Hyphal system monomitic; generative hyphae up to 4.0 µm wide, thin-walled, richly branched, septate, clamped, anatomoses frequent, crystalline encrustation present, especially on the subhymenial hyphae. Cystidia 25.0–35.0 × 2.5–3.5 µm, hypoid, somewhat subulate, thin-walled, with a basal clamp. Basidia 8.0–12.0 × 3.0–3.5 µm, clavate to sub-cylindrical, 4-sterigmate, with a basal clamp;

sterigmata up to 3.5 μm long. Basidiospores 3.0–4.5 × 1.5–2.5 μm, ellipsoid to sub-cylindrical, smooth, thin-walled, inamyloid, acyanophilous.

Distribution: Bhutan: Thimphu, Nawephu.

Collection examined: GSD 19365.

Substratum: On a decaying *Pinus* log.

Remarks: This species is characterized by hypochnoid to subpellicular fructification, thin-walled hyphoid cystidia, clavate to subcylindrical basidia, ellipsoid to subcylindrical basidiospores and clamped generative hyphae. It differs from *Fibulomyces muatbilis* (Bres.) Julich in the presence of cystidia.

Fibulomyces mutabilis (Bres.) Jülich, Willd. Beih. 7: 182, 1972. Plate 6.3d, Fig. 6.7a–d

Fructification resupinate, effused, thin, pellicular, loosely adnate; hymenial surface smooth, yellowish white; margins indeterminate, thinning, paler concolorous.

Hyphal system monomitic; generative hyphae 2–4 μm wide, branched, septate; basal hyphae thin to somewhat thick-walled, less branched; subhymenial hyphae thin-walled, somewhat narrower, crystalline encrustation present especially on subhymenial hyphae. Cystidia none. Basidia 9.6–19.3 × 2.8–3.4 μm, clavate to sub-cylindrical, 4-sterigmate, with a basal clamp; sterigmata up to 4.5 μm long. Basidiospores 2.8–4.5 × 2.0–2.3 μm, ellipsoid to sub-cylindrical, thin-walled, smooth, nonamyloid, acyanophilous, with or without guttules.

Distribution: H.P: Chamba- kalatope.

Collection examined: IBP 37019.

Substratum: On decaying gymnospermic log.

Remarks: This species is characterized by thin, pellicular, loosely adnate fructification; clamped generative hyphae in subiculum and subhymenium; clavate to sub-cylindrical basidia and ellipsoid to sub-cylindrical basidiospores and is a new record for Himalayas.

Hypochniciellum Hjortstam & Ryvarden,
Mycotaxon, 12(1):176, 1980.

Fructifications resupinate, arachinoid-pelliculose to atheloid, loosely adnate to separable, widely effused; hymenial surface cream yellow to yellow but fading on drying, smooth, discontinuous but becomes continuous at maturity. Hyphal system monomitic, clamped. Basidia clavate-cylindrical, 4-spored, sterigmata long. Basidiospores broadly ellipsoid to ovoid, minutely apiculate, smooth, non-amyloid but weakly dextrinoid.

Four species, widespread

Lit.: Legon (Mycologist 17: 41, 2003)

Type Species: *Leptosporomyces ovoideus* Jülich 1972.

Habitat: Wood

Himalayas: One

Hypochniciellum ovoideum (Jülich) Hjortstam & Ryvarden, Mycotaxon 12(1): 177 (1980) = *Leptosporomyces ovoideus* (Jülich) Telleria, in Telleria & Melo, Fl. Mycol. Iberica 1: 71 (1995). Plate 6.3e, Fig. 6.7e–h

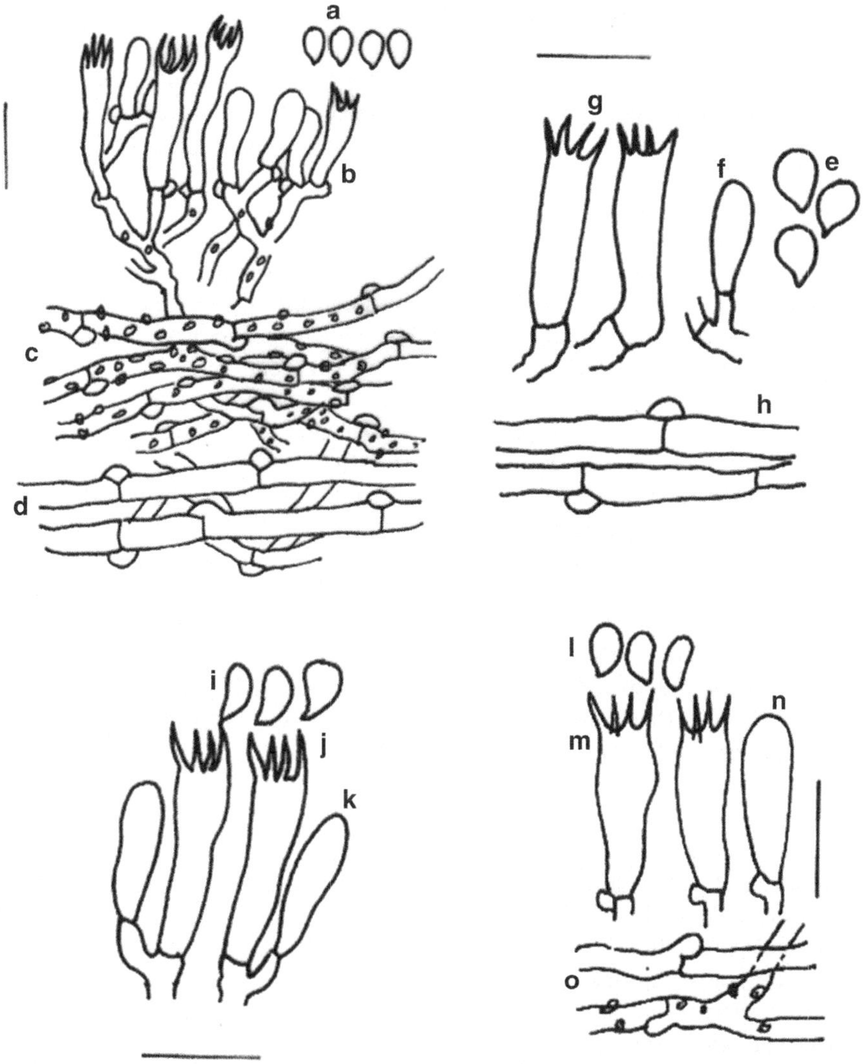

Fig. 6.7 (**a**–**d**) *Fibulomyces mutabilis* (**a**) Basidiospores, (**b**) Basidia, (**c**) Subhymenial hyphae, (**d**) Generative hyphae; (**e**–**h**) *Hypochniciellum ovoideum* (**e**) Basidiospores, (**f**) Basidioles, (**g**) Basidia, (**h**) Generative hyphae; (**i**–**k**) *Leptosporomyces adnatus* (**i**) Basidiospores, (**j**) Basidia, (**k**) Basidioles; (**l**–**o**) *Leptosporomyces galzinii* (**l**) Basidiospores, (**m**) Basidia, (**n**) Basidioles, (**o**) Generative hyphae

Fructifications resupinate, arachinoid-pelliculose to atheloid, loosely adnate to separable, widely effused, up to 300 µm thick in section; hymenial surface cream yellow to yellow but fading on drying, smooth, discontinuous but becomes continuous at maturity, often cracking finely and irregularly exposing the white subiculum; margin thinning, adnate, white to paler concolorous. Subiculum composed of loosely woven hyphae.

Hyphal system monomitic, hyphae 3–5.8 μm wide, branched at angles, septate, clamped, uninflated or rarely inflated up to 8.5 μm near the septa, the walls thin, subhyaline. Basidia 12–15 × 4–4.5 μm, clavate-cylindrical, 4-spored, sterigmata long. Basidiospores 3.7–4.2 × 2.5–3 μm, broadly ellipsoid to ovoid, minutely apiculate, the walls thin to slightly thick (up to 0.4 μm), subhyaline, smooth, non-amyloid but weakly dextrinoid.

Distribution: H.P.: Narkanda, Mahasu, Kullu.

Collection examined: SSR 5522, IBP 37021.

Substratum: On stump under conifers, decaying gymnospermic wood.

Remarks: The species is characterized by atheloid fructifications, cream yellow to yellow hymenial surface, thick-walled, broadly ellipsoid to ovoid basidiospores giving a weak dextrinoid reaction with Melzer's reagent.

Leptosporomyces Jülich, Wild. Beih.7: 192. 1972.

Fructifications resupinate, pelliculose to atheloid, loosely adnate; hymenial surface white to cream or yellow, smooth. Subiculum subhyaline in section, composed of loosely arranged hyphae. Hyphal system monomitic, thin-walled hyphae, clamped present almost all the septa. Cystidia and Gloeocystidia absent. Basidia clavate-cylindrical, short, 4-spored. Basidiospores small, globose to ellipsoid, thin-walled, subhyaline, smooth and non-amyloid.

Eleven species, widespread

Lit.: Greslebin & Rajchenberg (*N.Z. Jl.Bot.* **41**: 437, 2003; Patagonia)

Type Species: *Corticium galzinii* Bourd. 1972.

Habitat: Dead wood

Himalayas: Six

1. Basidia 4–sterigmate 2
1. Basidia 2–4 sterigmate .... *L. thindii*
2. Basidiospores globose .... *L. globosus*
2. Basidiospores ellipsoid or ellipsoid to ovoid 3
3. Basidiospores ellipsoid to ovoid .... *L. ovoideus*[a]
3. Basidiospores ellipsoid 4
4. Rhizomorphs present .... *L. adnatus*
4. Rhizomorphs absent 5
5. Hymenophore smooth to grandinioid, basidia clavate to subclavate, basidiospores 3.5–4.0 × up to 2.4 μm .... *L. galzinii*
5. Hymenophore smooth to tuberculate, basidia clavate to subclavate, basidiospores 5.3–7.0 × 2.9–4.4 μm .... *L. roseus var. microsporus*

[a]Extralimital not included in the text

Leptosporomyces adnatus (Rehill & B.K. Bakshi) S.S. Rattan, Biblthca Mycol. 60: 129, (1977) = *Corticium adnatum* Rehill & Bakshi, For. Bull., Dehra Dun, (n.s.) 242: 10. 1965. Fig. 6.7i–k

Fructifications resupinate, pelliculose to atheloid when young becoming loosely membranous with age, loosely adnate, widely effused; hymenial surface white when young turning cream yellow to yellowish brown with age, smooth, continuous; margin fibrillose to rhizomorphic, loosely adnate, white to paler concolorous. Subiculum composed of loosely woven hyphae.

Hyphal system monomitic, hyphae 2–3 (4) μm wide, branched at wide angles, septate, clamped, clamps prominent and almost at all septa, the walls thin to slightly thick, naked or finely impregnated with subhyaline crystals soluble in 3 % KOH solution. Rhizomorphs present, white, branched. Basidia 9.8–13.8 × 3.1–4.1 μm, clavate-cylindrical, 4-spored, sterigmata long. Basidiospores (3.5) 4–4.3 × 2–2.3 μm, ellipsoid, minutely apiculate, the walls thin, subhyaline, smooth, non-amyloid.

Distribution: H.P.: Narkanda; U.K: Chakrata.

Collection examined: SSR 5525, IBP 37257.

Substratum: On bark of *Taxus baccata, Abies pindrow, Cedrus deodara, Quercus dilatata.*

Remarks: The characteristic features of the species are atheloid to loosely membranous and adnate fructifications, presence of rhizomorphs and small ellipsoid basidiospores. It differs from *L. galzinii* which is characterized by easily separable, thin and fragile fructifications.

Leptosporomyces galzinii (Bourdot) Jülich, Willdenowia, Beih. 7: 192, 1972 = *Corticium galzinii* Bourdot, Rev. Sci. Bourd.Centr. Fr. 23(1): 11, 1910. Fig. 6.7l–o

Fructification resupinate, effused, loosely adnate; hymenial surface smooth to grandinoid, white to pale orange when fresh, greyish white with an orange on drying; margin thinning, paler concolorous to indeterminate.

Hyphal system monomitic; generative hyphae up to 2.4 wide, septate, thin-walled, clamped; basal hyphae thin to thick-walled; subhymenial hyphae highly branched, thin-walled. Cystidia absent. Basidia 8.7–13.0 × 3.5–5.1 μm, 4-sterigmate, clavate to subclavate, with basal clamp. Basidiospores 3.5–4.1 × 2.4 μm, apiculate, ellipsoid, thin-walled, smooth, inamyloid, acyanophilous.

Distribution: H.P.: Shimla, Tara Devi, Mandi- Manikaran, Kasol, Manali- Hidamba.

Collection examined: IBP 37259.

Substrate: On decaying stump of *C. deodara.*

Remarks: This species is characterized by yellow green fructification giving soft appearance. It is a new record for India/Himalayas.

Leptosporomyces globosus S.S. Rattan, Biblthca Mycol. 60: 269, 1977.

Fructifications resupinate, pelliculose to atheloid, loosely adnate, widely effused hymenial surface deep cream to cream yellow, smooth, continuous, not creviced; margin thinning, byssoid, loosely adnate, white to paler concolorous. Context composed of loosely woven hyphae.

Hyphal system monomitic, hyphae 2.1–3.1 μm wide, branched at wide angles, septate, clamped, clamps at almost all septa, the walls thin to firm (0.4 μm thick), subhyaline. Basidia 13.9–17.9 × 3.7–4.4 μm, clavate, 4-spored, sterigmata long. Basidiospores 2.4–3.1 μm in diameter, globose, minutely apiculate, the walls thin, subhyaline, smooth, non-amyloid, nondextrinoid.

Distribution: H.P.: Kullu, Pulga.

Collection examined: SSR 5687, IBP 37261, L 37262.

Substratum: On slash under conifers.

Remarks: This species is characterized by very small basidiospores which are globose. It seems to be closely related to *Leptosporomyces* sp. (no. 8650) described by Jülich (1972).

Leptosporomyces roseus Jülich, Willdenovia, Beih. 7: 208, 1972, var. *macrosporus* Prasher & Lalita var. nov. Plate 6.4a, Fig. 6.8d–f

Mycobank MB812327

Fructification resupinate, thin, loosely adnate, effused atheloid; hymenial surface smooth, at first yellowish-white then pale ochraceous with a tint of rose; margin determinate, with thin rhizomorph in young fructification.

Hyphal system monomitic; generative hyphae septate, clamped, thin-walled, 2–3.4 μm wide, straight and less branched in the subiculum, much branched in the subhymenium. Cystidia none. Basidia 8.5–12.5 3.25–4.25 μm, clavate, 4-sterigmate, with basal clamp; sterigmata up to 3 μm long. Basidiospores 3–4.4 × 1.8–2.4 μm, ellipsoid, smooth, thin-walled, non-amyloid, acyanophilous.

Etymology: Refers to the large sized spores.

Distribution: U.K.: Bhajpur- Nainital.

Collection examined: IBP 37551- PAN (Holotype), L 37556.

Substratum: On a decaying coniferous log.

Remarks: This species is closely related to *L. roseus* in all the morphological/anatomical details except for markedly smaller basidia 8.5–12.5 × 3.25–4.25 μm as compared to 12–17 × 3–4 μm in *L. roseus* as described by Eriksson & Ryvarden (1976). On the basis it is treated as a new variety.

Leptosporomyces thindii Prasher & Lalita sp. nov. Plate 6.3f, Fig. 6.8a–c

Mycobank MB812328

Fructification resupinate, effused, loosely adnate, thin athelioid; hymenial surface whitish, smooth to finally porulose; margin indeterminate.

Hyphal system monomitic; generative hyphae 2–3.5 μm wide, septate, clamped, thin-walled, less branched and straight in subiculum and much branched in the subhymenium. Cystidia none. Basidia 6.1–10.1 × 3.1–4.1 μm, short, clavate to subclavate, 2–4 sterigmata, with a basal clamp. Basidiospores 3–4 × 1.5–2 μm, ellipsoid to subcylindrical, smooth, thin-walled, non-amyloid, acyanophilous.

Etymology: In honour of late Prof. K. S. Thind.

Plate 6.4 (**a**) *Leptosporomyces roseus var. macrosporus.* (**b**) *Coniophora arida.* (**c**) *Coniophora cordensis.* (**d**) *Coniophora fusispora.* (**e**) *Coniophora olivacea.* (**f**) *Leucogyrophana mollusca*

Distribution: U.K.: Uttarkashi.

Collection examined: IBP37550, L 37555- PAN (Holotype).

Substratum: On decaying gymnospermic log.

Remarks: This species is closely related to *L. raunkiaeri* (M.P. Christ) Julich except for the smaller basidia (6–10×3.1–4 μm in *L. thindii* as compared to 8–15×3–4.5 μm in *L. raunkiaeri*) and basidiospores (3–4×1.5–2 μm in *L. thindii* as compared to 4–4.5×1.5–2.5 μm in *L. raunkiaeri*). On the basis it is treated as a new species.

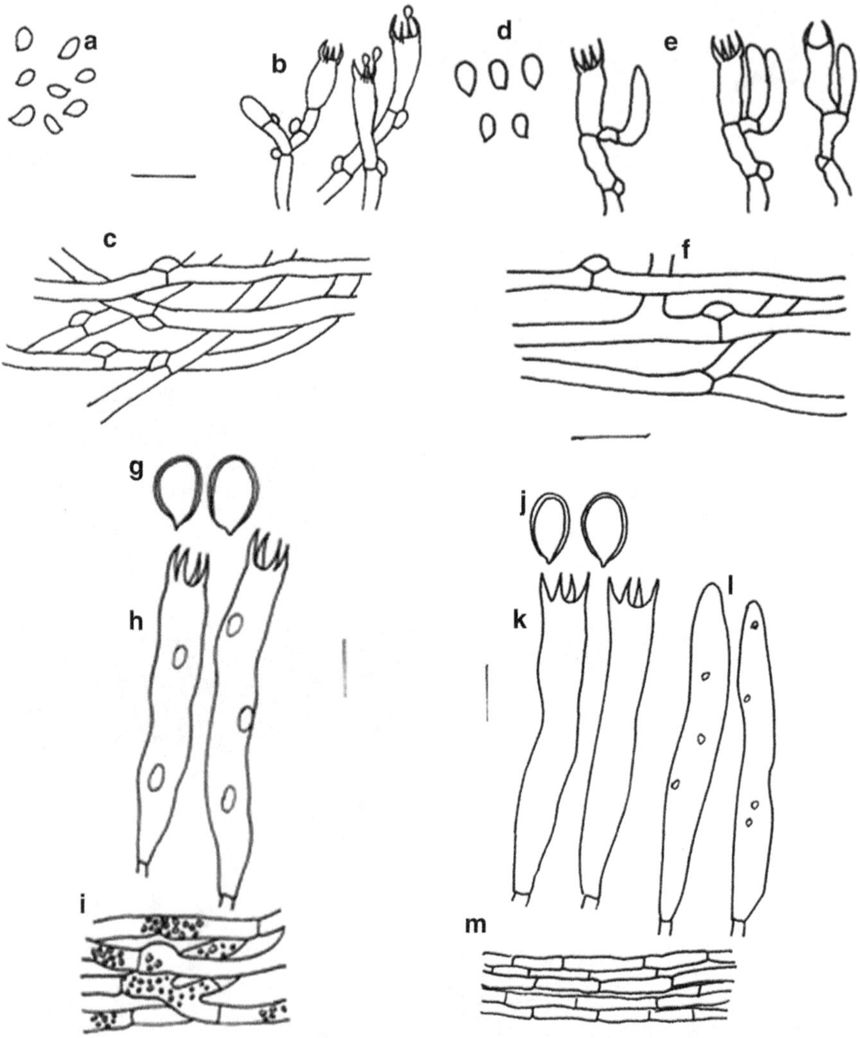

Fig. 6.8 (**a–c**) *Leptosporomyces thindii* (**a**) Basidiospores, (**b**) Basidia, (**c**) Generative hyphae; (**d–f**) *Leptosporomyces roseus* var. *macrosporus* (**d**) Basidiospores, (**e**) Basidia, (**f**) Generative hyphae; (**g–i**) *Coniophora arida* (**g**) Basidiospores, (**h**) Basidia, (**i**) Generative hyphae; (**j–m**) *Coniophora cordensis* (**j**) Basidiospores, (**k**) Basidia, (**l**) Cystidia, (**c**) Hyphal cordons

O-Boletales

Family- Coniophoraceae

Coniophora DC.,

Flore française, 6: 34, 1815

Fructification resupinate, somewhat fleshy to waxy-gelatinous; hymenial surface smooth to finely tuberculate or uneven, yellowish-brown to greenish-brown. Hyphal system monomitic or dimitic; generative hyphae branched, septate, with or without clamps, clamps single or sometimes on whorls, thin-walled. Cystidia absent. Gloeocystidia present or absent. Basidia utriform, 4-spored. Basidiospores ellipsoid to ovoid, thick-walled, smooth, yellowish-brown, cyanophilous, non-amyloid.

Twenty species, widespread.

Lit.: Lentz (Mycol. 49: 534, 1957; basidial development), Ginns (Op. bot. 61, 1982; world monogr.), Huckfeldt & Schmidt (Mycologist 20: 42, 2006; Key to strand-forming house-rot fungi).

Habitat: Dead wood.

Type species: *Coniophora membranacea* Dc. 1815.

Himalayas: Five

Key to species

1. Cystidial elements present 2
1. Cystidial elements absent 3
2. Cystidia septate .... *C. olivacea*
2. Cystidia not septate .... *C. cordensis*
3. Basidiospores navicular .... *C. fusispora*
3. Basidiospores ellipsoid to ovoid 4
4. Basidia utriform .... *C. puteana*
4. Basidia subclavate to subcylindrical .... *C. arida*

Coniophora arida (Fr.) P. Karst., Not. Sällsk. Fauna et Fl. Fenn. Förh. **9**: 370 (1868). Plate 6.4b, Fig. 6.8g–i

Fructification resupinate, subpelliculose to membranous, adnate, arising as small circular colonies which merge together and become effused; hymenial surface light brown to olivaceous brown, smooth, dry in appearance, not creviced; margin thinning, adnate, white to paler concolorous.

Hyphal system monomitic, hyphae 2–6 μm wide, septate, branched, thin-walled, clamps absent. Gloeocystidia absent. Basidia 55–90 × 8–11 μm, utriform, projecting out of the hymenium, 4-spored. Basidioles clavate 12–17.5 × 2–3.1 μm. Basidiospores 10.4–14 × 6.5–8.4 μm, walls thick, ellipsoid to ovoid, apiculate, smooth, cyanophilous, non-amyloid.

Distribution: H.P.: Dalhousie- Kalatope, Mahasu; Kullu; U.K.: Partapnagar-Tehri, Bhutan: Thimphu, Dochula; J& K: Pehalgam,

Collection examined: SSR 5773, 5489, IBP 37021.

Substratum: Stump of *Abies pindrow* and bark of *Cedrus deodara*.

Remarks: The main features of this species are smooth, dry fructification, simple septate hyphae. This differs from the *Coniophora puteana* in having thin and adnate fructification. Hymenial surface is smooth in *C. arida* as compared to corrugate-warted in *C. puteana*. The hyphae are without clamps where as in *C. puteana* a few clamps in whorls may be present. Earlier Rattan (1977) reported it from Himachal Pradesh and J&K. It is a new record for Uttarakhand.

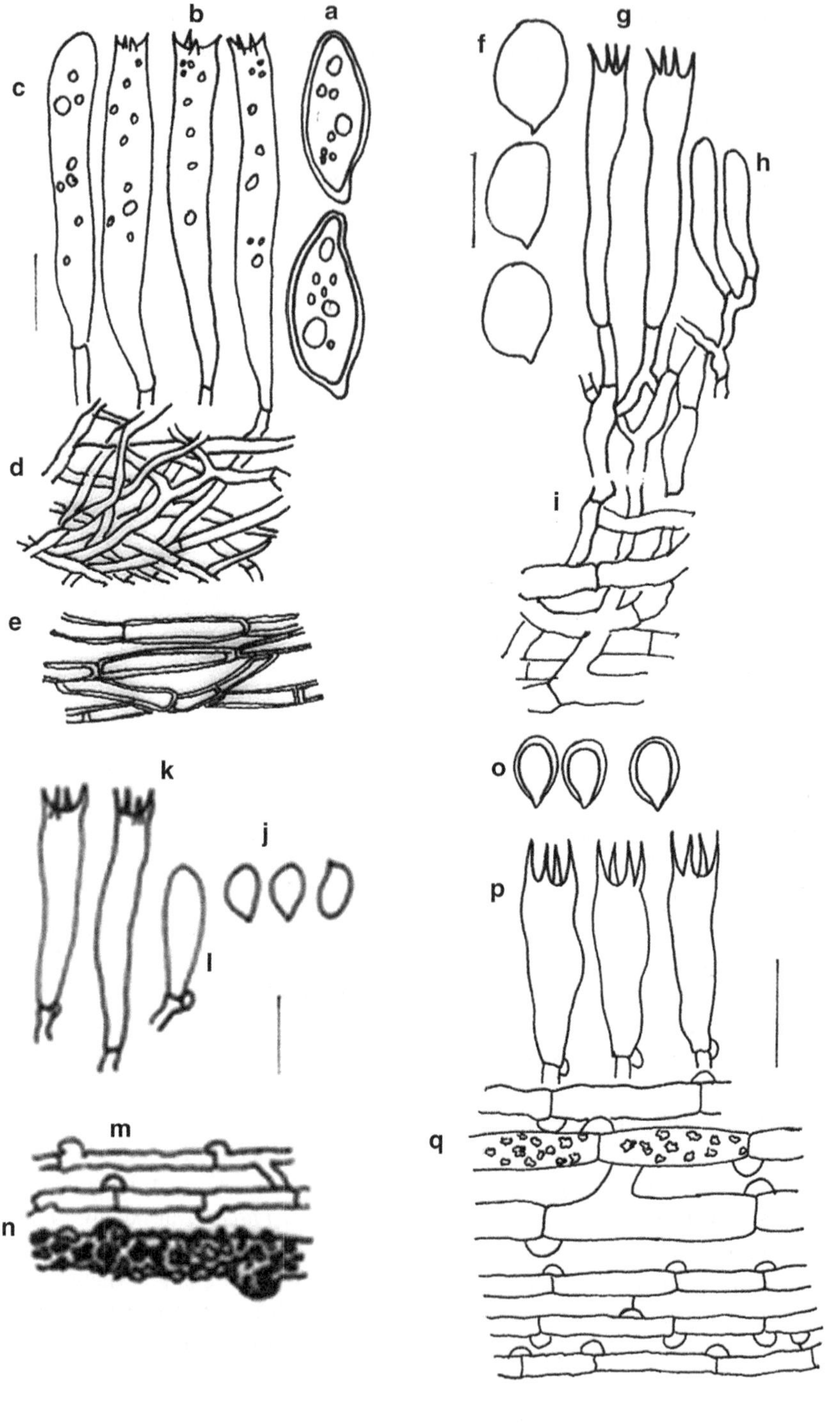
a
b
c
d
e
f
g
h
i
j
k
l
m
n
o
p
q

Coniophora cordensis S.S. Rattan, Biblthca Mycol. 60: 78, 1977. Plate 6.4c, Fig. 6.8j–m

Fructifications resupinate, pelliculose to submembranous, often arising in small circular colonies which may coalesce later, adnate, somewhat dry; hymenial surface yellow to yellowish brown, smooth to finely granulose; margin thinning, byssoid, yellow, adnate.

Hyphal system monomitic, hyphae 2–9 µm wide, branched, branches at wide angles, septate, clamps absent, slightly inflated between the septa, thin-walled, subhyaline, often covered with subhyaline crystals soluble in 3 % KOH sol. Gloeocystidia absent. Basidia 40–55×8.2–10 µm, utriform, 4-spored. Basidiospores 10–12.2×6.1–7.1 µm, ovoid, minutely apiculate, the walls thick, yellowish brown, cyanophilous, smooth, non-amyloid.

Distribution: H.P.: Mahasu- Khadrala, Narkanda, Shimla; Bhutan: Thimphu, Nawephu, Chankaphug.

Collection examined: GSD 19546, 19547, SSR: 5649, 5594, IBP 37022

Substratum: Wood, bark and logs of *Abies pindrow*.

Remarks: The characteristic features of this species are perriculose to submembranous fructifications, subhyaline to pale brown incrusted hyphae often organized in to prominent hyphal cordons and relatively short utriform basidia. It is close to *C. arida* but differs in having incrusted hyphae which are tinted brown and presence of hyphal cordons. *Coniophora betulae* is also related but differs in having pseudoparenchymatous subiculum and larger basidia.

Coniophora fusispora (Cooke & Ellis) Cooke, Syll. fung. (Abellini) 6: 650 (1889). Plate 6.4d, Fig. 6.9a–e

Fructification resupinate, loosely adnate, effused, up to 450 µm thick in section; hymenial surface even, yellowish-white to dark olivaceous-brown, margin thin, yellowish-white.

Hyphal system monomitic; generative hyphae up to 7.5 µm wide, without clamps; basal zone composed of loosely interwoven, thin to thick-walled, sparsely branched hyphae running almost parallel to the substrate; subhymenium composed of compactly packed to somewhat gelatinous hyphae and remains of old basidia. Basidia 58–118×10.0–14.5 µm, clavate at first, then elongated no basal

Fig. 6.9 (**a–e**) *Coniophora fusispora* a) Basidiospores, (**b**) Basidia, (**c**) Cystidia (**d**) Thin-walled generative hyphae, (**e**) Thick-walled generative hyphae; (**f–i**) *Coniophora puteana* (**f**) Basidiospores, (**g**) Basidia, (**h**) Basidioles, (**i**) Generative hyphae; (**j–n**) *Leucogyrophana mollusca* (**j**) Basidiospores, (**k**) Basidia, (**l**) Basidioles, (**m**) Thin walled generative hyphae, (**n**) Incrusted generative hyphae; (**o–q**) *Leucogyrophana olivascens* (**o**) Basidiospores, (**p**) Basidia, (**q**) Generative hyphae

clamp, 4-sterigmate. Basidiospores 15.9–22.3×6.4–8.1 μm, navicular-fusoid, smooth, thick-walled, cyanophillous with oily content, apiculus prominent.

Distribution: Bhutan: Thimphu, Paro, D'Dzong. Thimphu, on the way to Fajudin from Motithang; H.P.: Shimla, Chaupal, U. K.: Rishikesh.

Collection examined: GSD 19427, 19583, IBP 37263, L 37264.

Substratum: On decaying angiospermic and gymnospermic log.

Remarks: *Coniphora fusispora* is a very rare species and was reported by Cooke and Ellis (1879) as C*orticium fusisporum*, from N. America. In 1880, Cooke erected the subgenus *Coniophora*, which was later raised to the generic level. This species is marked by loosely attached fructification, generative hyphae without clamps and large, thick-walled, olivaceous brown, cyanophilous basidiospores. The collection resembles closely *C. fusispora* as described by Christiansen (1960), except for the smooth hyphae in comparison to smooth to crystal-incrusted in the type. It is being recorded from N.W. Himalayas for the first time.

Coniophora olivacea (Fr.) Karst., Finl. Basidsv. P 438, 1889 = *Hypochnus olivaceous* Fr., Obs. Mycol. 2: 282, 1818. Plate 6.4e

Fructification resupinate, effused, up to 150 μm thick in section, membranous; hymenial surface even, olivaceous-brown; margin thinning, fibrillose, whitish to paler concolorous.

Hyphal system monomitic; generative hyphae 2.5–6 μm wide, branched, septate, without clamps, basal zone formed of compactly arranged, thin-walled hyphae, running parallel to the host surface followed by a loosely interwoven zone of thin- to thick-walled, short celled hyphae encrusted with circular crystals, crystals dissolve in 3 % KOH sol; subhymenium composed of thin-walled hyphae; hymenium thickening, hyphal cordons present. Cystidia 72.5–102.5×5–6 μm, hyphoid, septate, olivaceous brown, thick walled, encrusted, immersed or sometimes projecting up to 80 μm out of the hymenium. Basidia 23.8–56.3×4.2–6.8 μm, clavate to subcylindrical, somewhat sinuous, 4-sterigmate, without a basal clamp; sterigmate up to 6.5 μm long. Basidiospores 6–10×3–6.3 μm ellipsoid to ovoid, smooth, thick-walled, inamyloid, cyanophilous, with one to many oil droplets.

Distribution: H. P.: Shimla.

Collection examined: IBP 39006.

Substratum: On logs of *C. deodara* and *P. roxburghii*.

Remarks: Fries (1818) described this species as *Hypochnus olivaceous* from Sweden. In 1828, he shifted it to *Thelophora*. However, Karsten (1889) erected a new genus *Coniophorella* to accommodate this species. Burt (1917) and Cunningham (1963) described it as a species of *Coniophora*. However, Christiansen (1960) and Talbot (1973) treated this species under the genus *Coniophorella* following Karsten. Dhingra (1992) described it as *Coniophorella olivacea* from Bhutan. Here this species is being first time reported for the first from India.

Coniophora puteana (Schum.) P. Karst., Not. Sällsk. Fauna et Fl. Fenn. Förh., II **6**(6): 370 (1868)=*Thelephora puteana* Schum.; Fr.,. Syst. Mycol. 1: 448, 1821. Fig. 6.9f–i

Fructification resupinate, membranous-fleshy, often arising as small circular colonies which coalesce later and become widely effused, loosely adnate; hymenial surface uneven, corrugated- warted, often minutely and concentrically zonate and radially sulcate, colour varies from the periphery towards the centre from white to cream, ochraceous, olivaceous, brownish olivaceous to dark brown, fleshy and slimy to touch when fresh; margin thinning, white, up to 1 cm wide, fimbricate, adnate.

Hyphal system monomitic, hyphae 3–10 µm wide, branched, septate, occasionally clamped, clamps single or in whorls, uninflated to slightly inflated, often collapsed in mature parts, thin-walled, subhyaline. Gloeocystidia absent. Basidia uniform, of projecting up to 35 µm out of the hymenium, 4-spored, sterigmata up to 6 µm long. Basidiospores 10–12×6–7.5 µm, broadly ellipsoid to ovoid, shortly apiculate, apiculus hyaline, the walls thick, light olivaceous brown, cyanophilous, non-amyloid.

Distribution: H.P.: Kullu, Manali, Dalhousie- Lakkarmandi; Mahasu- Narkanda, Kullu, Kothi.

Collection examination: SSR: 5154, 5629, IBP 37025.

Substratum: Coniferous wooden poles and stumps of *Abies pindrow*, logs of *Pinus roxburghii*, *Cedrus deodara.*

Remarks: This species is associated with dry rot and is commonly encountered on coniferous planks, logs and poles. The chief diagnostic features are fleshy nature, corrugate-warted hymenial surface and characteristic colour. The hyphae are subhyaline and simple septate but some are with single or whorled clamps.

Family- Hygrophoropsidaceae

Leucogyrophana Pouzar,
Ceská Mykologie, 12 (1): 32, 1958

Fructification resupinate, effuse, more or less atheloid, i.e. hymenial part forming a thin, continuous, easily detached layer, smooth or meruloid, white, yellow, orange, red; all hyphae libulate, subicular ones distinct, straight, sparsely branched, subhymenial hyphae dense, richly branched; cystidia lacking, or sometimes thin-walled, cylindrical cystidia present; basidia clavate, normally with 4 sterigmata; spores broadly or narrowly ellipsoid, smooth, thick-walled, cyanophilous in all species, in Melzer's it is unchanged, grey or yellow brown, not blue or violet as in the strictly amyloid species; spores of the generic type dextrinoid.

Thirteen species, widespread

Lit.: Jarosch & Besl (*Pl. Biol.* **3**: 443, 2001), Ginns (Can. J. Bot. 56, 1953, 1978, keys).

Type Species: *Merulius molluscus* Fr.1821

Habitat: Wood

Himalayas: Three

Key to species

1. Hymenophore meruloid 2
1. Hymenophore not meruloid,
   Basidiospores 4.4–5.4 × 3.0–4.1 µm...*L. olivascens*
2. Basidiospores 5–7 × 3.5–5.4 µm...*L. mollusca*
2. Basidiospores 3.5–4.4 × 2.2–2.7 µm..*L. thimphina*

Leucogyrophana mollusca (Fr.) Pouzar, Ceská Mykol. 12:33, 1958 = *Serpula mollusca* (Fr.) Donk. Persoonia 3: 209. 1964 = *Merulius molluscus* Fr., Syst. Mycol. 1:329, 1821. Plate 6.4f, Fig. 6.9j–n

Fructification resupinate, membranous to subceraceous, easily separable, often arising as small circular colonies which may coalesce later to form patches; hymenial surface orange, smooth when young but ridges appear at maturity; margin thinning, fibrillose, white to cream, sterile, separable. Subiculum composed of loosely woven hyphae in the basal part but these become compact to more or less agglutinated in the upper part.

Hyphal system monomitic, hyphae 3.5–7 µm wide, branched, septate, clamped at all septa, heavily incrusted, incrustations more prominent in the basal part of the context, the walls thin, subhyaline, often collapsed and difficult to discern in mature fructifications. Cystidia and gloeocystidia absent. Basidia 25–34.9 × 5.5–7.1 µm, clavate-cylindrical, 4-spored, sterigmata up to 5.5 µm long. Basidiospores 5–7 × 3.5–5.4 µm, broadly ellipsoid to ovoid, minutely apiculate, often clinging in groups and forming spore balls, the walls moderately thick, light yellow to yellowish brown, smooth, cyanophilous, subdextrinoid.

Distribution: H.P.: Kullu, Dalhousie- Lakkarmandi, Shimla, Mahasu- Narkanda, U. K.: Karanprayag- Chamoli, Chakrata, Mussoorie-Dehradun.

Collection examination: GSR 6026, IBP 37266, L 37265.

Substratum: Coniferous wooden poles and stumps of *Abies pindrow*.

Remarks: This species is characterized by the resupinate habit, monomitic hyphal system and small basidiospores which cling together and form spore bolls. Pouzar (1958) raised a new genus *Leucogyrophana* for this species. Parmasto (1968) treated this genus under Corticiaceae (subfamily Athelioideae). This arrangement was also accepted by Jülich (1971). The species is being reported for the first time from Uttarakhand.

Leucogyrophana olivascens (Berk. & M.A. Curtis) Ginns & Weresub, Mem. N. Y. bot. Gdn 28: 96 1976. Plate 6.5a, Fig. 6.9o–q

Fructification resupinate, delicately pelliculose to floccose, fragile, easily separable, often occurring in small colonies; hymenial surface light green to greenish brown, smooth to farinaceous; margin indeterminate, concolorous.

Hyphal system monomitic. Hyphae 1.4–6 µm wide, branched at wide angles, septate, clamped, thin-walled, subhyaline, distinct and often incrusted. Hyphal cordons present. Hymenium discontinuous, basidia and basidioles often arising in clusters forming pin-head cushions on hymenial surface. Gloeocystidia absent.

Plate 6.5 (**a**) *Leucogyrophana olivascens*. (**b**) *Serpula himantioides*. (**c**) *Serpula lachrymans*. (**d**) *Botryobasidium danicum*. (**e**) *Botryobasidium subcoronatum*. (**f**) *Botryobasidium vagum*

Basidia 12–15.8×5–5.8 µm, clavate to clavate cylindrical, 2–4 spored, sterigmata up to 3.5 µm long. Basidiospores 4.4–5.4×3–4.1 µm, broadly ellipsoid to ovoid, shortly apiculate, the walls thick, light greenish brown, smooth, cyanophilous, nonamyloid.

Distribution: Bhutan: Thimphu, Nawephu; H.P.: Shimla, Chaupal, Kullu- Jaggat suk.
Collection examined: GSD 19355, 19544, SSR 5752, IBP 43003.
Substratum: On decaying gymnospermic wood.
Remarks: This species is characterized by the delicately pelliculose to floccose fructifications, small size of basidia and basidiospores. *C. mustialesis* is very close to this species but the basidiospores are not cyanophilous in that species and instead turn blue in KOH sol.

Leucogyrophana thimphina Dhingra, In Plant Diversity in India: 481, 2004.
Fig. 6.10a–c

Fructification resupinate, loosely adnate, effused, soft when fresh, fragile on drying; hymenial surface light yellow to yellow when fresh, olivaceous on bruising and in the herbarium, meruloid; margins thinning, fibrillose, white when fresh, pale brownish on drying.

Hyphal system monomitic; generative hyphae septate, clamped, thin-walled; basal hyphae up to 5 μm wide, sparsely branched, loosely interwoven; subhymenial hyphae up to 2.5 μm wide, richly branched into a dense texture. Cystidia absent. Basidia 25–35 × 4.4–6 μm, narrowly clavate, 4-sterigmate, with a basal clamp; sterigmata up to 4.5 μm long. Basidiospores 3.5–4.4 × 2.2–2.7 μm, ellipsoid, subhyaline to tinted light yellow, smooth, thick-walled, cyanophilous, inamyloid, weakly dextrinoid, generally with one guttule.

Distribution: Bhutan: Thimphu, Dochula.
Collection examined: GSD 19566.
Substratum: On a decaying gymnospermic log.
Remarks: This species is an uncommon species in the Himalayas, reported only once from the type locality. It is closely related to *L. mollusca* except for the size of the basidia and basidiospores.

Family-Serpulaceae

Serpula (Pers.) Grey,
Nat. Arrang. Brit. Pl 1: 637. 1821.

Fructification resupinate to effused-reflexed or pileate, membranous to waxy gelatinous; hymenial surface reticulately folded to meruloid, varied in colour. Subiculum composed of loosely woven hyphae. Hyphal system monomitic or dimitic, generative hyphae with or without clamps. Cystidia or Gloeocystidia present or absent. Basidia clavate-cylindrical, 4-spored. Basidiospores ellipsoid to ovoid, thick-walled, yellowish brown to greenish brown, cyanophilous, non-amyloid.

Two species, widespread

Lit.: Cooke (*Mycol.* 49: 197, 1957; key), Huckfeldt & Schmidt (*Mycologist* 20: 42, 2006)
Type Species: *Merulius destruens* Pers., 1801.
Habitat: Dead wood
Himalaya: Two

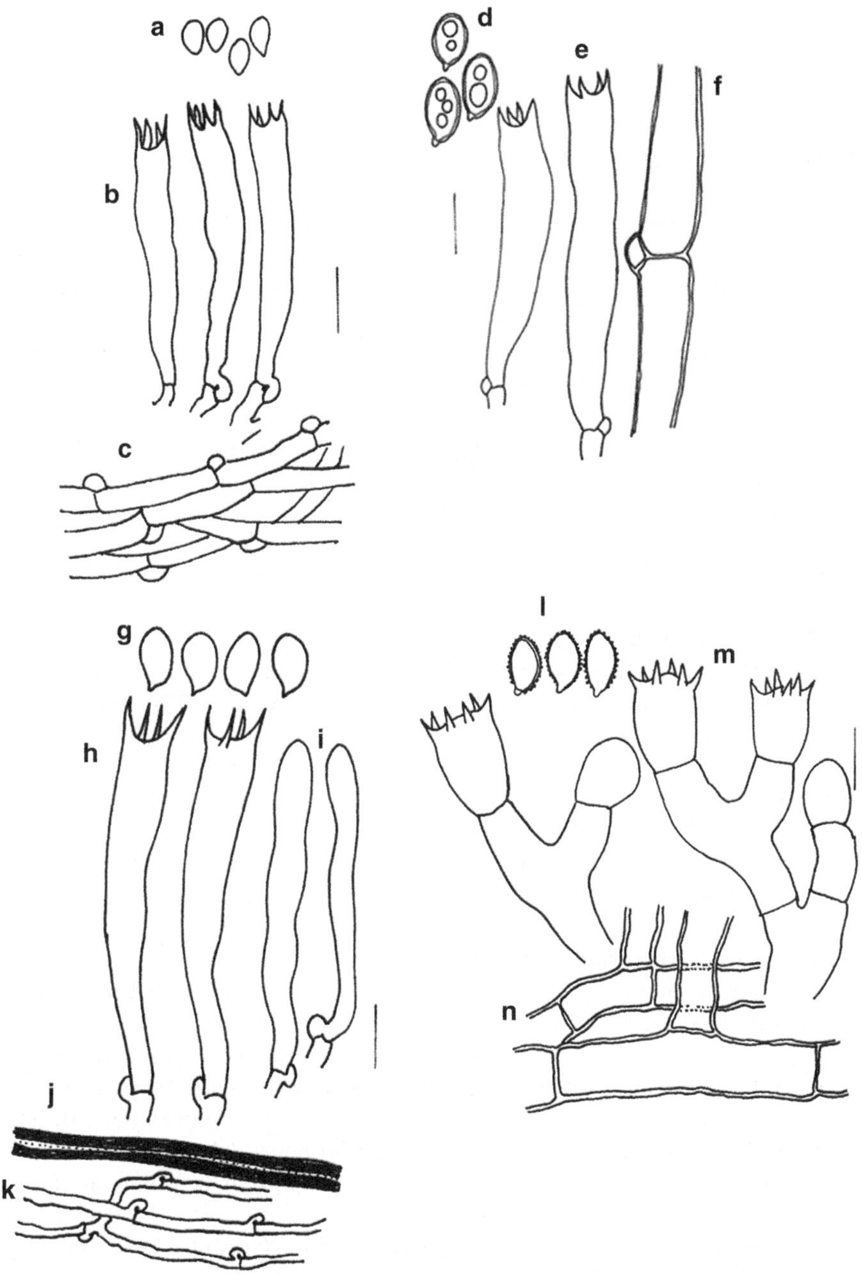

Fig. 6.10 (**a–c**) *Leucogyrophana thimphina* (**a**) Basidiospores, (**b**) Basidia, (**c**) Generative hyphae; (**d–f**) *Serpula himantioides* (**d**) Basidiospores, (**e**) Basidia, (**f**) Generative hyphae; (**g–k**) *Serpula lachrymans* (**g**) Basidiospores, (**h**) Basidia, (**i**) Cystidia, (**j**) Skeletal hyphae, (**k**) Generative hyphae; (**l–n**) *Botryobasidium asperulum* (**l**) Basidiospores, (**m**) Basidia, (**n**) Generative hyphae

Key to species

1. Hyphal system dimitic ....*S. lacrymans*
1. Hyphal system monomitic ....*S. himantioides*

Serpula himantioides (Fr.) P. Karst., Meddn Soc. Fauna Flora fenn. **11**: 21 (1885). Plate 6.5b, Fig. 6.10d–f

Fructification resupinate, membranous but fragile, loosely adnate, widely effused, hymenial surface light brown to brown, smooth when young, reticulation appearing at maturity, soon becoming convoluted and continue to grow so that at maturity hymenium becomes porose; pores angular, up to 700 µm deep; dissepiments 250–400 µm thick, fertile all over; margin up to 5 mm wide, thinning to fibrillose, sterile, smooth, white. Context composed of basal zone of repent hyphae and an intermediate zone of semi-erect hyphae.

Hyphal system monomitic, hyphae 3–6.5 µm wide, branched, septate, clamped. Cystidia and gloeocystidia absent. Basidia 35–40 × 6–9 µm, clavate to clavate-cylindrical, 4-spored. Basidiospores 9–10.5 × 5–6 µm, broadly ellipsoid, shortly apiculate, thick-walled, smooth, cyanophilous, inamyloid.

Distribution: A.P: West Kameng; West Bengal: Darjeeling; Bhutan: Thimphu; H.P.: Dalhousie- Lakkarmandi; Mahasu- Narkanda.

Collection examination: SSR: 6501, IBP 37678.

Substratum: Coniferous wooden poles and stumps of *Abies pindrow*.

Remarks: This is wide spread species in Himalayas. Rattan (1977) reported it for the first time from N.W Himalayas.

Serpula lachrymans (Wulfen) J. Schröt., Meddn Soc. Fauna Flora fenn.**11**: 21 (1885). Plate 6.5c, Fig. 6.10g–k

Fructifications effused-reflexed to distinctly pileate forms, resupinate forms widely effused, easily separable from substratum, margin thick. Upper surface smooth or concentrically sulcate, light cream to cream brown but usually appears rust brown; pore surface snuff brown or tobacco brown; margin obtuse.

Hyphal system dimitic; generative hyphae 1.4–7 µm wide, branched, septate, clamped, moderately thick-walled; skeletal hyphae 4–5.3 µm wide, unbranched to sparsely branched, aseptate, more abundant in the hard part of the context, the walls thick (1.4 µm) leaving little or no lumen, subhyaline;. Pores angular to irregular, often becoming irpiciform. Cystidia and gloeocystidia absent. Basidia 60–69.8 × 7–8.4 µm, clavate, 4-spored. Basidiospores 8.4–10.1 × 5.1–5.4 µm, broadly ellipsoid, the walls moderately thick (0.8 µm), light yellow to citrine, smooth, cyanophilous, nonamyloid.

Distribution: H.P: Dalhousie- Lakkarmandi, Mahasu- Narkanda; J&K: Gulmarg; U.K.: NDBR.

Collection examination: RSD 6728, 6667, IBP 37679.

Substratum: On log of *Abies pindrow*.

Remarks: This species is characterized by the pileate fructifications, dimitic hyphal system and large basidiospores. *Serpula himantoides* is similar but the fructifications in this species are mostly resupinate and the hyphal system is monomitic.

O- Cantharellales
Family- Botrybasidiaceae

Key to genera

1. Basidiospores echinulate.. *Botryohypochnus*
1. Basidiospores smooth .... *Botryobasidium*

Botryobasidium Donk,
Medded. Nedl. Mycol. Ver. 18–20: 116, 1931.

Fructification resupinate, loose, reticulate to hypochnoid; hyphal system monomitic; hyphae distinct, broad, septate, clamps present, branched at right angles; basal hyphae more or less thick-walled, Cystidia or Gloeocystidia absent. Basidia present in clusters, constricted, 4–8 sterigmate. Basidiospores vary in shape, smooth, hyaline or subhyaline, thin-walled, non-amyloid.

Fifty five species, widespread.

Lit.: Langer (*Biblthca Mycol.* **158**, 1994)

Type Species: *Corticium subcornatum* Höehn. & Litsch. 1907.

Habitat: Wood

Himalayas: Six

Key to species

1. Hyphae with clamps... *B. subcoronatum*
1. Hyphae without clamps...2
2. Basidiospores prominently warted... *B. asperulum*
2. Basidiospores smooth 3
3. Conidial stage present.. *B. candicans*
3. Conidial stage absent 4
4. Basidia urniform .... *B. subbotryosum*
4. Basidia not so..5
5. Basidia round to ellipsoid .... *B. danicum*
5. Basidia clavate to subcylindrical... *B. vagum*

Botryobasidium asperulum (D.P. Rogers) Boidin, Cahiers de La Maboké 8(1): 19 (1970). Fig. 6.10l–n

Fructification thin, reticulate to hypochnoid, yellowish-white to light ochraceous and of a loosely interwoven texture.

Hyphal system monomitic; generative hyphae up to 12 μm wide, without clamps, basal hyphae yellowish, thick-walled, sparsely branched, hymenial hyphae subhyaline, thin-walled, richly branched. Basidia 11.9–17.4 × 9.4–11.4 μm, ellipsoid to obovate, without basal clamp, generally 6-sterigmate. Basidiospores 6–10.5 × 4–6 μm, obliquely ovoid to fusiform-elliptical, somewhat thick-walled, prominently warted, non-amyloid.

Distribution: West Bengal: Darjeeling, Siliguri.
Collection examined: GSD 19215.
Substratum: On decaying angiospermic log.
Remarks: This collection resembles *B. asperulum* (Rogers) Boidin (Boidin 1970), in most of the characters except the size of the basidiospores which are larger in the former (6–10.5 × 4–6 µm in contrast to 4.2–6 × 3–4 µm).

Botryobasidium candicans J. Erikss., Sven. Bot. Tidskr. 52(1): 6 (1958). Fig. 6.11a–c

Fructifications resupinate, floccose to hypochnoid, loosely adnate, widely effused; hymenial surface deep cream to yellow-ochre; margin cottony, loosely adnate, white to paler concolorous.

Hyphal system monomitic; basal hyphae 6–10 µm wide, sparsely branched, branches at wide angles, septate, clamps absent, the walls thin to moderately thick (up to 0.7 µm), subhyaline to tinted yellow; subhymenial hyphae 4.5–7 µm wide, profusely branched, branches at right angle, ramified, septate, clamps absent, thin-walled. Conidia 10–15 × 7.5–8.5 µm, ovoid, arising apically in acropetal succession, in erect hyphoid conidiophores-similar to basal hyphae; the walls pale yellow, weakly dextrinoid, thick. Basidia 18 × 7 µm, uniform, 6-sterigmate, sterigmata short, often collapsing after spore discharge. Basidiospores 5–6 × 2.4–3 µm, ellipsoid to subfusiform, prominently apiculate, thin-walled, subhyaline, smooth, nonamyloid.

Distribution: H.P.: Shimla; J&K: Pehalgam.
Collection examined: SSR 5578, 5771, IBP 37557, 42003.
Substratum: On angiospermic twig.
Remarks: The characteristic features of this species are presence of conidial stage, absence of clamps and ellipsoid to subfusiform basidiospores.

Botryobasidium danicum J. Erikss. & Hjortstam, Friesia 9(1–2): 11 (1969).
Plate 6.5d, Fig. 6.11d–f

Fructification thin, reticulate to hypochnoid, greyish-white or with age somewhat yellowish and of a loosely inter-woven texture.

Hyphal system monomitic; generative hyphae without clamps, basal hyphae up to 10 µm wide, sparsely branched, thick-walled, hymenial branches comparatively narrower and thin-walled. Basidia 12–20.8 × 7.4–8.8 µm, round to ellipsoid, then obovate to subcylindrical, basal clamps absent, 6-sterigmate. Basidiospores 8–11.8 × 3–4 µm, navicular to subcylindrical, thin-walled, smooth, non-amyloid.

Distribution: Arunachal Pradesh: West Kameng, Bomdila, Shergaon, Basti side; West Bengal: Darjeeling.
Collection examined: GSD 19852, 19307.
Substratum: On a decaying gymnospermic stump, on a decaying *Cryptomeria japonica* branch.

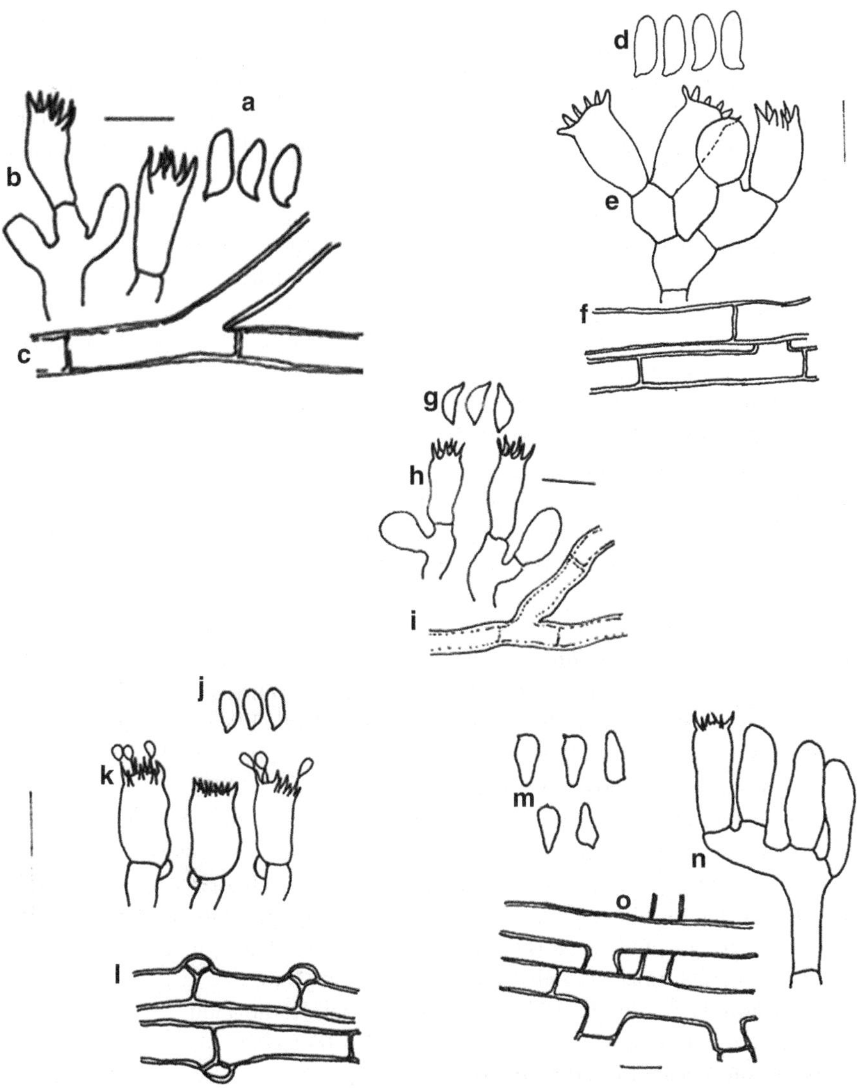

Fig. 6.11 (**a–c**) *Botryobasidium candicans* (**a**) Basidiospores, (**b**) Basidia, (**c**) Generative hyphae; (**d–f**) *Botryobasidium danicum* (**d**) Basidiospores, (**e**) Basidia, (**f**) Generative hyphae; (**g–i**) *Botryobasidium subbotryosum* (**g**) Basidiospores, (**h**) Basidia, (**i**) Generative hyphae; (**j–l**) *Botryobasidium subcoronatum* (**j**) Basidiospores, (**k**) Basidia, (**l**) Generative hyphae; (**m–o**) *Botryobasidium vagum* (**m**) Basidiospores, (**n**) Basidia, (**o**) Generative hyphae

Remarks: The description of the above mentioned species is in conformity with the type description except for minor variation in spore size. It is a new record Himalayas.

Botryobasidium subbotryosum S.S. Rattan [as '*subbotryosus*'], Biblthca Mycol. **60**: 210 (1977). Fig. 6.11g–i

Fructification resupinate, byssoid when young becoming loosely membranous with age, loosely adnate, widely effused; hymenial surface deep cream to pale yellow, smooth to somewhat farinose, discontinuous; margin thinning, loosely adnate, paler concolorous.

Hyphal system monomitic; hyphae 4–10 μm wide, branched at wide angles, septate, clamps absent, thin-walled, subhyaline to faintly tinted yellow. Basal hyphae are broader, more deeply tinted and less frequently branched than the subhymenial hyphae. Basidia 12.1–17.9 × 7–9 μm, urniform, 6-spored. Basidiospores 7–7.5 × 2.8–3.5 μm, ellipsoid to navicular with pointed ends, shortly apiculate, thin-walled, smooth, subhyaline, non-amyloid.

Distribution: H.P.: Kullu, Mahasu; Bhutan- Thimphu.

Collection examined: SSR 5696, 5569.

Substratum: On angiospermic twig.

Remarks: This species is close to *B. candicans* but differs in having slightly larger basidiospores and lacking conidial stage. *Botryobasidium botryosum* is also similar but possesses larger basidiospores (9–12 × 4.5–6 μm). It is a rare species in Himalayas.

Botryobasidium subcoronatum (Höhn. & Litsch.) Donk, Medd. Nederl. Mycol. Ver. 18–20: 117 (1931) = *Corticium subcornatum* Höhn. & Litsch., Sitz. Kais. Akad. Wiss. Wien, Math.-Nat. Klasse 116: 822. 1907. Plate 6.5e, Fig. 6.11j–l

Fructification thin, reticulate to hypochnoid, greyish-white or with age somewhat yellowish and of a loosely interwoven texture.

Hyphal system monomitic; generative hyphae without clamps, basal hyphae up to 10 μm wide, thick-walled, sparsely branched, hymenial branches comparatively narrower and thin-walled. Basidia 12.2–20.8 × 7.4–9 μm, first round to ellipsoid, than obovate to subcylindrical, without basal clamp, generally 6-sterigmate; sterigmata up to 4 μm long. Basidiospores 8.1–12.1 × 2.9–4.4 μm, narrowly navicular to subcylindrical, smooth, thin-walled, nonamyloid.

Distribution: H.P.: Shimla; J&K: Pehalgam; Nepal: Kathmandu.

Collection examined: SSR 5578, 5771, L 37027, IBP 42010.

Substratum: On angiospermic twigs.

Remarks: The above collections resembles *Botryobasidium danicum* Eriks. and Hjortst., in most of the characters except the size of the basidiospores.

Botryobasidium vagum (Berk. & M.A. Curtis) D.P. Rogers, Univ. Iowa Stud. nat. Hist. 17(1): 17 (1935). Plate 6.5f, Fig. 6.11m–o

Fructification thin, reticulate to hypochnoid, yellowish-white to pale-ochraceous and of loosely interwoven texture.

Hyphal system monomitic; generative hyphae branched at right angle, septate, without clamps; basal hyphae pale yellow, thick-walled, up to 8 μm wide. Basidia

15.0–22.0×8.0–10.9 µm, clavate to subcylindrical, rarely somewhat constricted, without basal clamp, generally 6-sterigmate. Basidiospores 8.0–10.0×3.9–4.9 µm, navicular, smooth, thin-walled, apiculate, non-amyloid.

Distribution: West Bengal: Darjeeling, Dhotrey; Bhutan: Thimphu, Chimakothi, Dochula; U.K.: Chakrata.

Collection examined: GSD 19289, 19494, 19573, IBP 37030, L 37003.

Substratum: On bark of a decaying *Cryptomeria japonica* stump, on decaying angiospermic stump, on a decaying gymnospermic log.

Remarks: This species was first proposed by Bresadola (1903) from Poland as *Cortricium botryosum*. Burt (1926) reported this species as a synonym of *Cortricium vagum* Berk. & Curt., on the basis of collection of various workers from N. America. Eriksson (1958) shifted this species to genus *Botryobasidium*. The himalayan collections resemble the type description. It is being reported from N.W. Himalayas for the first time.

***Botryohypochnus* Donk,**
Meded. Nederl. Mycol.
Vereen. 18–20: 118, 1931.

Fructification resupinate, floccose to arachinoid, hymenial surface cream coloured discontinuous. Subiculum composed of loosely woven hyphae. Hyphal system monomitic; generative hyphae septate, clamps absent, branched; basidia subcylindrical to obovate, 4-spored; basidiospores globose, echinulate, cyanophilous.

Type species: *Hypochnus isabellinus* J. Schröt, 1888.

Himalayas: One

Botryohypochnus isabellinus (Fr.) J. Eriksson, Sv. Bot. tidskr. 52: 2, 1958 = *Thelephora isabellinua* (Fr.), D. P. Rogers, Farlowia 50: 99 (1943). Fig. 6.12a, b

Fructification resupinate, loosely adnate, widely effused, floccose to hypochnoid; hymenial surface cream yellow, yellow-ochre to light brown or finally umber-brown, smooth but appears farinose under the lens, discontinuous; margin thinning or indeterminate, loosely adnate, paler concolorous.

Hyphal system monomitic; hyphae 5.5–15 µm wide, branched at right angles, septate, clamps absent, thin-walled, subhyaline to slightly tinted yellow. Basal hyphae are broader with thicker walls. Basidia 18–20×9–10 µm, 2–4 spored, subcylindrical. Basidiospores 8.5–10 µm in diameter, globose, thin-walled, pale yellow, echinulate, blunt, cyanophilous.

Distribution: H.P.: Chamba; Kullu; Shimla, Khajjiar; J&K: Pehalgam

Collection examined: GSD 19679, 19817, SSR 5770, IBP 37031.

Substratum: On a decaying angiospermic stump and logs, *Cedrus deodara*, *Abies pindrow*, *Pices simithiana*.

Remarks: The species is marked by the colour of the hymenial surface and shape & size of basidia and basidiospores. The species is abundantly found in North Western Himalayas.

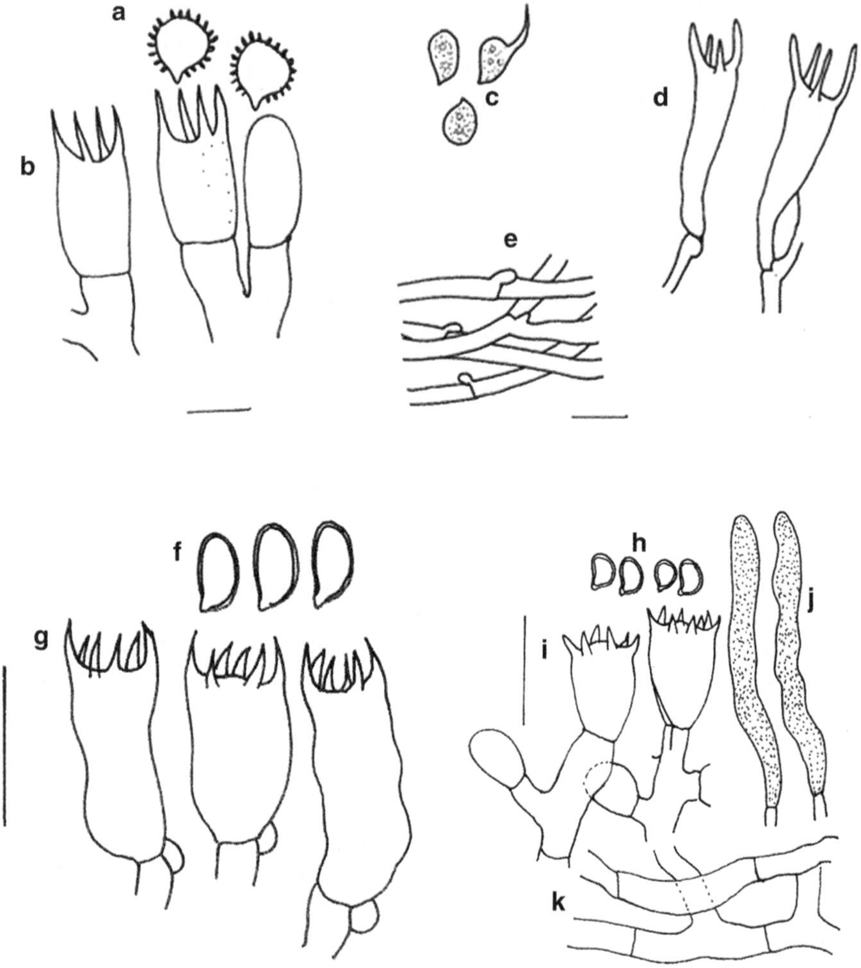

Fig. 6.12 (**a**, **b**) *Botryohypochnus isabellinus* (**a**) Basidiospores, (**b**) Basidia; (**c**–**e**) *Scotomyces subviolaceus* (**c**) Basidiospores, (**d**) Basidia, (**e**) Generative hyphae; (**f**, **g**) *Paullicorticium delicatissimum* (**f**) Basidiospores, (**g**) Basidia; (**h**–**k**) *Paullicorticium indicum* (**h**) Basidiospores, (**i**) Basidia, (**j**) Cystidia, (**k**) Generative hyphae

Family- Ceratobasidiaceae

Scotomyces Jülich,

Persoonia, 10(1):139, 1978.

Fructification annual, resupinate, effused, adnate, hypochnoid, waxy-gelatinous or ceraceous, thin film like, adnate; hymenial surface grey to black, smooth; margin indistinct. Hyphal system monomitic; hyphae thin to thick-walled, with or without clamps. Gloeocystidia or cystidia absent. Basidia clavate, thin-walled, 4-sterigmate, sterigmata long. Basidiospores subglobose to ovoid, thin-walled, smooth, subhyaline, nonamyloid.

Monotypic, widespread.
Lit.: Wojewoa (*Acta Mycologica* Warszawa **38**: 3, 2003; Poland)
Type Species: *Corticium fallax* G. Cunn. 1954.
Habitat: Wood
Himalayas: One

Scotomyces subviolaceus (Peck) Jülich, Persoonia 10(3): 334 (1979) = *Ceratobasidium subatratum* S.S. Rattan, Biblioth. Mycol. 60: 203 (1977). Plate 6.6a, Fig. 6.12c–e

Fructifications resupinate, floccose to loosely membranous when young becoming subceraceous to waxy gelatinous at maturity, often arising in small colonies; hymenial surface cream brown to brown to almost blackish brown, smooth but appears farinose; discontinuous, margin thinning, wide, loosely adnate, paler concolorous.

Hyphal system monomitic, hyphae 2.4–5.4 μm wide, branched at wide angles, septate, clamped at all septa, thin-walled, sub hyaline, frequently collapsing and gelatinizing in mature specimens, often becoming tinted to pale brown with age. Basidia often arising in clusters, 15–25 × 7–8.2 μm, clavate, 4-sterigmata. Basidiospores 6.4–7.2 × 4–5 μm, broadly ellipsoid to ovoid, thin-walled, smooth, non-amyloid.

Distribution: H.P: Manali; J&K: Pehalgam, U.K.: Badrinath, Vishnuprayag-Chamoli.

Collection examined: SSR 5671, 5772, IBP 37033, 42024.

Substratum: On stump, logs under conifers.

Remarks: The main features of this species are resupinate, browny black fructifications, presence of clamps and broadly ellipsoid to ovoid basidiospores. This is a new record for Uttarakhand.

Family- Hydnaceae

Key to genera

1. Basidia obovate to pyriform, 6-sterigmate, basidiospores ellipsoid .... *Paullicorticium*
1. Basidia urniform, 6–8 sterigmate, basidiospores ellipsoid to tetrahedral .... *Sistotrema*

Paullicorticium J. Eriksson

Symb. bot. Upsal. 16: 66, 1958.

Fructification resupinate, closely adnate, small and very thin, starting as a net of hyphae on the substrate, then interwining into a thin layer, ceraceous and perishable as wet, when dried pruinose-porulose under the lens; hyphae thin-walled, narrow, with or without clamps; no cystidia; basidia obconical-pyriform, with (4-)6-8 sterigmata; spores hyaline, thin-walled, allantoids, ellipsoid or navicular, non-amyloid, non-cyanophilous.

Five Species, widespread.

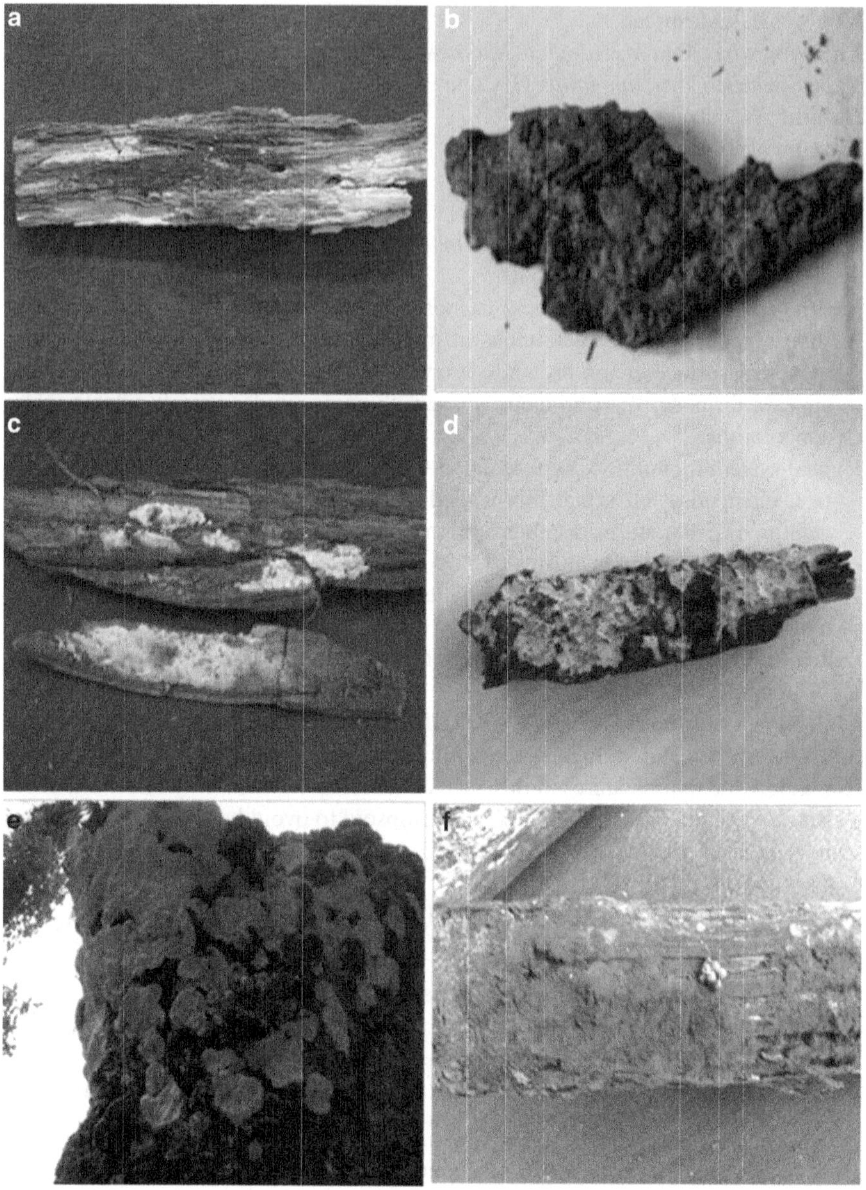

Plate 6.6 (**a**) *Scotomyces subviolaceus*. (**b**) *Paullicorticium indicum*. (**c**) *Sistotrema angustispora*. (**d**) *Dendrothele incrustans*. (**e**) *Dendrothele seriata*. (**f**) *Erythricium laetum*

Lit.: Liberta (*Brittonia* **13**: 219, 1962; key).
Type species: *Corticium pearsonii* Bourd.1922
Habitat: Wood
Himalayas: Two

Key to species

1. Generative hyphae with clamps .... *P. delicatissimum*
1. Generative hyphae without clamps .... *P. indicum*

Paullicorticium delicatissimum (Jacks.) Liberta, Brittonia **14**: 222 (1962) = *Corticium delicatissimum* Jacks., Can. Jour. Bot. 28: 722, 1950. Fig. 6.12f, g

Fructification resupinate, effused, loosely adnate, broom like; hymenial surface smooth, grey when fresh, greyish yellow on drying; margins thinning, paler concolorous to indeterminate.

Hyphal system monomitic; generative hyphae up to 3.1 µm wide, septate, branched, thin-walled, clamped; basal hyphae parallel to the substrate, branching less; subhymenial hyphae vertical, highly branched. Cystidia absent. Basidia 5.5–12.0 × 3.7–4.5 µm, 6-sterigmate with basal clamp, obovate to pyriform. Basidiospores 3.4–4.0 × 1.8–2.5 µm, ellipsoid to subcylindrical, thin-walled, smooth, acyanophilous, hyaline, inamyloid.

Distribution: H.P.: Solan.

Substrate: On decaying log of *Pinus roxburghii.*

Collection examined: IBP 37267.

Remarks: This species is characterized by thin basidiospores, small basidia and basidiospores. It was first reported from Sweden. It is a new record for Himalayas.

Paullicorticium indicum Dhingra, In Plant Diversity in India: 483, 2004. Plate 6.6b, Fig. 6.12h–k

Fructification resupinate, thin, adnate, when young subinvisible, then forming insignificant patches on the wood, greyish-white with a yellowish tint, in herbarium the surface appears reticulate or porulose.

Hyphal system monomitic; generative hyphae branched, septate, without clamps, thin-walled, up to 6.4 µm wide. Cystidia 45.0–60.0 × 5.0–7.0 µm, subcylindrical, somewhat sinuous, thin-walled, with yellowish resinous material. Basidia 7.4–12.0 × 6.4–8.9 µm, at first subglobose, then ovate, urniform or pyriform, without basal clamp, 6–8 sterigmate, sterigmata up to 4.0 µm long. Basidiospores 3.7–5.2 × 2.5–3.3 µm, ellipsoid to ovoid, smooth, somewhat thick-walled, inamyloid, acynophilous.

Distribution: West Bengal: Darjeeling, Tiger hill.

Collection examined: GSD 19255.

Substratum: On a decaying angiospermic stump.

Remarks: This species is characterized by thin fructification, generative hyphae without clamps, subcylindrical cystidia, ovate to urniform or pyriform, 6–8 sterigmate. Basidia and ellipsoid to ovoid, smooth, thick-walled basidiospores.

Sistotrema Fries. E.M.,
Systema Mycol. 1: 426, 1826.

Fructification resupinate, thin, arachnoid to membranous, often becoming subceraceous to ceraceius; hymenial surface smooth to odontoid or poroid. Subiculum subhyaline in section, not reacting with KOH sol. Hyphal system monomitic, hyphae branched, septate, clamps present, often ampullate near the septa, walls thin, subhyaline. Cystidia absent. Gloeocystidia present or absent Cystidia or Gloeocystidia absent. Basidia urniform or utriform, clavate, 4–8 spored. Basidiospores usually ellipsoid or subglobose, subhyaline, smooth, thin-walled, non-amyloid, acyanophilous.

Forty six species, widespread

Lit.: Hallenberg (*Mycotaxon* **21**: 389, 1984), Greslebin (*MR* **105**: 1392, 2001)

Type species: *Sistotrema confluens* Pers. 1794

Habitat: Dead wood

Himalayas: Seven

Key to species

1. Basidiospores tetrahedral .... *S. subtrigonospermum*
1. Basidiospores otherwise 2
2. Gloeocystidia present .... *S. sernanderi*
2. Gloeocystidia absent 3
3. Basidia 4–Sterigmate .... *S. brinkmannii*
3. Basidia 4–8 Sterigmate 4
4. Basidiospores lachrymiform .... *S. lachrymisporum*
4. Basidiospores not lachrymiform 5
5. Basidiospores more than 10 μm, navicular to subcylindrical .... *S. angustispora*
5. Basidiospores less than 5 μm, not navicular 6
6. Basidiospores narrowly ellipsoid to subcylindrical .... *S. binucleosporum*
6. Basidiospores ellipsoid .... *S. porulosum*

Sistotrema angustispora Dhingra, In Plant Diversity in India: 484, 2004.

Plate 6.6c, Fig. 6.13a–d

Fructification thin, at first reticulate to hypochnoid, with age more or less continuous; hymenial surface greyish-white when fresh, yellowish to ochraceous on drying, smooth; margin not differentiated.

Hyphal system monomitic; generative hyphae loosely interwoven, septate, clamped, basal hyphae up to 6.5 μm wide sparsely ramified, somewhat thick-walled, basidial branches up to 4.5 μm wide, comparatively thickly branched. Cystidia 55.0–135.0 × 7.0–8.0 μm, tubular with somewhat broadened base, thin-walled, encrusted with subhyaline granules, with a basal clamp, projecting up to 100 μm out of the hymenium. Basidia 18.0–28.2 × 8.0–10.1 μm, urniform, with a basal clamp, generally 6-sterigmata up to 4 μm long. Basidiospores 10.4–14.0 × 2.3–3.2 μm, narrowly navicular to subcylindrical, smooth thin-walled, inamyloid, acyanophilous, with many small oil droplets, often glued in groups of two or more.

Distribution: West Bengal: Darjeeling.

Collection examined: GSD 19233.

Substratum: On coniferous wood.

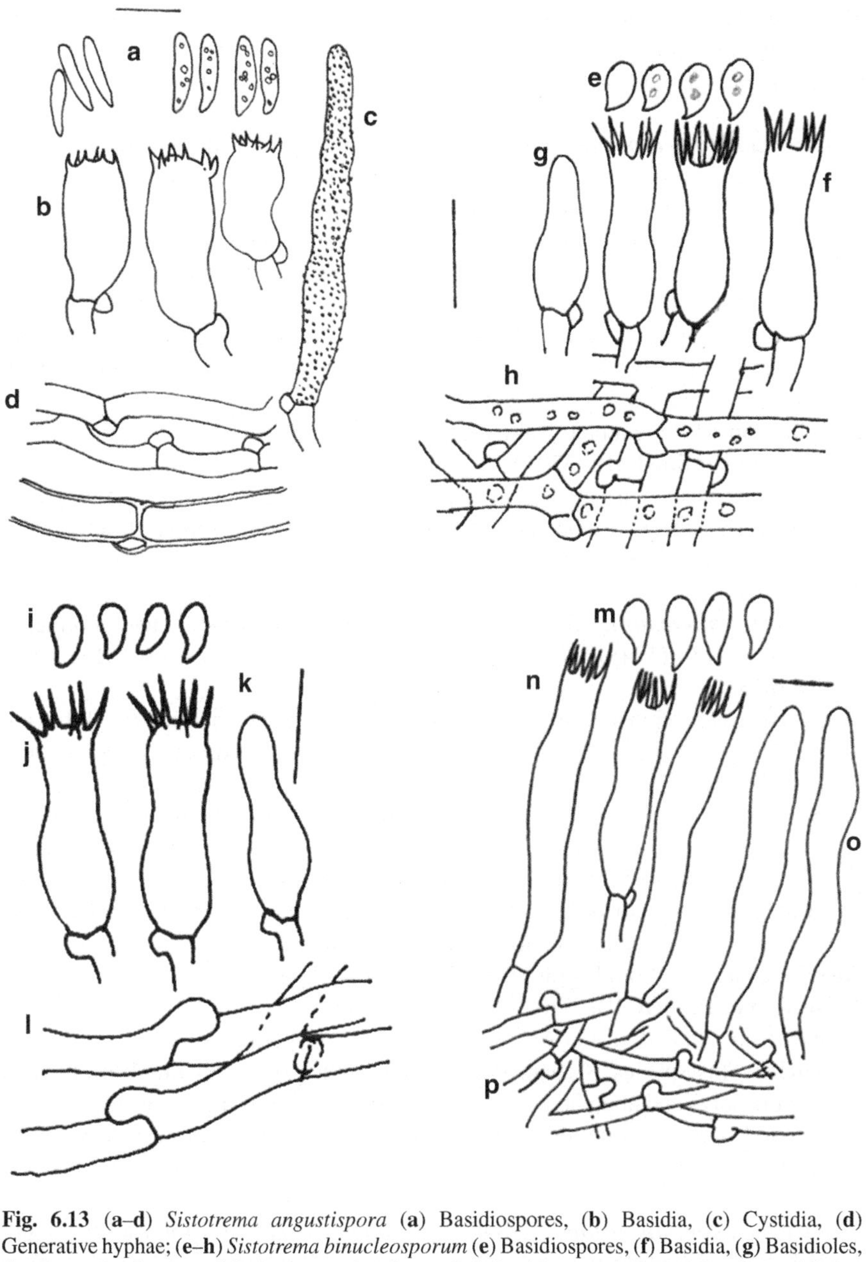

Fig. 6.13 (**a–d**) *Sistotrema angustispora* (**a**) Basidiospores, (**b**) Basidia, (**c**) Cystidia, (**d**) Generative hyphae; (**e–h**) *Sistotrema binucleosporum* (**e**) Basidiospores, (**f**) Basidia, (**g**) Basidioles, (**h**) Generative hyphae; (**i–l**) *Sistotrema brinkmannii* (**i**) Basidiospores, (**j**) Basidia, (**k**) Basidioles, (**l**) Generative hyphae; (**m–p**) *Sistotrema lachrymisporum* (**m**) Basidiospores, (**n**) Basidia, (**o**) Basidioles, (**p**) Generative hyphae

Remarks: The characteristic features of this species are peculiar granulation on the cystidia and narrowly navicular to subcylindrical basidiospores along with the absence of oily contents in the hyphae.

Sistotrema binucleosporum Hallenb., Mycotaxon 21: 409, 1984. Fig. 6.13e–h

Fructification resupinate, effused, adnate, very thin; hymenial surface smooth to rough, creamish to greyish white; margin fibrillose.

Hyphal system monomitic; generative hyphae 2.2–3.3 μm wide, thin-walled, septate, clamped; basal hyphae sparsely branched; subhymenial hyphae richly branched, densely united, short-celled. Basidia 12.5–17.5×4.4–5.1 μm, urniform, thin-walled, 6–8 sterigmata, with basal clamps, sterigmata 2.8 μm long. Basidiospores 3–4.5×1.6–2.2 μm, ellipsoid to subcylindrical, thin-walled, smooth, non-amyloid, acyanophilous.

Distribution: H.P.: Chamba- Dalhousie, Shimla, Narkanda.

Collection examined: IBP 37034.

Substratum: Under surface of gymnospermic log and on gymnospermic twigs.

Remarks: This species is characterized by very thin, adnate, porulose fructification and ellipsoid to subcylindrical basidiospores.

Sistotrema brinkmannii (Bres.) J. Erikss., K. Fysiogr. Sällsk. Lund. Förhandl. 18(no. 8): 134 (1948)=*Odontia brinkmannii* Bres., Ann. Mycol. 1: 88, 1903. Fig. 6.13i–l

Fructifications resupinate, effused, loosely adnate, hymenial surface rough having small aculei, greyish-white when fresh, becomes white on drying; margins irregularly thinning.

Hyphal system monomitic; generative hyphae up to 3 μm wide, branched, septate, with clamps at all septa, thin to thick-walled, crystalline abundant. Cystidia absent. Basidia 10.5–16×3.5–5 μm, uniform, generally with 6-sterigmata and a basal clamp. Basidiospores 3.4–4.5×1.9–3.1 μm, smooth, sub allantoid to ellipsoid, thin-walled, apiculate, non-amyloid, acyanophilous.

Distribution: H.P.: Shimla, Tara Devi, Narkanda.

Collection examined: IBP 37035.

Substratum: On bark of *Cedrus deodara*, cut gymnospermic wood.

Remarks: This species is characterized by presence of suballantoid to ellipsoid basidiospores and abundant crystals in the fructifications. Erikkson et al. (1984) reported it as a commonly occurring species in coniferous forest of Northern southern Europe.

Sistotrema lachrymisporum S.S. Rattan [as '*lachrymispora*'], Biblthca Mycol. **60**: 213 (1977). Fig. 6.13m–p

Fructification resupinate, membranous to subceraceous, adnate, widely effused, up to 250 μm thick in section; hymenial surface white when fresh becoming creamish on drying, smooth, not creviced or usually cracking minutely and irregularly on drying; margin thinning to fibrillose, adnate, white. Subiculum composed of basal zone of somewhat agglutinated repent hyphae and an upper zone of partly erect hyphae.

Hyphal system monomitic, hyphae 2.0–4.5 μm wide, branched, septate, clamped, clamps not at all septa, often irregularly inflated and collapsed and difficult to discern, the walls thin subhyaline. Gloeocystidia absent. Basidia (25) 50–55.1×6.4–7 μm, utriform, 4–6 sterigmate, sterigmata slender and up to 5.5 μm long. Basidiospores 8.5–11.5×3–5 μm, lachrymiform, with an obtuse apex and long curved base, shortly apiculate, often occurring in groups of four or six, thin-walled, hyaline, smooth, non-amyloid

Distribution: H.P.: Kullu.

Collection examined: SSR 5692.

Substratum: On coniferous wood.

Remarks: This species is marked by the shape and size of basidiospores. *Sistotrema hirschii* (Donk) Donk is very similar and may appear conspecific with it but the shape of the basidiospores is different (obovate to oblong with a slightly curved base). It is a rare species in Himalayas.

Sistotrema porulosum Hallenb., Mycotaxon 21: 407, 1984. Fig. 6.14a–d

Fructification resupinate, effused, closely adnate, thin; hymenial surface greyish white, porulose; margin indeterminate.

Hyphal system monomitic; generative hyphae 2.2–4 μm wide, septate, clamped, sometimes ampulate, with oily contents, subiculum very thin composed of slightly thick-walled hyphae; subhymenial hyphae thin-walled. Basidia 10.8–17.6×4–5.1 μm urniform, thin-walled, 6–8 sterigmate. Basidiospores 3.7–5.1×2.3–2.8 μm ellipsoid, thin-walled, smooth, minutely apiculate, non-amyloid, acyanophilous with one or more guttules.

Distribution: H.P.: Chamba- Banikhet.

Collection examined: IBP 37558

Substratum: On decaying angiospermic stump.

Remarks: This species is characterized by thin, adnate, porulose, greyish-white fructification and small, ellipsoid basidiospores. It is a rarely found species, and earlier reported only from Himachal Pradesh once (Dhingra et al., 2009).

Sistotrema sernanderi (Litsch.) Donk, *Fungus, Wageningen* **26**: 4 (1956)=*Gloeocystidium sernanderi* Litsch., Sv. Bot. Tidskr. 25: 437, 1931. Fig. 6.14e–h

Fructification resupinate, adnate, effused; hymenial surface rough, tuberculate, cream to creamish yellow when fresh, becomes pale yellow to pale, brown on drying; margin abrupt.

Hyphal system monomitic; generative hyphae up to 5.7 μm wide, branched, clamps at all septa, thin to thick-walled. Gloeocystidia 48.2–60×7.5–9.1 μm, thin-walled with oily content. Basidia 15.2–35.2×3.0–3.5 μm, uniform to cylindrical, basal clamp with 2- sterigmata. Basidiospores ellipsoid to sub allantoids, thin-walled, smooth, apiculate, non-amyloid, acyanophilous.

Distribution: H.P.: Shimla.

Collection examined: IBP 37558

Substratum: Decaying gymnospermic wood.

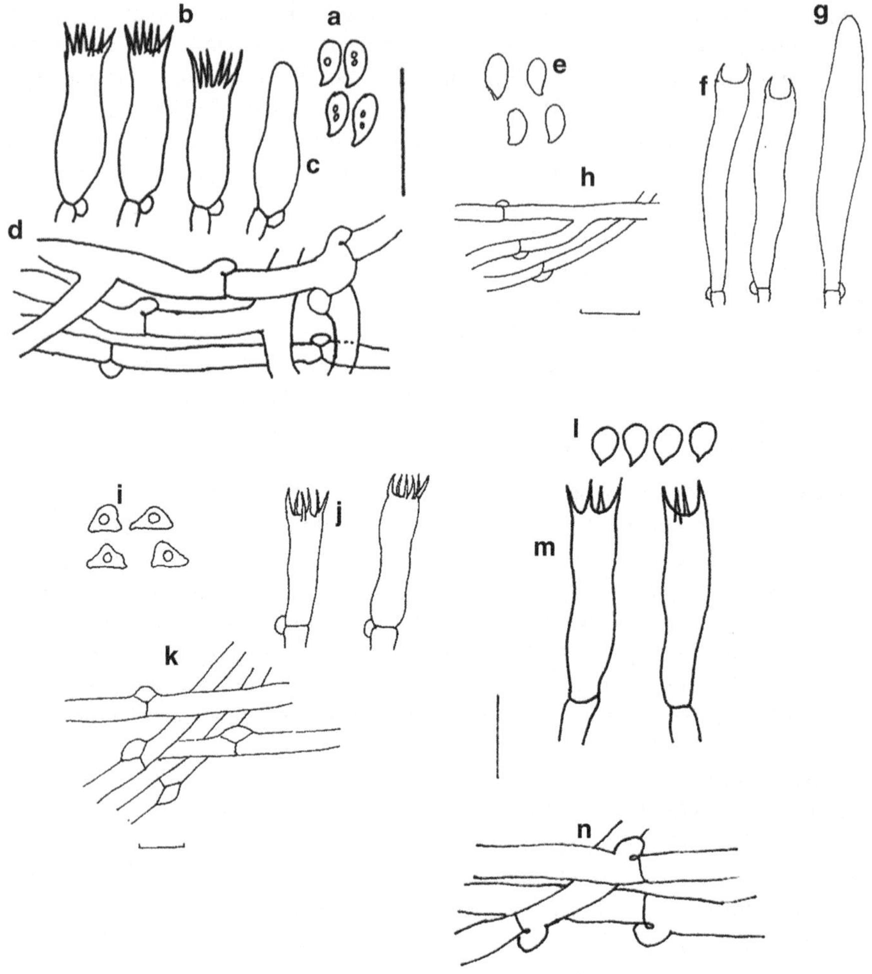

Fig. 6.14 (**a**–**d**) *Sistotrema porulosum* (**a**) Basidiospores, (**b**) Basidia, (**c**) Basidioles, (**d**) Generative hyphae; (**e**–**h**) *Sistotrema sernanderi* (**e**) Basidiospores, (**f**) Basidia, (**g**) Gloeocystidia, (**h**) Generative hyphae; (**i**–**k**) *Sistotrema subtrigonospermum* (**i**) Basidiospores, (**j**) Basidia, (**k**) Generative hyphae; (**l**–**n**) *Corticium confine* (**l**) Basidiospores, (**m**) Basidia, (**n**) Generative hyphae

Remarks: The species is marked by cylindrical to uniform basidia to narrowly ellipsoid basidiospores.

Sistotrema subtrigonospermum D.P. Rogers, Iowa St. Coll. J. Sci. **17**(1): 22 (1935). Fig. 6.14i–k

Fructification resupinate, adnate, effused, thin, just like a bloom on a host; hymenial surface almost smooth, discontinuous, greyish white, margins indeterminate.

Hyphal system monomitic; generative hyphae up to 3.0 µm wide, branched, septate, clamped, thin-walled. Basidia 14.1–23.3×4.6–5.4 µm, somewhat uniform, 6-sterigmate, with a basal clamps; sterigmata up to 6.2 µm long. Basidiospores 4.6–6.1×3.0–4.1 µm, tetrahedral, smooth, thin-walled, inamyloid, acyanophilous.

Distribution: H.P.: Kullu- Shoja, Shimla- Narkanda.

Collection examined: IBP 37560.

Substratum: On log of *Cedrus deodara* and *Pinus roxburghii.*

Remarks: *Sistotrema subtrigonospermum* is a very rarely found species, peculiar in having tetrahedral basidiospores. It is only recorded once from Himachal Pradesh.

O-Corticales
Family- Corticiaceae

Key to genera

1. Basidiocarps blue to violet.. *Pulcherricium*[a]
1. Basidiocarps creamish white to pale orange.. 2
2. Cystidial elements always absent.. 3
2. Cystidial elements present or absent.. 4
3. Basidiospore thick walled, hyaline, smooth .... *Erythricium*
3. Basidiospore thin-walled subhyaline smooth .... *Galzinia*
4. Cystidial elements present or absent,
   hyphae with or without clamps .... *Dendrothele*
4. Cystidial elements present, hyphae always with clamps 5
5. Basidiospores ellipsoid, to ovoid 6
5. Basidiospores globose to subglobose, thick-walled.. *Licrostroma*
6. Basidiospores ellipsoid, thin-walled .... *Laeticorticium*
6. Basidiospores ellipsoid to ovoid, thick walled.. *Corticium*

[a]Extra limital, Not included in this work

Corticium Pers.,
Neues Mag. Bot. 1:111, 1794.

Fructification resupinate to pileate, sessile to stipitate; hymenial surface smooth to tuberculate, hydnoid or poroid, brightly coloured. Subiculum composed of loosely or compactly arranged hyphae. Hyphal system monomitic; generative hyphae septate, clamped, swelling near the septa. Gloeocystidia or Cystidia present or absent. Basidia clavate to cylindrical, 4-spored. Basidiospores pale yellow, smooth to echinulate, nonamyloid, acyanophilous.

Twenty five species, widespread

Lit.: Eriksson & Ryvarden (Cortic. N. Eur. 4: 759, 1976; key 5 Eyr. spp.), Duhem & Michel (*Bull. Soc. Mycol. Fr.* **122**: 145, 2006; key s.str.), Larsson & Gilbertson (Norwi. Jl. Bot. 24: 99, 1977).

Type Species: *Corticium roseum* Pers. 1794

Habitat: Wood

Himalayas: One

Corticium confine Bourdot & Galzin, Bull. Soc. Mycol. Fr. 27(2): 260 (1911)=*Trechispora confinis* (Bourd. & Galz.) Liberta, Taxon 15: 318. 1966. Fig. 6.141–n

Fructification resupinate, arachnoid when young becoming submembranous on maturity, loosely adnate, often arising as small pin-head colonies which may coalesce later and extend; hymenial surface white, smooth to finely colliculose, discontinuous when young, may become continuous with age; margin thinning, fibrillose or occasionally pruinose, loosely adnate, white.

Hyphal system monomitic; hyphae 2–4.4 µm wide, branched at wide angles, septate, clamped, frequently with ampliform swellings near the septa, thin-walled, subhyaline. Cystidia absent. Basidia often occurring in clusters, 20–24.5×5–5.3 µm, short cylindrical to suburniform, 4-spored. Basidiospores 3.5–4.0×2.3–3.2 µm, broadly ellipsoid to ovoid, minutely apiculate, uniguttulate, thick-walled, subhyaline, smooth, nonamyloid.

Distribution: H.P.: Khadrala, Mahasu.

Collection examined: SSR 5652, L 37035, IBP 42039.

Substratum: On stump of *Abies pindrow*.

Remarks: The distinguishing features of this species are loose texture of the fructifications and small, broadly ellipsoid to ovoid, smooth basidiospores, short cylindrical or suburniform basidia which often occur in clusters.

Dendrothele Höhn. & Litsch.,
Sber. Akad. Wiss. Wien,
Math.-naturw. Ki., 116: 819, 1907.

Fructification resupinate, usually thin and orbicular, closely adnate, as dry crustose, smooth or with sparse papillae, white-greyish-ochraceous; hymenium comprising of basidia, cystidial organs, and numerous dendrohyphidia; hyphal system monomitic, hyphae thin-walled, profusely branched; cystidia normally few, clavate, with an apical appendix which is simple or branched as the dendrohyphidia; basidia clavate, comparably wide, with 2–4 sterigmata; dendrohyphidia numerous, profusely branched but difficult to observe as they are covered by crystals; spores ellipsoid to oblong, smooth, with somewhat thickened walls, cyanophilous, non-amyloid.

Thirty six species, widespread

Lit.: Legon et. al. (*Mycologist* **16**: 114, 2002), Nakasone (*Nova Hedwigia* **83**: 99, 2006).

Type Species: *Dendrothele papillosa* Höhn. & Litsch. 1907

Habitat: Wood

Himalayas: Three

Key to species

1. Hyphae with clamp connection .... *D. incrustans*
1. Hyphae without clamp connection 2
2. Cystidial elements present .... *D. alliacea*
2. Cystidial elements not present .... *D. seriata*

Dendrothele alliacea (Quél.) P.A. Lemke, Persoonia 3: 366, 1965 = *Corticium alliaceum* Quél., Compt. Rend. Assoc. France. Avanceum. Sci. 12: 505, 1884.
Fig. 6.15a–e

Fructification resupinate, effused, adnate; hymenial surface smooth to tuberculate, greyish white to orange white when fresh, cracked when drying; margin thinning, paler concolorous to indeterminate.

Hyphal system monomitic; generative hyphae up to 2.6 μm wide, thin-walled, clamps absent; basal hyphae parallel to the substrate; subhymenial hyphae highly branched. Dendrohyphidia appear in the hymenium having encrustations which dissolve in 3 % KOH. Cystidia 30.0–42.1 × 8–10.5 μm, fusiform, moniliform towards apex, thin-walled, basal clamp absent. Basidia 30–42.1 × 6–9 μm, clavate to subclavate, 2–4 sterigmate, basal clamp absent. Basidiospores 12.0–15.0 × 6.5–7.5 μm, broadly ellipsoid, thin to thick-walled, acyanophilous, smooth, inamyloid.

Distribution: H.P.: Dalhousie.

Collection examined: IBP 37036.

Substrate: On decaying log of angiosperm.

Remarks: This species is characterized by the presence of dendrohyphidia in the hymenium, clavate to subclavate basidia and ellipsoid basidiospores. It is differentiated from *D. acerina* by the shape of spores and thicker fructifications.

Dendrothele incrustans (P.A. Lemke) P.A. Lemke, Persoonia 3(3): 366 (1965) = *Aleurocorticium incrustans* P.A. Lemke, Can J. Bot. 42: 739, 1964. Plate 6.6d, Fig. 6.15f–i

Fructification resupinate, effused, adnate; hymenial surface smooth, cracked, greyish white when fresh, becoming white to grey later on drying; margin indeterminate.

Hyphal system monomitic; generative hyphae up to 2.3 μm wide, branched, septate, clamped, thin-walled; basal hyphae parallel to the substrate; subhymenial hyphae vertically arranged. Dendrohyphidia present, irregularly branched at the apex with encrustations and with basal clamp. Basidia 28–37 × 7–9 μm, suburniform to subclavate, 4-sterigmate, with basal clamp and oily contents; sterigmata up to 8 μm long. Basidiospores 8.5–10 × 6.5–9 μm, globose to subglobose, apiculate, smooth, thin-walled to somewhat thick-walled, inamyloid, acyanophilous, with oily contents.

Distribution: H.P: Shimla, Dalhousie.

Collection examined: IBP 37038, L 37039.

Substratum: On angiospermic twigs, cut stumps.

Remarks: It is characterized by globose basidiospores, dendrohyphidia in the hymenium, sub urniform to subclavate basidia and lack of cystidial elements.

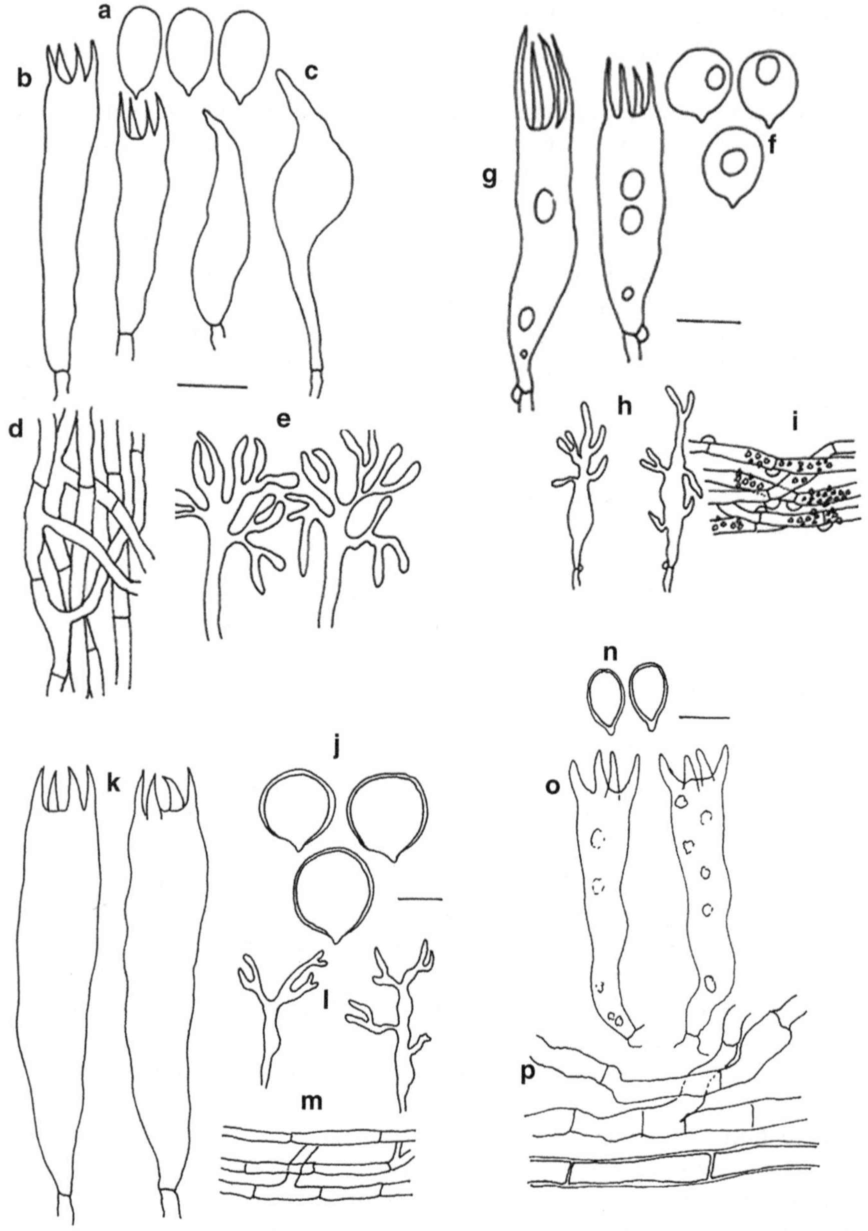

Fig. 6.15 (**a**–**e**) *Dendrothele alliacea* (**a**) Basidiospores, (**b**) Basidia, (**c**) Cystidia, (**d**) Generative hyphae, (**e**) Dendrohyphidia; (**f**–**i**) *Dendrothele incrustans* (**f**) Basidiospores, (**g**) Basidia, (**h**) Dendrohyphidia, (**i**) Generative hyphae, (**j**–**m**) *Dendrothele seriata* (**j**) Basidiospores, (**k**) Basidia, (**l**) Dendrohyphidia, (**m**) Generative hyphae; (**n**–**p**) *Erythricium laetum* (**n**) Basidiospores, (**o**) Basidia, (**p**) Generative hyphae

Dendrothele seriata (Berk. & M.A. Curtis) P.A. Lemke, Persoonia 3: 367 (1965) = *Stereum seriatum* Berk. & M.A. Curtis, J. Linn. Soc. Bot. 10:332, 1868. Plate 6.6e, Fig. 6.15j–m

Fructification resupinate, effused; hymenial surface smooth, cracked, greyish white when fresh, darkening on bruising, greyish white to pale brownish on drying; margin thinning.

Hyphal system monomitic; generative hyphae up to 3.5 μm wide, branched, septate, without clamps, thin-walled; basal hyphae parallel to the substrate; subhymenial hyphae vertical highly branched. Dendrohyphidia numerous, with irregular branches, abundant. Basidia 72.0–98.0 × 13.0–20.5 μm, clavate to subclavate, 2-4-sterigmate, without basal clamp and oily contents. Basidiospores 17.0–24.0 × 14.0–20.0 μm, globose to subglobose, apiculate, smooth, thick-walled, inamyloid, acyanophilous.

Distribution: H.P.: Shimla-Tara Devi, Narkanda, Manali.

Collection examined: IBP 37561, 42042, 42043.

Substratum: On bark of wooden log, broken branches.

Remarks: This species is marked by presence of dendrohyphidia, thick-walled basidiospores. It is being recorded for the first time from Himalayas.

Erythricium J. Erikss. & Hjortstam,
Svensk Botanisk Tidskrift, 64 (2): 165, 1970.

Fructification resupinate, effused, attached loosely to the substrate, forming membranous patches, surface reticulate-porus to smooth to slightly undulating, salmon-pink, fading to whitish when dry, margin whitish fibrillose. Consistency cottony, fibrous and soft. Hyphal system monomitic; generative hyphae septate, thin-walled, clamps absent. Cystidia absent. Basidia cylindric-clavate, sinuous, constricted, 4-sterigmate, basal clamp absent. Basidiospores ellipsoid to ovoid, with distinct apiculus, thick-walled, hyaline, smooth, cyanophilous.

Type Species: *Hyphoderma laetum* P. Karst.

Habitat: Underneath of mosses

Himalayas: One

Erythricium laetum (P. Karst.) J. Erikss. & Hjortst, Svensk Botanisk Tidskft, 64 (2: 166, 1970.) = *Corticium laetum* (Karst.) Bres., Annales Mycologici, 1(1): 94, 1903. Plate 6.6f, Fig. 6.15n–p

Fructification resupinate, effused, attached loosely to the substrate, forming membranous patches, surface reticulate-porus when young, smooth to slightly undulating, salmon-pink, fading to whitish when dry, margin whitish fibrillose. Consistency cottony, fibrous and soft.

Hyphal system monomitic; generative hyphae septate, thin-walled 4–16 μm across, clamps absent. Cystidia absent. Basidia cylindric-clavate, sinuous, constricted, 4-sterigmate, basal clamp absent, 38–50 × 9–12 μm. Basidiospores ellipsoid to ovoid, with distinct apiculus, thick-walled, hyaline, smooth, 11–13(15) × 6.5–7.5 μm, cyanophilous.

Distribution: H.P.: Shimla-Tara Devi, Narkanda, Manali
Collection examined: IBP 37038.
Substratum: On living branches of *Quercus* sp.
Remarks: The species is characterized by rhizomorphic margin, salmon pink, hymenial surface when fresh, absence of cystidia, and growing on living branches. It is a new generic record for Himalayas/India and first record from Asia.

Galzinia Bourdot,
C.R. Assoc. Fr. Avanc. Sc. 45: 577. 1922.

Fructification resupinate, membranous-ceraceous to subceraceous or wholly ceraceous; thin film like, adnate; hymenial surface smooth, variously coloured. Hyphal system monomitic, generative hyphae thin-walled, clamps present. Gloeocystidia or cystidia present or absent. Basidia cylindrical to urniform, proliferating, 4-spored. Basidiospores ellipsoid to cylindrical, subhyaline, smooth, nonamyloid, acyanophilous.
Nine species, widespread
Lit.: Rogers (Mycol. **36**: **70**, 1944), Boidin & Gills (*BSMF* **110**: **185**, 1994; key)
Type Species: *Galzinia pedicellata* Bourd. 1922
Habitat: Dead wood
Himalayas: Two
Key to species

1. Basidia clavate cylindrical, Basidiospores ellipsoid .... *G. ellipsospora*
1. Basidia clavate to suburniform, Basidiospores allantoid to suballantoid .... *G. incrustans*

Galzinia ellipsospora S.S. Rattan, Biblthca Mycol. **60**: 212 (1977). Fig. 6.16a–c
Fructification resupinate, membranous to ceraceous, adnate, often arising as small circular colonies which may coalesce later and become effused; hymenial surface cream brown to violaceous brown, smooth but appears farinose under the lens, not creviced; margin pruinose to fibrillose, adnate, white to paler concolorous. Subiculum composed of compactly arranged, somewhat agglutinated hyphae.
Hyphal system monomitic, hyphae 2–3 μm wide, branched, septate, clamped, thin-walled, subhyaline. Gloeocystidia or cystidia absent. Basidia15–20 × 3.4–4.0 μm, clavate-cylindrical, 4-spored. Basidiospores 3–4 × 1.6–2.1 μm, ellipsoid, minutely apiculate, thin-walled, subhyaline, smooth, nonamyloid.
Distribution: H.P: Shimla.
Collection examined: SSR 5550, IBP 37041, 42046.
Substratum: On stump, angiospermic cut logs.
Remarks: The chief features of this species are comparatively thicker fructification, cream brown to violaceous brown hymenial surface and shape and size of basidiospores. *Galzinia incrustans* (Höhn. & Litsch.) Parm., is very similar but possesses thinner fructification and slightly longer and allantoids basidiospores.

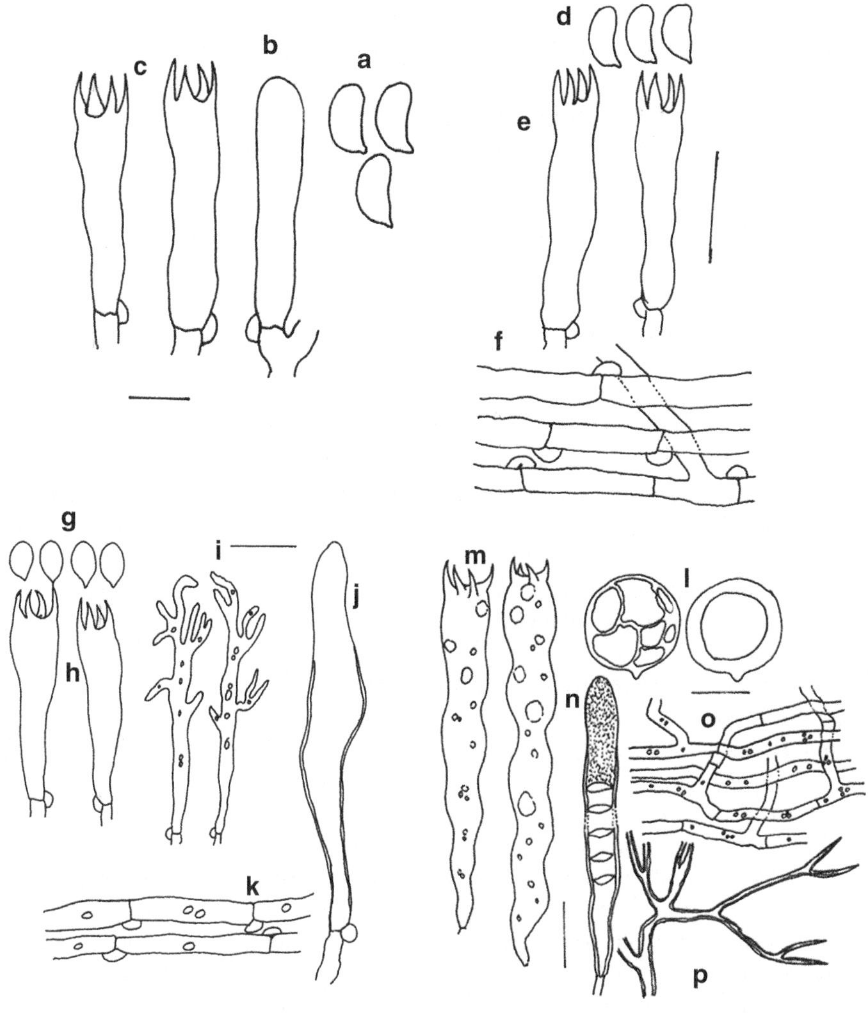

Fig. 6.16 (**a**–**c**) *Galzinia ellipsospora* (**a**) Basidiospores, (**b**) Basidioles, (**c**) Basidia; (**d**–**f**) *Galzinia incrustans* (**d**) Basidiospores, (**e**) Basidia, (**f**) Generative hyphae; (**g**–**k**) *Laeticorticium expallens* (**g**) Basidiospores, (**h**) Basidia, (**i**) Dendrohyphidia, (**j**) Gloeocystidia, (**k**) Generative hyphae; (**l**–**p**) *Licrostroma subgiganteum* (**l**) Basidiospores, (**m**) Basidia, (**n**) Cystidia, (**o**) Generative hyphae, (**p**) Dendrohyphidia

Galzinia incrustans (Höhn & Litsch.) Parmasto, Eesti NSV Tead. Akad. Toim., Biol. 14: 225, 1965 = *Corticium incrustans* Höhn & Litsch., Sber. Akad. Wiss. Wien, Math.- *Naturw. Kl. Abt.* 1:115: 1602. Fig. 6.16d–f

Fructification resupinate, effused, adnate; hymenial surface smooth , greyish white to orange white when fresh, orange white to brownish orange on drying; margin thinning, paler concolorous to indeterminate.

Hyphal system monomitic; generative hyphae up to 5.1 µm wide. Clamps present, septate, thin-walled; basal hyphae less branched; subhymenial hyphae vertical and highly branched. Cystidia absent. Basidia 9–21 × 3.8–4.6 µm, clavate to suburniform, 4-sterigmate with basal clamp. Basidiospores 4.5–6.0 × 2.6–3.5 µm, allantoids to suballantoid, thin-walled, apiculate, acyanophilous, smooth, inamyloid.

Distribution: H.P.: Dalhousie, Shimla.

Collection examined: IBP 37042, 42052.

Substratum: On bark and twigs of *Cedrus deodara* and *Picea* sp.

Remarks: The species is marked by clavate to suburniform constricted basidia and ellipsoid to allantoids basidiospores. It is being reported from Himalayas for the first time.

Laeticorticium Donk,
Fungus, **26**(1–4): 16 (1956)

Fructification resupinate, adnate, effused; hymenial surface smooth to tuberculate; hyphal system monomitic; generative hyphae septate, clamped; dendrohyphidia present in the hymenium. Cystidia present. Basidia clavate to subclavate, 4-sterigmate with basal clamp. Basidiospores broadly ellipsoid, apiculate, thin-walled, acyanophilous, inamyloid.

Wide spread

Type Species: *Corticium roseum* Persoon, 1794.

Habitat: Wood

Himalayas: One

Laeticorticium expallens (Bres.) Hjortstam, *Windahlia* **17**: 56 (1987) = *Corticium expallens* Bres., Annls mycol. 6:43, 1908. Plate 6.7a, Fig. 6.16g–k

Fructification resupinate, adnate, effused; hymenial surface smooth to tuberculate, orange white to brownish orange when fresh, pale orange to greyish orange to brownish orange on drying; margin thinning, whitish, paler concolorous to indeterminate.

Hyphal system monomitic; generative hyphae up to 4.0 µm wide, clamped, septate; basal hyphae parallel to the substrate, thick-walled; subhymenial hyphae thin-walled, highly branched. Dendrohyphidia present, highly branched with oil globules. Cystidia 50–64 × 7.5–9.0 µm, cylindrical, thin-walled at apex, smooth, having basal clamp. Basidia 20–34 × 5–7 µm, clavate to subclavate, 4-sterigmate with basal clamp. Basidiospores 5–7 × 3.5–4 µm, broadly ellipsoid, apiculate, thin-walled, acyanophilous, inamyloid.

Distribution: H.P.: Shimla-Tara Devi, Manali.

Collection examined: IBP 37044.

Substratum: Fallen twigs, cut stump.

Remarks: The species is characterized by the presence of dendrohyphidia, cylindrical cystidia, clavate to subclavate and broadly ellipsoid basidiospores and clavate cylindrical basidia with basal clamp.

Plate 6.7 (**a**) *Laeticorticium expallens*. (**b**) *Gloeophyllum abietinum*. (**c**) *Gloeophyllum carbonarium*. (**d**) *Gloeophyllum sepiarium*. (**e**) *Gloeophyllum subferrugineum*. (**f**) *Gloeophyllum striatum*

Licrostroma Lemke

Can. Jour. Bot. 42: 762, 1964.

Fructification resupinate, adnate, effused; hymenial surface smooth, continuous, cream-coloured to alutaceous on drying; hyphal system dimitic, generative hyphae septate, branched, without clamps, thin-walled to thick-walled; skeletal hyphae thick-walled, sparsely to rich branched, non dextrinoid. Pseudocystidia present, clavate to subclavate, flexuous. Basidia long, cylindric to narrowly clavate. Basidiospores large, globose to subglobose, smooth, nonamyloid.

One Species, North America, India.

Lit.: Lemke (CJB **42**: 762, 1964).

Type species: *Corticium subgiganteum* Berk. 1873.

Habitat: Wood

Himalayas: One

Licrostroma subgiganteum (Berk.) P.A. Lemke, Can. J. Bot. 42: 763 (1964).
Fig. 6.16l–p

Fructification initially small disciform colonies, which fuse together or become effused to form larger resupinate to substereoid patches, adnate; hymenial surface smooth to somewhat tuberculate, yellowish-white to light ochraceous when fresh, ochraceous later on; margin thinning, rarely inturned when fresh to always in turned on drying. Texture compact-membranous to ceraceous.

Hyphal system dimitic; generative hyphae up to 3 μm wide, branched, septate, without clamps, thin-walled, sometimes with oily contents; skeletal hyphae up to 4.0 μm wide, thick-walled. Pseudocystidia 128–350 × 14.0–22.0 μm, subcylindrical, flexuous, with rounded apices, thick-walled in the basal half, gradually thinning above, with oil contents giving positive reaction with sulfovanilline, secondary septa one to many. Basidia 102.0–168 × 15–22 μm, narrowly clavate, sinuous, thin- to thick-walled especially in the basal half, 4-sterigmate, normally without basal clamp. Basidiospores 16.0–22 × 16.0–20.0 μm, mostly globose, smooth, thick-walled, nonamyloid, acyanophilous.

Distribution: Arunachal Pradesh: West Kameng, Bomdila, Jamiri, about 6 km from Jamiri towards Nachephu; Bhutan: Thimphu, Bunakha.

Collection examined: GSD 19868, 19520.

Substratum: On a decaying angiospermic branches.

Remarks: This species was first described by Berkeley (1873) as *Corticium subgiganteum*. Massee (1889) shifted it to *Peniophora*. Hoehnel (1912) placed it under *Aleurodiscus*. Lemke (1964) erected the genus *Licrostroma* to accommodate this species. He considered *Aleurodiscus orientalis* Lloyd, and *A. reflexus* Yasuda, as its synonyms. The species has been recorded from Eastern and N.W. Himalayas. The specimens resembles *L. subgiganteum* as described by Lemke (1964) in most of the characters, except for somewhat larger size of pseudocystidia (128–350 × 14.0–22.0 μm, as compared to 100–250 × (8-)12-18(−23) μm) and basidia (102–168 × 15–22 μm in comparison to 70–100 × 13–15(−18) μm).

O- Gloeophyllales
Family- Gloeophyllaceae
Gloeophyllum Karst.

Bindr. Känn. Fiinl. Nat. Folk 37:79, 1882.

Fructification annual to perennial, resupinate to pileate, broadly attached, tough to woody; upper surface deep brown to greyish with age, glabrous; hymenophore poroid, daedaloid or lamellate; hyphal system dimitic to trimitic; generative hyphae septate, branched clamped; skeletal hyphae yellowish brown, sparsely branched; binding hyphae rare. Cystidia or cystidioles present or absent. Basidia clavate, 4-spored. Basidiospores smooth, cylindric, ellipsoid, thin-walled, negative in Melzer's reagent.

Thirteen species, widespread

Lit.: David & Fiasson (*Bull. Mens. Soc. Linn.* Lyon **46**: 304, 1977)

Habitat: Decayed Wood

Type Species: *Agaricus sepiaria* Wulfen, 1786

Himalaya: Five

Key to species

1. Fructification always on charred wood, hymenophore poroid .... *G. carbonarium*
1. Hymenophore lamellate to distinctly daedaleoid or with few pores; not on charred woods 2
2. Fructification light to dark reddish brown; spores 6–9 µm long; occurring on hardwoods 3
2. Fructification light yellowish to rusty brown; spores 6–13 µm long; occurring on conifers 4
3. Fructification fan shaped with a contracted base; glabrous; hymenophore breaking to dentate/hydnoid .... *G. striatum*
3. Fructification broadly attached; with scrupose protuberances; hymenophore .... *G. trabeum*[a]
4. Fructification yellowish brown, smooth, azonate; hymenophore distinctly lamellate; spores 6–8 µm long .... *G. subferrugineum*
4. Fructification pale to dark rusty brown, zonate, smooth or with coarse protuberances; spores 7–13 µm long 5
5. Fructification dark rusty brown, margin distinctly yellowish; hymenophore lamellate with few daedaleoid pores; lamellae 2–4 per mm .... *G. sepiarium*
5. Basidiocarps evenly brownish, pale with age; hymenophore with irregular to wavy lamellae; lamellae 1 per mm .... *G. abietinum*

[a]Extra limital not included in the text

Gloeophyllum abietinum (Bull.) Karst. P.A., Bidr. Känn. Finl. Nat. Folk 37: 80 (1882). Plate 6.7b, Fig. 6.17a–e

Fructifications annual to perennial, sessile, pileate, coriaceous when fresh, hard on drying, broadly attached, single or imbricate. Pileus dimidiate to broadly attached; upper surface flat, brown to greyish brown, tomentose, hyphae agglutinated to form small tufts, concentrically zonate; margin acute, wavy, concolorous or slightly darker. Pore surface dark brown, lamellate, lamellae 1 per mm. Context brown, soft coriaceous, homogenous, xanthochroic.

Hyphal system trimitic; generative hyphae hyaline, thin-walled, branched, septate, clamped, 2–4.5 µm in diameter; skeletal hyphae thick-walled, pale brown to rusty brown, sparsely branched, dominate the context and dissepiments, aseptate, 4–5.5 µm in diameter; binding hyphae subhyaline, thick-walled, much branched, few, up to 2.4 µm in diameter. Cystidia smooth, at first hyaline, later rusty brown, thick-walled, subulate, some apically encrusted, 20–40 × 4.4 µm. Basidiospores hyaline, thin-walled, smooth, cylindrical and non-amyloid, 7.8–11 × 3–3.4 µm.

Distribution: Bhutan-Thimphu, Dochula; A.P.: Bomdilla.

Collection examined: SSV 21391, 21414.

Substratum: On decaying coniferous log.

Remarks: Bagchee and Singh (1960) reported this species on *Pinus smithiana* and *Taxus baccata* from the N. W. Himalayas. The above mentioned collections are quite typical of the species and fit well in the description given by Thind and Dhanda (1980). The species is marked by annual, thin, flexible, dark greyish brown fructifications with soft tomentose upper surface; lamellate pore surface; few, smooth or apically encrusted, subulate cystidia; and hyaline smooth, cylindrical basidiospores.

Gloeophyllum carbonarium (Berk. & M.A. Curt.) Ryv., Mycotaxon 20(2): 334, 1984. Plate 6.7c, Fig. 6.17f–j

Fructification annual, resupinate, soft, flexible; upper surface grey; margin white, thin. Context brown, thin, fibrous; hymenial surface brown; pores angular.

Hyphal system trimitic; generative hyphae hyaline, thin-walled branched, septate, clamped, 2.4–3.0 µm in diameter; skeletal hyphae subhyaline, thick-walled, unbranched, aseptate 3.6–4.3 µm in diameter; binding hyphae subhyaline, thick-walled, branched 3.2–4.1 µm in diameter. Basidia clavate, 14.3–16.1 × 4.4–5.1 µm. Basidiospores hyaline, cylindrical 5.9–7.1 × 3.3–3.9 µm.

Distribution: H.P.: Kullu, Shimla.

Collection examined: IBP 37562.

Substratum: On decaying angiospermic log.

Remarks: This species is characterized by annual, soft, flexible fructification; trimitic hyphal system; clavate basidia and cylindrical hyaline basidiospores.

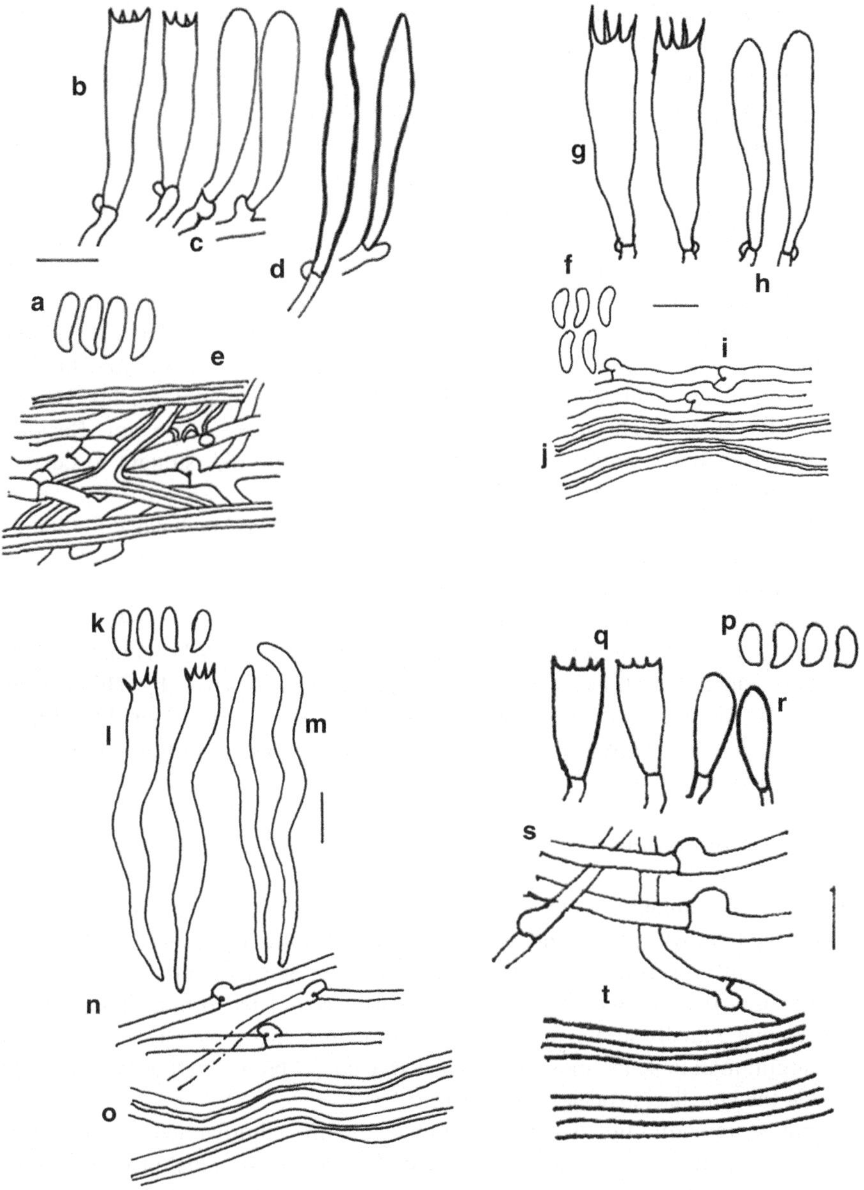

Fig. 6.17 (**a**–**e**) *Gloeophyllum abietinum* (**a**) Basidiospores, (**b**) Basidia, (**c**) Cystidia, (**d**) Gloeocystidia, (**e**) Generative hyphae and Skeletal hyphae; (**f**–**j**) *Gloeophyllum carbonarium* (**f**) Basidiospores, (**g**) Basidia, (**h**) Cystidia, (**i**) Generative hyphae, (**j**) Skeletal hyphae; (**k**–**o**) *Gloeophyllum sepiarium* (**k**) Basidiospores, (**l**) Basidia, (**m**) Cystidia, (**n**) Generative hyphae, (**o**) Skeletal hyphae; (**p**–**t**) *Gloeophyllum subferrugineum* (**p**) Basidiospores, (**q**) Basidia, (**r**) Basidioles, (**s**) Generative hyphae, (**t**) Skeletal hyphae

Gloeophyllum sepiarium (Wulfen) P. Karst., [as '*Gleophyllum*'] Bidr. Känn. Finl. Nat. Folk 37: 79 1882. Plate 6.7d, Fig. 6.17k–o

Fructification annual, sessile, solitary to imbricate, coriaceous when fresh, becoming firm and rigid on drying. Pileus, sessile, dimidiate, sometimes flabelliform, applanate; upper surface sepia coloured to snuff-brown, sometimes brownish yellow, creamish brown on drying, glabrous near the margin, lamellae 2–4 per mm; margin acute to blunt, concolorous or paler than the upper surface, entire. Pore surface lamellate, brown to dark-brown, straight or wavy, unbranched. Context brownish yellow to brown, homogenous, zonate, soft.

Hyphal system trimitic; generative hyphae hyaline, thin-walled, septate, branched, clamped, acyanophilous, 2.4–4.1 µm. in diameter; skeletal hyphae pale brown, thick-walled, aseptate, long, unbranched, acyanophilous, 2.8–4.5 µm; binding hyphae scanty, subhyaline, thick-walled, branched, acyanophilous, 2.0–3.0 µm. Basidia hyaline, clavate, 4-spored, about, 20.1–25.1 × 5.1–5.6 µm. Basidiospores, hyaline, thin-walled, smooth, cylindrical-ellipsoid, straight to curved, non-amyloid, 7.0–11.8 × 2.8–4.2 µm.

Distribution: U.K.: Mussoorie, Dehra Dun; A.P.- Bomdila; Meghalaya: Shillong.

Collection examined: SSV 21431, 21537, IBP 37045, 37046.

Substratum: On decaying deciduous log.

Remarks: The species was recorded from India as *Lenzites sepiaria* (Wulf.) Fr. by Bose (1928) from Shillong. It is quite common species in N.W. Himalaya. The above cited collection resembles with the description given by Thind et al. (1957). The species is characterized by sepia to snuff-brown coloured fructification with lamellate hymenophore, heavily encrusted, large cystidia, straight and curved basidiospores.

Gloeophyllum striatum (Fr.) Murrill, Bull. Torrey bot. Cl. 32(7): 370 (1905). Plate 6.7f, Fig. 6.18a–e

Fructifications annual, sessile, attached by broad or narrow lateral base, thin and flexible both when fresh and dry, solitary to imbricate. Pileus sessile, dimidiate or flabelliform, sometimes applanate; upper surface light brown to reddish-brown, or rusty-brown, soft tomentose near margin, later forming compact and even mat, radially striate, mostly zonate, sometimes inconspicuously zonate near margin; margin acute, mostly unbranched. Context brown, soft above, compact near the lamellae, azonate, xanthochroic.

Hyphal system dimitic; generative hyphae hyaline, thin-walled, occasionally thick-walled, septate, clamped, branched, contents staining in cotton blue, 2–4 µm in diameter; skeletal hyphae pale brown, thick-walled to almost solid, long, aseptate, rarely branched, 2–4.5 µm in diameter. Cystidia absent. Basidia hyaline, clavate or suburniform, 2–4 spored, 14–20 × 5.5–6.2 µm. Basidiospores hyaline, thin-walled, smooth, cylindric, slightly curved, non-amyloid, 7–11.2 × 2.7–3.4 µm.

Distribution: Assam: Naga Hills; Meghalaya: Shillong.

Collection examined: Dhanda 6936, SSR 6406.

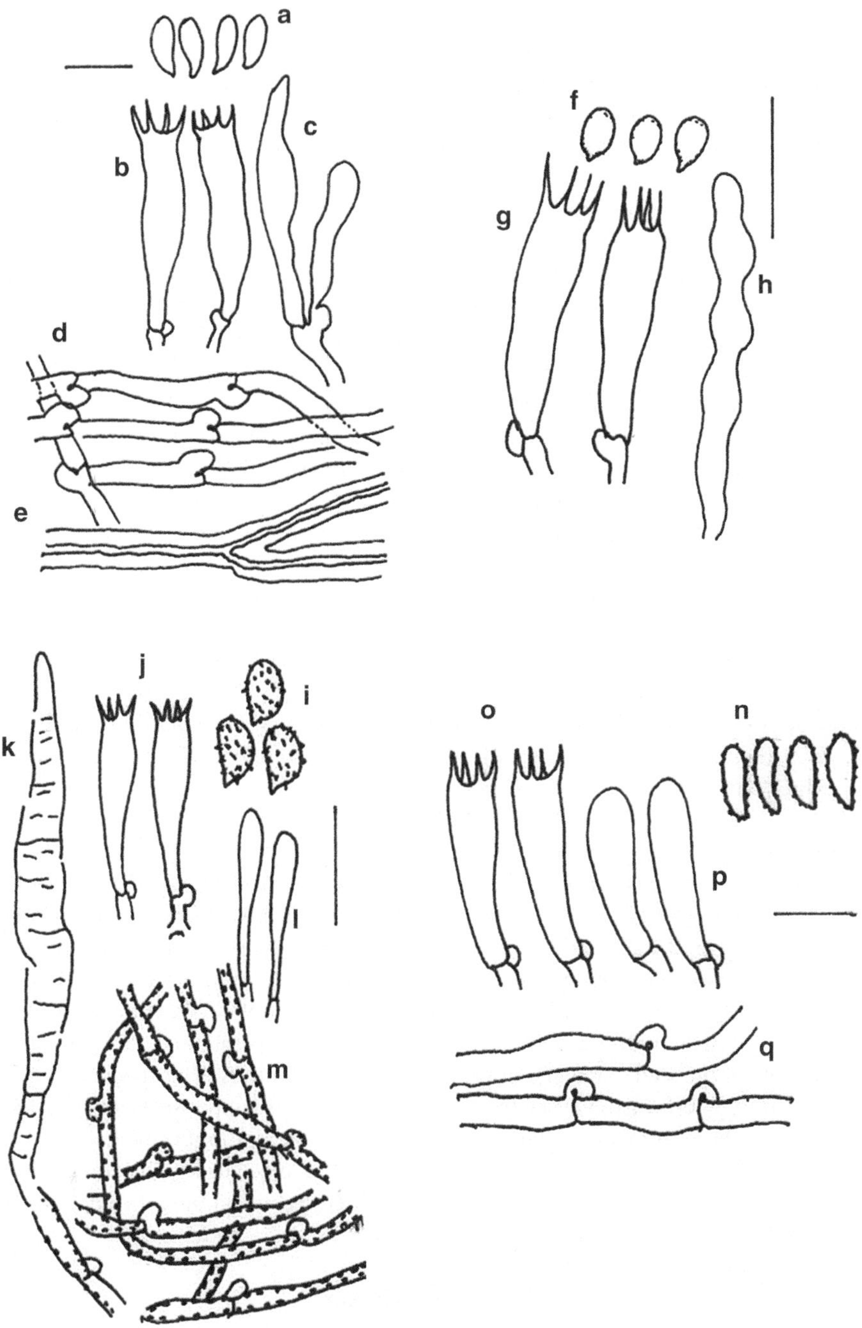

Fig. 6.18 (**a–e**) *Gloeophyllum striatum* (**a**) Basidiospores, (**b**) Basidia, (**c**) Cystidia, (**d**) Generative hyphae, (**e**) Skeletal hyphae; (**f–h**) *Dentipellis leptodon* (**f**) Basidiospores, (**g**) Basidia, (**h**) Gloeocystidia; (**i–m**) *Laxitextum bicolor* (**i**) Basidiospores, (**j**) Basidia, (**k**) Gloeocystidia, (**l**) Basidioles, (**m**) Generative hyphae; (**n–q**) *Kavinia alboviridis* (**n**) Basidiospores, (**o**) Basidia, (**p**) Basidioles, (**q**) Generative hyphae

Substratum: On decaying *Cedrus deodara, Pinus longifolia, Shorea robusta* log.

Remarks: The species is characterized by sessile, thin, solitary to imbricate fructifications with light brown to reddish-brown, compactly tomentose upper surface; lamellate hymenophore with closely packed lamellae; brown, xanthochroic, duplex context; and cylindric, slightly curved basidiospores.

Gloeophyllum subferrugineum (Berk.) Bond. & Sing. Annals Mycol. 39(1): 64, (1941). Plate 6.7e, Fig. 6.17p–t

Fructifications annual, sessile, pileate, solitary to imbricate, coriaceous when fresh, tough and rigid on drying, applanate, flabelliform to dimidiate, broadly attached; upper surface yellowish brown to golden brown, rusty brown with aging, tomentose, uneven, azonate to faintly concentrically zonate; margin concolorous with upper surface, entire, acute to obtuse, when acute slightly reflexed on drying. Pore surface pale brown to smoky brown, lamellate, lamellae thin and undulating, usually branched up to 10–12 per cm. Context yellowish brown to brown, fibrous, homogenous, xanthochroic.

Hyphal system dimitic; generative hyphae hyaline, thin-walled to slightly thick-walled, branched, septate, acyanophilous, 2–4 µm in diameter; skeletal hyphae brown, thick-walled to solid, aseptate, unbranched, 2.5–4.5 µm in diameter. Basidia hyaline, thin-walled, cylindric-clavate, up to 5.5 µm in diameter. Cystidia absent. Basidiospores hyaline, thin-walled, smooth, ellipsoid, non-amyloid, 6.8–8 × 2.5–3.5 µm.

Distribution: Bhutan-Chimakothi, Thimphu, Chankphug, Dochula, Begana, Paro, D' dzong, dochula to Wangdi ; A.P.-West Kameng-Bomdilla, Wang Basti, Twang, Three Km from to Dirang; U.K.: NDBR, Mussoorie.

Collection examined: SSV 21250, 21289, IBP 37047, 42054, L 37048.

Substratum: On decaying log of *Pinus roxburghii.*

Remarks: This species was first described from India by Berkeley (1854a) as *Lenzites subferruginea* Berk. on the basis of Dr. Hooker's collection from Khasi hills (Meghalaya). Bakshi (1971) also described it as *L. subferruginea* and mentioned its occurrence as common, usually on the logs of conifers in the temperate Himalayas. Dhanda (1977) collected it from different parts of N. W. Himalayas. The species is widely distributed and common in the temperate and sub temperate regions of the study area and always occur on the conifers. The species is characterized by brownish yellow fructifications; tomentose upper surface; lamellate hymenophore; brownish yellow, xanthochroic context; dimitic hyphal system; and hyaline, cylindric ellipsoid basidiospores. It is close to *S. sepiarium.* However, the latter differs in having 'sepia' coloured fructifications; trimitic hyphal system and encrusted subulate cystidia.

Family-Hericiaceae

Key to genera

1. Fructifications resupinate or rarely effused-reflexed............................*Dentipellis*
1. Fructification effused-reflexed to pileate ....*Laxitextum*

Dentipellis Donk,
Persoonia 2(2): 232. 1962.

Fructifications resupinate or rarely effused-reflexed, membranous; hymenial surface toothed. Subiculum subhyaline in section. Hyphal system monomitic, hyphae thin-walled, septate, clamped. Cystidia absent. Gloeocystidia present. Basidia clavate to cylindrical, 4-spored. Basidiospores globose to ellipsoid, amyloid, smooth to finely rough.

Four Species, widespread

Lit.: Ginns (*Windahlia* 16: 35, 1986)

Type Species: *Hydnum fragile* Pers. 1801.

Habitat: Dead wood

Himalayas: One

Dentipellis leptodon (Mont.) Maas Geest., Persoonia 7(4): 558 (1974) = *Dentipellis subseparaous* Khara & Rattan, Biblthca. Mycol. 60: 100, 1977. Fig. 6.18f–h

Fructification resupinate, membranous, somewhat fragile, separable, often arising as small circular colonies; hymenial surface white when fresh turning pale yellow to light yellowish brown on drying, toothed; margin thinning, separable, white to paler concolorous.

Hyphal system monomitic, hyphae 2–5 µm wide, branched, septate, clamped, thin-walled, subhyaline, distinct but more often collapsing with age, subhyaline. Gloeocystidia 3–5.6 µm broad, hypoid or subfusiform to cylindrical, sometimes beaded in appearance due to unequal inflations, thin-walled, immersed or rarely projecting slightly out of the hymenium. Basidia 11.8–15.8 × 3–4 µm, clavate-cylindrical, 4-spored. Basidiospores 2.9–4.4 × 2–3 µm, broadly ellipsoid to ovoid, minutely apiculate, thin-walled, smooth, amyloid.

Distribution: H.P.: Mahasu- Narkanda, Dharanghati.

Collection examined: HSK: 4029, 4034.

Substratum: Log under conifers and on bark of fallen logs.

Remarks: This species is quite distinctive because of basidiospores which are small in size with slightly roughened walls. *Dentipellis subseparans* (Peck) Donk is very similar but possesses larger basidiospores.

Laxitextum Lentz,
U.S. Dept. Agric. Monogr. **24**: 18 (1956) [1955]

Fructification effused-reflexed to pileate, spongy-fragile. Context devoid cuticle on the abhymenial side, composed of loosely woven hyphae. Hyphal system monomitic, subhyaline to tinted brown; generative hyphae subhyaline to tinted brown, thin-walled, clamped, septate. Cystidia absent. Gloeocystidia abundant. Basidia clavate, 4-spored. Basidiospores ellipsoid, subhyaline, finely echinulate, amyloid.

Three species, widespread

Lit.: Lentz (*Sydowia* 14: 123, 1960), Ginns & Freeman (*Biblthca Mycol.* 157, 1994).

Type species: *Thelephora bicolor* Pers. 1801

Habitat: Decay wood
Himalayas: One

Laxitextum bicolor (Pers.) Lentz, U.S. D.A. Agri. Monogr. 24: 19 (1955) = *Thelephora bicolor* Pers., Syn. Meth. Fung.: 568. 1801. Plate 6.8a, Fig. 6.18i–m

Fructification resupinate, effused-reflexed to pileate, soft, spongy when fresh becoming somewhat brittle and fragile on drying; pileus mostly solitary, flabelliform to conchate; upper surface brown; hymenial surface white in fresh specimen; margin acute, concolorous, rolling inward on drying. Subiculum brown, soft, spongy, composed of loosely woven hyphae.

Hyphal system monomitic; generative hyphae subhyaline to tinted brown, 2–4.5 µm branched, septate, clamped, encrusted septate, clamped, thin-walled; gloeocystidia numerous, thin-walled, 65–150 × 6–11 µm, fusiform to subcylindrical, empty or with subhyaline contents, thin-walled, immersed or projecting out of the hymenium. Basidia clavate, 4-sterigmate, 15–18 × 3–4 µm with basal clamp. Basidiospores globose to subglobose, subhyaline, finely echinulate, amyloid, 4.5–5 × 2.2–2.6 µm.

Distribution: H.P.: Dalhousie- Dunera; U.K.: Mussoorie- Kempty fall.

Collection examined: SSR: 5031, 19025, IBP 37049, L 37050.

Substratum: Fallen twigs under angiospermic forest.

Remarks: This species was first described by Banerjee (1935a, b) from Calcutta. Later, Thind and Rattan (1968) and Rattan (1977) reported it from the N.W. Himalayas. This species is characterized by effused-reflexed to pileate fructification, generative hyphae with clamps, presence of gloeocystidia, thin-walled, subulate cystidia, clavate, 4-sterigmate basidia and finally echinulate, amyloid basidiospores.

O- Gomphales

Family- Lentariaceae

Kavinia Pilát,
Stud. Bot. Čechslov. 1: 3. 1938.

Fructification resupinate, soft-fragile, widely effused; hymenial surface hydnoid, cream to reddish brown; margin thinning, white fibrillose. Subiculum subhyaline in section. Hyphal system monomitic, hyphae septate, clamped, teeth subulate, terrete. Cystidia and Gloeocystidia absent. Basidia clavate-cylindrical, 4-spored. Basidiospores ellipsoid, the walls minutely echinulate, cyanophilous, non-amyloid.

Five Species, widespread

Lit.: Eriksson & Ryvarden (Cortic. N. Europe. 4: 752, 1976), Legon (Mycologist 17: 42, 2003; UK)

Type Species: *Caldesiella sajanensis* Pilát 1936
Habitat: Dead wood
Himalaya: One

Plate 6.8 (**a**) *Laxitextum bilcolor.* (**b**) *Kavinia alboviridis.* (**c**) *Fomitiporia robusta.* (**d**) *Fuscoporia contigua*

Kavinia alboviridis (Morgan) Gilb. & Budington, J. Ariz. Acad. Sci. 6(2): 95 (1970) Plate 6.8b, Fig. 6.18n–q

Fructification resupinate, membranous, loosely adnate to separable, widely effused; hymenial surface hydnoid, cream when young turning cream brown to brown at maturity; margin thin, fibrilose to rhizomorphic, loosely adnate, white. Rhizomorphs present abundantly, white, branched. Teeth 5 mm long and 150 µm wide. Subiculum subhyaline, up to 300 µm thick in section, comprising of compactly arranged hyphae which are gelatinized.

Hyphal system monomitic, hyphae 1.9–8.4 μm wide, septate, thin-walled, clamps present, hyphae becomes inflated. Basidia 19.8–24.8 × 5–7 μm, clavate-cylindrical, 4-spored. Basidiospores 7.4–10 × 2.4–3.2 μm, ellipsoid thick-walled, smooth to slightly echinulate, brown, cyanophilous, non-amyloid.

Distribution: H.P.: Kullu, Manali; J&K: Bhadrwah.

Collection examination: HSK 4143, IBP 42049.

Substratum: On coniferous log.

Remarks: The species is widely distributed in the N.W. Himalaya and is easily identified as resupinate fructification with typical basidiospores.

O- Hymenochaetales
Family- Hymenochaetaceae

Key to genera:

1. Hymenial setae always present .... *Hymenochaete*
1. Hymenial setae present/absent 2
2. Hyphal system dimitic 3
2. Hyphal system monomitic 7
3. Basidiospores dextrinoid, strongly cyanophilous .... *Fomitiporia*
3. Basidiospores inamyloid, weakly cyanophilous 4
4. Generative hyphae usually encrusted .... *Fuscoporia*
4. Generative hyphae usually not encrusted 5
5. Context homogenous .... *Phellinus*
5. Context duplex 6
6. Setae absent, basidiospores faintly coloured .... *Phylloporia*
6. Setae arising from trama, basidiospores brownish .... *Porodaedalea*
7. Cystidia thick–walled with a capillary lumen, walls sub–hyaline ..... *Tubulicrinis*
7. Cystidia absent 8
8. Context homogenous .... *Inonotus*
8. Context duplex 9
9. Basidiospores more than 5 μm long .... *Onnia*
9. Basidiospores less than 5 μm long .... *Aurificaria*

Aurificaria D.A. Reid, Kew Bull 17(2): 278 (1963)

Fructification annual, sessile or substipitate, coriaceous when fresh, brittle on drying. Upper surface yellowish brown, tomentose near margin. Context golden-brown, homogenous. Hyphal system monomitic; generative hyphae pale brown to brown, septate, thin-walled, clamps absent. Basidia hyaline, clavate, 2–4 spored. Basidiospores pale brown to dark brown, broadly ellipsoid to subglobose.

Monotypic, widespread

Lit.: Reid (Kew Bull. 17: 278, 1963).

Type Species: *Polyporus indicus* Massee 1910.

Habitat: Dead wood

Himalayas: One

Aurificaria indica (Massee) D.A. Reid, Kew Bull. 17(2): 279 (1963). Fig. 6.19a–c
Fructification substipitate, solitary, coriaceous when fresh, becoming hard and somewhat brittle on drying. Pileus dimidiate to ligulate; upper surface yellowish brown, radially and irregularly striate, with narrow chestnut or dark brown and broad yellowish brown concentric zone; cuticle absent; hymenial surface light chocolate brown. Context yellowish brown, cinnamon in dried specimens, blackening in KOH solution, zoned, up to 1.5 cm in thickness.

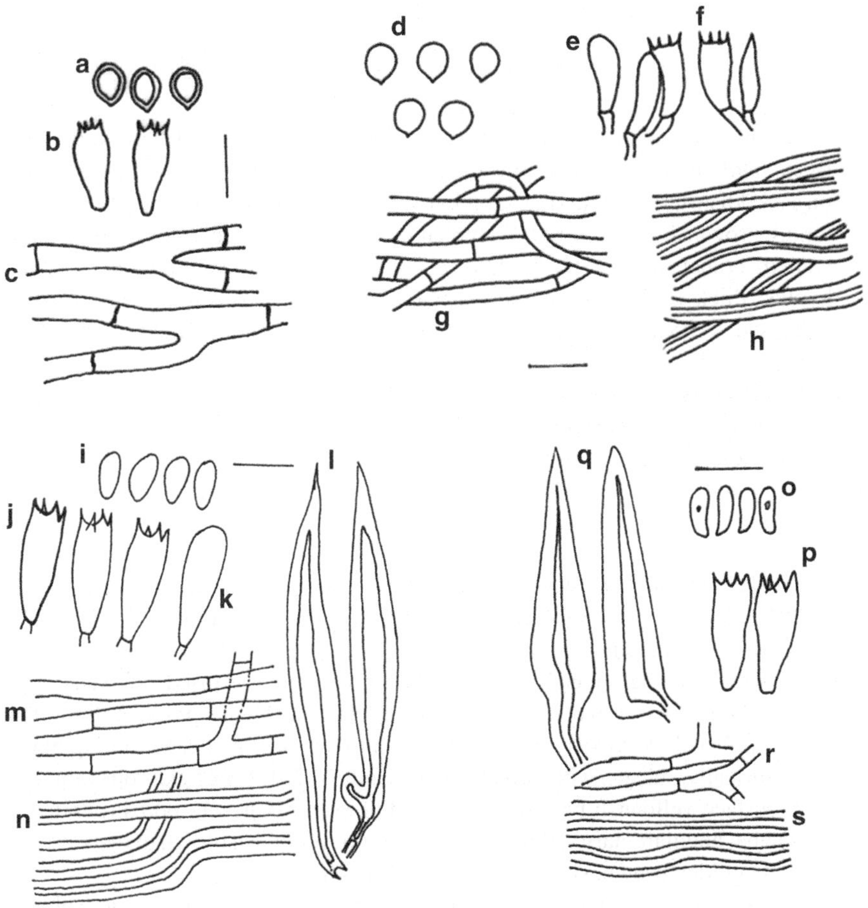

Fig. 6.19 (**a–c**) *Aurificaria indica* (**a**) Basidiospores, (**b**) Basidia, (**c**) Generative hyphae; (**d–h**) *Fomitiporia robusta* (**d**) Basidiospores, (**e**) Basidioles, (**f**) Basidia, (**g**) Generative hyphae, (**h**) Skeletal hyphae; (**i–n**) *Fuscoporia contigua* (**i**) Basidiospores, (**j**) Basidia, (**k**) Basidioles, (**l**) Setae, (**m**) Generative hyphae, (**n**) Skeletal hyphae; (**o–s**) *Fuscoporia ferrea* (**o**) Basidiospores, (**p**) Basidia, (**q**) Setae, (**r**) Generative hyphae, (**s**) Skeletal hyphae

Hyphal structure monomitic; generative hyphae yellowish brown, branched, septate, 4–8 μm diam., with thin or slightly thickened walls clamp connections absent. Basidia clavate, subhyaline, 4–6 μm diameters, bearing four short sterigmata. Spores dark amber brown, in reality very broadly elliptical, but appearing almost ovate or subglobose, uniguttulate, (−4)5-6 × (3.5-)4-4.5 μm.

Distribution: U.K.: Chakrata road, Dehra Dun.

Collection examined: 225

Substratum: On living deciduous tree.

Remarks: This species was reported from Mussoorie hills as *Polyporus indicus* var. *depauperatus* by Reid et al. 1959. It is a rare species.

Fomitiporia Murrill W.A.,
N. Amer. Flora 9(1):7, 1907.

Fructification perennial, sessile, solitary, sometimes effused-reflexed, soft when fresh, hard on drying. Pileus sessile, convex; upper surface yellowish brown and turns reddish-brown with age. Context yellowish-brown, homogenous, shining, tuning brownish black in KOH. Hyphal system dimitic; generative hyphae thin-walled, branched, septate, hyaline, clamps are absent, cyanophilous; skeletal hyphae thick-walled, sometimes septate, unbranched, acyanophilous. Basidia collapsed. Basidiospores subhyaline, smooth, thin-walled, apiculate, non-amyloid.

Eleven species, widespread

Lit.: Wagner & Fischer (*Mycol.* 94: 998, 2002).

Type Species: *Fomitiporia langloisii* Murrill.

Habitat: Wood

Himalayas: One

Fomitiporia robusta (P. Karst.) Fiasson & Niemelä, Karstenia 24(1): 25 (1984) = *Phellinus robustus* (Karst.) Bourd. & Galz. Hym. France P. 616, 1928. Plate 6.8c, Fig. 6.19d–h

Fructification perennial, pileate, sessile, solitary, sometimes effused-reflexed, tough and hard when fresh, woody hard on drying, heavy. Pileus ungulate, dimidiate; upper surface dark greyish brown, cinnamon to rust brown later becoming black, concentrically sulcate with broad round bands, 0.7–1.9 μm, glabrous, smooth. Pore surface yellowish brown to brown, dull even to uneven; pores rounded to somewhat angular; margin thick, round, smooth, yellowish brown to rusty brown. Context golden brown, shining, zonate, xanthochroic.

Hyphal system dimitic; generative hyphae thin-walled, branched, septate, subhyaline, clamps are absent, cyanophilous, 2.0–3.0 μm in diameter; skeletal hyphae thick-walled, sometimes septate, unbranched, acyanophilous, 2.8–5.6 μm in diameter. Cystidioles present, thin-walled, hyaline, ventricose, up to 3.7 μm wide. Setae absent. Basidia hyaline, thin-walled, clavate, 10.5–14.0 × 4.2–5.0 μm. Basidiospores light yellowish, smooth, 7–8.4(−9) × 6.0–8.0 μm thick-walled, globose to subglobose, apiculate, non-amyloid.

Distribution: A.P.- West Kameng, Shergaon; Meghalaya- Shillong, Mawphlang; U. K.: Chakrata.

Collection examined: SSV 21500, 21645, L 37308.

Substratum: On decaying angiospermic logs.

Remarks: The species was described by Bakshi (1971) as *Fomes robustus* P. Karst. as parasitic on 'Fir' from North Western Himalaya and South India. Later on it was described by Bakshi (1971) from N. W. Himalayas. It is being recorded from Uttarakhand for the first time.

Fuscoporia- Murrill,
W.A N. Amer. Flora 9(1):3, 1907.

Fructification resupinate, perennial, pileate or laterally stipitate, corky to woody hard, Pileus surface tomentose to velutinate, non-crusted. Context homogenous. Pores usually small. Hyphal system monomitic; generative hyphae at dissepiments edge or hymenium covered by crystals. Hymenial setae present in most species, straight, hooked in some species bearing mycelia setae; cystidioles present. Basidiospores cylindric, oblong-ellipsoid, broadly ellipsoid or subglobose, hyaline, thin-walled, smooth.

Widespread

Lit.: Donk (*Verh. K. Ned. Akad. Wet.* Tweede sect. 62:1, 1974).

Type Species: *Boletus ferruginosus* Sch. ex J. F. Gmel., 1792.

Habitat: Wood

Himalayas: Five

Key to species

1. Fructification resupinate.. 2
1. Fructification pileate.. 4
2. Mycelial setae present in the margin of fructification.. 3
2. Mycelial setae absent, basidiospores cylindrica ....1*F. ferrea*
3. Pores 7–8 per mm ....*F. ferruginosa*
3. Pores 2–3 per mm .... *F. contigua*
4. Fructifications sessile, broadly laterally attached.
   Setae swollen near the base..*F. torulosa*
4. Fructification solitary to imbricate, attached by narrow
   to broad lateral base. Setae not swollen near the base .... *F. senex*

Fuscoporia contigua (Pers.) G. Cunn., Bull. N.Z. Dept. Sci. Industr. Res., Pl. Dis. Div. 73: 4 (1948)=*Phellinus contiguus* (Pers. ex Fr.) Bourd. & Galz. Hym. France p. 624. 1928. Plate 6.8d, Fig. 6.19i–n

Fructification annual or perennial, resupinate, membranous-coriaceous, adnate, widely effused, up to 5 mm thick in section; pore surface yellowish brown to brown, rough, not creviced; margin thinning, more or less abrupt, paler concolorous. Tissues turning black with KOH sol. Pores obscurely stratose, pore mouth oval to angular but may become irpicoid with age due to unequal growth of dissepiments. Context yellowish brown in section, composed of narrow zone of hyphae.

Hyphal system dimitic; skeletal hyphae 2–3 µm wide, sparsely branched, aseptate, the walls light brown, thick (usually leaving little or no lumen); generative hyphae 1.5–2.5 µm wide, branched, septate, clamps absent, the walls subhyaline, thin to slightly thick. Setae 35–100×7–10 µm, subulate to subacicular, with acute apices, thick-walled, walls light brown to brown, immersed or slightly projecting into the pore cavity, immersed setae usually possess longer pedicels. Basidia 4–6 µm broad, clavate, 4-spored. Basidiospores 4.4–5×3–3.4 µm, broadly ellipsoid, minutely apiculate, the walls subhyaline, thin, smooth, nonamyloid.

Distribution: Mizoram-Kolasib; H.P.: Dalhousie- Khajjiar; Dharamsala- Yol camp, Rasan.

Collection examined: SSR 6119; Dhanda 6654; HSK 7542, 7543, 7544, IBP 42050, 42053.

Substratum: Stumps under mixed forest, Cut stump of *Pinus*.

Remarks: The characteristic features of this species are resupinate fructifications, large pores averaging 2–4 per mm, presence of setae and small, broadly ellipsoid basidiospores. This species occurs on *Pinus excelsa* and *Cedrus deodara* in N.W. Himalayas.

Fuscoporia ferrea (Pers.) G. Cunn., Bull. N.Z. Dept. Sci. Industr. Res., Pl. Dis. Div. 73: 7 (1948) = *Phellinus ferreus* (*Pers.*) Bourd. & Galz. Hym. Fr. P. 627, 1928.

Fig. 6.19o–s

Fructification perennial, resupinate, adnate, spongy hard when fresh, rigid and hard on drying, effused, up to 23 cm long, 4 cm wide and 0.8 cm thick, undulating and convex. Pore surface yellowish brown to greyish brown, black in older portions; pores round to subangular, regular 4–6 per mm, 110–168 µm in diameter dissepiments entire 55–68 µm thick, tubes stratified; margin fertile, smooth when fresh, rimose on drying, lighter or concolorous with pore surface, blackish on drying. Context fibrous, yellowish brown to brown, homogenous, xanthochroic, 0.6–1 mm thick.

Hyphal system dimitic; generative hyphae subhyaline to light yellow, thin-walled, branched, septate, cyanophilous, 1.8–2.7 µm in diam.; skeletal hyphae yellowish brown, thick-walled, 2.5–3.8 µm in diam. Setae abundant, projecting into the cavity, subulate with slightly swollen base, thick-walled, dark brown, apices straight and acute, 26.4–42×5.3–10.3 µm. Basidia hyaline, thin-walled, 4-spored, 8.9–13.9×3.7–5.4 µm. Basidiospores hyaline, thin-walled, smooth, cylindrical, minutely apiculate, non-amyloid, 5.7–8×2–2.5 µm.

Distribution: A.P.: West Kameng- Shergaon, Kalaktang.

Collection examined: SSV 21515.

Substratum: On decaying angiospermic twig.

Remarks: The description of the species resembles well with the description as given by Ryvarden and Johansen (1980). The species is marked by perennial, resupinate, rigid and hard pulvinate fructifications; larger pores 4–6 per mm; yellowish brown context; thick-walled, subulate setae; and hyaline thin-walled, cylindrical basidiospores. This species is being recorded for the first time from Himalayas.

Fuscoporia ferruginosa (Schrad.) Murrill [as '*ferruginosus*'], N. Amer. Fl. (New York) 9(1): 5 (1907)=*Phellinus ferruginosus* (Schrad. exFr.) Bourd. & Galz., Hym. Fr., P. 625, 1928. Plate 6.9a, Fig. 6.20a–f

Fructification annual or perennial, broadly effused, inseparable from the wood, corky; margin 'sayal brown', narrow, thin, sterile; subiculum brown, 0.15–0.2 mm thick; hymenial surface 'sayal brown', 'snuff brown', dull, pore tubes, pores circular to slightly angular, irpiciform when growing on vertical surface, pore wall entire.

Hyphal system dimitic; generative hyphae with simple septa, hyaline, thin-walled, 2–4 µm wide; skeletal hyphae pale yellow to light brown in KOH sol, thick-walled 2–5 µm diameter. Basidia clavate, 11–13×4.3–5.1 µm simple septate at base. Basidiospores hyaline, thin-walled, smooth, oblong, apiculate, 1 guttulate, 5.5–6.1×3–3.8 µm; setae present, dark brown, subulate, acuminate, abundant, 48–68(−102)×6.8–9.5 µm; hyphae (1) brown, thick-walled with wide to narrow lumen, unbranched, apparently aseptate, 2.1–3.7 µm diam., (2) setal hyphae with pointed ends, dark brown, 6–10 µm diam., (3) tramal hyphal hyaline, thin-walled, apparently unbranched, indistinctly septate.

Distribution: H.P.: Kullu- Kasole.

Collection examined: IBP 37051.

Substratum: Beam of Deodar cut log.

Remarks: This is rare species in the N.W. Himalayas and was reported by Bakshi (1971). It has been collected twice from Himachal Pradesh and Uttarakhand. It is closely related to *P. ferreus* but the latter differs in having narrower spores and setal hyphae.

Fuscoporia senex (Nees & Mont.) Ghob.-Nejh., in Ghobad-Nejhad & Dai, Mycotaxon 101: 208 (2007)=*Phellinus senex* (Nees & Mont.) Imaz. Bull. Govt. Forest Exp. Stn. Tokyo 57: 115, 1952. Plate 6.9b, Fig. 6.20g–k

Fructification perennial, pileate, solitary to imbricate, attached by narrow to broad lateral base, hard, light weight. Pileus flat or slightly convex, applanate, dimidiate or effused-reflexed, sometimes resupinate 3.5–7.5×1.5–6.4×1.3 cm; upper surface, brown to greyish brown, paler toward margin, fine tomentose, becoming glabrous, concentrically sulcate; margin thin to bluntly rounded, glabrous, brown. Pore surface brown, uneven; pores small, invisible with naked eye, round 7–10 per mm dissepiments equal 50–80 µm thick, tube stratified, separated by thin sterile layer, brown, up to 3.5 µm deep. Context fibrous, shiny, dark brown, darker than the tubes, homogeneous, xanthochroic.

Hyphal system dimitic, generative hyphae hyaline, thin-walled, branched, septate, clamps absent, 2–3 µm in diameter; skeletal hyphae yellow to brown, thick-walled to almost solid with narrow lumen, unbranched, aseptate, 2.4–3.8 µm wide in diameter. Setae subulate or ventricose, thick-walled, dark brown, swollen near the base, 22–35×7–9.5 µm. Basidia hyaline, thin- walled, clavate, 4-spored, 8–9.3×4–5 µm. Basidiospores hyaline, thin-walled to slightly thick-walled, smooth, hyaline to pale yellow, broadly ellipsoid, uniguttulate, 4–5.4×2.8–3.2 µm.

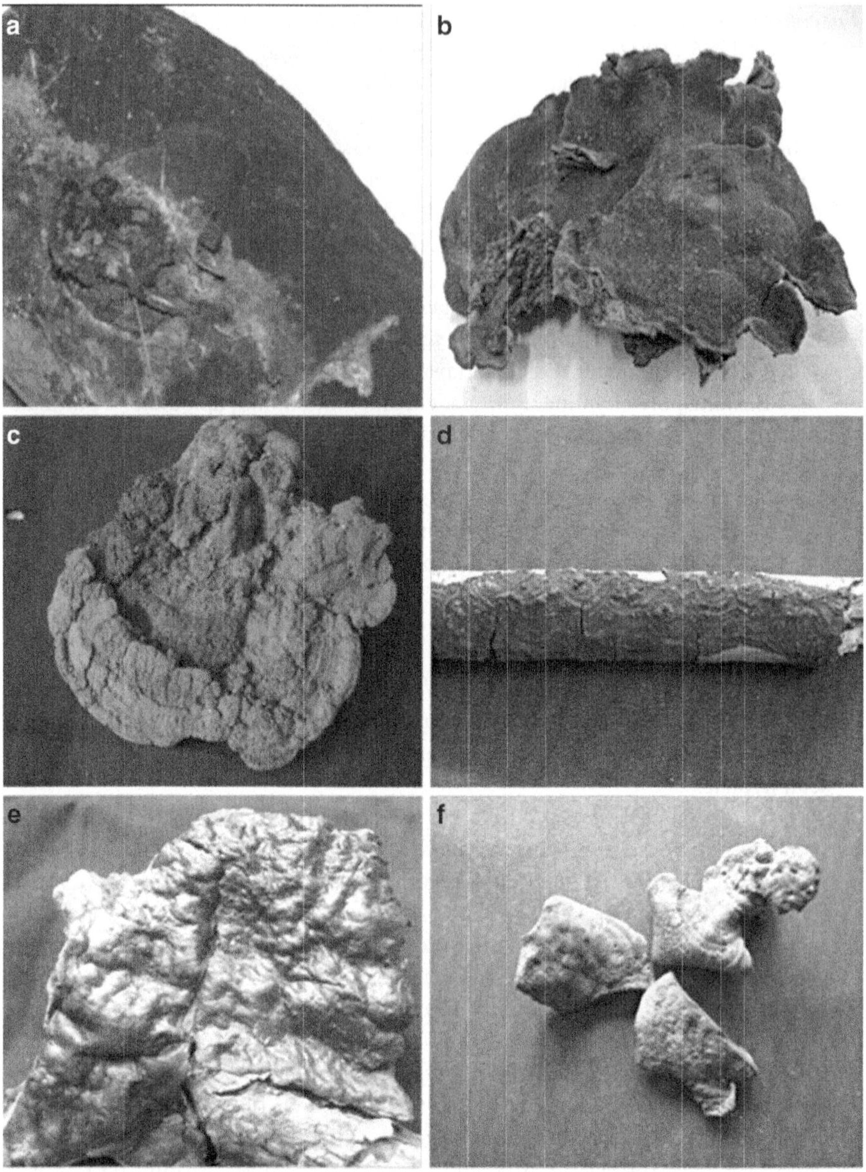

Plate 6.9 (**a**) *Fuscoporia ferruginosa*. (**b**) *Fuscoporia senex*. (**c**) *Fuscoporia torulosa*. (**d**) *Hymenochaete mougeotii*. (**e**) *Inonotus drydaeus*. (**f**) *Inonotus tabacinus*

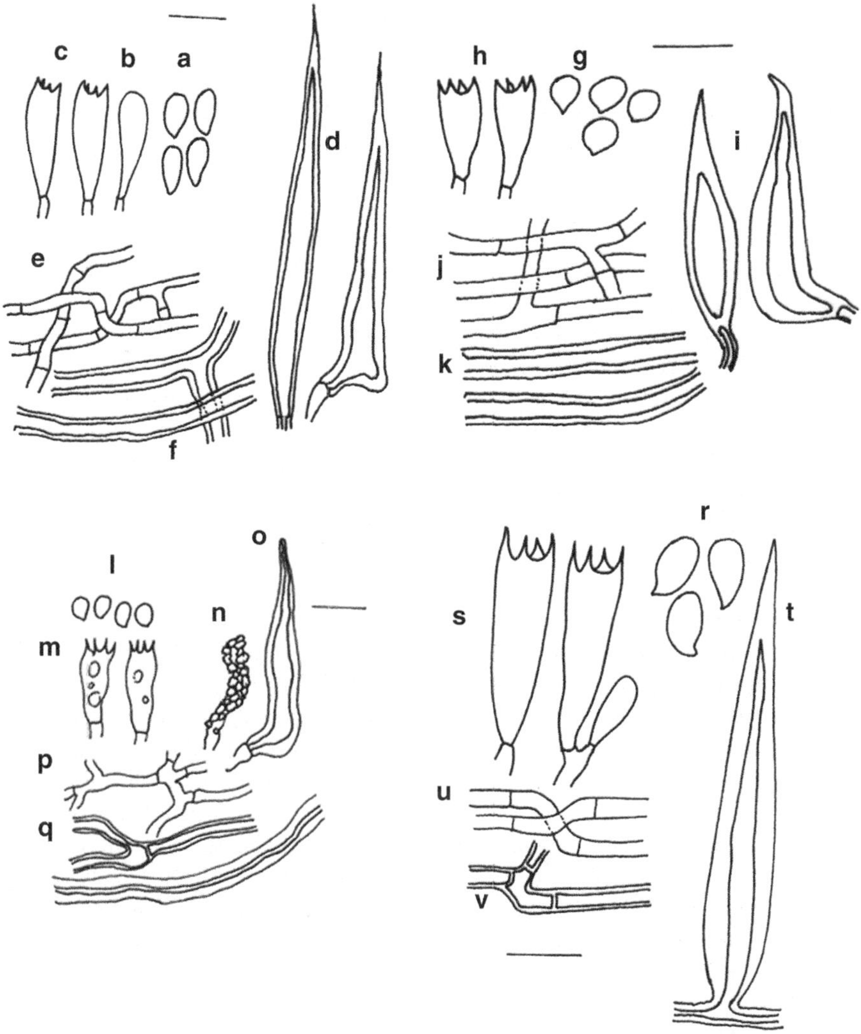

Fig. 6.20 (**a**–**f**) *Fuscoporia ferruginosa* (**a**) Basidiospores, (**b**) Basidioles, (**c**) Basidia, (**d**) Setae, (**e**) Generative hyphae, (**f**) Skeletal hyphae; (**g**–**k**) *Fuscoporia senex* (**g**) Basidiospores, (**h**) Basidia, (**i**) Cystidia, (**j**) Generative hyphae, (**k**) Skeletal hyphae; (**l**–**q**) *Fuscoporia torulosa* (**l**) Basidiospores, (**m**) Basidia, (**n**) Cystidia, (**o**) Setae, (**p**) Generative hyphae, (**q**) Skeletal hyphae; (**r**–**v**) *Hymenochaete fuscobadia* (**r**) Basidiospores, (**s**) Basidia, (**t**) Setae, (**u**, **v**) Generative hyphae

Distribution: A.P.: West Kameng, Dirang; Meghalaya: N.C. Hills, about 60 km from Garampani towards Haflong; Tripura: Agartala, Sepahijula; West Bengal: Darjeeling, Tiger Hill, Teesta; H.P.: Dalhousie- Panjpula.

Collection examined: SSV 21454, GSD 21069, IBP 37053, 42054.

Substratum: On decaying angiospermic logs, Bamboo, decaying logs of *Quercus*.

Remarks: This species was reported previously from India by various workers. The earliest record is by Berkeley (1856) who recorded it as *Polyporus senex* Nees & Mont. Based on Dr Hooker's collection from khasi hills (Meghalaya), Sikkim and Darjeeling. Subsequently it was reported by Theissen (1911), Bose (1921), Mitter and Tandon (1932), Banerjee (1947), Baghchee and Bakshi (1950) and Bakshi (1971) as *Fomes senex* (Nees and Mont.) Cooke. It is a widely distributed species in the Himalayas. All the Himalayan collections are quite typical of the species and closely resembles with the description given by Cunningham (1965) and Ryvarden & Johansen (1980). The species is marked by the resupinate to effused-reflexed to distinctly sessile fructifications; brown, greyish brown, concentrically sulcate upper surface; glacing pore surface; small 7–10 pores per mm; thick-walled, brown, subulate to ventricose setae; and hyaline to pale yellow, thick -walled, broadly ellipsoid basidiospores.

Fuscoporia torulosa (Pers.) T. Wagner & M. Fisch., Mycol. Res. 105(7): 780 (2001) = *Phellinus torulosus* (Pers.) Bourd. & Galz. Hym. Fr., P. 619, 1928. Plate 6.9c, Fig. 6.20l–q

Fructifiaction perennial, pileate, sessile, broadly laterally attached, sometimes effused-reflexed, solitary or imbricate, woody 14 × 6 × 5 cm in size. Pileus semicircular, dimidiate, flat to convex, triangular in section; upper surface yellowish brown when young, become rusty brown, with age concentrically sulcate with dark brown compact tomentum, not rimose; margin yellowish brown, round to acute entire or weakly lobed, sterile below. Pore surface brown to rusty brown, even to uneven dull; pores round 5–7 per mm, 40–148 μm in diameter; dissepiments entire; 65–120 μm thick; pore mouth velutinate; tubes distinctly stratified, brown, each layer up to 2.7 mm deep. Context yellowish brown, lower part hard and tough, upper part soft and corky, xanthochroic.

Hyphal system dimitic, generative hyphae hyaline to light yellow, thin-walled, branched, simple septate, 2–3 μm in diameter; skeletal hyphae yellow to brown, thick-walled, unbranched, aseptate, 2.4–3.6 μm wide in dissepiments. Setae abundant , projecting in to the cavity, subulate to subventricose, straight with pointed apices, dark brown, thick-walled, 24–38 × 6.4–9 μm. Basidia hyaline, clavate, 4-spored, up to 5 μm broad. Basidiospores hyaline, thin-walled, smooth, apiculate, uniguttulate, broadly ellipsoid, non-amyloid, 3.8–5.4 × 2.7–3.7 μm.

Distribution: A.P.-West Kameng; J&K: Batote; U.K.: Nainital; H.P.: Shimla, Solan.

Collection examined: SSV 21480, IBP 42068.

Substratum: Stumps under mixed forest, Cut stump of *Pinus*, angiospermic logs.

Remarks: It is a common species in N. W. Himalayas, and was first reported by Thind *et. al.* (1970) as *Fomes torulosus* (Pers.) Llyod, however the species is being reported for the first time from the Eastern Himalayas as well as from the

Himachal Pradesh. *Fuscoporia torulosa* closely resembles *F. senex*. However, the latter differs by thinner fructification; narrowly concentrically sulcate upper surface; smaller; and slightly shorter setae with swollen base. It is a new record for Eastern Himalayas.

Hymenochaete Lév.,
Ann. Sci. Nat., Bot. sér. 3 **5**: 150 (1846)

Fructification annual or perennial, resupinate to effused-reflexed to pileate, usually membranous-coriaceous or rarely hard and woody. Pileus with compact context and loose tomentum; hymenial surface smooth. Tissue darkening in KOH. Context yellowish brown to brown. Hyphal system monomitic or dimitic; generative hyphae simple-septate. Setae always present, usually abundant and occurring in overlapping layers and forming a distinct setal layer. Cystidia or Gloeocystidia absent. Basidia clavate-cylindrical to subclavate-cylindrical to subutriform, 4-spored. Basidiospores thin-walled, subhyaline, smooth, non-amyloid, acyanophilous of versatile shape.

One hundred ten species, widespread.

Lit.: Job (*Mycol. Helv.* 5: 1, 1990; temp. S. Hemisph. Spp.)

Type Species: *Helvella rubiginosa* Dicks., 1785.

Habitat: Decayed wood

Himalayas: Six

Key to species

1. Abhymenial cuticle present 2
1. Abhymenial cuticle absent 4
2. Hymenial surface red or deep red; paraphysoid hyphae with dendroid like process present at top; spores cylindrical to suballantoid .... *H. mougeotii*
2. Hymenial surface not red; paraphysoid hyphae absent; spores broadly ellipsoid 3
3. Basidiocarps coriaceous–tough, hard and woody on drying; spores 5–7 × 3–4.5 μm .... *H. rubiginosa*
3. Basidiocarps membranous–coriaceous, pliable on drying; spores 4.5–5.5 × 2–2.5 μm .... *H. luteobadia*
4. Hymenial surface deeply cracked into irregular isolated squares, at least in the resupinate portion 5
4. Not as above 6
5. Basidiocarps resupinate to effused–reflexed, brown to dark reddish brown; margin orange yellow; context with only one setigerous layer .... *H. tabacina*[a]
5. Basidiocarps resupinate ferruginous brown; margin yellowish brown; context stratified with usually two setigerous layers .... *H. semistupposa*
6. Basidiocarps pileate, thin, papery, remaining so on drying; abhymenial surface thickly to coarsely pubescent .... *H. rheicolor*[a]
6. Basidiocarps resupinate to pileate, somewhat hard and brittle on drying; abhymenial surface velvety to smooth 7
7. Setae densely crowded all over the setigerous layer .... *H. cacao*[a]

7. Setae arising from a distinct setal zone 8
8. Basidiocarps membranous–coriaceous; setae scanty and often covered by growing hyphae from the hymenium, giving them the pyramidal appearance; spores ellipsoid to subcylindrical, 6–7.5 × 3–4.5 μm .... *H. fuscobadia*
8. Basidiocarps subpelliculose to membranous; setae abundant and without any hyphal covering; spores ellipsoid, 5–7 × 3–3.5 μm (extra limital, not included in the text) .... *H. leonina*

[a]Extra limital not included in this work

Hymenochaete fuscobadia K.S.Thind & Adlakha, Trans. Brit. Mycol. Soc. 41(1): 133. 1958. Fig. 6.20r–v

Fructification annual, resupinate, membranous-coriaceous when fresh becoming somewhat brittle on drying, adnate, widely effused; hymenial surface ochre brown to somewhat greyish brown, smooth often cracking irregularly on drying; margin thinning, adnate but often curling away on drying, paler concolorous to concolorous. Context yellowish brown to ferruginous, composed of basal zone of somewhat parallel and compactly arranged hyphae and an upper zone of loosely woven, ascending hyphae supporting a pellicle-like hymenium, cuticle on the abhymenial side lacking.

Hyphal system monomitic, hyphae 2.5–5 μm wide, branched at wide angles, septate, clamps absent, thin-walled, pale yellow. Setae 75–110 × 7–11 μm, subulate, thick-walled, light brown to brown, often arising from the subhymenial zone and may project to 50 μm out of the hymenium. Hymenium and subhymenium compact and somewhat agglutinated forming a distinct pellicle-like layer up to 125 μm thick. Basidia 25 × 6 μm clavate cylindrical, 4-spored, sterigmata slender and up to 6 μm long. Basidiospores 6.5–7.5 × 3.8–4.2 μm, ellipsoid to subcylindrical with obtuse ends, minutely but distinctly apiculate, the walls thin, subhyaline, smooth, nonamyloid.

Distribution: Nepal: Daman, Kathmandu.

Collection examined: SSR 5581.

Substratum: On log under angiosperms.

Remarks: This species is quite rare and was known only from the type specimen which was collected from Mussoorie hills. It is very close to *H. leonine* but differs in having slightly larger basidiospores and more robust and adnate fructifications. Moreover, the setae are less abundant and generally covered by hyphae from the hymenium.

Hymenochaete leonina Berk. & Curt., J. Linn. Soc. Bot. 10(no.46): 334. 1868. Fig. 6.21a–e

Fructification annual, resupinate, subpelliculose to membranous, loosely adnate, somewhat fragile, often arising as small circular colonies which may coalesce and grow later, up to 400 μm thick in section; hymenial surface woody brown to greyish brown, smooth, rarely cracking irregularly on drying; margin thinning but occasionally indeterminate, byssoid, loosely adnate, bright orange-brown.

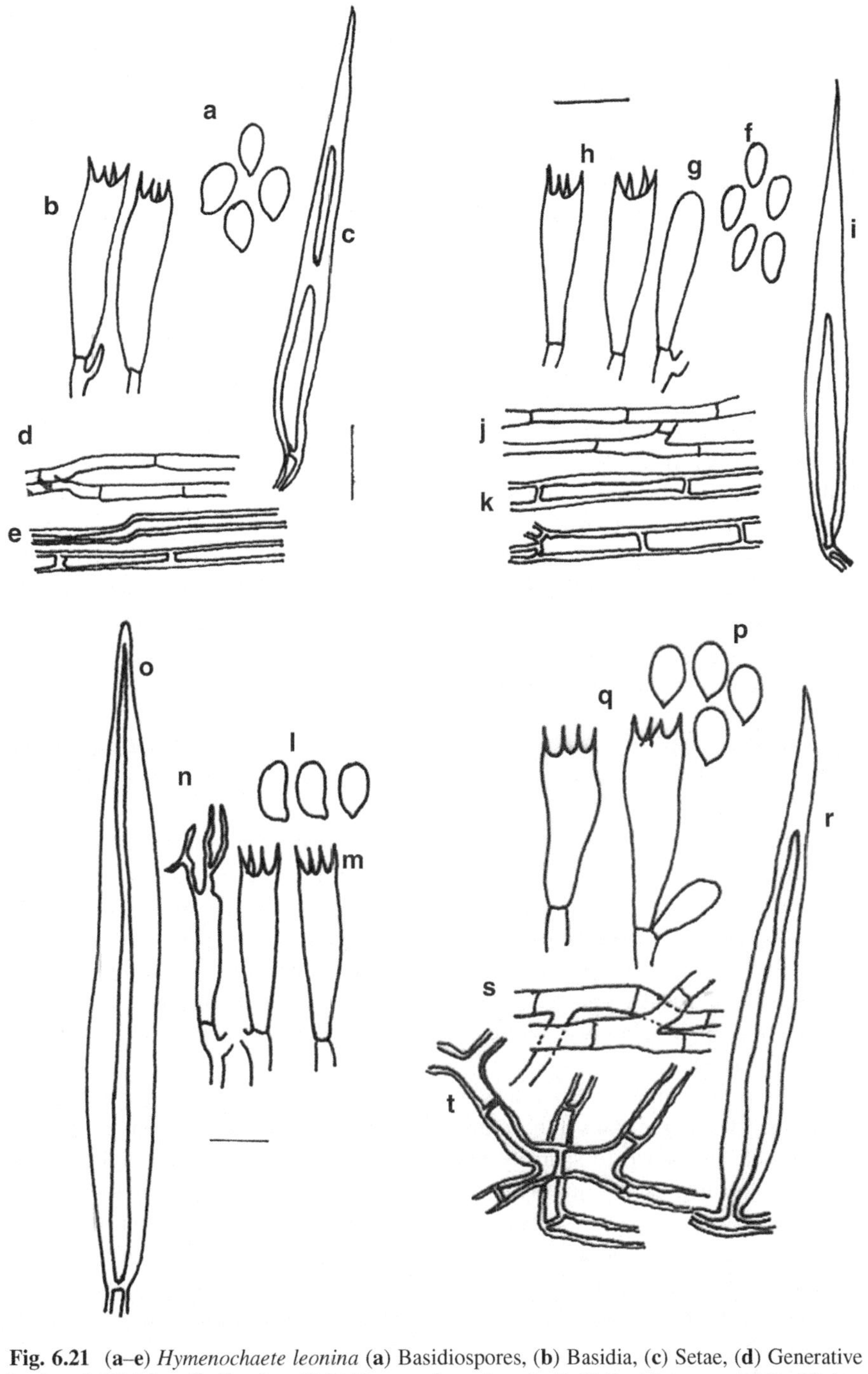

Fig. 6.21 (**a**–**e**) *Hymenochaete leonina* (**a**) Basidiospores, (**b**) Basidia, (**c**) Setae, (**d**) Generative hyphae, (**e**) Thick-walled hyphae; (**f**–**k**) *Hymenochaete luteobadia* (**f**) Basidiospores, (**g**) Basidioles, (**h**) Basidia, (**i**) Setae, (**j**) Thin-walled Generative hyphae, (**k**) Thick-walled generative hyphae; (**l**–**o**) *Hymenochaete mougeotii* (**l**) Basidiospores, (**m**) Basidia, (**n**) Dendrohyphidia, (**o**) Setae; (**p**–**t**) *Hymenochaete rubiginosa* (**p**) Basidiospores, (**q**) Basidia, (**r**) Setae, (**s**) Thin-walled generative hyphae, (**t**) Thick-walled generative hyphae

Context orange-brown in section, composed of basal zone of parallel, compactly arranged and more or less agglutinated hyphae and an upper zone of semi-erect and loosely woven hyphae.

Hyphal system monomitic, hyphae 1.6–4.1 µm wide, branched, septate, clamps absent, thin-walled, yellowish brown. Setae 60–110 (150) × 7–10 µm, subulate, thick-walled, brown, often arising from the subhymenial and may project up to 75 µm out of the hymenium. Hymenium and subhymenium compact and partly agglutinated forming a pellicle-like layer up to 80 µm thick. Basidia clavate-cylindrical to subutriform, 30 × 5.4 µm, 4-spored. Basidiospores 4.4–6 (7) × 3–3.2 µm, ellipsoid, minutely apiculate, thin-walled, subhyaline, smooth, non-amyloid.

Distribution: H.P: Shimla.

Collection examined: SSR 5571.

Substratum: On log under angiosperms.

Remarks: The characteristic features of this species are: resupinate and easily detachable fructifications, bright coloured context, long and narrow setae and ellipsoid basidiospores. The basal layer of the context is very compact and more or less agglutinated with the substratum. The hymenium and upper part of the context usually become separate from the basal layer, with little pressure and as a result the fructifications appear detachable.

Hymenochaete luteobadia (Fr.) Höhn. & Listch., Sber. Akad, Wiss, Math.-natur. Kl., Abt. 1, 116: 750 (1907) = *Thelephora luteoalba* Fr., Linnea 5: 526. 1830. Fig. 6.21f–k

Fructifications annual, effused reflexed to pileate, loosely adnate, membranous-coriaceous, widely effused. Pileus ranging from reflexed margin to sessile; hymenial surface smooth to sparsely tuberculate, radially sulcate and faintly concentrically zonate, zonations corresponding with those of upper surface, brown to olive brown; upper surface yellowish brown to greyish brown, strongly tomentose, concentrically zonate, zones formed by erect and appressed tomentum; margin acute to subobtuse, entire or rarely cleft, paler concolorous. Context yellowish brown.

Hyphal system dimitic; generative hyphae 2–3.9 µm wide, branched, septate, clamps absent, the walls thin, subhyaline to pale yellow; skeletal hyphae 2.4–3.4 µm wide, branched to sparsely branched, aseptate, thick-walled, light brown; tomentose hyphae 3–4 µm wide, light brown, moderately thick (up to 0.8 µm). Setae 40–60 × 6–6.8 µm, subulate, brown thick-walled, arising from the upper part of the context or subhymenium, immersed or projecting out of the hymenium, Basidia 18 × 5 µm, clavate to cylindrical, 4-spored. Basidiospores 5.4 × 2.4 µm, ellipsoid, minutely apiculate, the walls subhyaline, thin, smooth, nonamyloid.

Distribution: H.P.: Kullu.

Collection examined: SSR 5697, 5723, IBP 37056, L 37057.

Substratum: On twigs under mixed forests.

Remarks: The characteristic features of this species are pileate and membranous-coriaceous fructifications which remain pliable on drying, densely tomentose abhymenial surface, dimitic hyphal system and small setae.

Hymenochaete mougeotii (Fr.) Masee, J.Linn. Soc., Bot. 27: 111 (1890) = *Thelephora mougeotii* Fr., Elenchnus: 188. 1828. Plate 6.9d, Fig. 6.21l–o

Fructification annual, resupinate to effused-reflexed, membranous-coriaceous, adnate, arising as small colonies which later coalesce and become effused, up to 450 μm in section; hymenial surface smooth to tuberculate, deep red, tends to fade with age and become brownish red in old and withered patches; margin thinning, adnate, white to paler concolorous. Pileus when present mearly upturned margin, upper surface brown, tomentose and faintly concentrically zonate. Subiculum light brown in section, composed of compactly arranged parallel hyphae with a well developed cuticle (up to 20 μm thick) on the abhymenial side covering tomentum.

Hyphal system dimitic; generative hyphae 2–3 μm, branched, septate, clamps absent, the walls subhyaline, thin; skeletal hyphae 2–3.5 μm wide, sparsely branched, aseptate. Setae 60–90 × 7–10 μm, thick-walled, arising from the upper layers of the context, immersed or projecting out from hymenium paraphysate hyphae dendroid with few processessant at the top. Basidia 3–4.5 μm broad, clavate-cylindrical, 4-spored, sterigmata cylindrical. Basidiospores 5.4–6.4 × 2.7–3.1 μm, cylindrical to suballantoid, minutely apiculate, walls thin, smooth, subhyaline, non-amyloid.

Distribution: H.P.: Narkanda; J&K: Chandanwari; U.K: Mussoorie-Jabber Khet; Nepal: Gosainkhund.

Collection examined: SSR 5185, 5187, IBP 37058, L 37059.

Substratum: On twigs of *Quercus*.

Remarks: The species is usually in resupinate form but rarely becomes pileate. The fructification is marked by red colour hymenial surface and branched paraphysoid hyphae in the hymenium. It is found growing parasitically on *Pyrus* and number of angiospermic plants.

Hymenochaete rubiginosa (Dicks.) Leѷ., Ann. Sci. Nat., Bot., Sér. 3 5: 15. 1846. = *Helvella rubiginosa* Dicks., Fasc. Pl. Crypt. Brit. 1: 20. 1785. Fig. 6.21p–t

Fructifications perennial, effused reflexed to pileate, cartilaginous- tough when fresh become hard and woody on drying. Pileus up to 11 cm long, 5 cm broad and 2 mm thick, sessile or attached by narrow base, occasionally imbricate; hymenial surface chocolate brown to snuff brown, smooth to sparsely tuberculate or uneven or faintly concentrically sulcate, sulcations or zonations corresponding to those of upper surface; abhymenial surface dark brown to greyish brown; concentrically zonate, zones formed by erect and appressed tomentum; radially sulcate, with maturity the tomentum becomes compact and somewhat agglutinated and appears hard and bark like; margin acute, entire or rarely cleft, paler concolorous. Context light brown in section, composed of compact agglutinated hyphae, cuticle well developed on the abhymenial surface.

Hyphal system dimitic; skeletal hyphae 1.4–2.4 μm wide, unbranched, aseptate, the walls thick to almost solid, light brown; generative hyphae 1.5–2 μm wide, branched, septate, clamps absent, the walls thin, subhyaline. Tomentose hyphae similar to the skeletal hyphae. Setae 30.1–69.8 × 5–6.9 μm, subulate to subcylindrical, light brown, thick-walled, arising from the upper part of the context or subhymenium, immersed or projecting out of the hymenium. Basidia broadly clavate, 15–28 × 5–6 μm 4-sterigmate, hyaline to yellowish brown. Basidiospores 5–7 × 3.5–4.5 μm, ellipsoid, minutely apiculate, the walls subhyaline, thin, smooth, non-amyloid.

Distribution: U.K: Rishikesh.

Collection examined: SSR 5858.

Substratum: On stump of *Shorea robusta.*

Remarks: It is characterized by rigid fructifications with bright contrasting margin. This species is associated with heart- rot of *Shorea robusta* and is marked by the hard and woody fructifications with bark-like upper surface of pileus. It is a wide spread species in Himalayas. This species is new record for Uttarakhand.

Hymenochaete semistuposa Petch; Ann. Roy. Bot. Gdns. Perad. 9: 278. 1925.
Fig. 6.22a–c

Fructification annual or perennials, resupinate, membranous, inconspicuously stratose, each stratum up to 120 μm thick and separated by a layer of context, up to 300 μm thick in section, loosely adnate, widely effused; hymenial surface smooth to sparsely tuberculate, ferruginous to brown, rarely cracking irregularly on drying; margin thinning, byssoid, loosely adnate, yellowish brown. Subiculum yellowish-brown in section, composed of somewhat loosely woven hyphae supporting a compact and pellicle-like hymenium.

Hyphal system monomitic, hyphae (2.5) 3.4–5.4 μm wide, branched at wide angles and appear somewhat ramified, septate, clamps absent, the walls thin to slightly firm, light yellow. Setae 50–84 × 5–8 μm, subulate, thick-walled, brown, sometimes with furcate base, arising from all parts of the subiculum, immersed or projecting out of the hymenium. Basidia 16–20 × 4–5 μm, cylindrical to subutriform, 4-spored, sterigmata slender and up to 5 μm long. Basidiospores 4.2–5.2 × 1.5–2 μm narrowly ellipsoid to suballantoid, shortly apiculate, thin-walled, subhyaline, smooth, nonamyloid.

Distribution: H.P.: Narkanda.

Collection examined: SSR 5524, IBP 37680.

Substratum: On twigs of *Berberis sp.*

Remarks: This species is marked by the narrowly ellipsoid to suballantoid basidiospores and medium sized setae arising from all parts of the subiculum. The fructification possesses two strata which are separated by a narrow layer of the context. It is a rare species in the Himalayas.

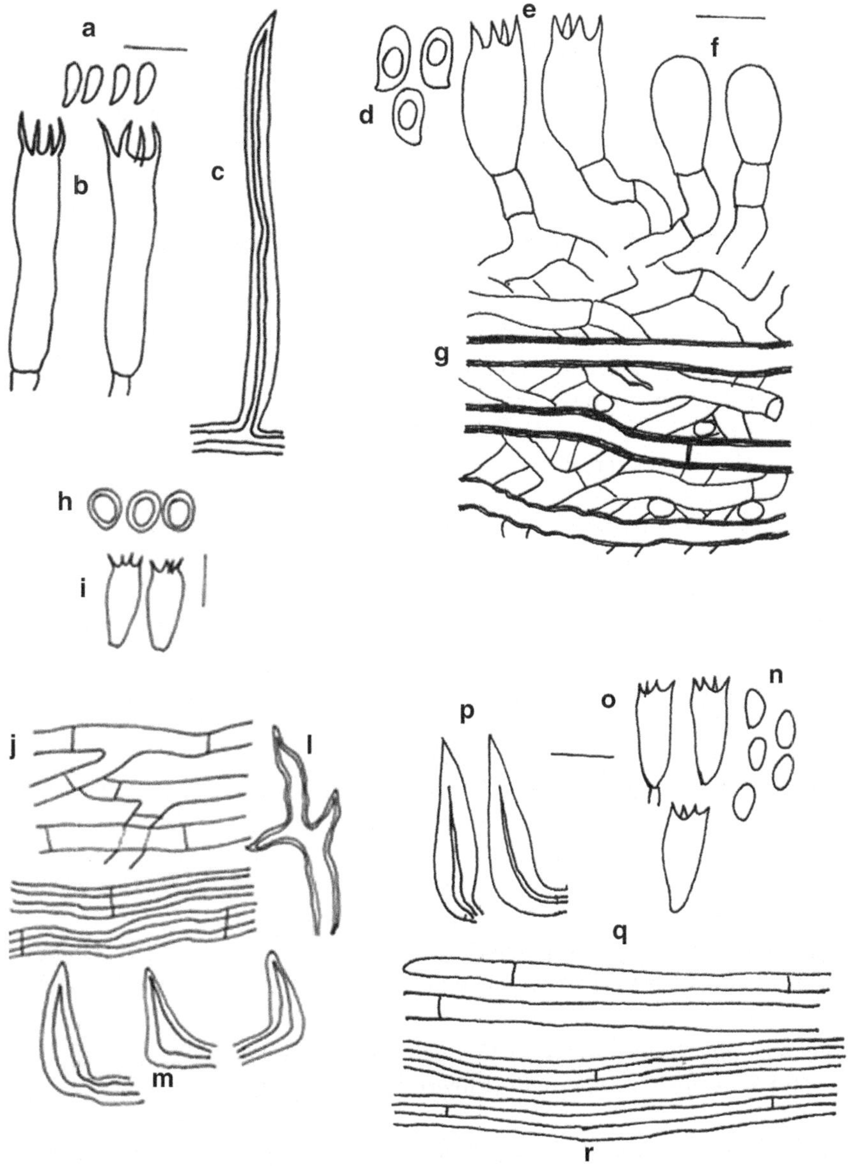

Fig. 6.22 (**a**–**c**) *Hymenochaete semistuposa* (**a**) Basidiospores, (**b**) Basidia, (**c**) Setae; (**d**–**g**) *Inonotus albertinii* (**d**) Basidiospores, (**e**) Basidia, (**f**) Basidioles, (**g**) Generative hyphae; (**h**–**m**) *Inonotus cuticularis* (**h**) Basidiospores, (**i**) Basidia, (**j**–**k**) Generative hyphae, (**l**) Pilosetae, (**m**) Setae; (**n**–**r**) *Inonotus tabacinus* (**n**) Basidiospores, (**o**) Basidia, (**p**) Setae, (**q**, **r**) Generative hyphae

Inonotus P.Karst.,
Medd. Soc. Fauna Fl. fenn. 5: 39, 1879.

Fructification annual, sessile, effused-reflexed or resupinate, solitary or imbricate agglomeration, soft, fleshy, corky when fresh, hard and woody on drying. Pileus sessile, dimidiate, applanate or convex; upper surface glabrous, tomentose to hirsute, uneven with thin or no cortext. Pore surface brown; pores round to angular, small to medium sized; tubes not stratose, brown in section. Context brown to dark brown, fibrous with compact hyphae, soft, homogenous, xanthochroic. Hyphal system monomitic; generative hyphae subhyaline to brown, thin-walled to slightly thick-walled, septate, branched, without clamps. Setae thick-walled, brown, subulate, ventricose, immersed in dissepiments or projecting out of hymenium. Cystidia absent. Basidia hyaline, clavate, 2–4 spored. Basidiospores hyaline, brown to rusty brown, smooth, thin-walled, ellipsoid to globose, non-amyloid.

Eighty Species, widespread.

Lit.: Peglar (*Trans. Brit. mycol. Soc.* 47: 175, 1964; KEY), Gilbertson N. Y. Bot. Garden 28: 67, 1976 Ryvarden (Syn. Fung. 15: 70., 2002; neotropical species)

Type species: *Boletus cuticularis* Bull., 1790.

Habitat: Dead wood

Himalayas: Four

Key to species

1. Setal hyphae present 3
1. Setal hyphae absent 2
2. Pores angular to irregular, 2–3 per mm, basidiospores ellipsoid, 5–7.4 × 3.8–4.4 µm .... *I. tenuicarinis*
2. Pores round to angular, 3–4 per mm, basidiospores subglobose, 6–8.5 × 5.2–7.2 µm .... *I. dryadeus*
3. Basidiospores hyaline, 3–5 × 2–3 µm .... *I. tabacinus*
3. Basidiospores hyaline to various shades of brown 4
4. Basidiospores deep brown in mass .... *I. albertinii*
4. Basidiospores sub– hyaline to light brown .... *I. cuticularis*

Inonotus albertinii (Lloyfmd) P.K. Buchanan & Ryvarden, Mycotaxon 31(1): 12 (1988) = *Coltricia albertinii* (Lloyd) Ryv. Kew Bull. 17: 277, 1963. Fig. 6.22d–g

Fructifications annual, lateral or centrally stipitate, solitary or several fructifications fused when they grow in clusters, soft coriaceous when fresh, brittle on drying. Pileus circular to infundibuliform, depressed in the centre; upper tomentose, concentrically zonate with brown and light brown zones; margin pale brown when fresh, become brown to rusty brown on drying, not incurved on drying, acute. Stipe short, central or lateral, expanded near the attachment of pileus, finely tomentose, solid, rusty brown when fresh, brown on drying, uneven; Pores surface greyish brown to brown when fresh, brown on drying, uneven, 2–4 per mm. dissepiments thin concolorous, pore tube in one layer, concolorous, up to 4 mm in section; pores angular, thin-walled. Context spongy, brown, homogenous, xanthochroic.

Hyphal system monomitic; generative hyphae hyaline to brown, thin to thick-walled, branched, septate, clamps absent. In dissepiments and sub hymenium, hyphae thin-walled to slightly thick-walled, 3–4.9 µm in diameter. Setal hyphae present in the dissepiments, embedded, run parallel with the pore wall, dark brown, thick-walled, straight and unbranched, 3–5.2 µm in diameter. Basidiospores hyaline to rusty brown, slightly thick-walled, smooth, ellipsoid, uniguttulate, cyanophilous, 6–8.9 × 4.4–5.1 µm.

Distribution: A.P.: West Kemeng, Bomidila.

Collection examined: SSV 21813.

Substratum: At the base of angiospermic stump, on the exposed roots, at the base of decaying stump.

Remarks: It appears to be rarely distributed species in the Himalayas. *Inonotus albertinii* has rusty black fructifications; thick-walled, brown setal hyphae, and rusty brown ellipsoid basidiospores. It is a new record for Himalayas.

Inonotus cuticularis (Bull.) P. Karst., Meddn Soc. Fauna Flora fenn. 5: 39 (1879) = *Polyporus cuticularis* (Bull.) Fr., Syst. mycol. (Lundae) 1: 363 (1821). Fig. 6.22h–m

Fructification annual, mostly imbricate, sometimes solitary, dimidiate, spongy and fleshy-tough when fresh, hard on drying. Pileus fan shaped, 3–6 × 5–10 × 0.7–1.5 cm, surface yellowish brown to dark brown, slightly darkening on drying, azonate, compactly woolly-tomentose, brown coloured setae which bear horn like processes or branches at the top are also present abundantly on the cuticle; margin acute or rounded, entire, or slightly wavy; hymenial surface plane or convex, yellowish brown, turning black brown on drying, even, margin whitish, darkening and becoming concolorous on drying dissepiments dentate. Pores not in strata, rounded, brown up to 5 mm deep, 3–5 mm. Context brown up to 1 cm thick of parallel hyphae.

Hyphal system monomitic both in context and trama, thin-walled to thick-walled; thick-walled hyphae long, abundant, yellowish brown, septate, branched, branching forked and divergent like a dichotomy, 4–8 µm wide, wall 0.4–3.3 µm thick, lumen 1.1–3.1 µm wide, hyphae hyaline, septate, branched, 4–5.6 µm wide, take stain, Setae sparse, deep brown, thick-walled, ventricose to subulate, 12–17.5 µm long and 4.8–8 µm broad at the base, projecting up to 4.8 µm beyond the hymenial surface. Hyphae supporting setae are concolorous and right angles with the latter. Basidia short oblong to clavate, subhyaline to light brown, 8–14.3 × 4–6.3 µm. 4-sterigmata, short. Basidiospores deep brown in mass, broadly ellipsoid, smooth, wall dark and thick, 7–9 × 4–5.5 µm.

Distribution: U.K: The Park, Mussoorie.

Collection examined: 206, 207.

Substratum: On living trees and stump of *Quercus incana*.

Remarks: The species is characterized by mostly imbricate fructifications, which are spongy watery and fleshy-tough when fresh, becoming hard or rigid on drying, yellowish brown to dark brown azonate, compactly woolly tomentose pileus with abundant setae on the cuticle, brown context, rounded pores and broadly ellipsoid, brown, basidiospores, 7–9 × 4–6 µm. The pores are reported as angular for *I. cuticularis* but these are typically rounded in the Mussoorie collection.

Inonotus dryadeus (Pers.) Murrill, North American Flora 9 (2): 86, 1908. = *Boletus dryadeus* Pers., Observationes mycologicae 2: 3, 1800. Plate 6.9e

Fructification annual, pileate, applanate, solitary to imbricate; pilei up to 30×20×4 cm; abhymenial surface smooth to tuberculate to very finely velutinate to glabrous, irregularly zonate, brown to dark brown with greyish orange tomentum; hymenial surface poroid, paler yellow to greyish brown, turning brown to dark brown on bruising; pores round to angular, 2–4 per mm; dissepiments thin, entire; context up to 2 cm thick, brown; pore tubes up to 2 cm long, concolorous with hymenial surface; margins obtuse, irregular to lobed, sterile up to 3 mm, yellowish white to concolorous with abhymenial surface.

Hyphal system monomitic. Generative hyphae up to 8.5 μm wide, branched, simple-septate, thin- to thick- walled, subhyaline to yellowish brown. Setal hyphae absent. Hymenial setae 32.0–96.0×6.0–20.0 μm, hooked, frequent in hymenial layer, thick-walled, projecting up to 20 μm. Basidia 11.0–18.0×5.2–7.8 μm, broadly clavate to subclavate, subhyaline, simple-septate at the base, 4-sterigmate; sterigmata up to 5.2 μm long. Basidiospores 6.0–8.5×5.2–7.2 μm, subglobose, smooth, yellowish brown, thick-walled, dextrinoid, acyanophilous.

Distribution: H.P.: Kullu, Mandi, Shimla.

Collections examined: IBP 39000, 42912, 42917.

Substratum: On stump of Q. *leucotrichophora*, on log of *C. deodara* and on stump of *R. arboretum*

Remarks: This species is characterized by its large sized fructifications, subglobose, dextrinoid basidiospores and strongly ventricose to hooked setae. It has earlier been reported from India by Bakshi (1971), Dhanda (1977) and Sharma (1985, 1995, 2000, 2012).

Inonotus tabacinus (Mont.) G. Cunn., Bull. N.Z. Dept. Sci. Industr. Res., Pl. Dis. Div. **78**: 3, 1948. Plate 6.9f, Fig. 6.22n–r

Fructification annual, occasionally biennial, lignicolous, sessile, solitary or imbricate, coriaceous, attached by broad or narrow lateral base; pileus sessile, applanate, flabelliform to dimidiate; upper surface ferruginous to dark brown, soft tomentose, concentrically zonate; margin acute, concolorous with upper surface, in turned on drying, entire, sterile below; pore surface dark brown, even, shining; pores rounded to angular. Context ferruginous or dark brown, xanthochroic, 1–3 mm thick, duplex, upper soft part composed of loose hyphae, lower hard part composed of compactly arranged hyphae.

Hyphal system monomitic; generative hyphae brown to dark brown, thin- to thick-walled, septate, clamps absent, sparsely branched, acyanophilous, 3–4.4 μm in diam.; setae abundant, 12–15×4–8 μm, thick-walled, dark brown, subulate or ventricose; cystidia absent; basidia hyaline, clavate, 10–14×4–6 μm; basidiospores hyaline, thin-walled, smooth, non-amyloid, ellipsoid, 3–5×2–3 μm.

Distribution: U.K.: NDBR, Dehra Dun (Lachhiwala forest), Rishikesh.

Collections examined: Hem 29094.

Substratum: On stump of *Shorea robusta*, on angiospermic wood.

Remarks: The diagnostic features of this species are annual, thin, applanate, dark brown, xanthochroic, duplex context; and hyaline, smooth, non-amyloid, ellipsoid basidiospores. The species is being reported for the first time from NDBR (U.K.).

Inonotus tenuicarinis Pegl. & Reid, Trans. Br. Mycol. Soc. 47(2): 173. 1964.
Fig. 6.23a–c

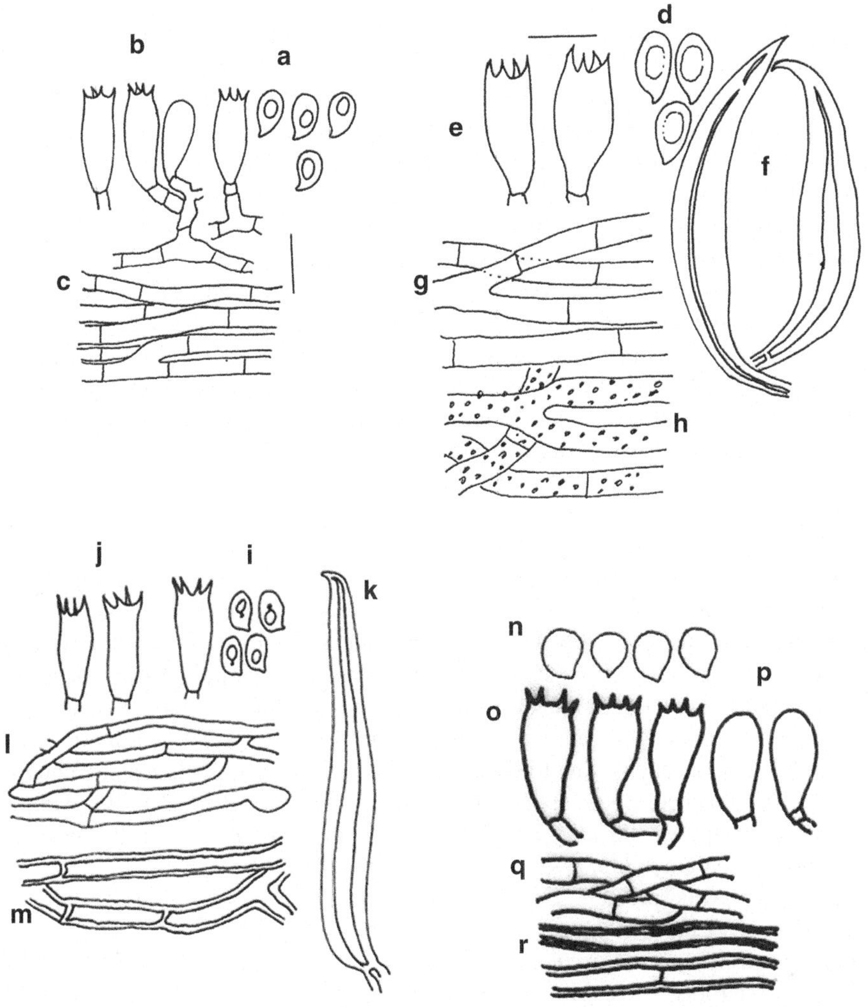

Fig. 6.23 (**a**–**c**) *Inonotus tenuicarinis* (**a**) Basidiospores, (**b**) Basidia and Basidioles, (**c**) Generative hyphae; (**d**–**h**) *Onnia circinata* (**d**) Basidiospores, (**e**) Basidia, (**f**) Setae, (**g**, **h**) Generative hyphae; (**i**–**m**) *Onnia tomentosa* (**i**) Basidiospores, (**j**) Basidia, (**k**) Setae, (**l**–**m**) Generative hyphae; (**n**–**r**) *Phellinus adamantinus* (**n**) Basidiospores, (**o**) Basidia, (**p**) Basidioles, (**q**–**r**) Generative hyphae

Fructification annual, pileate, broadly attached, solitary to accrete in to large agglomeration, soft and spongy when fresh, brittle on drying. Pileus sessile dimidiate to irregularly shaped, convex; upper surface brown, glabrous, concentrically zonate, radially striate, first soft becoming hard with age due to agglutination of cuticle; margin thin to thick, entire or slightly incised and wavy, sterile below, reddish brown to dark brown up to 1.8 mm wide , incurved on drying. Pore surface white when growing and young, brown with touch, yellowish brown to brown with age; pores angular to irregularly shaped, (1) 2–3 per mm. Pore mouth velutinate; pore tubes yellowish brown. Context yellowish brown, spongy, homogenous, xanthochroic up to 2 cm thick.

Hyphal system monomitic; generative hyphae hyaline to light brown, thin to slightly thick-walled, branched, septate, clamps absent, 2.4–9 μm in diameter. Subhymenial hyphae short septate. Hyphae thin to thick-walled in the context, 4.2–8 μm in diameter. Cystidia and setae absent, Basidiospores brown, smooth, slightly thick-walled, broadly ellipsoid, uni-guttulate, 5–7.4 × 3.8–4.4 μm.

Distribution: Meghalaya- Shillong.

Collection examined: SSV 21652.

Substratum: On decaying angiospermic log

Remarks: This is rarely distributed species. The Himalayan collections are typical of *Inonotus trenuicarinus*. The species is marked by annual, broadly arranged, soft spongy fructifications; reddish brown to greyish brown, radially striated upper surface; white to yellowish brown pore surface; large (1-)2-3 pores per mm; yellowish brown context; setae absent; brown, thick-walled, ellipsoid basidiospores. It is a new record for Himalayas.

Onnia P. Karst.,
Bidr. Känn. Finl. Nat. Folk 48: 326 (1889)

Fructification annual, stipitate, rarely sessile, brown, soft when fresh, slightly hard and brittle on drying. Stipe central or eccentric. Pileus stipitate, spathulate to flabelliform; upper surface brown, soft tomentose. Pore surface brown to dark brown, even, dull; pores round to irregular about 3–4 per mm. Tubes not stratified, brown in section. Context yellowish-brown to brown, xanthochroic, duplex, upper part soft and tomentose, lower part hard. Hyphal system monomitic; generative hyphae subhyaline to dark-brown, thin to thick-walled, septate, clamps absent, branched. Setae abundant, dark-brown, thick-walled. Basidia hyaline, clavate, 4-spored. Basidiospores ellipsoid, hyaline to pale brown, thin-walled, smooth, non-amyloid.

Five Species, widespread

Lit.: Wagner & Fischer (*Mycol.* 94: 998, 2002)

Type species: *Trametes circinatus* Fr. 1849.

Habitat: On roots

Himalayas: Two

Key to species

1. Duplex context consisting of well defined upper soft tomentose portion and lower hard and compact layer .... *O. circinata*
1. Duplex context consisting of upper not well defined tomentose layer and lower hard and compact layer .... *O. tomentosa*

Onnia circinata (Fr.) P. Karst., Bidr. Känn. Finl. Nat. Folk 48: 326 (1889). Fig. 6.23d–h

Fructification annual, centrally or eccentrically stipitate, soft tomentose and yellowish-brown with soft tomentose, azonate upper surface; large pores, 2–4 per mm; brown, thick, xanthochroic, duplex context, upper portion soft tomentose and well developed, lower portion hard and compact.

Hyphal system monomitic with hyaline to pale brown, septate, non-clamped generative hyphae 2–5.5 µm wide; Basidia 15–24×6–7 µm, 4-sterigmate; dark brown, thick-walled setae, mostly with hooked apices and hyaline to pale brown, smooth, ellipsoid, non-amyloid basidiospores, 5–6×3–4.1 µm.

Distribution: H.P.: Kullu, Shimla, Narkanda; J&K: Batote, Pehalgam.

Collection examined: SSR 6082, 6109, 6156.

Substratum: On rotten stump, on needles of *Cedrus deodara* and *Pinus excelsa* and exposed roots of *P. excelsa*.

Remarks: The species is characterized by yellowish-brown or golden-brown exposed well developed duplex context with conspicuous upper soft tomentose layer and abundant setae.

Onnia tomentosa (Fr.) P. Karst., Revue mycol., Toulouse 11(no. 47): 205 (1889) = *Coltricia tomentose* (Fr.) Murr. Bull. Torrey Bot. Club 31: 346, 1904. Fig. 6.23i–m

Fructification annual, stipitate to substipitate, terrestrial or growing on wood or roots, soft when fresh, somewhat hard and brittle on drying. Stipe lateral, cylindrical, solid, rusty-brown up to 4.8 cm long and 0.8 cm thick, merging into pileus above. Pileus spathulate to flabelliform; upper surface plane to convex, rusty-brown, soft tomentose, subzonate; margin concolorous with upper surface, acute, inturned on drying, entire or wavy, sterile below. Pore surface rusty-brown to dark-brown, even, dull; pores rounded to somewhat irregular about 3–4 per mm, dissepiments 40–120 µm thick equal apex velutinate. Context yellowish-brown, xanthochroic, duplex, upper soft tomentose layer not well developed up to 1 mm thick, lower layer hard and compact up to 2 mm thick.

Hyphal system monomitic; generative hyphae subhyaline to dark brown, thin to slightly thick-walled, septate, clamps absent, branched, 4–7 µm in diameter. Setae dark brown, thick-walled, subulate, with straight apices, immersed in the hymenium or projecting into pore cavity. Basidia hyaline, clavate, 4-spored up to 4.5 µm in diameter. Basidiospores hyaline to pale brown, thin-walled, smooth, ellipsoid, minutely apiculate, non-amyloid, 5–6.8×3–4.1 µm.

Distribution: A.P.: West Kameng, Rupa.

Collection examined: SSV 21828.

Substratum: At the base of angiospermic stump, on the exposed roots, at the base of decaying stump.

Remarks: This species was previously described from India by Bakshi (1955, 1971) as *Polyporus tomentose* Fr. from the temperate region all over the Himalayas and his description of *P. tomentose* includes *P. circinata* Fr. (= *Onnia circinata* (Fr.) P. Karst.) as a variety *circinatus* under it. *O. circinata* differs from *O. tomentosa* in having mostly hooked setae, central or eccentric stipe and well developed upper soft tomentose layer in duplex context. It is not a very common species in N.W, Himalayas. It is characterized by annual, stipitate fructification; xanthochroic, duplex context consisting of two indistinct layers, upper poorly developed tomentose layer; and dark brown, subulate setae with straight apices.

Phellinus Quél,
Enchir. Fung.: 172 (1886)

Fructifications resupinate to effused-reflexed or distinctly pileate, hard and woody, rarely coriaceous, usually very thick; hymenial surface poroid, yellowish brown to brown; margin usually thick and concolorous. Tissues permanently darkening or blackening in KOH sol. Context yellowish brown to ferruginous, homogenous. Hyphal system dimitic; generative hyphae always without clamps; skeletal hyphae yellowish brown to brown. Pores usually stratose. Cystidia or Gloeocystidia absent. Setae present or absent. Basidia clavate, 4-spored. Basidiospores subhyaline to brown, smooth, nonamyloid, acyanophilous.

About One hundred eighty species, widespread

Lit.: Ryvarden & Johansen (Prelim. Polyp. Fl. E. Afr.: 129, 1980; key 62 Afr. spp.), Larsson & Coldo Poulle (Syn. Fung. 3, 1990; world key), Ryvarden (Syn. Fung. 19: 229 pp., 2004)

Type Species: *Boletus igniarius* L.1753.

Habitat: Wood, bark of living angiospermous tree

Himalayas: Fourteen

Key to species

1. Spores hyaline 2
1. Spores coloured, pale brown to dark brown 8
2. Setae absent 3
2. Setae present 4
3. Pores 6–8 per mm*P. adamantius* ....
3. Pores 4–6 per mm*P. igniarius* ....
4. Both setae and setal hyphae present*P. ferruginosus* .... [a]
4. Only setae present, setal hyphae absent 5
5. Context homogeneous 6
5. Context duplex, marked into upper and lower layers 7
6. Basidiospores broadly ellipsoid, 4–5 × 2.4–2.8 μm .... *P. gilvus*
6. Basidiospores subglobose to ellipsoid, 3.5–5.6 × 2.5–3.5 μm .... *P. nilgheriensis*

7. Upper surface compactly tomentose, concentrically sulcate; pores 5–6 per mm; context 2 cm thick .... *P. torulosus*[a]
7. Fructification annual; upper surface soft tomentose, later glabrous, faintly zonate; pores 3–4 per mm; context up to 1 mm thick .... *P. xeranticus*
8. Setae absent 9
8. Setae present 13
9. Pores up to 6 per mm 10
9. Pores small, more than 6 per mm 11
10. Basidia 10–14 × 6–7.6 μm , Basidiospores 5–6 × 4–5 μm .... *P. merrillii*
10. Basidia about 9 × 4.1 μm, Basidiospores 4.1–5.4× 3.5–3.9 μm .... *P. grenadensis*
11. Pores rounded, 6–8 per mm .... *P. fastuosus*
11. Tubes distinctly stratified; pores 7–10 per mm 12
12. Pores 6–8 per mm; basidiospores dark brown .... *P. caryophylli*
12. Pores 7–10; basidiospores pale rusty brown .... *P. allardii*
13. Pores large, 3–4 per mm .... *P. pini*[a]
13. Pores small, more than 5 per mm 14
14. Fructification effused reflexed; upper surface not rimose .... *P. johnsonianus*
14. Fructification applanate; upper surface rimose 15
15. Context concentrically zonate with thin black lines; basidiospores subhyaline to pale golden brown 16
15. Context not concentrically zonate; basidiospores dark brown .... *P. sanfordii*
16. Basidiospores 2.8–4.5 × 2.8–3.8 μm .... *P. linteus*
16. Basidiospore 4–5.0 × 3.8–4 μm .... *P. setulosus*

[a]Extra limital, not included in the text

Phellinus adamantinus (Berk.) Ryv., Norw. J. Bot. 19:234,1972. Plate 6.10a, Fig. 6.23n–r

Fructification perennial, pileate, solitary, sessile, attached by a narrow base, woody hard. Pileus applanate to dimidiate, sessile up to 8 × 4 × 2 cm; upper surface yellowish brown to dark brown, grey with age, tomentose, glabrous later, concentrically sulcate, dark brown to dark greyish brown, crust present; margin yellowish brown acute to blunt, entire, sterile below. Pore surface yellowish brown to greyish brown, even; pores small, round 6–8 per mm; pore mouth finely velutinate. Context yellowish brown to dark brown, azonate, homogenous, separated with thin, black and shining crust above, xanthochroic up to 7.5 mm thick.

Hyphal system dimitic; generative hyphae hyaline to subhyaline, branched, septate, clamps absent, cyanophilous, 2.1–4 μm in diameter; skeletal hyphae pale brown, thick-walled, aseptate, sparsely branched, dominate in the dissepiments and context, cyanophilous, 3.4–6.4 μm in diameter. Cystidia and setae absent. Basidia hyaline, clavate, 4-spored, up to 5 μm in diameter. Basidiospores hyaline, thin-walled, smooth, broadly ellipsoid to subglobose, non-amyloid, 3–4 × 2–3 μm.

Distribution: Bhutan- Bunakha, Thimphu.

Collection examined: SSV 21278, 21288.

Substratum: On decaying stump in mixed forest.

Plate 6.10 (**a**) *Phellinus adamantinus*. (**b**) *Phellinus allardii*. (**c**) *Phellinus caryophylli*. (**d**) *Phellinus fastuosus*. (**e**) *Phellinus gilvus*. (**f**) *Phellinus igniarius*

Remarks: *Phellinus adamantinus* is distinguished by perennial, woody hard fructification; concentrically sulcate, dark brown to dark greyish brown crust on upper surface and hyaline, thin-walled, broadly ellipsoid to subglobose, small basidiospores. *P. adamantinus* is close to *P. rhytipholeus* (Mont.) Ryv. However, latter differs in having light brown upper surface and larger thick-walled and globose 4–5 μm basidiospores. It has already been recorded from Himalayas.

Phellinus allardii (Bres.) S. Ahmad, Basidiomyc. W. Pakist. **6**: 57 (1972).
Plate 6.10b, Fig. 6.24a–d

Fructification perennial, effused-reflexed to pileate, sessile, solitary or imbricate, broadly attached, hard and medium weight. Pileus sessile semicircular or applanate, ungulate, convex, triquetrous in section 2.5–8 × 1.5–3 × 2 cm; upper surface reddish, rusty brown to greyish black, fine tomentose when young, later becomes glabrous, concentrically sulcate, slightly rimose on drying; margin reddish brown, thin, acute. Pore surface reddish brown to greyish brown, glancing in incident light, new pore layer develops frequently, younger parts distinctly brighter than the older parts; pores round, small 7–10 per mm; tubes distinctly to indistinctly stratified, reddish brown. Context reddish brown to brown, fibrous, azonate, homogenous, xanthochroic with a black crust above it.

Hyphal system dimitic; generative hyphae hyaline to pale yellow, thin-walled, branched, septate, clamps absent, 1.8–3 μm in diameter; skeletal hyphae golden to rusty brown, thick-walled unbranched, aseptate, 2.1–4.7 μm in diameter. Setae and cystidia absent, Basidia collapsed on drying. Basidiospores pale rusty brown, thin-walled, smooth, broadly ellipsoid, minutely apiculate, aguttulate, 4–4.9 × 2.4–3.5 μm.

Distribution: A.P.: West Kameng; Bhutan; U.K.: NDBR.

Collection examined: SSV 21459, 21275, IBP 42067.

Substratum: On decaying angiospermic logs, *Rhododendron* logs.

Remarks: Bakshi (1971) reported this species as *Fomes allardii* Bres. and mentioned it as very common on roots, stumps and trunks of living Indian oaks in the temperate region all over the western Himalayas. *Phellinus allardii* is close to *P. caryophylli*. However, latter has 7–9 pores per mm; and light brown to brown, slightly thick-walled, 3.6–4.5 × 2.8–3 μm basidiospores.

Phellinus caryophylli (Racib.) G. Cunn., Bull. N.Z. Dept. Sci. Industr. Res., Pl. Dis. Div. **164**: 238 (1965) = *Fomes caryophylli* (Racib.) Bres., Annls mycol. 10(5): 498 (1912). Plate 6.10c, Fig. 6.24e–i

Fructification perennial, sessile effused-reflexed to pileate, solitary to imbricate, hard, woody, dimidiate; upper surface dark brown, lighter towards margin, sulcate with ridges; hymenial surface brown, pores small, angular, pore tubes stratified; margin brown, blunt, obtuse. Pore surface yellowish brown to brown, even, smooth, glancing; pores small, angular, 7–9 per mm, dissepiments 32–50 μm thick; pore mouth velvety, tubes indistinctly stratified, each layer up to 2 mm deep. Context brown, fibrous 3.2 mm thick.

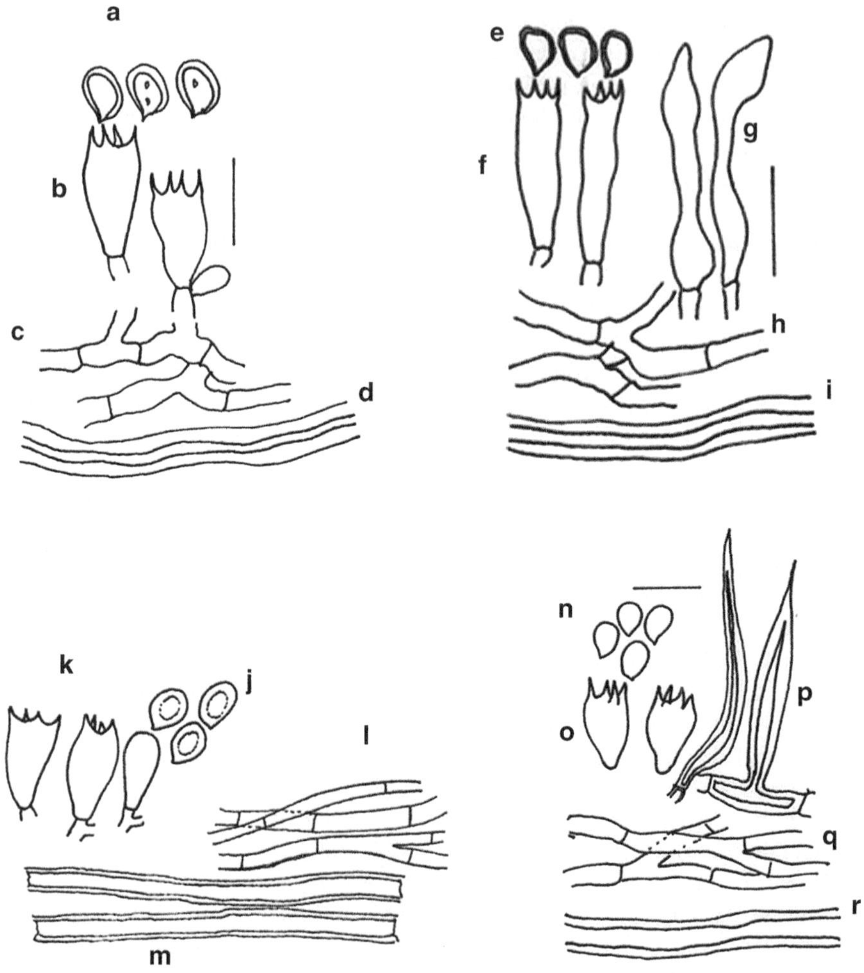

Fig. 6.24 (**a**–**d**) *Phellinus allardii* (**a**) Basidiospores, (**b**) Basidia, (**c**) Generative hyphae, (**d**) Skeletal hyphae; (**e**–**i**) *Phellinus caryophylii* (**e**) Basidiospores, (**f**) Basidia, (**g**) Cystidia, (**h**) Generative hyphae, (**i**) Skeletal hyphae; (**j**–**m**) *Phellinus fastuosus* (**j**) Basidiospores, (**k**) Basidia, (**l**) Generative hyphae, (**m**) Skeletal hyphae; (**n**–**r**) *Phellinus gilvus* (**n**) Basidiospores, (**o**) Basidia, (**p**) Setae, (**q**) Generative hyphae, (**r**) Skeletal hyphae

Hyphal system dimitic; generative hyphae hyaline, branched, septate, clamps absent; skeletal hyphae brown, thick-walled, unbranched, aseptate with narrow lumen; setae absent. Basidia clavate, 10.4–12×4–5.2 µm. Basidiospores light brown, smooth, ellipsoid 4×2.5 µm.

Distribution: Bhutan: Thimphu; A.P.-West Kameng; H.P.: Dalhousie- Panjpulla, Chamba- Thali; Shimla- The Glen, Kullu- Pulga; Manali- Hidamba Devi, Dharamsala- Forsyth ganj.

Collection examined: SSR 6092, 6093, IBP 42921, 42927.

Substratum: stumps and fallen twigs under angiospermic forest and mixed forest, burnt logs, dead wood, roots of *Rhododendron*, *Quercus semecarpifolia*, base and stumps of *Quercus incana*, *Cedrus deodara*, *Juglans regia*.

Remarks: The species has been reported earlier by Bagchee (1950, 1961) on *Shorea robusta*, *Anogenesis latifolia*, *Syzigium* sp. Thind et al. 1970 described it as *Fomes caryophylli*. It causes sap and heart rot of Sal. The species is characterized by perennial, effused reflexed with dark brown, concentrically sulcate, velvety upper surface; stratified tubes, absence of setae, dark brown, smooth, ellipsoid basidiospores. It is close to *P. pectinatus* (Klot) Quél. But the latter has large, ungulate fructifications with brown smooth upper surface and duplex context.

Phellinus fastuosus (Lév.) Ryv., Norw. J. Bot. 19(3 & 4): 234, 1972. Plate 6.10d, Fig. 6.24j–m

Fructification perennial, sessile, solitary or imbricate, rigid, applanate, somewhat convex; pileus surface yellowish brown to dark brown, compactly tomentose, concentrically sulcate, velvety, rimose; margin acute to blunt, rigid, entire, concolorous with pileus surface; hymenial surface yellowish brown to brown, even; pores rounded to angular; pore tubes finely velutinate, stratified, brown. Context yellowish brown, azonate, homogenous, xanthochroic, separated by a thin black line from tomentose on upper surface.

Hyphal system dimitic; generative hyphae thin-walled, hyaline, septate, clamped, branched, 2–3 μm in diam.; skeletal hyphae yellowish brown, thick-walled, rarely branched, aseptate or pseudoseptate, acyanophilous, 3–7 μm in diam.; setae absent. Basidia hyaline, clavate, 4-sterigmate, 8.6–12.7 × 6–7 μm. Basidiospores brown to dark brown, ovoid to subglobose, smooth, thin-walled, nonamyloid, 5–6 × 4.5–5 μm.

Distribution: U.K: NDBR.

Collection examined: Hem 28247.

Substrate: on gymnospermic tree.

Remarks: The species is characterized by perennial, solitary or imbricate fructification; hymenial surface yellowish brown; hyphal system dimitic; basidia clavate, 4-sterigmate; ovoid to subglobose, smooth basidiospores.

Phellinus gilvus (Schwein.) Pat., Essai Tax. Hyménomyc. (Lons-le-Saunier): 82 (1900) = *Phellinus scruposus* (Fr.) G.H.Cunn., Bull. N.Z. Dep. Scient. Ind. Res. 164: 230, 1965. Plate 6.10e, Fig. 6.24n–r

Fructification annual to perennial, sessile, imbricate, corky when fresh, hard on drying. Pileus applanate, effused-reflexed; upper surface brown, reddish brown to rusty brown near the base, concentrically sulcate, scrupose with coarse erect tufts; margin thin, incurved on drying, dentate. Pore surface brown, even, dull; pores small round, 5–8 per mm; dissepiments equal, 32–50 μm thick; pore mouth velutinate; tubes in one layer. Context golden brown, fibrous, homogenous, xanthochroic.

Hyphal system dimitic; generative hyphae thin-walled, branched, septate, hyaline, clamps are absent, 2.3–3.5 μm in diameter; skeletal hyphae thick-walled, aseptate, unbranched, 3–4.6 μm in diameter. Setae abundant, thick-walled, subulate to subventricose, apices acute projecting beyond hymenium. Basidia thin-walled, hyaline, clavate, 4-spored, 9.6–13 × 3.8–4.2 μm. Basidiospores hyaline, smooth, thin-walled, broadly ellipsoid, 4–5 × 2.4–2.8 μm.

Distribution: Meghalaya- Shillong; H.P: Chamba- Thali, Dharamsala- Dadh, Sidhbari; Shimla- Narkanda, Dalhousie- Jandrighat springs, Manali- Gojra.

Collection examined: SSV 21061, 21653, SSR 6119, Dhanda 6871, L 37064, IBP 42069.

Substratum: On decaying fencing log.

Remarks: The species was first reported by Berkley from India (1856) as *Polyporus scruposus* Fr. Based on Dr. Hooker's collection. It is a very common species in N.W. Himalaya. The species is characterized by annual, sessile solitary or imbricate fructification with yellowish brown to reddish brown tomentose upper surface, blunt margin; brown pore surface, minute pores 6–8 per mm, subulate setae.

Phellinus grenadensis (Murr.) Ryv., Norw. J. Bot. 19: 234, 1972. Fig. 6.25a–d

Fructification perennial, sessile, imbricate, hard and woody both when fresh as well as dry, light in weight. Pilei sessile, ungulate; upper surface dark brown, older portions turning brownish-black, closely concentrically sulcate, glabrous; margin blunt to obtuse, yellowish brown or concolorous with upper surface, entire, sterile below. Pore surface yellowish brown to dark brown, even, dull; pores rounded. Context yellowish-brown, homogenous, azonate, limited on the upper surface by black, thin and shining crust, xanthochroic; tubes distinctly stratified, separated by context in between, brown in section.

Hyphal system dimitic; generative hyphae hyaline, thin-walled, septate, clamps absent, sparsely branched, acyanophilous, 1.5–2.8 μm in diameter; skeletal hyphae pale brown to dark brown, slightly thick-walled, aseptate, mostly unbranched, acyanophilous, 1.4–4.2 μm in diameter. Setae absent. Basidia hyaline, clavate, 4-spored, about 9 × 4.1 μm. Basidiospores pale brown to brown, thin-walled, smooth, ovoid to subglobose, 4.1–5.4 × 3.5–3.9 μm.

Distribution: H.P.: Dharamshala.

Collection examined: I.P.S Khurana 7549.

Substratum: On cut stump under angiospermic forest.

Remarks: This collection resembles *P. grenadensis* in all respects which is marked by perennial, hard and woody fructifications with dark brown, closely concentrically sulcate upper surface; obtuse margin; and brown, ovoid to subglobose basidiospores.

Phellinus igniarius (L.) Quél., Enchir. Fung. (Paris): 177 (1886) = *Phellinus trivialis* (Bres.). Plate 6.10f, Fig. 6.25e–i

Fructification perennial, effused-reflexed to conchate, sessile, imbricate, hard and woody both when fresh and dry. Pileus sessile to effused-reflexed,

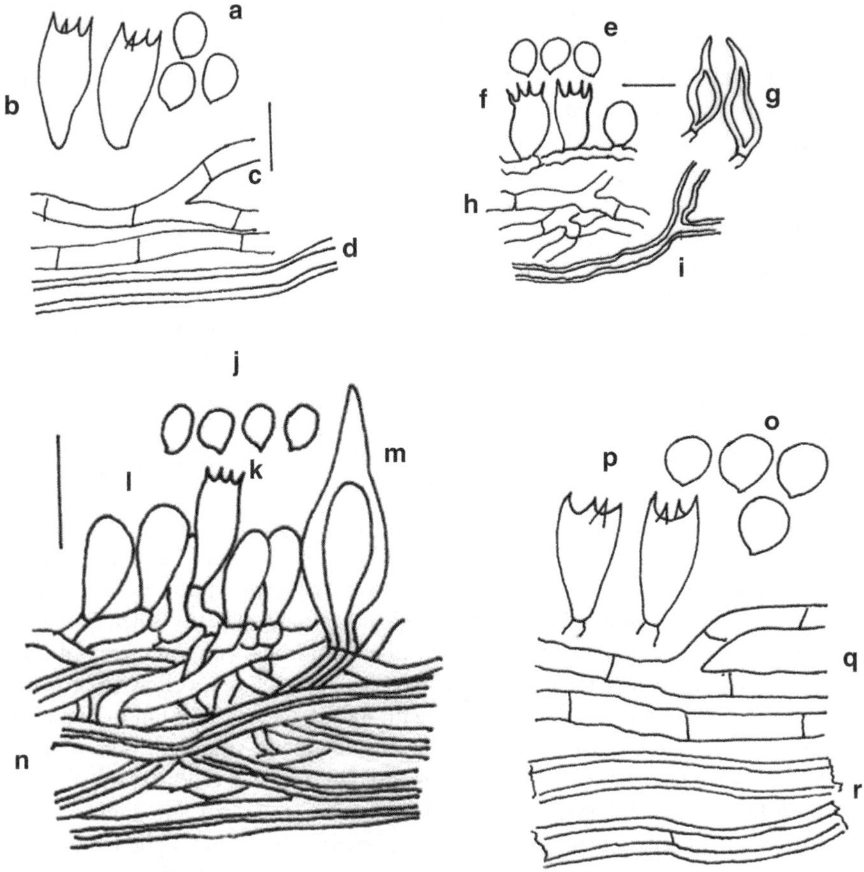

Fig. 6.25 (**a–d**) *Phellinus grenadensis* (**a**) Basidiospores, (**b**) Basidia, (**c**) Generative hyphae, (**d**) Skeletal hyphae; (**e–i**) *Phellinus ignarius* (**e**) Basidiospores, (**f**) Basidia, (**g**) Setae, (**h**) Generative hyphae, (**i**) Skeletal hyphae; (**j–n**) *Phellinus johnsonianus* (**j**) Basidiospores, (**k**) Basidia, (**l**) Basidioles (**m**) Setae, (**n**) Skeletal hyphae; (**o–r**) *Phellinus merrillii* (**o**) Basidiospores, (**p**) Basidia, (**q–r**) Generative hyphae

1.5–3 × 3–6 × 0.5–1.5 cm upper surface brown, older portion turning dark brown to greyish black, not rimose, concentrically sulcate, pubescent; margin blunt or obtuse, concolorous with upper surface, entire, somewhat wavy, sterile below, sterile border up to 3 mm wide. Pore surface yellowish brown to dark brown, even, dull; pores rounded, 100–150 μm in diameter, about 4–6 per mm, pore mouth finely, velutinate; dissepiments 60–180 μm thick, equal, entire, composed of very compact parallel hyphae. Context yellowish-brown, azonate, homogenous, limited on upper surface by thin black line, xanthochroic, up to 4 mm thick; tubes indistinctly stratified brown in section, up to 5 mm deep.

Hyphal system dimitic; generative hyphae hyaline to pale brown, thin-walled, septate, clamps absent, branched, acyanophilous, 2–3 µm in diameter; skeletal hyphae yellowish brown, thick-walled, aseptate to sometimes pseudoseptate, mostly unbranched, acyanophilous, 3–6.3 µm in diameter. Setae absent. Basidia hyaline, clavate, 4-spored, up to 6.3 µm in diameter. Basidiospores hyaline, thin-walled, smooth, some uniguttulate, some apiculate, dextrinoid, globose to subglobose, 5–6.3 × 5.3–5.6 µm.

Distribution: Haryana: Surajpur.

Collection examined: Bhatnagar 7314, IBP 37065.

Substratum: On dead wood of *Salix sp*.

Remarks: The species is characterized by perennial, effused-reflexed, imbricate, concentrically sulcate fructification; dark brown pore surface; dimitic hyphal system; hyaline, globose to subglobose, dextrinoid basidiospores.

Phellinus johnsonianus (Murr.) Ryv., Norw. J. Bot. 19:234, 1972. Plate 6.11a, Fig. 6.25j–n

Fructification perennial, effused-reflexed to pileate, solitary to imbricate, broadly attached, woody hard. Pileus sessile, dimidiate to applanate, conchate, flat to convex, up to 4 cm wide 6.5 cm broad, and 1 cm thick; upper surface reddish brown, brown to smoky brown, fine tomentose, concentrically sulcate; margin even, smooth, obtuse, yellowish brown to concolorous with the upper surface, sterile to fertile below, up to 0.6 mm wide. Pore surface even, yellowish brown to brown; pores small, angular, invisible with naked eye, 62–112 µm in diameter; dissepiment equal, 28–77 µm thick; tubes indistinctly stratified, brown, up to 2.3 mm deep in each layer. Context firm, golden brown to reddish brown with a thick dark brown cuticle, homogenous, xanthochroic, up to 4 mm thick.

Hyphal system dimitic; generative hyphae hyaline to yellow, thin-walled, branched, septate, 2.2–3 µm in diameter; skeletal hyphae dominate the dissepiment and context, yellow to dark brown, aseptate, unbranched, 2.6–4.2 µm in diameter. Setae present, rare, thick-walled, brown, ventricose, acute, 13–23.5 × 6–9 µm. Basidia thin-walled, hyaline, clavate, up to 4.6 µm wide. Basidiospores pale yellow to brown, slightly thick-walled, smooth, broadly ellipsoid to subglobose, minutely apiculate, 3.5–4.0 × 2.9–3.4 µm.

Distribution: U.K.: NDBR- Reni, Lata, Juma.

Collection examined: SSV 21583, Hem 28222, 28246, 28279.

Substratum: On decaying angiospermic logs.

Remarks: In India, this species was described for the first time by Thind et al. (1970) as *Fomes johnsonianus* Murr. *Phellinus johnsonianus* is close to *P. dependens* (Murr.) Ryv. However, latter has pendent, ungulate fructifications with tapering base; smaller setae 12–18 × 5–8 µm; and pale reddish brown basidiospores. *Phellinus johnsonianus* is marked by effused- reflexed yellowish brown fructifications; ventricose, thick-walled setae; and pale yellow to brown, ellipsoid to globose basidiospores. It is a new record for NDBR (U.K.)

Plate 6.11 (**a**) *Phellinus johnsonianus*. (**b**) *Phellinus linteus*. (**c**) *Phellinus merrillii*. (**d**) *Phellinus nilgheriensis*. (**e**) *Phellinus sanfordii*. (**f**) *Phellinus xeranticus*

Phellinus linteus (Berk. & M.A. Curtis) Teng, Chung-kuo Ti Chen-chun, [Fungi of China]: 762 (1963). Plate 6.11b

Fructification perennial, sessile, imbricate, sometimes effused-reflexed, soft when fresh, hard on drying. Pileus sessile, dimidiate to semi-circular, applanate; upper surface dark reddish dark chestnut -brown, concentrically zoned and sulcate, later glabrous and grey to black from the base, concentrically and radially cracked, frequently covered with mosses from the base. Pore surface yellowish brown to reddish-brown; margin acute to rounded, entire, velutinate, paler; tubes slightly paler than context; pores round to weakly angular. Context yellowish-brown to dark brown, fibrous, homogenous, hard, concentrically zonate.

Hyphal system dimitic; generative hyphae thin-walled, branched, septate, hyaline, clamps are absent, cyanophilous, up to 2.8 μm in diameter; skeletal hyphae thick-walled, aseptate, unbranched, acyanophilous, 2.8–4.2 μm in diameter. Setae brown, abundant, thick-walled, with straight pointed apices. Basidia sub-hyaline, thin-walled, smooth, clavate, 2-4-spored, about 9.9–12.2 × 6–7.2 broad. Basidiospores golden yellow to rusty brown, smooth, thin to slightly thick-walled, apiculate, non-amyloid, ovoid to subglobose, 4.6–5.5 × 3.5–4.6 μm.

Distribution: U.K.: NDBR.

Collection examined: Hem 28958.

Substratum: On decaying gymnospermic tree.

Remarks: The species was described by Bakshi (1971) as *Fomes linteus* (Berk. & Curt.) Cook. from Western Himalaya and South India. It is not a very common species in N.W. Himalaya. The above cited collections are identical with the description given by Bakshi (1971).

Phellinus merrillii (Murr.) Ryv. Norw. J. Bot. **19** : 234, 1972. Plate 6.11c, Fig. 6.25o–r

Fructification perennial, sessile, solitary, pileate, broadly attached, woody hard when dry; pileus ungulate to conchate, up to 10 × 9 × 6 cm near the base; upper surface dark yellowish brown to reddish brown, matted tomentose to coarsely scrupose, concentrically sulcate in broad to narrow zones, in age becoming glabrous and blackish from the base, sometimes covered with mosses, more or less weakly cracked, crust as a distinct layer only in older and glabrous specimens. Context up to 3 cm thick, shiny, golden brown, hard, woody, concentrically zonate, mostly dense and resinous on drying; margin round, entire, yellowish brown, velutinate to tomentose in actively growing specimens; pore surface dark cinnamon to dark purplish brown, glancing; tubes indistinctly stratified, up to 8 mm thick, yellowish brown; pores round, regular, 4–5 per mm, dissepiments entire and thick.

Hyphal system dimitic, generative hyphae simple septate, thin to slightly thick-walled, hyaline to pale yellow, 2–3.5 μm wide; skeletal hyphae thick-walled, dark brown; hymenial setae absent. Basidia 10–14 × 6–7.6 μm, broadly clavate. Basidiospores ovoid to subglobose, pale yellow to rusty brown, 5–6 × 4–5 μm in diameter.

Distribution: U.K.: NDBR.

Collection examined: Hem 28189, IBP 37067
Substrate: On angiospermic tree, on gymnospermic wood.
Remarks: The species is characterized by thick, sulcate, matted pileus, absence of setae and subglobose coloured spores.

Phellinus nilgheriensis (Mont.) G. H. Cunn., *Bull. N.Z. Dep. Scient. Ind. Res.* **164**: 226, 1965. Plate 6.11d, Fig. 6.26a–d

Fructification biennial or perennial, sessile, imbricate, sometimes effused-reflexed, soft when fresh, hard on drying. Pileus sessile, applanate to conchate; upper surface brown to reddish-brown densely tomentose near margin. Pore surface golden brown to brown, even, shinning; pores rounded; margin paler concolorous with upper surface, blunt, sterile below. Pore surface golden brown to brown, even, shiny, sometimes cracking on drying. Pores rounded to angular, about 6–8 per mm, pore mouth finely velutnate, tubes stratified, golden-brown to brown in section, up to 1 cm deep, dissepiments 14–68 µm, thick, equal. Context brown, homogenous, azonate xanthochroic, limited on the upper surface by hard, black and shiny crust.

Hyphal system dimitic; generative hyphae thin-walled, branched, septate, hyaline, clamps are absent, cyanophilous, 2.1–2.8 µm in diameter; skeletal hyphae thick-walled, aseptate, unbranched, a cyanophilous, 2.8–4.8 µm in diameter. Setae brown, abundant, thick-walled, with straight apices. Basidia hyaline, smooth, clavate, 2-4-spored, about 5.6–9.7 × 3.5–4.2 µm broad. Basidiospores hyaline, smooth, thin-walled, apiculate, non-amyloid, subglobose to ellipsoid, 3.5–5.6 × 2.5–3.5 µm.

Distribution: H.P.: Shimla- Baghi, Narkanda; Dalhousie- Baloon, Khajjiar.
Collection examined: Dhanda 6279, 6452, IBP 42072, 42073.
Substratum: Bark of living angiosperms, *Juglans regia*, decaying log of *Pinus excelsa*, stumps of *Taxus baccata* and angiospermic stumps.
Remarks: The species was first described by Montagne (1842) from Nilgiri hills (Tamil Nadu), South India. It is not a very common species in N.W. Himalaya. The above cited collections are identical with the description given by Bakshi (1971).

Phellinus sanfordii (Lloyd) Ryvarden, Norw. Jl Bot. 19: 235 (1972) = *Fomes sanfordii* Lloyd, Mycol. Writ. 4 (Synopsis of the Genus *Fomes*): 258 (1915). Plate 6.11e, Fig. 6.26e–i

Fructification perennial, new year's growth is formed below the older portion, imbricate, dimidiate, hard and woody. Pileus applanate, convex, 3–10 × 7–13 × 1.5–3.7 cm, surface dark brown, previous year's portion turns black, rough, concentrically and abundantly zonate, rimose with age, cuticle absent, hairy, hairs fine and abundant, fall off from the older portion, velvety when young. Margin entire to wavy, sometimes slightly cleft at places, acute, yellowish brown. Hymenial surface deep brown, even, margin lighter coloured, sterile. Dissepiments not toothed. Pores stratose, different layers indistinct, rounded, brown in section, straight, 2 mm–1 cm deep, 106–143 µm in diameter, or 4–5 per mm. dissepiments 26–50 µm thick, equal, edges concolorous. Context brown, about 2 mm thick.

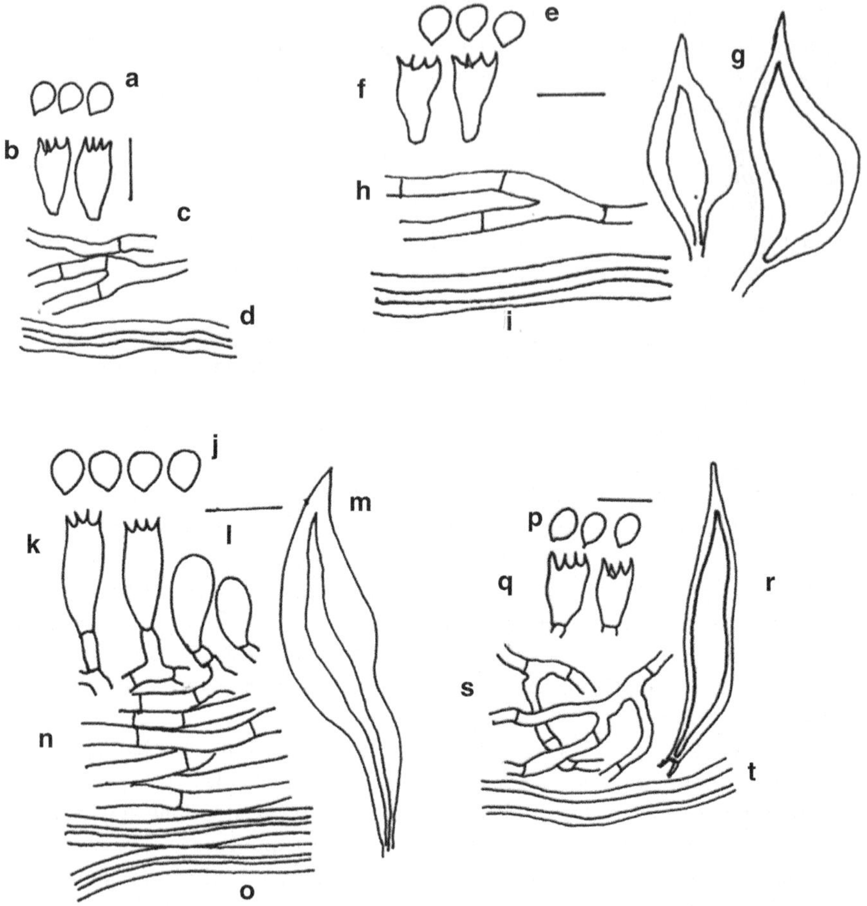

Fig. 6.26 (**a**–**d**) *Phellinus nilgheriensis* (**a**) Basidiospores, (**b**) Basidia, (**c**) Generative hyphae, (**d**) Skeletal hyphae; (**e**–**i**) *Phellinus sanfordii* (**e**) Basidiospores, (**f**) Basidia, (**g**) Setae, (**h**) Generative hyphae, (**i**) Skeletal hyphae; (**j**–**o**) *Phellinus setulosus* (**j**) Basidiospores, (**k**) Basidia, (**l**) Basidioles, (**m**) Setae, (**n**) Generative hyphae, (**o**) Skeletal hyphae; (**p**–**t**) *Phellinus xeranticus* (**p**) Basidiospores, (**q**) Basidia, (**r**) Setae, (**s**) Generative hyphae, (**t**) Skeletal hyphae

Hyphal system dimitic; generative hyphae subhyaline, branched, septate, 1.6–2.4 μm wide, thin-walled, takes stain; skeletal hyphae long, sulphur yellow in colour, unbranched, aseptate, some pseudo-septa also present, 1.5–2.8 μm wide, thick-walled, walls 0.4–0.8 μm thick, do not take stain; Setae not abundant, ventricose, deep brown, 12–23 × 6–8 μm, projecting up to 14 μm beyond the hymenial layer. Basidia honeycomb type, clavate, light brown, 3.0–4.6 μm in diameter. Basidiospores brown, subglobose or ovoid, smooth, 2.4–3.2 × 2.4–2.8 μm.

Distribution: U.K.: Mussoorie, H.P.: Kullu.

Collection examined: 213, IBP 42080, 42081.

Substratum: On living deciduous tree.

Remarks: This species is characterized by its applanate, larger fructifications, zonate pileus becoming rimose with age, dimitic hyphae, and presence of setae.

Phellinus setulosus (Lloyd) Imaz., Bull. Tokoyo Sci. Mus. 6: 104, 1943 = *Fomes setulosus* Lloyd, Mycol. Writ. 4 (Synopsis of the Genus *Fomes*): 243 (1915).

Fig. 6.26j–o

Fructification perennial, pileate, solitary, rarely 2 or 3 fructifications grow closely, sessile, attached by narrow to broad base, rigid and hard on drying. Pileus applanate to ungulate, up to 25 cm long, 22 cm wide and 9 cm thick near the base; upper surface brown to reddish brown to rusty brown, finely tomentose, concentrically zonate, glabrous with age; margin thick, entire, light brown, lighter than the upper surface, rounded, sterile below, up to 5 mm wide; pore surface yellowish brown to reddish brown, even, shiny; pores small, round to subangular, 6–8 per mm; dissepiment thin, 34–83 μm thick; tubes continuous, indistinct into different strata. Context reddish brown, fibrous, homogeneous, xanthochroic, up to 2 cm thick.

Hyphal system dimitic, generative hyphae hyaline to pale yellow, thin-walled, branched, septate, without clamps, 1.8–3.5 μm wide; skeletal hyphae light brown to dark brown, dominate in dissepiment and context, thick-walled, 2.5–4.5 μm in diameter. Setae numerous, dark brown, subulate, ventricose to variable shaped, 20–35 × 8–11 μm. Basidiospores golden yellow to brown, smooth, slightly thick-walled, subglobose to broadly ellipsoid, nonamyloid, 4–5.0 × 3.8–4 μm.

Distribution: A.P.: West Kameng, Bomdilla.

Collection examined: SSV 21444.

Substratum: On decaying angiospermic logs.

Remarks: This species is characterized by perennial, ungulate, rigid and hard fructifications with fine tomentose distinct cuticle; brown to reddish brown upper surface; obtuse margin; pores 6–8 per mm; ventricose, with swollen base, thick-walled, brown setae; and golden brown, slightly thick-walled, subglobose basidiospores. The Himalayan collections, however, possesses slightly smaller basidiospores than reported for east African collection (5–7 × 4–6 μm).

Phellinus xeranticus (Berk.) Peglar., Kew. Bull. 21: 44, 1967 = *Phellinus cereus* (Berk.) Ryv. Norwegian Jl. Bot. 19: 234, 1972. Plate 6.11f, Fig. 6.26p–t

Fructification annual, resupinate, adnate, not separable, spreading irregularly on the substratum, widely effused, up to 4–20 cm, up to 2 mm thick. Margin thinning or abrupt, yellowish brown, myceloid, narrowly sterile. Pore surface yellowish brown to brown, uneven, dull; pores poorly developed in these collections, rounded, about 4–6 per mm; pore mouth finely velutinate. Context brown, thin, homogenous, xanthochroic; tubes not stratified, brown in section, up to 2 mm deep.

Hyphal system dimitic; generative hyphae hyaline, thin-walled, branched, septate, clamps absent, staining in cotton blue, 1.4–3 μm in diameter; skeletal hyphae pale brown to golden brown to dark brown, thin to thick-walled, up to 1.5 μm thick, aseptate, branched, acyanophilous, 2.4–5.3 μm in diameter. Setae dark brown, subulate, thick-walled, straight, pointed apices, 35–49 × 7–9.8 μm. Basidia hyaline, thin-walled, clavate, 4-spored, up to 5.6 μm in diameter. Basidiospores hyaline, thin-walled, smooth, minutely apiculate, non-amyloid, ellipsoid, 4.0–4.8 × 2.0–2.8 μm.

Distribution: H.P.: Dalhousie- Jandrighat springs, Baloon; Shimla- The Glen, Tara Devi. Bhutan, Nepal; Meghalaya- Shillong, West Bengal: Darjeeling

Collection examined: IPSK 7530, R 6441, 6445, 6465, 6725; D 6729, 6890.

Substratum: On angiospermic stumps of *Cedrus deodara* and *Quercus* twigs.

Remarks: This species was described for the first time by Berkeley (1854b) as *Polyporus* (*Resupinatus*) *cereus* Berk. based on Dr. Hooker's collection made from dead wood, high valleys, East Nepal.

Phylloporia Murrill,
Torreya 4: 141(1904)

Fructification annual, effused-reflexed to sessile, imbricate, dimidiate, soft when fresh, hard and woody on drying. Pileus cinnamon to yellowish brown to dark reddish brown, tomentose and spongy; pore surface brown to reddish brown; tubes concolorous with pore surface; pores angular to round; context light to dark brown, separated from a thick tomentum by a thin black line; hyphal system monomitic; hyphae light to golden brown, simple septate; hymenial setae absent; spores ellipsoid, 4–5 × 3–4.5 μm, light yellowish to brownish, slightly thick-walled, on living or dead branches of angiospermous trees.

Seven species, widespread

Lit.: Ryvarden & Johanson (Prelim, Polyp. Fl. E. Africa: 230, 198, Key), Wagner & Ryvarden (*Mycol. Progr*. 1: 105, 2002; phylogeny and taxonomy).

Type Species: *Phylloporia parasitica* Murrill., 1904

Habitat: Wood/Living trees

Himalayas: Three

Key to species

1. Hyphal system monomitic 2
1. Hyphal system dimitic .... *P. ribis*
2. Pores 5–6 per mm .... *P. pectinata*
2. Pores 7–9 per mm ....*P. spathulata*

Phylloporia pectinata (Klotzsch) Ryvarden, Syn. Fung. (Oslo) 5: 196 (1991) = *Fomes pectinatus* (Klotzsch) Gillet, Grevillea 14(no. 69): 20 (1885). Fig. 6.27a–d

Fructification perennial, imbricate, dimidiate, coriaceous when fresh, hard and woody on drying. Pileus sessile, usually convex to compressed, ungulate, 2–6 × 3–10 × up to 1 cm, surface dark brown, rough, abundantly and concentrically sulcate, slightly cracked or not, cuticle absent, hairy, hairs small, soft and fine. Margin entire to wavy, acute, concolorous; hymenial surface brown, older surface turns dark brown, even, smooth and velvety to touch; margin lighter coloured, becoming more or less concolorous on drying, sterile. Dissepiments not toothed. Pores stratose, rounded, 1–2 mm long each season, straight, 58–108 μm in diameter or 5–6 per mm. Dissepiments 32–90 μm thick, equal, edges concolorous, at apex velutinate. Context brown, about 1 mm thick.

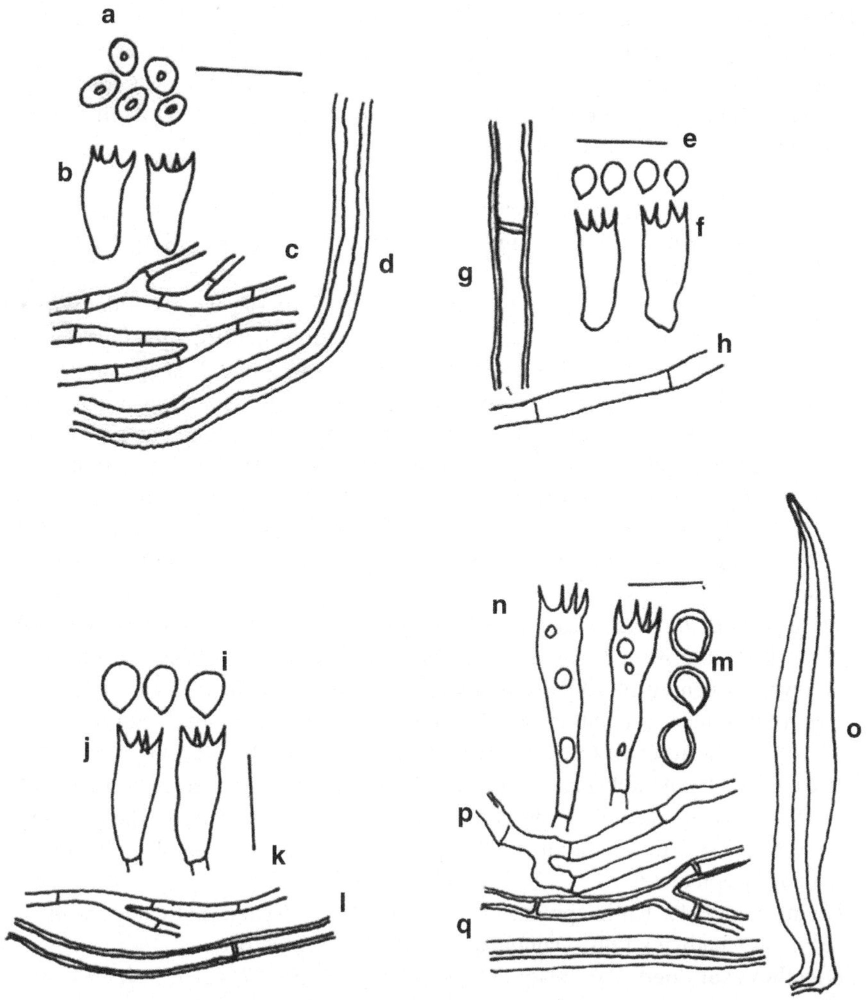

Fig. 6.27 (**a–d**) *Phylloporia pectinata* (**a**) Basidiospores, (**b**) Basidia, (**c**) Generative hyphae, (**d**) Skeletal hyphae; (**e–h**) *Phylloporia ribis* (**e**) Basidiospores, (**f**) Basidia, (**h**) Generative hyphae, (**h**) Skeletal hyphae; (**i–l**) *Phylloporia spathulata* (**i**) Basidiospores, (**j**) Basidia, (**k–l**) Generative hyphae; (**m–q**) *Porodaedalea pini* (**m**) Basidiospores, (**n**) Basidia, (**o**) Setae, (**p**) Generative hyphae, (**q**) Skeletal hyphae

Hyphae monomitic both in the context and trama; hyphae light yellowish brown to deep yellowish brown, or reddish brown, unbranched, rarely branched, pseudo-septate, 2–4.5 μm wide, thick-walled, walls 0.6–1.2 μm thick, do not take stain. Basidia clavate type, oblong, subhyaline, 3.2–5.1 μm in diameter. Sterigmata 4, thin and short. Basidiospores hyaline, subglobose, smooth, 2–3.0 × 1.5–2.5 μm.

Distribution: U.K.: Dehra Dun; H.P.: Dharamsala- Siwalik hills.

Collection examined: 212.

Substratum: Cut stumps and tree of *Murraya exotica*, on living deciduous tree.

Remarks: This species is characterized by the perennial, imbricate fructifications, abundantly sulcate, ungulate, hairy pileus, brown hymenium, stratose pores, brown context, monomitic hyphae, and hyaline, subglobose, smooth basidiospores, 2–3.2 × 1.6–2.8 µm in size.

Phylloporia ribis (Schumach.) Ryvarden, Polyp. N. Eur. (Oslo) 2: 371 (1978) = *Phellinus ribis* (Schum. exFr.) Quél., Ench. Fung., P. 173, 1886. Plate 6.12a, Fig. 6.27e–h

Fructification annual to perennial, imbricate, corky to sub-woody. Pileus up to 5 × 12 × 1.5 cm, effused-reflexed to pileate, sometimes attached by a narrow lateral base, applanate to subconchate to sometimes irregular to shape; upper surface brown near the margin (New Year's growth) but turns brownish-black with age, concentrically zonate to sulcate, softly tomentose in young parts but tomentum withers away in older parts exposing the darker layer below; margin acute, often lobed, paler concolorous; pore surface yellow to yellowish-brown, even, glabrous. Context up to 1.5 cm thick, brown, turning black when touched with KOH, duplex, lower part hard and separated by a dark line from the upper soft part.

Hyphal system dimitic; generative hyphae 2–5 µm wide, subhyaline, sparsely branched, thin-walled, septate, clamps absent; skeletal hyphae 3–6 µm wide, brown, darkening in KOH, unbranched to rarely branched, thick-walled with capillary lumen, aseptate. Pores in strata, each stratum up to 2 mm deep each year, circular to oval, 60–125 µm in diameter, averaging 7–9 per mm; dissepiments 25–80 µm thick, equal, concolorous with the context. Setae absent. Basidia 4–5 µm broad, clavate, 4-spored, sterigmata up to 1.5 µm long. Basidiospores 2.6–3.0 × 2.4–3 µm, ovoid to subglobose, light brown with a dark wall, smooth, minutely apiculate.

Distribution: H.P.: Kullu- pulga, Dharamsala- Mcleod ganj, Yol camp, Sidhbari, Shimla- Chadwick falls, Dalhousie- Khajjiar, U.K.: NDBR.

Collection examined: RW 6046; R 6150; D 6277, 6655, 6745, 6753; K 7541.

Substratum: living angiospermic trees, stumps of *Quercus incana*, *Cedrus deodara*, Buried wood and cut stumps of *Pyrus patio*.

Remarks: The characteristic features of this species are perennial nature, brown context darkening in KOH, narrow pores (7–9 per mm), dimitic hyphal system, absence of setae and duplex context. *Phellinus pectinatus* (Klotz.) Quél. is also similar but differs in having homogenous context and small fructifications with fine sulcations.

Phylloporia spathulata (Hook.) Ryvarden, Syn. Fung. (Oslo) 5: 196 (1991). Fig. 6.27i-l

Fructification annual, laterally stipitate, soft coriaceous when fresh, azonate, light brown to brown; margin acute, sterile below, concolorous, entire; pore surface concolorous with the upper surface, smooth, shiny and soft waxy. Context light brown, darkening in KOH, duplex: lower part (adjoining the tubes) is hard and lighter coloured while the upper part is spongy-soft and deeper in colour.

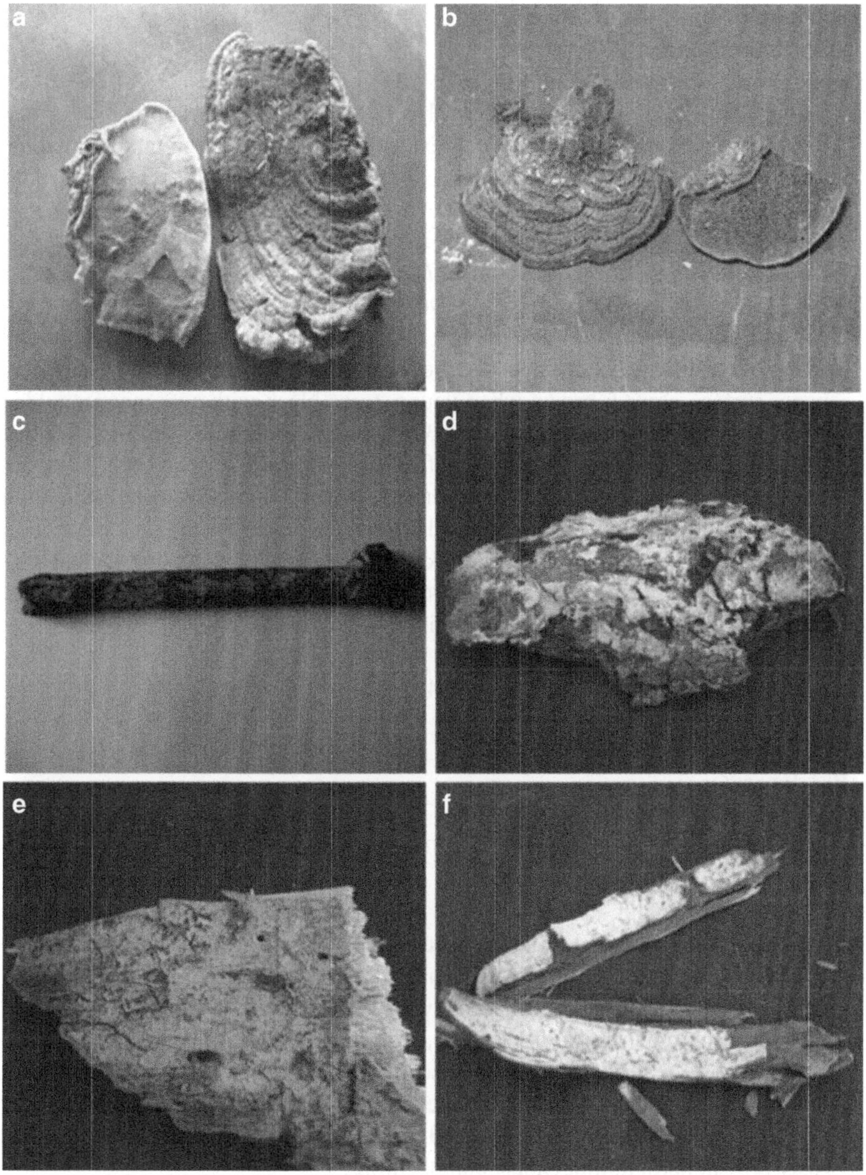

Plate 6.12 (**a**) *Phylloporia ribis*. (**b**) *Porodaedalea pini*. (**c**) *Tubulicrinis glebulosus*. (**d**) *Peniophorella praetermissa*. (**e**) *Peniophorella pubera*. (**f**) *Peniophorella rude*

Hyphal system monomitic, hyphae 3–6 μm wide, light brown, thin-to moderately thick-walled, sparsely branched, septate, clamps absent. Pores not in strata, up to 750 μm deep, circular to oval, 60–100 μm in diameter, averaging 7–9 per mm; dissepiments 30–90 μm thick, equal, concolorous. Setae absent. Basidia 4–4.2 μm broad, clavate, 4-spored. Basidiospores 2.6–3.0 × 2.3–2.7 μm, subglobose, light brown, smooth.

Distribution: H.P.: Kullu, Manali.
Collection examined: RW 6090.
Substratum: On angiospermic stump.
Remarks: This species is characterized by the small stipitate fructifications often reviving for 2 or 3 years, brown context, darkening in KOH, monomitic hyphal system, narrow pores (7–9 per mm) and small, subglobose basidiospores (2.8–3.2 × 2.3–2.8 µm).

Porodaedalea **Murrill,**
W.A. 1905. 367.

Fructification perennial, effused-reflexed to pileate, solitary to imbricate, hard and woody, broadly attached. Pileus sessile, conchate to convex, becomes ungulate; upper surface greyish brown when young, turns black with age; margin golden brown to greyish brown, obtuse, up to 3 mm wide. Pore surface yellowish brown; pores angular, tubes indistinctly stratified. Context rusty brown, hard, homogenous, xanthochroic. Hyphal system dimitic; generative hyphae subhyaline, thin to slightly thick-walled, unbranched, aseptate. Setae abundant, ventricose, enlarged at the base, thick-walled, dark brown. Basidia hyaline, thin-walled, clavate, 4-spored. Basidiospores hyaline to subhyaline, thin to slightly thick-walled, smooth, subglobose to broadly ellipsoid, non-amyloid.
Ten species, wide spread
Lit.: Fiasson & Niemalä (Karstenia 24:14, 1984).
Habitat: Decayed wood
Type species: *Boletus pini* Brot. 1804.
Himalayas: One

Porodaedalea pini (Brot.) Murrill, Bull. Torrey bot. Club 32(7): 367(1905) = *Phellinus pini* (Fr.) Ames., Ann. Mycol. 11: 246, 1913. Plate 6.12b, Fig. 6.27m–q
Fructification perennial, effused-reflexed to pileate, solitary to imbricate, hard and woody, broadly attached. Pileus sessile, conchate to convex, becomes ungulate; upper surface greyish brown when young, turns black with age, narrow to broadly concentrically sulcate, fine velutinate when young, later becomes rough, with a distinct black crush towards the base, rimose on drying, only the margin and zones just within remain fine tomentose; margin golden brown to greyish brown, obtuse, up to 3 mm wide. Pore surface yellowish brown; pores angular, 3–4 per mm; dissepiment equal, 65–244 µm thick; tubes indistinctly stratified. Context rusty brown, hard, homogenous, xanthochroic.
Hyphal system dimitic; generative hyphae subhyaline, thin to slightly thick-walled, unbranched, aseptate, 2.2–3.5 µm in diameter. Setae abundant, ventricose, enlarged at the base, thick-walled, dark brown, apices straight, 18.5–35 × 6–11 µm. Basidia hyaline, thin-walled, clavate, 4-spored, 10–13 × 4.8–5.2 µm. Basidiospores hyaline to subhyaline, thin to slightly thick-walled, smooth, subglobose to broadly ellipsoid, non-amyloid, 3.5–4.5 × 3.5–4 µm.
Distribution: Bhutan- Thimphu; A.P.: West kameng, Bomdilla; Meghalaya: Cherrapunji, H.P.; Dalhousie, Dharamshala, Kinnaur, Kullu, Shimla.
Collection examined: SSV 21324, SSV 21353, IBP 42092.

Substratum: On decaying *Pinus* logs, on living angiospermic stump.

Remarks: This species was previously recorded from India by Bagchee and Bakshi (1950) as *Fomes pini* (Thore: Pers.) Lloyd, and mentioned it as a serious parasite on 'blue pine' all over its range of distribution in the temperate region of Western Himalayas. Thind et al. (1957) described *Fomes pini* in detail. The species is marked by perennial, effused-reflexed to ungulate fructification; greyish brown to black, closely concentrically zonate upper surface; thick-walled, dark brown, ventricose, setae; and hyaline to brown, globose basidiospores.

Tubulicrinis **Donk,**
Fungus 26: 13. 1956.

Fructifications resupinate, mealy floccose to membranous, adnate; hymenial surface smooth but appears hispid under the lens, pale coloured, discontinuous to continous. Subiculum composed of loosely or compactly woven hyphae. Hyphal system monomitic, hyphae often collapsing or agglutinating and becoming indistinct, rarely distinct, branched, septate, clamps present or absent, the walls thin or moderately thick, subhyaline. Gloeocystidia absent. Cystidia with a rooting base, thick-walled with a capillary lumen, lumen uniform or may broaden abruptly near the apex, the walls subhyaline and soluble in 3 % KOH. Basidia clavate-cylindrical, 4-spored. Basidiospores globose to cylindrical or allantoid, the walls thin, subhyaline, smooth and nonamyloid. Or allantoid, the walls thin, subhyaline, smooth and nonamyloid.

Thirty one species, widespread.

Lit.: Hayashi (*Bull. Govt For. Exp. Stn* **260**, 1974), Hjortstam (*Windahlia* **24**: 1, 1998).

Type Species: *Tubulicrinis glebulosus* (Fr.) Donk., 1956.

Habitat: Dead wood

Himalayas: Five

Key to species

1. Lycocystidia conspicuous, umbrella like cap of incrustation present 2
1. Lycocystidia absent. 3
2. Lycocystidia 42–78 × 6–8 μm, slender subulate to obclavate .... *T. hamatus*
2. Lycocystidia 68–95 × 6–8 μm, cylindrical, wide in the middle and narrowing towards neck… .... *T. strangulates*
3. Cystidia cylindrical with obtuse apex, lumen capillary type but expands abruptly in to a cylindrical bulb near the apex .... *T. glebulosus*
3. Cystidia with attenuated and pointed apex. 4
4. Lumen capillary type, not expanding at the top .... *T. chaetophora*
4. Lumen capillary type and expanding in to an acute conical apex .... *T. subulatus*

Tubulicrinis chaetophorus (Höhn.) Donk, Fungus 26: 14. 1956. Fig. 6.28a–d

Fructification resupinate, loosely adnate, widely effused, 48–298 μm thick in section; hymenial surface white to cream when fresh ochre on drying, porose-reticulate to farinose to pilose microscopically, discontinuous not creviced or sometimes irregularly cracked on drying; margin thinning, loosely adnate, white

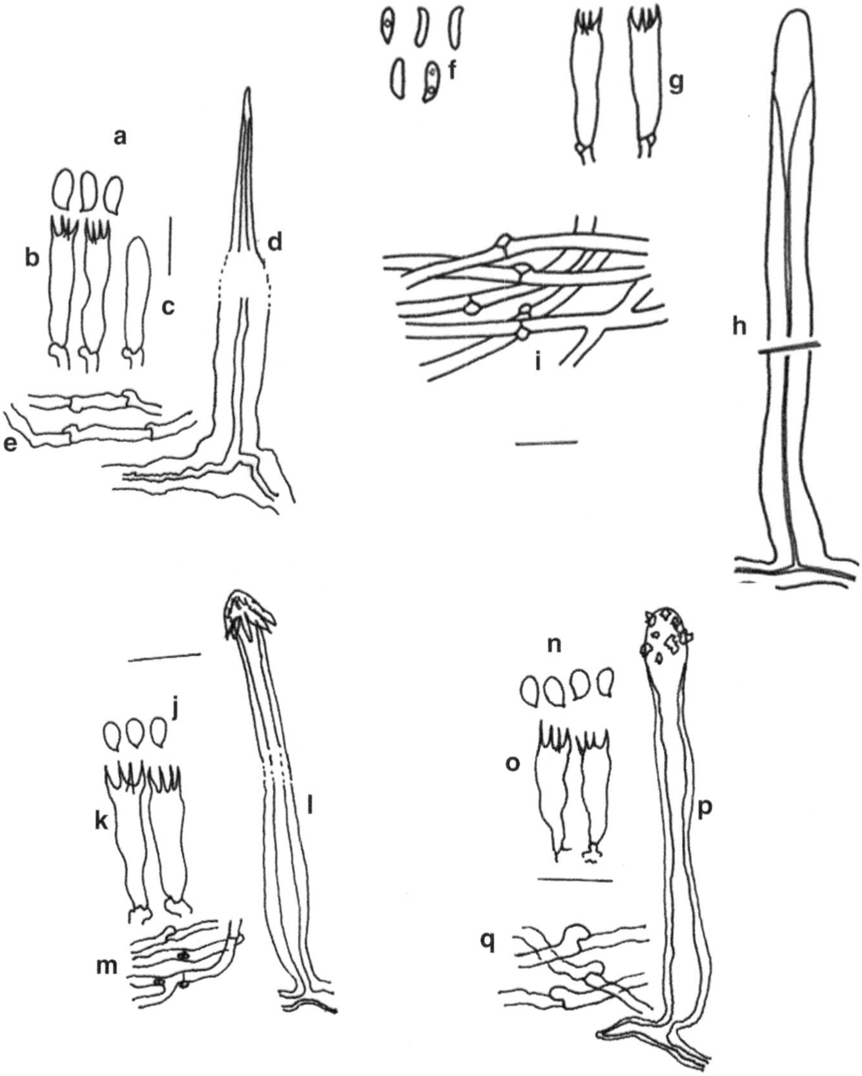

Fig. 6.28 (**a–d**) *Tubulicrinis chaetophorus* (**a**) Basidiospores, (**b**) Basidia, (**c**) Basidioles, (**d**) Cystidia, (**e**) Generative hyphae; (**f–i**) *Tubulicrinis glebulosus* (**f**) Basidiospores, (**g**) Basidia, (**h**) Cystidia, (**i**) Generative hyphae; (**j–m**) *Tubulicrinis hamatus* (**j**) Basidiospores, (**k**) Basidia, (**l**) Cystidia, (**m**) Generative hyphae; (**n–q**) *Tubulicrinis strangulates* (**n**) Basidiospores, (**o**) Basidia, (**p**) Cystidia, (**q**) Generative hyphae

to paler concolorous. Subiculum subhyaline in section, composed of loosely woven hyphae.

Hyphal system monomitic; generative hyphae septate, branched, clamped, thin to thick-walled, 1.5–3.5 μm wide. Cystidia conical with a broad rooting base, 45–175×6.5–11.5 μm, thick-walled, subhyaline, amyloid, soluble in 3 % KOH sol, lumen capillary and not expanding at the top. Basidia clavate to cylindrical, sometimes pleurobasidiate, 4-spored, sterigmata up to 4.0 µm long. Basidiospores ellipsoid, shortly apiculate, thin-walled, smooth, 5–8 × 3–3.5 µm, subhyaline, non-amyloid.

Distribution: H.P.: Chamba- Khijjiar- Dalhousie- Kalatope; Shimla- Chail; J&K: Sanasar.

Collection examined: SSR: 5137, 5152, 5583.

Substratum: stumps of *Cedrus deodara* and *Abies pindrow*, stumps under conifers.

Remarks: The species is characterized by cystidia having pointed apices and amyloid walls. The lumen is narrow and utriform. Some of the basidia are pleurobasidiate.

Tubulicrinis glebulosus (Fr.) Donk [as '*glebulosa*'], Fungus, Wageningen 26(1–4): 14 (1956) = *Tubulicrinis gracillima* (Ell. & Ev.) Cunn., N.Z. Dept. Sci. Ind. Res. Bull. 145: 141, 1963 = *Peniophora gracillima* Ell. & Ev. ex Rog. & Jacks., Farlowia 1 p. 317, 1943. Plate 6.12c, Fig. 6.28f–i

Fructification membranous, resupinate, adnate widely effused; hymenial surface cream to cream brown, smooth to farinose, continuous, not creviced; margin thinning, adnate, concolorous.

Hyphal system monomitic, generative hyphae 1.5–3.6 µm, branched, septate clamped, thin to thick-walled, subhyaline. Cystidia (65–100 × 7.5–9 µm) cylindrical with obtuse apices, thick-walled, lumen capillary but expands abruptly near the apex into a cylindrical bulb; Basidia subcylindrical to narrowly clavate (13.5–25 × 3.5–5 µm), 4-spored with basal clamp. Basidiospores cylindrical to suballantoid, 6–8 × 1.5–2 µm, minutely apiculate, thin-walled, smooth, thin-walled, non-amyloid, acyanophilous.

Distribution: A.P.: West Kameng, Bomdila, West Bengal: Darjeeling; H.P.: Manali, Kinnaur; U.K.: Dehra Dun.

Collection examined: GSD 19771, 19212, IBP 37069, L 37070.

Substratum: On decaying angiospermic log, twigs and broken branches.

Remarks: This species is characterized by cylindrical cystidia with narrow lumen broadening at the top and cylindrical to suballantoid basidiospores. Rehill & Bakshi (1965) reported it for the first time on the basis of a single collection from Dehra dun (North-Western Himalayas). Dhingra et al. (2011) recorded it from Eastern Himalayas. It is now reported from Himachal Pradesh as well as from more localities in Uttarakhand.

Tubulicrinis hamatus (H.S. Jacks.) Donk [as '*hamata*'], Fungus, Wageningen 26(1–4): 14 (1956). Fig. 6.28j–m

Fructification resupinate, effused, thin, loosely adnate, up to 143 µm thick in section; hymenial surface porulose to continuous, greyish white; margin indeterminate, thinning.

Hyphal system monomitic; generative hyphae up to 3.2 µm wide, thin-walled, septate, clamped; subicular hyphae usually straight, branched, parallel to the substrate; hymenial hyphae profusely branched, compactly arranged in to dense texture. Lycocystidia 42–78 × 6–8 µm, slender subulate to obclavate with a conspicuous umbrella like cap of incrustation, thick-walled with a wider lumen at the base, weakly amyloid. Basidia 18–22×5–6.3 µm, subclavate, thin-walled, 4-sterigmate, with a basal clamp; sterigmata up to 3.7 µm long. Basidiospores 4.5–6×3–3.8 µm, ellipsoid, thin-walled smooth, non-amyloid.

Distribution: H.P.: Dalhousie, Manali- Hidamba, reserve forest; U.K.: Mussoorie.

Collection examined: IBP 37071.

Substratum: Decaying gymnospermic stump, *Cedrus deodara* stump, on cut *Pinus* stump.

Remarks: *Tubulicrinis hamatus* is easily recognized by the lycocystidia with a conspicuous umbrella-like cap of encrustation and ellipsoid basidiospores.

Tubulicrinis strangulates Larss. & Hjortst., Mycotaxon 26: 438, 1986. Fig. 6.28n–q

Fructification resupinate, effused, loosely adnate, up to 300 µm thick in section; hymenial surface porulose when young becoming continuous with maturity, white to pale ochraceous.

Hyphal system monomitic; generative hyphae 2–2.8 µm wide, branched, septate, clamped, weakly amyloid; subhymenial hyphae narrower than basal hyphae. Lycocystidia 68–95×6.0–8.0 µm, cylindrical, wide in the middle part, narrowing towards the neck, then widened to an obtuse apex, generally encrusted apically with crystalline matter, often exclusively around the neck; capillary lumen ending gradually or more commonly abrupt in a thin-walled capitulate tip, amyloid. Basidia 12.0–15.0×3.5–4.0 µm, subclavate, thin-walled, non-amyloid, 4-sterigmate, with a basal clamp; sterigmata up to 4.0 µm long. Basidiospores 4.0–6.0×2.0–2.6 µm. Basidiospores 4.0–6.0×2.0–2.6 µm, subreniform, adaxial side concave, thin-walled, smooth, apiculate, acyanophilous, non-amyloid, with oily contents.

Distribution: H.P.: Chamba- Kalatope.

Collection examined: IBP 37072.

Substratum: Decaying stump, wood.

Remarks: This species is characterized by its amyloid cystidia with a neck encrustation, subreniform basidiospores and fairly thin-walled, inamyloid basidia. It is recorded for the first time from Uttarakhand (N.W. Himalayas).

Tubulicrinis subulatus (Bourdot & Galzin) Donk [as 'subulata'], Fungus, Wageningen 26: 14 (1956). Fig. 6.29a–d

Fructification resupinate, adnate, loosely membranous to membranous, effused, up to 145 µm thick in section; hymenial surface cream when fresh turning ochraceous on drying, smooth to somewhat pilose, continuous, not creviced but sometimes may crack irregularly and aerolately in thicker parts; margin indeterminate, loosely adnate, paler concolorous. Context subhyaline in section, comprised of loosely woven hyphae. Hyphae wholly collapsed in older parts and are difficult to discern.

Hyphal system monomitic, hyphae 2–3 µm wide, branched, septate, clamped, thin-walled, subhyaline. Cystidia 49–92×7.8–11 µm, conical with a broad rooting base and pointed apice, smooth to sometimes impregnated with subhyaline crys-

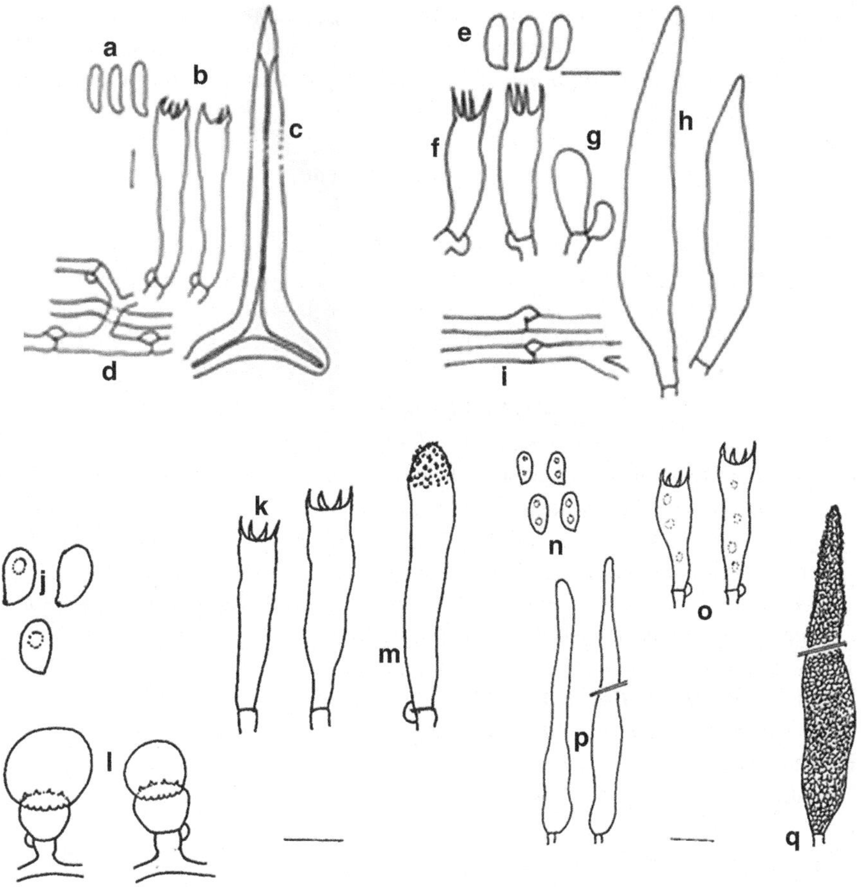

Fig. 6.29 (**a–d**)*Tubulicrinis subulatus* (**a**) Basidiospores, (**b**) Basidia, (**c**) Cystidia, (**d**) Generative hyphae; (**e–i**) *Peniophorella pallida* (**e**) Basidiospores, (**f**) Basidia, (**g**) Basidioles, (**h**) Cystidia, (**i**) Generative hyphae; (**j–m**) *Peniophorella praetermissa* (**j**) Basidiospores, (**k**) Basidia, (**l**) Stephanocystidia (**m**) Gloeocystidia; (**n–q**) *Peniophorella pubera* (**n**) Basidiospores, (**o**) Basidia, (**p**) Lamprocystidia, (**q**) Gloeocystidia

tals in the apical part, arising from different parts of the context, immersed or projecting up to 45 μm, the walls thick, soluble in 10 % KOH sol., nonamyloid, lumen capillary but broadens near the apex forming an acute conical bulb (15–20 × 4–4.6 μm). Basidia 16–22 × 3.4–4.5 μm, clavate-cylindrical, 4-spored, strigmata up to 4.2 μm long. Basidiospores 5.2–7.8 × 1.5–2 μm, cylindrical to narrowly ellipsoid, minutely apiculate, thin-walled, nonamyloid.

Distribution: H.P.: Kullu- Rohtang; J&K: Pehalgam; Bhutan: Thimphu, Begana.

Collection examined: SSR: 5729, IBP 37074.

Substratum: stumps under a mixed forest.

Remarks: This species is characterized by loosely membranous fructifications, the shape and size of basidiospores, and conical cystidia with pointed apices. The lumen is capillary through-out but widens abruptly near the top forming an apical bulb in cystidia.

Family-Rickenellaceae

Key to genera

1. Sterile elements of two types consisting of halocystidia and stellate cystidia present .... *Resinicium*
1. Sterile elements of three types consisting of gloeocystidia, stephanocystidia capitate, cystidia present........ *Peniophorella*

Peniophorella P. Karst., Bidr. Kann. Finl. Nat.Folk. 48, 1889.

Fructification resupinate, ceraceous when wet, whitish or yellowish; pore surface smooth to tuberculate, white to yellowish. Hyphal system monomitic; generative hyphae branched, clamps at all the septa. Cystidia present in most of the species, leptocystidia, metuloids or gloeocystidia. Basidia narrowly to broadly clavate, 4-sterigmate, and basal clamp. Basidiospores ellipsoid to allantoids, thin-walled, with oily contents, acyanophilous, non-amyloid.

Thirty five, widespread

Lit.: Larsson (*Mycol. Res.* 111, 2007)

Type Species: *Thelephora pubera* Fr., 1828

Habitat: Wood

Himalayas: Five

Key to species

1. Hymenial surface odontoid .... *P. rude*
1. Hymenial surface smooth to tuberculate........ 2
2. Cystidial elements of three types .... *P. praetermissa*
2. Cystidia elements of two types 3
3. Encrusted cystidial elements present .... *P. puberum*
3. Encrusted cystidial elements absent........ 4
4. Hyphae with bladder like echinulate cells present........ *P. pallida*
4. Hyphae without such contents .... *P. microtsugae*

Peniophorella microtsugae Prasher & Lalita sp. nov. Plate 6.13a, Fig. 6.30e–i
Mycobank MB812329

Etymology: Signifying the smaller size of the cystidia and basidiospores.

Fructification resupinate, adnate, effused, thin; hymenial surface smooth, greyish-white to pale ochraceous, under the lens porose and marked with brown dots; margin not well-marked.

Hyphal system monomitic; generative hyphae 1.6–3 μm wide, branched, septate, clamped, thin-walled; basal zone not well-differentiated, composed of irregularly interwoven hyphae; texture dense in the subhymenial part. Cystidia of two types: (i) leptocystidia 20–35 × 4.5–5.5 μm, generally fusiform with mostly acute

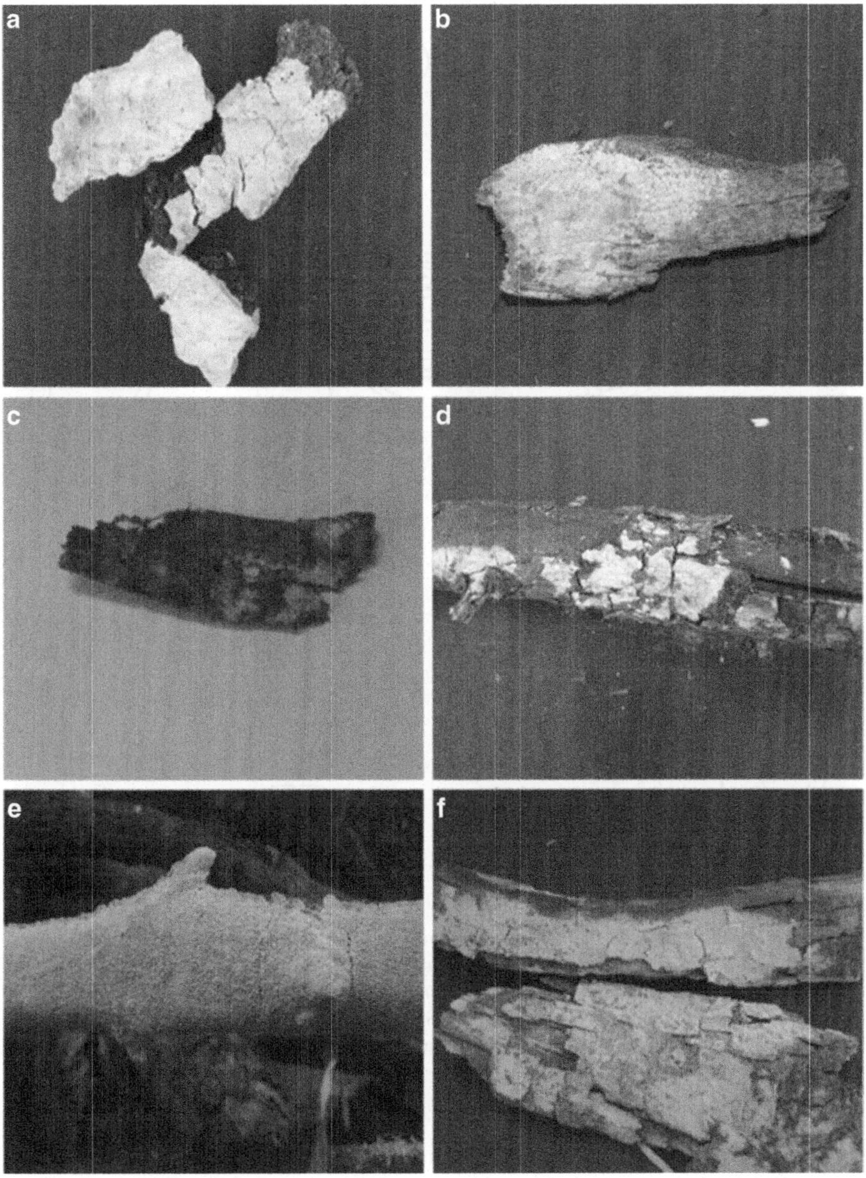

Plate 6.13 (**a**) *Peniophorella microtsugae sp. nov.* (**b**) *Sidera lenis var. minutispora.* (**c**) *Alutaceodontia alutacea.* (**d**) *Hyphodontia alienata.* (**e**) *Hyphodontia arguta* (**f**) *Hyphodontia aspera*

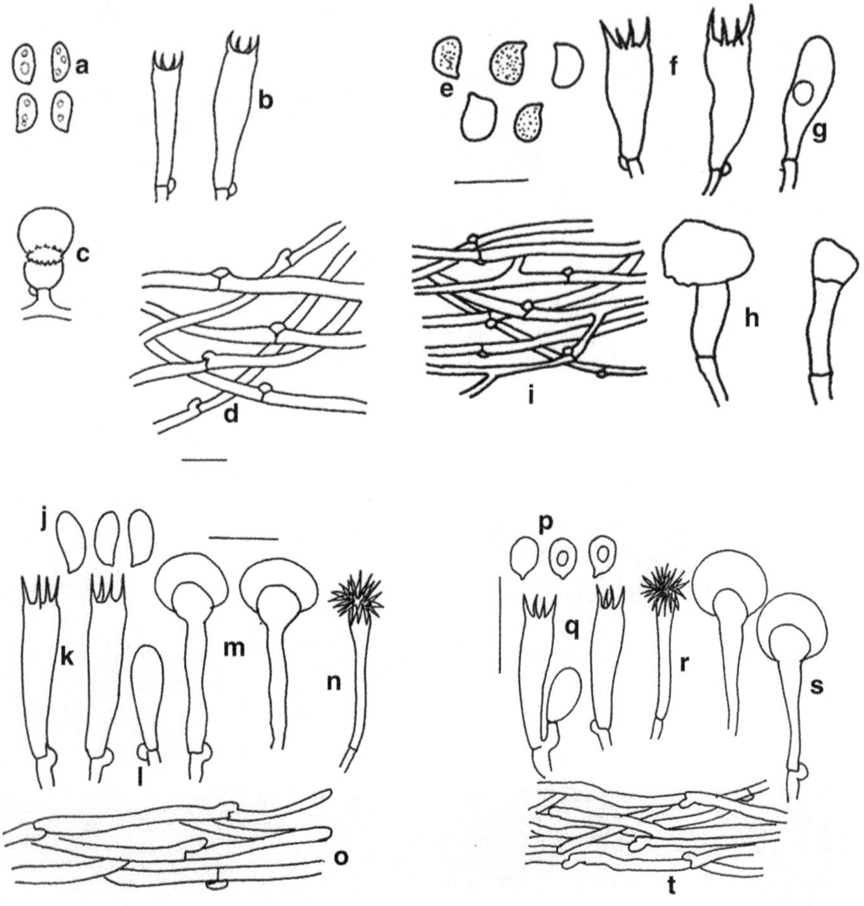

Fig. 6.30 (**a–d**) *Peniophorella rude* (**a**) Basidiospores, (**b**) Basidia, (**c**) Stephanocystidia, (**d**) Generative hyphae; (**e–i**) *Peniophorella microtsugae* (**e**) Basidiospores, (**f**) Basidia, (**g**) Basidioles, (**h**) Cystidia, (**i**) Generative hyphae; (**j–o**) *Resinicium bicolor* (**j**) Basidiospores, (**k**) Basidia, (**l**) Basidioles, (**m**) Vesicular type cystidia, (**n**) Stellate cystidia, (**o**) Generative hyphae; (**p–t**) *Resinicium friabile* (**p**) Basidiospores, (**q**) Basidia, (**r**) Stellate cystidia, (**s**) Vesicular type cystidia, (**t**) Generative hyphae

or rarely obtuse apices, thin-walled, slightly projecting out of the hymenium (ii) 20–35 × 3.5–7(–21) µm, more or less enclosed, capitate cystidia, apically excreting a brown, resinous, amorphous matter. Basidia 17.5–27.5 × 5.5–7 µm, clavate, somewhat constricted or sinuous, with numerous oil drops and a basal clamp, 4-sterigmate; sterigmata up to 5 µm long. Basidiospores 6–7.5 × 3.5–5.0 µm, ellipsoid, smooth, thin-walled, non-amyloid, acyanophilous with many oil drops or irregular oily bodies.

Distribution: U.K.: Badrinath.

Collection examined: IBP 37077- PAN (Holotype), 37078, L 37681.

Substratum: On a decayed stump.

Remarks: The above mentioned collection is close to *P. tsugae* in gross morphological details, as described by Eriksson and Strid (1975), except for the markedly smaller size of the cystidia (22–35 × 4.5–6 µm as compared to 50–80 × 6–9 µm) and smaller basidiospores (6–8 × 3.5–5 µm in comparison to 8–10 × 4–5 µm). On this basis it is treated as a new species.

Peniophorella pallida (Bres.) K.H. Larss., Mycol. Res. **111**(2): 192 (2007) = *Hyphoderma pallidum* (Bres.) Donk, Fungus 27: 15, 1957 = *Corticium pallidum* Bres., Fung.Trid. 2: 59. 1898. Fig. 6.29e–i

Fructification resupinate, membranous, adnate, often occurring in small colonies which may extend up to 4 × 2 cm, up to 75 µm thick in section; hymenial surface whitish brown to greyish brown, smooth but farinose under the lens, continuous, not creviced; margin indeterminate, adnate, concolorous.

Hyphal system monomitic, hyphae 2.5–4 µm wide, branched at wide angles, septate, clamped, the walls subhyaline, thin to slightly firm, tending to collapse or agglutinate in mature specimens. Cystidia absent. Gloeocystidia 40–70 × 7–9.5 µm, cylindrical to subfusiform with acute to obtuse apices, subhyaline with deeply staining granular contents, thin-walled, immersed or projecting slightly out of the hymenium, immersed gloeocystidia are often covered with brownish or umber coloured resinous mass which disappears in 3 % KOH sol. Basidia 18–20 × 5–6 µm, subutriform, 4-spored, sterigmata slender and up to 6 µm long. Basidiospores 7–9.5 × 2.8–3.2 µm, ellipsoid-cylindrical to suballantoid with obtuse ends, minutely apiculate, the walls thin-walled, subhyaline, smooth, nonamyloid.

Distribution: H.P.: Dalhousie.

Collection examined: SSR 5779.

Substratum: On wood under conifers.

Remarks: This species is marked by thin and whitish brown to greyish brown fructifications, cylindrical to sub fusiform gloeocystidia and size of basidiospores.

Peniophorella praetermissa (P. Karst.) K.H. Larss., Mycol. Res. **111**(2): 192 (2007) = *Hyphoderma praetermissum* (Karst.) J. Eriksson & Strid, in J. Eriksson and Ryv., Cort. N. Europe 3 p.505, 1975. Plat 6.135 D, Fig. 6.29j–m

Fructification resupinate, membranous to membranous ceraceous, hard on drying, effused, adnate; hymenial surface cream to pale cream yellow, smooth to farinose, continuous; margin thinning, concolorous, adnate.

Hyphal system monomitic; generative hyphae 1.5–2.6(4.5) µm wide, septate, branched, thin-walled, clamps present. Stephanocystidia present, 12–18 × 8–12 µm, two-celled, thin-walled. Gloeocystidia 34–68 × 6–12 µm, conical to subcylindrical with narrow obtuse apex, empty or with subhyaline, thin to thick-walled, immersed or projecting out of the hymenium. Basidia subutriform to utriform, 25–34 × 4–6 µm, 4-spored, sterigmata up to 5 µm long. Basidiospores broadly ellipsoid to curved cylindrical, thin-walled, minutely apiculate, subhyaline, smooth, non-amyloid.

Distribution: A.P.: West Kameng, Bomdila; Bhutan: Thimphu, Ha.; H.P.: Dalhousie, U.K.: Rishikesh, Mussoorie, Nainital; J&K: Gulmarg.

Collection examined: GSD 19752, 19647, SSR 5122, 5850, IBP 37076, 42094.

Substratum: On decaying angiospermic and gymnospermic logs and branches.

Remarks: The species is characterized by membranous to ceraceous fructification becoming horny on drying, smooth hymenial surface, clamped generative hyphae; two types of cystidial elements often constricted, 4-spored basidia with oily contents and a basal clamp, ellipsoid to suballantoid, smooth, thin-walled, non-amyloid, acyanophilous basidiospores with numerous oil drops or granular oily contents.

Peniophorella pubera (Fr.) P. Karst., Bidr. Känn. Finl. Nat. Folk 48: 427 (1889)=*Hyphoderma puberum* (Fr.) Wallr., Fl. crypt. Germ.: 576, 1833. Plate 6.12e, Fig. 6.29n–q

Fructification resupinate, membranous-ceraceous to subceraceous, adnate, widely effused, up to 400 μm thick in section; hymenial surface cream to deep cream or cream yellow, smooth to finely tuberculate when examined under lens, continuous, not creviced or rarely cracking irregularly in thicker parts on drying; margin thinning, adnate, white to paler concolorous.

Hyphal system monomitic; generative hyphae clamped; thick-walled, 2–4.6 μm wide, branched, septate, lamprocystidia (60–120×12–18 μm)conical to subfusiform, immersed or projecting, subhyaline, thick, impregnated with abundant subhyaline crystals. Gloeocystidia 65–80×7–10 μm, cylindrical to flexuous, immersed or occasionally may project slightly out of the hymenium, thin-walled, subhyaline. Basidia subclavate to clavate, often constricted, 4-sterigmate (25–35×6–7.8 μm); cylindrical to suburiform. Basidiospores 6–9×3–4.2 μm, ellipsoid-cylindrical to narrowly ellipsoid, thin-walled, non-amyloid, acyanophilous.

Distribution: A.P.: West Kameng, Bomdila; West Bengal: Darjeeling; Bhutan: Thimphu; H.P.: Dalhosie, Jandrighat, The Glen, Shimla, Narkanda, Mahasu, Kalatop; J&K: Batote, Gulmarg; U.K.: Nainital, Hemkunt.

Collection examined: IBP 37072, 42097, L 37073.

Substratum: On decaying stump of *C. japonica,* on a decaying *Pinus* log and decaying gymnospermic wood.

Remarks: Rattan (1977) reported it from the N. W. Himalayas on the basis of a single collection. All the specimens resemble closely the description of *H. pallidum* as given by Rattan (1977) and Eriksson and Ryvarden (1975). It is a fairly common species in the Himalayas.

Peniophorella rude (Bres.) K.H. Larss., Mycol. Res. **111**(2): 192 (2007)= *Hyphoderma rude* Hjortst. & Ryv., Mycotaxon 10(2): 275, 1980=*Odontia rudis* Bres., Ann. mycol. 18(1/3): 42, (1920). Plate 6.12f, Fig. 6.30a–d

Fructification resupinate, adnate, effused, up to 360 μm thick in section; hymenial surface yellowish-white to pale yellow, smooth to tuberculate or odontoid; aculei small, conical; margin thinning, generally indeterminate.

Hyphal system monomitic; generative hyphae branched, septate, clamped; basal hyphae loosely interwoven, long-celled; subhymenial hyphae densely packed, short-celled; in the centre of the aculei are present thick-walled hyphae, some rosettes of crystalline matter present at the apices of the hyphae. Cystidia of two types: (i) leptocystidia 30–60 × 8–12 µm, tubular, widened at the base, thin-walled, smooth, enclosed; (ii) stephanocystidia up to 20–14 µm, subglobose, in the middle surrounded by a whorl of some teeth, with a basal clamp. Basidia 25–30 × 5.4–7.3 µm, clavate to subclavate, often somewhat constricted, 4-spored, with a basal clamp; sterigmata up to 7 µm long. Basidiospores 6.3–9 × 4–5.2 µm, ellipsoid, smooth, thin-walled, non-amyloid, acyanophilous, with oil drops or granular oily contents.

Distribution: Arunachal Pradesh: West Kameng, Bomdila, Shergaon; H.P.: Shimla-Tara Devi, Keylong; Manali, Palampur; U. K.: Vishnuprayag-Chamoli.

Collection examined: IBP 37074, 37075, 42099, 42102, L 37075.

Substratum: On decaying bark of an angiospermic log, twigs, bark of angiospermic log.

Remarks: This species is characterized by odontoid hymenial surface. It is closely related to *P. praetermissum*, but differs on the basis of odontoid hymenial surface and absence of capitates, projecting cystidia. *Peniophorella rude* is a rare species It is a new record for N.W. Himalayas.

Resinicium Parm.,
Consp. Syst. Cort.: 97. 1968.

Fructifications resupinate, membranous-ceraceous becoming hard and crust-like on drying, adnate; hymenial surface finely tuberculate to distinctly toothed. Teeth usually small, irregular and sparse. Subiculum subhyaline in section, subceraceous, somewhat fibrous in young specimens. Hyphal system monomitic, hyphae sparsely branched, septate, clamped, clamps not at all septa, walls subhyaline, thin to thick-walled, often collapsing and agglutinating. Gloecystidia absent. Cystidia of two types: capitate with radiating crystals at the top and vesicular type. Basidia clavate, 4-spored. Basidiospores ellipsoid to subcylindrical, walls thin, subhyaline, smooth, nonamyloid.

Five Species, widespread.

Lit.: Parmasto (*Consp. System. Cortciaceae.*: 97, 1968).

Type Species: *Hydnum bicolour* Alb. & Schwein., 1805.

Habitat: Dead wood

Himalayas: Two

Key to species

1. Basidiospores ellipsoid, 4.5–7.2 × 2.5–3 µm .... *R. bicolor*
1. Basidiospores subcylindrical 5–7 × 3.2–4.5 µm .... *R. friable*

Resinicium bicolor (Alb. & Schwein.) Parm., Consp. Syst. Cort. p. 98. 1968 = *Hydnum bicolor* Alb. & Schwein., Consp. Fung. (Leipzig):m270, 1805. Fig. 6.30j–o

Fructifications resupinate, membranous-ceraceous to sub ceraceous, adnate, widely effused, covering the underside of the substratum, up to 700 µm thick in section (including the length of spine); hymenial surface smooth to finely tuberculate or sparsely toothed, white to cream to cream-grey, continuous, not creviced; margin thinning, byssoid, adnate, concolorous. Teeth ranging from small papillate to 450 μm long, scattered or rarely gregarious, subulate to cylindrical with acute to subacute apices but more often malformed and poorly developed. Context subhyaline in section, somewhat fibrous in young specimens but becomes subceraceous with age.

Hyphal system monomitic, hyphae 1.5–3(4.4) μm wide, branched, septate, clamped, clamps not at all septa, the walls thin to slightly thick, subhyaline, often collapsing and agglutinating and difficult to discern. Cystidia are of two types: (i) stellate type, 10–20 × 3–4 μm., clavate with a slightly swollen apex, often crowned with stellate crystals, arising from the subhymenium, not immersed but always projecting out of the hymenium, the walls thin, subhyaline; (ii) vesicular type with swollen apical vesicle up to 15 μm broad, supported on the stalk, thin-walled, subhyaline, often collapsing in KOH sol mounts and difficult to discern (best seen in water mounts). Basidia 15–20 × 3–4 μm, clavate, 4-spored. Basidiospores 4.5–7.2 × 2.5–3.4 μm, ellipsoid, minutely apiculate, the walls thin, subhyaline, smooth, nonamyloid.

Distribution: H.P.: Kullu, Shimla; J&K: Sanasar; U.K.: Nainital; Nepal: Kathmandu.

Collection examined: HSK 4006, SSR 4333, L 37079, IBP 42109.

Substratum: On angiospermic log, on log of *Abies pindrow* &*Cedrus deodara*.

Remarks: This is one the commonest species growing under conifers and is easily distinguished by subceraceous texture and two type of cystidia. The presence of stellate cystidia with a crown of radiating crystals is quite characteristics of the species.

Resinicium friabile Hjortstam & Melo, Mycotaxon 65: 324, 1997. Fig. 6.30p–t

Fructification resupinate, effused, adnate up to 380 μm thick in section; hymenial surface smooth to tuberculate, orange white to pale orange to greyish orange; margin thinning, paler concolorous to indeterminate.

Hyphal system monomitic; generative hyphae up to 3.0 μm wide, clamped, septate, thin-walled; basal hyphae parallel to the substrate, branched; subhymenial hyphae vertical, profusely branched. Cystidia of two types: Halocystidia thin-walled with apical bladder and with basal clamp, stellate cystidia thin-walled with star like cluster of crystals at the apex with basal clamp. Basidia 10.8–18.0 × 3.5–4.5 μm, 4-sterigmata with basal. Basidiospores 5–7 × 3.2–4.5 μm, apiculate, subcylindrical, thin-walled, acyanophilous, smooth, inamyloid.

Distribution: H.P.: Shimla.

Collection examined: IBP 37080.

Remarks: This species is characterized by subcylindrical basidiospores. It is being recorded from Himalayas for the first time.

Family- Repetobasidiaceae

Sidera Miettinen & K.H. Larss.,
Mycol. Progress 10(2): 136.2011.

Fructification annual to sometimes reviving, resupinate, separable, without any distinct taste. Margin white to creamish, thinning to narrow fimbriate. Pore surface white to creamish when fresh, darker and cracked on drying, glancing; pores small, round to angular; dissepiments equal; tubes in one layer. Context white, soft, thin, soft, homogenous, non-xanthochroic. Hyphal system dimitic; generative hyphae hyaline, thin-walled, branched, septate, clamped, cyanophilous. skeletal hyphae subhyaline, thick-walled, branched, aseptate, acyanophilous. Cystidia absent, cystidioles as branched hyphae with crystal incrustations at their tips project into the pore cavity. Basidia thin-walled, hyaline, clavate, 4-spored. Basidiospores hyaline, thin-walled, smooth, cylindrical to allantoids, non-amyloid.

Type Species: *Sidera lenis* (P. Karst.) and Larss., Miettinen 2011

Habitat: Wood

Himalayas: Two

Key to species

1 Basidiospores cylindrical to allantoid, 2.8–3.0 × 1.4 μm .... *S. lenis* var. *minutispora*

1 Basidiospores allantoid, 4.5–7 × 1.5–2 μm .... *S. lunata*

Sidera lenis (P. Karst.) Miettinen, in Miettinen and Larsson, Mycol. Progr. 10(2): 136 (2011) var. *minutispora* Prasher & Lalita var nov. Plate 6.13b, Fig. 6.31a–e

Mycobank MB812330

Fructification annual to sometimes reviving, resupinate, widely effused up to 14 × 5 cm, up to 3 mm thick, separable, without any distinct taste; margin white to creamish, thin to narrow, fimbriate. Pore surface white to creamish, creamish-brown at places, glancing, even, soft to touch, cracking when dry; pores round to angular, 60–170 μm in diameter, 5–8 per mm, pore mouths finely velutinate; dissepiments 25–40 μm thick, equal. Context white, soft, thin; tubes creamish in section, up to 3 mm long. Tissue does not change color in KOH sol.

Hyphal system dimitic; generative hyphae hyaline, thin-walled, branched, septate, clamped, cyanophilous, 1.4–2.1 μm in diameter; skeletal hyphae subhyaline, thick-walled, wall up to 1 μm thick, branched, aseptate, acyanophilous, 1.4–2.8 μm in diameter. Cystidia absent, cystidioles as branched hyphae with crystal incrustations at their tips project into the pore cavity. Basidia thin-walled, hyaline, clavate, 4-spored, 12–16 × 4.5–6 μm. Basidiospores (only few spores seen) hyaline, thin-walled, smooth, cylindrical to allantoid, 2.8–3.0 × 1.4 μm.

Etymology: refers to smaller spores than the type.

Distribution: U.K: Chakrata, Dehra Dun.

Collection examined: IBP 37563- PAN (Holotype), 42107, L 37564.

Substratum: On under surface of burnt *Pinus* log.

Remarks: The above collections closely resemble *S. lenis* except that the spores are smaller (2.8–3.0 μm) than those mentioned for the species (3.5–4 μm) in Lowe's manual (1966). On this basis it is considered as a new variety.

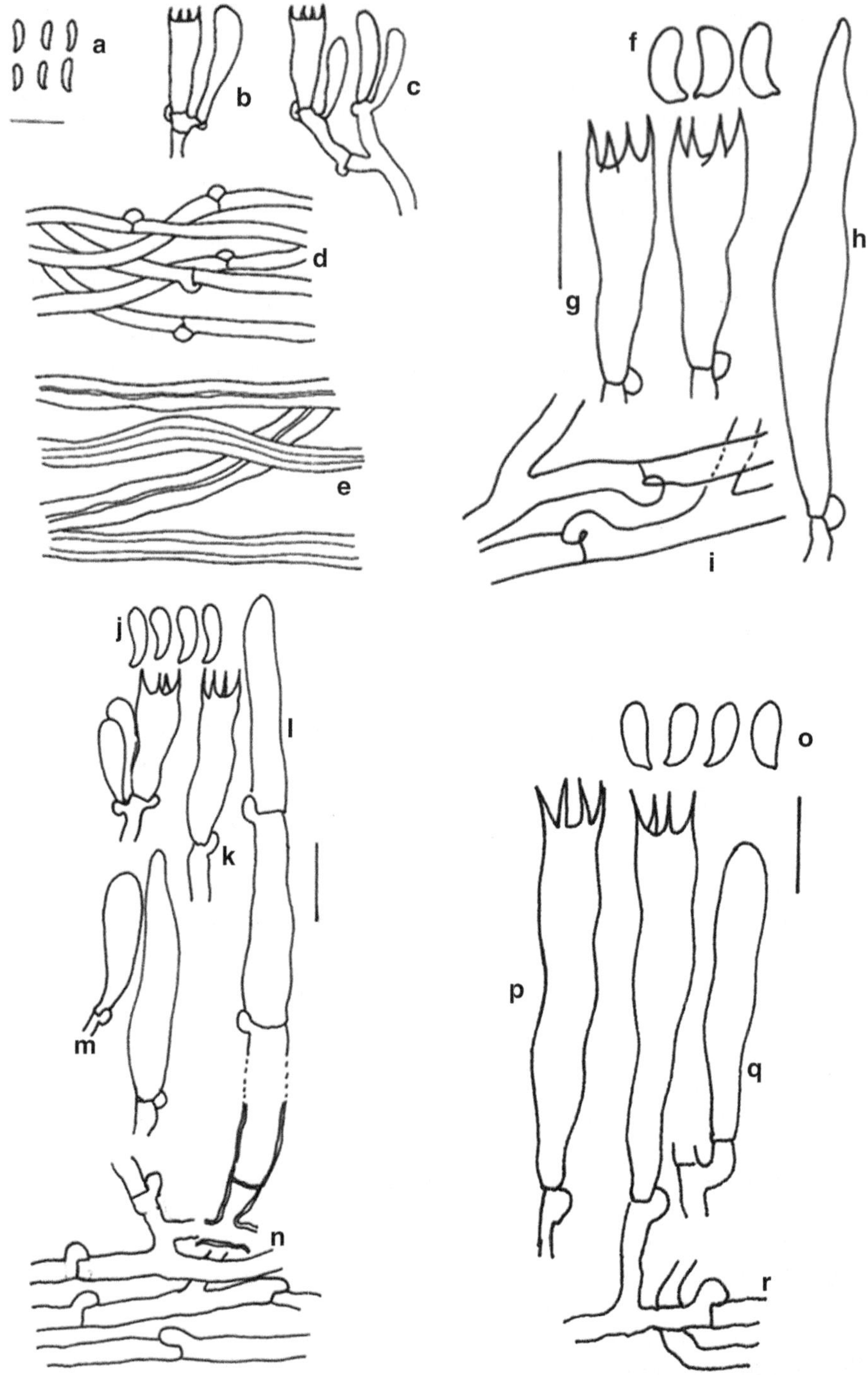

Fig. 6.31 (**a–e**) *Sidera lenis* var. *minutispora* (**a**) Basidiospores, (**b–c**) Basidia and Basidioles, (**d**) Generative hyphae, (**e**) Skeletal hyphae, (**f–i**) *Sidera lunata* (**f**) Basidiospores, (**g**) Basidia, (**h**) Cystidia, (**i**) Generative hyphae; (**j–n**) *Alutaceodontia alutacea* (**j**) Basidiospores, (**k**) Basidia, (**l**) Cystidia, (**m**) Leptocystidia, (**n**) Generative hyphae; (**o–r**) *Basidioradulum tuberculatum* (**o**) Basidiospores, (**p**) Basidia, (**q**) Cystidia, (**r**) Generative hyphae

Sidera lunata (Romell ex Bourdot & Galzin) K. H. Larss., in Miettinen & Larsson, Mycol. Progr. **10**(2): 136 (2011)=*Athelopsis lunata* (Romell ex Boudot & Galzin) Parmasto, Consp. System. Corticiac.: 43, 1968=*Grandiana lunata* Romell ex Boudot & Galzin, Hymenomyc. De France: 410, 1928. Fig. 6.31f–i

Fructification resupinate, effused, loosely adnate; hymenial surface grandinoid, greyish white to yellowish; margin thinning, concolorous to indeterminate.

Hyphal system monomitic; generative hyphae up to 3.3 µm, septate, clamped; basal hyphae thick-walled; subhymenial thin-walled, highly branched. Cystidia 30–54×3.2–4.6 µm, smooth, with basal clamp, project out of hymenium, thin-walled. Basidia 10–22×3–4 µm, clavate to subclavate, 4-sterigmate with basal clamp. Basidiospores 4.5–7.0×1.5–2.4 µm, apiculate, allantoids, smooth, thin-walled, inamylod, acyanophilous.

Distribution: H.P.: Manali, Sundernagar.

Collection examined: L 37082, 37083, IBP 37084.

Substratum: On decaying wood and rotten twigs.

Remarks: The characteristic feature of this species are fusiform cystidia, suclavate basidia and allantoids basidiospores.

Family-Schizoporaceae

Key to genera

1. Basidiospores allantoids, hymenial surface yellowish when fresh to ochraceous, flocculose to odontoid with small aculei .... *Alutaceodontia*
1. Basidiospores not of above combination 2
2. Basidiospores cylindrical to oblong or pip-shaped .... *Basidioradulum*
2. Basidiospores globose to ellipsoid 3
3. Hyphal system monomitic 4
3. Hyphal system dimitic .... *Schizopora*
4. Cystidia present 5
4. Cystidia absent .... *Xylodon*
5. Clamps present .... *Hyphodontia*
5. Clamps absent .... *Oxyporus*

Alutaceodontia (Parmasto) Hjortstam & Ryvarden, Syn. Fung. 15:8, 2002.

Fructification resupinate, adnate, effused, thin; hymenial surface yellowish when fresh to sordidly ochraceous in herbarium, flocculose to odontoid with small aculei; margin indistinct. Hyphal system monomitic; generative hyphae branched at wide angles, septate, clamped, cyanophilous. Cystidia present. Basidia subclavate with a suburniform constriction, 4-sterigmate, with a basal clamp. Basidiospores allantoid, smooth, thin-walled, non-amyloid, acyanophilous.

Monotypic, widespread

Lit.: Hjortstam & Ryvarden (*Syn. Fung.* **15**: 8, 2002)

Type Species: *Hydnum alutaceum* Fr. 1821

Habitat: Wood

Himalayas: One

Alutaceodontia alutacea (Fr.) Hjortstam & Ryvarden, Syn. Fung. (Oslo) 15: 8 (2002)=*Hyphodontia alutacea* (Fr.) J. Eriksson, Symb. bot. ups. 16(1): 104, 1958. Plate 6.13c, Fig. 6.31j–n

Fructification resupinate, adnate, effused, thin; hymenial surface yellowish initially ochraceous later on, flocculose becoming more or less odontoid with small (less than 1 mm long) aculei towards maturity; margin indistinct.

Hyphal system monomitic; generative hyphae branched at wide angles, septate, clamped, cyanophilous, up to 3–8 μm wide; subhymenial hyphae densely, thin-walled; basal hyphae and hyphae in the centre of the aculei somewhat thick-walled. Cystidia 85–128×4.5–7.2 μm, hyphoid, somewhat sinuous, thin-walled, enclosed or projecting, sometimes with a clamped septum. Thin-walled, subcylindrical leptocystidia also present in between the basidia, 18–35×5–5.5 μm. Basidia 9–13×4.5–5.0 μm, subclavate with a suburniform constriction, with a basal clamp, 4-sterigmate; sterigmata up to 4 μm long. Basidiospores 6–8×1.5–3 μm, allantoid, smooth, thin-walled, non-amyloid, acyanophilous.

Distribution: Bhutan: Thimphu, Begana.

Collection examined: GSD 19625.

Substratum: On a decaying gymnospermic log, on gymnospermic wood.

Remarks: This species is characterized by hyphoid cystidia, small, subclavate basidia with a suburniform constriction and allantoid basidiospores.

Basidioradulum Nobles,
Mycologia 59(2): 192. 1967.
Emend. Parm., Consp. Syst. Cort. p. 112. 1968.

Fructifications resupinate, membranous to rarely subceraceous, adnate, usually thick; hymenial surface smooth to tuberculate or distinctly raduloid. Hyphal system monomitic, hyphae usually thin-walled, septate, clamped at all septa, usually distinct. Cystidia present or absent. Gloeocystidia absent. Basidia clavate-cylindrical to subutriform, 4-spored. Basidiospores cylindrical to oblong or pip-shaped, thin-walled, smooth, subhyaline, nonamyloid and acyanophilous.

Two species, widespread

Lit.: Legon (*Mycologist* **19**: 81, 2005)

Type species: *Hydnum radula* Fr. 1818

Habitat:1as: One

Basidioradulum tuberculatum (Berk. & M.A. Curtis) Hjortstam, Mycotaxon 54: 183 (1995)=*Phlebia albida* Fr., Monogr. Hymenomyc. Svec. (Upsaliae) 2(2): 280 (1863). Fig. 6.31o–r

Fructification resupinate, ceraceous when fresh becoming rigid on drying, widely effused; hymenial surface light pink to orange or pale flesh coloured when fresh but fades on drying; margin thinning, fibrillose, white, wide. Subiculum composed of compactly arranged more or less ceraceous hyphae.

Hyphal system monomitic; generative hyphae 1.5–3.5 μm wide, septate, branched, clamps present, clamps prominent, walls thin to thick-walled. Cystidia absent. Basidia clavate, sterigmate, 30–44×4.4–5.4 μm, 4-spored. Basidiospores narrowly ellipsoid, thin-walled, minutely apiculate, subhyaline, smooth, 6–6.4×2.4–3.0 μm, non-amyloid.

Distribution: H.P.: Mahasu, Narkanda; U.K.: Chakrata, Mussoorie.

Collection examined: SSR 5342, 5412, IBP 37682.

Substratum: On log of *Rhododendron arboretum, Quercus semicarpifolia* and *Picea smithiana*

Remarks: The species is common and characterized by reticulate folded hymenial surface, lack of cystidia and large basidiospores. The margin is very wide and white in colour.

Hyphodontia Erikss.,
Symb. bot. upsal. 16(no. 1): 101. 1958.

Fructifications resupinate, floccose to membranous; hymenial surface usually pale coloured, smooth to tuberculate or distinctly toothed. Context subhyaline in section, composed of somewhat loosely woven hyphae. Hyphal system monomitic, hyphae branched, branches at wide angles and ramified, septate, clamped or rarely without clamps, thin to firm or slightly thick-walled, subhyaline, nondextrinoid but usually take a deep yellow colour with Melzer's reagent, mostly distinct but in some species they may collapse especially in the subhymenial zone and become indistinct. Cystidia (leptocystidia) present or absent. Gloeocystidia absent but in some species it is difficult to differentiate between lepocystidia and gleocoystidia. Basidia small utriform, 4-spored, rarely two spored. Basidiospores small, globose to ellipsoid, cylindrical or suballantoid, smooth, nonamyloid.

Sixty four species, widespread

Lit.: Wersub (*CJB* 39: 1475, 1961), Wu (*Mycol* 93: 1019, 2001; new spp.)

Type Species: *Hyphodontia pallidula* (Bers.) Erikss 1958.

Habitat: Dead wood

Himalayas: Eighteen (Table 6.1)

Hyphodontia abieticola (Bourdot & Galzin) J. Erikss., Symb. bot. upsal. 16(1): 84, 1958 = *Odontia barba-jovis* subsp. *abieticola* Bourdot & Galzin, Hymenomyc. De France: 426, 1928. Fig. 6.32a–d

Fructification resupinate, effused, adnate, up to 350 μm thick in section; hymenial surface smooth to odontoid, aculei dense, conical, creamish white to pale ochraceous initially, becoming pale yellow to greyish yellow to brownish grey on drying; margins indeterminate or thinning, paler concolorous.

Hyphal system monomitic; generative hyphae, branched, clamped; basal hyphae up to 4.5 μm wide, thick-walled, running parallel to the substratum, loosely arranged; subhymenial hyphae up to 3.0 μm wide, thin-walled vertical and

Table 6.1 Synopsis of *Hyphodontia* species

Name	Fructification	Hymenium	Cystidia	Basidia	Basidiospore
H. abieticola	Resupinate, effused adnate	Smooth to odontoid, creamish white to pale	Thick walled except in apical part 115–190 × 5–6.5 μm	Subclavate to subcylindrical 17.5–20 × 3–5 μm	Ellipsoid 5.2–6 × 2.5–3.6 μm
H. alienata	Resupinate, adnate effused	Smooth, creamish white to pale yellow	Thick walled at the base 105–135 × 2.8–5.4 μm	Subclavate to subcylindrical 17.5–20 × 3–5 μm	Ellipsoid 4.6–6.2 × 2.7–3.2 μm
H. altaica	Resupinate, floccose, membranous	Smooth to finely tuberculate, creamish yellow	Lagenocystidia 20–34 μm and Leptocystidia 40–69 × 4.5–5 μm present	Subutriform 12–17 × 4–5 μm	Ellipsoid 4.2–5.7 × 2.2–2.6 μm
H. alutaria	Resupinate, pelliculose, submembranous	Smooth to finely tuberculate, cream to creamish yellow	Lagenocystidia 20–35 × 3.5–4.5 μm, Leptocystidia 40–70 × 4–5 μm	Subutriform 12–17 × 4–4.4 μm	Ellipsoid 4.2–5.3 × 2.2–2.6 μm
H. arguta	Resupinate, floccose, membranous	Odontoid to distinctly toothed; creamish to ochre	Lagenocystidia 20–30 × 2.5–4 μm, Leptocystidia 20–31 × 3–3.6 μm	Clavate cylindrical to sub utriform 15–17 × 4–4.4 μm	Ellipsoid to ovoid 4–4.6 × 3–3.6 μm
H. aspera	Resupinate, floccose, membranous	Smooth to finely granulose, white to cream	Leptocystidia or hyphidia of variable length present	Subutriform 14–17 × 3.5–4.5 μm	Ellipsoid to broadly ellipsoid 4.5–6 × 3.5–4.5 μm
H. barbajovis	Resupinate, effused	Smooth to odontoid, creamish to yellowish grey	Cystidia with secondary septa 118–160 × 4.8–7.2 μm	Subclavate 12–16.5 × 4–5.2 μm	Broadly ellipsoid 4–6 × 3.5–4 μm
H. caulicystidiata	Resupinate, adnate, effused, thin	Smooth to porose-floccose, yellowish grey to pale ochraceous	Cystidia subcylindrical to subfusiform with stalk 55–83 × 9–12.2 μm	Clavate to subclavate to sinuous	Ellipsoid 2.4–3.2 μm
H. crustosa	Resupinate, effused, adnate	Smooth to odontoid; greyish white to light yellow to orange	Only subulate hyphae with resinous encrustations present	Clavate to subclavate to subcylindrical 13–25 × 3–4.6 μm	Ellipsoid 4.5–6 × 3.5–4.2 μm

H. efibulata	Resupinate, soft membranous, adnate	Smooth to finely tuberculate or aculeate, white to cream to ochre	Cystidia cylindrical to subfusiform with encrusted apical part forming a globose head. 85–100 × 4.5–6 μm	Utriform 12–14 × 4–4.5 μm	Ellipsoid to cylindrical 3.5–4.2 × 2–2.4 μm
H. hastata	Resupinate, effused, adnate	Smooth to tuberculate, whitish orange to pale orange	Moniliform cystidia and subulate cystidia present	Clavate 14–20 × 3.5–4.5 μm	Ellipsoid 4–5.2 × 2–3 μm
H. juniperi	Resupinate, effused, adnate	Smooth to tuberculate, greyish white to creamish	Cystidia subcylindrical to fusiform, 28–35 × 6–7.2	Clavate 10–22 × 5.4–5.8 μm	Ellipsoid 5.5–7.4 × 3.5–4 μm
H. pallidula	Resupinate, widely effused	Smooth to finely tuberculate, cream to ochre	Leptocystidia cylindrical to flexuous	Subutriform 15–18 × 4–5 μm	Ellipsoid 3.5–4.5 × 2.2–3 μm
H. papilosa	Resupinate, floccose to submembranous	Granulose to toothed; cream to yellow	Axial bundle of hyphae look like leptocystidia	Clavate to subutriform 10–15 × 3.5–4.5 μm	Ellipsoid to subcylindrical 4.3–5 × 2.8–3.4 μm
H. propinqua	Resupinate, effused, adnate	Smooth to finely tuberculate, creamish white to yellow	Hypoid cystidia and lagenocystidia present	Subcylindrical 14.5–20 × 5–5.5 μm	Ellipsoid 5.2–8.4 × 4–5 μm
H. nespori	Resupinate, submembrano-us to subceraceous	Odontoid with conical aculei, creamish to pale ochraceous	Not differentiated	Subcylindrical 10–18 × 3.5–4.5 μm	Ellipsoid to subcylindrical 4.4–5.3 × 2.2–3.2 μm
H. spathulata	Resupinate, membranous to ceraceous	Axial bundle of hyphae with impregnation appear like leptocystidia	Axial bundle of hyphae with impregnation appear like leptocystidia	Utriform 12–15 × 3.5–4.4 μm	Ellipsoid to ovoid 4–5 × 3–3.5 μm
H. stipata	Resupinate, floccose to loosely membranous	Axial bundle of hyphae with impregnation appear like leptocystidia	Axial bundle of hyphae with impregnation appear like leptocystidia	Subutriform 15–18 × 3–4.5 μm	Ellipsoid to ovoid 4–5.5–3–4 μm
H. sambuci	Resupinate, effused	Cystidia of variable shape with apical encrustations	Cystidia of various shape with apical encrustations	Subclavate to subcylindrical 12–24 × 3.5–6 μm	Ellipsoid 5–63.5–4 μm

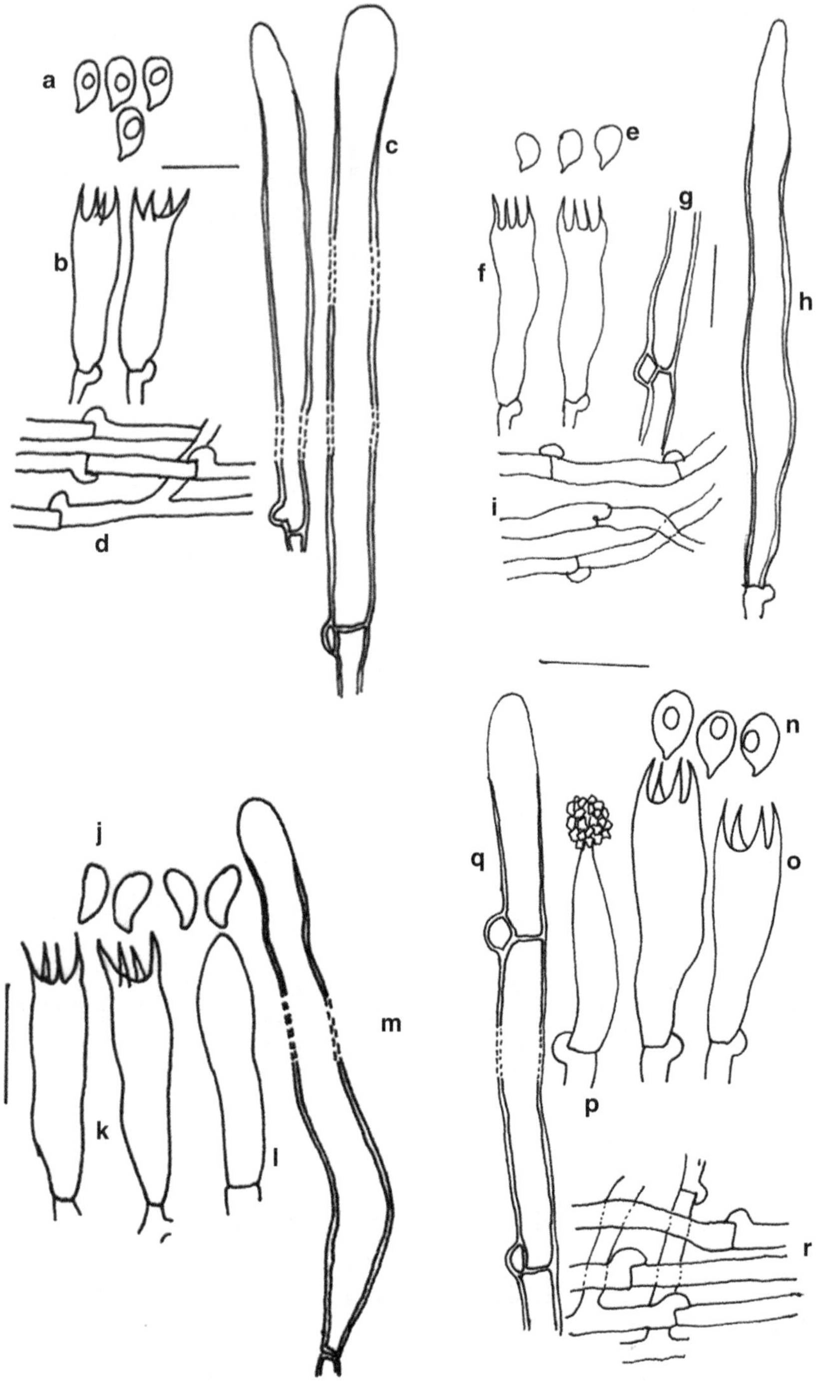

Fig. 6.32 (**a–d**) *Hyphodionia abieticola* (**a**) Basidiospores, (**b**) Basidia, (**c**) Cystidia, (**d**) Generative hyphae; (**e-i**) *Hyphodionia alienate* (**e**) Basidiospores, (**f**) Basidia, (**g**) Thick-walled Generative hyphae, (**h**) Cystidia, (**i**) Thin-walled generative hyphae; (**j-m**) *Hyphodontia altaica* (**j**) Basidiospores, (**k**) Basidia, (**l**) Lagenocystidia, (**m**) Leptocystidia; (**n-r**) *Hyphodontia alutaria* (**n**) Basidiospores, (**o**) Basidia, (**p**) Lagenocystidia, (**q**) Leptocystidia, (**r**) Generative hyphae

compactly arranged. Cystidia 115.0–190.0×5.0–6.5 µm, thick-walled except in the apical part, with basal clamp, projecting up to 80 µm out of the hymenium. Basidia 17.5–20.0×3.0–5.0 µm, subclavate to subcylindrical, somewhat constricted, 4-sterigmate, with basal clamp; sterigmata up to 3.2 µm wide. Basidiospores 5.2–6.0×2.5–3.0 µm, ellipsoid, apiculate, smooth, thin-walled, inamyloid, acyanophilous, with oil droplets.

Distribution: H.P.: Shimla, Narkanda, Dalhousie, Manali.

Collection examined: IBP 37086, 42122.

Substratum: On stump of *Cedrus deodara,* on decaying gymnospermic wood.

Remarks: This species is characterized by smooth to odontoid fructifications, tubular cystidia and ellipsoid basidiospores. It is being reported for the first time from Himalayas.

Hyphodontia alienata (S. Lundell) J. Erikss., Symb. bot. Upsal. 16: 104, 1958=*Peniophora alienta* S. Lundell, Fungi Exsiccati Suecici 21–22: 28, 1941. Plate 6.13d, Fig. 6.32e–i

Fructification resupinate, adnate, effused; hymenial surface smooth, creamish white to pale yellow initially, becoming yellowish brown on drying; margin thinning, paler concolorous to indeterminate.

Hyphal system monomitic; generative hyphae up to 4.8 µm wide, clamped, septate; basal hyphae parallel to substrate, thin to thick-walled, less branched; subhymenial hyphae vertical, thin-walled, highly branched. Cystidia 105.0–135.0×2.8–5.4 µm, thick-walled at the base, basal clamp present at the secondary septa projects out of the hymenium. Basidia 11.5–18.6×2.6–4.3 µm, 4-sterigmate, clavate to subclavate with basal clamp. Basidiospores 4.6–6.2×2.7–3.2 µm, apiculate, ellipsoid, smooth, thin-walled, acyanophilous, inamyloid.

Distribution: H.P.: Manali-reserve forest.

Collection examined: IBP 37088, 37089

Remarks: The species is characterized by the presence of tubular thick-walled cystidia, clavate to subclavate, 4-sterigmate basidia and ellipsoid basidiospores. Ranadive et al. (2011) reported it from Western Ghats. It is recorded for the first time from Himalayas.

Hyphodontia altaica Parm., Consp. System. Corticiac.: 211. 1968. Fig. 6.32j–m

Fructifications resupinate, floccose-membranous, adnate, widely effused; hymenial surface cream to cream yellow, smooth to finely tuberculate, discontinuous and appear porose-reticulate under the hand lens, not creviced; margin thinning, adnate, concolorous.

Hyphal system monomitic, hyphae 2–4 µm wide, branched at wide angles and ramified, septate, clamped, the walls thin, subhyaline. Lagenocystidia abundant, 20–34×3.4–4 µm, subulate with acute and draw out apex, arising from the hymenial layer and may project to 15 µm out of the hymenium, subhyaline, slightly thick-walled, characteristically impregnated with subhyaline crystals in the apical part forming a crystal-coated apex up to 12 µm long and 4.5 µm broad.

Leptocystidia scanty, 40–69 × 4.5–5 µm, cylindrical with obtuse apices, aseptate or occasionally with one or two septa, arising from the upper layers of context and may project to 25 µm out of the hymenium, the walls thin to slightly thick especially in the basal part, subhyaline, not incrusted. Basidia 12–17 × 4–5 µm, subutriform, 4-spored, sterigmata slender and up to 4 µm long. Basidiospores 4.2–5.7 × 2.2–2.6 µm, ellipsoid, minutely apiculate, the walls thin, subhyaline, smooth, nonamyloid.

Distribution: H.P.: Narkanda, Mahasu.

Collection examined: SSR 4336, 5677.

Substratum: On bark of *Abies pindrow*, on stump of *Cedrus deodara.*

Remarks: The above collection is typical of the species except for smaller cystidia. This species belongs to *H. subalutacea* complex where it is marked by almost smooth hymenial surface and size of basidiospores.

Hyphodontia alutaria (Burt.) Erikss., Symb. bot. upsal. 16: 104. 1958 = *Peniophora alutauria* Burt, Ann. Mo. Bot. Gdn. 12:231. 1925. Fig. 6.32n–r

Fructification resupinate, pelliculose-membraneous to submembranous, adnate, widely effused, up to 150 µm thick in section; hymenial surface cream to cream yellow, smooth to finely tuberculate, discontinuous and appears porose-reticulate under the hand lens, not creviced; margin thinning, adnate, concolorous.

Hyphal system monomitic, hyphae 2–4 µm wide, branched at wide angles and ramified, septate, clamped, thin-walled, subhyaline. Lagenocystidia abundant, 20–35 × 3.5–4 µm, subulate with acute and drawn out apex, arising from the hymenial layer and may project to 15 µm out of the hymenium, subhyaline, slightly thick-walled, characteristically impregnated with subhyaline crystals in the apical part forming a crystals-coated apex up to 12 µm long and 4.5 µm. Leptocystidia scanty, 40–70 × 4–5 µm, cylindrical with obtuse apices, aseptate or occasionally with one or two septa, arising from the upper layers of the context and may project to 25 µm out of the hymenium, the walls thin to slightly thick especially in the basal part, subhyaline, not incrusted. Basidia 12–17 × 4–4.4 µm, subutriform, 4-spored, sterigmata slender and up to 4 µm long. Basidiospores 4.2–5.3 × 2.2–2.6 µm, ellipsoid, minutely apiculate, thin-walled, subhyaline, smooth, non-amyloid.

Distribution: H.P.: Narkanda, Kullu; J&K: Pehalgam.

Collection examined: SSR 5609.

Substratum: On bark of *Cedrus deodara.*

Remarks: The characteristic features of this species are submembranous fructifications, smooth hymenial surface, presence of abundant lagenocystidia and cylindrical Leptocystidia. *Hyphodontia pallidula* is very similar but that lacks lagenocystidia or these are very rare.

Hyphodontia arguta (Fr.) Erikss., Symb. bot. upsal. 16: 104. 1958 = *Hydnum argutum* Fr., Syst. Mycol. **1**: 424. 1821. Plate 6.13e, Fig. 6.33a–e

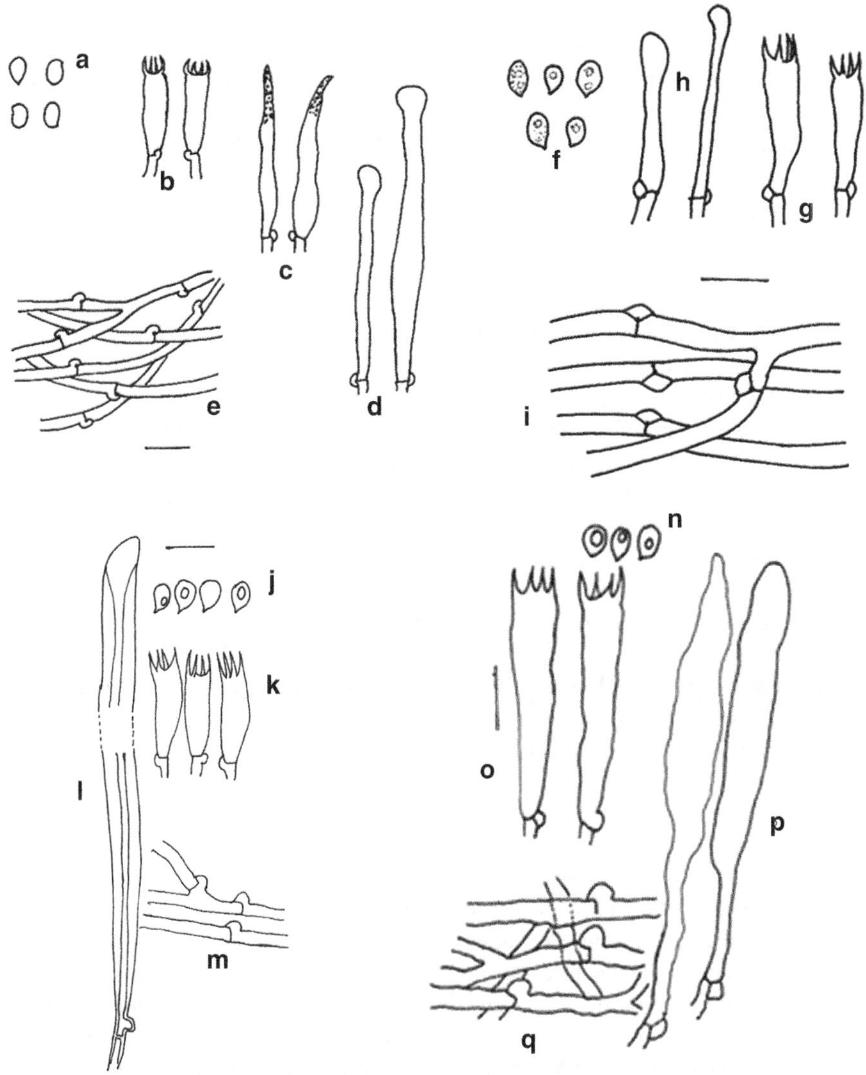

Fig. 6.33 (**a**–**e**) *Hyphodontia arguta* (**a**) Basidiospores, (**b**) Basidia, (**c**) Lagenocystidia, (**d**) Leptocystidia, (**e**) Generative hyphae; (**f**–**i**) *Hyphodontia aspera* (**f**) Basidiospores, (**g**) Basidia, (**h**) Cystidia, (**i**) Generative hyphae (**j**–**m**) *Hyphodontia barbajovis* (**j**) Basidiospores, (**k**) Basidia, (**l**) Cystidia, (**m**) Generative hyphae; (**n**–**q**) *Hyphodontia caulicystidiata* (**n**) Basidiospores, (**o**) Basidia, (**p**) Cystidia, (**q**) Generative hyphae

Fructifications resupinate, floccose to floccose-membranous but become somewhat brittle on drying, loosely adnate, widely effused, up to 250 µm thick (extruding the length of spine) in section; hymenial surface cream yellow to light ochre, finely odontoid to distinctly toothed, discontinuous, not creviced; margin thinning, arachnoid, loosely adnate, paler concolorous. Teeth up to 1.5 mm long, subulate to cylindrical with acute apices. Context composed of loosely woven hyphae but these become compact and parallel in the trama of the spine.

Hyphal system monomitic, hyphae 2–4.5 µm wide, branched at wide angles and ramified, septate, clamped, the walls thin to firm, subhyaline. Lagenocystidia 20–30 × 2.5–4 µm, subulate, apex pointed and characteristically impregnated with crystals, thin-walled, subhyaline, arising from the hymenial layer and projecting to 10 µm or rarely immersed. Leptocystidia 20–31 × 3–3.6 µm, cylindrical, apex capitate surrounded by a thin-walled vesicle (up to 6.5 µm broad) which often collapses in KOH sol but is best seen in water mounts. Basidia 15–17 × 4–4.4 µm, clavate-cylindrical to subutriform, 4-spored. Basidiospores 4–4.6 × 3–3.6 µm, broadly ellipsoid to ovoid, minutely apiculate, the walls thin subhyaline, smooth, nonamyloid.

Distribution: H.P.: Shimla- Narkanda, Mahasu, Kullu, Chamba, Dalhousie; U.K.: Hemkunt; J&K: Sanasar.

Collection examined: HSK 4024, 4272, 4155, IBP 37091, 37092, L 37093.

Substratum: On base of *Abies pindrow*, on stump of *Cedrus deodara*.

Remarks: This is one of the commonest species occurring mostly on conifers and is marked by distinctly hydnoid fructifications and presence of two types of cystidia (lagenocystidia and capitate leptocystidia).

Hyphodontia aspera (Fr.) Erikss., Symb. bot. upsal. 16(1): 104. 1958 = *Grandinia aspera* Fr., Hym.Eur.: 627. 1874. Plate 6.13f. Fig. 6.33f–i

Fructification resupinate, loosely adnate, widely effused, up to 200 µm thick in section, palliculose to submembranous; hymenial surface white to cream, granules sparse, smooth to finely granulose; margin thinning, loosely adnate, white to paler concolorous. Context subhyaline in section, composed of loosely woven hyphae.

Hyphal system monomitic; generative hyphae branched, septate, clamped, thin to thick-walled, 3–4 µm wide, subhyaline. Leptocystidia (Hyphidia) 3–4 µm wide, of variable length, cylindrical to hyphoid, with a tendency to form capitates head, thin-walled, subhyaline, arising from subhymenium or upper layer of context, projecting out of the hymenium. Basidia subutriform 4.5–6 × 3.5–4.5 µm, 4-sterigmate, with a basal clamp. Basidiospores 4.5–6 × 3.5–4.5 ellipsoid to broadly ellipsoid, smooth, thin-walled, non-amyloid, acyanophilous.

Distribution: U.K.: Cheena peak, Nainital, Mossy fall.

Collection examined: HSK 4282, GSD 19433, IBP 37094, 37095.

Substratum: On angiospermic twig, on gymnospermous wood.

Remarks: The species is characterized by resupinate, loosely adnate, widely effused, fructification; hymenial surface white to cream; hyphal system monomitic; generative hyphae branched, septate, clamped, thin to thick-walled, 3–4 µm wide, subhyaline. The above collection resembles closely the description of *H. aspera* as given by Eriksson and Ryvarden (1976) and Rattan (1977).

Hyphodontia barbajovis (Bull.) J. Erikss., Symb. bot. upsal. 16(1): 104, 1958 = *Hydnum barbajovis* Bull., Herb. Fr. 11: 481, 1791. Fig. 6.33j–m

Fructification resupinate, adnate, effused; hymenial surface smooth to odontoid, creamish white when fresh, becoming yellowish to greyish yellow on drying; margin thinning; paler concolorous to indeterminate.

Hyphal system monomitic; generative hyphae up to 3.5 μm wide, clamped, septate; basal hyphae parallel to the substrate, thin to thick-walled. Cystidia 118.0–160 × 4.8–7.2 μm, thick-walled at the base, cylindrical, secondary septa present. Basidia 12.0–16.5 × 4.1–5.2 μm, 4-sterigmate, subclavate with basal clamp. Basidiospores 4.2–6.0 × 3.5–4.0 μm, apiculate, broadly ellipsoid, thin-walled, inamyloid, smooth, acyanophilous.

Distribution: H.P.: Manali- Hadimba, reserve forest.

Collection examined: IBP 37096.

Substratum: On twigs of *Cedrus deodara.*

Remarks: This species is characterized by smooth to odontoid hymenial surface, cylindrical cystidia with apical walls and thick walls elsewhere and ellipsoid basidiospores. It is a new record for Himalayas.

Hyphodontia caulicystidiata Dhingra, J.Indian. bot. Soc. 84: 120, 2005. Fig. 6.33n–q

Fructifications resupinate, adnate, effused, soft, thin; hymenial surface smooth to porose-floccose under lens, whitish to yellowish-grey to pale ochraceous; margins indistinct.

Hyphal system monomitic; generative hyphae branched at wide angles, septate, clamped, up to 3.4 μm wide; basal hyphae loosely interwoven, somewhat thick-walled, subhymenial hyphae denser and thin-walled. Cystidia 55–83 × 9–12.2 μm, numerous, subcylindrical to subfusiform with a distinct stalk, smooth, somewhat thick-walled at the base, gradually thinning above, cyanophilous. Basidia 25–35 × 5–6.3 μm, clavate to subclavate, somewhat sinuous, with oily contents and a basal clamp, 4-sterigmate; sterigmata up to 5 μm long. Basidiospores 3.5–5 × 2.4–3.2 μm, ellipsoid, smooth, thin-walled, inamyloid, acyanophilous, with one to many oil drops.

Distribution: H.P.: Kullu, Dharamshala.

Collection examined: GSD 19262, IBP 37660.

Substratum: On angiospermous wood.

Remarks: The species is characterized by having resupinate, soft membranous fructification; hymenial surface smooth to floccose; hyphal system monomitic, clamped; cystidia numerous; basidia clavate to subclavate; ellipsoid, smooth basidiospores. It is a new record for N. W. Himalayas.

Hyphodontia crustosa (Pers.) J. Erikss., Symb. bot. upsal. 16: 104, 1958. Plate 6.14a, Fig. 6.34a–c

Fructifications resupinate, membranous to membranous-ceraceous, adnate, widely effused, up to 150 μm thick in section; hymenial surface finely toothed, cream yellow, continuous, not or rarely cracking on drying; margin determinate to indeterminate, adnate, paler concolorous. Context subhyaline in section, composed

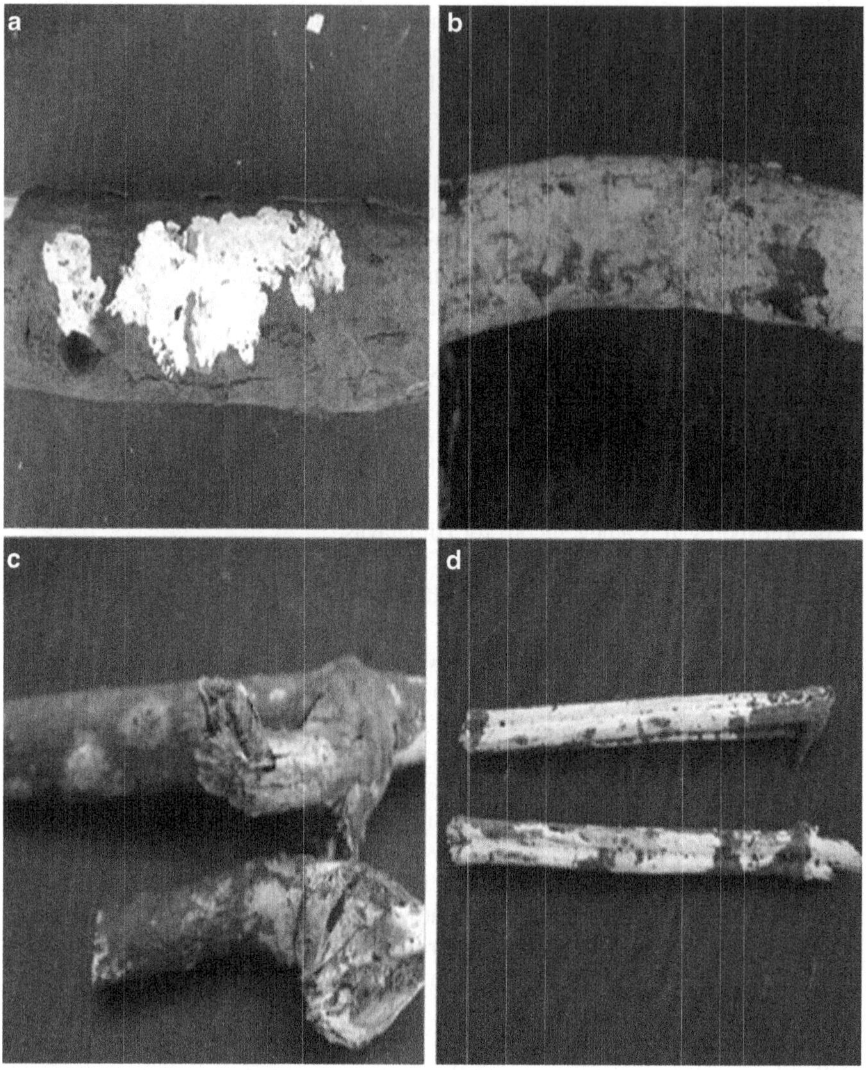

Plate 6.14 (**a**) *Hyphodontia crustosa*. (**b**) *Hyphodontia juniperi*. (**c**) *Hyphodontia nespori*. (**d**) *Hyphodontia pallidula*

of compactly arranged hyphae, impregnated with abundant crystalline matter which often obscures hyphal details.

Hyphal system monomitic, hyphae 2–4 μm wide, branched at wide angles and ramified, septate, clamped, the walls thin, subhyaline, often collapsed and difficult to discern. Teeth gregarious, ranging from small aculei to 200 μm long, subulate with acute apices. Leptocystidia absent but some subulate hyphae may be present,

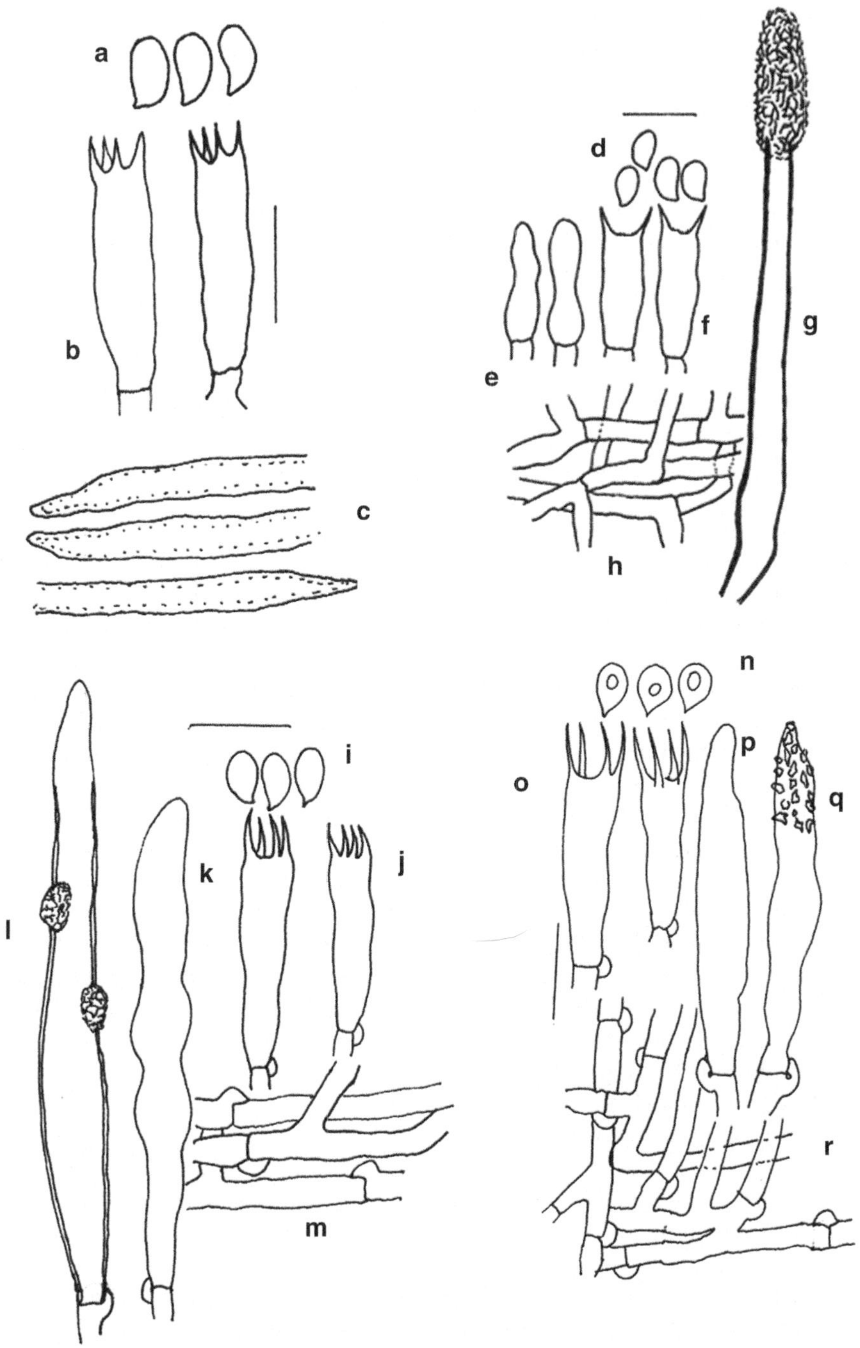

Fig. 6.34 (**a**–**c**) *Hyphodontia crustosa* (**a**) Basidiospores, (**b**) Basidia, (**c**) Cystidia; (**d**–**h**) *Hyphodontia efibulata* (**d**) Basidiospores, (**e**) Basidioles, (**f**) Basidia, (**g**) Cystidia, (**h**) Generative hyphae; (**i**–**m**) *Hyphodontia hastata* (**i**) Basidiospores, (**j**) Basidia, (**k**–**l**) Cystidia, (**m**) Generative hyphae; (**n**–**r**) *Hyphodontia juniperi* (**n**) Basidiospores, (**o**) Basidia, (**p**–**q**) Cystidia, (**r**) Generative hyphae

immersed or slightly projecting out of the hymenium. Basidia 15–20 × 3.5–4.5 μm, subutriform, 4-spored, sterigmata slender and up to 4 μm long. Basidiospores 5.5–6.5 × 2.8–3.5 μm, cylindrical, minutely apiculate, the walls thin, subhyaline, smooth, nonamyloid.

Distribution: U.K.: Hemkunt, Mussoorie.

Collection examined: HSK 4271, 4290.

Substratum: On coniferous logs, on twigs of *Rhododendron.*

Remarks: The characteristic features of this species are membranous-ceraceous fructifications, small teeth or aculei and cylindrical basidiospores. Leptocystidia are absent but some subulate hyphae may be present in the hymenium, immersed or slightly projecting.

Hyphodontia efibulata Erikss. & Hjortst.Svensk bot.Tidskr. 63(2): 226 (1969). Fig. 6.34d–h

Fructifications resupinate, soft-membranous, adnate, widely effused, up to 600 μm thick in section; hymenial surface smooth to finely tuberculate or aculeate, continuous but sometimes reticulate-porose under the lens, not or rarely creviced irregularly especially on drying, white to cream when young becoming cream yellow to ochre on drying; margin variable, thinning to abrupt or indistinct, adnate, paler concolorous. Tubercles minute, acute to hemispherical and gregarious, fertile all over.

Hyphal system monomitic, hyphae 2.5–5.4 μm wide, branched at wide angles and ramified, septate, clamps absent, uninflated but usually with a tendency to become inflated, the walls thin, subhyaline, distinct except in the subhymenial zone where these may become somewhat agglutinated. Cystidia 85–100(125) × 4.5–6(7.5) μm, cylindrical to subfusiform with obtuse apices, arising from the upper part of the context and may project to 45 μm out of the hymenium, the walls subhyaline, slightly thick (up to 0.8 μm) especially in the basal part, naked but more often impregnated with subhyaline crystals in the apical part forming a capitate or globose head up to 25 μm long and 8 μm broad. Basidia 12–14 × 4.0–4.5 μm, utriform, 2-spored, sterigmata up to 4.5 μm long. Basidiospores 3.5–4.2(5.5) × 2–2.4 μm, ellipsoid to ellipsoid-cylindrical, minutely apiculate, thin-walled, smooth, subhyaline, nonamyloid.

Distribution: H.P.: Kullu.

Collection examined: SSR 4354, IBP 42923.

Substratum: On wood of *Picea smithiana.*

Remarks: This species is marked by the tuberculate hymenial surface and absence of clamps on the hyphae. It resembles *H. efibulata* in most of the respects except that the basidiospores are slightly longer in *H. efibulata*.

Hyphodontia hastata (Litsch.) J. Erikss., Symb. bot. upsal. 16(1): 104 (1958) = *Peniophora hastata* Litsch., Öst. bot. Z. 77(2):130(1928). Fig. 6.34i–m

Fructification resupinate, effused, adnate, up to 210 μm thick in section; hymenium surface smooth to tuberculate; whitish orange to pale orange when fresh, yellow to brownish orange on drying; margin thinning, paler concolorous to indeterminate.

Hyphal system monomitic; generative hyphae up to 3.1 µm wide, clamped, septate; subhymenial hyphae thin-walled, highly branched. Cystidia of two types: Monoliform cystidia 38–75×6.2–8 µm thick-walled with basal clamp and subulate cystidia up to 52×6 µm, thick-walled with resinous encrustation which dissolve in 2 % KOH sol. Basidia 14–20×3.5–4.5, 4-sterigmate, clavate, with basal clamp. Basidiospores 4–5×2–3 apiculate, smooth, thin-walled, acyanophilous, inamyloid.

Distribution: H.P.: Solan, Chail.

Collection examined: IBP 37101, 37102.

Substratum: On coniferous wood and twigs.

Remarks: The species is characterized by the presence of numerous subulate cystidia along with moniliform cystidia and ellipsoid basidiospores.

Hyphodontia juniperi (Bourdot & Galzin) J. Erikss. & Hjortst., Cortic. N. Eur. (Oslo) 4: 666 (1976)=*Corticium serum* var. *juniperi* Bourdot & Galzin, Bull. Soc. mycol. Fr. 27(2): 246, 1911. Plate 6.14b, Fig. 6.34n–r

Fructification resupinate, effused, adnate, up to 320 µm thick in section; hymenial surface smooth to tuberculate, greyish white to creamish white when fresh, yellowish white to pale yellow on drying; margin thinning, paler concolorous to indeterminate.

Hyphal system monomitic; generative hyphae up to 2.8 µm wide, clamped, septate; basal hyphae parallel to the substrate; subhymenial hyphae thin-walled, highly branched. Cystidia 28.0–35.0×6.0–7.2 µm, thin-walled, subcylindrical to subfusiform, with basal clamp, thin-walled. Basidia 10.0–22.0×5.4–5.8 µm, 4-sterigmate, clavate, with basal clamp. Basidiospores 5.5–7.4×3.5–4.0 µm, thin-walled, apiculate, ellipsoid to broadly ellipsoid, inamyloid.

Distribution: H.P.: Sunder Nagar, Mandi, Bilaspur.

Collection examined: IBP 37103, 42117.

Substratum: On rotten sticks.

Remarks: This species is characterized by clavate, subcylindrical to fusiform cystidia, 4-spored basidia and ellipsoid and broadly ellipsoid, thin-walled, uniguttulate basidiospores.

Hyphodontia nespori (Bres.) J. Eriksson & Hjortst., in Eriksson & Ryv., Cort. N.Europe 4: 665, 1976=*Odontia nespori* Bres., Ann. mycol. 18(1/2): 43, 1920. Plate 6.14c, Fig. 6.35m–q

Fructification resupinate, adnate, effused, submembranous to subcrustaceous; hymenial surface odontoid, with small, conical aculei, apically fimbriate by projecting hyphae, creamish white to pale ochraceous; margin somewhat determinate.

Hyphal system monomitic; generative hyphae branched, septate, clamped, somewhat thick-walled, irregularly interwoven; subhymenial hyphae thinner and densely packed; projecting hyphae in the aculei enlarged, somewhat thick-walled. Cystidia not clearly differentiated. Basidia 10–18×3.5–4.0 µm, subcylindrical, 4-sterigmate, with a basal clamp. Basidiospores 4.4–5.3×2.2–3.2 µm, ellipsoid to subcylindrical, smooth, thin-walled, non-amyloid, acyanophilous.

Distribution: Arunachal Pradesh: West Kameng, Bomdila, U.K.: Dehradun.
Collection examined: GSD 19745, L 37319.
Substratum: On decaying angiospermic twigs.
Remarks: The species is characterized by having odontoid fructifications with small, dense aculei, wider, somewhat thick-walled projecting hyphae of the aculei, few capitates hyphae, subcylindrical with a suburniform constriction basidia and ellipsoid to subcylindrical basidiospores. The collection resembles closely the description given by Eriksson and Hjortstam (1976). The species is being recorded for the first time from Uttarakhand and N. W. Himalayas.

Hyphodontia pallidula (Bres.) Erikss., Symb. bot. upsal. 16(1): 104. 1958 = *Gonatobotrys pallidula* Bres., Ann. Mycol. 1(2): 127. 1903. Plate 6.14d, Fig. 6.35a–e

Fructification resupinate, adnate, widely effused, up to 180 μm thick in section, membranous; hymenial surface cream yellow to ochre, continuous, smooth to finely tuberculate; margin thinning or indeterminate, adnate, concolorous.

Hyphal system monomitic; generative hyphae branched at wide angles, septate, clamps are present, thin to thick-walled, 2–4 μm wide. Leptocystidia cylindrical to flexuous, arising from the upper layer of the subiculum, projecting out of the hymenium, walls thin to thick-walled, uncrusted or rarely crusted. Basidia subutriform, 4-spored, 15–18 × 4–5 μm. Basidiospores 3.5–4.5 × 2.2–3 μm ellipsoid to broadly ellipsoid, minutely apiculate, thin-walled, subhyaline, smooth, nonamyloid.

Distribution: H.P.: Khajjiar, Chamba; U.K.: Rishikesh; J&K: Patnitop.
Collection examined: SSR 5141, 4379, IBP 37104, 37105.
Substratum: On bark of *Abies pindrow*, on stump of *Cedrus deodara*.
Remarks: The characteristic feature of the species is smooth hymenial surface, presence of capitate leptocystidia and small ellipsoid basidiospores.

Hyphodontia papilosa (Fr.) Erikss., Symb. bot. upsal. 16(1): 104. 1958 = *Thelephora papilosa* Fr., Elench. fung. 1: 212. 1828. Fig. 6.35f–g

Fructification resupinate, floccose to submembranous, loosely adnate, widely effused, up to 500 μm thick in section; hymenial surface granulose to finely toothed, deep cream to cream yellow, continuous, not crevised; margin thinning, byssoid, loosely adnate, white to paler concolorous. Subiculum subhyaline in section, composed of loosely woven hyphae.

Hyphal system monomitic, hyphae 2–4.5 μm wide, branched at wide angles and ramified, septate, clamped, the walls thin, subhyaline. Teeth gregarious, subulate to subcylindrical with acute to penicillate apices, terrete, concolorous. Leptocytidia absent but the axial bundle of hyphae projecting from teeth apices become somewhat specialized and look like leptocytidia, up to 5 μm broad and often impregnated with small, discrete, spherical crystals. Basidia 10–15 × 3.5–4 μm, clavate-cylindrical to subutriform, 4-spored. Basidiospores 4.3–5 × 2.8–3.4 μm, ellipsoid to subcylindrical, minutely apiculate, the walls thin, subhyaline, smooth, non-amyloid.

Distribution: J&K: Bhadarwah; Nepal: Gosainkund (extra limital).

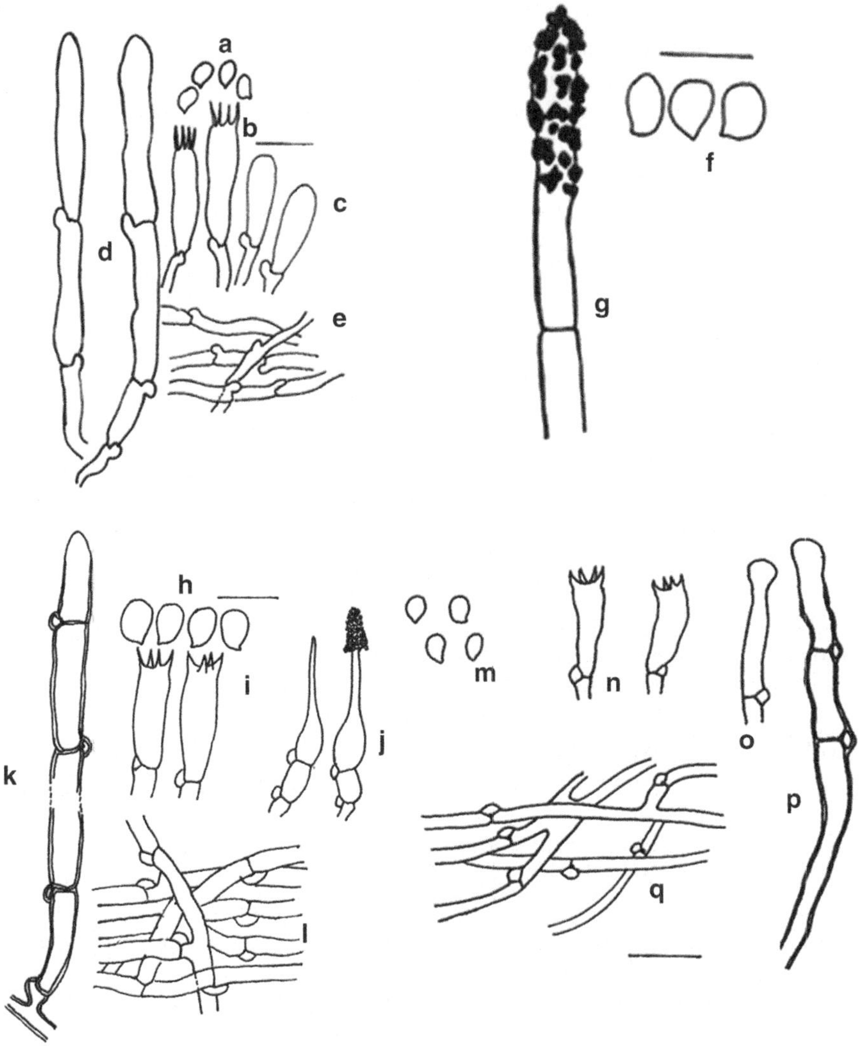

Fig. 6.35 (**a**–**e**) *Hyphodontia pallidula* (**a**) Basidiospores, (**b**) Basidia, (**c**) Basidioles, (**d**) Cystidia, (**e**) Generative hyphae; (**f**–**g**) *Hyphodontia papilosa* (f) Basidiospores, (**g**) Cystidia, (**h**–**l**) *Hyphodontia propinqua* (**h**) Basidiospores, (**i**) Basidia, (**j**) lagenocystidia, (**k**) Cystidia, (**l**) Generative hyphae; (**m**–**q**) *Hyphodontia nespori* (**m**) Basidiospores, (**n**) Basidia, (**o**–**p**) Cystidia, (**q**) Generative hyphae

Collection examined: HSK 4169, 4299.

Substratum: On bark of *Cedrus deodara.*

Remarks: This species is characterized by the floccose to soft membranous fructifications, granulose to finely toothed hymenial surface and ellipsoid to sub-cylindrical basidiospores. True leptocystidia are absent but the axial bundle of hyphae projecting from teeth apices become somewhat specialized and look like leptocystidia.

Hyphodontia propinqua Hjortst., Mycotaxon 17: 553, 1983. Fig. 6.35h–l

Fructification thin, resupinate, adnate, effused, at first porose, with time continuous, subceraceous; hymenial surface creamish-white to yellowish, smooth to finely tuberculate, pilose under lens by the projecting cystidia; margin not well differentiated.

Hyphal system monomitic; generative hyphae 2–3.5 μm wide, branched at wide angles, septate, clamped, cyanophilous; basal hyphae loosely interwoven, somewhat thick-walled; subhymenial hyphae thin-walled, densely interwoven. Cystidia of two kinds, (i) 48–120×5–6.2 mm, hyphoid, somewhat thick-walled, septate, clamped, cyanophilous, apically obtuse, projecting up to 50 mm out of hymenium, number of septa varies from one to several; (ii) 18–25×3.5–4.5 mm, generally immersed lagenocystidia which consist of hyphal ends, abruptly ending in needle like apices, provided with a characteristic encrustation. Basidia 14.5–20×5–5.5 mm, subcylindrical, with a basal clamp, cyanophilous, 4-sterigmate; sterigmata up to 4 mm long. Basidiospores 5.2–8.4×4–5 mm, ellipsoid to broadly ellipsoid, smooth, thin-walled, non-amyloid, acyanophilous, usually uniguttulate.

Distribution: West Bengal: Darjeeling, H.P.: Chamba.

Collection examined: GSD 19267, IBP 37106, L 37107.

Substratum: On angiospermous wood.

Remarks: The above specimen resembles in all respects the type description. It is a new record for N.W. Himalayas.

Hyphodontia sambuci (Pers.) J. Erikss., Symb. bot. upsal. 16(no. 1): 104 (1958)=*Hyphoderma sambuci* (Pers.) Jülich, Persoonia 8(1): 80, 1974. Plate 6.15a, Fig. 6.36a–f

Fructification resupinate, adnate, effused, up to 105 μm thick in section, subceraceous when fresh, cracking and of chalky consistency on drying; hymenial surface smooth to somewhat tuberculate, white to creamish-white or yellowish-white; margin thinning, pruinose in the periphery, mostly determinate in mature fructification.

Hyphal system monomitic; generative hyphae branched, septate, clamped, 2.5–3.5 μm wide; basal hyphae somewhat thick-walled, loosely interwoven; subhymenial hyphae thin-walled, densely united, predominantly in vertical direction, covered with lots of crystals. Cystidial elements frequent in the hymenium, of vertical shape of variable shape either capitate or somewhat tapering, generally with apical encrustation, 26–70×4.5–6.3 μm. Basidia 12–24×3.5–6 μm, subclavate to subcylindrical, 4-sterigmate, with a basal clamp. Basidiospores 5–6×3.5–4 μm, ellipsoid to broadly ellipsoid, smooth, thin to somewhat thick-walled, non-amyloid, weakly cyanophilous when thick-walled.

Distribution: A. P.: West Kameng, Bomdila; Bhutan: Thimphu, Chimakothi.

Collection examined: GSD 19753, 19503.

Substratum: On decaying angiospermic branches.

Remarks: This species is characterized by whitish, chalky fructification, clamped generative hyphae, projecting tapering cystidia and ellipsoid to broadly ellipsoid, thin to somewhat thick-walled basidiospores with oily content.

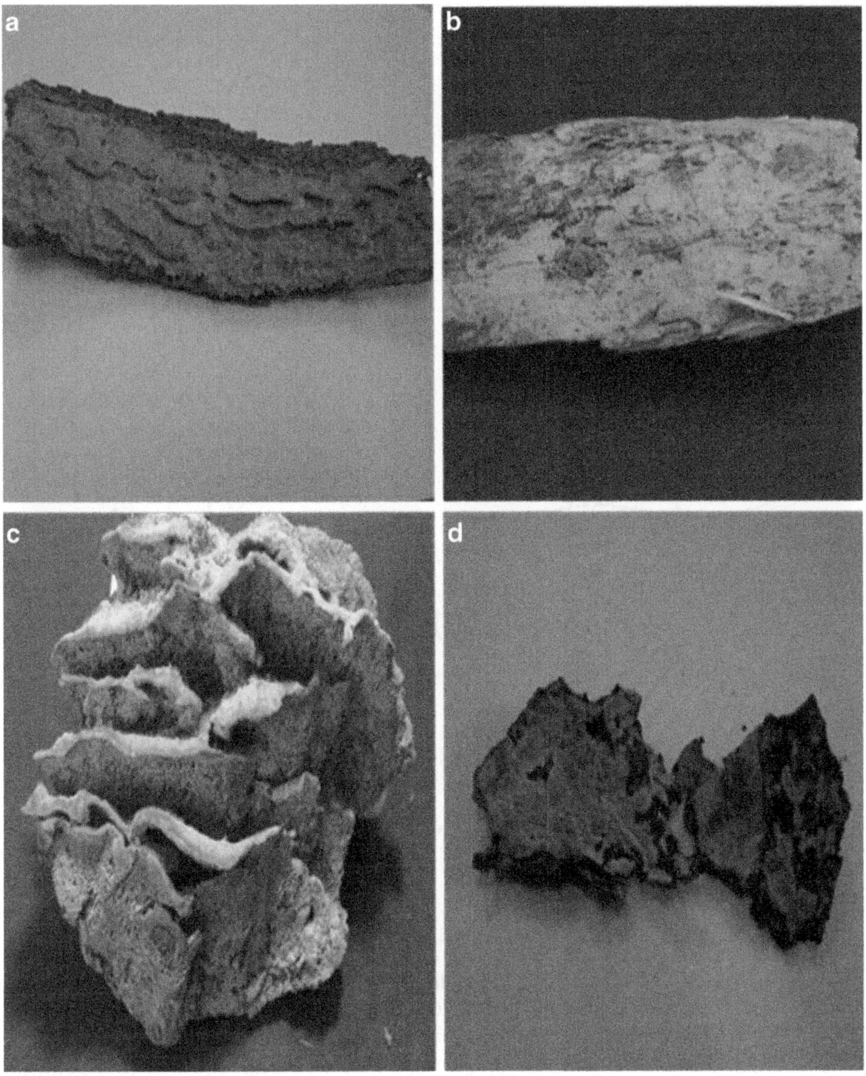

Plate 6.15 (**a**) *Hyphodontia sambuci*. (**b**) *Oxyporus corticola*. (**c**) *Oxyporus ravidus*. (**d**) *Schizopora flavipora*

Hyphodontia spathulata (Schard.) Parm., Consp. Syst. Cort.: 123. 1968 = *Hydnum spathulatum* Schard., Spicil. Fl. Germ. 1: 178 (1794). Fig. 6.36g–j

Fructification resupinate, membranous to membranous-ceraceous creamish when young becoming somewhat horny and brittle on drying, adnate, widely effused; hymenial surface distinctly toothed, cream to cream yellow or pale ochraceous, not or rarely cracking irregularly on drying; margin thinning, byssoid, adnate, white to paler concolorous. Teeth gregarious, subulate to subcylindrical and

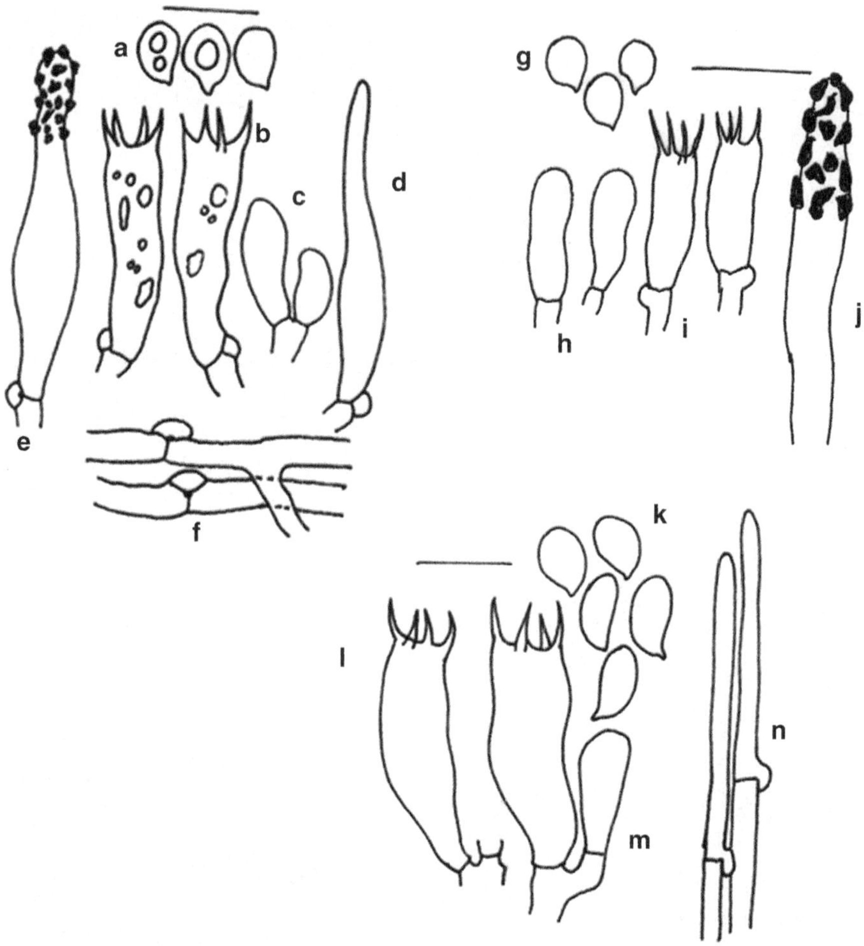

Fig. 6.36 (**a**–**f**) *Hyphodontia sambuci* (**a**) Basidiospores, (**b**) Basidia, (**c**) Basidioles, (**d**) Cystidia, (**e**) Encrusted cystidia, (**f**) Generative hyphae; (**g**–**j**) *Hyphodontia spathulata* (**g**) Basidiospores, (**h**) Basidioles, (**i**) Basidia, (**j**) Encrusted cystidia; (**k**–**n**) *Hyphodontia stipata* (**k**) Basidiospores, (**l**) Basidia, (**m**) Basidia with basidioles, (**n**) Generative hyphae

terrete in young specimens but often become flattened, confluent with age and appear irpicoid superficially, apices acute to mostly penicillate due to the projecting hyphae. Subiculum composed of compactly arranged, somewhat agglutinated hyphae.

Hyphal system monomitic, hyphae 2–3.5 µm wide, branched at wide angles and ramified, septate, clamped, the walls thin to slightly firm, subhyaline, distinct but more often agglutinated especially in the trama of the spine. Leptocytidia absent but the axial bundle of hyphae projecting from teeth apex become, impregnated with discreet crystals and look-like leutocystidia. Paraphysoid hyphae are also sometimes present in the hymenium. Basidia 12-15×3.5–4.4 µm, utriform,

4-spored, sterigmata slender and up to 3.5 μm long. Basidiospores 4–5 × 3–3.5 μm, broadly ellipsoid to ovoid, minutely apiculate, the walls thin, subhyaline, smooth, non-amyloid.

Distribution: H.P.: Shimla; J&K: Patnitop; U.K.: Mussoorie; Extra limital: Nepal: Gosainkund.

Collection examined: HSK 4011, 4149.

Substratum: On wood of *Cedrus deodara,* on log of *Betula utilis.*

Remarks: The characteristic features of this species are toothed to irpiciform hymenial surface and broadly ellipsoid to ovoid basidiospores. The teeth apices are penicillate due to the projecting axial bundle of hyphae. These are often covered with discrete crystals and sometimes look like leptocystidia but true leptocystidia are absent in this species. The membranous-ceraceous texture is distinctive feature which is also shared by *H. crustosa* and serves to distinguish both these species from *H. papillosa* and *H. stipata.*

Hyphodontia stipata (Fr.) Gilb., In Petersen, Evol. High. Basidio.: 300 (1971) = *Hydnum stipatum* Fr., Syst. Mycol.1: 425.1821. Fig. 6.36k–n

Fructification resupinate, floccose to loosely membranous, loosely adnate to separable, widely effused; hymenial surface finely odontoid to distinctly toothed, cream yellow to light ochre, discontinuous, not crevised; margin thinning, byssoid, easily separable, cream to paler concolorous. Teeth gregarious, terrete, ranging from small aculei (up to 300 μm long), to 1 mm long, subulate to cylindrical with acute apices, concolorous. Context subhyaline in section, composed of loosely woven hyphae.

Hyphal system monomitic, hyphae 2–3.5 μm wide, branched at wide angles and ramified, septate, clamped, the walls thin to slightly firm, subhyaline. Leptocytidia absent but the axial bundle of hyphae projecting from teeth apex appear like leptocystidia. Basidia 15–18 × 3–4.5 μm, subutriform, 4-spored, sterigmata straight. Basidiospores 4–5.5 × 3–4 μm, broadly ellipsoid to ovoid, minutely apiculate, thin-walled, subhyaline, smooth, non-amyloid.

Distribution: U.K: Mussoorie.

Collection examined: HSK 4234, SSR 4380, IBP 42123.

Substratum: On bark *Salix tetrasperma.*

Remarks: This species is characterized by the floccose fructifications, densely toothed hymenial surface and broadly ellipsoid to ovoid basidiospores. It is common on *Shorea robusta* or other angiospermic substrata especially in the Himalayan foothills. It is close to *H. papillosa* but the teeth and basidiospores are much smaller in *H. papillosa* then *H. stipata*. The basidiospores are ellipsoid to subcylindrical in *H. papillosa* in comparison to ellipsoid to ovoid in *H. stipata.*

Oxyporus (Bourd. & Galz.) Donk
Revis. Nederi. Heterobasid. 2: 202, 1933

Fructification annual or perennial, sessile to resupinate, whitish or pale coloured. Pileus effused-reflexed to sessile. Pore surface whitish to creamish. Context white to creamish. Hyphal system monomitic; generative hyphae thin to thick-walled, septate, clamps absent. Cystidia present, cylindrical. Basidia hyaline, clavate, 4-spored. Basidiospores hyaline, thin-walled, smooth, ellipsoid, non-amyloid.

Ten species, widespread

Lit.: Gilbertson & Ryvarden (*N. Amer. Polypores.* Vol. **2** *Megasporoporia-Wrightoporia:* 437, 1987)

Type Species: *Polyporus connatus* Weinm., 1826

Habitat: Decayed wood

Himalayas: Four

Key to species

1. Hymenophore reddish to yellowish brown or pink to buff, grey 2
1. Hymenophore otherwise white ochraceous, greyish to greenish at base 3
2. Cystidia long, club-shaped; basidiospores oblong-ellipsoid, 5.1–7.0 × 2.0 .... *O. vellereus*[a]
2. Cystidia clavate, basidiospores cylindric to ellipsoid, 6–8.8 × 3–4 μm .... *O. cervinogilus*
3. Cystidia usually smooth, cylindrical or irregular in shape, basidiospores yellowish and thick-walled........................ *O. mollissimus*[a]
3. Cystidia encrusted 4
4. Cystidia with yellowish resinous excretions at the apex, fusiform....... *O. ravidus*
4. Cystidia coarsely encrusted........................ 5
5. Cystidia of one type, all encrusted 6
5. Cystidia of two types, ventricose and encrusted, smooth to hyphoid .... *O. corticola*
6. Fructification annual, widely effused, cystidia clavate to ventricose .... *O. latemarginatus*[a]
6. Fructification perennial, pileate cystidia hyaline to yellowish .... *O. populinus*

[a]Extra limital, not included in the text

Oxyporus cervinogilvus (Jungh.) Ryv., Norw. Jl. Bot. 20:3, 1973 = *Polyporus cervinogilvus* Jungh., Verh. Batav. Genootsch. Kunst. Wetensch. 17:45, 1838.

Fig. 6.37a–c

Fructification annual, resupinate to effused-reflexed, solitary or imbricate, somewhat incurved, rigid when dry; margins papery thin, acute, sterile, often slightly lobed or incised, pale brown to yellowish-white; upper surface brown or straw colored often slightly pale pink, finely tomentose to coarsely hirsute, faintly zoned. Context brown or straw colored, fibrous, coriaceous to corky; hymenial surface concolorous with the context, often slightly ochraceous to buff with pale pink tint, pores regular or angular nearly irpicoid towards the older regions extending up to edges, 1–2 per mm, pore tubes straight or slightly oblique up to 2 mm long. Cystidia hyaline, thick-walled, clavate, coarsely encrusted at apex, 20–35 × 5–9 μm, hymenial in origin.

Hyphal system monomitic; generative hyphae simple septate, branched, hyaline to pale yellow, thin to thick-walled, lumina broad, occasionally swelled, 2.5–5.7 μm wide. Basidia 4-sterigmate, 16.0–20.5 × 5.5–6.5 μm. Basidiospores hyaline, smooth, thin-walled, cylindrical to oblong, ellipsoid, 6–8.6 × 3–4.5 μm.

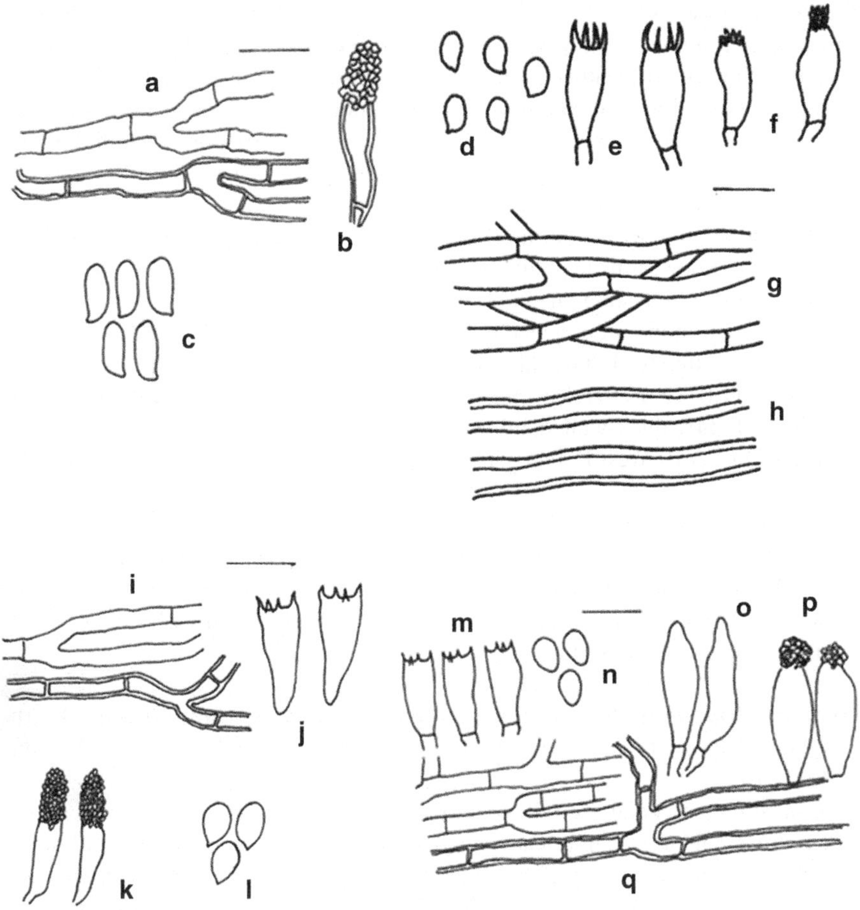

Fig. 6.37 (**a**–**c**) *Oxyporus cervinogilvus* (**a**) Generative hyphae, (**b**) Encrusted cystidia, (**c**) Basidiospores; (**d**–**h**) *Oxyporus corticola* (**d**) Basidiospores, (**e**) Basidia, (**f**) Cystidia, (**g**) Generative hyphae, (**h**) Skeletal hyphae; (**i**–**l**) *Oxyporus populinus* (**i**) Generative hyphae, (**j**) Basidia, (**k**) Encrusted cystidia, (**l**) Basidiospores; (**m**–**q**) *Oxyporus ravidus* (**m**) Basidia, (**n**) Basidiospores, (**o**) Basidioles, (**p**) Encrusted cystidia, (**q**) Generative hyphae

Distribution: U.K.: Dehra Dun; Assam: Lokhra hills.
Collection examined: IBP 42125.
Substratum: On decaying angiospermic log of *Dalbergia paniculata.*
Remarks: The collection resembles very much *Coriolopsis zeylanicus* but distinguishing characters are: *C. zeylanicus* has trimitic hyphal system with clamped generative hyphae, longer cylindrical basidiospore and it lacks cystidia, while *O. cervinogilvus* has a monomitic hyphal system with simple septate generative hyphae, ellipsoid basidiospores and encrusted cystidia.

Oxyporus corticola (Fr.) Ryv., Persoonia 7: 19, 1972. Plate 6.15b, Fig. 6.37d–h

Fructification annual, rarely perennial, resupinate or widely effused, context thin cream fibrous, soft leathery when dry; margins narrow, finely fimbricate. Hymenial surface white to cream turning light cream brown on drying, pores circular to angular or irregular and partly elongated, 2–4 per mm, pore wall thin, fimbriate, sometimes lacerate, pore tube up to 5.5 mm long. Cystidia of two kinds- (i) hyaline, thin-walled, elongated or cylindrical, with an apical crown of crystals, 12–20 × 4–7 μm, hymenial in origin; (ii) smooth, hyaline, thin to slightly thick-walled, cylindrical to hyphoid, apical end rounded to tapering, 20–24.5 × 6–15 μm, originating in subhymenial layer.

Hyphal system monomitic; generative hyphae simple septate, branched, hyaline to subhyaline, thin to slightly thick-walled, 2.0–4.5 μm wide. Basidia 4-sterigmate, clavate 11.0–12.0 × 5.0–6.2 μm. Basidiospores hyaline to pale yellow, smooth, thin to slightly thick-walled, ellipsoid, 4.5–6.0 × 3.6–4.0 μm.

Distribution: A.P.: West Kameng, about 5 km from Shergaon towards Rupa; U.K.: Nainital, Mussoorie.

Collection examined: SSV 21504, IBP 37019, L 37110.

Substratum: On decaying angiospermic log, also occurring on *Abies pindrow.*

Remarks: It is characterized by annual, resupinate, soft fructification; white pore surface; large 2–4 per mm pores; monomitic hyphal system with simple septate hyphae, encrusted cystidia.

Oxyporus populinus (Schum.) Donk, Meddn. Bot. Mus. Univ. Utrecht. 9: 204, 1933 = *Polyporus populinus* (Schum.) Fr., Syst. Mycol.1: 367, 1821. Fig. 6.37i–l

Fructification perennial, sessile or effused-reflexed, mostly imbricate, less frequently singular or as imbricate pilei from a common base, pileus mostly broadly attached, semi-circular to elongated, often irregular, bony hard when dry; margin acute; upper surface cream to ochraceous, sometimes blackish, azonate, intricately villose, later glabrous and dull or even slightly scrupose in old parts. Context cream to ochre, soft, mostly thin, 2–4 μm deep; hymenial surface whitish to pale ferruginous with a distinct sterile and finely velutinate margin, shiny, pores circular to angular, sometimes labelliform in mature conditions, 4–5 per mm, pore tube thin and papery, whitish, stratified, each stratum 1–4 mm thick, totally up to 4 cm deep at the base. Cystidia hyaline to yellowish, thin- to slightly thick-walled, clavate to club-like, coarsely encrusted at the apex, 10–21 × 3–5.5 μm, subhymenial or tramal in origin.

Hyphal system monomitic; generative hyphae simple septate, branched, hyaline to brownish, thin- to thick-walled, 2.5–4.3 μm wide. Basidia clavate, 4-sterigmate, 9–21 × 5–6 μm. Basidiospores hyaline to pale yellow, thin- to slightly thick-walled, ellipsoid to subglobose smooth, 3.5–4.6 × 3–4.5 μm.

Distribution: U.K.: Dehra Dun, Mussoorie.

Collection examined: IBP 42127.

Substratum: On decaying angiospermic log on living tress of *Mallotus philippinensis* Muell.

Remarks: The species is characterized by perennial, sessile fructification with the tube layers separated by thin layers of context.

Oxyporus ravidus (Fr.) Bond. & Sing., Ann. Mycol. 39(1): 63, 1941 = *Polyporus ravida* Fr., Epicr. Syst. Mycol.: 475 (1838) = *Trametes ravida* (Fr.) Pilat., Atl. Champ. Polyp. 272, 1939. Plate 6.15c, Fig. 6.37m–q

Fructification annual, sessile, effused-reflexed or resupinate, often imbricate, sometimes laterally fused with the substratum and covering the large portion of the dead logs, attached by narrow or broad base, flexible when fresh and rigid and corky on drying; margin thin, acute, sterile, whitish when fresh, later becoming light brown to straw yellow slightly involute or incurved inwards on drying; upper surface white to cream when fresh, straw yellow when old, sometimes dark brown crust appearing at some places, uneven with distinct cracking, nearly azonate to indistinctly zonate near the margin, strongly tomentose in isolated nodular areas or strigose or even fibrillose-setose, somewhat radiate-rubose, finally becoming glabrous, at some places sometimes covered with mosses. Context homogenous, white to cream, flexible to coriaceous when fresh, straw yellow to pale brown and corky after drying, sometimes with dark brown or black narrow lines separating the tomentum from the context; hymenial surface wide when fresh, finally light brown or straw yellow coloured, pores more or less circular to angular or nearly lamellar, 2–3 per mm, pore tubes mostly in one, less frequently 2–3 layers, short, white when fresh, drying to straw yellow, up to 3 mm long. Cystidia are of two types: (1) fusiform or rarely cylindrical, hyaline, thin-walled with yellow resinous excretions at the apex, 10–15 × 5–7 μm. (2) Hyaline, thin-walled, wide cylindric to ventricose, smooth, 15.2–25.5 × 4–7.5 μm.

Hyphal system monomitic; generative hyphae hyaline to subhyaline, thin- to thick-walled, branched with prominent simple septa, sometimes collapsed, 1.5–6 μm wide. Basidia 4-sterigmate, 9–12 × 5–8 μm. Basidiospores hyaline thin-walled, smooth, oblong ellipsoid, 5–6.5 × 3–4 μm. Cystidia abundant, projecting beyond the hymenium.

Distribution: U.K.: Dehra Dun, Chakrata, Kumaun; H.P.: Kullu, Manali.

Collection examined: IBP 42119, 42121.

Substratum: On decaying angiospermic logs.

Remarks: The species is characterized by having annual, sessile, effused-reflexed or resupinate fructification; monomitic hyphal system; 4-sterigmate basidia; thin-walled smooth, oblong ellipsoid basidiospores.

Schizopora Velen. Českă Houby 4–5: p. 638, 1922 Emend. Donk-Persoonia 5: 76, 1967.

Fructification annual to biennial, resupinate, adnate to separable; margin sterile with or without rhizomorph. Context creamish, thin, homogenous, non-xanthochroic; pores round, angular, split pores oblong or labyrinthine. Pore surface whitish, cream to light brown. Hyphal system dimitic; generative hyphae thin to thick-walled, clamped, branched non- inflating frequently ending into a vesicle; skeletal hyphae thick-walled, aseptate, fusiform at the tip, not penetrating into the hymenium, cyanophilous. Leptocystidia and lamprocystidia present. Basidia utriform, clavate, 4-spored, cyanophilous. Basidiospores allantoids to cylindrical, hyaline, thin-walled, amyloid.

Four Species, widespread

Lit.: Ryvarden & Johansen (*Prelim. Polyp. Fl. E. Afr.:* 315, 1980)

Habitat: Dead Wood

Type species: *Polyporus laciniatus* Velen., 1922.

Himalaya: Two

Key to species

1. Basidiospores thin-walled, hyaline, smooth, non-amyloid, globose to subglobose, minutely apiculate, 3.2–4.4 × 2.8–3.5 μm....... *S. flavipora*
1. Basidiospores oblong-ellipsoid, smooth, hyaline, thin-walled, apiculate, 4.1–5.6 × 3.0–4.2 μm .... *S. paradoxa*

Schizopora flavipora (Cooke) Rejv. Bull. Soc. Mycol. Fr. 51: 383, 1935, – Grzyby 2: 53, 1965. Plate 6.15d

Fructification perennial, resupinate, effused, coriaceous when fresh, becoming brittle on drying, spreading up to 12 × 5 cm. Margin adnate, thin, up to 3 mm wide, lighter than pore surface, pale creamish. Pore surface rimose, light yellowish orange when fresh, wood coloured after drying; pores round to somewhat angular, 92–208 μm in diameter 9–13 per mm; pore mouth velutinate; dissepiment, entire,42–97 μm thick; tubes stratose, creamish in section. Context cream coloured, xanthochroic, homogeneous 0.3 mm thick.

Hyphal system dimitic; generative hyphae hyphodontoid, flexuose, thin-to thick-walled, cyanophilous, branched, septate, clamped, 2.3–3.5 μm in diameter. In dissepiment, hyphae 1.8–2.8 μm in diameter; skeletal hyphae subhyaline, thick-walled, cyanophilous, flexuose, 124–410 μm long and up to 4.6 μm in diameter. Hyphae frequently and terminally into a vesicle, 6.5–7.8 μm in diameter. Sometimes vesicle prolonged into a hypha on the opposite side. Cystidia abundant, ventricose, 14–17.5 × 4.2–5.4 μm. Basidia collapsed on drying. Basidiospores thin-walled, hyaline, smooth, non-amyloid, globose to subglobose, minutely apiculate, 3.2–4.4 × 2.8–3.5 μm.

Distribution: A. P.: West Kameng, New Bomdila.

Collection examined: SSV 21799.

Substratum: On decaying angiospermic log.

Remarks: This species is marked by effused, wood coloured fructifications; dimitic hyphal system with thick-walled generative hyphae; terminal vesicles; ventricose cystidia and globose to subglobose basidiospores. *Schizopora phellinoides* is quite similar to *S. flavipora* (Schrad.: Fr.) Donk. However, the latter differs in having annual fructifications; larger angular to irpicoid to hydnoid pores; and larger ellipsoid basidiospores.

Schizopora paradoxa (Schard.) Donk, Persoonia 5(1): 76. 104. 1967 = *Hydnum paradoxum* Schrad. Spicil. Gern. 1: 179, 1794 = *Poria versipora* (Pers.) Sacc., Syll. Fung. 6: 311(1988). Plate 6.16a, Fig. 6.38a–g

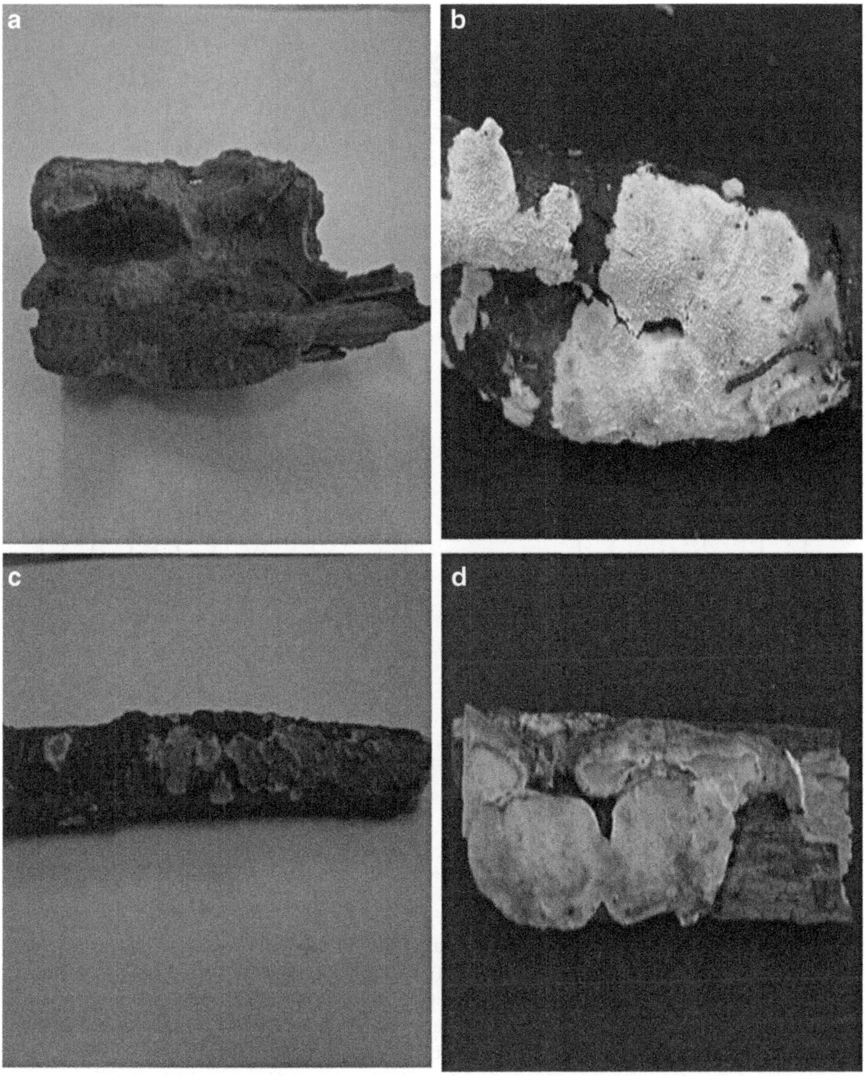

Plate 6.16 (**a**) *Schizopora paradoxa*. (**b**) *Xylodon rimmossima*. (**c**) *Antrodia albida*. (**d**) *Antrodia serialis*

Fructification annual, rarely biennial, resupinate, widely effused, brittle when dry; margin cream; context pinkish buff, less than 1 mm thick; hymenial surface pinkish buff, pores angular, irregular, often toothed with age, pore tubes up to 1 mm long.

Hyphal system dimitic; generative hyphae thin-walled to thick-walled, clamped, branched, hyaline, some characteristically swelled at the apex, frequently encrusted particularly at the dissepiment ends, 1.5–4.0 μm wide; skeletal hyphae

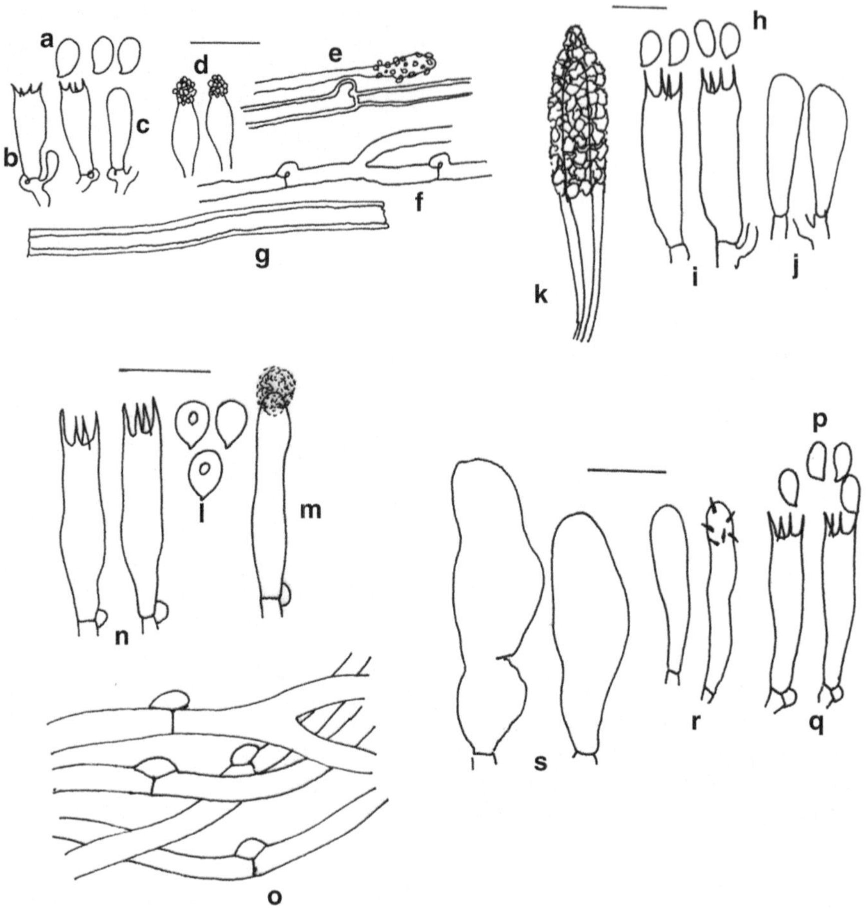

Fig. 6.38 (**a**–**g**) *Schizopora paradoxa* (**a**) Basidiospores, (**b**) Basidia, (**c**) Basidioles, (**d**–**e**) Encrusted Cystidia, (**f**) Generative hyphae, (**g**) Skeletal hyphae; (**h**-**k**) *Xylodon pruni* (**h**) Basidiospores, (**i**) Basidia, (**j**) Basidioles, (**k**) Encrusted cystidia, (**l**-**o**) *Xylodon rimosissimus* (**l**) Basidiospores, (**m**) Cystidia, (**n**) Basidia, (**o**) Generative hyphae; (**p**-**s**) *Cystostereum murrayi* (**p**) Basidiospores, (**q**) Basidia, (**r**) Cystidia, (**s**) Gloeocystidia

hyaline, thick-walled, aseptate, 3.0–4.5 μm wide. Cystidiole abundant, with acute or swollen apex, also may have resinous or crystalline encrustations at the apex. Basidia 4-sterigmate, 12.0–15.0×3.5–4.0 μm. Basidiospores oblong-ellipsoid, smooth, hyaline, thin-walled, apiculate, 4.1–5.6×3.0–4.2 μm.

Distribution: U.K: Dehra Dun; H.P.: Kullu.

Collection examined: IBP 37111, 37112.

Substratum: On logs.

Remarks: This species is characterized by the presence of encrusted hyphae and hyphae with terminal swellings.

Xylodon (Pers.) Grey,
Nat. Arr. Brit. Pl. (London) 1: 649(1821)

Fructifications resupinate, adnate, confluent, hymenophore smooth to grandinoid, tuberculate, raduloid or distinctly hydnoid, in some species poroid. Hyphal system monomitic, clamped, septate, thin to thick-walled, cyanophilous. Cystidia present, conical to more commonly captate rarely septate smooth or incrusted. Basidia suburniform, thin-walled or basally thickened, 4-spored and a basal clamp. Basidiospores globose to ellipsoid, cylindrical or allantoids, smooth, thin to thick-walled, cyanophilous.

Lit.: Donk (*Taxon* **12:** 113, 1963)

Type Species: *Xylodon quercinus* (Pers.) Grey 1821

Habitat: Wood

Himalayas: Two

Key to species

1. Basidia subcylindrical, subclavate,
basidiospores ellipsoid to broadly ellipsoid.................................. *X. rimosissima*
1. Basidia clavate, subclavate, basidiospores ellipsoid................................*X. pruni*

Xylodon pruni (Lasch.) Hjortstam & Ryvarden, Syn. Fung. (Oslo) 23: 100 (2007) = *Phanerochaete pruni* (Litsch.) Rattan, Biblthca Mycol. 60: 258(1977) = *Odontia pruni* Lasch. In Rabenh., Fungi Eur. Exsicc. No. 1915. Fig. 6.38h–k

Fructifications resupinate, membranous-ceraceous to wholly ceraceous becoming horny and brittle on drying, adnate, widely effused; hymenial surface deep cream to yellow ochre, densely tuberculate to minutely toothed, not creviced to rarely cracking in thicker parts on drying; margin thinning to more or less abrupt, adnate, white to paler concolorous. Teeth up to 250 μm long and 150 μm broad at the base, subulate to cylindrical, apices usually tufted, concolorous. Context subceraceous to ceraceous.

Hyphal system monomitic, hyphae 2.5–4 μm wide, branched, septate, clamps absent, thin- walled, subhyaline, often collapsing and agglutinating as discern. Cystidia 8–12 μm broad, usually long and cylindrical, the walls moderately thick, subhyaline, heavily incrusted, usually occurring as axial bundle in the trama of spine. Basidia 20–25 × 5–6 μm, clavate-cylindrical, 4-spored, sterigmata up to 4 μm long. Basidiospores 6–7.5 × 3.2–4.5 μm, ellipsoid to broadly ellipsoid, minutely apiculate, the walls thin, subhyaline, smooth, non-amyloid.

Distribution: J&K: Bhadarwah.

Collection examined: HSK 4158.

Substratum: On fallen twigs.

Remarks: This species is characterized by the odontoid hymenial surface, presence of simple-septate hyphae and lamprocystidia forming an axial bundle in the trama of the spine.

Xylodon rimosissimus (Peck) Hjortstam & Ryvarden, Syn. Fung. (Oslo) 26: 39 (2009) = *Odontia rimosissima* Peck, Ann. Rep. N. Y. St. Mus. 50(1): 114, 1897. Plate 6.16b, Fig. 6.38l–o

Fructification resupinate, effused, up to 320 µm thick in section; hymenial surface smooth to odontoid; greyish white to pale orange when fresh, pale yellow to pale orange on drying; margin thinning, paler concolorous to indeterminate.

Hyphal system monomitic; generative hyphae up to 4.0 µm wide, clamped, septate, hyphal ends in the aculei encrusted; basal hyphae parallel to the substrate, thick-walled; subhymenial hyphae thin to thick-walled, highly branched. Cystidia absent. Secondary septa present with encrustation at the tip. Basidia 17.0–25.0 × 3.1–4.3 µm subclavate to subcylindrical, 4-sterigmate with basal clamp. Basidiospores 5.6–7.0 × 3.0–4.3 µm apiculate, thin-walled, smooth, ellipsoid to broadly ellipsoid, inamyloid, acyanophilous.

Distribution: H.P.: Manali, Manikaran.

Collection examined: IBP 37112.

Substratum: On log of angiospermic tree.

Remarks: This species is characterized by ellipsoid to broadly ellipsoid, acyanophilous basidiospores, 4-spored basidia and absence of cystidia. This species is retained as *X. rimossimus* instead of *Hyphodontia rimossima* (Peck) Gilb. after Bernicchia & Gorjon (2010) due to absence of lagenocystidia, tubular and encrusted cystidia.

O- Polyporales

Family- Cystostereaceae

Cystostereum Pouz.,
Česká Mykol. 13(1):18. 1959.

Fructifications effused-reflexed to totally resupinate, perennial, often stratose, membranous-ceraceous to somewhat corky or woody; hymenial surface pale coloured, smooth to densely and prominently tuberculate, often cracking deeply and irregularly on drying. Hyphal system dimitic; generative hyphae subhyaline and clamped; skeletal hyphae subhyaline. Cystidia absent. Gloeocystidia present. Basidia clavate, 4-spores. Basidiospores thin-walled subhyaline, smooth, nonamyloid, acyanophilous.

Six species, widespread

Lit.: Hallenberg & Ryvarden (*Mycotaxon* **2:** 135, 1975).

Type Species: *Cystostereum murrayi* Berk. & Curt. 1869.

Habitat: Wood

Himalayas: One

Cystostereum murrayi (Berk. & Curt.) Pouz. Česká Mykol. 13(1): 18. 1959 = *Thelephora murrayi* Berk. & Curt., J. Linn. Soc. Bot. 10: 329(1868) [1869].
Fig. 6.38p–s

Fructification resupinate to effused-reflexed, annual to perennial, adnate, membranous-ceraceous when fresh but turn hard and brittle on drying, often arising as small colonies which may become widely effused due to confluence and growth; hymenial surface white to cream, smooth but become strongly tuberculate with age, tubercles abundant and gregarious, continuous but become areolately and irregularly creviced on drying; margin thick, adnate, concolorous. Pileus poorly developed and often represented by merely upturned margin; upper surface brownish-black to almost black, concentrically sulcate, uneven and bark-like. Context subhyaline to light yellow, composed of compactly arranged more or less agglutinated hyphae enclosing abundant gleoecystidia. Sometimes gleoecystidia collapse and disintegrate leaving empty space in the context.

Hyphal system dimitic; generative hyphae 1.5–2.4 µm wide, branched, septate, clamped, the walls thin, subhyaline; skeletal hyphae 1.5–2.5 µm wide, sparsely branched, aseptate, the walls subhyaline to tinted yellow, thick with little or no lumen; Gleoecystidia 30–53 × 7–12 µm, cylindrical to clavate, abundant and usually arranged in overlapping rows, empty or with pale yellow granular contents, the walls thin, subhyaline, distinct in young specimens but often collapsing in mature parts. Basidia clavate-cylindrical, averaging 20 × 3 µm, 4-spored. Basidiospores 4–4.4 × 1.8–2.1 µm, ellipsoid to slightly pip-shaped, minutely apiculate, thin-walled, subhyaline, smooth, non-amyloid.

Distribution: H.P.: Dalhousie- Lakkarmandi, Mahasu- Narkanda.

Collection examination: SSR: 5154, 5629.

Substratum: Coniferous wooden poles and stumps of *Abies pindrow.*

Remarks: The characteristic features of this species are tuberculate hymenial surface, dimitic hyphal system, presence of abundant gleoecystidia and nonamyloid basidiospores. The gleoecystidia are clavate to cylindrical but some may appear globose to subglobose depending upon their orientation in the slide.

Family Fomitopsidaceae

Key to genera

1. Generative hyphae septate.. 2
1. Generative hyphae clamped.. 7
2. Gloeoplerous hyphae present.. *Phaeolus*
2. Gloeoplerous hyphae absent.. 3
3. Hyphal system monomitic, cystidia present.. *Pycnoporellus*
3. Hyphal system dimitic with skeleton–binding hyphae.. 4
4. Cystidia present.. *Dacryobolus*
4. cystidia absent.. 5
5. Fructification resupinate.. 6
5. Fructification sessile or substipitate.. *Laetiporus*
6. Fructification thin membranous soft fleshy
   when fresh, fragile on drying.. *Anomoporia*
6. Fructification fragile when fresh, brittle on drying.. *Antrodia*
7. Fructification hard and tough, annual to perennial,
   hyphal system di/trimitic.. 8
7. Fructification soft, annual, hyphal system usually monomitic.. 9
8. Pores angular to daedaleoid to lamellate.. *Daedalea*
8. Pores round to angular.. *Fomitopsis*
9. Basidiospores amyloid to dextrinoid, slightly thick–walled.. *Parmastomyces*
9. Basidiospores not so.. *Postia*

Antrodia P. Karst., Emend. Donk
Medd. Soc. Fauna Fl. fenn. 5: 40, 1879.

Fructification annual to reviving, resupinate to effused-reflexed, adnate, soft to tough; upper surface azonate, white to cream coloured. Pore surface white to creamish brown, pores round to angular to irregular, glancing, tubes not stratified. Context cream to light brown, thin, compact, homogenous coriaceous. Hyphal system dimitic; generative hyphae hyaline, branched, thin-walled, septate, clamped; skeletal hyphae hyaline, thick-walled to solid, aseptate, abundant. Cystidia and setae absent. Cystidioles xanthochroic. Basidia hyaline, clavate, 4-spored. Basidiospores cylindrical to allantoids, hyaline, thin-walled, smooth, acyanophilous, non-amyloid.

Forty Six Species, widespread

Lit.: Donk (*Persoonia* 4: 339, 1966), Lombard (*Mycol.* 82:185, 1990)

Type Species: *Polyporus serpens* Fr. 1818.

Habitat: wood

Himalayas: Five

Key to species

1. Hyphal system trimitic .... *A. serialis*
1. Hyphal system dimitic 2
2. Basidiospores ellipsoid 3
2. Basidiospores allantoid to cylindrical 4
3. Basidiospores cylindric, ellipsoid; 8–12 × 3–4.5 µm .... *A. albida*
3. Basidiospores ellipsoid; 4–6 × 2.5–3.3 µm .... *A. gossypium*
4. Basidiospores allantoid to cylindrical; 4–5 × 1–1.8 µm .... *A. xantha*
4. Basidiospores cylindric; 4.2–5.5 × 1.2–1.8 µm .... *A. sinuosa*

Antrodia albida (Fr.) Donk, Persoonia 4(3): 339 (1966) = *Daedalea sepium* (Berk.) Ravenel, Fung. Carol. Exs. 1: no. 21, 1855. Plate 6.16c, Fig. 6.39a–c

Fructification annual, resupinate to effused-reflexed, with small distinct pilei or sometimes close pilei, confluent to form extended pilei along the wood, imbricate, coriaceous when fresh, tough and corky on drying; upper surface white to cream with pinkish tinge; margin white very narrow. Pore white to cream with pinkish tint finely tomentose, faintly zonate. Pore surface white to creamish; pores round to angular, 2–3 per mm, dissepiments even, 40–120 µm thick; tubes in one layer, up to 7 mm deep in section. Context white, soft, homogenous, coriaceous.

Hyphal system dimitic; generative hyphae hyaline, thin-walled, septate, branched, clamped, cyanophilous, 2.0–4.3 µm in diameter; skeletal hyphae hyaline to subhyaline, thick-walled to solid, aseptate, unbranched, 3.0–5.2 µm in diameter. Cystidia absent. Basidia clavate, 4-spored, up to 7.5 µm broad, Basidiospores hyaline, thin-walled, smooth, cyanophilous, cylindric- ellipsoid, non-amyloid, 8.0–12.5 × 3.0–4.5 µm.

Distribution: Bhutan- Chimakothi; U.K.: Chakrata.

Collection examined: SSV 21260, IBP 37114, L 37115.

Substratum: On stump of *Quercus dilate* and *Q. semecarpifolia.*

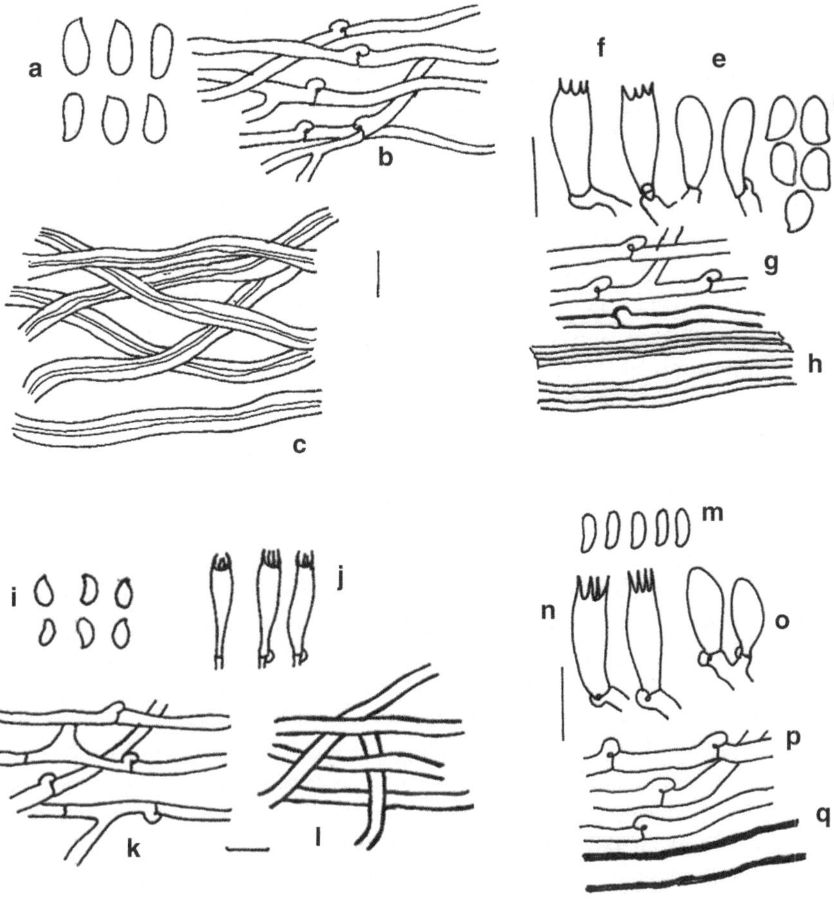

Fig. 6.39 (**a**–**c**) *Antrodia albida* (**a**) Basidiospores, (**b**) Generative hyphae, (**c**) Skeletal hyphae; (**d**–**h**) *Antrodia gossypium* (**d**) Basidiospores, (**e**) Basidioles, (**f**) Basidia, (**g**) Generative hyphae, (**h**) Skeletal hyphae; (**i**–**l**) *Antrodia serialis* (**i**) Basidiospores, (**j**) Basidia, (**k**) Generative hyphae, (**l**) Skeletal hyphae; (**m**–**q**) *Antrodia sinuosa* (**m**) Basidiospores, (**n**) Basidia, (**o**) Basidioles, (**p**) Generative hyphae, (**q**) Skeletal hyphae

Remarks: This species was first recorded from India by Bakshi et al. (1958) as *Trametes sepium* Berk. The above cited collection are quite typical of the species and resembles with description given by Bakshi (1971) characterized by effused reflexed to sessile fructification, white to pinkish pore surface; smooth cylindric-ellipsoid basidiospores.

Antrodia gossypium (Speg.) Ryv., Norw. J. Bot. 20: 8, 1973**.** Fig. 6.39d–h
Fructification annual, resupinate, adnate, soft when fresh, brittle on drying, effused, up to 8 × 2 × 2.6 cm. Margin white, thinning, narrow to wide, myceloid. Pore surface white to creamish, light brown on drying, uneven, dull; pores angular,

3–4(5) per mm, tubes in one layer, gelatinous, soft, brittle on drying, white up to 1.8 mm deep. Context white to creamish, soft, thin, homogenous, non-xanthochroic, up to 0.5 mm thick.

Hyphal system dimitic; generative hyphae dominate in the dissepiment, hyaline to subhyaline, thin- to thick-walled, branched, septate, clamped, irregular, cyanophilous, 2.5–6.0 µm in diameter; skeletal hyphae hyaline, thick-walled, straight, rarely branched, aseptate, cyanophilous, 3.2–4 µm in diameter. Cystidia and cystidioles absent. Basidia hyaline, thin-walled, clavate, 4-spored, cyanophilous, 12–13 × 4.5–5.2 µm. Basidiospores hyaline, thin-walled, smooth, broadly ellipsoid, apiculate, non-amyloid, 4–6 × 2.5–3.3 µm.

Distribution: A.P.-West Kameng, Bomdilla, New Bomdilla; H.P.: Shimla.

Collection examined: SSV 21821, Dhanda 6533, IBP 37116.

Substratum: On log of *Abies pindrow* **.**

Remarks: The species is characterized by annual, resupinate, adnate, soft fructifications; white pore surface; white margin; angular, 3–4 (−6) pores per mm; thin, white, 0.5 mm thick context; dimitic hyphal system with thin- to thick-walled generative hyphae; and hyaline, smooth, ellipsoid, 4–6 × 2.5–3.5 µm basidiospores. The species was first reported from India by Thind and Dhanda (1978) on the basis of single collection Dhanda 6533 from Shimla. This species appears to be uncommon in the study area since exhaustive fungal explorations yielded only one collection.

Antrodia serialis (Fr.) Donk, Persoonia 4: 339: 1966 = *Daedalea serialis* (Fr.) Aoshima, Trans. Mycol. Soc. Japan 8(1): 2, 1967. Plate 6.16d, Fig. 6.39i–l

Fructification resupinate to effused-reflexed, coriaceous, tough at maturity; pileus 1 × 1–5 × 0.3–1 cm, flat or slightly convex above and concave below, frequently imbricate and laterally confluent; upper surface of pileus ochraceous to pale brown, indistinctly zonate, slightly pubescent; margin thin or thick, acute or obtuse, slightly undulate, white or pale; context white, coriaceous, 1 mm or less thick; pore surface smooth, usually white to light brown pore tubes white, 2–8 mm long, pore mouth white or nearly so, circular to angular, in old specimens often becoming dentate, 2–3 per mm.

Hyphal system trimitic; generative hyphae hyaline, 2–4 µm wide, clamped, thin- to thick-walled, a few with irregularly thickened walls and more common in the trama than in the context; skeletal hyphae hyaline to subhyaline, thick-walled to solid, unbranched, 2.5–3 µm wide, predominating everywhere; binding hyphae scarce, hyaline to subhyaline, thick-walled to solid, shortly branched, occurring particularly in the basal part of the pileus, 2–4 µm wide. Basidia narrow-clavate, 16–20 × 4–6 µm. Basidiospores hyaline, thin-walled, short cylindric, apiculate, 6.4–8.4 × 2.4–3.4 µm.

Distribution: H.P.: Mahasu, Shimla, Kullu; J&K: Pehalgam.

Collection examined: RSD 6549, 6686, IBP 37117, L 37118, 37119.

Substratum: On stump of *Abies pindrow,* on stump of *Pinus excelsa.*

Remarks: The species is characterized by tough fructification than most of the species.

Antrodia sinuosa (Fr.) Karst, Medd. Soc. Fauna. Fl. fenn. 6: 10,1881. Fig. 6.39m–q

Fructification annual, resupinate, widely effused, 5–32×5×4–5,soft and corky when fresh, hard on drying, separable from the wood when fresh, margin white, thin to absent. Pore surface cream to yellowish when fresh, brown on drying; pores angular initially, later becoming elongate and irregularly sinuate,1–3 per mm,; dissepiment 98–180 μm thick, dentate; tubes in one layer white to light yellowish when fresh, concolorous with pores surface on drying. Context thin, cream, homogenous, slightly xanthochroic.

Hyphal system dimitic; generative hyphae hyaline, thin-walled to slightly thick-walled, branched, septate, clamped, faintly cyanophilous, 1.7–3.5 μm in diameter; skeletal hyphae subhyaline, thick-walled, aseptate, unbranched, acyanophilous, 3–5 μm in diameter. Cystidia and cystidioles absent. Basidia clavate, 12–15×4–5 μm. Basidiospores hyaline, thin-walled, smooth, cylindrical, slightly curved, nonamyloid, 4.2–5.5×1.2–1.8 μm.

Distribution: A.P.: West Kameng.

Collection examined: SSV 21506.

Substratum: On Wood.

Remarks: The species is characterized by annual, brown resupinate, widely effused fructifications; angular to sinuate pores; dentate dissepiments; dimitic hyphal system, with thin- to slightly thick-walled generative hyphae; thick-walled, subhyaline skeletal hyphae, smooth and thin-walled, cylindrical, 4.2–5.5×1.2–1.8 μm basidiospores. *Antrodia sinuosa* is close to *A. albida* but later differs by white fructifications; and larger cylindric-ellipsoid, 8–12.5×3–4.5 μm, basidiospores. Another closely related species *A. cervina* is distinguished by bigger pores (2–4 per mm), and larger basidiospores. It is a new record for Himalayas/India.

Antrodia xantha (Fr.) Ryv., Norw. Jl. Bot. 20: 8, 1973=*Daedalea xantha* (Fr.) Roy & De, Mycotaxon 61: 421 (1997)=*Polyporus xantha* Fr., Obs. 1 L 128, 1815=*Poria xantha* (Fr.) Cke., Grevillea 14: 112,1886. Plate 6.17a, Fig. 6.40a–d

Fructification annual, resupinate, soft, fragile when fresh, brittle on drying, easily separable from the wood when fresh, broadly effused, smooth when fresh, cracking freely into a characteristic pattern of small segments on drying; margin white, very small to almost absent. Pore surface pale yellow when fresh, fades to cream on drying; pores round to split, 5–6 per mm; pore tubes yellow. Context white, homogenous, non-xanthochroic, non-amyloid.

Hyphal system dimitic; generative hyphae hyaline, thin-walled, branched, septate, clamped, cyanophilous, 2.0–4.1 μm in diameter; skeletal hyphae hyaline to subhyaline, thick-walled, aseptate, unbranched, run parallel in dissepiment, weakly amyloid, gelatinized with concentrate KOH sol and become irregular and swollen, 3.2–5.9 μm in diameter. Cystidia absent but tapering cystidioles present. Basidia clavate, 4-spored, cyanophilous, up to 4.4 μm broad. Basidiospores hyaline, thin-walled, allantoid to cylindrical, nonamyloid, 4–5.0×1–1.8 μm.

Distribution: Bhutan: Thimphu, Utselpong; Begana; A.P.-West Kameng-Rupa, Shergaon; H.P.: Shimla; U. K.: Dehradun.

Plate 6.17 (**a**) *Antrodia xantha*. (**b**) *Daedalea dochmia*. (**c**) *Daedalea gollanii*. (**d**) *Daedalea quercina*. (**e**) *Dacryobolus karstenii*. (**f**) *Laetiporus sulphureus*

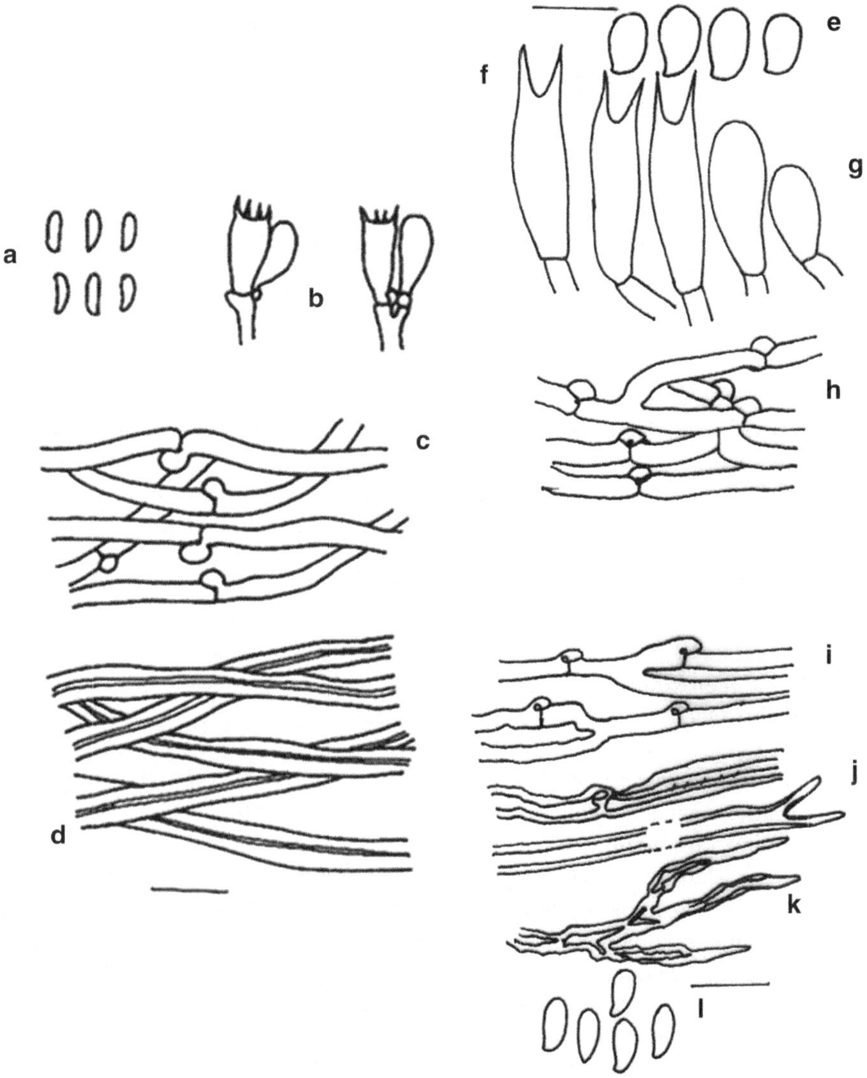

Fig. 6.40 (**a**–**d**) *Antrodia xantha* (**a**) Basidiospores, (**b**) Basidia, (**c**) Generative hyphae, (**d**) Skeletal hyphae; (**e**–**h**) *Anomoporia bombycina* (**e**) Basidiospores, (**f**) Basidia, (**g**) Basidioles, (**h**) Generative hyphae; (**i**–**l**) *Daedalea dickinsii* (**i**) Thin-walled generative hyphae, (**j**) Skeletal hyphae, (**k**) Binding hyphae, (**l**) Basidiospores

Collection examined: SSV 21521, IBP, 37120, 37121, L 22080.
Substratum: On decaying brunt log of *Pinus*.
Remarks: Annual, resupinate, widely effused, soft, friable fructifications with smooth when fresh and freely crack on drying, pale yellow pore surface; medium, 5–6 pores per mm; dimitic hyphal system with weakly amyloid skeletal hyphae, and hyaline, allantoid to cylindric basidiospores are the distinguishing features of *A. xantha*. It is common species in the N. W. Himalayas (Bakshi 1958, 1971; Dhanda 1977). This species is new record for Uttarakhand.

Anomoporia Pouzar,
Česká Mykol. 20: 172, 1966.

Fructification resupinate, annual effused, soft when fresh. Pore surface white, light yellow; grey to light violet or pale brown; pore thin-walled, medium round to angular, 2–4 per mm; margin wide and floccose, running out in the rhizomorphs, white to pale yellow, non-xanthochroic. Context thin, white to pale yellow. Hyphal system monomitic; generative hyphae hyaline, thin-walled, branched, clamped, amyloid. Cystidia absent. Basidia clavate, thin-walled, 2–4 spored. Basidiospores ellipsoid to oval, hyaline, smooth, thin-walled, amyloid.
Eight Species, widespread
Lit.: Pouzar (*Česká Mykol.* **20**: 172, 1966.)
Type species: *Polyporus bombycinus* Fr., 1825.
Habitat: Rotten Wood
Himalayas: One

Anomoporia bombycina Fr., Pouzar., Česká Mykol. 20: 172 (1966). Fig. 6.40e–h
Fructification annual, resupinate, thin-membranous, soft fleshy when fresh, fragile on drying, effused up to 6 cm, adnate, separable near the margin; margin white with greyish tint, thin, narrow, translucent. Pore surface white, light grey to light yellow with green tint when fresh, light brown on drying. Pores large, irregular, round, angular to elongated, 1–2 per mm; dissepiment equal, 35–54 µm thick; tubes in one layer, concolorous with the pore surface, up to 0.9 mm deep. Context very thin, white, homogenous, non-xanthochroic.
Hyphal system monomitic; generative hyphae hyaline, thin-walled, branched, septate, clamps very common, hyphae filled with light brown mass, acyanophilous, 2.2–4.2 µm in diameter. Cystidia and cystidioles absent. Basidia hyaline, thin-walled, clavate, 2–4 spored, 17.4–20 × 7–8.1 µm; sterigmata prominent. Basidiospores hyaline, thin-walled, smooth, globose to broadly ellipsoid, uniguttulate, amyloid, 6–8.6 × 4–5.5 µm.
Distribution: Bhutan- Thimphu.
Collection examined: SSV 21351.
Substratum: On decaying Wood.
Remarks: This species has been recorded for the first time from Himalayas. It appears to be of rare distribution in the Himalayas. *Anomoporia bombycina* is marked by resupinate, thin fructifications; white with green tinted pore surface; large, 1–2 pores per mm; monomitic hyphal system, filled with brown mass, thin-walled, clamped generative hyphae; and amyloid basidiospores.

Daedalea Pers.,
Syn. meth. fung.2: 500, 1801.

Fructification annual to perennial, sessile, broadly attached dimidiate, effused-reflexed, hard and woody on drying; upper surface smooth, tomentose to hirsute, concentrically zonate and sulcate. Pore surface ochraceous to dark brown; pores typically daedaloid with radially elongated pores, anastomosing freely, sometimes labyrinthiform or lamellar or sinuous, even poroid at places. Context white to creamish or yellowish brown, homogenous, non-xanthochroic or slightly darkening in KOH. Hyphal system trimitic; generative hyphae hyaline, thin-walled, septate, clamped, branched; skeletal hyphae thick-walled with narrow lumen; binding hyphae subhyaline, thick-walled, much branched. Cystidia absent or present. Basidia hyaline, clavate, 4-spored. Basidiospores hyaline, cylindric or ellipsoid, non-amyloid.

Seven Species, widespread.

Lit.: Legon (*Mycologist* 19: 44, 2005)

Type Species: *Agaricus quercinus* L., 1753.

Habitat: Dead Wood

Himalaya: Seven

Key to species

1. Fructification annual, sessile, imbricate 2
1. Fructification pileate, dimidiate 5
2. Hymenial surface plane, deep orange or deep brownish orange........... *D. gollanii*
2. Hymenial surface cream to yellow, pinkish brown/dark brown.......................... 3
3. Hymenial surface cream to yellow, pores irregular 1–2 per mm .... *D. flavida*
3. Hymenial surface pinkish brown/dark brown 4
4. Basidiospores 5–6 × 6.2 × 2.1 µm .... *D. dicknisii*
4. Basidiospores 9–11 × 3–4 µm.. *D. imponens*
5. Pores circular to hexagonal .... *D. sulcata*
5. Pores regular to irregular, daedaleoid 6
6. Basidiospores cylindrical, hyaline, 5–6 × 2.5–3.4 µm......................... *D. quercina*
6. Basidiospores oblong-elliptical, smooth, 4–6.5 × 2–2.5 µm .... *D. dochima*

Daedalea dickinsii Yasuda, Bot. Mag., Tokyo 36: 127, 1923 [1922] = *Trametes dickinsii* Berk. ex Cke., Greviellea 19(no. 92): 100, 1891. Fig. 6.40i–l

Fructification sessile, reflexed, imbricate, applanate, coriaceous, becoming rigid when dry, usually 5–20 × 3–15 × 2–3 cm or much larger, margin thin; upper surface brown, becoming blackish with age, glabrous, usually zoned, rough with small raised areas; hymenial surface pinkish brown, pores regular or irregular, round to angular, tending to be labrynthiform, 1–3 per mm, pore wall somewhat thick, pore tubes brown, up to 2 cm or longer, penetrating to different depths in context, cavities white, empty or filled with white lint-like material, detachable in form of thread. Context light to dark pinkish brown, soft, corky;

Hyphal system monomitic; hyphae (i) pale brown, thick-walled, lumen large or small, unbranched, 2.4–5 µm broad, common and (ii) hyaline, thin-walled, or slightly thick-walled, branched, septate with clamps, 1.5–3.8 µm broad, few.

Basidia clavate, 7–9.4×4 μm with 4-sterigmata. Basidiospores hyaline, thin-walled, cylindric, apiculate, 5.6–6.2×2.1–2.7 μm.

Distribution: Assam: Lokhra hills; U.K.: Chakrata, Mussoorie.

Collection examined: L 42129, 42131.

Substratum: On stump of *Quercus dilate, Q. semecarpifolia.*

Remarks: This species is very common on stumps of Oak in Chakrata (N.W. Himalayas).

Daedalea dochmia (Berk. & Br.) T. Hatt., Mycoscience **46**(5): 307(2005) = *Polyporus dochmius* Berk. et Br., Linn. Soc. Bot. J. 14:50(1873) [1857] = *Fomes dochmius* (Berk. et. Br.) Cke., Grevillea 14: 17, 1885. Plate 6.17b

Fructification perennial, sessile, conchate to applanate usually solitary, sometimes imbricate, somewhat woody, up to 20×15×2.5 cm; upper surface brownish grey when young, later greyish black to black with distinct crust, with sulcate zones and radial cracks; margin pale yellow, finely tomentose when young; context pinkish, fibrous, up to 1 cm thick; hymenial surface pink to cream, mostly smooth, pore circular, 6–8 per mm, pore tubes pink to cream, stratified, up to 3 mm long in each layer. Cystidia somewhat fusoid, thin-walled to slightly thick-walled, clamped at the base, minutely encrusted at the apex, 7.5–9×2–3 μm.

Hyphal system trimitic; generative hyphae hyaline, thin-walled to thick-walled, clamps, 2–4 μm wide; skeletal hyphae thick-walled to solid, tortuous, subhyaline to yellowish, 2–6 μm wide; binding hyphae subhyaline to yellowish, moderately branched, subsolid to solid, 2–4 μm wide. Basidia hyaline, thin-walled, clavate, 4-sterigmate, 10–15×4–5 μm. Basidiospores hyaline, thin-walled, oblong-elliptical, smooth, 4.0–6.5×2–2.5 μm.

Distribution: U.K.: NDBR.

Collection examined: Hem 28627.

Substratum: On decaying angiospermic log.

Remarks: This species is easy to recognise because of the greyish to black pileus with numerous radial cracks and the pinkish to buff pore surface. It is a new record for N. W. Himalayas and Uttarakhand.

Daedalea flavida Lév., Annls Sci. Nat., Bot., sér. 3 2: 198 (1844) = *Lenzites adustus* Mass. Kew Bull. 11: 250, 1910. Fig. 6.41f–k

Fructification annual to perennial, solitary to imbricate, sessile, attached by broad lateral base, light weight, corky coriaceous when fresh, hard on drying. Pileus sessile, applanate, dimidiate, convex, up to 12 cm long, 8 cm broad and 3.5 cm thick; upper surface white to cream light brown on drying, concentrically zonate, finely tomentose, later glabrous; margin concolorous with upper surface, blunt, entire. Pore surface cream light to light yellow, even to uneven, dull; pores irregular, somewhat elongated near the margin, large, 1–2 per mm; dissepiments 150–240 μm thick; tubes in one layer, cream, up to 1.5 cm deep in section. Context white, azonate, homogenous, non-xanthochroic, up to 1.5 cm thick.

Hyphal system trimitic; generative hyphae hyaline, thin-walled, septate, branched, clamped, cyanophilous, 2–3.5 μm in diameter; skeletal hyphae subhyaline, thick-walled to solid with narrow lumen, aseptate, rarely branched, acyanophilous, 2–6 μm in diameter; binding hyphae hyaline, thick-walled, much branched, aseptate, acyanophilous, 2–4 μm in diameter. Cystidia absent. Basidia collapsed on drying. Basidiospores thin-walled, smooth, hyaline, cylindric-ellipsoid, non-amyloid, 3.5–4.4 × 1.5–1.8 μm.

Distribution: Assam: Khasi Hills; H.P.: Kullu; W.B.: Calcutta; A.P.: Jamiri-Buragaon Road.

Collection examined: SSV 21565, 21834.

Substrate: On angiospermic log.

Remarks: This species is characterized by annual to perennial, broadly attached, hard and rigid fructification; white to creamish, concentrically zonate upper surface; white context; and hyaline, cylindric-ellipsoid basidiospores.

Daedalea gollanii Massee, Bull. Misc. Inf., Kew: 217, 1908. Plate 6.17c, Fig. 6.41a–e

Fructification annual, imbricate, dimidiate, coriaceous when fresh, hard on drying. Pileus fan shaped, plane, sessile, surface light brown, coloured, color not changing on drying, glabrous, concentrically zonate, cuticle absent, hair absent; margin acute, entire, lighter colored, strongly incurved on drying; hymenial surface plane, deep orange or deep brownish orange, young pilie with whit or cream colored hymenium, even, margin concolorous, with narrow sterile margin, 0.5–1 mm wide. Pores not in strata, mostly daedaloid to irregular, a few rounded, cream colored in section, up to 5 mm deep, straight, near margin, 187–420 μm broad or 1–2 per mm; dissepiments 75–345 μm thick, slightly toothed, unequal, of interwoven hyphae, at apex velutinate. Context cream colored, 0.5–3 mm thick, of interwoven hyphae.

Hyphae dimitic both in context and trama. Generative hyphae hyaline, septate, thin-walled, branched, clamped, clamps abundant, 2–3.5 μm wide, take stain; skeletal hyphae subhyaline, aseptate, unbranched or sometimes sparsely branched, without clamps, 2.4–4.7 μm broad, thick-walled, wall 0.6–1.8 μm thick, acyanophilous;. Psuedocystidia represented by abundant projecting skeletal hyphae. Basidia clavate 8–14.5 × 2.4–3.0 μm, sterigmata-4, mostly straight, 1.4–2.4 μm long. Basidiospores hyaline, broadly ellipsoid, smooth, eguttulate, 3.2–4.8 × 1.6–2.5 μm.

Distribution: U.K.: Mussoorie, Nanital.

Collection examined: IBP 42143, 42147.

Substratum: On stump of deciduous trees.

Remarks: This species is characterized by glabrous, lighter colored, imbricate fructifications, deeper colored hymenium, daedaloid to irregular pores, and broadly ellipsoid small basidiospores. Pseudocystidia, represented by projecting hyphae, are also present.

Daedalea imponens Ces., Atti Accad. Sci. fis. mat. Napoli 8(8): 7 (1879) = *Gloeophyllum imponens* (Ces.) Teng., Fungi from China, p. 760–761, 1963.
Fig. 6.41l–o

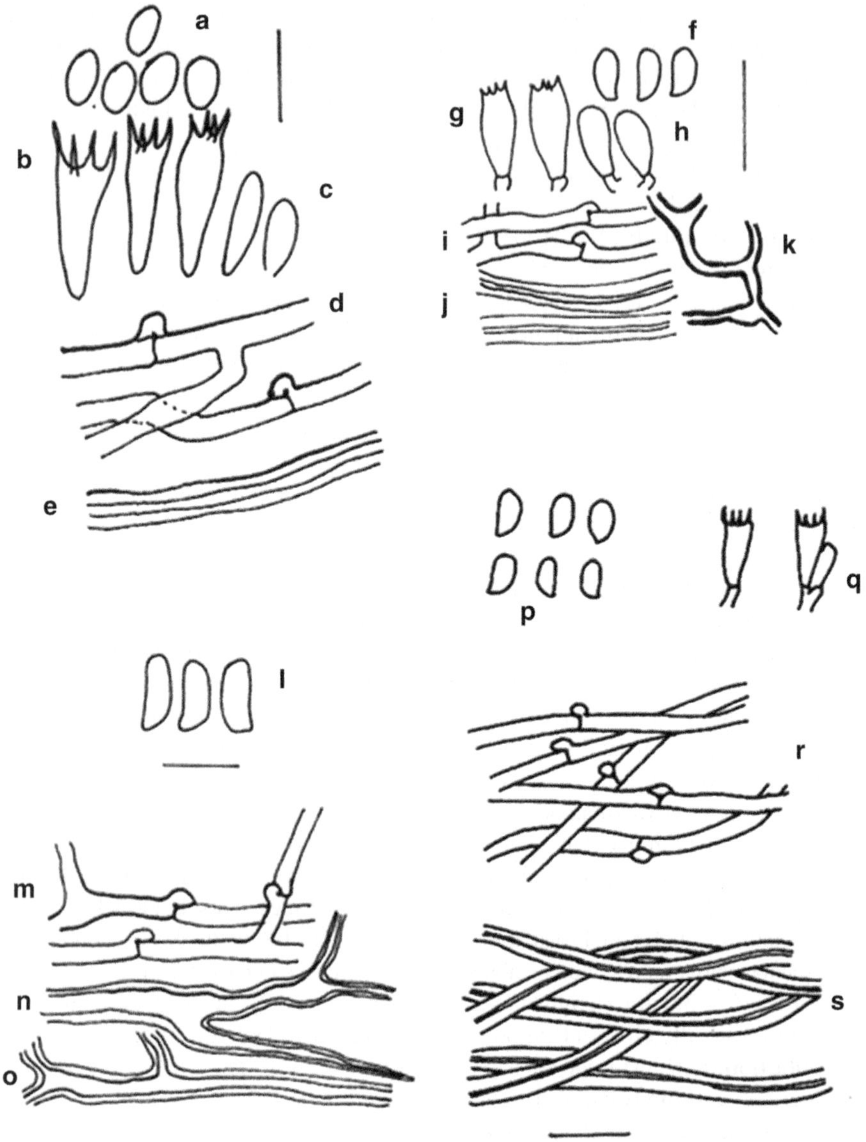

Fig. 6.41 (**a–e**) *Daedalea gollanii* (**a**) Basidiospores, (**b**) Basidia, (**c**) Basidioles, (**d**) Generative hyphae, (**e**) Skeletal hyphae; (**f–k**) *Daedalea flavida* (**f**) Basidiospores, (**g**) Basidia, (**h**) Basidioles, (**i**) Generative hyphae, (**j**) Skeletal hyphae, (**k**) Binding hyphae; (**l–o**) *Daedalea imponens* (**l**) Basidiospores, (**m**) Generative hyphae, (**n**) Binding hyphae, (**o**) Skeletal hyphae; (**p–s**) *Daedalea quercina* (**p**) Basidiospores, (**q**) Basidia, (**r**) Generative hyphae, (**s**) Skeletal hyphae

Fructifications annual, pileate, sessile, solitary to imbricate, coriaceous when fresh, tough on drying, light weight, laterally attached. Pileus sessile, dimidiate to flabelliformis; upper surface olive brown, dark brown to blackish brown at the base, concentrically zonate, glabrous, irregularly wrinkled on drying; margin thin, light grey, wavy, fertile below; pore surface cream colored when fresh, becomes darker on drying; lamellate with forked lamellae, and partly poroid with elongated rounded-angular pores with thick walls, lamellae 0.4–1.2 mm thick, 3–6 mm deep. Context fibrous, homogenous, xanthochroic.

Hyphal system trimitic; generative hyphae hyaline to subhyaline, thin-walled, branched, clamped, cyanophilous, 1.7–3.4 μm in diam; skeletal hyphae pale yellow, thick-walled, aseptate, unbranched, dominating the context and dissepiments, 3–8.4 μm in diameter; binding hyphae subhyaline, thick-walled, moderately branched, aseptate, acyanophilous, 2.2–3.7 μm wide. Cystidia fusoid to cylindrical, blunt, thick-walled, 35–60 × 4.6–6 μm. Basidia collapse on drying. Basidiospores thin-walled, hyaline, smooth, cylindrical, 9.6–11.5 × 3–4.0 μm wide.

Distribution: Meghalya: Nongpoh.

Collection examined: GSD 21148.

Substratum: On decaying gymnospermous log.

Remarks: The species is characterized by annual, coriaceous, laterally attached fructifications; dimidiate to flabelliformis pilei; olive brown to blackish brown, concentrically zonate upper surface; creamish pore surface, lamellate with forked lamellae and partly poroid with elongated pores; trimitic hyphal system; and hyaline, cylindrical basidiospores.

Daedalea quercina (L.) Pers., Syn. Meth. Fung. (Gottingen) 2: 500(1801). Plate 6.17d, Fig. 6.41p–s

Fructification annual, sometimes reviving, sessile, attached by broad base, hard and coriaceous, light in weight, mostly solitary, occasionally imbricate. Pileus sessile, applanate, dimidiate; upper surface greyish brown to yellowish-brown near margin, greyish brown towards base, very minutely tomentose near margin, later glabrous, concentrically zonate near margin; margin acute or subobtuse, yellowish-brown, entire. Pore surface yellowish-brown to light brown, even, dull; pores somewhat regular near margin, later mostly becoming irregular and daedaloid, 1–2 per mm; dissepiments 95–260 μm thick; tubes in one layer, yellowish brown up to 3.5 μm deep in section. Context pallid brown, homogenous, darkening in KOH sol. Cystidia absent.

Hyphal system trimitic; generative hyphae hyaline, thin-walled, septate, clamped, branched, 3–4.3 μm in diameter; skeletal hyphae pale brown, thick-walled, wall up to 1.5 μm thick, long, aseptate, rarely branched, 3–4.2 μm in diam. Basidia clavate 20–27 × 6–7.5 μm, 4-sterigmate with basal clamp, very difficult to find even in fresh specimens. Basidiospores cylindrical, hyaline, thin-walled, smooth, negative Melzer's reagent, 5–6 × 2.5–3.4 μm.

Distribution: Bhutan-Chimakothi, Chukha, Thimphu, Dochula; A.P.: West Kameng, Bomdilla, Wang Basti; U.K.: Nanital.

Collection examined: SSV 21251, IBP 37129, 37130.
Substratum: On decaying angiospermic stumps.
Remarks: This species was reported previously from India by Bose (1919) and Banerjee (1947) from Calcutta (W. Bengal). Bakshi (1971) listed it under "Species inadequately described" and recorded its occurrence from Calcutta. The species is characterized by having annual, hard and coriaceous, imbricate fructification; hymenial surface greyish brown to yellowish brown; hyphal system trimitic; basidia clavate, 4-sterigmate; cylindrical, hyaline, smooth basidiospores. This species is new record for Uttarakhand.

Daedalea sulcata (Berk.) Ryv., Norw. Jl. Bot. 24: 213–230, 1977=*Hexagonia sulcate* Berk., London J. Bot. 6: 510, 1847. Fig. 6.42a–e

Fructification pileate, dimidiate to appalante with broad base, imbricate, adjacent pilei may fuse to give perennial appearance, coriaceous to corky when fresh, hard when dry; pileus surface deep brown to blackish, yellowish brown near the margin, crusty, concentrically sulcate, rimose when old, radially cracked in old and dried specimens, surface at the base of old specimens may be warty and covered with mosses; margin even, thick, entire, sterile up to 3 mm below. Context pinkish when fresh, brown when dried, coriaceous to corky and hard; hymenial surface characteristically, pinkish to brick red when fresh, yellowish brown to brown on drying; pores almost circular to typically hexagonal, sometimes irregular, 0.8–4.5 cm in diameter, pore wall thin to thick, pore tubes concolorous with hymenial surface, straight to oblique, 0.4–2.0 cm long.

Hyphal system trimitic; generative hyphae with clamp connection, branched, of various forms; (i) hyaline, thin-walled, some irregularly wide, many in collapsed condition, 1.2–5.3 μm wide; (ii) hyaline, thick-walled to subsolid, occasionally showing broken clamp connections at one end and characteristic thickened projections on side walls, frequently encrusted, 1.5–4.0 μm wide, often with wall irregularly thickened, 1.5–5.2(−7.5) μm wide (iii) subhyaline, thick-walled, few, found in the skin like areas on the pileus bearing both clamp connection and pseudosepta, repeatedly branched with short branches and intricately interwoven to form a kind of thin plectenchymatic tissue found in the skin like areas of pileus surface. Hyphae 1.0–2.4 μm wide; skeletal hyphae subhyaline, thick-walled to solid, abundant, frequently with the collapsed thin-walled clamps connection at the basal end and pseudosepta at apical end, usually unbranched, occasionally apically branched with 1–4 short tapering branches, 1.5–4.5 μm wide; binding hyphae hyaline, slightly to fairly thick-walled, distantly and sparsely branched, few and found only in the pore field and trama, 1.5–2.5 μm wide. Basidia narrow clavate, 13.0–23.5 × 3.5–5.0 μm, 4-sterigmate. Cystidia hyaline, slightly thick-walled, cylindrical to clavate, strongly encrusted with hyaline crystals over most of the length, abundant and projecting beyond the hymenial level, up to 9 μm wide. Cystidioles hyaline, thick-walled, 7.8–28.0 × 2.4–3.8 μm. Basidiospores hyaline, thin-walled, slightly allantoids, 4.6–6.0 × 1.5–2.5 μm.
Distribution: West Bengal: Jalpaiguri, Northern India; U.K.: NDBR.
Collection examined: IBP 37685, 42159.
Substratum: On stump of *Shorea robusta* and *Albizzia stipulata.*

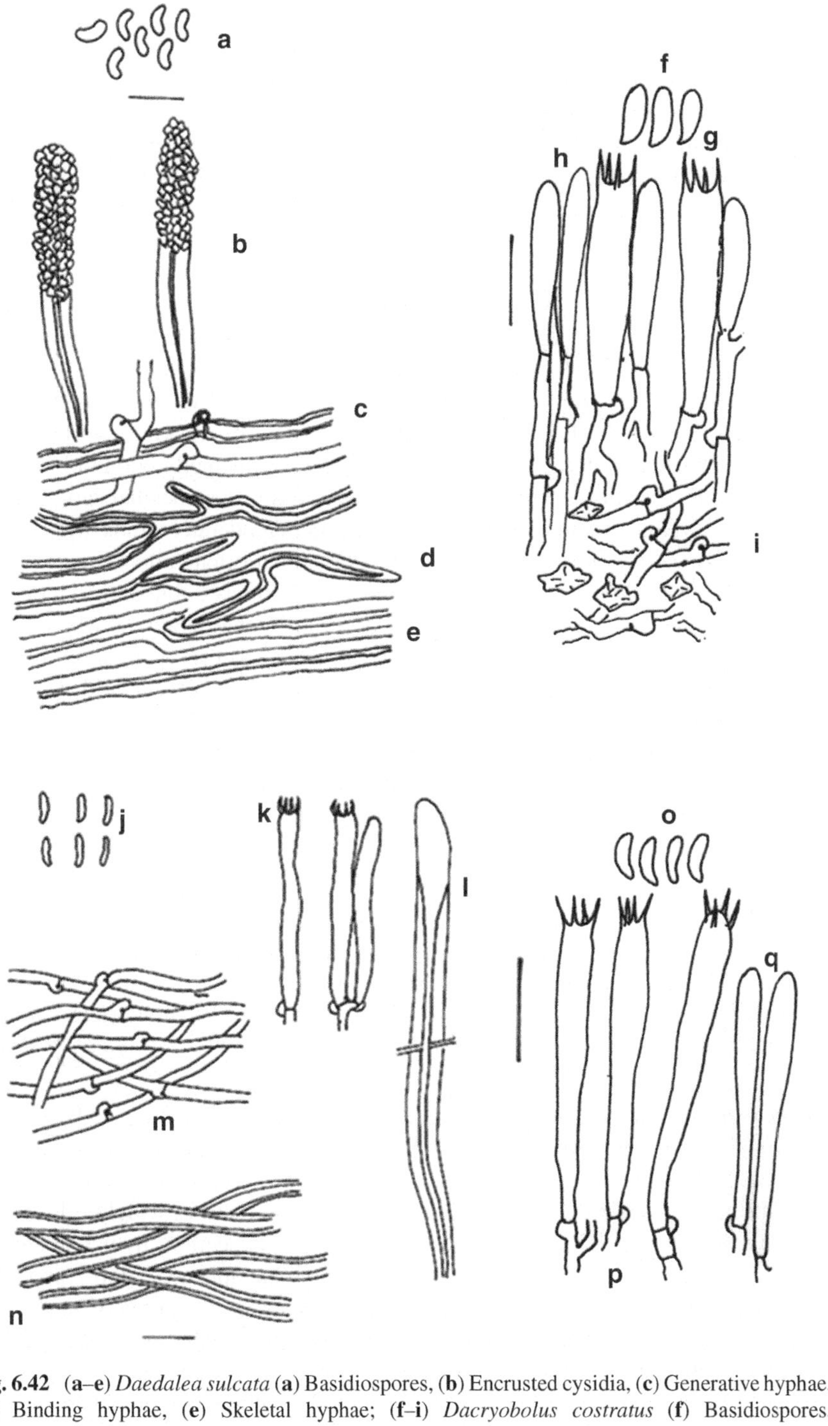

Fig. 6.42 (**a–e**) *Daedalea sulcata* (**a**) Basidiospores, (**b**) Encrusted cysidia, (**c**) Generative hyphae, (**d**) Binding hyphae, (**e**) Skeletal hyphae; (**f–i**) *Dacryobolus costratus* (**f**) Basidiospores, (**g**) Basidia, (**h**) Cystidia, (**i**) Generative hyphae; (**j–n**) *Dacryobolus karstenii* (**j**) Basidiospores, (**k**) Basidia, (**l**) Cystidia, (**m**) Generative hyphae, (**n**) Skeletal hyphae; (**o–q**) *Dacryobolus sudans* (**o**) Basidiospores, (**p**) Basidia, (**q**) Cystidia

Remarks: The species is characterized by having appalante to dimidiate, pileate fructification with broad base, imbricate, hard when dry; hymenial surface characteristically, pinkish to brick red when fresh, yellowish brown to brown on drying; pores almost circular to typically hexagonal, sometimes irregular, Hyphal system trimitic; basidia clavate, 4-sterigmate; cystidia hyaline and hyaline, allantoids basidiospores.

Dacryobolus Fr.,
Summaveg. Scand; Section Post. (Stockholm): 2.
404, 1849, Emend. Overw., Z. Pilzk. 31: 40. 1965.

Fructifications resupinate, membranous-ceraceous to wholly ceraceous or waxy-gelatinous when fresh but often shrinking, irregularly cracking and becoming hard and brittle on drying, smooth to tuberculate or distinctly odontoid. Context subhyaline in section, composed of compactly arranged, somewhat agglutinating hyphae. Hyphal system monomitic, hyphae thin to moderately thick-walled, septate, clamped. Cystidia when present cylindrical but often collapse and become distorted in KOH sol., thin-walled to moderately thick, subhyaline, smooth or often exuding resinous mass which may occur in patches along the whole length of it or may accumulate at the tip of the teeth like a wax drop in hydnoid species. Basidia clavate, 4-spored. Basidiospores cylindrical to suballantoid, the walls thin, subhyaline, smooth, nonamyloid.

Five Species, widespread

Lit.: Manjon *et.al.* (*An. Jard. Bot. Madr.* **40**: 297, 1984)

Type Species: *Hydnum sudans* Alb. & Schwein, 1805.

Habitat: Dead wood

Himalayas: Three

Key to species

1. Hymenial surface odontiod with exudations at the top of the teeth....... *D. sudans*
1. Hymenial surface smooth, without exudations 2
2. Cystidia present.. *D. karstenii*
2. Cystidia absent..*D. costratus*

Dacryobolus costratus (Rehill & Bakshi) Rattan, Biblthca. Mycol. 60: 279(1977) = *Corticium costratum* Rehill & Bakshi, Ind. For. Bull. (n.s.) 242: 12. 1965. Fig. 6.42f–i

Fructifications resupinate, membranous-ceraceous to crustose, widely effused but more often restricted, adnate, up to 450 μm thick in section; hymenial surface brown but turns whitish brown on drying, smooth but appears farinose under the lens, not creviced; margin abrupt, cliff-like, adnate, concolorous. Context subhyaline in section, stratose, composed of compactly arranged somewhat agglutinated hyphae enclosing abundant umber coloured amorphs matter.

Hyphal system monomitic, hyphae 2–2.5 μm wide, subhyaline, branched, septate, clamped, thin-walled, often collapsing and difficult to discern. Cytidia absent but some basidioles become prominent and look like cystidioles. Basidia 20–25 × 3.5–5 μm, clavate, 4-spored, starigmata up to 4.5 μm long. Basidiospores 5–7 × 2–2.5 μm, suballantoid, subhyaline, thin-walled, smooth, nonamyloid, shortly apiculate.

Distribution: U.K.: Chakrata.
Collection examined: SSR 7201, IBP 37131.
Substratum: On log of *Picea smithiana*
Remarks: This species is characterized by the membranous-ceraceous texture, presence of abundant crystals in context and small suballantoid basidiospores. The exact affinity of this species is uncertain but it comes near *Dacryobolus* in micromorphology. The presence of abundant amorphous matter (umber coloured) in the context is reminiscent of such matter present along the length of cystidia in *D. karstenii* and on teeth apex in *D. sudans*.

Dacryobolus karstenii (Bers.) Overw. ex. Parm., Consp. Syst. Cort.: 98. 1968 = *Stereum karstenii* Bers., Atti. I.R. Acad. Agiati. Rovereto III, 3(1): 109. 1897. Plate 6.17e, Fig. 6.42j–n

Fructification resupinate, membranous to submembranous, fleshy when fresh and becomes hard on drying arising as pin head colonies; hymenial surface pale yellow to yellowish brown, continuous, smooth, thicker specimens usually crade deeply and irregularly on drying; margin thick, more or less abrupt, adnate, concolorous. Context composed of compactly arranged, agglutinated hyphae.

Hyphal system monomitic; generative hyphae septate, branched, clamped, thin-walled, 1.5–4.5 μm wide. Cystidia present 100–350 × 5–8 μm, cylindrical, not radicate, arising from different parts of the context, immersed or projecting up to 35 μm out of the hymenium, walls subhyaline, thick, forming a capillary like lumen throughout their length except at the apex where it widens gradually, wall collapsing 5 % KOH sol, exuding subhyaline crystals along whole length of cystidia. Basidia clavate, 30–40 × 2.5–3.0 μm, 4-spored. Basidiospores narrowly ellipsoid to suballantoid, minutely apiculate, thin-walled, subhyaline, smooth, non-amyloid.

Distribution: H.P.: Kullu; J&K: Patnitop, U.K.: Rudarprayag.
Collection examined: SSR 5120, 5228, IBP 37132, 4216 3.
Substratum: On stump of *Cedurs deodara, Abies pindrow, Pinus excelsa* .
Remarks: The species is characterized by thick fleshy fructification when fresh but becomes hard on drying, smooth hymenial surface, presence of cystidia having incrustation occurring in patches. Sometimes the margin becomes loose and rolls inwards on drying as a result of which the fructifications appear stereoids.

Dacryobolus sudans (Alb. Schein) Fr., Summa veg. Scand. 404. 1849.
Fig. 6.42o–q

Fructifications resupinate, membranous-ceraceous to wholly ceraceous, often shrinking on drying, adnate, widely effused; hymenial surface distinctly toothed, cream brown to light brown, continuous; margin fibrillose, adnate, white to paler concolorous. Teeth gregarious, subulate, terrete when fresh but tends to coalesce on drying, apices opalescent due to the accumulation of resinous material. Context composed of compactly arranged more or less agglutinated hyphae.

Hyphal system monomitic, hyphae 1.3–4.2 μm wide, branched, septate, clamped, the walls subhyaline, thin to slightly thick (up to 1.5 μm). Cystidia 2.5–4.8 μm broad, of variable length, cylindrical or hyphoid, often occurring in clusters and projecting from the apices of the spines, the walls thin, subhyaline, often collapsing in 10 % KOH sol., exuding resinous material which flows towards the tip and accumulate at teeth apices giving them the typical appearance. Basidia 15–21 × 2–3.3 μm, clavate, 4-spored. Basidiospores 4.5–6 × 1.3–1.6 μm, narrowly ellipsoid to suballantoid, minutely apiculate, thin-walled, hyaline, smooth, nonamyloid.

Distribution: H.P.: Narkanda.

Collection examined: HSK 4209.

Substratum: On log under conifers & *Quercus incana.*

Remarks: This species is easily recognized by the waxy texture, toothed hymenial surface and presence of an opalescent drop of viscid material at spine apices.

Fomitopsis Karst.,
Medd. Soc.Fl. Fauna 6: 9, 1881.

Fructification annual to perennial, pileate, sessile to effused reflexed, large, tough to woody hard. Pileus velvate becomes smooth. Pores small round and entire. Pore surface cream, brownish-grey to pink. Context cream pale, brown to pink. Hyphal system trimitic/dimitic; generative hyphae hyaline, thin-walled, clamped; skeletal hyphae and binding hyphae hyaline to pale yellow. Cystidia present or absent. Basidiospores subglobose to cylindrical to ellipsoid, hyaline, smooth on amyloid.

Thirty Species, widespread

Lit.: Kim *et.al.* (Mycol. **97**: 812, 2005; phylogeny)

Type Species: *Boletus pinicola* Swartz., 1810

Habitat: Dead wood

Himalayas: Five

Key to species

1. Hymenium and context reddish/black 2
1. Hymenium and context white to tan 4
2. Fructification perennial 3
2. Fructification annual, pileus applanate, conchate or ungulate, surface glabrous to tomentose, pores 6–8 per mm .... *F. rubida*
3. Fructification small, context 6 cm thick, pores 1–2 per mm, basidiospores 4.4–7.3 × 3.5–4.4 μm .... *F. rufolaccata*
3. Fructification large, context up to 7 cm thick, pores 3–5 per mm, basidiospores 5.5–7.0 × 2.0–2.5 μm .... *F. rosea*
4. Fructification columnar, surface sulcate, gloeoporus hyphae present, basidiospores not more than 5.5 μm long .... *F. officinalis*[a]
4. Fructification not columnar and sulcate, gloeopleuros hyphae absent, basidiospores large more than 6 μm long 5
5. Fructification applanate to ungulate, surface greyish to blackish brown, sometimes laccate, cystidia present, hyphoid, basidiospores ellipsoid–cylindric .... *F. pinicola*
5. Fructification applanate, surface white to pale buff, smooth, cystidia absent, basidiospores cylindric .... *F. palustris*

[a]Extra limital, not included in the text

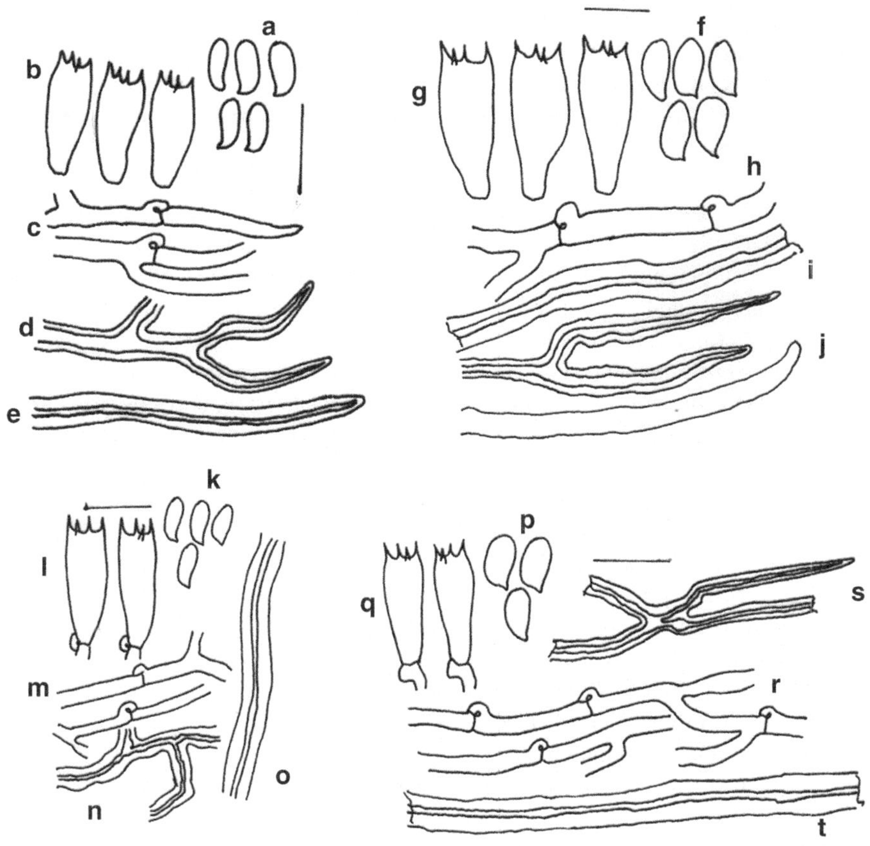

Fig. 6.43 (**a**–**e**) *Fomitopsis palustris* (**a**) Basidiospores, (**b**) Basidia, (**c**) Generative hyphae, (**d**) Binding hyphae, (**e**) Skeletal hyphae; (**f**–**j**) *Fomitopsis pinicola* (**f**) Basidiospores, (**g**) Basidia, (**h**) Generative hyphae, (**i**) Skeletal hyphae, (**j**) Binding hyphae, (**k**–**o**) *Fomitopsis rosea* (**k**) Basidiospores, (**l**) Basidia, (**m**) Generative hyphae, (**n**) Binding hyphae, (**o**) Skeletal hyphae; (**p**–**t**) *Fomitopsis rubida* (**p**) Basidiospores, (**q**) Basidia, (**r**) Generative hyphae, (**s**) Binding hyphae, (**t**) Skeletal hyphae

Fomitopsis palustris (Berk. & Curt.) Gilbn. & Ryv., Mycotaxon 22(2): 364, 1985 = *Polyporus palustris* Berk. & Curt., Grev. 1: 51, 1872. Fig. 6.43a–e

Fructification perennial, sessile to effused reflexed, dimidiate, solitary or imbricate, individual pilei applanate, watery when fresh, drying rigid and corky; upper surface white, becoming pale buff on age, minutely tomentose to glabrous, smooth to faintly zonate; hymenial surface white cream when fresh, turning pale buff and often greyish on drying, pores circular to angular, 3–4 per mm. margin acute, inturned when dry. Context cream, somewhat corky, 0.5–2.0 cm thick.

Hyphal system trimitic; generative hyphae clamped hyaline, thin to thick-walled to solid, often broken at clamps, 2.6–4 μm wide; skeletal hyphae hyaline thick-walled, 3.0–5.6 μm wide; binding hyphae rare, hyaline, shortly branched, 3–5.2 μm wide. Basidia 4-sterigmate, clavate 13–19 × 5–7 μm. Basidiospores cylindric, hyaline thin-walled, 6.0–8.0 × 2.0–3.0 μm.

Distribution: U.K.: Dehra Dun, Mussoorie.
Collection examined: IBP 42701
Substratum: On decaying angiospermic log.
Remarks: This species has been reported and described from Uttarakhand earlier by (Roy & de 1996).

Fomitopsis pinicola (Sw.) Karst., Meddn. Soc. Fauna Flora fenn. 6: 9 (1881) = *Polyporus pinicola* (Sw.) Fr., Syst. Mycol. 1: 372, 1821. Fig. 6.43f–j

Fructification perennial, sessile, rarely resupinate to effused- reflexed, woody, applanate to ungulate; upper surface is with reddish brown resinous layer when young becoming glabrous later and sometimes laccate, greyish to black brown, smooth to sulcate; context cream to woody brown, 1–3 cm or more thick; hymenial surface brownish darkening where bruised when fresh, pores usually circular, 4–5 per mm, pore tube usually stuffed with hyphae, stratified, 5–10 mm long in each layer.

Hyphal system trimitic; generative hyphae hyaline, thin-walled and clamped, 3–5.5 µm wide; skeletal hyphae hyaline, thick-walled, up to 7 µm wide. Binding hyphae rare, thick-walled, hyaline, up to 5 µm wide. Basidia short clavate, 16.0–22.2 × 6.5–8.2 µm. Basidiospores hyaline, thin-walled, smooth, ellipsoid-cylindric, 6–8 × 3–4 µm. Cystidia hyphoid, hyaline, 2.5–4.0 µm wide, projecting above basidial layer.

Distribution: U.K.: Chakrata, Nainital; W. B.: Darjeeling.
Collection examined: Dhanda 6791, L 37135, 37136.
Substratum: On dead conifers.
Remarks: It is one of the most common species in the coniferous forests.

Fomitopsis rosea (Alb. & Schw. ex Fr.) Karst., Meddn. Soc. Fauna Flora fenn. 6: 91 (1881) = *Polyporus roseus* (Alb. &Schw.) Fr.,Observ. Mycol. 2: 260 (1818). Fig. 6.43k–o

Fructification perennial, solitary, ungulate, rarely imbricate, woody; Pileus convex to angulate, up to 20 × 11 × 7.5 cm upper surface dark brown to black, glabrous, concentrically sulcate, often rimose with age, initially pink to grey pink, purplish brown to black on ageing; margin acute to obtuse, sterile below, entire or rarely cleft, pink to rosy in colour; pore surface pink when young, becoming pinkish brown to dark brown with age. Context up to 2.5 cm thick, pink to pink brown, soft corky, azonate; hymenial surface pink when young, pinkish brown to brown with age, pores circular to angular, 3–5 per mm.

Hyphal system trimitic; generative hyphae rare, hyaline, clamped, 2.0–3.6 µm wide; skeletal hyphae abundant, thick-walled with narrow lumina 2.5–5.2 µm wide; binding hyphae rare with short branches, subsolid to solid, 2.6–4.4 µm wide. Basidia 4- sterigmate, 14.0–16.0 × 4.6–5.6 µm. Basidiospores cylindric, hyaline, thin-walled, 5.5–7.3 × 2.0–2.6 µm.

Distribution: H.P.: Kullu, Narkanda, Mahasu; U.K.: Nainital, Mussoorie; J&K: Khillanmarg.
Collection examined: D 6028.

Substratum: On decaying *Cedrus deodara, Pinus excelsa* log *Abies.*

Remarks: The species is characterized by its perennial nature, convex to ungulate fructification, pink to pinkish brown context and margin, trimitic hyphal system and small cylindrical, hyaline basidiospores.

Fomitopsis rubida (Berk.) Roy & De, [as *rubidus*], Mycotaxon 60: 317, 1996 = *Polyporus rubidus* Berk., Lond. J. Bot. 6:500, 1847. Fig. 6.43p–t

Fructifications usually annual, rarely biennial or so, sessile or effused-reflexed, corky, becoming woody on age, dimidiate to semi-circular, applanate, conchate or even ungulate, imbricate; upper surface cinnamon buff to pinkish brown, at first minutely tomentose, glabrous, zonate; margin thin, slightly incurved. Context light pinkish brown, fibrous, 1.0–2.5 mm thick; hymenial surface light pinkish brown to pale cinnamom, pores small, circular to subangular, 6–8 per mm, pores tubes up to 3 mm long.

Hyphal system trimitic; generative hyphae hyaline, thin-walled, clamped, 1.5–2.6 µm wide; skeletal hyphal hyaline or nearly so, thick-walled to sub solid, straight or flexous, 2.5–4.6 µm wide; binding hyphae moderately branched, hyaline, thick-walled to solid, 2.5–4.6 µm wide. Basidia clavate, 4- sterigmate, 10–15 × 3–5 µm. Basidiospores hyaline, thin-walled, smooth, cylindric-ellipsoid, apiculate, 5.0–7.5 × 3.5–4.0 µm.

Distribution: U.K.: Dehra Dun; W.B.: Calcutta.

Collection examined: IBP 37137, L 37138.

Substratum: On decaying angiospermic log.

Remarks: The species is characterized by its light pink, corky to woody fructification. It causes brown cuboidal rot in heart wood of living trees and dead wood.

Fomitopsis rufolaccata (Bose) Dhanda [as *'rufolaccatus'*], in Thind & Dhanda, Indian Phytopath. 33(3): 386 (1981) [1980] = *Fomes rufolaccatus* Bose, Annls mycol. 19(1/2): 129, 1921. Fig. 6.44a–d

Fructification perennial, sessile, mostly solitary, sometimes imbricate, hard and woody, of medium weight. Pileus ungulate, sometimes applantate, 6–17 × 11–22 × 5–16 cm; upper surface reddish-brown to black, resinous, woody grey when old, glabrous, azonate, rough, rimose in very old places; margin pale yellow to concolorous with upper surface, blunt to obtuse, sterile below. Pore surface pale yellow to brownish, dull, uneven; pores rounded to somewhat angular, sometimes irregular, 380–949 µm in diameter, 1–2 per mm, pore mouth even, velutinate. Tubes indistinctly stratified, tube layers separated by thick context in between, creamish brown in section; dissepiments 81–11 µm thick, equal, of compact interwoven hyphae. Context pale yellow, soft homogenous, azonate, slightly darkening in KOH sol, up to 6 cm thick.

Hyphal system trimitic; generative hyphae hyaline, thin-walled, septate, clamped, branched, cyanophilous, 2.5–3.4 µm in diameter; skeletal hyphae subhyaline, thick-walled to solid, long, rarely branched, aseptate, acyanophilous, 3.3–8.4 µm in diameter; binding hyphae subhyaline, thick-walled to solid, much branched, aseptate, acyanophilous, 2.5–4.6 µm in diameter. Setae absent. Cystidia absent.

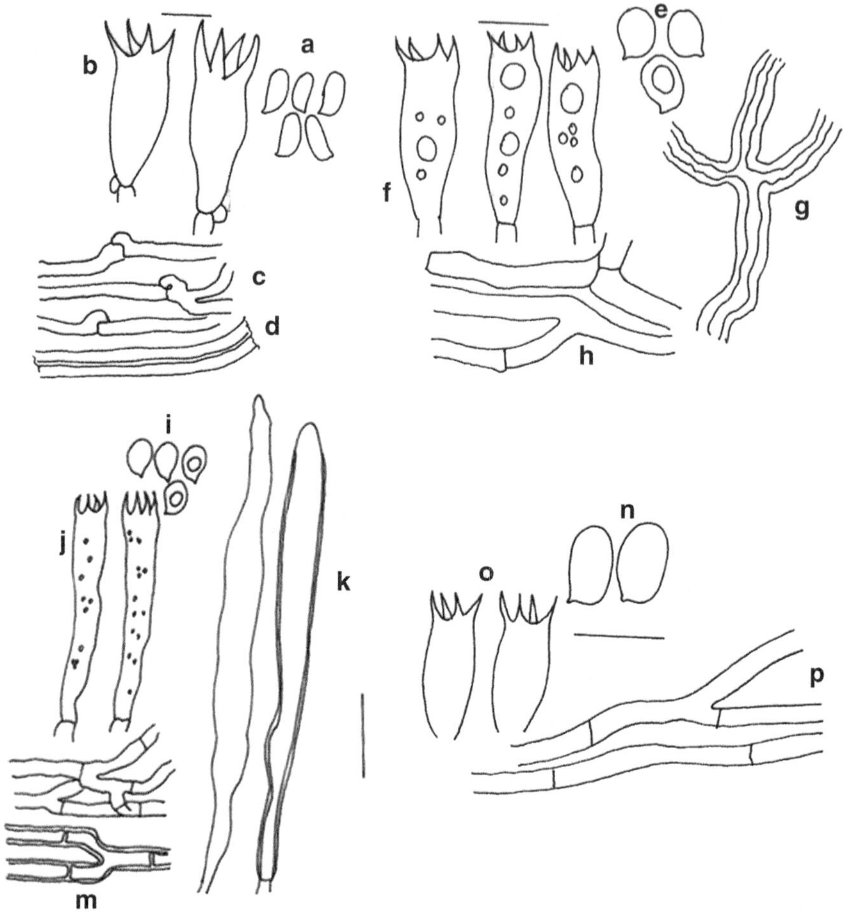

Fig. 6.44 (**a–d**) *Fomitopsis rufolaccata* (**a**) Basidiospores, (**b**) Basidia, (**c**) Generative hyphae, (**d**) Skeletal hyphae; (**e–g**) *Laetiporus sulphureus* (**e**) Basidiospores, (**f**) Basidia, (**g**) Binding hyphae, (**h**) Generative hyphae; (**i–m**) *Parmastomyces corticola* (**i**) Basidiospores, (**j**) Basidia, (**k**) Cystidia, (**l**) Thin-walled generative hyphae, (**m**) Thick-walled generative hyphae; (**n–p**) *Phaeolus schweinitzii*; (**n**) Basidiospores, (**o**) Basidia, (**p**) Generative hyphae

Basidia hyaline, thin-walled, clavate, 4-spored, up to 7.3 μm in diameter; sterigmata thin, straight, up to 5.5 μm long. Basidiospores hyaline, thin to slightly thick-walled, smooth, non-amyloid, broad ellipsoid to sub globose, 4.4–7.3 × 3.5–4.4 μm.

Distribution: H.P.: Dalhousie- Khajjiar; Manali- Kothi; Kullu- Pulga.

Collection examined: SSR 6442, D 6637.

Substratum: *Cedrus deodara* and logs of *Abies pindrow.*

Remarks: It is characterized by large, perennial fruiting bodies with resinous and sticky upper surface owing to raddish- laccate exudation; large, somewhat angular pores, 1–2 per mm; pale yellow context; trimitic hyphal system and hyaline, smooth basidiospores.

Laetiporus Murr.,
Bull.Torrey bot. Club. 31(11): 607, 1904.

Fructification annual, yellowish-orange, sessile soft fleshy when fresh, drying rigid; upper surface yellowish-brown, even. Context white to creamish, chalky brittle when dry. Hyphal system dimitic; generative hyphae hyaline, thin-walled, septate, clamps absent, branched; binding hyphae thick-walled, much branched. Cystidia absent. Basidia hyaline, clavate, 2–4 spored. Basidiospores hyaline, smooth, broadly ellipsoid to ovoid.

Five species, widespread

Lit.: Burdsall & Banik (*Harvard Pap. Bot.* **6**:43, 2001; N. Am.)

Type Species: *Laetiporus speciosus* Battarra ex Murr.1904

Habitat: Dead wood

Himalayas: One

Laetiporus sulphureus (Bull.) Murr., Annl. Mycol. 18(1/3): 51(1920) = *Polyporus sulphurous* (Bull.) Fr., Syst. Mycol.1: 357, 1821. Plate 6.17f, Fig. 6.44e–h

Fructification annual, sessile, with a narrowed base or substipitate, stipe lateral, numerous pilei arising from a common stem in imbricate clusters, flesh and brittle when fresh, usually 5.0–20.0 × 5.0–25.0 × 0.5–2.5 cm; upper surface of pileus sulphur yellow to orange, fading to almost white with age on drying, smooth, glabrous; margin thick, undulate, rounded; context light yellow, fleshy when fresh, soft and fragile when dry, 0.5–1.8 cm thick; hymenial surface sulphur yellow when fresh, brownish in drying, pores angular 3–4 per mm, pore tubes light yellow, straight or oblique, 1–3 mm long.

Hyphal system dimitic; generative hyphae hyaline, thin-walled, simple septate, rarely branched, 6–13 μm wide; binding hyphae thick-walled, hyaline, aseptate, much branched and ramifying, 3.5–14.5 μm wide. Basidia 4- sterigmate, 15–18 × 4–6 μm. Basidiospores subcylindric, hyaline, smooth, 5.0–7.0 × 4.0–5.5 μm.

Distribution: Meghalaya: Shillong; H.P.: Bashahr; U.K.: Chakrata, Kumaun, NDBR- Vishnupryag; Assam: Cheerapunji, J&K: Sonamarg.

Collection examined: Hem 28606, IBP 37132.

Substratum: On decaying angiospermic logs, oaks and logs of conifers, occurring as a parasite on toon in Dehra Dun causing heart rot.

Remarks: The species is characterized by sulphur yellow pileal surface with bright orange and pink flesh scattered when fresh.

Parmastomyces Kotl. & Pouzar,
Feddes Regni. Veg. 69: 138(1964)

Fructification annual, resupinate to effused-reflexed or sessile, white to light brown, soft when fresh, fragile when dry; pores angular. Context duplex, with a dense dark gelatinous layer next to the tubes and a white, soft-fibrous layer next to the substrate; hyphal system monomitic; generative hyphae with clamps; Cystidia absent. Basidiospores cylindrical, smooth, hyaline, slightly thick-walled, dextrinoid in Melzer's reagent; on conifers and hardwood.

Seven species, widespread
Lit.: Kotlaba & Pouzar (*Reprium nov. Spec. Regni. Veg.* **69;** 138, 1964).
Type Species: *Tyromyces kravtzevianus* Bondartsev & Parmasto 1957
Habitat: Wood
Himalayas: One

Parmastomyces corticola Corner [as *'corticicola'*], Beih. Nova Hedwigia 96: 96 (1989) = *Perenniporia corticola*(Corner) Decock, Mycologia 93(4): m776, 2001.
Fig. 6.44i–m
Fructification annual, widely effused, inseparable soft leathery when dry; subiculum white, very thin, inconspicuous; hymenial surface 'pale ochraceous buff', 'light ochraceous buff' or a shade darker, dull, pore tube up to 2 mm long, pores round to angular or irregular, 2–4 per mm, pore wall thin, fimbriate, sometimes lacerate.
Hyphal system monomitic; hyphae hyaline to 'light buff' under the microscope, thick-walled, branched, simple septate, 2.6–5.7 μm diameter, tramal hyphae similar to subiculum hyphae. Cystidia conspicuous, abundant, capitate, elongate, diameter head up to 9.3 μm, also present in the subhymenium. Basidia clavate, 5.0–6.5 μm broad. Basidiospores hyaline, smooth, ellipsoid, 6.1–7 × 3.6–4 μm;
Distribution: J&K: Gulmarg.
Collection examined: RSD 6666.
Substratum: On stump of *Abies pindrow.*
Remarks: The species is characterized by inseparable soft-leathery fructification, thick-walled dextrinoid basidiospores.

Phaeolus (Pat.) Pat.,
N. Egsai. Taxonomique: 86, (1900)

Fructification annual, sessile, imbricate. Pileus applanate, watery and spongy when fresh, drying rigid and brittle; upper surface rusty-brown to dark brown. Pore surface brown to dark brown; pores large and irregular. Tubes not stratified, dark brown in section. Context soft, succulent, rusty-yellow. Hyphal system monomitic; generative hyphae light yellowish-brown to reddish-brown, septate, clamps absent. Cystidia pale brown, cylindric-clavate. Basidia clavate, 4-spored. Basidiospores hyaline, thin-walled, smooth, broadly ellipsoid.
Two species, widespread
Lit.: Simpson & May (Australas. Pl. Path. 31: 99, 2002; Australia).
Type Species: *Polyporus schweinitzii* Fr.1821
Habitat: Dead wood
Himalayas: One

Phaeolus schweinitzii (Fr.) Pat., Essai Tax. Hyménomyc. (Lons-le-Saunier): 86 (1900) = *Polyporus schweinitzii* Fr., Syst. mycol. (Lundae) 1: 351 (1821).
Fig. 6.44n–p
Fructification sessile or stipitate, spongy and softy-corky when fresh, becoming rigid and brittle on drying. Stipe present or absent, when present up to 8 × 4 cm,

cylindrical solid, central to eccentral, yellowish-brown, tomentose. Pileus imbricate, applanate to flabelliform, 10–41 × 24–30 × 1–3 cm; upper surface yellowish-brown to rusty brown, strigose, sometimes tomentose to nearly glabrous, azonate to concentrically zonate; margin acute to obtuse, concolorous to paler concolorous, entire, turning inward on drying; pore surface yellow to greenish-yellow in young specimens but turns yellowish-brown in older ones. Context yellowish-brown to rusty-brown, softy and spongy when fresh, becoming firm but fragile on drying. All parts turn cherry-red and then black when touched with KOH sol.

Hyphal system monomitic; hyphae 3.7–11 μm wide, light yellow-brown to pale brown, ribbon-like, thin to moderately thick-walled, branched, septate, clamps absent. Pores up to 4 mm deep decurrent, pores circular to oval but become irregular and large due to the breakdown of dissepiments with age, averaging 1–3 per mm; dissepiments 52–200(453) μm thick, concolorous with the context, even to dentate, finely velutinate. Basidia 12.4–15.3 × 5–6.8 μm, clavate, 4-spored. Basidiospores 6–11.5 × 4–6.8 μm, subhyaline to light yellow individually but appear yellow in mass, broadly ellipsoid, smooth, non-amyloid, shortly apiculate.

Distribution: H.P.: Shimla.

Collection examined: SSR 6016, IBP 37133.

Substratum: On stump.

Remarks: It is characterized by soft, sessile to stipitate fructification, yellowish-brown context turning cherry-red and then black with KOH sol, monomitic hyphal system and large broadly ellipsoid basidiospores.

Postia Fr.

Hymen. Europe p. 586,1874.

Fructification annual, resupinate to pileate, soft, fleshy when fresh, white, reddish orange; upper surface tomentose to glabrous. Pore surface white to cream. Context white, cream to brown, succulent when fresh. Hyphal system monomitic; generative hyaline, thin to thick-walled with narrow lumen, clamped, branched. Cystidia absent or present. Basidia clavate, hyaline, 2–4 spored. Basidiospores hyaline, thin-walled, smooth, allantoids to ellipsoid, non-amyloid.

Thirty species, widespread

Lit.: Yaon et al. (*FEMS Microbiol. Lett.* 242: 109, 2005; *Postia caesia* complex).

Type Species: *Polyporus lacteus* Fr., 1821.

Habitat: Dead wood

Himalayas: Seven (Table 6.2)

Postia caesia (Schrad.) P. Karst., Revue mycol., Toulouse 3: 360 (1881) = *Tyromyces caesius* (Schrad.) Murr., North Am. Flora 9: 34, 1907.

Plate 6.18a, Fig. 6.45g–j

Table 6.2 Synopsis of *Postia* species

Species	Fructification type	Abhymenial surface nature/ color and after bruising	Pores (per mm)	Hyphal system	Cystidia(c)/ cystidioles(cy)	Basidia	Basidiospores
P. caesia	Annual, pileate, sessile to effused reflexed, imbricate	Greyish to bluish Tomentose to strigose	Round/angular, 3–4	Monomitic, gen. Hyphae 2.1–4 μm, thin-walled in hymenium and trama, thick in context	c/cy Absent	14.6–17.6 × 6 –7.8 μm, Clavate to cylindrical, clamped at base, sterigmata, 6 μm	5.5–7 × 1.5–2.7 μm, Cylindrical to suballantoid, inamyloid, acyanophyllous
P. ceriflua	Pileate, solitary, pendent with basal attachment	Smooth to tuberculate to finally scrupose	Round to angular 3–5	2.6–4.6 μm wide thin to thick walled, monomitic	cy present, 14–20 × 4.6–5.2 μm, fusoid	4.0–5.5 μm, Clavate to sinuous, sterigmata 2 μm, clamped at base	3.5–4.6 × 1.2–2.0 μm, Inamyloid, acyanophyllous
P. guttulata	Pileate, imbricate, dimidiate to applanate	Brownish on bruising Smooth with shallow, circular depression, azonate	Round to angular, 6–7	6 μm	cy 14–19 × 3.9–5.2 μm	Clavate to subclavate, 14 × 6 μm, sterigmata 3.3 μm	4.6–6 × 2.6–3.3 μm, ellipsoid to suballantoid, guttulate
P. sericeomollis	Resupinate	Membranous	Round to angular 3–4	Cyanophilous 1.4–5.6 μm	c Present capitates encrusted	Clavate, 2–4 spored, 11–14 × 4.2–5.6 μm,	2.8–4.2 × 1.6–2.1 μm, ellipsoid to broadly ellipsoid

P. leucomella	Sessile, dimidiate, effused reflexed or imbricate	Radiate, fibrillose to minutely tomentose as azonate to lightly zonate	3–5 circular, lateral, irregular, angular often sinuous	2–6 μm some encrusted	c/cy absent	Clavate, 15–22 × 4.5 μm	4–5 × 1.5–2 μm, cylindric to allantoids
P. mappa	Effused, adnate, resupinate	Azonate, soft fleshy	Angular to irregular 2–4	1.8–4.1 μm	c/cy absent	Clavate, 14–16.5 × 4–5.5 μm	6–8.2 × 2.0–2.7 μm, cylindrical ellipsoid to ellipsoid, slightly curved
P. undosus	Friable, sessile, reflexed, pileate, single to imbricate dimidiate	Concentrically zonate, wrinkled on drying.	Angular, isodiametric	1.8–5 μm	c/cy absent	Clavate, 9–16.5 × 3.8–4.2 μm	4.2–6 × 1–1.8 μm, Cylindrical to allantoid

Plate 6.18 (**a**) *Postia caesia*. (**b**) *Postia ceriflua*. (**c**) *Postia mappa*. (**d**) *Ganoderma applanatum*

Fructification annual, effused-reflexed, adnate, reflexed part from a small pileus along the margin, soft when fresh, brittle on drying, effused, up to 10 cm × 2 cm × 5 mm, usually adjacent pilei laterally fused to form a long fructification. Pileus small, effused-reflexed, 8.5 mm wide and 4.8 mm thick, pore layer decurrent along the upper edge of reflexed part; upper surface white to cream, darken on drying, smooth, glabrous, azonate, irregularly wrinkled, shrink on drying; margin thin, acute, smooth, wavy, fertile below, incurved on drying.

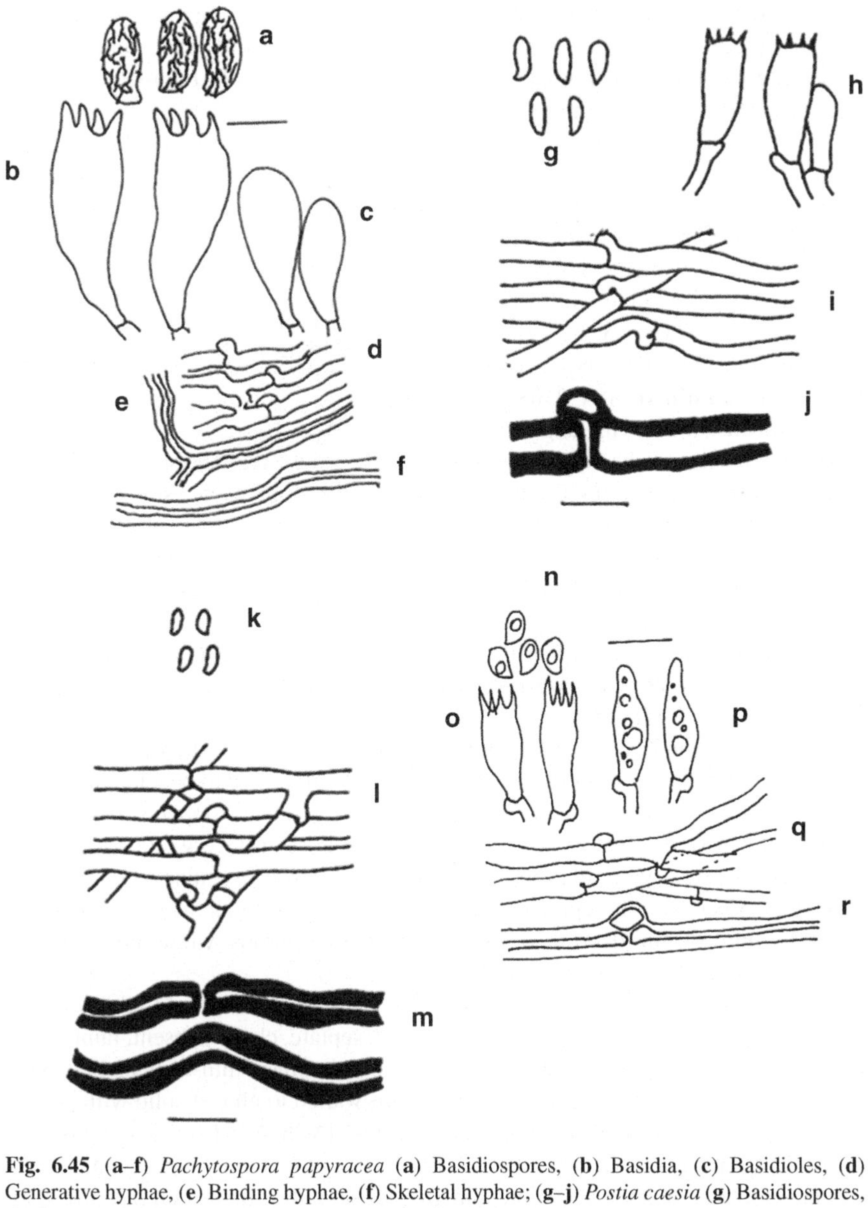

Fig. 6.45 (**a–f**) *Pachytospora papyracea* (**a**) Basidiospores, (**b**) Basidia, (**c**) Basidioles, (**d**) Generative hyphae, (**e**) Binding hyphae, (**f**) Skeletal hyphae; (**g–j**) *Postia caesia* (**g**) Basidiospores, (**h**) Basidia, (**i**) Thin-walled generative hyphae, (**j**) Thick-walled generative hyphae; (**k–m**) *Postia ceriflua* (**k**) Basidiospores, (**l**) Thin-walled generative hyphae, (**m**) Thick-walled generative hyphae, (**n–r**) *Postia guttulata* (**n**) Basidiospores, (**o**) Basidia, (**p**) Cystidia, (**q**) Thin-walled generative hyphae, (**r**) Thick-walled generative hyphae

Pore surface white when fresh, cream on drying; pores angular, 3–4 per mm; pores mouth velutinate; dissepiments entire, 40–185 μm thick; tubes in one layer, concolorous with upper surface. Context white, soft, homogenous, non-xanthochroic, up to 2 mm thick.

Hyphal system monomitic; generative hyphae hyaline, thin to thick-walled, branched, septate, clamps present, hyphae thin-walled in dissepiments, cyanophilous, 2.1–4 μm in diameter. Context hyphae thin to thick-walled, faintly cyanophilous, 2.5–7.0 μm in diameter. Cystidia and cystidioles absent. Basidia hyaline, thin-walled, clavate, 4-spored, cyanophilous, 14.6–17.6 × 6–7.8 μm. Basidiospores hyaline, thin-walled, smooth, cylindric-ellipsoid to ellipsoid, non-amyloid, 5.5–7.0 × 1.5–2.7 μm.

Distribution: N.W. Himalayas, Tripura: Sepahijula; U.K.: Rudraprayag, Nainital.

Collection examined: SSV 21089, L 37328.

Substratum: On decaying angiospermic stump.

Remarks: It seems to be uncommon species in the study area. The species is characterized by annual, effused-reflexed, soft fructification; white, smooth, azonate upper surface; white pore surface; large, 3–4 pores per mm; monomitic hyphal system with thin to thick-walled, clamped generative hyphae; and hyaline, cylindrical-ellipsoid to ellipsoid basidiospores. It is a new record for Uttarakhand.

Postia ceriflua (Berk. & M.A. Curtis) Jülich, Persoonia 11(4): 423 (1982) = *Tyromyces cerifluus* (Berk. & Curt.) Murr., North Am. Flora 9: 33, 1907.

Plate 6.18b, Fig. 6.45k–m

Fructification annual, pendent, broadly attached, soft, fleshy when fresh, hard on drying, up to 8 × 5 × 1 cm in size, single or close pilei unite to become imbricate. Pileus circular to semi-circular or dimidiate, applanate; upper surface white when fresh sterile below, up to 2.8 mm wide. Pore surface white when fresh, ochraceous on drying, uneven; pores small, angular to irregular on uneven surface, 3–5 per mm; dissepiments thin, even, 30–85 μm thick; tubes in one layer, white, up to 4 mm deep in section. Context soft, fibrous, white, homogenous, non-xanthochroic, up to 6.8 mm thick.

Hyphal system monomitic; generative hyphae hyaline, thin-walled to thick-walled to almost solid with narrow lumen, branched, septate, clamps absent, faintly cyanophilous. Dissepiments and sub hymenium hyphae, thin-walled, branched, 2.6–4.6 μm in diameter. Context hyphae thin-walled to almost solid with narrow lumen, rarely branched, solid hyphae gelatinised with KOH sol, 3–6 μm diameter. Cystidia absent. Basidia clavate, 4-spored, cyanophilous, up to 4–5.5 μm in diameter. Basidiospores hyaline, thin-walled, smooth cylindric-ellipsoid to ellipsoid, non-amyloid, 3.5–4.6 × 1.2–2.0 μm.

Distribution: Bhutan-Thimphu: Bunakha, U.K.: Mussoorie, Lal Tibba-Dehradun.

Collection examined: SSV 21286, L 37329.

Substratum: On decaying *Pinus* log.

Remarks: The species is characterized by annual, pendent, broadly attached, soft, fleshy white fructification; medium, angular, three to five pores per mm; white context; monomitic hyphal system with thin to thick-walled to solid, clamped generative hyphae; and hyaline, thin-walled, cylindrical-ellipsoid to ellipsoid basidiospores. It is a new record for India/Himalayas as well as Bhutan.

Postia guttulata (Peck ex Sacc.) Jülich, Persoonia 11(4): 423 (1982) = *Tyromyces guttulatus* (Peck.) Murr., North Am. Flora 9: 31, 1907. Fig. 6.45n–r

Fructification annual, soft, pileate, stipitate, growing singly. Pileus reniform or flabelliform, soft and friable when fresh, hard and brittle on drying; upper surface cream to yellow when fresh, brown on drying, glabrous, irregularly wrinkled, azonate, small excretory drops present on the thin pellicle; margin thin to almost absent, lobed, pale yellow, regular. The pileus tapers to form a contracted stipe, up to 4 cm long and up to 2 cm broad, white to cream colored. Pore surface white when fresh, cream when touched; pores angular, 4–6 per mm; pore tubes white when fresh; pore mouth velutinate; dissepiment concolorous with the pore surface, up to 50–85 μm wide. Context white, homogenous, non-xanthochroic up to 5 mm thick.

Hyphal system monomitic; generative hyphae hyaline, branched, septate, clamped, thin- walled to slightly thick-walled, 1.8–4.6 μm wide in dissepiment. In context, hyphae thick-walled to almost solid with narrow lumen, 3–7.6 μm wide. Cystidia absent, subulate cystidioles present. Basidia cylindric clavate to clavate, 4-spored, cyanophilous, 12–17.5 × 4.5–5.5 μm; sterigmata up to 3 μm long. Basidiospores hyaline, thin-walled, smooth, ellipsoid to oblong, ellipsoid, uniguttulate, 4.6–5.3 × 2–2.5(−3) μm.

Distribution: A.P.: Bomdilla; H.P: Shimla; U.K.: NDBR.

Collection examined: SSV 21756, Dhanda 6554, IBP 37139, L 37140.

Substratum: On decaying angiospermic stump.

Remarks: The species is marked by its short stipitate flabelliform fructifications; yellowish upper surface with small excretory liquid drops; white hymenial surface with pinkish tinge; monomitic hyphal system with thin to almost solid generative hyphae; and ellipsoid to broadly ellipsoid basidiospores. It is a common species in the Himachal Pradesh, from where it was first reported by Thind and Dhanda (1978). However, it is uncommon in the Eastern Himalayas. The collection resembles *T. guttulatus* in all respects except that the basidiospores are slightly bigger 4.5–5.2 × 2–2.5(−3) μm, which are mentioned to be 3.5–4.5 × 1.7–2.1 μm by Thind and Dhanda (1980). It is a new record for Uttarakhand and Eastern Himalayas.

Postia leucomallella (Murrill) Jülich, Persoonia 11(4): 423 (1982) = *Postia fragilis* (Fr.) Jülich, sesu auct; fide Checklist of Basidiomycota of Great Britian & Ireland (2005). Fig. 6.46a–e

Fructification annual, sessile, dimidiate, resupinate or effused-reflexed, light in weight, often imbricate, soft when fresh, becoming coriaceous and brittle when dry; upper surface white, becoming pinkish brown to reddish brown, radiate-fibrillose, minutely tomentose, azonate or lightly zonate; margin thin, undulate, often involute on drying. Context white to light brown, 2–7 mm thick, fibrous; hymenial surface white when fresh, pinkish brown to rusty brown on drying,

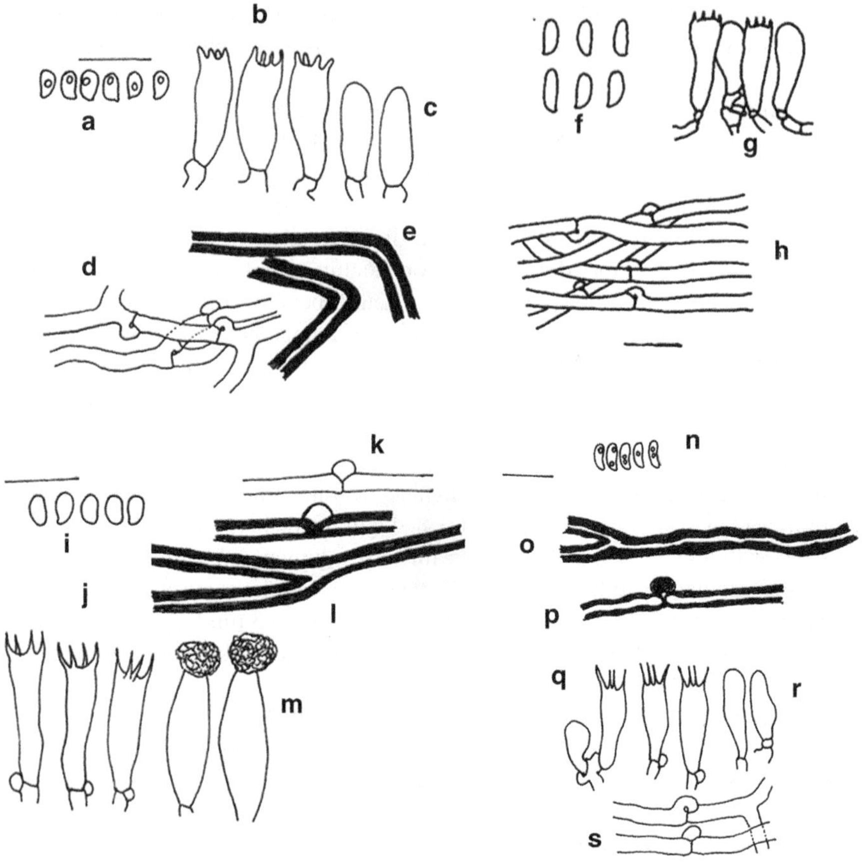

Fig. 6.46 (**a**–**e**) *Postia leucomallella* (**a**) Basidiospores, (**b**) Basidia, (**c**) Basidioles, (**d**) Generative hyphae, (**e**) Skeletal hyphae; (**f**–**h**) *Postia mappa* (**f**) Basidiospores, (**g**) Basidia, (**h**) Generative hyphae; (**i**–**m**) *Postia sericeomollis* (**i**) Basidiospores, (**j**) Basidia, (**k**) Generative hyphae, (**l**) Skeletal hyphae, (**m**) Cystidia; (**n**–**s**) *Postia undosa* (**n**) Basidiospores, (**o**) Skeletal hyphae, (**p**) Thick-walled generative hyphae, (**q**) Basidia, (**r**) Basidioles, (**s**) Thin-walled generative hyphae

pore tube 2–7 mm long; pores initially circular, later irregular, angular, often sinuous, 3–5 per mm.

Hyphal system monomitic; generative hyphae hyaline to brown, thin- to slightly thick-walled, branched, clamped, with walls sometimes encrusted, 2–6 µm wide. Basidia 4-sterigmate, 15–22 × 4–5 µm. Basidiospores hyaline, thin-walled, smooth, cylindric to allantoid, 4.0–5.0 × 1.5–2.0 µm.

Distribution: U.K.: Chakrata; H.P.: Shimla- Khadrala.

Collection examined: Dhanda 6535, 6568, 6918, 6986, 6987.

Substratum: On decaying *Cedrus deodara* log.

Remarks: The species is characterized by annual, resupinate to effused reflexed, light in weight, brittle fructification; monomitic hyphal system, smooth, cylindric to allantoids basidiospores.

Postia mappa (Overh. & J. Lowe) M.J. Larsen & Lombard, Mycotaxon 26: 272 (1986) = *Tyromyces mappa* (Overh. & Lowe) Ryv. Norw. J. Bot. 20: 10, 1973. Plate 6.18c, Fig. 6.46f–h

Fructification annual, resupinate, effused, adnate, soft, fleshy when fresh, fragile on drying, succulent when fresh, shrink on drying, up to 6 cm long and 6 mm thick; margin thin, white, fleshy, rhizomorphic. Pore surface white when fresh, cream to brown on drying, uneven; pores angular to irregular on slopping substrate; dissepiment even, 70–195 μm thick; tubes in one layer, white, up to 3 mm deep in section Context thin, white, homogenous, non-xanthochroic, up to 135 μm thick.

Hyphal system monomitic; generative hyphae thin-walled, branched, septate, clamps prominent, cyanophilous, 1.8–4.1 μm in diameter. Cystidia absent. Basidia clavate, thin-walled, 4-spored, cyanophilous, 14–16.5 × 4–5.5 μm. Basidiospores hyaline, thin-walled, smooth, cylindric ellipsoid to ellipsoid, slightly curved, non-amyloid, 6–8.2 × 2–2.7 μm.

Distribution: A.P.-West Kameng, Bomdilla-on Tawang road, U.K.: Badrinath- Chamoli.

Collection examined: SSV 21601.

Substratum: On under surface of decaying angiospermic stump.

Remarks: The species is characterized by its resupinate, soft, fleshy, succulent, white fructifications; 2–4 per mm pores; monomitic hyphal system with prominent clamps; and cylindric-ellipsoid 6–8 × 2–2.5 μm basidiospores. This collection fits well in the description given by Ryvarden (1978), where the basidiospores are slightly bigger, 6.5–10 × 2.5–3 μm. This species is being reported here for the first time from Indian Himalayas.

Postia sericeomollis (Romell) Jülich, Persoonia 11(4): 423 (1982) = *Tyromyces sericeo-mollis* (Rom.) Bond., Annls. Mycol., 39:52, 1941. Fig. 6.46i–m

Fructification annual, resupinate, arising as small patches which coalesce later becoming effused, 2–9 × 1.5–5 cm and up to 2 mm thick, separable, often membranous; margin white to creamish, thinning, arachnoid, wide; pore surface white to pale creamish, uneven; pores rounded to angular, about 3–4 per mm; dissepiment 68–95 μm thick, equal, apices finely velutinate. Context white, soft, thin; tubes soft when fresh, hard and brittle when dry, not well formed, extending mostly along the slope up to 4 mm long. Tissue not changing color in KOH sol.

Hyphal system monomitic; generative hyphae hyaline to subhyaline, thin- to slightly thick-walled, branched, septate, clamped, cyanophilous, 1.4–5.6 μm wide. Cystidia present, subhyaline, thin- to slightly thick-walled, capitate-incrusted, embedded to slightly projecting beyond the hymenium, 14–18.3 × 4.2–5.8 μm. Basidia hyaline, clavate, thin-walled, 2–4 spored, 11.5–14 × 4.2–5.8 μm, sterigmata short to long. Basidiospores hyaline, thin-walled, smooth, ellipsoid to broadly ellipsoid, non-amyloid, 3.0–4.5 × 1.6–2.3 μm.

Distribution: H.P.: Kullu.

Collection examined: Dhanda 6599.

Substratum: On stump of *Cedrus deodara*.

Remarks: It is very rare in the N. W. Himalayas and could be gathered only once. The species is marked by resupinate fructifications with monomitic hyphal system, clamped generative hyphae and capitates-encrusted cystidia.

Postia undosa (Peck) Jülich, Persoonia 11(4): 424 (1982) = *Tyromyces undosus* (Peck.) Murr., North Am. Flora: 34, 1907. Fig. 6.46n–s

Fructification annual, soft, friable, easily separable from wood, sessile, reflexed-pileate, single to imbricate with numerous pilei with narrow distance on a decurrent Pileus broadly attached, dimidiate, individual, pilei soft, friable; margin white, acute, regular, sterile below, soft, up to 1.8 mm wide; pore surface, up to 8 × 2 cm; upper surface white when fresh, cream on drying, hairy, becoming glabrous in older fructifications, faintly concentrically zonate, on drying irregularly wrinkled, entire; pores angular, isodiametric, round, 3–4 per mm; tubes in one layer, white up to 5.5 mm deep in section. Context white, soft, homogenous, non-xanthochroic.

Hyphal system monomitic; generative hyphae hyaline, thin- to thick-walled, branched, septate, clamped, 2.0–5.2 μm in diameter. Thin-walled to thick-walled hyphae present in dissepiment and context. Basidia clavate, cyanophilous, thin-walled, 4-spored, 9.2–16.7 × 3.8–4.5 μm. Cystidia absent. Basidiospores hyaline, thin-walled, cylindrical to allantoids, acyanophilous, non-amyloid, 4.2–6.2 × 1–2 μm.

Distribution: A.P.-West Kameng, Bomdilla, 5 km from Bomdilla towards Munna.

Collection examined: SSV 21782.

Substratum: On angiospermic decaying log.

Remarks: This species is characterized by narrow, reflexed-pileate, imbricate, soft, white fructifications; faintly zonate upper surface; isodiametric pores; thin to thick-walled generative hyphae; and hyaline, thin-walled, cylindrical to allantoid, thin, 1–1.8 μm wide basidiospores. It is new record for Himalayas/India.

Family- Ganodermataceae

Ganoderma Karst.,

Revue Mycol., Toulouse **3(9)**: 17, 1881

Fructification annual or perennial, pileate, centrally or laterally stipitate or sessile, dimidiate to flabelliform, hard, woody and heavy or corky and light. Upper surface covered with a coloured, glossy distinct cortext, brown to deep purplish, sometimes as if varnished, smooth. Pore surface entire, ochraceous to brown; pores small to medium, round or slightly angular; tubes often stratified. Stipe when present, glossy with distinct cortext, yellowish, reddish to deep purplish in colour. Context white to brown, corky fibrous to hard and woody non-xanthochroic. Hyphal system trimitic; generative hyphae hyaline, thin-walled, septated, with clamps; skeletal hyphae pale brown to brown, thick-walled, aseptate, long; binding hyphae pale brown to brown, thick-walled, much branched. Cystidia and setae absent. Basidia hyaline, clavate, 4-spored. Basidiospores truncate, large, distinctly double walled, outer wall (exosporium) hyaline, thin and smooth; inner wall (endosporium), brown, thick, ornamented, echinulate.

Fifty species, wide spread
Lit.: Ryvarden (*Mykol* **92**:180, 2000)
Habitat: At the base of trunk
Himalayas: Five
Key to species

1. Pileus astipitate 5
1. Pileus stipitate 2
2. Pileus orangish brown, abhymenial surface weakly laccate .... *G. lucidum*
2. Pileus reddish brown, abhymenial surface strongly laccate 3
3. Hyphal system dimitic .... *G. multiplicatum*
3. Hyphal system trimitic 4
4. Basidiospores verrucose elliptic 10–13 × 7–9 μm .... *G. resinaceum*
4. Basidiospores globose; 6.5 × 4.5 μm .... *G. sessiliformae*
5. Basidiospores 7.0–7.6 × 4.0–4.8 μm .... *G. applanatum*
5. Basidiospores 7.5–9.5 × 5–8.5 .... *G. philippii*[a]

[a]Extra limital, not included in the text

Ganoderma applanatum Karst., Hym. Eur.: 143, 1887. Plate 6.18d, Fig. 6.47a–e
Fructification perennial, sessile, broadly attached, single or imbricate, hard, stipitate, woody. Pileus applanate to ungulate, flat to convex, 5–19 × 11–29 × 4–6 cm thick; upper surface brown to greyish brown, broad centrically sulcate, smooth; margin acute, concolorous with upper surface; pore surface white when fresh, discoloured when touched, light brown to brown with aging; pores round, entire, 4–6 per mm; tubes concolorous with the pore surface, often with white mycelium, distinctly stratified. Context cinnamon brown, thinner than the tube layer, homogenous, non-xanthochroic.
Hyphal system trimitic; generative hyphae clamped, thin-walled, hyaline, branched, cyanophilous, 2–3 μm in diameter; skeletal hyphae light brown, thick-walled, occasionally branched, 3–5 μm in diam.; binding hyphae subhyaline to light yellowish brown, thick-walled, branched, 3–4 μm in diameter. Basidia hyaline, clavate, 4-sterigmate, 12.8–19.0 × 5.3–7.5 μm. Basidiospores ganodermatoid, brown ornamented, ovoid-ellipsoid, truncate at one end, 6.6–8.8 × 4.5–6.0 μm.
Distribution: Bhutan: Chukha; Thimphu, Nawephu; A.P.: West Kameng, New Bomdila; U.K.: NDBR, Nanital.
Collection examined: Hem 28600, 28623, SSV 21263, 21295, L 37142.
Substratum: on angiospermic tree, on gymnospermic tree.
Remarks: The species is characterized by perennial, applanate, woody fructification, brown to blackish, thick hard crust present on the upper surface; yellowish brown context; trimitic hyphal system and ganodermoid basidiospores.

Ganoderma lucidum (Curt.:Fr.) Karst., Rev. Mycol. **3(9)**: 17, 1881. Plate 6.19a, Fig. 6.47f–j
Fructification annual to perennial, short stipitate, corky and coriaceous when fresh, hard and light weight on drying; pileus reniform to flabelliform, 5–13 × 1.8–2.8 cm; upper surface shiny with laccate crust, reddish-brown, smooth, azonate to faintly zonate; hymenial surface cream coloured; context; pore surface white

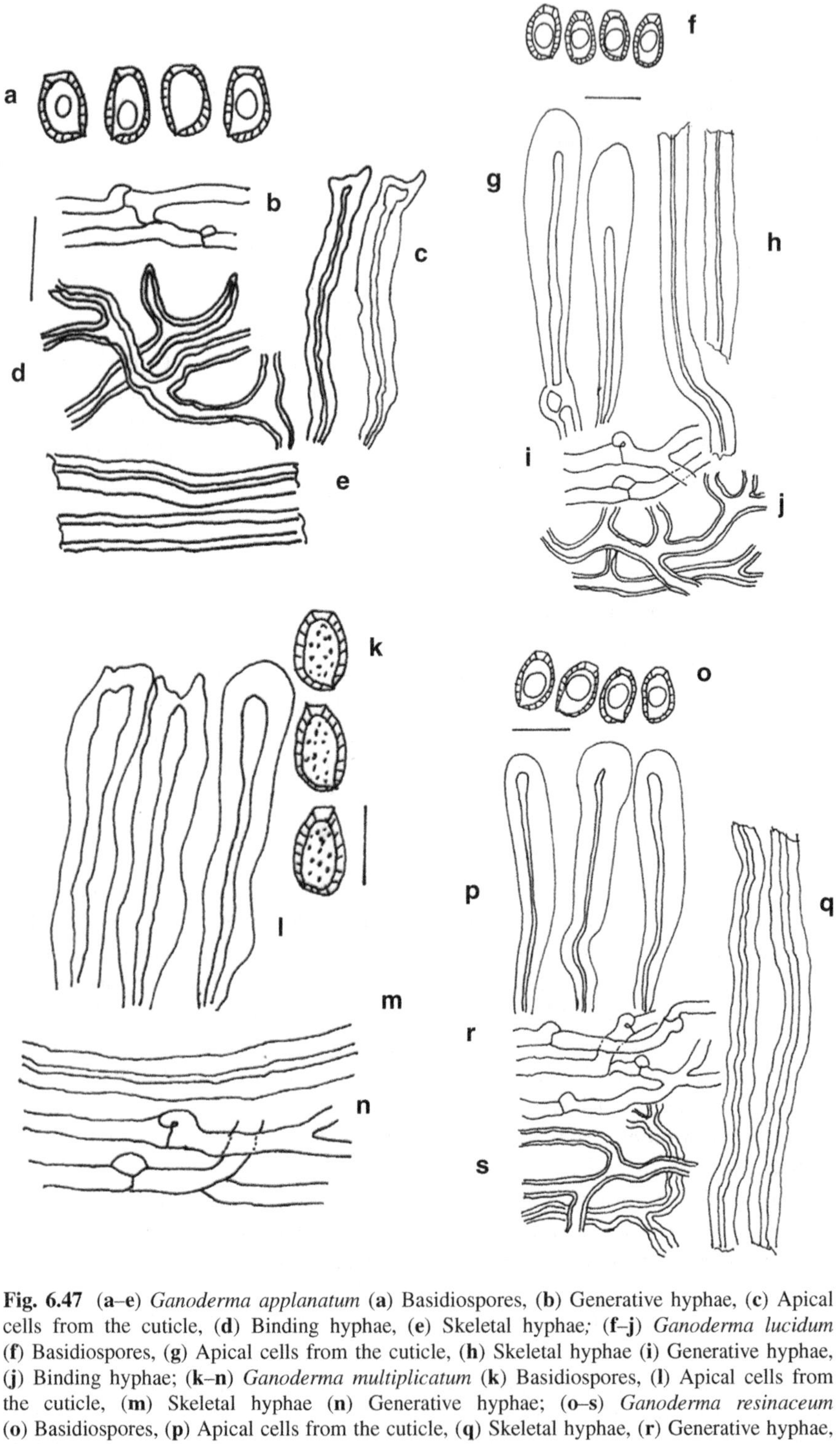

Fig. 6.47 (**a–e**) *Ganoderma applanatum* (**a**) Basidiospores, (**b**) Generative hyphae, (**c**) Apical cells from the cuticle, (**d**) Binding hyphae, (**e**) Skeletal hyphae; (**f–j**) *Ganoderma lucidum* (**f**) Basidiospores, (**g**) Apical cells from the cuticle, (**h**) Skeletal hyphae (**i**) Generative hyphae, (**j**) Binding hyphae; (**k–n**) *Ganoderma multiplicatum* (**k**) Basidiospores, (**l**) Apical cells from the cuticle, (**m**) Skeletal hyphae (**n**) Generative hyphae; (**o–s**) *Ganoderma resinaceum* (**o**) Basidiospores, (**p**) Apical cells from the cuticle, (**q**) Skeletal hyphae, (**r**) Generative hyphae, (**s**) Binding hyphae

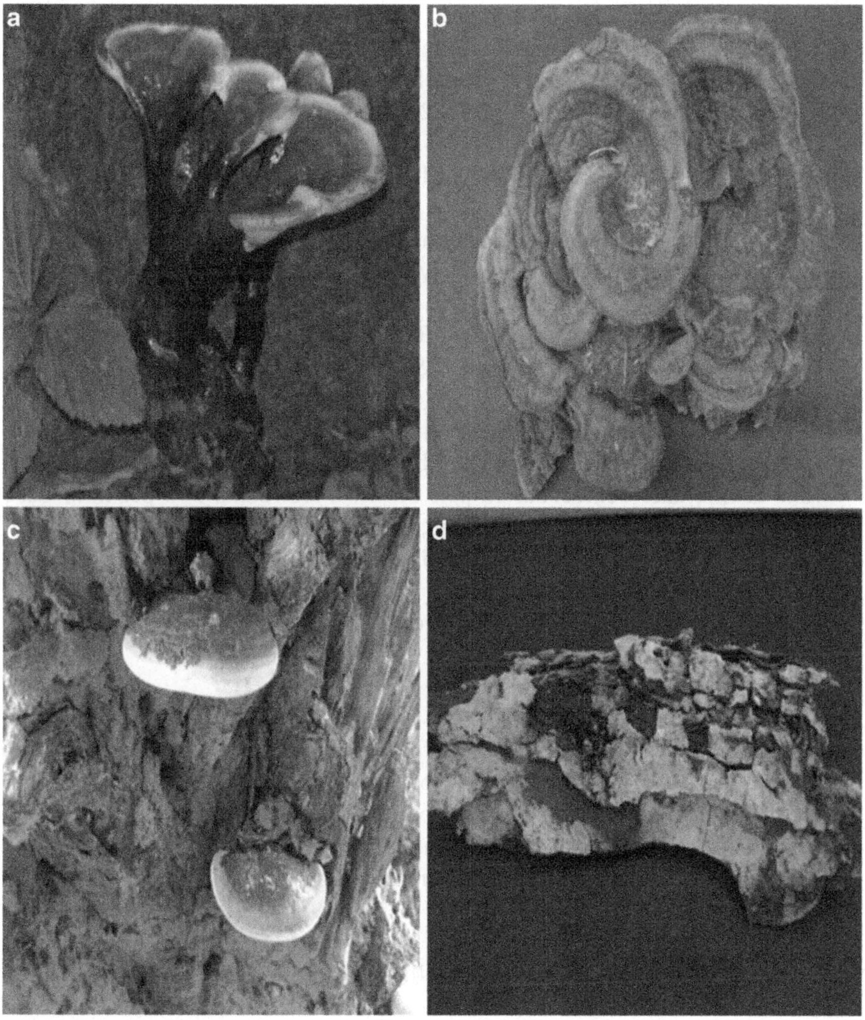

Plate 6.19 (**a**) *Ganoderma lucidum*. (**b**) *Ganoderma multiplicatum*. (**c**) *Ganoderma resinaceum*. (**d**) *Intextomyces contiguus*

to cream, turning brown when touched, darken on drying; pores small, round, 4–6 per mm, brown; pore tubes 6–7 µm long. Context white to cream, fibrous, homogenous, non-xanthochroic.

Hyphal system trimitic; generative hyphae thin-walled, hyaline, clamped, branched, cyanophilous, 3.3–2.3 µm in diameter; skeletal hyphae light brown, thick-walled, lumen narrow, sparsely branched, 1.8–5.0 µm in diameter; binding hyphae pale-brown, thick-walled, much branched, 3.5–5.0 µm in diam. Cystidia absent. Basidia hyaline, clavate, 4-sterigmate, 10.6–17.2 × 5.3–7 µm. Basidiospores ganodermatoid, truncate at one end, brown, ornamented, ovoid-ellipsoid, truncate, 8–11.5 × 6.8–8 µm.

Distribution: Bhutan: Chimakothi; A.P.: West Kameng, Bomdilla, Wang Basti; Manipur: Imphal; Meghalaya: Shillong; Mizorum: Aizwal; Nagaland: Kohima; Tripura: Agartala; West Bengal: Darjeeling; U.K.: NDBR.

Collection examined: Hem 28187, SSV 21257, 21261, 21576, 21593.

Substratum: On angiospermic tree.

Remarks: The species is characterized by perennial, broadly attached, stipitate fructification; upper surface brown to greyish brown; hyphal system trimitic; basidia clavate, 4-sterigmate; ganodermatoid, ovoid-ellipsoid basidiospores. The species is the most common and widely distributed species.

Ganoderma multiplicatum (Mont.) Pat., *Bull. Soc. Mycol. Fr.* 5:74, 1889.
= *Polyporus multiplicatus* Mont. *Ann. Sci. Nat. Bot. Ser.* 41:128, 1854. Fig. 6.47k–n

Fructification perennial, pileate, stipitate, dimidiate, laccate 17–18 × 11–5 × 1.5–2 cm. Upper surface concentrically sulcate, brown of chestnut. Pore surface creamy white at first later ochraceous to pale brown. Pore round, 5–8 per mm. Tube 3 mm thick snuff brown. Context 5 mm, snuff brown, shiny. Cutis type diverticulate, 28–30.8 × 14 µm. Hyphal system dimitic, generative hyphae 3.8 µm diameter, thin-walled, hyaline with clamp connection; skeletal hyphae 5.8–7.5 µm diameter, thick walled, yellowish green colour. Basidiospore 7–9 × 5–7 µm, brown ellipsoid, truncate.

Distribution: H. P.: Dalhousie.

Collection examined: L 39010.

Substratum: On dead stump of *Holoptelea intergrifolia.*

Remarks: The species is characterized by perennial, pileate, stipitate, dimidiate fructification. Upper surface concentrically sulcate; pore surface creamy white. Earlier it was reported from Venezuela, Brazil, China, New Guinea and Egypt. From India, it has been reported from Maharashtra. It is a new record for Himalayas.

Ganoderma resinaceum Boud., in Patouillard, Bull. Soc. Mycol. Fr. 5: 72 (1890) [1889]. Plate 6.19c, Fig. 6.47o–s

Fructification broadly attached to the substrate, upper undulating, smooth, when fresh hard on drying, red-brown in colour; margin inflated, yellowish to orange-yellow.

Hyphal system trimitic; generative hyphae septate, thin-walled, clamps present, 2–3.5 µm; skeletal hyphae thick-walled up to 6 µm; binding hyphae branched, thick-walled 1.5–4 µm. Basidia 4-sterigmate with basal clamp absent., 15 × 8 µm. Basidiospores verrucose, elliptical, 10–13 × 7–9 µm with pores.

Distribution: Bhutan: Thimphu, Newephu.

Collection examined: GSD 19550.

Substratum: On a decaying angiospermic log.

Remarks: The species is characterized by broadly attached, smooth fructification; trimitic hyphal system; verrucose to elliptical basidiospores.

Ganoderma sessiliforme Murrill, Bull. N. Y. Bot. Garden 8: 149, 1912.

Fructification stipitate, dimidiate, hard like wood; upper surface reddish brown, brittle; margin dark brown, thick; pore surface yellowish white. Hyphal system trimitic; generative hyphae thin-walled, septate, clamps present; skeletal hyphae thick-walled, 5 μm in diameter; binding hyphae thick-walled, yellow in colour. Basidiospore globose, 6.5×4–5 μm.

Distribution: U.K.: Dehradun.

Collection examined: IBP 37144, L 37146, 37147.

Substrate: On decayed wood.

Remarks: The species is characterized by hard woody reddish brown fructification; dimitic hyphal system, clamped hyphae; globose basidiospores.

Family- Hyphodermataceae

Intextomyces J. Erikss. & Ryv.,
Cort. N. Europe 4: 735, 1976.

Fructification resupinate, effused, adnate, whitish, greyish or rose grey; ceraceous when wet, hard when dry; hymenium at first smooth, then more or less tuberculate or even odontoid; margin determinate, mostly fertile throughout; hyphae of the fructification thin-walled, indistinct, richly and irregularly branched and interwoven into dense texture; cystidia absent; basidia suburniform, basally continuing into a narrow, stalk-like hypha, which penetrates the fructification or the substrate if the fructification is very young; basidiospores ellipsoid or ovoid, often subangulate, thick walled, smooth, cyanophilous, not amyloid.

Four Species, widespread.

Lit.: Eriksson & Ryvarden (*Cortic. N. Europ.* **4**: 735, 1976).

Type species: *Corticium contiguum* Karst.

Habitat: Dead and decaying Wood, Bark.

Himalayas: One

Intextomyces contiguous (Karst.) J. Erikss. & Ryv., Cort.N. Europe 4: 735 (1976)=*Corticium contiguum* Karst., Soc. F. Fl. Fenn. Acta 2: 1 p.39, 1881. Plate 6.19d, Fig. 6.48a–d

Fructification resupinate, effused, adnate, perennial, up to 300 μm thick in section, at first very thin but thickening with time, hymenial surface smooth or when vigorously growing more or less tuberculate or even odontoid, continous but when dried often cracked by shrinkage, ceraceous and greyish white when alive and wet, hard when dry, calcareously whitish or rose grey; margin mostly determinate.

Hyphal system monomitic; hyphae mostly indistinct, thin-walled, fibulate, 1–3.5 μm wide, richly ramified densely interwoven into a texture, composed of two hyphal elements (1) plasma-rich, very thin-walled, irregularly sinuous hyphae, penetrating the fruitbody, mostly in vertical direction (2) densely agglutinated hyphae, poor in protoplasm and together with remains of basidia and spores forming a pseudoparenchymatic texture; subicular and subhymenial layers not distinguishable; whole of the hyphal tissue is covered with some crystalline material in the shape

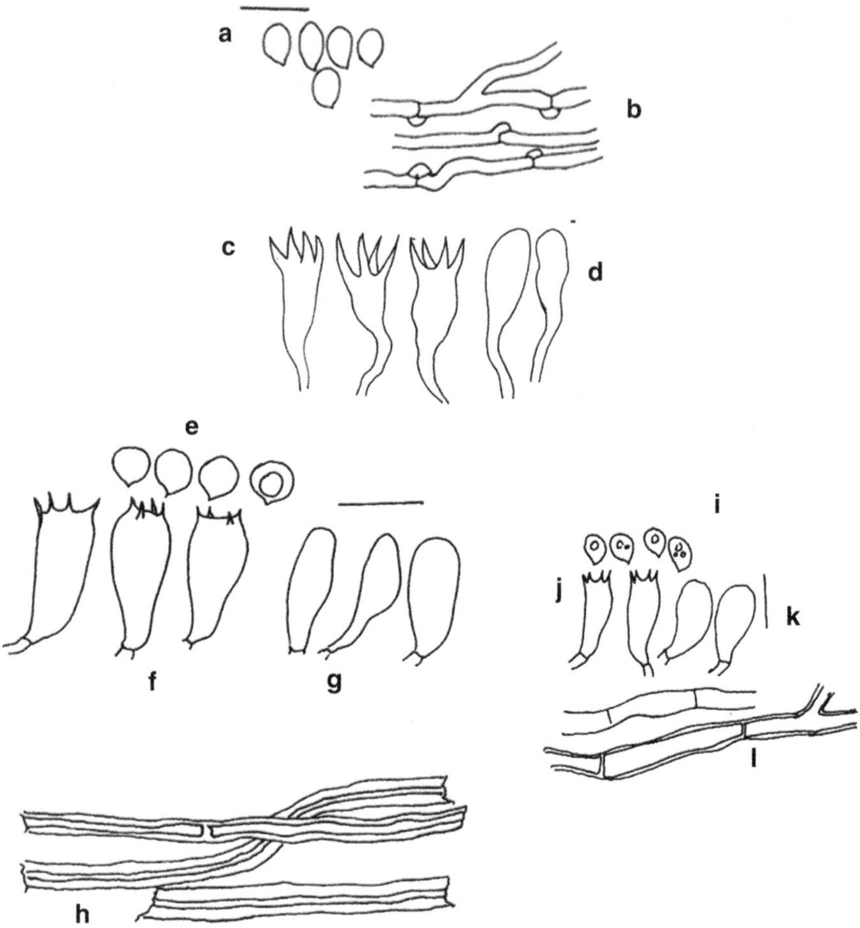

Fig. 6.48 (**a**–**d**) *Intextomyces contiguus* (**a**) Basidiospores, (**b**) Generative hyphae, (**c**) Basidia, (**d**) Basidioles, (**e**–**h**) *Physisporinus rivulosus* (**e**) Basidiospores, (**f**) Basidia, (**g**) Basidioles, (**h**) Generative hyphae; (**i**–**l**) *Rigidoporus crocatus* (**i**) Basidiospores, (**j**) Basidia, (**k**) Basidioles, (**l**) Generative hyphae

of rods, hyphae enclosed in the substrate more distinct and regular, 2–3 μm wide, branched at approximately right angles. Cystidia none. Basidia start as swellings at the end of penetrating hyphae, when mature short cylindrical with suburniform constriction at the base abruptly narrowed into the bearing hyphae, no clamps at the base of the true basidium, but the bearing hyphae is fibulate; the epibasidial part 10–20.5 × 4–6(−8.5) μm; 4-sterigmate, rather long and stout, at first cylindrical and obtuse, later conical, somewhat curved, 6–8 μm long. Basidiospores ellipsoid or ovoid, often subangulate, smooth, thick-walled, cyanophilous, non-amyloid, 4.5–5(−6) × 3.5–4.3 μm.

Distribution: Bhutan: Thimphu, Newephu.
Collection examined: GSD 19550.
Substratum: On a decaying angiospermic log.
Remarks: The species is characterized by hard fructification, dense, ceraceous texture, peculiar basidia and thick-walled, cyanophilous and subangular basidiospores.

Family-Meripileaceae

Key to genera

1. Fructification annual, resupinate, often arising as small circular colonies which coalesce to form large patches .... *Physisporinus*
1. Fructification annual to perennial, resupinate, effused-reflexed to pileate .... *Rigidoporus*

Physisporinus P. Karst.,
Bidr. Känn. Finl. Nat. Folk 48: 324 (1889)

Fructification annual, resupinate, often arising as small circular colonies which coalesce to form large patches, effused, adnate; margin determinate, white, membranous, narrow, sterile. Pore surface white, drying creamish-white dull; pores oval to angular. Context white, hard, thin of agglutinated interwoven hyphae Hyphal system monomitic; generative hyphae hyaline to subhyaline, thin to slightly thick-walled, branched, septate, clamped, cyanophilous. Cystidioles not abundant subhyaline. Basidia hyaline, thin-walled, clavate. Basidiospores obovoid to subglobose, hyaline, thin-walled, smooth non-amyloid.

Two species, Europe, Uganda, India/Himalayas.

Lit.: Ipulet & Ryvarden (Syn. Fung. 20: 87, 2005; Uganda), Gilbertson & Ryvarden (N.Amer. Polyp. 2:628, 1987).
Type Species: *Poria vitrea* Pers., 1795.
Habitat: Wood
Himalayas: One

Physisporinus rivulosus (Berk. & M.A. Curtis) Ryvarden [as *'rivulosa'*], Mycotaxon 20(2): 353 (1984) = *Poria rivulosa* (Berk. & M.A. Curtis) Cooke, Grevillea 14(no. 72): 109 (1886). Fig. 6.48e–h

Fructification annual, resupinate, often arising as small circular colonies which coalesce later to form large patches, effused, 10 × 3 × 2 cm, adnate, inseparable; taste mild; margin determinate, white, adnate when fresh loosening from substratum on drying, membranous, narrow. Pore surface white, drying creamish-white dull; pores oval to angular, becoming somewhat irregular due to breakdown of dissepiments, 4–6 per mm; dissepiments up to 135 μm thick, equal, hard, composed of parallel agglutinated hyphae; apices velutinate. Context white, hard, thin of agglutinated interwoven hyphae up to 200 μm thick; tubes not stratose, hard, white to creamish in section, up to 1.5 mm deep. Tissue not darkening in KOH sol.

Hyphal system monomitic; generative hyphae hyaline to subhyaline, thin to slightly thick-walled, branched, septate, clamped, cyanophilous, 1.5–4.5 μm in diameter. Cystidioles not abundant subhyaline, fusoid, 4–7.3 μm in diameter. Basidia hyaline, thin-walled, clavate, 2–4 spored, up to 8.5 μm in diameter; sterigmata short to long. Basidiospores hyaline, thin-walled, smooth non-amyloid, obovoid to subglobose, some uniguttulate, 4.3–6.4×4.3–5.6 μm.

Distribution: H.P.: Kullu- Rangri, Manali, Shimla.

Collection examined: RW 6084, L 42924, 42929.

Substratum: stumps of *Cedrus deodara.*

Remarks: The species is marked by annual resupinate, separable and bitter fructification with white pore surface, drying creamish-white; angular pores, 3–5 per mm; white, non-xanthochroic context, monomitic hyphal system; fusoid cystidia; and hyaline, smooth, non-amyloid, ovoid to subglobose basidiospores.

Rigidoporus Murr.,
Bull. Torrey bot. club 32(9): 478, 1905.

Fructification annual to perennial, resupinate, effused-reflexed to pileate, soft when fresh becomes hard on drying; Pileus effused-reflexed to sessile, applanate to convex; upper surface glabrous, smooth, uneven, concentrically. Pore surface white to pinkish. Pores small, circular, angular to irregular; tubes in single layer or in many layers when stratified, separated by thin layers of context. Context white, pale cream to light coloured, soft coriaceous homogenous, non xanthochroic thin to thick. Cystidia encrusted or smooth present or absent. Hyphal system monimitic; generative hyaline to subhyaline to yellow, thin to thick-walled, branched, septate, clamps absent. Basidia clavate, 4-spored, thin-walled. Basidiospores subglobose to globose, thin-walled, nonamyloid.

Forty Species, widespread

Lit.: Ryvarden & Johansen *(Prelim. Polyp. Fl. E. Afr.: 537, 1980)*, Corner (Beih. Nova. Hedw. 86: 152, 1987; key Malaysia spp.), Gilbertson and Ryvarden (N. Amer. Polyp. 2:693, 1987; key temp. spp.)

Type Species: *Polyporus micromegas* Mont.1842

Habitat: Dead Wood

Himalayas: Five

Key to species

1. Fructification pileate 2
1. Fructification resupinate........ 5
2. Pileus pinkish white to pale buff; basidiospores 8–9.5 μm........ *R. ulmarius*
2. Pileus reddish black to orange; basidiospores up to 6 μm 3
3. Cystidia club-like, encrusted........ *R. lineatus*
3. Cystidia ventricose-mucronate, not encrusted 4
4. Pore surface reddish black to brownish .... *R. fusco-lineatus*[a]
4. Pileus orange red........ *R. microporus*
5. Cystidia absent........ *R. crocatus*
5. Cystidia obliquely projecting, strongly encrusted .... *R. vinctus*

[a]Extra limital, not included in the text

Rigidoporus crocatus (Pat.) Ryvarden, Occ. Pap. Farlow Herb. Crypt. Bot.**18**: 13 (1983)=*Poria nigrescense* Bres., Acad. Rover. Agiati Atti 3: 83. 1897= *Rigidoporus nigrescense* (Bers.) Donk, Persoonia 4: 341. 1966. Fig. 6.48i–l

Fructification resupinate, annual, separable, membranous-crustaceous when fresh becoming hard and brittle on drying, widely effused, subiculum up to 1 mm thick light ochraceous buff turning darker at the margin; pore surface white to cream when fresh becoming cream yellow to yellowish brown or pinkish buff on drying, not creviced; margin thinning to thick, loosely adnate but often curls up on drying, concolorous. Pores round to angular, not stratose, up to 4 mm deep; pore mouths rounded to angular. Context composed of compactly arranged more or less agglutinated hyphae.

Hyphal system monomitic, hyphae (2)4–5.5 μm wide, branched, septate, clamps absent, the walls subhyaline, thin to thick with capillary lumen. Cystidia or gloeocystidia absent. Basidia not observed. Basidiospores 3.5–5.6×3–4.7 μm, subglobose, minutely apiculate, the walls thin, subhyaline, smooth, nonamyloid.

Distribution: U.K.: Chakrata- Dehra Dun.

Collection examined: IBP 37149, L 37150.

Substratum: On stump of angiospermic wood.

Remarks: It is marked by the resupinate, hard and brittle fructifications, white to cream pore surface which darkens considerably on drying, moderately thick-walled, simple-septate hyphae which become compact and agglutinated forming pseudoparenchymatous tissue and subglobose basidiospores.

Rigidoporus lineatus (Pers.) Ryvarden, Norw. Jl. Bot. 19: 236 (1972)=*Rigidoporus zonalis* (Berk.) Imaz., Bull. Govt. For. Exp. Stn. 57:119 (1952)=*Polyporus zonalis* Berk., Ann. Nat. Hist. 10:375(1842). Fig. 6.49a–e

Fructification usually annual, sometimes reviving in the second season, sessile or effused-reflexed, solitary or imbricate, with a broad or narrow base, dimidiate or flabelliform, leathery when fresh, rigid on drying, 1.2–7.5×1.3–5.3×0.3–0.6 μm; upper surface pinkish buff, concentrically zonate, with different shades of brown, finally tomentose to glabrous, showing radiating wrinkles on drying; margin acute, strongly incurved when dry. Context white or pale buff, fibrous; hymenial surface white to pinkish, with a silky lusture, pore circular to somewhat angular, 8–9 per mm, pore tubes up to 3 mm long.

Hyphal system dimitic; generative hyphae hyaline, thin-walled to slightly thick-walled, simple septate, branched, 3.0–4.5 μm wide; skeletal hyphae hyaline, unbranched, slightly thick-walled, showing distant septa or highly thick-walled showing occluded lumina, with apex round, twisted or slightly swollen, 4.5–7.5 μm wide. Pseudoparenchymatous cells hyaline, thick-walled and transparent, occurring in the lower part of the context. Acanthophyses hyaline, thin-walled, 5.6–8.5 μm wide, found at the growing apex of dissepiments. Basidia hyaline, clavate or subclavate, 10.0–14.3×5.5–7.0 μm, 4-sterigmate. Basidiospores hyaline, thin-walled, globose or subglobose, apiculate, 3.2–4.5 μm in diameter. Cystidia hyaline, formed by heavy encrustation at the apical portion of subsolid to solid skeletal hyphae, 5.0–9.8 μm.

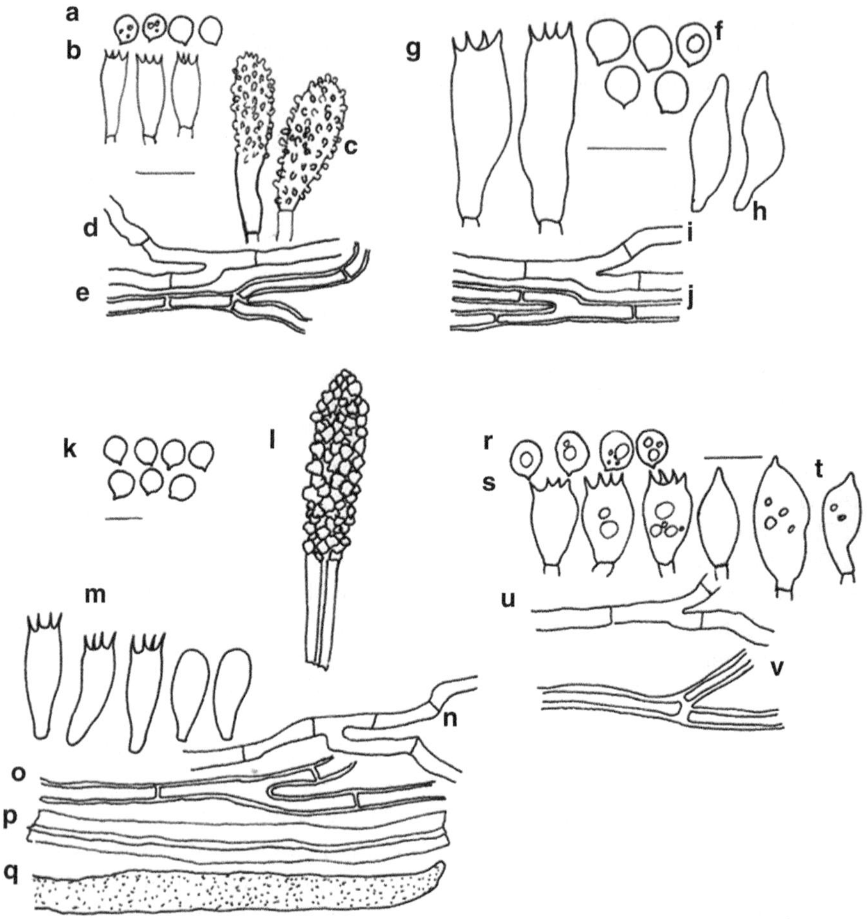

Fig. 6.49 (**a–e**) *Rigidoporus lineatus* (**a**) Basidiospores, (**b**) Basidia, (**c**) Mucronate cystidia, (**d**, **e**) Generative hyphae, (**f–j**) *Rigidoporus microporus* (**f**) Basidiospores, (**g**) Basidia, (**h**) Mucronate cystidia, (**i**) Thin-walled generative hyphae, (**j**) Thick-walled generative hyphae; (**k–q**) *Rigidoporus vinctus* (**k**) Basidiospores, (**l**) Encrusted cystidia, (**m**) Basidia, (**n**) Thin-walled generative hyphae, (**o**) Thick-walled generative hyphae, (**p**) Skeletal hyphae, (**q**) Gloeopleuros hyphae; (**r–v**) *Rigidoporus ulmarius* (**r**) Basidiospores, (**s**) Basidia, (**t**) Cystidia, (**u**) Thin-walled generative hyphae, (**v**) Thick-walled generative hyphae

Distribution: U.K.: Chakrata, Nainital; W.B.: Darjeeling.
Collection examined: L 42301.
Substratum: On decaying *Cedrus deodara* log.
Remarks: The species is close to *Oxyporus lignosus* from which it differs in having pinkish upper surface, white pore tubes, encrusted cystidia and acanthophyses. In *O. lignosus*, upper surface of fructification is reddish brown or yellowish brown, pore tubes reddish brown, and cystidia mucronate.

Rigidoporus microporus (Sw.) Overeem, Icon. Fung. Malay.5: 1 (1924) = *Oxyporus lignosus* (Kl.) Roy & De, Mycotaxon 67: 406 (1998). Fig. 6.49f–j

Fructification annual, rarely perennial, attached by broad base, rarely resupinate to effused-reflexed, applanate to dimidiate, usually imbricate, leathery when fresh, hard and rigid on drying, up to 2 cm thick; margin fade yellow, thin or rounded, usually strongly incurved; upper surface reddish brown to yellowish brown, tomentose when young concentrically zonate and with a thin crust, becoming glabrous on drying. Context cream to wood coloured, corky, up to 1 cm thick; hymenial surface reddish brown; pores circular to angular, 6–9 per mm, pores tubes up to 6 mm long.

Hyphal system monomitic; generative hyphae simple septate, sparsely branched, septa close or distant, hyaline to yellowish, thin to slightly thick-walled, 2.5–4.8 μm wide. Cystidia hyaline, thin-walled, mucronate and variable in shape, bulbous at the base, 10.8–14.2 × 3.8–6.2 μm, hymenial or subhymenial in origin. Basidia clavate, 4-sterigmate, 9–12.5 × 7–8.5 μm. Basidiospores hyaline, thin-walled, smooth, globose, 3.6–4.6 × 3.5–4.3 μm.

Distribution: U.K.: Dehra Dun; W.B.: Darjeeling.

Collection examined: L 42304.

Substratum: On log of *Shorea robusta & Ficus* sp.

Remarks: The species is characterized by annual to perennial, resupinate to effused-reflexed, dimidiate fructification; monomitic hyphal system; globose basidiospores. The species is close to *Rigidoporus zonalis.*

Rigidoporus ulmarius (Sowerby) Imazeki, Bull. Gov. Forest Exp. St. Tokyo 57: 119 (1952) = *Fomes ulmarius* Fr., Hymenomyc. eur. (Upsaliae): 683 (1874). Plate 6.20b, Fig. 6.49r–v

Fructification perennial, pileate, broadly-attached, effused reflexed to applanate, consistency tough and corky when fresh, woody hard when dry. Pileus 11–16 × 5–11 × 5–6 μm, appalante, whitish to ochraceous or dirty wood-coloured in old specimens, first tomentose but soon glabrous, dull, azonate or with concentric, broad sulcate zones, margin usually thick and rounded. Pore surface orange to pinkish brown when fresh, clay pink to buff when dry, pores round to angular, 5–8 per mm, tubes tough to resinous to consistency when fresh, very hard when dry, tubes mostly distinctly stratified with light, ochraceous context layers, which, however, may be lacking. Context cream to wood-coloured, often with zones reflecting the different growth stages.

Hyphal system monomitic; generative hyphae with simple septa, in the subhymenium hyaline, thin-walled and collapsed in dry specimens, 2–4.1 μm wide, in the trama thick-walled with distinct lumen 3–5 μm wide, in the context even more thick-walled and up to 8 μm wide. Cystidioles scattered in the hymenium, ventricose, subulate, slightly projecting above the basidia, thin-walled, hyaline in the parts collapsed in dry specimens, 15–25 × 7–9 μm. Basidia 5–7.5 μm broad, clavate, 4-spored. Basidiospores globose, hyaline, thin to slightly thick-walled, hyaline to pale yellowish, when mature 7–9.3 × 6.5–8.4 μm in diameter.

Distribution: U.K.: Mussoorie.

Plate 6.20 (**a**) *Rigidoporus vinctus*. (**b**) *Rigidoporus ulmarius*. (**c**) *Bjerkandera adusta*. (**d**) *Crustoderma dryinum*. (**e**) *Flavodon flavus*

Collection examined: IBP 42309.

Substratum: On trunks of living tree of *Rhododendron* sp.

Remarks: The species is characterized by white to cream coloured fructification, monomitic hyphal system, absence of clamps, presence of thin-walled cystidia and globose to subglobose basidiospores.

Rigidoporus vinctus (Berk.) Ryvarden, Norw. Jl. Bot.19 (2): 143 (1973) = *Poria vincta* (Berk.) Cooke, Grevillea 14: 110. 1886. = *Polyporus vinctus* Berk., Ann. Mag. Nat. Hist. 9: 196. 1852. Plate 6.20a, Fig. 6.49k–q

Fructification annual to perennial, resupinate, widely effused, adnate; subiculum pale brown, floccose, up to 0.5 mm thick, membranous but becomes hard and brittle on drying; margin white to cream; hymenial surface pale ochraceous buff to light pinkish ochraceous, pores circular to angular to oblique, 8–12 per mm, pore wall thin, pore layer stratified, up to 1 mm thick in each layer.

Hyphal system dimitic, generative hyphae hyaline, simple septate, branched, thin-to slightly thick-walled, 1.5–4.0 µm wide. Skeletal hyphae aseptate, unbranched, hyaline to subhyaline, thick-walled to solid, 2.6–6.5 µm wide. Gloeopleurous hyphae simple septate, unbranched, hyaline, thin-walled and with dens contents taking deep stain in cotton blue, 3.6–6.0 µm wide. Thin- to thick-walled pseudoparenchymatous cells present in the subiculum. Cystidia of two kinds-(i) hyaline, thin-walled, variable in shape, with tapering ends, unincrusted, 13–16 × 7–10 µm, hymenial in origin; (ii) hyaline, cylindrical, thick-walled, heavily encrusted in the apical parts, subhymenial in origin and penetrating into the pore mouths, 20–80 × 5–19 µm. Basidia short clavate, 4-sterigmate, 8.0–12.5 × 4.0–6.0 µm. Basidiospores hyaline, thin-walled, subglobose to globose, smooth, 3.0–5.2 × 2.6–4.0 µm.

Distribution: Meghalaya; U.K.: Dehra Dun.

Collection examined: SSV 21051, 21059; IBP 37153.

Substratum: On decaying angiospermic log.

Remarks: The species is characterized by annual, effused, adnate fructification; light pinkish brown pore surface; monomitic hyphal system with simple septate generative hyphae; apically encrusted cystidia; and hyaline, ovoid to subglobose basidiospores

Family-Meruliaceae

Key to genera

1. Fructification resupinate.. 2
1. Fructification pileate 12
2. Fructification resupinate effused.. *Hypochnicium*
2. Fructification ceraceous to membranous 3
3. Hymenium smooth/loosely interwoven hyphae.. 4
3. Hymenium smooth to tuberculate.. 5
4. Hymenium surface smooth .... *Crustoderma*
4. Hymenium surface of loosely interwoven hyphae.. *Conohypha*
5. Hyphal system monomitic 6
5. Hyphal system dimitic 9
6. Basidia utriform, medium to large.. *Hyphoderma*
6. Basidia clavate to subclavate.. 7
7. Basidiospore ellipsoid to suballantoid 8
7. Basidiospore ellipsoid to subglobose.. *Sarcodontia*
8. Basidiospores subcylindrical .... *Crustoderma*

8. Basidiospores ellipsoid to suballantoid..*Phlebia*
9. Fructification resupinate, effused, reflexed... 10
9. Fructification resupinate, membranous to coriaceous....................................... 11
10. Pores angular to irregular, 7–10 per mm...*Flaviporus*
10. Pores round to irregular, 2–3 per mm ....*Irpex*
11. Pores angular to oval, 3–7 per mm .... *Junghunia*
11. Hymenial surface toothed ....*Radulodon*
12. Upper surface creamish brown, pubescent ....*Abortiporus*
12. Upper surface of variously coloured... 13
13. Context duplex ....*Bjerkendera*
13. Context subhyaline.. 14
14. Hymenial surface cream to light brown, smooth to hydnoid......... *Cabalodontia*
14. Hymenial surface first poroid then irpicoid 15
15. Cystidia present, encrusted ....*Flavodon*
15. Cystidia absent.. 16
16. Pores small, round to angular.. *Gloeoporus*
16. Pores not small... 17
17. Cystidia clamped...*Gyrophanopsis*
17. Cystidia without clamps.. 18
18. Teeth subulate or subcylindrical with acute apices............................ *Mycoacia*
18. Teeth absent 19
19. Cystidia conical, encrusted at upper part with obtuse apex ....*Scopuloides*
19. Cystidia arising as prolongations of skeletal hyphae or embedded in trama or project out of the hymenium or teeth apices .... *Steccherinum*

Abortiporus Murrill,

Bull. Torrey. bot. club 31(8): 421, 1904.

Fructification annual, sessile to subsessile or short stipiate. Pileus dimidiate or spathulate, solitary or imbricate; upper surface creamish-brown, pubescent, villose or glabrous without context azonate. Pore surface uneven, creamish-brown to brown; pores irregular, mostly radially elongated. Tubes not stratified. Context creamish, duplex; upper layer soft and spongy, lower layer hard and coriaceous. Hyphal system monomitic; generative hyphae hyaline, thin to thick-walled, clamped, septate. Cystidia present or absent. Basidia clavate, 2–4 spored. Basidiospores hyaline, ellipsoid, subglobose to globose, smooth.

Three species, widespread

Lit.: Ryvarden (*Syn. Fung.* 5: 104, 1991).

Type Species: *Boletus distortus* Schwein.1822.

Habitat: Wood

Himalayas: One

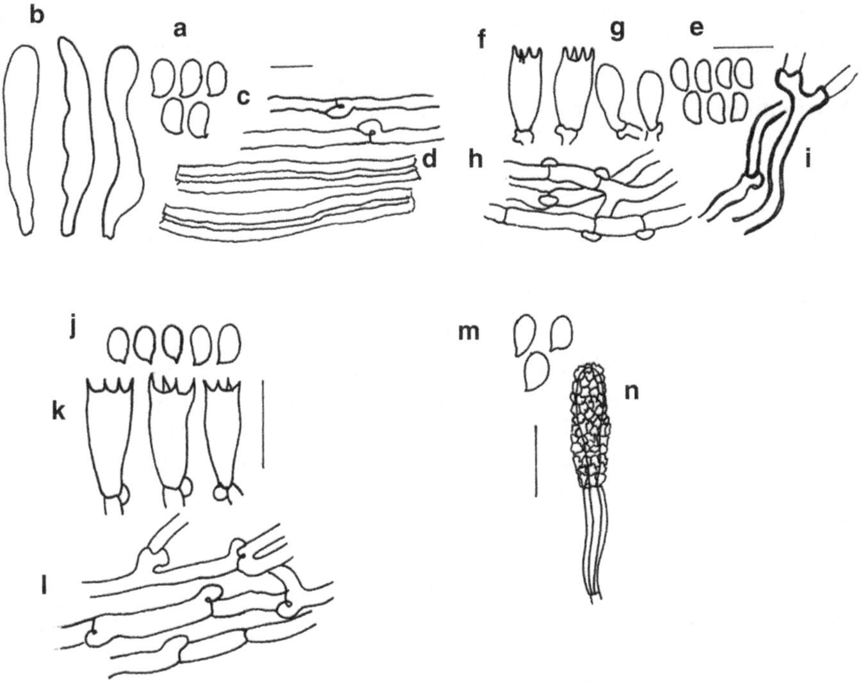

Fig. 6.50 (**a–d**) *Abortiporus biennis* (**a**) Basidiospores, (**b**) Basidia, (**c**) Generative hyphae, (**d**) Skeletal hyphae; (**e–i**) *Bjerkendra adusta* (**e**) Basidiospores, (**f**) Basidia, (**g**) Basidioles, (**h**) Generative hyphae, (**i**) Thick-walled generative hyphae; (**j–l**) *Bjerkendra fumosa* (**j**) Basidiospores, (**k**) Basidia, (**l**) Generative hyphae; (**m–n**) *Cabalodontia queletii* (**m**) Basidiospores, (**n**) Encrusted Cystidia

Abortiporus biennis (Bull.) Singer, Mycologia 36(1): 68 (1944). Fig. 6.50a–d

Fructification annual, substipitate, mostly solitary, rarely imbricate, coriaceous and leathery when fresh becoming tough and brittle on drying. Pileus up to 8 × 8 cm and up to 0.7 cm thick at the base, laterally stipitate, flabelliform, often narrowed at the base forming a small vertex; upper surface light ochraceous to pale reddish-brown when fresh, fading to cream to cream-ochre on drying, smooth to finely tomentose, devoid of cuticle; margin becoming thin, acute, even and concolorous; pore surface concolorous with the upper surface, irpiciform. Context white, drying pale ochraceous, duplex: upper part soft and loosely woven while lower part is hard and compact.

Hyphal system monomitic, hyphae 2.7–5.7 μm wide, hyaline, thin to thick-walled, some with capillary lumen, branched, septate, clamped, clamps prominent. Pores decurrent, often torn and irregular, irpiciform, sometimes elongating radially in a lamellar pattern; dissepiments 45–120 μm upper surface, irpiciform.

Context white, pale on drying, concolorous with the context. Basidia 15–20.5×5–6.5 µm, irpiciform, clavate, 4-spored. Basidiospores 4.5–5.5× 3.5–4 µm, ovoid, hyaline, smooth, shortly apiculate.

Distribution: H.P.: Dalhousie; U.K.: Dehradun.

Collection examined: IBP 37158, 37159.

Substratum: On stump of *Pinus roxbughii.*

Remarks: The characteristic features of this species are substipitate fructification, absence of cuticle, monomitic hyphal system and irpiciform to lamellate pores.

Bjerkandera Karst.,
Medd. Soc. Fauna Fl. fenn. 5:38, 1879.

Fructification annual, sessile, effused-reflexed to pileate, soft coriaceous when fresh, brittle on drying, upper surface creamish to grey, tomentose, azonate. Context creamish, duplex. Hyphal system monomitic; generative hyphae hyaline, thin to thick-walled, septate, clamped. Cystidia absent. Basidia hyaline, clavate, 2–4 spored. Basidiospores hyaline, smooth, thin-walled, ellipsoid, non-amyloid.

Two Species, widespread

Lit.: Ryvarden & Gilbertson *(Europ. Polyp.* **1**: 168, 1993)

Type Species: *Boletus adustus* Willd. 1787

Habitat: Dead Wood

Himalayas: Two

Key to species:

1. Upper surface pale, tomentose & azonate,
   finely tomentose when young .... *B. adusta*
1. Upper surface creamish grey to greyish brown,
   tomentose to glabrous, azonate to weakly zonate................................. *B. fumosa*

Bjerkandera adusta (Fr.) Karst., Medd. Soc. Fauna Fl. fenn. 5: 38, 1879.

Plate 6.20c, Fig. 6.50e–i

Fructification annual, lignicolous, sessile, occasionally effused, thin, imbricate; upper surface pale, tomentose & azonate, finely tomentose when young; margin white to cream reddish grey to blackish brown when touched and on drying, thin, greyish, inturned on drying; Pileus 6×3 cm and 7 mm thick; pore surface grey to greyish black when fresh, darken on drying; pores 6–8 per mm, round to angular to irregular, 4–6 per mm; dissepiments thin, even. Context pale or creamish, demarcated by a thin dark line from the tubes, non-xanthochroic, upper part is white; lower layer is blackish grey and joining the tubes.

Hyphal system monomitic; generative hyphae hyaline to subhyaline, thin to thick-walled, septate, clamped, branched, non-amyloid, acyanophilous, 3–6.5 µm in diameter. Cystidia none. Basidia hyaline, clavate, 2–4 sterigmate, 12–15.5×5–9.5 µm. Basidiospores hyaline, ellipsoid, non-amyloid, acyanophilous, 5–8.2×2–3.2 µm.

Distribution: Bhutan: Paro; Nepal: Kathmandu; U.K.: Chakrata-Dehra Dun; H.P.: Narkanda; A.P: West Kameng; Manipur: Imphal; Meghalaya: Shillong; Mizoram: Aizwal; Nagaland: Kohima; Tripura: Agartala.

Collection examined: SSV 21362, 21398, IBP 37160, L 37161.
Substratum: On under surface of decaying angiospermic log.
Remarks: The species is characterized by annual, sessile, effused, imbricate fructification; cream upper surface; grey pore surface; context separated by a dark brown line from the tubes; monomitic hyphal system and hyaline, ellipsoid basidiospores.

Bjerkandera fumosa (Pers.: Fr.) Karst., Medd. Soc. Fauna Fl. fenn. 5: 38, 1879. Fig. 6.50j–l

Fructification annual, sessile, effused-reflexed to pileate, imbricate with fused pilei, soft coriaceous when fresh, brittle on drying, initially tomentose, later glabrous, azonate, smooth. Upper surface creamish grey to greyish brown, tomentose to glabrous, azonate to weakly zonate; margin thin, undulating, concolorous with upper surface, entire, sterile below. Pore surface brown to greyish brown, darken on drying, even to uneven, dull. Context white or cream, more or less fibrous when fresh, corky on drying, separated from the white context by a dark brown line; pores circular to angular, becoming irpiciform at places, 2–4 per mm.

Hyphal system monomitic; generative hyphae with abundant clamp connections, hyaline thin- to thick-walled to solid, branched, septate, often showing broken clamp at one end, acyanophilous, non-amyloid, 3–6.5 µm wide, loosely arranged in the context and much agglutinated in the trama due to formation of many short side branches. Cystidia absent. Basidia subclavate, hyaline, 4-sterigmate, up to 4 µm. Basidiospores short cylindric, hyaline, smooth, 4.2–6.7 × 2.0–3.8 µm.

Distribution: Bhutan: Paro, U.K.: Dehra Dun.
Collection examined: SSV 21376, L 42932.
Substratum: On decaying angiospermic log.
Remarks: The species is characterized by having thinner pilei; brownish grey pore surface; smaller, 4–6 pores mm; and slightly bigger 4–6 × 2.2–2.8 µm basidiospores. It is a common species in N.W. Himalayas.

Cabalodontia M. Piątek,
Polish. Bot. J. 49(1):2 2004

Fructifications resupinate, annual to perennial, ceraceous, adnate, widely effused; hymenial surface cream to light brown, smooth to hydnoid. Subiculum subhyaline in section, composed of compactly arranged, more or less agglutinated and ceraceous hyphae. Hyphal system monomitic, hyphae branched, septate, clamps septate, tortuous, collapsed and difficult to discern individually, the walls thin, hyaline. Cystidia absent. Basidia clavate with tapering base, 4-spored. Basidiospores suballantoid, thin-walled, subhyaline, smooth, non-amyloid.

Five species, widespread
Lit.: Piątek (*Polish Botanical Journal* **49:** 2, 2004)
Type Species: *Cabalodontia queletii* (Bourdot & Galzin) M. Piatek 2004
Habitat: Wood
Himalayas: Two

Key to species

1. Cystidia absent; basidia clavate to clavate-cylindrical;
   basidiospores ovoid, 4–5.2 × 3–3.5 μm ....*C. queletii*
1. Cystidia present; basidia clavate with tapering base;
   basidiospores suballantoid, 5–5.8 × 1.5–2 μm............................... *C. subcretacea*

Cabalodontia queletii (Bourdot & Galzin) Piątek, Polish Botanical Journal 49(1): 3 (2004) = *Metulodontia queletii* (Bourd. & Galz.) Parm., Consp. Syst. Cort.: 118. 1968 = *Odontia queletii* Bourd. & Galz., Bull. Soc. mycol. France 30(3): 270. 1912.

Fig. 6.50m–n

Fructifications resupinate, membranous- ceraceous to subceraceous becoming hard and somewhat brittle on drying, adnate, often arising in small colonies which may coalesce latter and become widely effused, up to 1 mm thick in section; hymenial surface cream to pale ochraceous, smooth to finely and densely tuberculate or distinctly toothed, continuous, not creviced; margin thinning, fibrillose, adnate, white to paler concolorous, subulate to cylindrical, rarely flattened, apices acute. Subiculum composed of compactly arranged and somewhat agglutinated hyphae.

Hyphal system monomitic, hyphae (2)2.8–5 μm wide, branched, septate, clamped, the walls thin to slightly thick, subhyaline. Cystidia (lamprocystidia) 40–100 × 6–14 μm. fusiform to subcylindrical, often arranged imbricately in the hymenium, the walls subhyaline, thick (up to 2.5 μm), heavily impregnated with subhyaline crystals specially in the upper 2/3 part. Basidia 14.3–18.3 × 4–5.2 μm, clavate to clavate- cylindrical, 4-spored. Basidiospores 4–5.2 × 3–3.5 μm, ovoid, minutely, apiculate, the walls thin, subhyaline, smooth, non-amyloid, acyanophilous.

Distribution: H.P.: Kufri, Shimla, U.K.: Nainital.

Collection examined: HSK 4197, 4228, IBP 37162.

Substratum: On angiospermic twig, on log of *Quercus incana*.

Remarks: The characteristic features of this species are subceraceous texture, odontoid, hymenial surface, presence or lamprocystidia and small ovoid basidiospores. The lamprocystidia are usually arranged imbricately in the axis of spine.

Cabalodontia subcretacea (Litsch.) Piątek, Polish Botanical Journal 49(1): 3 (2004) = *Phlebia subcretacea* (Litsch.) Christ. Dansk bot. Arkv. 19: 165, 1960= *Corticium subcretaceum* Litsch., Ostr. bot. Zeitschr. 88: 110. 1939. Fig. 6.51a–c

Fructifications resupinate, ceraceous, adnate, widely effused; hymenial surface cream to cream yellow, smooth to somewhat rough finely tuberculate, continuous, not creviced; margin thinning to somewhat cottony, loosely adnate, white. Context composed of compactly arranged, more or less agglutinated and ceraceous hyphae, demarcation in to basal zone and upper zone not clear.

Hyphal system monomitic; hyphae 1–2 μm wide, branched, septate, clamps septate, often unevenly inflated up to 3 μm wide, tortuous, collapsed and difficult to

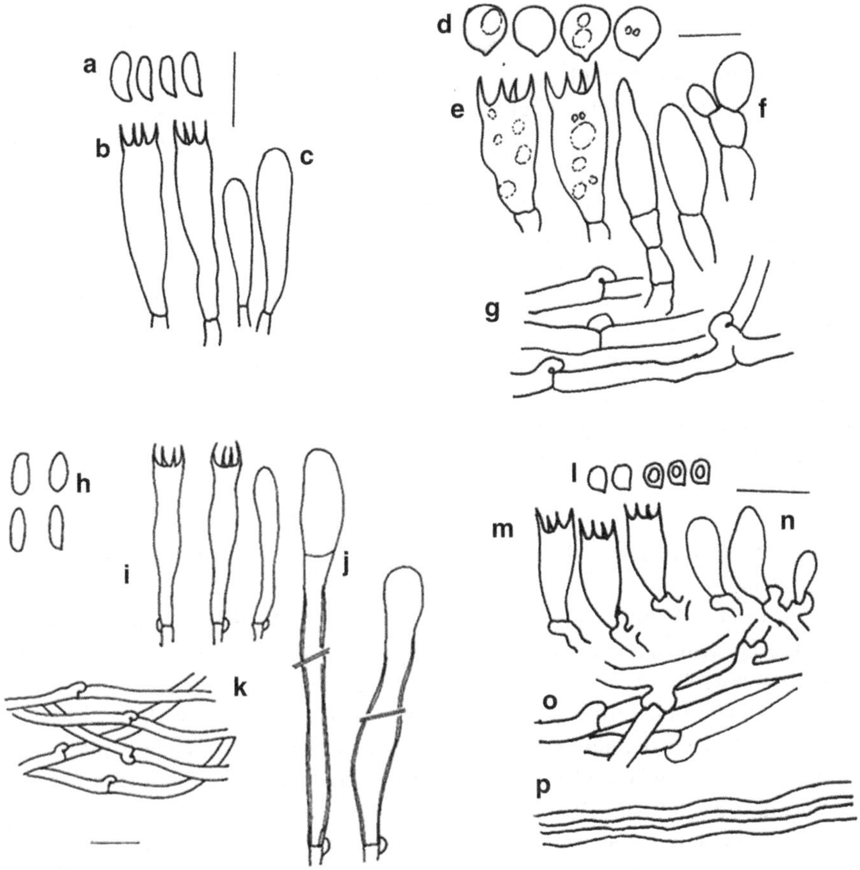

Fig. 6.51 (**a**–**c**) *Cabalodontia subcretacea* (**a**) Basidiospores, (**b**) Basidia, (**c**) Basidioles; (**d**–**g**) *Conohypha grandispora* (**d**) Basidiospores, (**e**) Basidia, (**f**) Basidioles, (**g**) Generative hyphae; (**h**–**k**) *Crustoderma dryinum* (**h**) Basidiospores, (**i**) Basidia, (**j**) Cystidia, (**k**) Generative hyphae; (**l**–**p**) *Flaviporus hydrophilus* (**l**) Basidiospores, (**m**) Basidia, (**n**) Basidioles, (**o**) Generative hyphae, (**p**) Skeletal hyphae

discern individually, the walls thin, hyaline. Cystidia absent. Basidia 20–25.3 × 4–5.6 μm, clavate with tapering base, 4-spored. Basidiospores 5.5–8 × 1.5–2.0 μm, suballantoid, minutely apiculate, the walls thin, subhyaline, smooth, non-amyloid.

Distribution: H.P.: Mahasu

Collection examined: SSR 5868.

Substratum: On stump under conifers.

Remarks: This species is characterized by a smooth, cream to cream- yellow hymenial surface, and comparatively longer and suballantoid basidiospores. It differs from *Phlebia livida* in having smaller basidiospores and darker hymenial surface.

Conohypha Jülich,
Persoonia 8: 303, 1975.

Fructification resupinate, effused, hyphochnoid or membranous; hymenial surface, very thin, greyish white to light grey or yellowish- white, of loosely woven texture, forming a greyish bloom on the substrate; margins not differentiated. Hyphal system monomitic; generative hyphae thin-walled, clamped, generally branching from the top of the cells. Basidia subcylindrical, often constricted, thin-walled, with basal clamp, 4-sterigmate, with oily contents. Basidiospores broadly, ellipsoid to ovoid, smooth thin-walled to somewhat thick-walled, inamyloid, acyanophilous, with a single large oil drop or many smaller oil drops.

Two species, world-wide

Lit.: Jülich (*Persoonia* **8**: 303, 1975), Wojewoda (*Acta Mycologica Warszawa* **38**: 3, 2003; Poland).

Type species: *Corticium albocremeum* Höhn & Litsch, 1908

Habitat: Wood

Himalayas: One

Conohypha grandispora Dhingra, In Plant Diversity in India: 478, 2004.
Fig. 6.51d–g

Fructification very thin, hypochnoid, greyish white to light grey or yellowish- white, of loosely woven texture, forming a greyish bloom on the substrate; margins not differentiated.

Hyphal system monomitic; generative hyphae thin-walled, clamped; basal hyphae up to 5.7 μm wide, almost parallel to the substrate; subhymenial hyphae composed of short and broadened cells, generally branching from the top of the cells. Basidia 15–26.8 × 8.2–9.3 μm, subcylindrical, often constricted, thin-walled, with basal clamp, 4-sterigmate, with oily contents; sterigmata up to 11 μm long, Basidiospores 7.2 × 11.2 × 6.0–7.5 μm, broadly, ellipsoid to ovoid, smooth thin-walled to somewhat thick-walled, inamyloid, acyanophilous, with a single large oil drop or many smaller oil drops.

Distribution: West Bengal: Darjeeling.

Collection examined: GSD 19286, 19323.

Substratum: On a burnt, decaying stump.

Remarks: The species is characterized by having membranous fructification, bigger, broadly ellipsoid to ovoid thick-walled basidiospores having very long sterigmata.

Crustoderma Parm.,
Consp. Syst. Cort.: 87, 1968.

Fructification resupinate, ceraceous to crustaceous, effused, closely attached to the substrate, margin mostly abrupt, sometimes finely fibrillose; hymenium mostly smooth, first yellowish, then ochraceous to cinnamon coloured. Hyphal system monomitic; hyphae yellowish, then ochraceous to cinnamon coloured, densely agglutinated, somewhat thick-walled, with clamps and richly branched. Cystidia cylindrical, obtuse, basally with thickened walls, sometimes with one or several adventitious septa. Basidia narrowly clavate to almost cylindrical, at first thin walled but in age somewhat thickened with 4 thin sterigmata. Spores narrowly ellipsoid to subcylindrical, nonamyloid, strongly stained by cotton blue.

Fourteen Species, world-wide

Lit.: Nakasone (*Mycol.* **76**: 40, 1984), Gilbertson & Nakasone (*Mycol.* **95**: 467, 2003; key).

Type species: *Corticium dryinum* Berk. & M.A. Curtis, 1873

Habitat: Wood

Himalayas: One

Crustoderma dryinum (Berk. & Curt.) Parm., Consp. Syst. Cort.: 88(1968) = *Corticium dryinum* Berk. & Curt., Grevillea 1 p.179, 1873. Plate 6.20d, Fig. 6.51h–k

Fructification resupinate, adnate, effused, ceraceous when alive, membranous to crustaceous on drying; hymenial surface smooth, velvety microscopically due to the projecting cystidia, light yellow to pale ochraceous; margin abrupt or undifferentiated.

Hyphal system monomitic; generative hyphae branched, septate, clamped 3–5 µm wide; basal hyphae somewhat thick-walled; subhymenial hyphae thin-walled, mainly vertical, compactly packed to agglutinated. Cystidia cylindrical to subcylindrical, thick-walled with basal clamp, often secondarily septate, 72–162 × 9–13 µm. Basidia 24.2–34.3 × 5–5.5 µm, clavate to subcylindrical, 4-sterigmate, with a basal clamp. Basidiospores 7–8.7 × 3–3.7 µm, broadly ellipsoid to subglobose, thin to thick-walled, non-amyloid, cyanophilous.

Distribution: A. P.: West Kameng, Bomdila, Wang Basti; U. K.: Mussoorie, Nanital.

Collection examined: GSD 19700, IBP 37164.

Substratum: On a burnt stump, cut stump and logs.

Remarks: The species is marked by its yellow coloured fructification, clamped hyphae, cylindrical cystidia, clavate to subcylindrical, 4-spored basidia and thin to slightly thick-walled, cyanophilous basidiospores. This is a new record for N. W. Himalayas.

Flaviporus Murrill,

Bull. Torrey bot. Club 32(7): 360(1905)

Fructification annual, effused-reflexed to resupinate; pileus dark brown when fresh, glabrous; pore surface vivid chrome yellow; pores tiny invisible to naked eye; tubes concolorous. Context thin and yellow. Hyphal system dimitic; generative hyphae clamped, septate, branched; skeletal hyphae abundant, unbranched. Cystidia present. Basidia 4-sterigmate. Basidiospores ellipsoid, smooth, negative in Melzer's reagent, shorter than 5 µm.

Thirteen species, widespread

Lit.: Ginns (*Can. J. Bot.* 58: 1578, 1980)

Type Species: *Polyporus rufoflavus* Berk. & M.A. Curtis 1869.

Habitat: Wood

Himalayas: One

Flaviporus hydrophilus (Berk. & M.A. Curtis) Ginns, Can. J. Bot. 58(14): 1583(1980) = *Tyromyces hydrophilous* (Berk. & Curt.) Lowe, Mycotaxon 2: 46, 1975. Fig. 6.51l–p

Fructification annual, pileate, soft, fleshy, succulent, solitary to imbricate, sessile. Pileus applanate, dimidiate to spathulate with a distinctly tapering base, 7.5 × 0.5 × 13 cm; upper surface white to light orange initially changes to brown the orange to reddish brown; margin thin, acute, fleshy, irregular, incurved on drying. Pore surface wide when fresh, on drying ochraceous with pinkish tinge; pores small, angular to irregular, 7–10 per mm, 65–105 μm in diameter, difficult to observe with naked eye, on drying, pores contract and cracked to form split pores; pore mouth smooth; pore tubes strongly agglutinate after drying; dissepiment entire, whitish. Context succulent, homogenous, white when fresh, white orange to slightly darker with age, acyanophilous, non-xanthochroic, up to 5 mm thick.

Hyphal system monomitic; generative hyphae hyaline, thin to thick-walled to almost solid with narrow lumen, branched, septate, clamped, faintly cyanophilous, 2.2–6 μm in diameter. Hyphae parallel, 2.2–4.5 μm in diameter in the dissepiments. The context consists of thick-walled to almost solid skelerified generative hyphae with narrow lumen, very rarely clamped, straight, rarely branched, 3–6 μm in diameter. Cystidia absent. Basidia clavate, 4-spored, thin-walled, cyanophilous, up to 4.5 μm in diameter. Basidiospores hyaline, thin-walled, smooth, broadly ellipsoids, uniguttulate, acyanophilous, non-amyloid, 2.8–3.2 × 2–2.8 μm.

Distribution: Bhutan: Thimphu.

Collection examined: SSV 21255.

Substratum: On decaying angiospermic log.

Remarks: The species is characterized by annual, fleshy, succulent imbricate, dimidiate to spathulate fructification which are white to light orange on the upper surface; white with tinted pore surface, thin-walled to solid skelerified generative hyphae. It is a new record for Himalayas.

Flavodon Ryvarden, Norw. J. of Bot., 20(1): 3, 1973.

Fructification annual, resupinate to pileate; pileus tomentose, yellowish grey. Context bright, sulphur yellow; hymenophore first poroid, later irpicoid, sulphur yellow. Hyphal system dimitic; generative hyphae simple septate, cystidia encrusted, formed of terminal ends of skeletal hyphae. Basidiospores broadly ellipsoid, smooth, hyaline, acanthophyses present.

Two species, wide spread

Type Species: *Irpex flavus* Kl. 1833

Himalayas: One

Flavodon flavus (Klotzsch) Ryvarden, Norw. Jl Bot. 20(1): 3 (1973) = *Irpex flavus* (Kl.) Ryv., Norw. J. Bot. 20: 3. 1973. Plate 6.20e, Fig. 6.52a–f

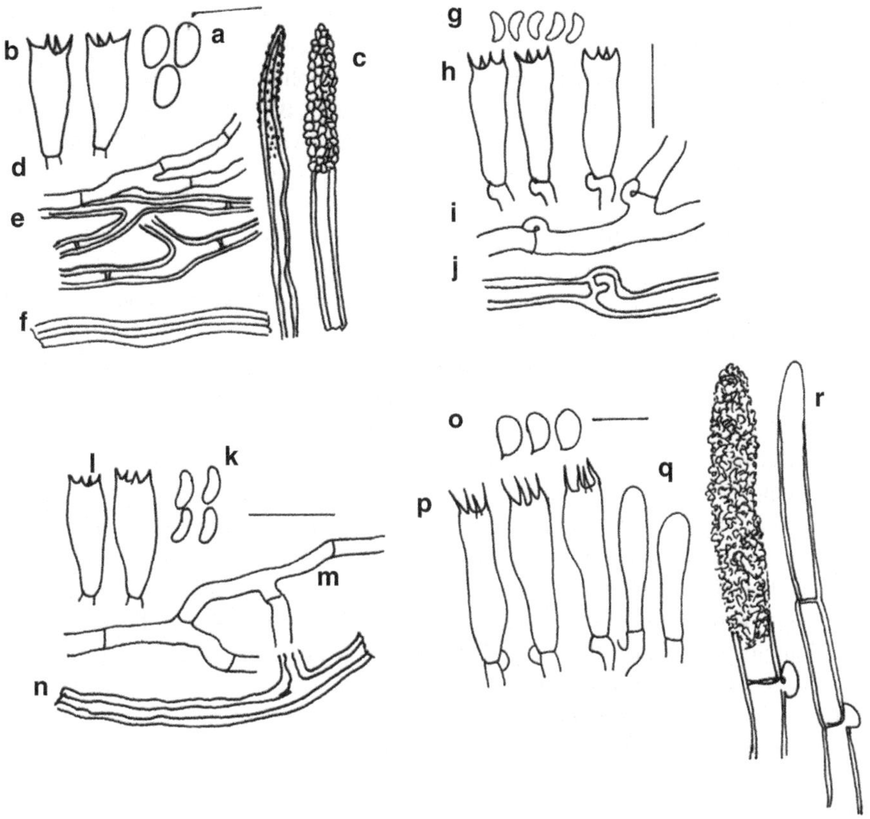

Fig. 6.52 (**a–f**) *Flavodon flavus* (**a**) Basidiospores, (**b**) Basidia, (**c**) Encrusted cystidia, (**d**) Thin-walled generative hyphae, (**e**) Thick-walled generative hyphae, (**f**) Skeletal hyphae; (**g–j**) *Gloeoporus dichorus* (**g**) Basidiospores, (**h**) Basidia, (**i**) Thin-walled generative hyphae (**j**) Thick-walled generative hyphae; (**k–n**) *Gloeoporus thelephoroides* (**k**) Basidiospores, (**l**) Basidia, (**m**) Generative hyphae, (**n**) Skeletal hyphae; (**o–r**) *Gyrophanopsis polonensis* (**o**) Basidiospores, (**p**) Basidia, (**q**) Basidioles, (**r**) Encrusted cystidia

Fructification annual, effused-reflexed or resupinate, imbricate, reflexed, portion dimidiate to applanate, corky to leathery; upper surface pinkish buff to light yellowish brown, sometimes becoming black at the base, matted tomentose, concentrically zonate, soft to touch; pileus tomentose, unzoned or in narrow concentric zones. Context yellowish brown, fibrous, up to 2 mm thick, hymenial surface sulphur yellow when fresh, cream coloured to yellowish brown with age, spiny to irpiciform, poriod near the margin, pores 1–2 per mm, pore tubes up to 4 mm long.

Hyphal system dimitic; generative hyphae hyaline, simple septate, 2–5 μm, branched; skeletal hyphae hyaline, up to 7 μm thick-walled, dominating in the context. Cystidia cylindrical to subclavate, capitate incrusted, thick-walled, 3.5–

5.5 μm at the apex; hyphae of two types: (i) pale yellow to sulphur yellow with lumen usually obliterate or narrow, unbranched, aseptate or rarely simple septate, 3–6 μm diameter and (ii) nearly hyaline to sulphur yellow, thin to slightly thick-walled, branched, septate, 2–4 μm diameter. Basidia clavate, 4.5–5.5 μm broad. Basidiospores hyaline, short ellipsoid, thin-walled, 4–5 × 1.8–2.8 μm.

Distribution: U.K.: Saharanpur; Nainital, Mussoorie, Pauri; West Bengal: Shantiniketan.

Collection examined: IBP 37168, L 37170.

Substratum: On decaying angiospermic log, stumps.

Remarks: The species is characterized by having yellowish colour, hydnoid hymenophore; monomitic hyphal system.

Gloeoporus Mont.,
in Sagra, Historia fisica, polirica y natural
de la islea de Cuba 2: 385 (1842) [1838–1842]

Fructification annual, resupinate to pileate; upper surface, white to greyish and tomentose; pore surface pinkish to white orange to bay or reddish; pores small, round to angular with continuous layer of basidia over dissepiments tube layer gelatinous in fresh condition dense to cartilaginous when dry, darker and denser than the white and cottony, context. Hyphal system monomitic; generative hyphae with simple septate, clamped. Cystidia present or absent. Basidiospores allantoids to cylindrical, thin-walled, smooth, negative in Melzer's reagent.

Twenty six species, widespread.

Lit.: Coelho et al. (*Mycol.* **98**: 821, 2006; Brazil)

Type Species: *Gloeoporus conchoides* Mont.1842

Habitat: Dead wood

Himalayas: Two

1. Generative hyphae clamped in the context .... *G. dichorus*
1. Generative hyphae non clamped .... *G. thelephoroides*

Gloeoporus dichorus (Fr.) Bres., Hedwigia 53: 74 (1913) = *Polyporus dichorus* Fr., Observ. Mycol. (Haviniae) 1: 125 (1815). Fig. 6.52g–j

Fructification annual, resupinate to effused-reflexed, often imbricate, pileus white to yellowish and coriaceous when fresh, on drying cream to ochraceous and rigid, pileus surface at first finely tomentose, later smooth or hispid, concentrically zonate in different shades; margin acute. Context duplex, upper part white, corky, separated from the lower part of hymenium by narrow dark line; hymenial surface waxy, first reddish purple, becoming reddish brown when old, poroid, pores circular to angular, 5–7 per mm, pore tubes shallow, hymenial surface separable from pileus as a thin elastic layer.

Hyphal system monomitic; generative hyphae in the context hyaline, clamped, well branched, thin or thick-walled, 3–6 μm wide, in the resinous zone above the pore tubes strongly agglutinated; tramal hyphae thin-walled, 1–3 μm wide, loosely interwoven with in a gelatinous matrix. Basidia hyaline, 4-sterigmate,

12.2–15.3 × 2–4.3 µm. Basidiospores hyaline, thin-walled, smooth, allantoid, 3.5–5.2 × 0.7–1.5 µm.

Distribution: widespread throughout the Himalayas, H.P.: Shimla, Narkanda.

Collection examined: L 42312

Substratum: On decaying angiospermic log

Remarks: The characteristic features of the species is resupinate to effused-reflexed pileus with tomentose surface, waxy reddish hymenophore, tube layer being separated from white context by a dark line, monomitic hyphal system with clamped hyphae and allantoid basidiospores.

Gloeoporus thelephoroides (Hook.) G. Cunn., Bull. N. Z. Dept. Sci. Industr. Res., Pl. Dis. Div. 164: 111 (1965) = *Gloeoporus conchoids* Mont. in Sagra, Historia fisica, polirica y natural de la islea de Cuba 2: 385 (1842) [1838–1842]. Fig. 6.52k–n

Fructification annual, resupinate to pileate, often imbricate, soft when fresh, brittle when dry, pileus thin, at times laterally connate; surface finely tomentose when young, later smooth, azonate to slightly zonate, white to cream to yellowish or greyish cream; margin thin, acute, sharp and undulating. Context duplex, upper part white to cream, separated from the lower thin part by dark layer, the lower part of context gelatinous when fresh, and waxy when dry; pore surface with tint of flesh colour, pinkish brown when old, separable from the context as thin parchment like layer, pores small, circular round to subangular, 6–8 per mm, pore tubes up to 0.3 mm long.

Hyphal system monomitic; generative hyphae hyaline, simple septate, in the context thick-walled to almost solid, moderately branched, 3.8–5.8 µm wide, in the waxy zone above the pore tubes strong agglutinated, in the trama thin-walled, mostly collapsed and up to 3.5 µm wide. Basidia subclavate, 4-sterigmate, 10–13.2 × 4–5.2 µm. Basidiospores hyaline, thin- walled, smooth, 3–5–5.0 × 1.0–1.5 µm.

Distribution: W.B.: Jalpaiguri, Darjeeling.

Collection examined: GSR 2039.

Substratum: On decaying angiospermic log.

Remarks: *Gleoporus conchoides* has been described by Fidalgo & Fidalgo (1968) as having clamp bearing generative hyphae. This fungus is monomitic with simple septate generative hyphae, as also observed by Gilbertson and Ryvarden (1986).

Gyrophanopsis Jülich,
Persoonia 10(3): 329, 1979.

Fructification resupinate, effused, adnate, white, smooth, tuberculate or irregularly odontoid; margin varied. Hyphal system monomitic; generative hyphae thin to thick-walled, smooth. Cystidia present or absent. Basidia subutriform, 4-spored, basidia with basal clamp. Basidiospores ellipsoid to subglobose, smooth, echinulate, cyanophilous.

Two species, widespread

Lit.: Stalpers & Buchanan (N.Z. Jl Bot. 29: 333, 1991)

Type Species: *Pellicularia zealandica* G. Cunn. 1953.

Habitat: Wood
Himalayas: One

Gyrophanopsis polonensis (Bres.) Stalpers & P.K. Buchanan, N.Z. Jl Bot. 29(3): 333 (1991) = *Hypochnicium polonense* (Bres.) Strid, Wahlenb. 1: 68, 1975.
Plate 6.21a, Fig. 6.52o–r

Fructification resupinate, floccose to subpelliculose, adnate, effused, soft and floccose, pilose under lens by the projecting cystidia; hymenial surface more or less tuberculate, white to cream, discontinuous, not creviced, smooth on drying; margin thinning, loosely adnate, paler concolorous. Context subhyaline in section, composed of loosely arranged hyphae.

Hyphal system monomitic; generative hyphae branched, 3.7–6.2 μm wide, septate, clamped, thin to thick-walled. Cystidia septate, 6–11.2 μm broad, subhyaline, thin to thick-walled, naked or finely impregnated with subhyaline crystals especially in the upper part, arising from all levels of the context, clamped covered with grainy crust of crystals, with fine encrustation. Gloeocystidia absent. Basidia cylindrical, 25.2–30.3 × 5.2–6.2 μm, subutriform 4-sterigmate, with a basal clamp. Basidiospores broadly ellipsoid to subglobose, 7–8.4 × 4.5–5.2 μm, verruculose, thin-walled, minutely apiculate, non-amyloid, cyanophilous.

Distribution: H.P.: Dalhousie-Lakkar Mandi, Shimla-Narkanda, Mahasu, Bagi; Kullu.

Collection examined: SSR 5101, 5335, 5623.

Substratum: On decaying angiospermic twigs, on a decaying stump of *Cryptomeria japonica* and logs of *Abies pindrow.*

Remarks: The species is characterized by having floccose to subpelliculose fructification and presence of septate and clamped cystidia. It is a fairly common species in the Himalayas. The collection resembles the description given by Thind and Rattan (1970) and Eriksson and Ryvarden (1976).

Hyphoderma Wallr.,

Fl. crypt. Germ. (Norimbergae) 2: 576,1833.

Fructifications resupinate, floccose-membranous to membranous-ceraceous; hymenial surface smooth or rarely tuberculate, mostly white or cream or occasionally pale coloured, discontinuous or continuous. Hyphal system monomitic, hyphae subhyaline, thin-walled, clamped, often collapsing or rarely distinct. Cystidia or Gloecocystidia present or absent. Basidia subutriform to utriform, 4-spored, medium or large sized. Basidiospores broadly ellipsoid to subcylindric with a tendency to become sausage-shaped or curved, but may be larger in some species, the walls thin, subhyaline, smooth, nonamyloid, acyanophilous.

Thirty five species, world-wide

Lit.: Donk (*Fungus Wageningen* **27**: 13, 1957), Larsson (*Mycol. Res.* **111**: 186, 2007; phylogeny).

Type species: *Hyphoderma setigerum* (Fr.) Donk
Habitat: Dead wood
Himalayas: Ten (Table 6.3)

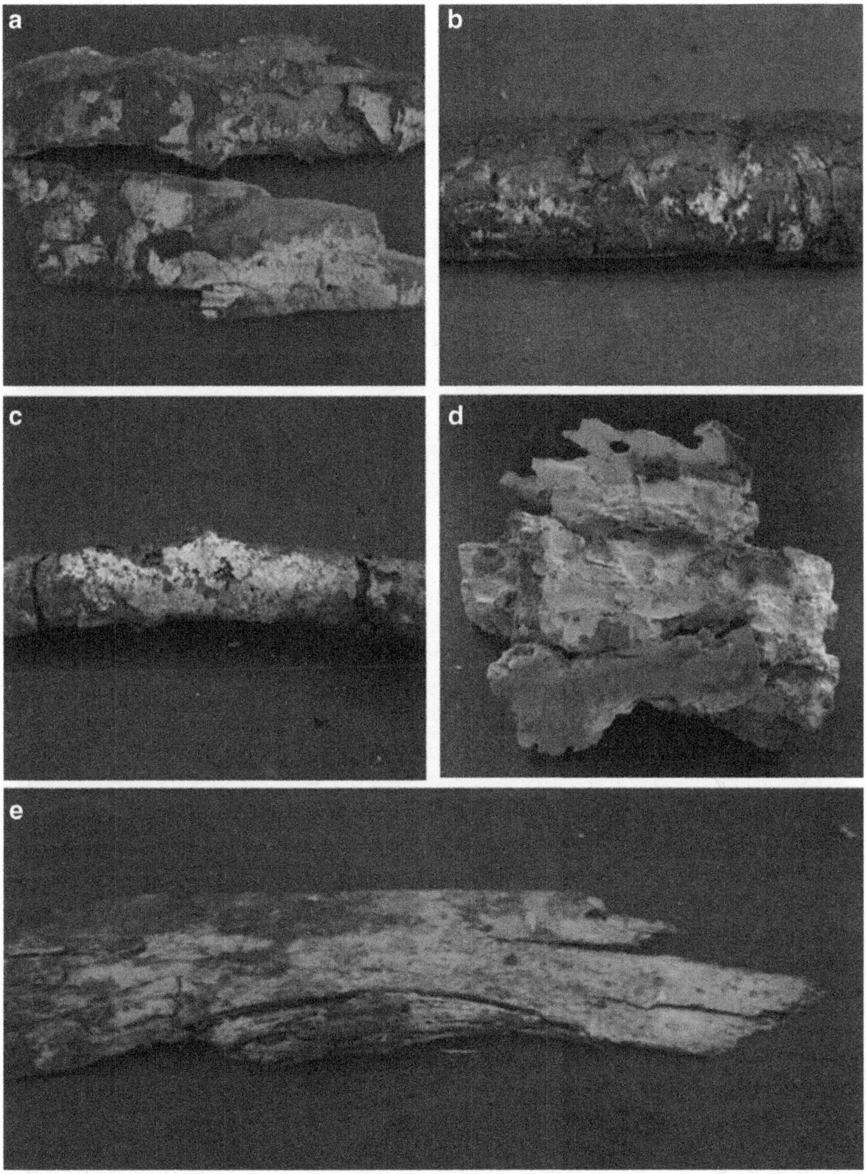

Plate 6.21 (**a**) *Gyrophanopsis polonensis*. (**b**) *Hyphoderma argillaceum*. (**c**) *Hyphoderma setigerum*. (**d**) *Hyphochnicium caucasium*. (**e**) *Hypochnicium geogenium*

Table 6.3 Synopsis of different species of *Hyphoderma*

Species	Fructification	Hymenium	Cystidia	Basidiospore
H. argillaceum	Resupinate, floccose to subpelliculose, widely effused	White to cream becoming ochraceous	Leptocystidia 70–120 × 10–15 μm tubular, with a brown excretion at the top	Ellipsoid, smooth, thin-walled, non-amyloid, acyanophilous 7–10 × 4.5–6 μm
H. clarusproprietas	Resupinate, adnate, thin ceraceous	Pinkish white, to yellowish grey, tuberculate	Subcylindrical, 87–160 × 10–12 μm, thinning above	Ellipsoid, smooth, thin-walled, non-amyloid, acyanophilous 6.6 × 9.2 × 4.6–6.2 μm
H. macedonicum	Resupinate, effused, adnate	White to ochraceous; smooth to tuberculate	Cystidia wide at base, tapering at top with brown resinous matter 72–80 × 7.5–10 μm	Allantoid to suballantoid, smooth, thin-walled, non-amyloid, acyanophilous 6.2–8 × 2.5–3 μm
H. occidentale	Resupinate floccose-pelliculose, loosely adnate	White, smooth to tuberculate	Gleocystidia with a capitate apex 40–70 × 5–7 μm	Ellipsoid cylindrical to cylindrical with obtuse ends, smooth, thin-walled, non-amyloid, acyanophilous 8.5–12 × 4–5 μm
H. parvisporum	Resupinate, effused, pellicular athelioid	Pale yellowish smooth to rough	Clavate 34–75 × 5.7–9.6 μm	Subglobose to globose, smooth, thin-walled, non-amyloid, acyanophilous, 5.7–7.4 × 5.1–6.2 μm
H. setigerum	Resupinate, subpelliculose to submembranous adnate	Leptocystidia, cylindrical to hyphoid, 3–5 septate, impregnated in apical half	White to deep cream, smooth to tuberculate to toothed	Cylindrical with rounded curved ends, smooth, thin-walled, non-amyloid, acyanophilous 8–10 × 3–4 μm
H. sibiricum	Resupinate, effused, in small patches	White to cream, smooth	Absent	Ellipsoid, smooth, thin-walled, non-amyloid, acyanophilous, 7–9.2 × 4.1–5.3 μm
H. sikkimum	Resupinate, effused, adnate, ceraceous	Greyish white to cream; smooth	Leptocystidia, 50–75 × 7–9 μm Lamprocystidia 42–120 × 12–18 μm bladder like structure also present	Ellipsoid to suballantoid, smooth, thin-walled, non-amyloid, acyanophilous, 11–15 × 4.5–7.5 μm
H. sporulosum	Resupinate, effused ceraceous	Whitish to yellowish with orange tint	Cystidia, sub cylindrical, 35–70 × 6–10 μm encrusted, septate	Ellipsoid, smooth, thin-walled, non-amyloid, acyanophilous, 4.5–6.2 × 3–4.2 μm
H. singularibasidium	Resupinate, loosely adnate	Greyish white to yellowish white, rough to farinaceous	Absent	Ellipsoid, smooth, thin-walled, non-amyloid, acyanophilous, 4.5–6.2 × 3–4.2 μm

Hyphoderma argillaceum (Bres.) Donk, Fungus 27: 14, 1957 = *Corticium argillaceum* Bres., Fung. trid. 2: 63, 1898. Plate 6.21b, Fig. 6.53a–d

Fructification resupinate, floccose to subpelliculose loosely adnate, widely effused, up to 150 μm thick in section, porose-reticulate; hymenial surface white to cream becoming light ochraceous on drying, smooth but finely pilose under the lens, discontinuous, not creviced; margin thinning somewhat pruinose, loosely adnate, white to paler concolorous.

Hyphal system monomitic; generative hyphae branched, septate, clamped, thin-walled, hyphae 3–4.5 μm wide. Cystidia (leptocystidia) 70–120 × 10.0–15.0 μm, tubular, subulate to subconical, the walls subhyaline, thin to moderately thick-walled, projecting a greater part of the length out of the hymenium, generally with a brown or reddish-brown excretion. Basidia 21.3–26 × 5.3–8.3 μm, clavate to subclavate with a suburniform constriction, 4-sterigmate, with a basal clamp. Basidiospores 7.0–10.0 × 4.5–6.0 μm, ellipsoid with oblique apiculus, smooth, thin-walled, non-amyloid, acyanophilous.

Distribution: Arunachal Pradesh: West Kameng, Bomdila, Shergaon; West Bengal: Darjeeling, Sanchal Lake; Bhutan: Thimphu, Nawephu, Begana, Paro, D'Dzong. H.P.: Dalhousie, J&K: Pehalgam

Collection examined: GSD 19825, 19221, 19545, L 42934.

Substratum: On decaying angiospermic branch.

Remarks: This species is characterized by thin and floccose fructifications and characteristic leptocystidia. This species was first reported from India by Thind and Rattan (1970) from N.W. Himalayas. Above collection resemble closely the description of *H. argillaceum* as given by Thind and Rattan (1970) and Eriksson and Ryvarden (1975), with minor variation in size of basidiospores.

Hyphoderma claruspropríetas Dhingra, Plant Science Research in India: 205, 1989. Fig. 6.53e–i

Fructification resupinate, adnate, effused, thin, ceraceous; hymenial surface pinkish-white when fresh, yellowish grey in the herbarium, more or less tuberculate, smoothening on drying; margin not well marked.

Hyphal system monomitic; generative hyphae septate, clamped, up to 3–5.3 μm wide; basal hyphae somewhat thick-walled, sparsely branched, almost parallel to the substrate; subhymenial hyphae much branched, thin-walled, at right angles to the substrate. Cystidia 87.0–160.0 × 10.0–12.0 μm, numerous, subcylindrical, basally somewhat thick-walled, gradually thinning above, projecting out of the hymenium. Basidia 24–30 × 7–8 μm, clavate to subcylindrical, with oily contents and a basal clamp, 4-spored. Basidiospores 6.6–9.2 × 4.6–6.2 μm, ellipsoid, smooth, thin-walled, non-amyloid, acyanophilous, with one large guttule or many small oil drops.

Distribution: West Bengal: Darjeeling, Sanchal Lake, Haryana: Panchkula.

Collection examined: GSD 19239. IBP 37172 (Extra limital).

Substratum: On a decayed angiospermic branch.

Remarks: This species has been collected only once from the Himalayas. The author has once again collected it from a place other than the study area.

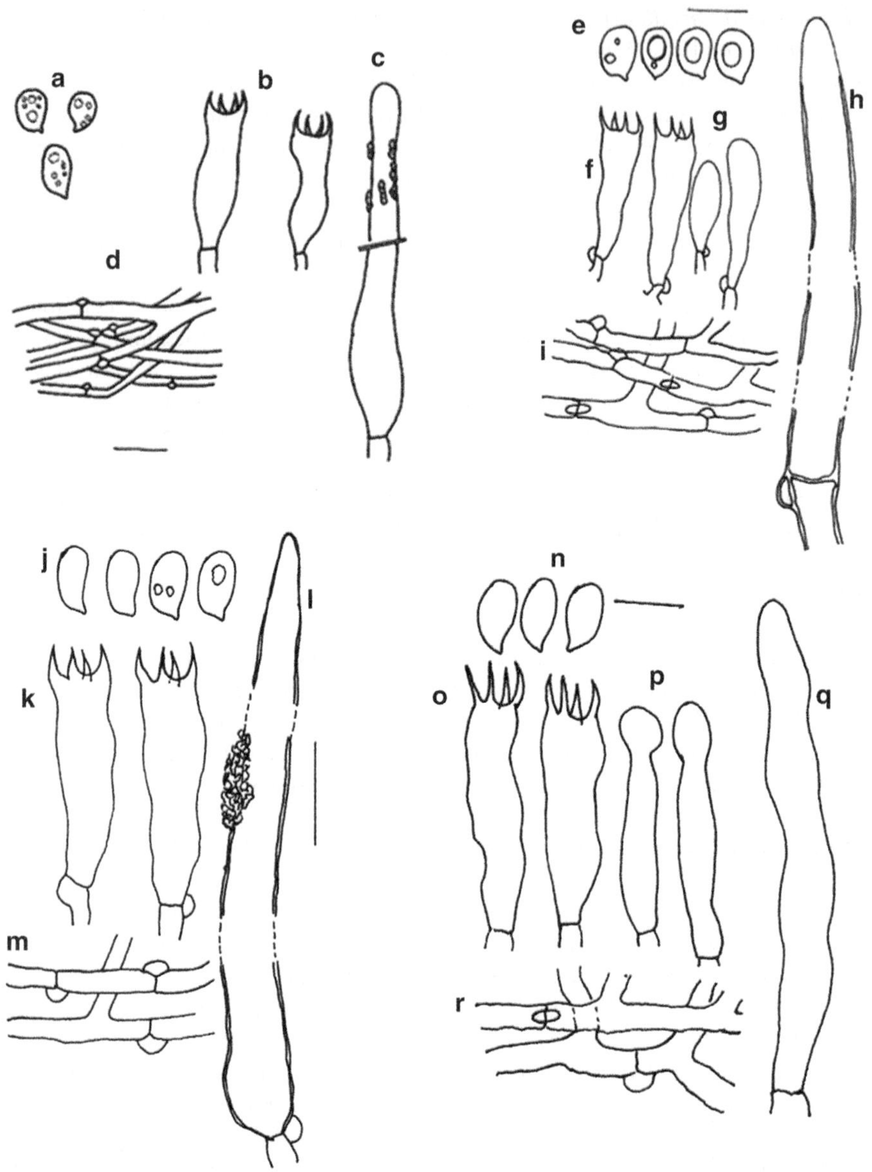

Fig. 6.53 (**a**–**d**) *Hyphoderma argillaceum* (**a**) Basidiospores, (**b**) Basidia, (**c**) Cystidia, (**d**) Generative hyphae; (**e**–**i**) *Hyphoderma clarusproprietas* (**e**) Basidiospores, (**f**) Basidia, (**g**) Basidioles, (**h**) Cystidia, (**i**) Generative hyphae; (**j**–**m**) *Hyphoderma macedonicum* (**j**) Basidiospores, (**k**) Basidia, (**l**) Cystidia, (**m**) Generative hyphae; (**n**–**r**) *Hyphoderma occidentale* (**n**) Basidiospores, (**o**) Basidia, (**p**, **q**) Cystidia, (**r**) Generative hyphae

Hyphoderma macedonicum (Litsch.) Donk, Fungus, Wageningen 27: 15, 1957 = *Gloeocystidium macedonicum* Litsch., Glasnik (Bull.) Soc. Scient. Skoplje 18(6): 181, 1938. Fig. 6.53j–m

Fructification resupinate, effused, adnate, up to 270 µm thick in section; hymenial surface smooth to tuberculate, white when fresh, becoming ochraceous white on drying; margins thinning, paler concolorous to indeterminate.

Hyphal system monomitic; generative hyphae up to 3.5 µm wide, branched, septate, clamped, thin-walled; basal hyphae irregularly interwoven, parallel to the substrate; subhymenial hyphae denser and vertical. Cystidia 72.0–80.0 × 7.5–10.0 µm, widen at the base, tapering at the apex, thin-walled, covered with brownish resinous matter, with basal clamp. Basidia 20–26.0 × 4.5–6.0 µm, subclavate, 4-sterigmate, with basal clamp; sterigmata up to 4.5 µm long. Basidiospores 6.2–8.0 × 2.5–3.0 µm, allantoids to sub allantoid, apiculate, thin-walled, smooth, non-amyloid, acyanophilous, with oil droplets.

Distribution: H.P.: Shimla.

Substratum: On stump of *Cedrus deodara.*

Collection examined: IBP 37174.

Remarks: This species is close to *H. argillaceum* with projecting cystidia with reddish brown amorphous matter, and allantoids to suballantoid basidiospores. However, it is a new record for Himalayas.

Hyphoderma occidentale (D.P. Rogers) Boidin & Gilles, Cryptog. Mycol. 15(2): 138 (1994) = *Hyphoderma subdefinitum* Erikss. & Strid. In Erikss. & Ryv., Cort. North Europe 3: 539. 1975. Fig. 6.53n–q

Fructifications resupinate, floccose-pelliculose to pelliculose, loosely adnate, arising in small orbicular colonies which may grow later but usually remain restricted, up to 120 µm thick in section; hymenial surface white, smooth to finely tuberculate, discontinuous, somewhat porose-reticulate under the lens, not creviced; margins thinning to more or less determinate, loosely adnate, white. Context subhyaline in section, composed of loosely woven hyphae.

Hyphal system monomitic; generative hyphae 2.5–4.5 µm wide, branched, septate, clamped, the walls thin subhyaline. Gloeocystidia 40–70 × 5–7 µm, cylindrical with obtuse apices, usually uniform but occasionally become constricted below the apex forming a capitate apex, thin-walled, subhyaline, empty or with subhyaline granular contents, immersed or rarely projecting up to 15 µm out of the hymenium. Basidia 25–30 × 7–8.5 µm, subutriform or typically utriform, 4-spored, sterigmata strongly curved, up to 7.5 µm long. Basidiospores 8.5–9.5(12) × 4–5 µm, ellipsoid cylindrical to cylindrical with obtuse ends and slightly depressed on the adaxial side appearing more or less suballantoid, minutely apiculate, the walls thin, subhyaline, smooth, non-amyloid.

Distribution: J&K: Gulmarg.

Collection examined: SSR 4382.

Substratum: On stump of *Abies pindrow.*

Remarks: This species is characterized by floccose-pelliculose fructifications with smooth to tuberculate hymenial surface and presence of cylindrical gloeocystidia. It belongs to *H. roseocremeum* group (fide Eriksson and Ryvarden 1975) where it is marked by the shape and size of the basidiospores.

Hyphoderma parvisporum Avn. P. Singh, Priyanka, Dhingra & Singla, as [*parvispora*], Mycotaxon 111: 71 (2010). Fig. 6.54a–c

Basidiospores resupinate, adnate, effused, thin, pellicular, almost athelioid; hymenial surface smooth, pale yellowish; margins indeterminately thinning.

Hyphal system monomitic; generative hyphae up to 3.3 μm wide, branched, septate, clamped; basal hyphae somewhat thick-walled, encrusted; subhymenial hyphae thin-walled, without encrustation. Basidia 11.0–16.2×3.0–4.0 μm, clavate, basally stalked, 4-sterigmate, with a basal clamp; sterigmata up to 4.5 μm long. Basidiospores 4.5–4.9×1.7–2.0 μm, narrowly ellipsoid thin-walled, smooth, inamyloid, acyanophilous.

Distribution: H.P.: Dalhousie, Shimla- Tara Devi.

Collection examined: L 42317.

Substratum: On a decayed gymnospermous wood.

Remarks: The species has close affinity with *H. capitatum* J. Erikss. and A. Strid. It differs markedly from *H. capitatum* in the size of basidiospores and the shape of the cystidia; subcylindrical to subfusiform cystidia in *H. parvisporum* in comparison to capitate cystidia in *H. capitatum*.

Hyphoderma setigerum (Fr.) Donk, Fungus, Wageningen 27: 15 (1957)=*Thelephora setigera* Fr., Elench. fung. 1: 208. 1828. Plate 6.21c, Fig. 6.54d–h

Fructification resupinate, subpelliculose to submembranous, arise in small colonies which may coalesce and grow later but always remain restricted, adnate, effused, smooth to somewhat odontoid or tuberculate, white to deep cream coloured, more or less continuous; margin thinning, adnate, white to paler concolorous. Hymenial surface smooth to tuberculate or toothed, white to deep cream, more or less continous; margin thinning, adnate, white to paler concolorous. Teeth small up to 250 μm long, conical to subulate with acute apices.

Hyphal system monomitic; generative hyphae branched, septate, clamped, thin to somewhat thick-walled, sparsely branched, 2.5–4 μm wide. Cystidia 75–200×7.5–10 μm, hyphoid, somewhat thin to thick-walled, subhyaline, clamped, smooth to encrusted, projecting. Gloeocystidia absent. Basidia 29–32.5×5.5–7 μm, subclavate to clavate, 4-sterigmate, with a basal clamp. Basidiospores 7–11.7×3–4.8 μm, narrowly ellipsoid to subcylindrical, smooth, thin-walled, non-amyloid, acyanophilous with several oil drops.

Distribution: Arunachal Pradesh: West Kameng, Bomdila, Shergaon, Basti side, Bhutan: Thimphu, Chimakothi, Bunakha; H.P.: Shimla; U.K.: Mussoorie- Kempty fall, Ranikhet; J&K: Pehalgam.

Collection examined: SSR 4330, HSK 4263, GSD 19691, IBP 37176, 37177.

Substratum: On stump and bark of *Cedrus deodara,* angiosperm bark, twigs.

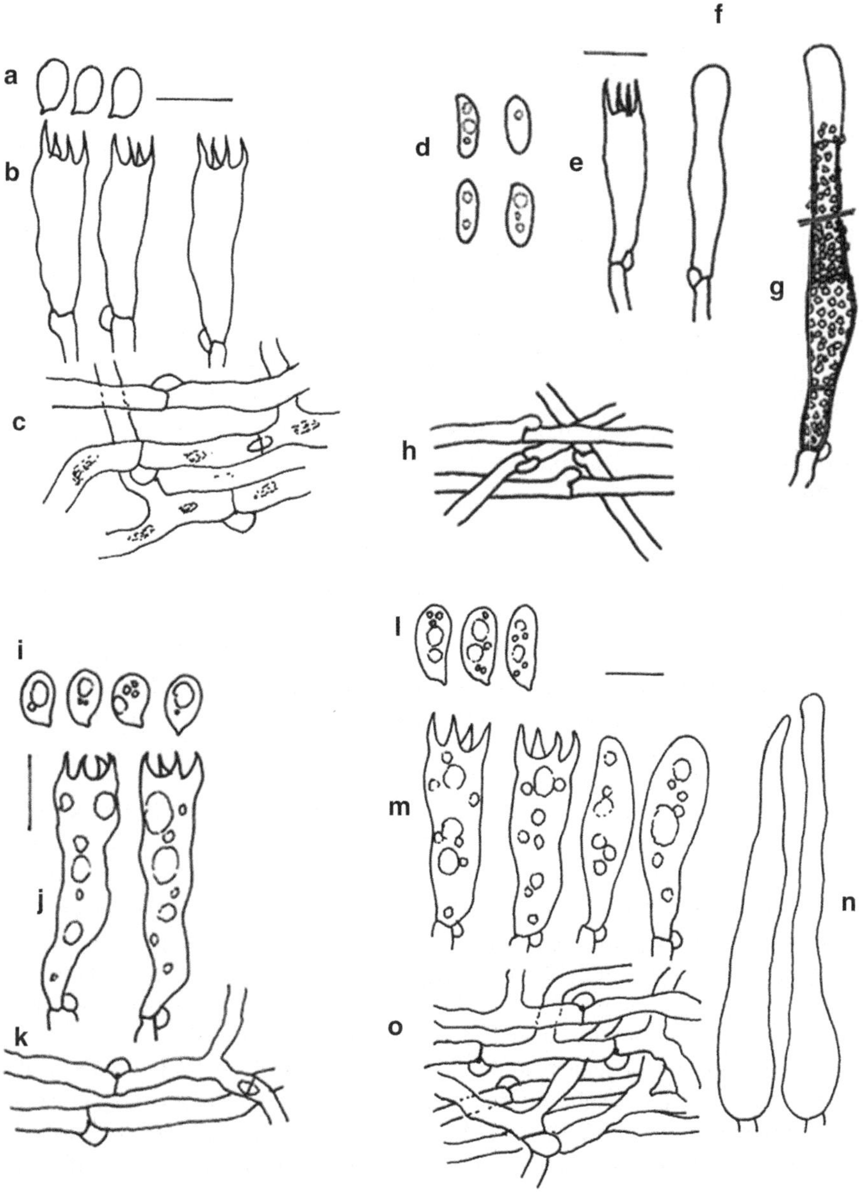

Fig. 6.54 (**a–c**) *Hyphoderma parvisporum* (**a**) Basidiospores, (**b**) Basidia, (**c**) Generative hyphae; (**d–h**) *Hyphoderma setigerum* (**d**) Basidiospores, (**e**) Basidia, (**f**) Basidioles, (**g**) Encrusted cystidia, (**h**) Generative hyphae; (**i–k**) *Hyphoderma sibiricum* (**i**) Basidiospores, (**j**) Basidia, (**k**) Generative hyphae; (**l–o**) *Hyphoderma sikkimium* (**l**) Basidiospores, (**m**) Basidia, (**n**) Cystidia, (**o**) Generative hyphae

Remarks: This is a widely distributed species, first reported from India by Thind and Rattan (1970) from N.W. Himalayas. The above collection resembles the description given by Rattan (1977) and Eriksson and Ryvarden (1975).

Hyphoderma sibiricum (Parm.) John Eriksson. & Strid, In John Erikss. & Ryrv., Cort. N. Europe 3: 535, 1975 = *Radulomyces sibiricus* Parm., Consp. Syst. Cort.: 223, 1968. Fig. 6.54i–k

Fructification resupinate, effused, smooth, thin, in small patches, white when fresh, to cream-coloured on drying; margin not differentiated, sometimes fibrillose.

Hyphal system monomitic; hyphae septate, clamps present at all septa, texture dense in the subhymenium, branched, thin-walled, 2–3.2 μm wide. Cystidia absent. Basidia cylindrical, sinuous, 25–35.5 × 5.0–7.3 μm, 4-sterigmate, non-amyloid. Basidiospores ellipsoid, thin-walled, smooth, 7.0–9.2 × 4.1–5.3 μm with numerous oil drops in the protoplasm.

Distribution: Bhutan: Thimphu.

Collection examined: GSD 19467.

Substratum: On decaying stump of *Cryptomeria japonica,* on a decaying *Pinus* log.

Remarks: *Hyphoderma sibricum* is closely related to *H. cremeoalbum* but is differentiated by the smaller spores and basidia.

Hyphoderma sikkimium Dhingra, Pl. sci. res. in India: 201 (1989). Fig. 6.54l–o

Fructification resupinate, adnate, effused, thin, up to 0.3 mm in section, ceraceous; hymenial surface greyish-white to cream-colored, smooth; margins thinning, pruinose in young fructification.

Hyphal system monomitic; generative hyphae up to 3.2 μm wide, branched, septate, clamped; basal hyphae loosely interwoven, thin- to somewhat thick-walled; subhymenial hyphae thin-walled, densely interwoven. Cystidia of two types: (i) 50–75.2 × 7–9 μm, tubular, basally somewhat widened at the base, often somewhat constricted, thin-walled, leptocystidia, generally enclosed in the hymenium. (ii) 42–120 × 12–18.2 μm, generally subfusiform, encrusted lamprocystidia, projecting up to 50 μm out of the hymenium. Some bladder-like, thin-walled structures are also observed, which probably can be the initial stages in the development of leptocystidia. Basidia 25–35.3 × 8.5–10.2 μm, subclavate, often constricted in a suburniform manner, with oily contents and a basal clamp, 4-sterigmate. Basidiospores 11–15.2 × 4.5–6(7.5) μm, ellipsoid to narrowly ellipsoid to suballantoid, smooth, thin-walled, non-amyloid, acyanophilous, often with many oil drops.

Distribution: Sikkim-Gangtok; H.P.: Chamba-Dalhousie.

Collection examined: GSD 19349, L 37683.

Substratum: On decaying stump of *Cryptomeria japonica,* on a decaying *Pinus* log.

Remarks: The species is characterized by ceraceous fructification and two types of cystidia and ellipsoid to suballantoid basidiospores. It is a new record for N. W. Himalayas.

Hyphoderma singularibasidium Dhingra, Avn. P. Singh & Singla, Mycotaxon 108: 197 (2009). Fig. 6.55a–d

Fructifications resupinate, loosely adnate, thin; hymenial surface rough, farinaceous under the lens, greyish-white to yellowish white; margins indeterminately thinning out.

Hyphal system monomitic; generative hyphae up to 8 μm wide, thin- to somewhat thick-walled, clamped; basal hyphae running parallel to the substrate, less branched, with large cells; subhymenial hyphae much branched, branches arising from the clamps. Basidia 17–35.2 × 7.0–9.3 μm, clavate to sub-clavate but constricted in the middle, thin-walled, with a peculiar unilateral outgrowth arising from the middle of the basidium, 4-sterigmate, with a basal clamp, with oily contents; sterisgmata up to 7.4 μm long. Basidiospores 7.4–9.2 × 4.5–5.8 μm, broadly ellipsoid, thin- to somewhat thick-walled, smooth, inamyloid, acyanophilous, with a large oil drop or many smaller oil droplets.

Distribution: H.P.: Chamba, Manali, Shimla.

Collection examined: L 42940.

Substratum: Decayed gymnospermous log.

Remarks: This species is characterized by lateral outgrowth from the middle of basidium, pointing in apical direction.

Hyphoderma sporulosum Dhingra, Pl. sci. res. in India: 201 (1989). Fig. 6.55e–h

Fructification resupinate, adnate, effused, thin in section, ceraceous; hymenial surface smooth to finely tuberculate, whitish initially becoming, light yellow with an orange tint at maturity; margins thinning, whitish, somewhat fibrillose.

Hyphal system monomitic; generative hyphae up to 3.5 μm wide, branched, septate, clamped; basal hyphae thin- to somewhat thick-walled, loosely interwoven; subhymenial composed of thin-walled, densely interwined hyphae. Cystidia 35–70.2 × 6–10 μm, subcylindrical, with one or more constrictions, apically obtuse, thin-walled, initially becoming thick-walled later on, encrusted, especially in the upper half, with one or more secondary septa. Basidiospores 4.5–6.2 × 3–4.2 μm, ellipsoid to broadly ellipsoid, smooth, thin-walled, non-amyloid, acyanophilous.

Distribution: West Bengal: Darjeeling; Bhutan- Thimphu, Bunakha.

Collection examined: GSD 19107, 19526

Substratum: On decaying stump.

Remarks: This species is characterized by subcylindrical, constricted, apically obtuse, thick-walled, encrusted cystidia; subulate to clavate, sinous and spored basidia and ellipsoid to broadly ellipsoid spores. It is differentiated from *H. puberum* on the basis of cystidia and smaller spores.

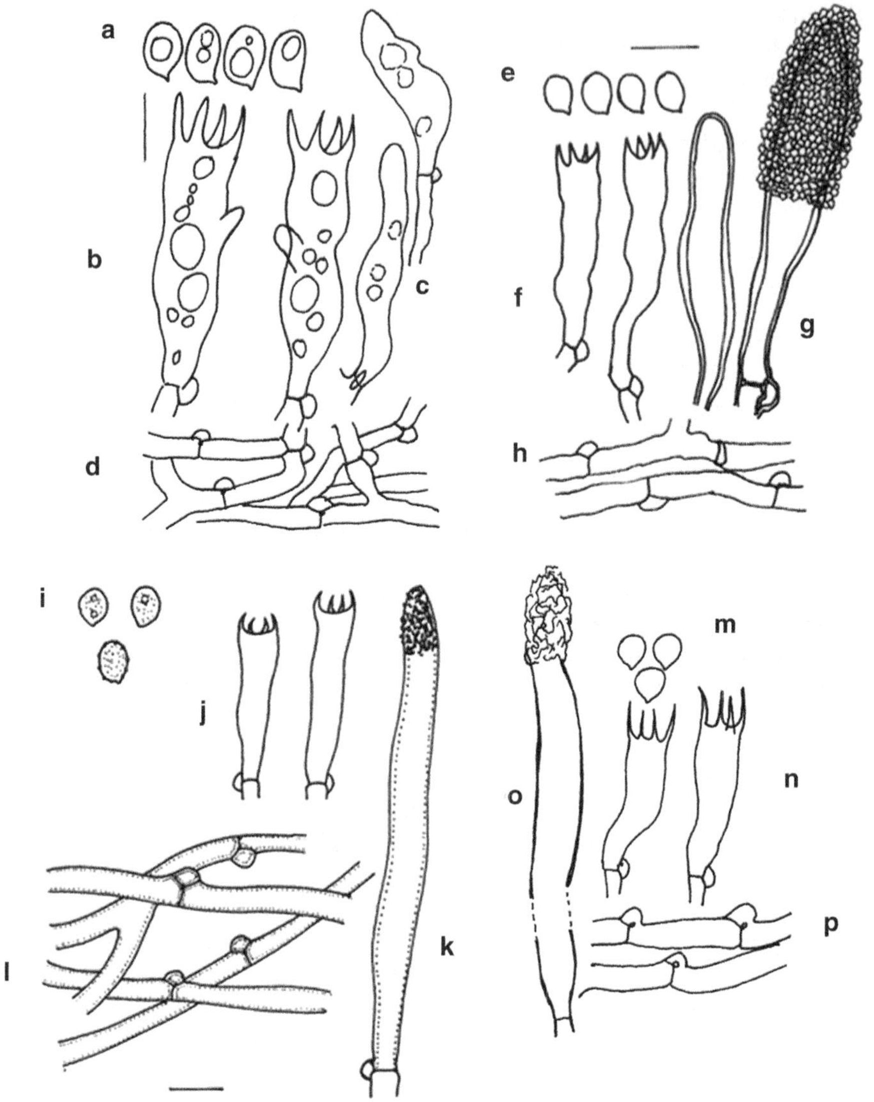

Fig. 6.55 (**a**–**d**) *Hyphoderma singularibasidium* (**a**) Basidiospores, (**b**) Basidia, (**c**) Cystidia, (**d**) Generative hyphae; (**e**–**h**) *Hyphoderma sporulosum* (**e**) Basidiospores, (**f**) Basidia, (**g**) Encrusted cystidia, (**h**) Generative hyphae (**i**–**k**) *Hypochnicium caucasicum* (**i**) Basidiospores, (**j**) Basidia, (**k**) Cystidia, (**l**) Generative hyphae; (**m**–**p**) *Hypochnicium cystidiatum* (**m**) Basidiospores, (**n**) Basidia, (**o**) Cystidia, (**p**) Generative hyphae

Hypochnicium J. Erikss.,
Symb. bot. upsal. 16(no. 1): 100. 1958.

Fructification resupinate, effused, adnate, white, yellowish or reddish, glabrous or pilose with projecting cystidia, smooth, tuberculate or irregularly odontoid; margin varying, usually not differentiated; hyphal system monomitic; hyphae thin to more or less thick-walled, fibulate, hyaline; cystidia present or absent, if present projecting or enclosed, thin or thick-walled, smooth or with a thin crust of crystals, with homogenous contents except in one species which has sulfocystidia, rich in oildrops; basidia more or less suburniform, normally with 4-sterigmata and basal clamp; spores ellipsoid to subglobose, with thickened, cyanophilous walls, smooth, verruculose or echinulate.

Twenty species, widespread

Lit.: Nilsson & Hallenberg (*Mycol.* 95: 54, 2003; phylogeny *Hypochanicium punctulatum* complex)

Type Species: *Thelephora bombycina* Sommerf. 1826.

Habitat: Dead wood

Himalayas: Seven (Table 6.4)

Hypochnicium caucasicum Parm., Eesti NSV Tead. Akad. Toim. Biol. seer. 16(4): 385, 1967. Plate 6.21d, Fig. 6.55i–k

Fructification resupinate, adnate, effused up to 280 μm thick in section, porose first, then continous, yellowish white to pale yellow; hymenial surface more or less tuberculate, smooth on drying; margin not well marked.

Hyphal system monomitic; generative hyphae branched, septate, clamped; basal hyphae up to 7.8 μm wide, irregularly branched, thick-walled; subhymenial hyphae thinner, thin-walled, vertical, branched into a dense texture. Cystidia 58–122 × 8.0–12.0 μm, enclosed to somewhat projecting, thin to somewhat thick-walled, apically encrusted, with a basal clamp. Basidia 22–30 × 6.0–7.8 μm, subclavate, 4-sterigmate, with a basal clamp. Basidiospores 5.5–7.8 × 4.5–5.3 μm, broadly ellipsoid to subglobose, verruculose, thick-walled, non-amyloid, cyanophilous, usually with one guttule.

Distribution: Meghalaya: Shillong, Barapani; U.K.: Mussoorie, Lal tibba.

Collection examined: GSD 19011, IBP 37178.

Substratum: On decaying bark of a *Pinus* stump, dead coniferous wood.

Remarks: This species has been reported earlier by Dhingra et al. (2011) from Eastern Himalayas. The species is characterized by thick-walled, randomly branched basal hyphae – a character which differentiate it from *H. puntulatum.* This collection resembles closely the measurements and figure of *H. cauasium* as given by Eriksson and Ryvarden (1976) except for the presence of apical encrustation on cystidia. It is the second record of this species and first from N.W. Himalayas.

Table 6.4 Synopsis of *Hypochnicium* species

Name	Fructification	Hymenium	Cystidia	Basidiospore
H. caucasicum	Resupinate, adnate, effused, porose to continuous	Yellowish white to pale yellow; tuberculate	Apically encrusted; 52–122 × 8–12 μm	Ellipsoid to subglobose, thick walled, verrucose, cyanophilous, 5.5–7.8 × 4.5–5.3 μm
H. cystidiatum	Resupinate, floccose, membranous, adnate,	Cream yellow to light yellow; smooth to tuberculate	Apically encrusted; 70–105 × 10–18 μm	Subglobose to ovoid, slightly thick- walled, echinulate weakly cyanophilous 5.5–7 × 4.5–5.6 μm
H. geoginium	Resupinate, effused, ceraceous to membranous	White to yellowish white; smooth to tuberculate	Not encrusted	Broadly ellipsoid, thick-walled, cyanophilous, 6–7.7 × 4.5–5.6 μm
H. longicystidiosum	Resupinate, arachnoid to floccose, loosely adnate	White, smooth totuberculate	Leptocystidia with encrustations in patches in upper part 175–230 × 4.75 μm	Globose, thin-walled subhyaline, smooth, 4.8–5 μm
H. lundellii	Resupinate, pelliculose to membranous	White to cream, tuberculate	Absent	Ellipsoid, thick-walled, cyanophilous, 5.5–6.8 × 4.2–5.2 μm
H. punctulatum	Resupinate, pelliculose to membranous, adnate	White to cream rarely cream yellow	Gloeocystidia 60–135 × 8–11 μm	Ellipsoid to ovoid or subglobose thin- to thick-walled, cyanophilous, 6–7.3 × 5–6 μm
H. sphaerosporum	Resupinate, membranous, adnate, effused to subceraceous	White to cream yellow; smooth to tuberculate	Gloeocystidia thin-walled, not encrusted	Globose to subglobose, sub hyaline, smooth, thick-walled, weakly cyanophilous

Hypochnicium cystidiatum Boid. & Gill., Cah. Mabok. 9(2): 90. 1971. Fig. 6.55m–p

Fructifications resupinate, floccose to floccose-membranous, adnate, arising in small colonies; hymenial surface cream yellow to light yellow, smooth to finely tuberculate, tubercles few and sparse, reticulate-porose to somewhat pilose under a hand lens, discontinuous to somewhat continuous, not creviced; margin thinning, somewhat fibrillose, adnate, white to paler concolorous. Subiculum subhyaline to pale yellow, somewhat fibrous.

Hyphal system monomitic, hyphae 2.7–6 μm wide, branched at wide angles, branches often proliferating from the clamps, and often ramified, closely septate, clamps at all septa and quite prominent, the walls thin to slightly thick (up to 1 μm), subhyaline. Cystidia 70–105.2×10–17.5 μm, conical to subfusiform, with obtuse apices, arising from the upper part of context and scarcely projecting out of the hymenium, the walls are usually thick (up to 1 μm) in the basal part but become progressively thin towards the apices, subhyaline, naked or may be finely impregnated with subhyaline crystals near the top, incrustations may extend to 25 μm. Gloeocystidia 5–7 μm broad, cylindrical to flexuous, immersed, often, thin-walled, subhyaline, empty or sometimes with granular contents staining deeply with phloxine. Basidia 20–25×5.5–7 μm, clavate-cylindrical, 4-spored, sterigmata up to 3.5 μm long. Basidiospores 5.5–7×4.5–5.8 μm, subglobose to ovoid, shortly apiculate, the walls subhyaline, slightly thick, tend to swell in KOH solution, finely but distinctly echinulate, non-amyloid, weakly cyanophilous.

Distribution: H.P.: Shimla

Collection examined: SSR 4321, IBP 37175.

Substratum: On stump under angiosperms.

Remarks: This species is characterized by the cystidia which resemble those of *Hypoderma argillaceum* in gross morphology. It is a rare species in the Himalayas.

Hypochnicium geogenium (Bres.) J. Eriksson, Symb. bot. ups. 16(no. 1): 101(1958) = *Corticium goegenium* Bres., Ann. mycol. 1 p.98, 1903.

Plate 6.21e, Fig. 6.56a–e

Fructification resupinate, effused, ceraceous when fresh, membranous on drying; hymenial surface smooth to tuberculate when fresh, smoothing on drying, white to yellowish white; margin not well differentiated. Hyphal system monomitic; generative hyphae richly and irregularly branched, septate, clamped, thin to thick-walled; basal hyphae up to 5.0 μm wide. Cystidia subcylindrical, thin-walled, with basal clamp, projecting up to 58 μm out of the hymenium. Basidia 20.0–31.6×5.0–7.2 μm, clavate, 4-sterigmate, with a basal clamp. Basidiospores 6.0–7.7×4.5–5.6 μm, broadly ellipsoid, thick-walled, non-amyloid, cyanophilous.

Distribution: West Bengal: Darjeeling; Bhutan: Thimphu, Dochula.

Collection examined: GSD 19121, 19419, IBP 37178, 37179.

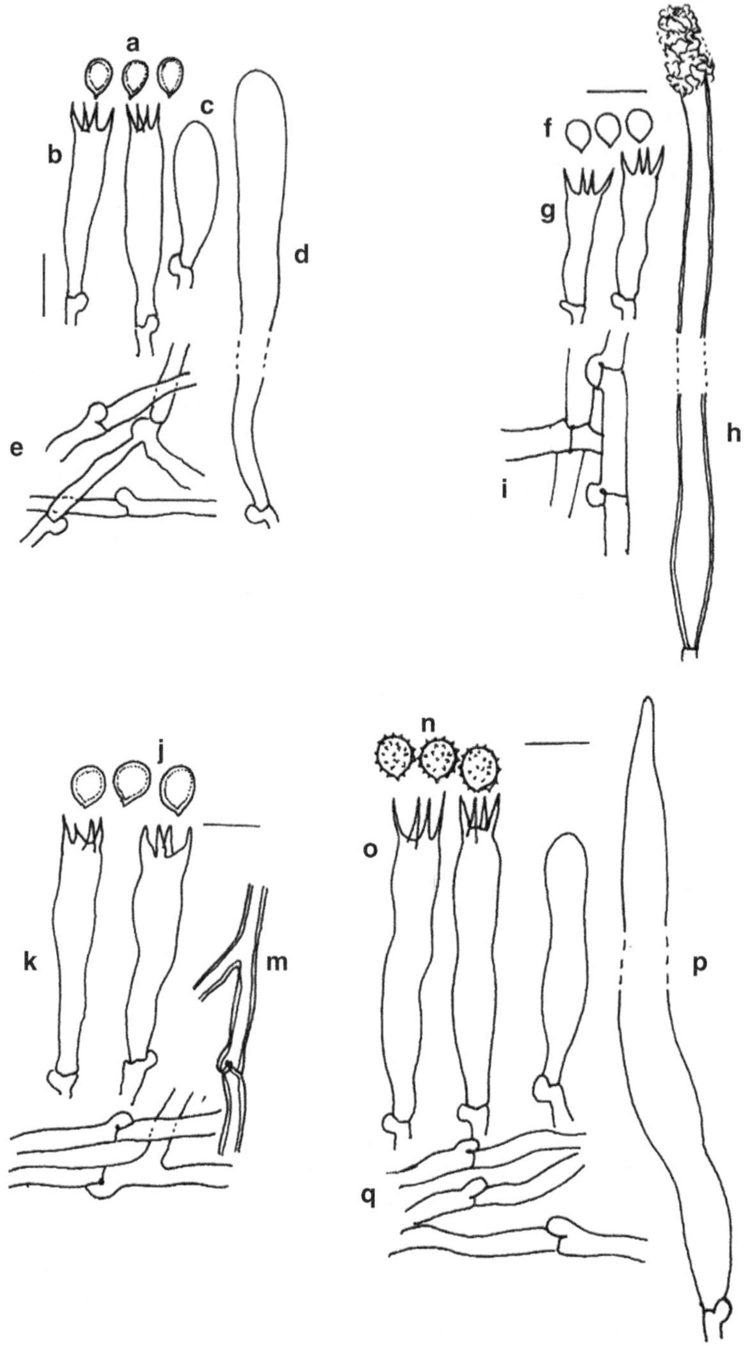

Fig. 6.56 (**a–e**) *Hypochnicium geogenium* (**a**) Basidiospores, (**b**) Basidia, (**c–d**) Cystidia, (**e**) Generative hyphae; (**f–i**) *Hypochnicium longicystidiosum* (**f**) Basidiospores, (**g**) Basidia, (**h**) Cystidia, (**i**) Generative hyphae; (**j–m**) *Hypochnicium lundellii* (**j**) Basidiospores, (**k**) Basidia, (**l**) Thin-walled generative hyphae (**m**) Thick walled generative hyphae (**n–q**) *Hypochnicium punctulatum* (**n**) Basidiospores, (**o**) Basidia, (**p**) Cystidia, (**q**) Generative hyphae

Substratum: On a decaying angiospermic log, on a decaying log of *Cryptomeria japonica.*

Remarks: This species was first reported by Bresadola (1903) from N. America as *Corticium geogenium.* Eriksson (1958) shifted it to *Hypochnicium.* The Himalayan collection resembles *H. geogenium* as described by Eriksson and Ryvarden (1976) except for somewhat broader basidiospores It has been earlier recorded from E. Himalayas (Dhingra et al. 2011). However, it is a new record for N.W. Himalayas

Hypochnicium longicystidiosum (S.S. Rattan) Hjortstam & Ryvarden, Mycotaxon 20(1): 135 (1984)=*Hyphodontia longicystidiosa* S.S. Rattan, Biblthca Mycol 60: 340(1977). Fig. 6.56f–i

Fructifications resupinate, arachnoid to floccose, loosely adnate, arising in small colonies; hymenial surface white, smooth to tuberculate, tubercles common and irregular, discontinuous, not creviced; margin indistinct, adnate, white to paler concolorous. Subiculum composed of somewhat compactly arranged hyphae.

Hyphal system monomitic, hyphae 2.6–4.6 µm wide, branched at wide angles and ramified, septate, clamped, the walls thin, subhyaline, distinct in young specimens but tend to collapse in mature ones. Leptocystidia 175–228×4–7.5 µm, subulate-cylindrical to hyphoid with obtuse apices, subhyaline, the walls thin to slightly thick (up to 1.5 µm) especially in the basal part but it becomes progressively thin towards the apices, incrusted to finely incrusted with subhyaline crystals, incrustations present in patches or uniformly in the apical part, arising from the basal part of the context and may project to 90 µm out of the hymenium. Basidia 16–20.3×5.5–8 µm, clavate-cylindrical to subutriform, 4-spored. Basidiospores 4.8–5 µm in diameter, globose, minutely apiculate, the walls thin, subhyaline, smooth, nonamyloid.

Distribution: H.P.: Manali, Kullu.

Collection examined: SSR 4377, IBP 37181.

Substratum: On twigs of *Cedrus deodara,* on gymnospermic wood.

Remarks: The characteristic features of this species are arachnoid to floccose fructifications, tuberculate hymenial surface, presence of very long cystidia and globose basidiospores.

Hypochnicium lundellii (Bourd.) Erikss. Symb. bot. upsal. 16 (no.1): 101. 1958. = *Corticium lundellii* Bourd. In Erikss., Svensk. Bot. Tidskr. 43: 56. 1949.

Plate 6.22a, Fig. 6.56j–m

Fructification resupinate, pelliculose-membranous to membranous, adnate, arising as small colonies; hymenial surface white to cream, more or less tuberculate, tubercles sparse but prominent and blunt, continous to discontinuous, finely creviced on drying; margin thinning, fibrillose, adnate, white to paler concolorous. Subiculum composed of loosely woven hyphae in the basal part but these become more compactly arranged and somewhat ceraceous in the upper part.

Plate 6.22 (**a**) *Hypochnicium lundellii*. (**b**) *Hypochnicium punctulatum*. (**c**) *Hypochnicium sphaerosporum*. (**d**) *Irpex consors*. (**e**) *Irpex lacteus*

Hyphal system monomitic, branched at wide angles and ramified, closely septate, often proliferating in to new branches, the walls thin, subhyaline; generative hyphae thin to thick-walled, septate, clamped. Cystidia absent. Basidia 23–30.9 × 4.7–6.2 µm, clavate to subclavate, 4-sterigmate, with a basal clamp. Basidiospores 5.5–6.8 × 4.2–5.2 µm, broadly ellipsoid, thick-walled, non-amyloid, cyanophilous.

Distribution: H.P.: Shimla-Glen, Manali.

Collection examined: SSR 5552, IBP 42947.

Substratum: On stump under mixed forest.

Remarks: It is a rarely found species in Himalayas and was first reported by Rattan (1977) from N.W. Himalayas. The above collection resembles closely the description of *H. lundellii* as given by Erikssion and Ryvarden (1976) and Rattan (1977). It is the second record of this species from Himalayas.

Hypochnicium punctulatum (Cooke) Erikss., Symb. bot. upsal. 16(no. 1): 101. 1958. = *Corticium punctulatum* Cooke, Grevillea 6: 132. 1878.

Plate 6.22b, Fig. 6.56n–q

Fructification resupinate, pelliculose-membranous to soft membranous, adnate, often arising as small circular colonies which may coalesce later and become effused; hymenial surface white to cream or rarely cream yellow, smooth to distinctly tuberculate, tubercles sparse, few and hemispherical, continous, not creviced; margin thinning, adnate, concolorous. Subiculum subhyaline in section, composed of some what loosely interwoven hyphae.

Hyphal system monomitic, generative hyphae 3–6 µm, branched at wide angles, branches, proliferating from clamps and ramified, septate, clamped, basal hyphae thick-walled and more loosely interwoven but become compact towards hymenium. Cystidia absent. Gloeocystidia 60–135.2 × 8–11.2 µm, enclosed to somewhat projecting. Basidia 24.2–30.2 × 5.5–7.3 µm, clavate to subclavate, 4-sterigmate, with a basal clamp. Basidiospores 6.0–7.3 × 5–6 µm, broadly ellipsoid to ovoid or subglobose, verruculose, thin to thick-walled, non-amyloid, cyanophilous.

Distribution: H.P.: Kullu, Shimla, Dalhousie- Kalatop; U.K.: Mussoorie, J&K: Gulmarg.

Collection examined: GSR 5032, SSR 5410, IBP 37183, 37184, L 37185, 37186.

Substratum: On log of *Abies pindrow*, on bark of *Cedrus deodara, Quercus incana*.

Remarks: This species is characterized by the texture and colour of the fructification, presence of gloeocystidia and finely echinulate basidiospores.

Hypochnicium sphaerosporum (Hoehn. & Litsch.) Erikss., Symb. bot. upsal. 16(no. 1): 101. 1958 = *Peniophora sphaerospora* Hoehn. & Litsch., K. Acad. Wiss. Wien. Math.-Nat. Klasse Sitzber. 115: 1559. 1906. Plate 6.22c, Fig. 6.57a–d

Fructification resupinate, membranous, adnate, widely effused, at first porose then continuous, subceraceous; hymenial surface smooth to rarely tuberculate, tubercles sparse and small, white to cream yellow, continuous or creviced irregularly; margin thinning, adnate, white to paler concolorous. Subiculum subhyaline in section, composed of somewhat compactly arranged hyphae.

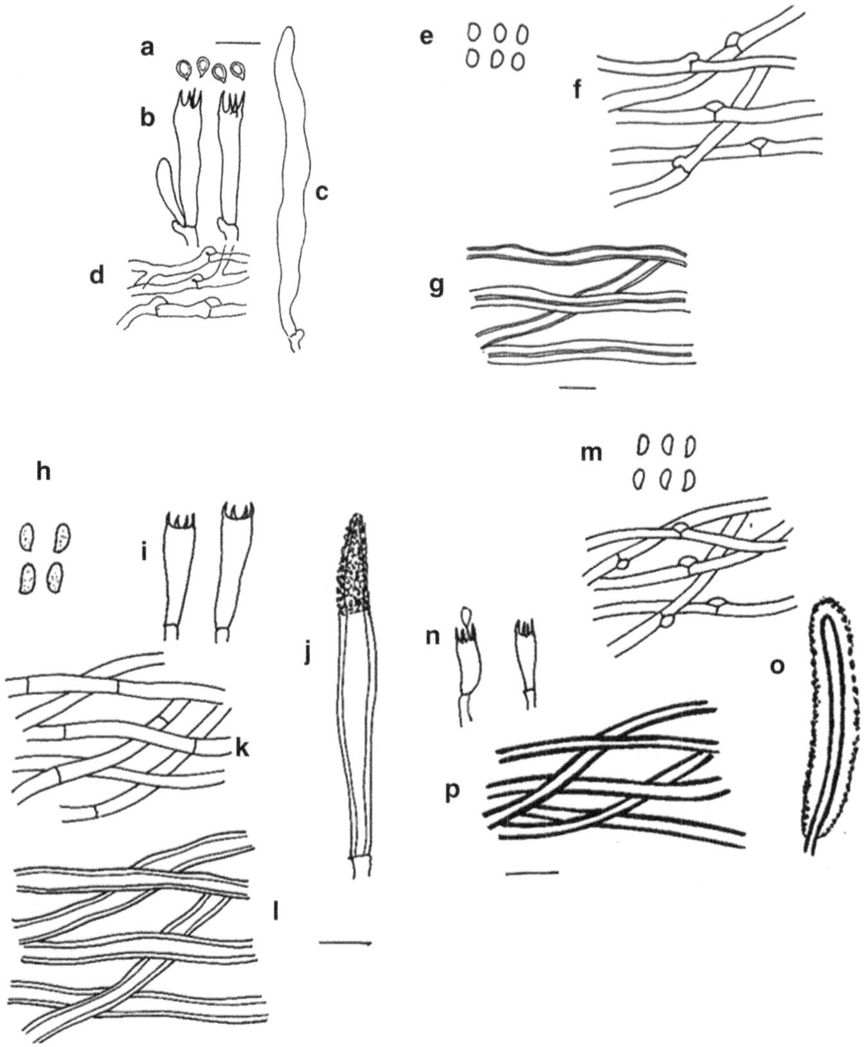

Fig. 6.57 (**a**–**d**) *Hypochnicium sphaerosporum* (**a**) Basidiospores, (**b**) Basidia, (**c**) Cystidia, (**d**) Generative hyphae; (**e**–**g**) *Irpex consors* (**e**) Basidiospores, (**f**) Generative hyphae, (**g**) Skeletal hyphae; (**h**–**l**) *Irpex lacteus* (**h**) Basidiospores, (**i**) Basidia, (**j**) Encrusted Cystidia, (**k**) Generative hyphae, (**l**) Skeletal hyphae; (**m**–**p**) *Junghuhnia collabens* (**m**) Basidiospores, (**n**) Basidia, (**o**) Cystidia, (**p**) Generative hyphae

Hyphal system monomitic; hyphae 2–5 μm wide, branched at wide angles and closely ramified, thin-walled, septate, clamped. Gloeocystidia cylindrical to subfusiform, 8.5–11.0 μm wide, thin-walled. Basidia 26–35×6.5–8.5 μm, subclavate, 4-sterigmate, with a basal clamp. Basidiospores 4.8–6.8 μm in diameter, globose to subglobose, subhyaline, smooth, thick-walled, non-amyloid, weakly cyanophilous.

Distribution: H.P.: Narkanda, Khardrala, Manali
Collection examined: SSR 5633, 5766.
Substratum: On stump of *Pinus excelsa, Abies pindrow*.
Remarks: This species was reported from Himalayas by Rattan (1977) from Himachal Pradesh N.W. Himalayas. The above collections are quite typical of *H. sphaerosporum* as described by Rattan (1977) and Eriksson and Ryvarden (1976).

Irpex Fr.,
Syst. orb. veg. 1: 81, 1825;

Fructification annual, pileate to resupinate, sessile, soft coriaceous when fresh; upper surface glabrous or tomentose, azonate to zonate. Context white to creamish. Hyphal system dimitic; generative hyphae hyaline, thin-walled, septate, with or without clamps; skeletal hyphae thick-walled with narrow lumen, aseptate. Cystidia present or absent. Basidia hyaline, thin-walled, 4-spored, clavate. Basidiospores allantoids, smooth, non-amyloid, acyanophilous.
Six species, widespread
Lit.: Mass Geesteranus (*Persoonia* **7**: 443, 1974), Buzina et al. (*J. Clin. Microbiol.* 43: 2009, 2005)
Type Species: *Sistrotrema lacteum* Fr. 1818
Habitat: Dead wood
Himalaya: Three
Key to species

1. Upper surface creamish brown to white to light buff, minutely tomentose, later becomes glabrous, azonate to inconspicuously concentrically zonate 2
1. Upper surface pinkish buff, matted tomentose, faintly azonate ....*I. vellereus*
2. Pore surface yellowish brown, mostly irpicoid, poroid near margin ..... *I. consors*
2. Pore surface white to pinkish when fresh, grey when dried ....*I. lacteus*

Irpex consors Berk., J. Linn. Soc. Bot. 16(no. 89): 51. 1878 [1877].
Plate 6.22d, Fig. 6.57e–g

Fructification annual, largely resupinate, widely effused-reflexed, brittle on drying; Pileus effused reflexed, confluent, forming wider pilei, imbricate; upper surface creamish brown, minutely tomentose, later becomes glabrous, azonate to inconspicuously concentrically zonate; margin acute, concolorous with upper surface, wavy, in turned on drying, fertile below. Pore surface yellowish brown, mostly irpicoid, poroid near margin, uneven, dull; pores round to irregular near the margin, sinuous, 1–3 per mm, irpiciform later. Spines not stratose yellowish brown, 2.8–5.5 mm long, subulate. Context creamish, homogenous, non-xanthochroic, up to 1.8 mm thick.

Hyphal system dimitic; generative hyphae hyaline, thin-walled to thick-walled, septate, branched, clamped, acyanophilous, 2.5–6.5 μm. in diameter; skeletal hyphae subhyaline, thick-walled to solid, long, sparsely branched, aseptate, non-amyloid, acyanophilous, 2.5–6.5 µm. Cystidia absent. Basidia hyaline, clavate, 4-spored, about, 12.7–14×4.7–6.1 µm. Basidiospores, hyaline, thin-walled, smooth, apiculate non-amyloid, broadly ellipsoid, 4.3–6.4×2.9–4.3 µm.

Distribution: H.P.: Dalhousie- Banikhet; Shimla- The Glen, U.K.: Karanprayag-Chamoli.

Collection examined: SSR 6444; D6446, IBP 37186.

Substratum: Stump of *Quercus*, dead wood

Remarks: This species was recorded from India by Bakshi (1971) from the Western Himalaya. The above cited collection resembles with the description given by Dhanda (1977). The species is characterized by sessile fructification with concentrically zonate upper surface.

Irpex lacteus (Fr.) Fr., Elench. Fung.1: 142, 1828. Plate 6.22e, Fig. 6.57h–l

Fructification annual, usually effused-reflexed or resupinate, pilei often imbricate, thin and coriaceous; upper surface white to light buff, adpressed tomentose to velutinate, azonate to faintly zonate; margin thin, entire, fertile, incurved. Pore surface white to pinkish when fresh, grey when dried, pores 2–3 per mm near the margin, angular to irregular, often irpiciform. Context white to pale, fibrous; entire, fertile, incurved.

Hyphal system dimitic; generative hyphae hyaline, thin-walled to thick-walled, non-xanthochroic, hyaline, simple septate, clamps absent, branched, 2.7–5.1 µm wide; skeletal hyphae hyaline, thick-walled to solid, 2.7–5.6 µm wide, unbranched, acyanophilous. Cystidia conspicuous, formed by heavy encrustations on apical ends cylindrical, often projecting in the pore cavity, 5.3–12.9 µm wide. Basidia clavate, 4-sterigmate, 18.3–20.3×4.7–5.7 µm, thin-walled. Basidiospores clavate, cylindric to cylindric-ellipsoid, hyaline, thin-walled, smooth, 4.5–6.5×2.5–3.2 µm, non-amyloid.

Distribution: Bhutan: Thimphu, Nawephu, H.P.: Kullu, Kinnaur U.K.: Dehra Dun, Mussoorie.

Collection examined: SSV 21298, IBP 37187, 37188, L 37189

Substratum: On decaying *Pinus roxburghii* log, on gymnospermic wood.

Remarks: This species is characterized by effused-reflexed coriaceous fructification; dimitic hyphal system, encrusted cystidia and thin-walled cylindric ellipsoid spores.

Irpex vellereus Berk. & Broome, J. Linn. Soc., Bot. 14(2): 61 (1875). Plate 6.22f

Fructification resupinate, effused reflexed, leathery to corky, imbricate up to 3.4×1.3×0,5 µm; upper surface pinkish buff, matted tomentose, faintly azonate, margin thin. Context cream coloured to pinkish buff fibrous; hymenial surface pinkish buff, brown in old specimens, irpiciform, poroid near the margin; pores 1–2 per mm.

Hyphal system monomitic; hyphae hyaline thin to thick-walled, branched, simple septate, 4–6.4 µm in diameter. Cystidia cylindrical to subclavate, slightly encrusted up to 6.4 µm broad; Basidia clavate, up to 5.5 µm broad. Basidospores hyaline, ellipsoid to ovoid, thin-walled, 4.1–5.8×2.5–3.5 µm.

Distribution: U.K.: Mussoorie, Tehri; H.P: Mandi
Collection examined: IBP 37190, L 37191.
Substratum: On decaying angiospermic log, log of *Shorea robusta*
Remarks: This species is characterized by resupinate to effused-reflexed, imbricate fructification; hymenial surface pinkish buff; ellipsoid to ovoid basidiospores.

Junghuhnia **Corda,**
Anl. Stud. Mycol.:195, 1842.
Emend. Ryv., Persoonia 7:17–21, 1972.

Fructifications resupinate, membranous, coriaceous to membranous, becoming somewhat horny on drying, widely effused; pore surface cream pink to light pink, glaucous, not creviced; margin thinning to more or less abrupt, fibrillose, adnate, white to paler concolorous. Context composed of compactly arranged hyphae, subhyaline to pink pale in section. Hyphal system dimitic; skeletal hyphae, subhyaline to tinted yellow, thick, sometimes leaving little or no lumen; generative hyphae, branched, septate, clamped, thin-walled, subhyaline. Pores not stratose, angular to oval; dissepiments somewhat hairy at the edges. Cystidia clavate to clavate-cylindrical, the walls subhyaline, thick leaving little or no lumen, heavily impregnated with subhyaline crystals especially in the upper part, arising from the skeletal hyphae and may remain immersed in the context, dissepiments or project out of the hymenium. Gloeocystidia absent. Basidia small, clavate-cylindrical, 4-spored. Basidiospores variable in shape, minutely apiculate, the walls thin, subhyaline, smooth, nonamyloid, acyanophilous.

Twenty species, widespread

Lit.: Ryvarden (*Persoonia* 7: 17, 1972), Ipulet & Ryvarden (*Syn. Fung.* 20: 87, 2005; Uganda)

Type Species: *Laschia crustacea* Jungh. 1838

Habitat: Dead wood

Himalayas: Four

Key to species

1. Fructification resupinate, adnate, widely effused.. 2
1. Fructification pileate to effused reflexed.. *J. zonata*
2. Basidiospores smooth to allantoids.. *J. collabens*
2. Basidiospores smooth to cylindrical.. 3
3. Basidiospores 3.6–5.0 × 1.6–1.9 µm.. *J. luteoalba*
3. Basidiospores 3.2–4.5 × 1.2–2 µm.. *J. nitida*

Junghuhnia collabens (Fr.) Ryvarden, Persoonia 7(1): 18 (1972) = *Irpex collabens* (Fr.) Kotir & Saaren., Polish Bot. J. 47(2): 104, 2002 = *Polyporus collabens* Fr., Hym. Europ., p. 572, 1874 = *Poria rixosa* Karst., Rev. Mycol. 3,9: 19, 1881. Plate 6.23a, Fig. 6.57m–p

Fructification annual, largely resupinate, widely effused reflexed, adnate to separable. Pileus effused reflexed, confluent, forming wider pilei, imbricate; upper surface white to creamish, drying creamish brown, glabrous near the margin; margin thin

Plate 6.23 (**a**) *Junghuhnia collabens*. (**b**) *Junghuhnia luteoalba*. (**c**) *Junghuhnia nitida*. (**d**) *Junghuhnia zonata*. (**e**) *Phlebia livida* (**f**) *Phlebia radiata*

to abrupt, adnate, creamish to pinkish brown. Pore surface pinkish to pinkish-brown, even, drying somewhat darker; pores rounded to angular. Context subhyaline to light pink to brown, thin firm.

Hyphal system dimitic; generative hyphae hyaline, thin-walled, septate, branched, clamped, cyanophilous, 2.1–4.1 μm. in diameter; skeletal hyphae subhyaline, thick-walled to highly thick-walled, sparsely branched, acyanophilous, Pores stratose, each stratum up to 1 mm thick; pore mouth round to angular, 60–150 μm in diameter, 1–6 per mm; dissepiments 25–75 μm thick, equal, hairy. Cystidia abundant, subhyaline, clavate to clavate-cylindrical, thick-walled with narrow lumen, highly encrusted, encrusted, 34–60×7–13 μm arising from the skeletal hyphae and remain immersed in the trama or projecting up to 25 μm out of the hymenium. Basidia hyaline, clavate, 4-spored, about, 10.2–12.1×3.6–5.1 μm. Basidiospores, hyaline, thin-walled, smooth, cylindrical, straight to curved, non-amyloid, 2.9–4.5×1.4–2.2 μm.

Distribution: H.P.: Kullu, U.K.: Chakrata.

Collection examined: SSR 6101, 6169. SSV 21303, IBP 37688

Substratum: On decaying log of *Abies pindrow, Pinus excelsa.*

Remarks: This species is characterized by resupinate fructification; heavily encrusted, large cystidia; straight and curved basidiospores.

Junghuhnia luteoalba (Karst.) Ryv., Persoonia 7: 18. 1972=*Physisporus luteoalbus* Karst., Rev. Mycol. 9:10. 1887=*Poria luteoalba* (Karst.) Sacc., Syll. Fung. 6: 299. 1888. Plate 6.23b, Fig. 6.58a–d

Fructifications annual to often reviving, resupinate, adnate to mostly widely effused long along the wood, soft when fresh, hard on drying, adanate to separable near the margin; margin entire, lighter concolorous with pore surface. Pore surface pale brown to brownish orange; pores medium sized, angular, 3–5 per mm; tubes in one layer, light orange. Context thin, light orange, more homogenous, non-xanthochroic.

Hyphal system dimitic; generative hyphae hyaline, thin- walled, branched, aseptate, clamped, cyanophilous, 1.9–3.1 μm in diameter; skeletal hyphae subhyaline, thick-walled straight, aseptate, unbranched, acyanophilous, 2.5–4.2 μm in diameter. Cystidia numerous, embedded or projecting up beyond the hymenium in to pore cavity, especially encrusted, long, 30.2–65.2×7.2–17.1 μm. Basidia clavate, 4-spored, cyanophilous, up to 5 μm broad. Basidiospores hyaline, thin-walled, smooth, cylindric, slightly curved non-amyloid, 3.6–5.0×1.6–1.9 μm.

Distribution: U.K.: Chakrata, Mussoorie, Nanital; Bhutan-Thimphu; Manipur-Ukhrul.

Collection examined: SSV 21288, 21259, IBP 42945.

Substratum: On angiospermic twig.

Remarks: This species was first described by Bakshi (1971) as *Poria lutea-alba* (Karst.) Sacc., occurring on the logs of *Cedrus deodara* from Chakrata (U.K.). However, this species seems to be widely distributed in the Himalayas since above cited collections were made from widely separate localities, including Bhutan and from Manipur. The species is marked by annual, resupinate, widely effused fructifications with pale brown pore surface; thin, yellowish context, encrusted cystidia; and hyaline, cylindric, slightly curved basidiospores. It is a new record for Eastern Himalayas.

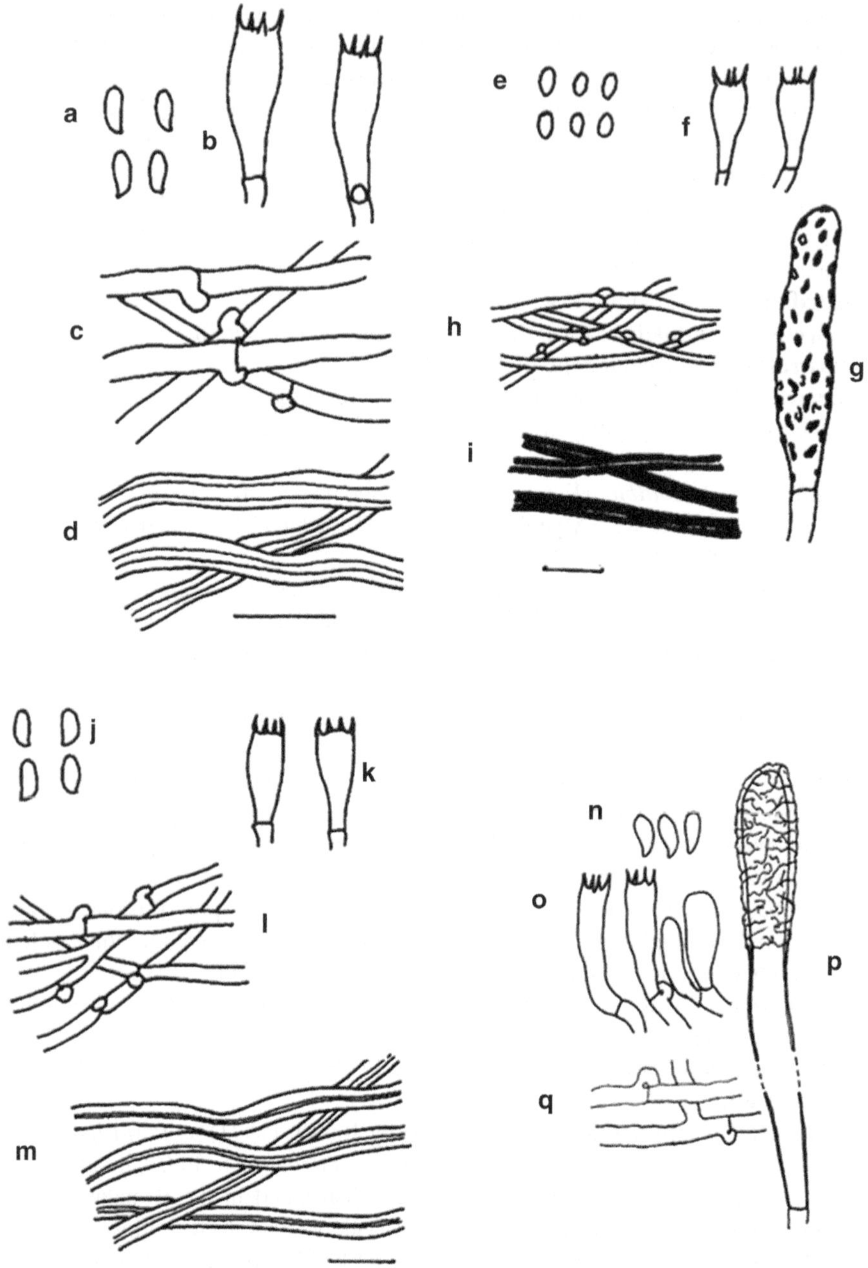

Fig. 6.58 (**a**–**d**) *Junghuhnia luteoalba* (**a**) Basidiospores, (**b**) Basidia, (**c**) Generative hyphae, (**d**) Skeletal hyphae; (**e**–**i**) *Junghuhnia nitida* (**e**) Basidiospores, (**f**) Basidia, (**g**) Cystidia, (**h**) Generative hyphae, (**i**) Skeletal hyphae; (**j**–**m**) *Junghuhnia zonata* (**j**) Basidiospores, (**k**) Basidia, (**l**) Generative hyphae, (**m**) Skeletal hyphae; (**n**–**q**) *Mycoacia fuscoatra* (**n**) Basidiospores, (**o**) Basidia, (**p**) Encrusted cystidia, (**q**) Generative hyphae

Junghuhnia nitida (Pers.) Ryv., Persoonia 7(1): 18. 1972=*Polyporus nitidus* (Pers.) Fr.,Observ. Mycol. 2: 262, 1818=*Poria eupora* (Karst.) Cooke, Grevillea 14: 110. 1886. Plate 6.23c, Fig. 6.58e–i

Fructifications annual to often reviving, resupinate to adnate, separable, usually arise as small circular patches which later become effused; margin creamish, thin, pubescent to fimbricate. Pore surface even, creamish to pale pinkish; pores rounded to angular, averaging 5–7 per mm; dissepiments 30–85 µm, thick, equal, apices velutinate, tubes non-stratose, rigid and tough on drying, pinkish brown, up to 1.5 mm deep. Context creamish, soft, thin, homogenous, not darkening in KOH.

Hyphal system dimitic; generative hyphae hyaline, thin-walled, branched, septate, clamped, cyanophilous, 1.9–2.89 µm in diameter; skeletal hyphae subhyaline, thick-walled to highly thick-walled with narrow lumen, walls up to 1.5 µm thick, rarely branched, aseptate, acyanophilous, 2.2–4.4 µm in diameter, hyphal pegs present, up to 10 µm in diameter projecting into the pore cavity. Cystidia abundant, clavate to clavate cylindrical, subhyaline, thick-walled, highly encrusted all over, 32–70×7.5–15 µm immersed as well as projecting into the pore cavity. Basidia hyaline, cyanophilous, thin-walled, clavate,2–4 spored, 8–12×3–4.5 µm, sterigmata up to 3 µm long. Basidiospores hyaline, 3.2–4.5×1.2–2 µm.

Distribution: U.K.: Nainital, Chakrata, Mussoorie; H.P.: Shimla- Narkanda; Bhutan-Thimphu

Collection examined: MPS 6290. SSV 21287, 21314, IBP 37192, L 37193.

Substratum: On log under a mixed forest.

Remarks: This species has been previously described from India as *Poria eupora* Karst. on *Picea morinda* Link by Anonymous (1950), Bagchee and Singh (1960). Bakshi (1971) described it as *P. eupora* (Karst.) Cooke from Chakrata forest division (U.K.) Thind and Dhanda (1980) described it in detail. The species is characterized by resupinate, pale pinkish, annual fructifications; dimitic hyphal system; heavily encrusted cystidia; and small ellipsoid basidiospores.

Junghuhnia zonata (Bres.) Ryvarden, Persoonia 7(1): 18 (1972)=*Irpex zonatus*(Bres.) Saaren. & Kotir., in Kotiranta & Saarenoksa, Polish Bot. J. 47(2): 107(2002). Plate 6.23d, Fig. 6.58j–m

Fructifications annual, pileate to effused reflexed, soft coraceous when fresh, hard on drying, imbricate, many pilei arise from the same base. Pileus dimidiate, applanate, flabelliform to irregularly shaped, conchate; upper surface white to cream, faintly concentrically zonate, finely tomentose, later glabrous, radially wrinkled on drying; margin smooth, acute, concolorous with upper surface. Pore surface white to cream, uneven; pores angular, 3–4 per mm, distinct pores only present near the margin, at older portion the pores torn in to spines; tubes white, up to 1.5 mm deep. Context white, soft homogenous, non-xanthochroic, up 2.4 mm thick.

Hyphal system dimitic, generative hyphae hyaline, thin-walled, branched, septate, clamped, cyanophilous, 2.9–4.6 µm in diameter; skeletal hyphae hyaline, thick-walled, unbranched, acyanophilous, 2.5–3.8 µm in diameter. Cystidia absent.

Basidia clavate, hyaline, 4-spored, cyanophilous up to 8 μm in diameter. Basidiospores hyaline, thin-walled, smooth, ellipsoid, 4.3–5.9 × 2.3–3.9 μm.

Distribution: Bhutan: Thimphu, Dochula; Sikkim

Collection examined: SSV 21311.

Substratum: On decaying angiospermic stump.

Remarks: The species is characterized by annual, effused reflexed, inbricate fructifications; cream yellow, glabrous, concentrically zonate upper surface; creamish brown pore surfaces; pores at first angular, soon torn in top spines; dimitic hyphal system; cystidia absent; hyaline, smooth, ellipsoid basidiospores.

Mycoacia Donk,
Med. Nederi. Mycol. Ver., 18–20: 150 (1931)

Fructifications resupinate, membranous-ceraceous to subceraceous but turns rigid and somewhat brittle on drying, adnate; hymenial surface toothed, yellow-brown to dark brown. KOH reaction tissues not or slightly darkening in 3 % sol. Teeth subulate or subcylindrical with acute apices without any axial bundles of cystidia or cystioles. Subiculum yellow-brown in section, composed of compactly arranged more or less agglutinated hyphae. Hyphal system monomitic, hyphae branched, septate, clamps absent or occasionally present, the walls thin to moderately thick, often tinted brown. Cystidia generally absent or may occur as thin-walled lepocystidia. Gloeocystidia absent. Basidia clavate, 4-spored. Basidiospores ellipsoid to suballantoid, thin-walled, subhyaline, nonamyloid.

Fifteen species, widespread

Lit.: Ragab (*Mycol.* **43:** 459, 1951), Hjortstam & Ryvarden (*Syn. Fung.* **18:** 20, 2004).

Type Species: *Hydnum fuscoatrum* Fr. 1879

Habitat: Dead wood

Himalaya: Two

Key to species

1. Basidiospores 4–4.7 × 2.3–3.9 μm, ellipsoid, minutely apiculate .... *M. fuscoatra*
1. Basidiospores 3.6–4(4.2) × 1.6–2.0 μm, narrowly ellipsoid to suballantoid .... *M. stenodon*

Mycoacia fuscoatra (Fr.) Donk, Med. Nederl. Mycol. Ver. 20: 152. 1931 = *Hydnum fuscoatrum Novit. fl.* svec. 2: 39, 1814. Fig. 6.58n–q

Fructifications resupinate, membranous-ceraceous to wholly ceraceous becoming hard and brittle drying, adnate, widely effused; hymenial surface distinctly toothed, yellowish brown to light brown, continuous, not creviced; margin thinning to more or less abrupt, adnate, concolorous. Teeth up to 1 mm long, subulate with acute apices, terrete when fresh but often clustering on drying. Subiculum composed of compactly arranged to almost agglutinated hyphae.

Hyphal system monomitic, hyphae 2–4.2 μm wide, branched, septate, clamped, the walls thin, subhyaline to tinted yellow, often collapsing and difficult to discern.

Cystidia rather scanty, clavate-cylindrical, immersed in trama of spine or projecting slightly out of the hymenium, the walls slightly thick, subhyaline to tinted yellow, smooth or finely impregnated with subhyaline crystals. Basidia 15.2–18.2 × 3.5–4.2 µm, clavate-cylindrical, 4-spored. Basidiospores 4–4.7 × 2.3–3.9 µm, ellipsoid, minutely apiculate, the walls thin, subhyaline, smooth, non-amyloid.

Distribution: U.K.: Hemkunt.

Collection examined: HSK 4270.

Substratum: On log under conifers.

Remarks: This species is characterized by the presence of thin-walled cystidia and clamped hyphae, presence of subulate teeth with acute apices, terrete when fresh but often clustering on drying.

Mycoacia stenodon (Pers.) Donk, Med. Nederl. Mycol. Ver.18–20: 151. 1931. = *Hydnum stedon* Pers., Mycol eur. II, 188. 1825. Fig. 6.59a–c

Fructifications resupinate, membranous to membranous ceraceous but turns hard and brittle on drying, adnate, widely effused up to 2 mm thick; hymenial surface distinctly toothed, yellow brown to dark brown to almost brownish black with age, continuous, not crevised; margin thinning, adnate paler concolorous. Teeth up to 2 mm long, subulate with acute apices, terrete when fresh but often clustering on drying. Subiculum yellowish brown, composed of compactly arranged more or less agglutinated hyphae.

Hyphal system monomitic; hyphae 1.7–3.7 µm wide, branched, septate, clamped, clamps prominent and almost at all septa, thin-walled, subhyaline to finely tinted yellow. Cystidia absent. Basidia 12.2–18.2 × 3.2–4.2 µm, clavate-cylindrical, 4-spored. Basidiospores 3.6–4(4.2) × 1.6–2.0 µm, narrowly ellipsoid to suballantoid, minutely apiculate, the walls thin, subhyaline, smooth, nonamyloid.

Distribution: U.K.: Chakrata. Bhutan-Thimphu, Manipur-Ukhrul.

Collection examined: SSV 21288, SSV 21259

Substratum: On angiospermic twig.

Remarks: The characteristic features of this species are clamped hyphae, lack of cystidia and small narrowly ellipsoid to suballantoid basidiospores.

Phlebia Fr.,
Systema Mycol. 1: 426, 1821.
emend. Donk, Fungus 27: 8. 1957.

Fructifications resupinate, membranous-ceraceous to wholly ceraceous, becoming rigid or horny on drying, loosely adnate to separable, often effused widely; hymenial surface smooth to tuberculate or distinctly toothed, sometimes with reticulate folds or ridges which usually disappear on drying. Subiculum composed of compactly arranged, subceraceous or ceraceous hyphae which are difficult to discern individually. Hyphal system monomitic, hyphae with or without clamps. Cystidia present but usually absent. Gloeocystidia absent. Basidia usually clavate with a tapering base, 4-spored. Basidiospores ellipsoid to suballantoid, usually less than 6 µm long, thin-walled, subhyaline, smooth, nonamyloid, acyanophilous.

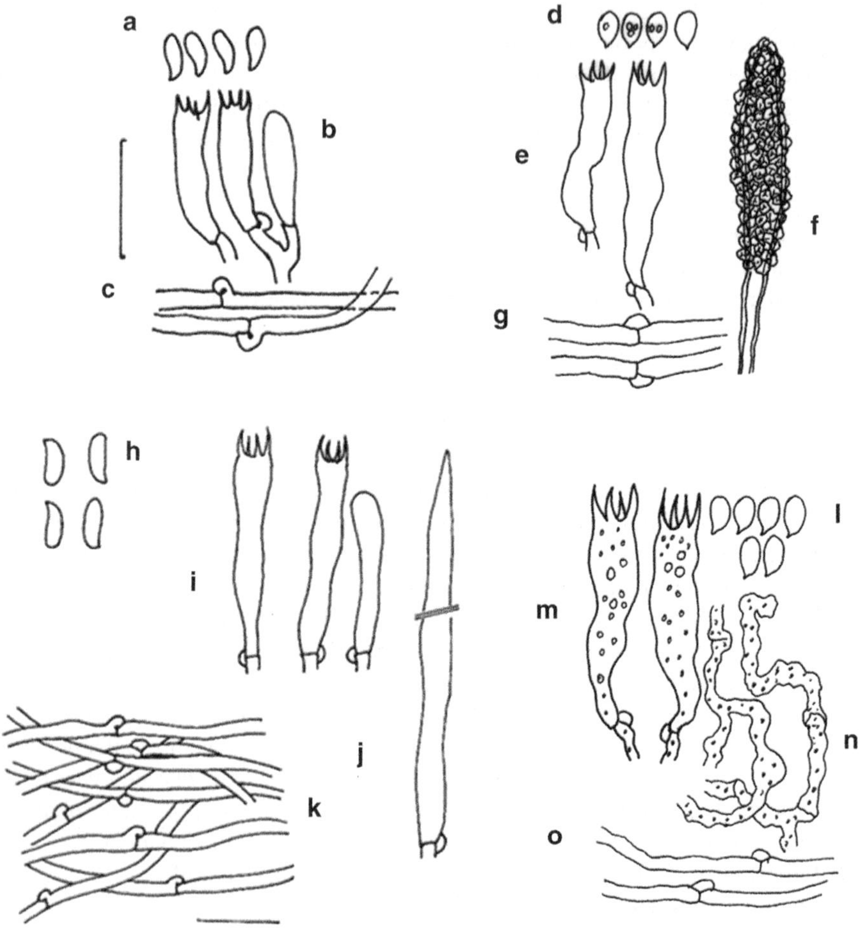

Fig. 6.59 (**a**–**c**) *Mycoacia stenodon* (**a**) Basidiospores, (**b**) Basidia, (**c**) Generative hyphae; (**d**–**g**) *Phlebia crassisubiculata* (**d**) Basidiospores, (**e**) Basidia, (**f**) Encrusted cystidia, (**g**) Generative hyphae*;* (**h**–**k**) *Phlebia livida* (**h**) Basidiospores, (**i**) Basidia, (**j**) Cystidia, (**k**) Generative hyphae; (**l**–**o**) *Phlebia interjacenoids* (**l**) Basidiospores, (**m**) Basidia, (**n**) Generative hyphae in context, (**o**) Generative hyphae

Lit.: Cooke (*Mycol.* **48:** 386, 1956), Spirin & Zmitrovich (*Novosti Sistematiki Nizshikh Nov. Sist. Niz. Rast.* **37:** 166, 2004)
Type Species: *Phlebia radiata* Fr. 1821
Habitat: Dead wood
Himalayas: Twelve (Table 6.5)

Table 6.5 Synopsis of *Phlebia* species

Name	Hymenial surface colour/type	Cystidia type	Basidiospores
P. crassisubiculata	Cream yellow, darkening in 3 % KOH; smooth to pubescent	Pseudoseptate, thick walled, 23–80 × 7–10 μm	Ellipsoid, smooth, thin-walled, inamyloid, acyanophilous 5.3–6.9 × 2.9–4.7 μm
P. livida	Reddish brown to greyish brown; smooth to tuberculate to odontoid	Absent	Subclylindrical to suballantoid, thin-walled, smooth, inamyloid, acyanophilous; 3.2–4.1 × 1.8–2.0 μm
P. interjacenoides	Whitish; smooth with cracks in mature part	Absent	Ellipsoid to suballantoid, smooth, thin-walled, inamyloid, acyanophilous; 4.2–6.1 × 2.2–3.2 μm
P. kamengii	Yellowish white to greyish yellow; reticulately folded to almost poroid	Narrowly clavate, thin-walled, embedded in subhymenium 35–65 × 7–11 μm	Broadly ellipsoid, smooth, thin-walled, inamyloid, acyanophilous, with 1- more oil droplets; 4.6–6.6 × 3.1–4.6 μm
P. microspora	Yellowish white to pale yellow; odontoid	Cystidia fusiform thin-walled to thick-walled; encrusted in the upper half; 30–60 × 4.7–6 μm	Broadly ellipsoid to ovoid, smooth, thin-walled, inamyloid, acyanophilous 3.3–4.6 × 2.4–3.5 μm
P. radiate	Reddish brown to flesh coloured; reticulately folded	Cystidia clavate, horizontal/ vertical; 65–100 × 8–11 μm	Suballantoid, smooth, thin-walled, inamyloid, acyanophilous; 4.0–5.0 × 1.5–2.5 μm
P. rufa	Pale yellowish to brownish; reticulately folded	Cystidia narrowly clavate; 30–50 × 6–9 μm	Suballantoid, smooth, thin-walled, inamyloid, acyanophilous; 4.3–6.1 × 2.1–2.6 μm
P. sagregata	Whitish to brownish; smooth	Subfusiform tosubclavate; 45–60 × 6–7.2 μm	Subellipsoid to suballantoid, smooth, thin-walled, inamyloid, acyanophilous; 4.8–6.0 × 2.4–3.0 μm
P. subserialis	Cream yellow to yellow brown; smooth to sulcate	Leptocystidia, impregnated at the tip; 44–74 × 4.8–8.1 μm	Ellipsoid cylindrical to suballantoid, smooth, thin-walled, inamyloid, acyanophilous; 4.5–6.4 × 2.0–2.4 μm
P. subochracea	Yellow to yellowish brown; toothed	Absent	Ellipsoid, suballantoid, smooth, thin-walled, inamyloid, acyanophilous; 4.6–5.6 × 3.0–3.3 μm
P. thindii	Yellowish white to pale yellow; smooth to tuberculate	Absent	Ellipsoid, suballantoid, smooth, thin-walled, inamyloid, acyanophilous; 5.2–7.6 × 2.5–3.5 μm
P. singularis	Pale yellow to pestle yellow to ochraceous	Cystidia subcylindrical often widening in the apical part, 50–85 × 5–9 μm	Ellipsoid to obovate, suballantoid, smooth, thin-walled, inamyloid, acyanophilous; 4.6–6.1 × 2–3.6 μm

Phlebia crassisubiculata Singh Avn. P., Priyanka, Dhingra & Singla, Mycotaxon 112: 21–24. Fig. 6.59d–g

Fructification resupinate, arising as small colonies which may coalesce later on become effused, adnate; hymenial surface smooth to finely pubescent under lens due to projecting cystidia, creamy yellow, darkening in 3 % KOH sol; margins abrupt or indeterminately thinning, paler concolorous.

Hyphal system monomitic; generative hyphae up to 4 μm wide, branched, septate, clamped, thin- to somewhat thick-walled. Cystidia 23.2–80.2 × 7.0–10.4 μm, thick-walled, enclosed to somewhat projecting, pseudo-septa may be present. Basidia 23.2–40.3 × 5.1–6.4 μm, subclavate, thin-to somewhat thick-walled, sterigmata, with a basal clamp; sterigmata up to 5.1 μm long. Basidiospores 5.3–6.9 × 2.9–4.7 μm, ellipsoid, smooth, thin-walled, inamyloid, acyanophilous, with oil droplets.

Distribution: H.P.: Chamba, 2 km from Dalhousie in direction to Banikhet, Shimla- Glen.

Collection examined: IBP37185.

Substratum: On underside of gymnospermous stump.

Remarks: This species is characterized by resupinate, effused fructification; hymenial surface creamy yellow; margin abrupt; hyphal system monomitic, hyphae clamped, septate and ellipsoid basidiospores.

Phlebia interjacenoides Dhingra, J. Ind. Bot. Soc. 84: 116, 2005. Fig. 6.59l–o

Fructification resupinate, closely adnate, effused up to 160 μm thick in section, ceraceous-subgelatinous when fresh, crustaceous-corneous on drying; hymenial surface smooth, continuous, cracks developing in mature parts on drying, whitish; margins indistinct.

Hyphal system monomitic; generative hyphae up to 2.8 μm wide, thin-walled, septate, clamped, richly branched and interwined, often gelatinized mature parts; basal zone not well differentiated, only a few straight, sparsely branched hyphae observed next to substrate; subhymenial hyphae winding, with irregular constrictions and dilations, and with numerous oil drops. Cystidia absent. Basidia 18.2–30.1 × 4.8–6.2 μm, clavate, basally narrowing into hyphae like part, 4-sterigmate, with a basal clamp; sterigmata up to 6 μm long. Basidiospores 4.2–6.1 × 2.2–3.2 μm, ellipsoid to suballantoid, smooth, thin-walled, inamyloid, acyanophilous.

Distribution: Bhutan; India: U.K.-Mussoorie.

Collection examined: GSD 19628, IBP 37197.

Substratum: On gymnospermous wood.

Remarks: The species is characterized by ceraceous-subgelatinous to crustaceous-corneous fructification; hymenial surface smooth cracks develop in mature parts on drying, hyphal system monomitic, cystidia absent, ellipsoid to suballantoid basidiospores. It is a new record for N. W. Himalayas/India.

Phlebia livida (Pers.) Bres., Atti, Accad. Agiati. Rovereto 3: 105. 1897.
Plate 6.23e, Fig. 6.59h–k

Fructification resupinate, closely adnate, ceraceous to subgelatinous when fresh and living, firm membranous to corneous on drying adnate but tends to become loose on drying, widely effused; hymenial surface reddish brown to greyish brown, abundantly sporulating parts, rarely cracking irregularly in thicker parts when rapidly dried; margin thinning to more or less abrupt loosely adnate, paler concolorous, smooth, tuberculate, phlebioid, odontoid. Subiculum composed of a basal zone of compactly arranged, slightly thick-walled. Repent hyphae and upper zone of slightly thin-walled more or less agglutinated semi-erect hyphae, in mature parts the context becomes wholly ceraceous and hyphae are difficult to discern.

Hyphal system monomitic; generative hyphae 2.5–4.8 μm wide, branched, septate, clamped, thin-walled to moderately thick, subhyaline basal hyphae thick-walled. Cystidia absent. Basidia subclavate, 4-sterigmate, 15–25 × 3.2–4.2 μm with a basal clamp. Basidiospores smooth, 3.2–4.1 × 1.8–2.0 μm, subcylindrical to suballantoid, thin-walled, non-amyloid, acyanophilous.

Distribution: H.P.: Narakanda.

Collection examined: SSR 5357, 5648, IBP 37194.

Substratum: On log of *Abies pindrow, Pinus excels.*

Remarks: It is a common species in Himalayas. The above collection resembles *Phebia livida* as described by Rattan (1977).

Phebia microspora Dhingra, J. Ind. Bot. Soc. 84:114, 2005. Fig. 6.60e–h

Fructification resupinate, adnate, effused, thin; hymenial surface yellowish-white to pale yellow when fresh, pale ochraceous after drying, odontoid with dense; margins thinning, paler concolorous.

Hyphal system monomitic; generative hyphae up to 3.7 μm wide, branched, septate, clamped, thin to somewhat thick-walled, densely united into a conglutinate tissue both in the subiculum and subhymenium. Cystidia 30.1–60.2 × 4.7–6.1 μm, numerous, especially in the aculei, generally fusiform, thin-walled when young to thick-walled on maturity, encrusted in the apical half, often secondarily septate. Basidia 10.2–15.1 × 3.3–4.6 μm, clavate to subclavate, generally 4-sterigmate, with a basal clamp; sterigmata up to 3.5 μm long. Basidiospores 3.6 μm long. Basidiospores 3.3–4.6 × 2.4–3.5 μm, broadly ellipsoid to ovoid, smooth, thin-walled, inamyloid, acyanophilous.

Distribution: Bhutan.

Collection examined: GSD 19203.

Substratum: On angiospermous wood.

Remarks: This species is close to *Phlebia queletii* (Bourd. & Galz.) Christ. in having odontoid fructification and the resemblance in hyphal and cystidial elements. But it differs from it in having smaller basidia (10.2–15.1 × 3.3–4.6 μm in *P. microspora* as compared to 15–25 × 4–5 μm in *P. queletii*) and smaller basidiospores (3.3–4.6 × 2.4–3.5 μm in *P. microspora* as compared to 5–6 × 3–3.5 μm in *P. queletii*).

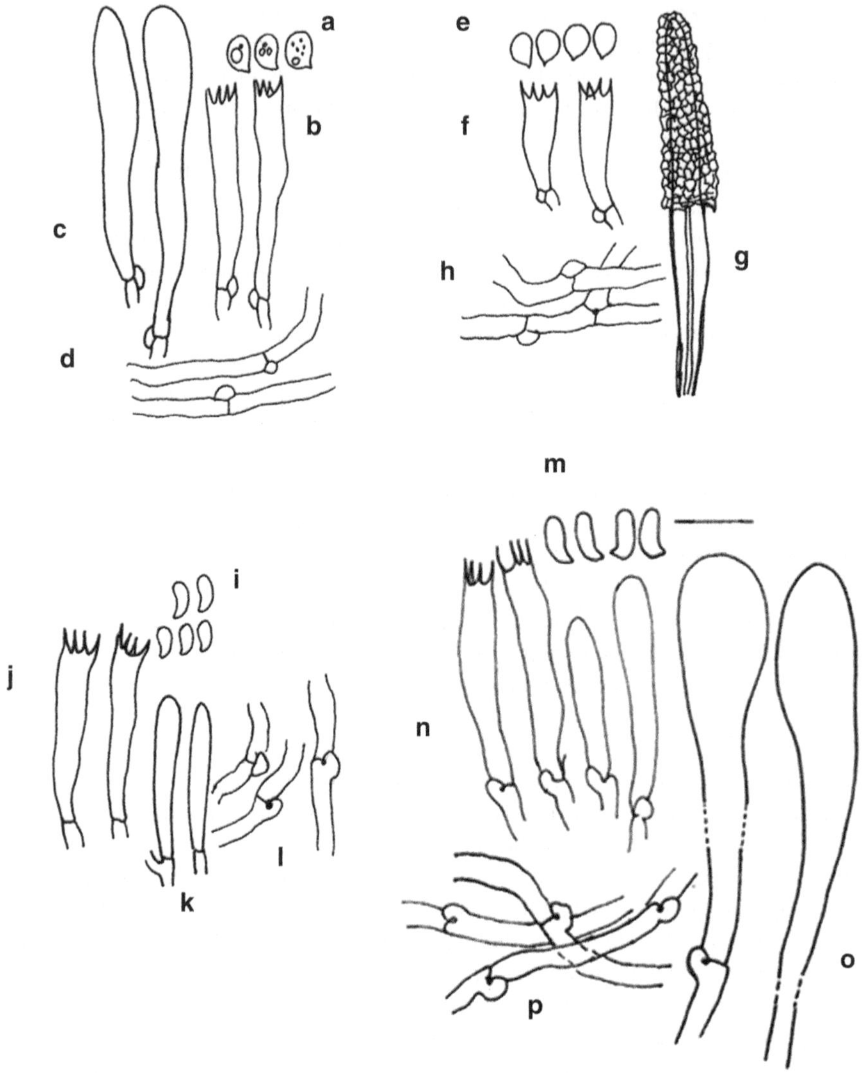

Fig. 6.60 (**a–d**) *Phlebia kamengii* (**a**) Basidiospores, (**b**) Basidia, (**c**) Cystidia, (**d**) Generative hyphae; (**e–h**) *P. microspora* (**e**) Basidiospores, (**f**) Basidia, (**g**) Encrusted cystidia, (**h**) Generative hyphae; (**i–l**) *P. radiata* (**i**) Basidiospores, (**j**) Basidia, (**k**) Cystidia, (**l**) Generative hyphae; (**m–p**) *P. rufa* (**m**) Basidiospores, (**n**) Basidia, (**o**) Cystidia, (**p**) Generative hyphae

Phlebia radiata Fr., Syst. Mycol. 1: 427.1821. Plate 6.23f, Fig. 6.60i–l

Fructification resupinate, ceraceous when fresh turning brittle and horny on drying, loosely adnate, arising in small circular colonies which coalesce; radially-irregularly plicate, reddish brown to flesh coloured; hymenial surface irregularly reticulate-folded; margin more or less abrupt, usually rolling away from the substratum on drying. Subiculum composed of compactly arranged more or less agglutinated and ceraceous hyphae.

Hyphal system monomitic; generative hyphae 3.0–4.5 μm wide richly branched, clamped, thin-walled. Cystidia present horizontal or vertical cystidia 65–100 × 8–11 μm narrowly clavate. Basidia 25.1–35.1 × 4.1–4.2 μm, narrowly clavate, 4-sterigmate, with a basal clamp, arranged in a dense palisade and smooth. Basidiospores suballantoid, thin-walled, nonamyloid, acyanophilous 4.0–5.0 × 1.5–2.25 μm.

Distribution: H.P.: Shimla, Dalhousie.

Collection examined: SSR 5557, 5653, IBP 37198.

Substratum: On stump of *Junglens regia.*

Remarks: Rattan (1977) reported this species for the first time India on the basis of two collections from the N.W. Himalayas. The characteristic features of this species are reticulately-folded, reddish brown; hymenial surface and small suballantoid basidiospores.

Phlebia rufa (Fr.) M.P. Christ., Dansk. bot. Arkv. 19(2): 164, 1960 = *Merulius rufus* Pers. Syn. meth. Fung. 2: 498, 1801. Plate 6.24a, Fig. 6.60m–p

Fructification resupinate, closely adnate, effused, ceraceous-fleshy when fresh, corneous on drying; hymenial surface pale yellowish to brownish, darker on drying, reticulately folded to almost poroid; margin thinning, determinate, paler concolorous.

Hyphal system monomitic; generative hyphae clamped; subicular hyphae 3.2–4.6 μm wide, somewhat thick-walled, horizontal, 2–3 μm wide, thin-walled, densely united, vertical in the subhymenium, gelatinised zone between subiculum and subhymenium. Cystidia 30.1–50.1 × 6.1–9.1 μm, clavate or narrowly clavate, thin-walled, horizontal in transition layer and vertical in subhymenium. Basidia 22.2–30.1 × 3.6–4.8 μm, narrowly clavate, 4-sterigmate, with a basal clamp. Basidiospores 4.3–6.1 × 2.1–2.6 μm, smooth, suballantoid, thin-walled, non-amyloid, acyanophilous.

Distribution: Meghalaya: Shillong, Risa colony; H.P.: Kullu.

Collection examined: GSD 19002, IBP 37195.

Substratum: On a decaying angiospermic log, angiospermic twigs.

Remarks: This species is characterized by reticulately folded to almost poroid brown coloured hymenium, generative hyphae with clamp and suballantoid spores. It resembles *Phlebia radiata* anatomically but the later differs in having radially folded, orange-red hymenial surface. The above collection resembles the description of *Phlebia rufa* given by Eriksson et al. (1981). It has been recorded from Eastern Himalayas by Dhingra et al. (2011). However, it is being recorded from Himachal Pradesh/N. W. Himalayas for the first time.

Phlebia segregata (Bourdot & Galzin) Parmasto, Eesti NSV Tead. Akad. Toim., Biot. Seer 16 (4): 393, 1967 = *Peniophora segregata* Bourdot & Galzin, Hymenomyc. De France: 284, 1928. Fig. 6.61a–d

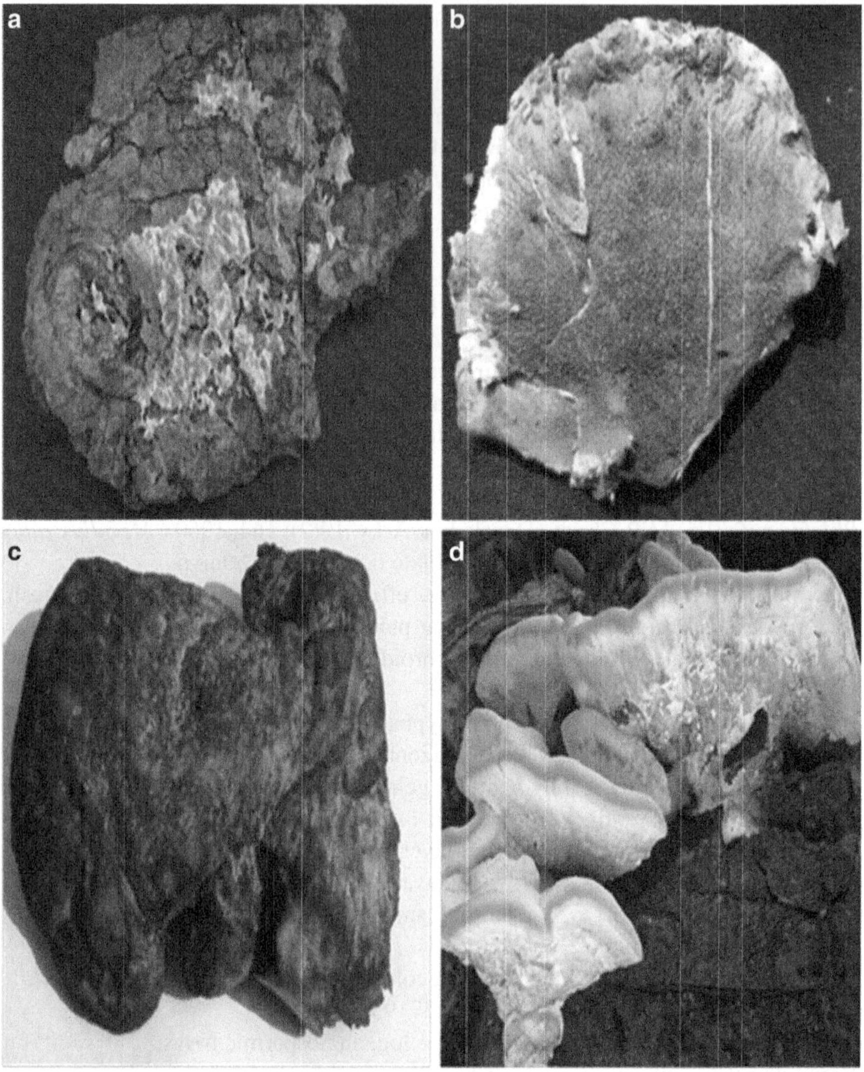

Plate 6.24 (**a**) *Phlebia rufa*. (**b**) *Sarcodontia delectans*. (**c**) *Sarcodontia spumea*. (**d**) *Antrodiella semisupina*

Fructification resupinate, adnate, ceraceous; hymenial surface smooth whitish brownish; margins indeterminate to abrupt.

Hyphal system monomitic; generative hyphae up to 3.4 μm wide, branched, septate, clamped, basal hyphae thick-walled, less branched, long-celled, horizontal; subhymenial hyphae thin-walled, much branched, short-celled. Cystidia 45–60 × 6–7.2 μm, numerous, subfusiform to subulate with tapering apex,

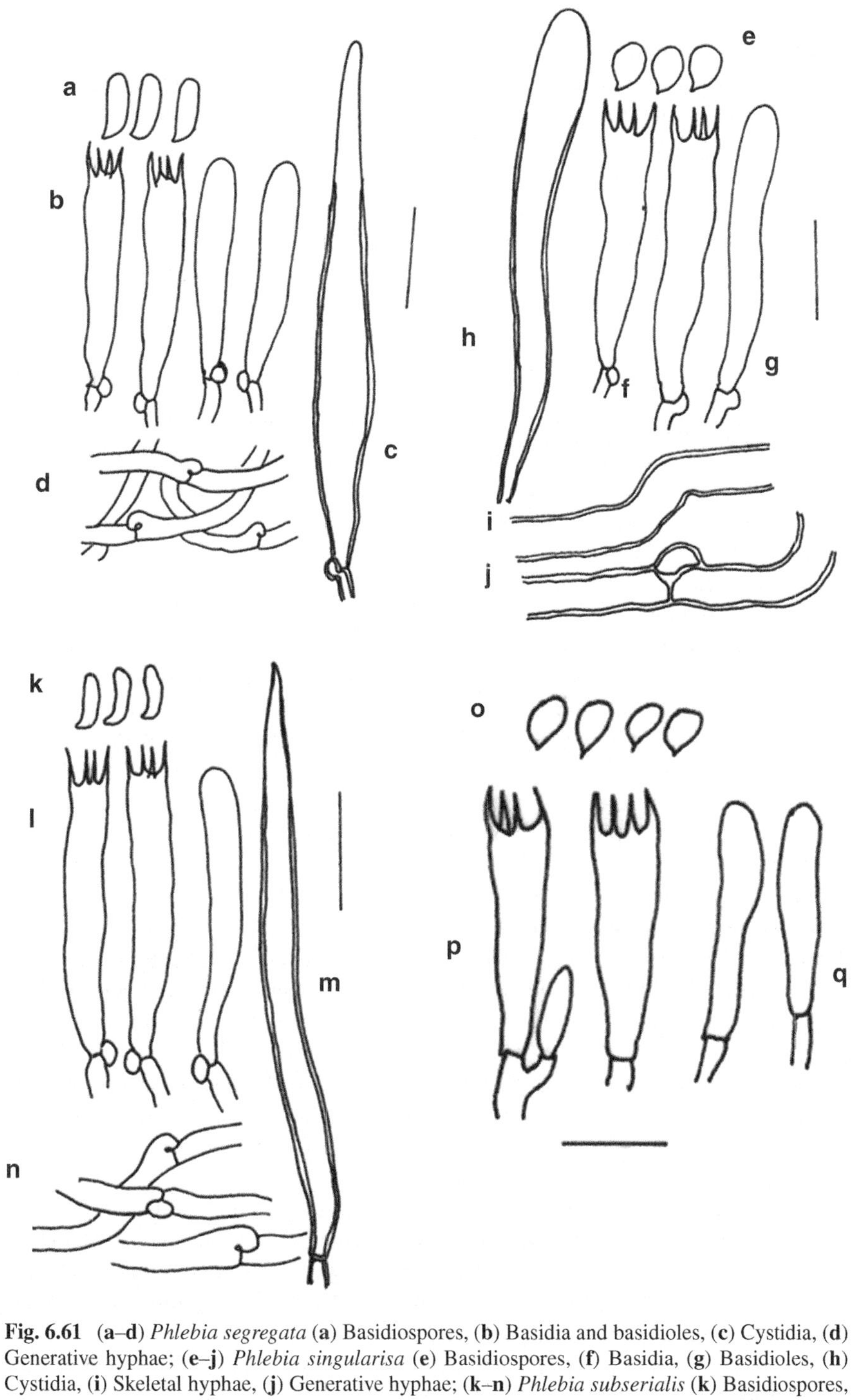

Fig. 6.61 (**a**–**d**) *Phlebia segregata* (**a**) Basidiospores, (**b**) Basidia and basidioles, (**c**) Cystidia, (**d**) Generative hyphae; (**e**–**j**) *Phlebia singularisa* (**e**) Basidiospores, (**f**) Basidia, (**g**) Basidioles, (**h**) Cystidia, (**i**) Skeletal hyphae, (**j**) Generative hyphae; (**k**–**n**) *Phlebia subserialis* (**k**) Basidiospores, (**l**) Basidia, (**m**) Cystidia, (**n**) Generative hyphae; (**o**–**p**) *Phlebia subochracea* (**o**) Basidiospores, (**p**) Basidia, (**q**) Basidioles

thick-walled, few cystidia branched near the base; projecting up to 30 μm out of the hymenium. Basidia 20.0–26.0×4.5–5.0 μm, clavate to subclavate, 4-sterigmate, with basal clamp, sterigmata up to 3.0 μm long. Basidiospores 4.8–6.0×2.4–3.0 μm, subellipsoid to suballantoid, apiculate, smooth, thin-walled, non-amyloid, acyanophilous.

Distribution: H.P.-Shimla; U. K.: Karanprayag-Chamoli.

Collection examined: IBP 37200.

Substratum: On a decaying angiospermic log, on gymnospermic wood.

Remarks: This species is distinguished by the presence of smooth to finely tuberculate fructification, projecting subfusiform to subulate cystidia, clavate, clavate to subclavate basidia and subellipsoid to suballantoid thin-walled basidiospores. This species is a new record for Uttarakhand.

Phlebia singularisa Dhingra, J. Ind. Bot. Soc. 84:114, 2005. Fig. 6.61e–j

Fructification resupinate, loosely adnate, loosening in margins on drying, effused, membranous-ceraceous when fresh, horny on drying; hymenial surface tuberculate when fresh, almost smooth when dried, pale yellow to pastel-yellow when alive, pale ochraceous in herbarium; abhymenial surface yellowish-white, rough; margin abrupt to thinning, white, fibrillose.

Hyphal system monomitic; generative hyphae septate, clamped; basal composed of up to 6.0 μm wide, thick-walled, less branched, compactly interwoven hyphae; subhymenial hyphae up to 3.3 μm wide, richly branched and interwoven into a dense to almost agglutinated texture. Cystidia 50.1–85.2×5.0–9.0 μm, subcylindrical, often widening in the apical part, thin-to slightly thick-walled, projecting out of the hymenium. Basidia 18.1–27.6×4.6–6.1 μm, clavate to subclavate, 4-sterigmate, with a basal clamp. Basidiospores 4.6–6.1×2–3.6 μm, ellipsoid to obovate, smooth, thin-walled, inamyloid, acyanophilous.

Distribution: Bhutan.

Collection examined: GSD 19612.

Substratum: On gymnospermous wood.

Remarks: This species is characterized by loosely adnate fructification with fibrillose whitish margin, clamped generative hyphae, subcylindrical cystidia, 4-sterigmate, clavate to subclavate basidia and ellipsoid to subfusiform spores.

Phlebia subochracea (Alb. & Schwein.) J. Erikss. & Ryvarden, Cortic. N. Eur. (Oslo) 4: 873 (1976) = *Mycoacia subochraceae* (Alb. & Schw.) Parm., Est: NSV Tead. Akad. Toim. Biol. Seer 16(4): 388, 1967 = *Grandinia subochracea* Bres., Hedwigia 33: 206. 1894. Fig. 6.61o–p

Fructification resupinate, membranous to subceraceous but turn somewhat hard and brittle on drying, adnate, widely effused; hymenial surface distinctly toothed, yellow to yellowish brown, continuous, not creviced; margin thinning to more or less indeterminate, adnate, paler concolorous. Teeth up to 1.5 μm long, subulate to subcylindrical with acute apices, terrete when fresh but often clustering on drying. Subiculum composed of compactly arranged more or less agglutinated hyphae.

Hyphal system monomitic, hyphae 2.5–5.5 µm wide, branched, septate, clamps absent, the walls thin to slightly thick, subhyaline to tinted yellow–brown. Cystidia absent. Basidia 20–30.2 × 5.6–6 µm, clavate-cylindrical, 4-spored, sterigmata up to 4.5 µm long. Basidiospores 4.6–5.6 × 3–3.3 µm, ellipsoid, minutely apiculate, thin-walled, subhyaline, smooth, nonamyloid.

Distribution: H.P.: Shimla, Chamba, Manali.

Collection examined: HSK 4017, 4119, IBP 37196.

Substratum: On angiospermic stump of *Quercus incana.*

Remarks: This species is characterized by the simple-septate hyphae, lack of cystidia and the shape and size of basidiospores.

Phlebia subserialis (Bourd. & Galz.) Donk, Fungus Wageningen 27: 12.1957 = *Corticium subseriale* Bourd. & Galz., Hym. de France p. 219. 1928 (1927).
Fig. 6.61k–n

Fructification resupinate, membranous-ceraceous becoming rigid and brittle on drying, separable, widely effused; hymenial surface cream yellow to yellow-brown, smooth to somewhat uneven and wavy to sulcate, not creviced; margin determinate to abrupt, loosely adnate but rolls away from the substratum on drying, concolorous. Subiculum composed of compactly arranged, more or less agglutinated hyphae, distinction in to the basal zone and upper zone not well demarcated.

Hyphal system monomitic, hyphae 3–6 µm wide, branched, septate, clamped, clamps not at all septa, the walls thin to slightly thick, subhyaline, often collapsing. Cystidia (Leptocystidia) 44–74 × 6.8–8.1 µm, fusiform, arising from the subhymenial layer and may project out of the hymenium, apices subacute to acute, usually with a long drawn out tip, thin to slightly thick-walled, subhyaline, smooth or finely impregnated with subhyaline crystals. Basidia 30.1–45.1 × 4.6–5.6 µm, clavate, 4-spored. Basidiospores 4.5–6.4 × 2–2.6 µm, ellipsoid-cylindrical to suballantoid, minutely apiculate, thin-walled, subhyaline, smooth, nonamyloid.

Distribution: H.P.: Shimla, Kullu, Manali.

Collection examined: SSR 5554, IBP 37203.

Substratum: On wood of *Cedrus deodara.*

Remarks: This species is characterized by the smooth and light coloured hymenial surface, presence of thin-walled leptocystidia with characteristic long drawn out apices. The hymenial surface is wavy to sulcate but not folded-reticulate.

Phlebia thindii Dhingra, J. Ind. Bot. Soc. 84: 115, 2005. Fig. 6.62a–d

Fructification resupinate, adnate, effused, ceraceous when fresh, crustaceous on drying; hymenial surface smooth to somewhat tuberculate, continuous but cracking transversely in the older parts on drying, yellowish-white to pale yellow; margin indeterminate, thinning out into a pruinose periphery or abrupt.

Hyphal system monomitic; generative up to 3.0 µm wide, thin-walled, septate, clamped; subiculum very narrow, composed of compactly packed to agglutinated hyphae; subhymenium of densely interwoven, semi erect hyphae.

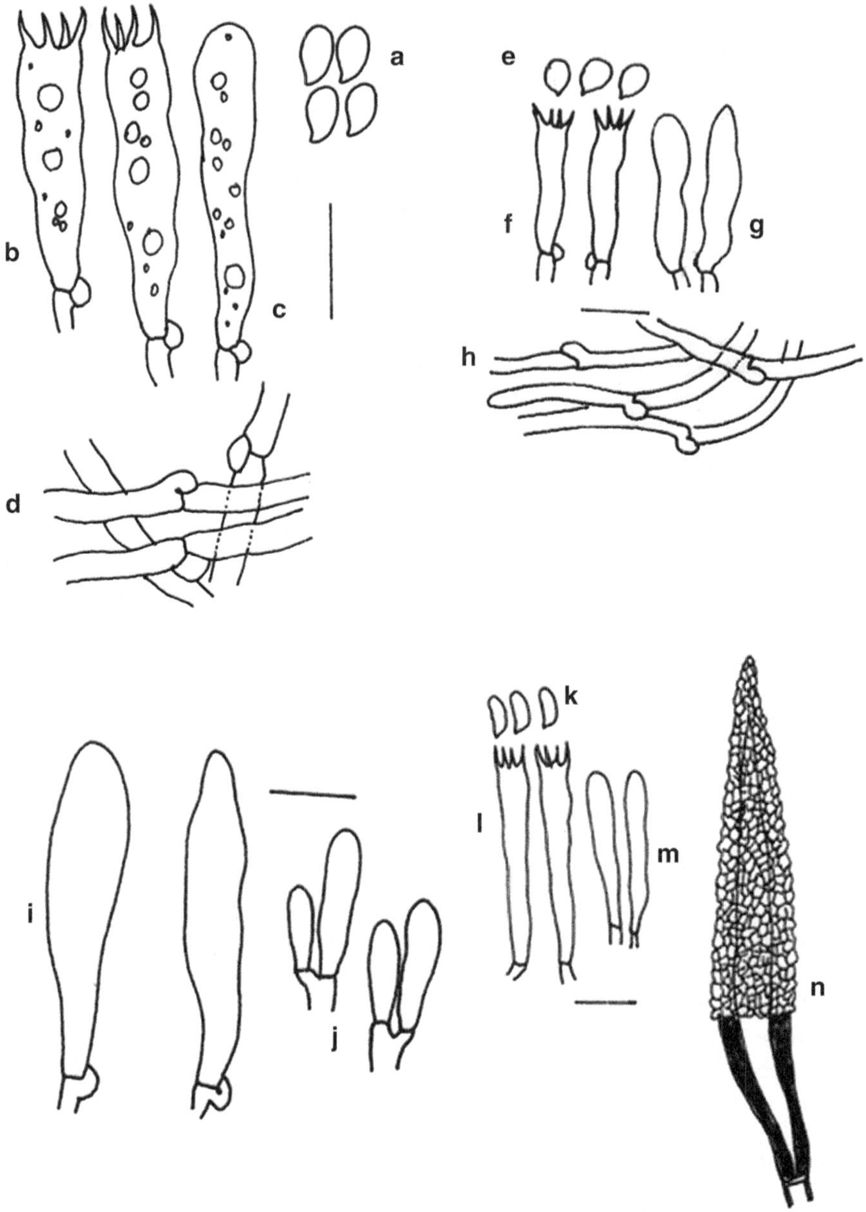

Fig. 6.62 (**a**–**d**) *Phlebia thindii* (**a**) Basidiospores, (**b**) Basidia, (**c**) Basidioles, (**d**) Generative hyphae; (**e**–**h**) *Radulodon americanus* (**e**) Basidiospores, (**f**) Basidia, (**g**) Cystidia, (**h**) Generative hyphae; (**i**–**j**) *Radulodon erikssonii* (**i**) Gloeocystidia, (**j**) Basidioles; (**k**–**n**) *Scopuloides hydnoides* (**k**) Basidiospores, (**l**) Basidia, (**m**) Basidioles, (**n**) Encrusted cystidia

Cystidia absent. Basidia 17.6–24.1×6.1–7.2 μm, clavate, narrowed into a stalk like part at the base, 4-sterigmate, with a basal clamp. Basidiospores 5.2–7.6×2.5–3.5 μm, ellipsoid, smooth, thin-walled, inamyloid, acyanophilous, with oily contents.

Distribution: West Bengal: Darjeeling- Tiger hill.

Collection examined: GSD 19305.

Substratum: On angiospermous wood.

Remarks: This species is characterized by yellowish fructification, generative hyphae with clamps, clavate basidia, narrowing into stalk like structure at the base and ellipsoid basidiospores. It is closely related to *P. griseoflavescens* (Litsch.) Eriksson & Hjortst. However, the latter is marked by whitish pale fructification, larger basidia and longer and narrower basidiospores.

Radulodon Ryvarden,
Can. J. Bot. 50: 2073. 1972.

Fructifications resupinate, membranous to subceraceous, adnate, effused; hymenial surface toothed, creamish yellow to yellowish-brown. Teeth subulate to flattened, terrete initially often clustered and becoming hard and somewhat brittle on drying. Subiculum composed of compactly arranged, agglutinated hyphae. Hyphal system monomitic, hyphae normally thin-walled, with or without clamps. Cystidia absent. Gloeocystidia present or absent. Basidia clavate-cylindrical, 4-spored. Basidiospores globose to subglobose, the walls thin, subhyaline, smooth, nonamyloid, acyanophilous.

Nine Species, widespread

Lit.: Hjortstam et al., (*Kew Bull.* **45:** 303, 1990, key.), Ryvarden (Česka Mykol 30:38, 1976 key.)

Type species: *Radulodon americanus* Ryvarden, 1972.

Habitat: Dead wood

Himalayas: Two

Key to species

1. Gloeocystidia absent; hymenial teeth with flattened apices............*R. americanus*
1. Gloeocystidia present; hymenial teeth with acute apices..................*R. erkissonii*

Radulodon americanus Ryvarden, Can. J. Bot. 50: 2074. 1972. Fig. 6.62e–h

Fructification resupinate, membranous to subceraceous when fresh becoming hard and somewhat brittle on drying, adnate, widely effused; hymenial surface toothed, cream yellow to ochraceous not creviced; margin thinning, adnate, paler concolorous. Teeth 1–5.2 mm long, scattered to occasionally gregarious, subulate to flattened with fimbricate apices, terrete when fresh but often clustered on drying. Subiculum subhyaline in section, composed of compactly arranged more or less agglutinated hyphae.

Hyphal system monomitic, hyphae 2–3.5 μm wide, branched, septate, clamped, thin-walled, subhyaline. Gloeocystidia absent. Basidia 20.5–30×4–5.2 μm, clavate to clavate-cylindrical, 4-spored, sterigmata up to 4.5 μm long. Basidiospores globose to subglobose 4.6–5.5 × 3.5–4.5 μm, minutely apiculate, thin-walled, subhyaline, smooth, nonamyloid, acyanophilous.

Distribution: H.P.: Dalhousie, Shimla, Manali; J&K: Pehalgam.

Collection examined: SSR 4392, 4391, IBP 37662.

Substratum: On log under conifers, cut coniferous stumps.

Remarks: This species is characterized by the presence of clamps on the hyphae, absence of gloeocystidia and subglobose basidiospores.

Radulodon erikssonii Ryvarden, Can. J. Bot. 50: 2075. 1972. Fig. 6.62i–j

Fructifications resupinate, membranous to subceraceous, becoming hard and brittle on drying, adnate, widely effused; hymenial surface cream yellow to ochraceous, toothed, not creviced; margin thinning, adnate, concolorous. Teeth 1–5 mm long, scattered to gregarious, subulate, apices acute or rarely fimbricate, terrete when fresh but often clustered and becoming brittle on drying. Subiculum composed of compactly arranged, almost agglutinated hyphae.

Hyphal system monomitic, hyphae 2–4.6 μm wide, branched, septate, clamped, thin-walled, subhyaline. Gloeocystidia 32–40 × 7–12.5 μm, subfusiform to clavate-cylindrical, immersed, thin-walled, subhyaline, often collapsing in mature specimens and difficult to discern. Basidia 20.5–25.0 × 5–6.2 μm, clavate-cylindrical, 4-spored. Basidiospores 4.5–5.5 μm across, globose to rarely subglobose, minutely apiculate, the walls thin, subhyaline, smooth, nonamyloid, acyanophilous.

Distribution: H.P.: Narkanda.

Collection examined: SSR 4402, HSK 4223, IBP 37689.

Substratum: On log under angiosperms.

Remarks: This species is characterized by the presence of clamps, gloeocystidia and predominantly globose basidiospores.

Scopuloides (Massee) Höhn. & Litsch.,
Wiesner Festschrift 57,58,1908.

Fructification resupinate, adnate, totally attached, odontoid; hyphal system monomitic; subiculum thin, composed of parallel arranged hyphae, thin-walled or with slightly thickened walls, subhymenial layer rather dense but not firm; all hyphae without clamps; cystidia numerous, conical, thick-walled, encrusted in the upper part; projecting hyphal ends present, usually septate and encrusted, with obtuse apex; basidia small, slightly thickened towards the base, normally with 4-sterigmata, without basal clamp; spores smooth, thin-walled, in the type species short, allantoid, not amyloid, not dextrinoid, not cyanophilous.

Five species, widespread

Lit.: Boidin et al. (*Cryptog. Mycol.* **14:** 195, 1993; key Eur. Spp.)

Type Species: *Peniophora hydnoides* Cooke & Massee. 1888.

Habitat: Wood

Himalayas: One

Scopuloides hydnoides (Cooke & Massee) Hjortstam & Ryvarden, Mycotaxon 9(2): 509 (1979) = *Phlebia hydnoides* (Cooke. & Mass.) Christ. Dansk.bot. Arkv. 19(2): 175. 1960 = *Peniophora hydnoides* Cooke & Massee, in Cooke, Grevillea 16(no. 79): 77(1888). Fig. 6.62k–n

Fructifications resupinate, membranous-ceraceous to ceraceous when fresh becoming horny and brittle on drying, loose adnate but tends to curl away from the substratum on drying, widely effused, up to 1 mm thick in section; hymenial surface cream to cream yellow, smooth to finely tuberculate or distinctly odontoid, warts scattered to gregarious, up to 480 µm long, subulate, continuous, not creviced or finely cracking irregularly on drying; margin thinning to more or less abrupt, fibrillose, loosely adnate, white to paler concolorous. Subiculum subhyaline in section, composed of a basal zone of somewhat loosely woven distinct hyphae and upper zone of compactly arranged more or less agglutinated and ceraceous hyphae.

Hyphal system monomitic, hyphae 2.5–5.2 µm wide, branched, septate, clamps absent, walls thin to moderately thick, subhyaline. Cystidia 38–92 × 7.8–14 µm, fusiform to subcylindrical with acute apices, immersed or projecting out of the hymenium, walls thick (up to 2 µm), subhyaline, heavily impregnated with white crystalline matter. Leptocystidia hyphoid, confined to the axil of teeth only but sometimes may project out of the teeth apices in the bundles, morphologically similar or slightly wider than the axial hyphae. Basidia 30–40 × 3.5–4.5 µm, clavate, 4-spored. Basidiospores 4.5–5.6 × 2–2.6 µm, narrow ellipsoid to suballantoid, minutely apiculate, the walls thin, subhyaline, smooth, nonamyloid.

Distribution: H.P.: Kullu, Rohtang, Gojra, Dalhousie,

Collection examined: HSK 4054, SSR 4374, IBP 37206.

Substratum: On stump of *Cedrus deodara, Abies pindrow.*

Remarks: This species is characterized by the odontoid hymenial surface and presence of two types of cystidia (lamprocystidia and leptocystidia). The teeth are usually few and scattered and some fructification may even appear smooth. It was recorded by Rattan (1977) as *Phlebia hydnoides* from Himachal Pradesh.

Sarcodontia Schulzer,
in Schulzer, Kanitz and Knapp
Verh. Zool. Bot. Ges. Wien 16 (Abh.): 41 (1866).

Fructification resupinate or on vertical substrate forming nodes with fertile undersides; hymenophore of long, conical aculei; hymenophore in the living state ceraceous, when dried rather hard; hyphal system monomitic, all hyphae with clamps, generative hyphae thin-walled, subicular hyphae very thick-walled; cystidia absent; basidia clavate, often sinuous or constricted, 4 sterigmate; spores ellipsoid to subglobose, thick-walled, not amyloid, not cyanophilous.

Two species, widespread

Lit.: Stalpers (*Folia Cryptog. Estonica* 33: 133, 1998).

Type Species: *Sarcodontia mali* Schulzer, 1866

Habitat: Wood

Himalayas: Three

Key to species

1. Dissipiments lacerate and appear toothed .... *S. pachydon*
1. Dissipiments not so 2
2. Basidia up to 6.2 μm in diameter, pores angular in shape, 3–5 per mm .... *S. spumea*
2. Basidia up to 6 μm in diameter, pores angular to irregular, splitting on drying, 1–2 per mm .... *S. delectans*

Sarcodontia delectans (Peck) Spirin, Mycena 1(1): 64–71 (2001)=*Spongipellis delectans* (Peck.) Murr., North Am. Fl. 9(1): 38, 1907. Plate 6.24b, Fig. 6.63a–d

Fructification annual, pileate, soft fleshy, corky on drying, light in weight, solitary to imbricate, sessile, applanate, convex, semi-circular up to 20×12×4.5 cm in size; upper surface cream when fresh, brown on drying, tomentose hyphae agglutinated to form small protuberances of tufted hyphae azonate, thinning abruptly, concolorous with the upper surface, sterile below, up to 2.8 mm wide. Pore surface white to cream when fresh, becoming brown on drying, even, shining, pores angular to irregular, splitting on drying, 1–2 per mm; dissepiments equal 65–125 μm thick; pore mouth velutinate; margin even, acute. Context cream, soft, spongy thick; tubes in one layer concolorous with pore surface.

Hyphal system monomitic; generative hyphae hyaline, thin to thick-walled, branched, septate, clamped; thin-walled hyphae 2.5–3.0 μm in diameter; thick-walled with narrow lumen 3.9–4.0 μm in diameter. Cystidia absent. Basidia hyaline, clavate 13–16.5×4–6 μm. Basidiospores hyaline, thick-walled, smooth, minutely apiculate, subglobose to ovoid, uniguttulate, non-amyloid, 4.4–6.2×4.8–6.1 μm.

Distribution: U.K.: Nainital; Bhutan: Thimphu; H.P.: Shimla, Solan, Bilaspur.

Collection examined: SSR 4347, IBP 37207, 37208.

Substratum: Stumps under a mixed forest, cut bamboo stumps.

Remarks: The species is characterized by having sessile to pileate fructification; upper surface cream; hyphal system monomitic; generative hyphae clamped, thin to thick-walled; basidia clavate, 4-sterigmate; smooth, subglobose to ovoid basidiospores.

Sarcodontia pachyodon (Pers.) Spirin, Mycena 1(1): 64–71 (2001)=*Irpiciporus pachydon* (Pers.) Kotl. & Pouz., Ceska Mykol. 11: 156, 1957. Fig. 6.63e–f

Fructification resupinate, membranous to ceraceous when fresh becoming hard and brittle on drying, adnate, widely effused, up to 5 mm thick in section; hymenial surface poroid when young but dissepiments soon grow unequally, become lacerate and appear toothed, cream brown when fresh becoming ochre-brown on drying, not creviced; margin indeterminate, adnate, concolorous. Teeth represented by elongated and lacerated dissepiments, quite irregular, usually up to 5 mm long and 300 μm broad at the base, subulate to cylindrical but occasionally flattened and irregular, apices acute, sterile.

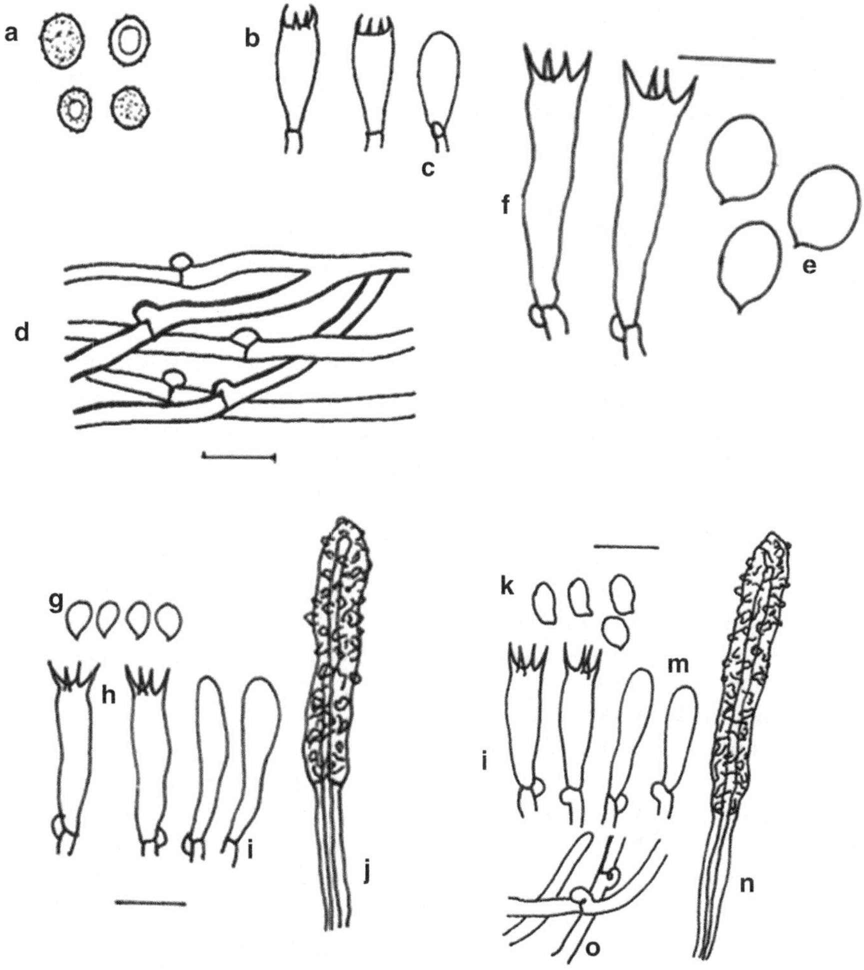

Fig. 6.63 (**a–d**) *Sarcodontia delectans* (**a**) Basidiospores, (**b**) Basidia, (**c**) Basidioles, (**d**) Generative hyphae; (**e–f**) *Sarcodontia pachyodon* (**e**) Basidiospores, (**f**) Basidia; (**g–j**) *Steccherinum ciliolatum* (**g**) Basidiospores, (**h**) Basidia, (**i**) Basidioles, (**j**) Encrusted cystidia; (**k–o**) *Steccharinum fimbriatum* (**k**) Basidiospores, (**l**) Basidia, (**m**) Basidioles, (**n**) Encrusted cystidia, (**o**) Generative hyphae

Hyphal system monomitic, hyphae 2–4.5 μm wide, sparsely branched, septate, rarely clamped, thin to thick-walled, subhyaline, distinct but occasionally agglutinating forming loose hyphal strands. Hymenium dense palisade of basidia and basidioles, wholly ceraceous and agglutinating. Basidia 4–5 μm broad, 4-spored. Basidiospores 5–7.8 μm in diameter, globose to subglobose, minutely apiculate, thin-walled, subhyaline, smooth, nonamyloid, acyanophilous.

Distribution: H.P.: Mahasu- Sarahan, Sundernagar; J&K: Pehalgam.

Collection examined: K: 4033, IBP 37208.

Substratum: logs under mixed forest.

Remarks: This species is characterized by irpicoid hymenial surface, absence of cystidia and globose to subglobose basidiospores.

Sarcodontia spumea (Sowerby: Fr.) Pat., Ess. Tax.: 84, 1900 = *Polyporus spumens* Sowerby: Fr., Systema Mycologicum 1: 358, 1821. Plate 6.24c

Fructification annual, pileate, sessile, solitary; pileus 9.0 × 6.5 × 1.0 cm, dimidiate to applanate, fleshy and soft when fresh, hard and brittle on drying; abhymenial surface smooth to tomentose, orange white to pale orange to greyish orange when fresh, pale orange to greyish orange to light brown to brown on dying; hymenial surface poroid, orange white to greyish orange to brownish orange when fresh, somewhat darkening on drying; pores round to angular, 2–3 per mm; dissepiments thin, entire; context duplex, upper part, loose, orange white, lower part dense, pale orange, up to 8 mm thick; pore tubes up to 3 mm long, concolorous with hymenial surface; margins obtuse, sterile up to 3 mm, pale orange.

Hyphal system monomitic; generative hyphae up to 6.5 μm wide, branched, septate, clamped, thin- to thick- walled, subhyaline. Cystidia absent. Basidia 22.0–28.0 × 7.8–9.1 μm, clavate, 4-sterigmate, subhyaline, clamped at the base; sterigmata up to 5.2 μm long. Basidiospores 6.5–7.8 × 4.6–6.5 μm, subglobose to broadly ellipsoid, smooth, thick-walled, subhyaline, cyanophilous, inamyloid.

Distribution: H.P.: Kullu.

Collection examined: IBP 39001.

Substratum: On angiospermous wood.

Remarks: This species is characterized by piletae, sessile, solitary, dimidiate to appalante, fleshy and soft when fresh, hard and brittle on drying fructification, round to angular pores, and thick–walled, subglobose to broadly ellipsoid, cyanophilous basidiospores. It has earlier been reported from A.P. It is being reported for the first time from N. W. Himalayas.

Steccherinum Grey,
Nat. Arrang. Br. Pl.1: 651. 1821.

Fructifications resupinate, effused-reflexed to subpileate to pileate, membranous-coriaceous; hymenial surface odontoid to hydnoid, always cream with shade of pink, brown or violaceous, glabrous and shiny. Pileus ranging from merely upturned margin to distinct and well developed, sessile or umbonate. Context tough but pliable. Hyphal system dimitic; skeletal hyphae subhyaline, with or without clamps. Cystidia arising as the prolongations of skeletal hyphae which may be embedded in the trama of spine or project out of the hymenium or teeth apices, thick-walled, subhyaline, generally impregnated with subhyaline crystals. Gloeocystidia absent. Basidia clavate to clavate-cylindrical, 4-spored. Basidiospores broadly ellipsoid or ovoid to subcylindrical, thin-walled, smooth, subhyaline, nonamyloid, acyanophilous.

Thirty three species, widespread

Lit.: Yuan & Dai (*Mycotaxon* 93: 173, 2005)
Habitat: Dead wood
Type Species: *Hydnum ochraceum* (Pers. ex J.F. Gmel, 1792)
Himalaya: Four
Key to species

1. Fructification resupinate; aculei up to 1 mm long 2
1. Fructification effused- reflexed to pileate, aculei up to 3 mm long 3
2. Margin with abundant rhizomorphs;
   basidiocarps easily separable; aculei up to 0.3 mm long .... *S. fimbriatum*
2. Margin fibrillose without rhizomorphs;
   basidiocarps loosely adnate; aculei up to 1 mm long .... *S. ciliolatum*
3. Hymenial surface cream to light pinkish brown;
   spores 3–4 × 1.6–2.5 µm .... *S. ochraceum*
3. Hymenial surface pinkish brown to orange red;
   spores 4–4.5 × 2.5–3 µm .... *S. laeticolor*

Steccherinum ciliolatum (Berk.& Curt.) Gilb.& Bud., Journ. Arizona Acad. Sci. **6**: 97. 1970. Fig. 6.63g–j

Fructifications resupinate, membranous to fibrous, loosely adnate, often arising in small circular colonies which become confluent later on up to 140 µm in section; hymenial surface cream to cream-brown, distinctly toothed, not creviced; margin thinning, indistinct to fibrillose, adnate, paler concolorous. Teeth gregarious up to 1.2 mm long, subulate to flattened, terrete, apices acute or serrate to finely hispid.

Hyphal system dimitic; generative hyphae 2.4–5 µm wide, branched, septate, clamped, walls hyaline, thin to slightly thick; skeletal hyphae 2.6–5.6 µm wide, sparsely branched, aseptate or occasionally with few retraction septa, walls subhyaline to pale yellow, thick with little or no lumen. Cystidia 5–7.5 µm broad, of variable length being the prolongations of the skeletal hyphae, clavate to cylindrical, subhyaline, thick-walled, finely impregnated with subhyaline crystals especially in the apical half, often immersed in the trama of spine or projecting out of the hymenium or teeth apices. Gloeocystidia absent. Basidia 12.5–15.5 × 4–5.3 µm, clavate, 4-spored, sterigmata up to 3 µm long. Basidiospores 3.5–4.1 × 2.6–3.2 µm, broadly ellipsoid to ovoid, minutely apiculate, thin-walled, subhyaline, smooth, nonamyloid, acyanophilous.

Distribution: H.P.: Shimla- Glen, Annandale; Chamba- Khijjiar.
Collection examined: HSK: 4016, 4020, 4105; SSR: 4323, IBP 37212.
Substratum: Twigs and wood under mixed forest, angiospermic stumps.
Remarks: This species is characterized by the resupinate fructifications, with cream to cream brown hymenial surface, absence of rhizomorphs and small broadly ellipsoids to ovoid basidiospores. It is close to *S. fimbriatum* but differs in lacking the rhizomorphs and possessing longer teeth and broader basidiospores.

Steccharinum fimbriatum (Pers.) Erikss., Symb. bot. upsal. 16:134. 1958= *Hydnum fimbriatum* Pers. Syst. Mycol. 1:421, 1821. Fig. 6.63k–o

Fructifications resupinate, membranous to membranous- coriaceous, separable, effused; hymenial surface cream with a shade of pink brown when fresh but fades on drying, papiliose to distinctly toothed, not creviced; margin thinning, fibrillose to rhizomorphic, loosely adnate, paler concolorous. Rhizomorphs abundant, often running appressed to the substratum, branched, creamish pink. Teeth gregarious up to 280 μm long, subulate to cylindrical, apices acute to fimbricate due to projecting cystidia, terrete when young but usually coalesce at the base at maturity.

Hyphal system dimitic; generative hyphae 1.6–3.6 μm wide, branched, thin, subhyaline, septate, clamped, clamps at all septa; skeletal hyphae 2.6–4.7 μm wide, sparsely branched, aseptate or occasionally with few retraction septa, clamps absent, walls subhyaline, thick with little or no lumen. Cystidia 6–9.5 μm broad, of variable length, cylindrical, subhyaline, thick-walled, incrusted heavily especially in the apical part, immersed or may project to 25 μm out of the hymenium. Gloeocystidia absent. Basidia 12.4–16.2×4–4.7 μm, clavate, 4-spored. Basidiospores 3.6–4.7×1.5–2.0 μm, broadly ellipsoid, minutely apiculate, thin-walled, smooth, subhyaline, nonamyloid, acyanophilous.

Distribution: H.P.: Dalhousie, Kullu; J&K: Bhadarwah.

Collection examined: HSK: 4118, 4163, IBP 37215.

Substratum: On angiospermic stumps.

Remarks: The species is characterized by the creamish to pinkish brown color of the hymenial surface, presence of rhizomorph, small teeth and narrow ellipsoid basidiospores. It has already been recorded from number of places from N. W. Himalayas.

Steccherinum laeticolor (Berk. & Curt.) Banker, Mycologia 4: 316. 1912=*Hydnum laeticolor* Berk. & Curt., in Berk., Grevillea 1: 99. 1873. Fig. 6.64a–c

Fructifications resupinate to effused-reflexed, loosely adnate, membranous- coriaceous to fibrous, often arising as small circular colonies; hymenial surface pinkish brown or orange red, usually brightly colored, distinctly toothed, not creviced, margin thinning, fibrillose and adnate in resupinate part but becomes more or less thick, abrupt and upturned in effused-reflexed forms. Teeth gregarious, up to 2.4 μm long, terrete and subulate when young but become flattened and irregular at maturity, giving an irpicoid appearance, apices acute, sterile, concolorous.

Hyphal system dimitic; generative hyphae 2.6–4.5 μm wide, branched, septate, clamped, the walls thin, subhyaline; skeletal hyphae 2.4–5.5 μm wide, unbranched to sparsely branched, aseptate, the walls subhyaline, thick with little or no lumen;. Cystidia 8–12 μm broad, of variable length, arising as the prolongations of the skeletal hyphae, cylindrical to subclavate, thick-walled, subhyaline, impregnated with subhyaline crystals in the apical part, immersed in the trama of spine or may project out of the hymenium, not or rarely projecting out of the teeth apices. Gloeocystidia absent. Basidia 4.5–5.8 μm broad clavate, 4-spored. Basidiospores 4.0–4.6×2.6–3 μm, broadly ellipsoid to ovoid, minutely apiculate, the walls thin, subhyaline, smooth, non-amyloid.

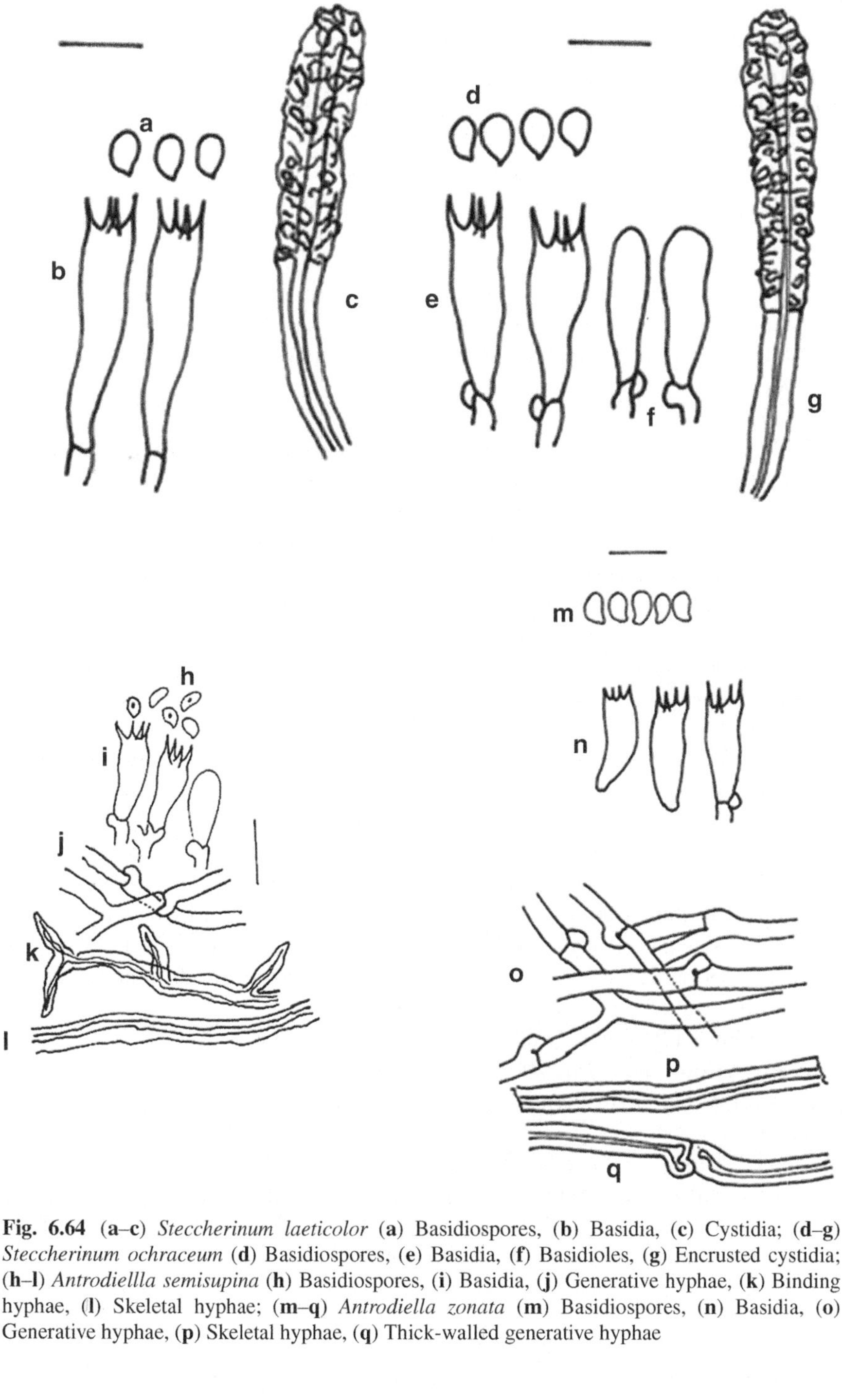

Fig. 6.64 (**a**–**c**) *Steccherinum laeticolor* (**a**) Basidiospores, (**b**) Basidia, (**c**) Cystidia; (**d**–**g**) *Steccherinum ochraceum* (**d**) Basidiospores, (**e**) Basidia, (**f**) Basidioles, (**g**) Encrusted cystidia; (**h**–**l**) *Antrodiellla semisupina* (**h**) Basidiospores, (**i**) Basidia, (**j**) Generative hyphae, (**k**) Binding hyphae, (**l**) Skeletal hyphae; (**m**–**q**) *Antrodiella zonata* (**m**) Basidiospores, (**n**) Basidia, (**o**) Generative hyphae, (**p**) Skeletal hyphae, (**q**) Thick-walled generative hyphae

Distribution: H.P: Kullu, Chamba-Dalhousie.

Collection examined: HSK 4005, IBP 37213.

Substratum: On angiospermic twig.

Remarks: This species is characterized by the toothed to irpicoid configuration and bright colour of the hymenial surface. *S. ochraceum* is very similar but possess lighter color fructifications and slightly smaller basidiospores. Moreover, the teeth are subulate and terrete while these often coalesce and become irpicoid in *S. laeticolor*.

Steccherinum ochraceum (Pers.) Grey, Nat. Arrang. Brit. P1. 1: 651. 1821 = *Hydnum ochraceum* Pers. in Gmelina Systema Naturae, Edn 13 2(2): 1440(1792) = *Steccherinum rhois* (Schw.) Banker, Mem. Torrey Bot. C1. 12: 12. 1906.

Fig. 6.64d–g

Fructifications resupinate to effused- reflexed or distinctly pileate, membranous-coriaceous when fresh becoming rigid but pliable on drying, adnate, often arising as small colonies which later become confluent; hymenial surface cream brown to pinkish brown, distinctly toothed, not crevised; margin thinning, wide, loosely adnate, white to paler concolorous. Pileus ranging from merely upturned margin to distinct, sessile or pendulus and attached by a narrow base or umbo; upper surface cream to cream brown, concentrically zonate, zones of erect and appressed tomentum, soft and woolly. Teeth gregarious subulate or rarely becoming flattened, terrete, apices acute or finely penicillate due to the projecting cystidia.

Hyphal system dimitic; generative hyphae 2.6–4.5 μm wide, branched, septate, clamped, thin-walled, subhyaline; skeletal hyphae 2–4.5 μm wide, unbranched to separately branched, aseptate, the walls subhyaline, thick leaving little or no lumen;. Cystidia 6–9.8 μm broad, variable in length, prolongations of skeletal hyphae, cylindrical to subclavate, thick-walled, subhyaline, impregnated, in the apical half, immersed in trama of spine or projecting out of hymenium or teeth apices. Gloeocystidia absent. Basidia 12.2–20.5 × 3–5.2 μm, clavate, 4 spored. Basidiospores 2.5–3.5 × 2–3.2 μm, broadly ellipsoid to ovoid, minutely apiculate, thin-walled, subhyaline, smooth, non-amyloid.

Distribution: H.P: Manali, Kullu;U.K: Nainital, Mussoorie.

Collection examined: HSK 4008, 4232, IBP 37214.

Substratum: On angiospermic wood, on female cone of *Picea smithiana*, dead decaying wood.

Remarks: This species is characterized by the pinkish brown hymenial surface with distinct and regular white margin, presence of subulate and terrete spines. The fructifications varies from resupinate to pileate. The pileus may range from merely upturned margin to sessile or pendulus.

Family-Phanerochaetaceae

Key to genera

1. Fructification smooth to tuberculate having peg like projections........*Phlebiopsis*
1. Fructification smooth to tuberculate 2
2. Hyphal system dimitic 3
2. Hyphal system monomitic 4
3. Cystidia absent... *Antrodiella*
3. Cystidia present.. *Porostereum*
4. Hyphal clamps present.. *Ceriporiopsis*
4. Hyphal clamps absent 5
5. Pseudocystidia present, hyphoid arising from basal hyphae..... *Candelabrochaete*
5. Not as above 6
6. Cystidia absent .... *Byssomerulius*
6. Cystidia present... 7
7. Basidiospores smooth ellipsoid... *Rhizochaete*
7. Basidiospores ellipsoid to allantoid 8
8. Hymenial surface whitish to yellowish orange brown.................. *Phanerochaete*
8. Hymenial surface white yellow, cinnamon, green or purple................. *Ceriporia*

Antrodiella Ryvarden & I. Johans.,
A prelim. Polyp. Flor. Of E. Afri:256,1980

Fructification annual, resupinate to pileate, mostly sessile, imbricate fructification, creamish to creamish yellow, zonate, waxy soft when fresh, dense and hard on drying; margin acute; pore surface light ochraceous to straw coloured when dry; pore entire and small, tubes concolorous with the pore surface. Hyphal system dimitic; generative hyphae clamped; skeletal hyphae thick-walled, usually unbranched. Cystidia absent. Basidiospores ellipsoid, hyaline, smooth, nonamyloid.

Fifty species, widespread

Lit.: Dai (*Mycotaxon* 89: 389, 2004; China).

Type Species: *Polyporus semisupinus* Berk. & M.A. Curtis 1872

Habitat: Wood

Himalayas: Two

Key to species

1. Hyphal system trimitic; basidiospores 2.5–3.5 μm.......................... *A. semisupina*
1. Hyphal system dimitic; basidiospores 4.3–5.4 × 2.5–2.9 μm................. *A. zonata*

Antrodiella semisupina (Berk. & Curt.) Ryv. & Johansen, Prelim. Polyp. Fl. E. Afr. (Oslo): 261(1980) = *Antrodiasemisupina* (Betk. & Curt.) Ryv., N. Amer. Polyp. Vol. 1 *Abortiporus*– Lindtneri (Oslo) 1: 88 (1976). Plate 6.24d, Fig. 6.64h–l

Fructification resupinate to pileate usually small, when pileate often imbricate with many narrow pilei often fused laterally over an effused pore surface, up to 5 mm thick at the base, and 3 cm wide, resupinate specimens rarely above 4 mm thick, tough when fresh, hard and resinous, often slightly translucent when dry;

subiculum white to cream, up to 2 mm thick; margin narrow and white; upper surface cream in colour, wavy, smooth; margin thin, narrow and white; pores present at the underside also in cream coloured; pore surface white to light straw coloured or ochraceous; pores rounded-angular.

Hyphal system dimitic-trimitic; generative hyphae septate, clamps absent, thin to thick-walled, 1.5–3 µm across; skeletal hyphae thick-walled 2–6 µm; binding hyphae thick-walled, highly branches showing nodules, 1–3.5 µm. Cystidia absent. Basidia clavate, 4-sterigmate, basal clamps present, 10–15×4.5 µm. Basidiospores elliptical, hyaline, smooth, 2.5–3.5 µm, non-amyloid.

Distribution: U.K.: Rudraprayag.

Collection examined: IBP 37216.

Substrate: On deciduous wood, on coniferous wood

Remarks: This species is characterized by resupinate to pileate fructification, trimitic hyphal system; small pores; oblong basidiospores.

Antrodiella zonata (Berk.) Ryvarden, Boln Soc. argent. Bot. 28(1–4): 228 (1992)=*Irpex zonatus* Berk., Hooker J. Bot. 6: 168, 1854. Fig. 6.64m–q

Fructification annual, mostly sessile, imbricate fructification, creamish to creamish yellow, zonate; margin acute.

Hyphal system dimitic; generative hyphae clamped; skeletal hyphae thick-walled. Basidiospores ellipsoid, 4.3–5.4×2.5–2.9 µm, hyaline, smooth, nonamyloid.

Distribution: H.P.: Dalhousie- Lakkar mandi, Khajjiar, Kalatope, Panjpulla, Lover's walk, Jandrighat, Banikhet, Shimla- Mashobra, The Glen, Chadwick falls, Narkanda; Manali- Gojra.

Collection examined: SSR 6110, 6118, 6443, D 6198, 6216.

Substratum: Stumps of *Cedrus deodara, Picea smithiana, Quercus, Abies pindrow, Q. semicarpifolia* and *Q. incana.* Stumps under mixed forest and angiospermic forest.

Remarks: This species is characterized by having annual, sessile fructification; dimitic hyphal system; ellipsoid, smooth basidiospores.

Byssomerulius Parm.,
Izv. Akad. Navk Estonsk. SSR,
Ser. Biol. 16: 383(1967)

Fructification resupinate to pileate, hymenium smooth, to merulioid, white initially, changing to yellowish or light brown later on; margin white, subhymenium thickening with age. Hyphal system monomitic, hyphae thin to slightly thick-walled without clamps. Basidia narrowly clavate, with 4-sterigmata with basal clamp. Basidiospores subcylindrical, smooth and nonamyloid.

Eight species, world-wide

Lit.: Parmasto (*Consp. System. Corticiac*, 1968), Zimtrovich (*Micol. Fitopatol.***35**: 9, 2001).

Type species: *Thelephora corium* Pers. 1801

Habitat: Wood

Himalayas: One

Byssomerulius corium (Fr.) Parm., Easti NSV Tead. Akad. Toimet, Biol. 16: 383, 1967 = *Merulius corium* Fr., Elench. Fung. p.58, 1828. Plate 6.25a, Fig. 6.65a–e

Fructification resupinate, with reflexed margin, adnate, effused; abhymenial surface dirty white, tomentose; hymenial surface smooth, to meruloid, creamish-white to brownish-orange to brownish; margin irregular, thinning, adnate in young to reflexed.

Hyphal system monomitic; generative hyphae branched, septate, without clamps; basal zone composed of 3–5 µm wide, somewhat thick-walled, loosely interwoven hyphae; subhymenial hyphae 2.6–3.6 µm wide, densely interwoven encrusted, basal hyphae not encrusted and thickened. Cystidia absent. Basidia 18.5–30.5 × 4.5–5.6 µm, narrowly clavate, 4-sterigmate, without a basal clamp. Basidiospores 4.5–6.5 × 2.5–3 µm, subcylindrical, smooth, thin-walled, non-amyloid, acyanophilous.

Distribution: A.P.: West Kameng, Bomdila, Jamiri, Manipur: Ukhrul, Meghalaya: Shillong; Bhutan: Thimphu, Begana; H.P.: Shimla- Narkanda

Collection examined: GSD 19800, 19892, 19067, IBP 37217.

Substratum: On decaying angiospermic twigs, on decaying *Rubus* twigs.

Remarks: This species is easily recognised by meruloid hymenial surface, thin to somewhat thick-walled, densely interwoven, generative hyphae without clamps, narrowly clavate, 4-spored basidia and subcylindrical, smooth, thin-walled, non-amyloid, acyanophilous basidiospores. This species was first reported from India by Bakshi (1971) from N.W. Himalayas on the basis of a single collection.

Candelabrochaete Boid., Cahiers Maboke 8: 1:24, 1970.

Fructification resupinate, effused byssoid to submembranaceous, smooth to more or less odontoid, of fragile consistency in the dry state. Hyphal system monomitic; hyphae thin to thick-walled, short-celled, branched at wide angles, without clamps, basal hyphae with thickened (double) walls; cystidia (pseudocystidia) cylindrical more or less projecting at first thin-walled becoming thick-walled later on, often with secondary simple septa.; basidia sub-cylindrical, constricted, sub-urniform, often with linear repetition, several times renewed, 4-sterigmate. Basidiospores oblong to ellipsoid, thin-walled, smooth, non-amyloid, noncyanophilous.

Eleven Species, world-wide

Lit.: Hjortstam (*Mycotaxon* **56**: 451, 1995).

Type species: *Candelabrochaete africana* Boid. 1970

Habitat: Wood

Himalayas: One

Candelabrochaete himalayana Dhingra, In Plant Diversity in India: 477, 2004. Fig. 6.65f–i

Fructification resupinate, effused, fragile when dried; hymenial surface golden brown to light brown, turning purplish on putting a drop of 3 % KOH sol, smooth, velutinous under lens by projecting pseudocystidia; margins not differentiated.

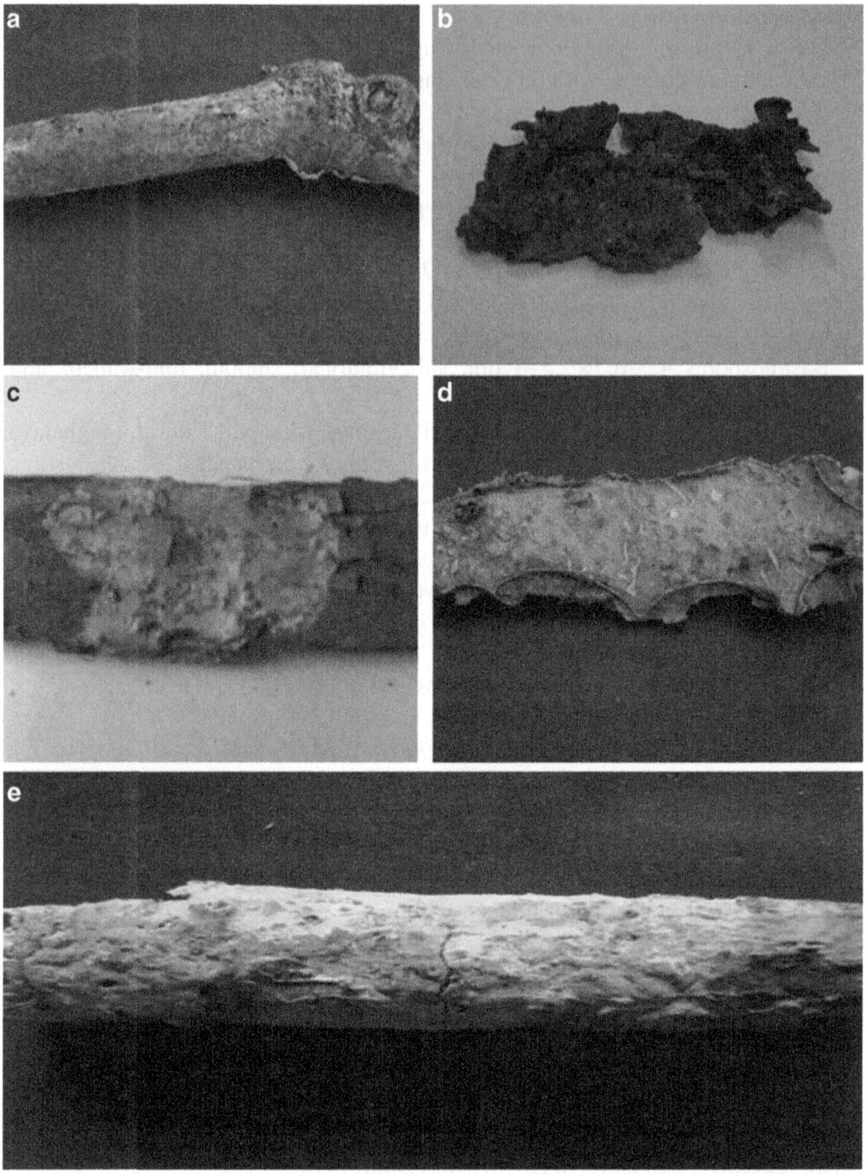

Plate 6.25 (**a**) *Byssomerulius corium*. (**b**) *Ceriporia viridans*. (**c**) *Phanerochaete galactites*. (**d**) *Phanerochaete laevis*. (**e**) *Phanerochaete sordida*

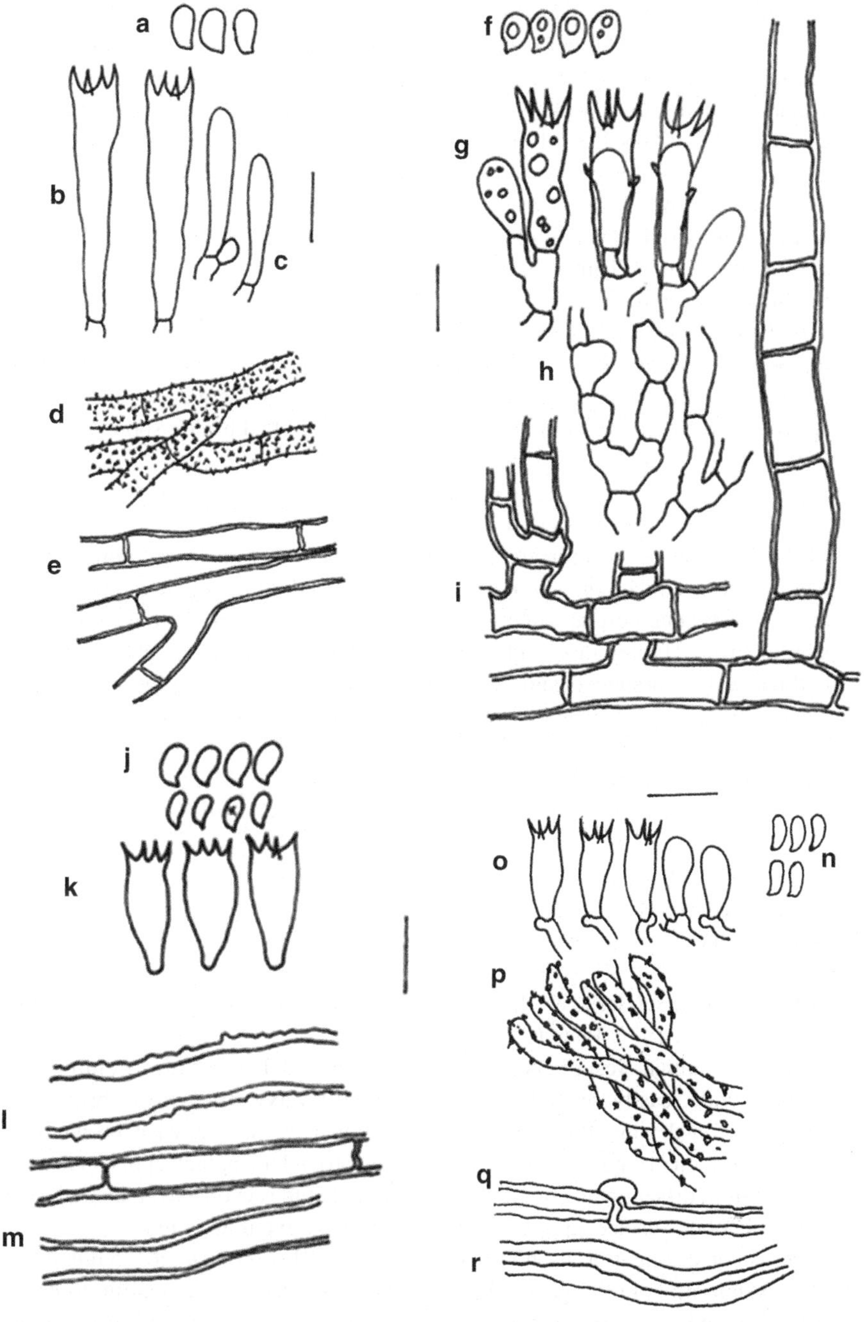

Fig. 6.65 (**a**–**e**) *Byssomerulius corium* (**a**) Basidiospores, (**b**) Basidia, (**c**) Basidioles, (**d**–**e**) Generative hyphae; (**f**–**i**) *Candelabrochaete himalayana* (**f**) Basidiospores, (**g**) Basidia, (**h**–**i**) Generative hyphae; (**j**–**l**) *Ceriporia viridans* (**j**) Basidiospores, (**k**) Basidia, (**l**) Skeletal hyphae, (**m**) Generative hyphae; (**n**–**p**) *Ceriporiopsis gilvescens* (**n**) Basidiospores, (**o**) Basidia, (**p**) hyphae with crystals, (**q**) Generative hyphae, (**r**) Skeletal hyphae

Hyphal system monomitic; generative hyphae branched at wide angles, short-celled, without clamps; basal hyphae up to 10 µm wide, sparsely branched, thick-walled. Pseudocystidia 60–.0–170.0 × 7.5–10.5 µm, hyphoid arising from the basal hyphae, septate, thick-walled, projecting up to 100 µm out of the hymenium. Basidia 14.2–22.6 × 4.2–6.1 µm, subcylindrical, apically widened, without a basal clamp, 4-sterigmate, linear repetition frequent. Basidiospores 4.5–6.5 × 3.0–4.5 µm, ellipsoid to broadly ellipsoid, smooth, thin-to slightly thick-walled, inamyloid, acyanophilous, with one to many oil drops.

Distribution: Bhutan: Thimphu, Paro, D'Dzong.

Collection examined: GSD 19430.

Substratum: On a decaying gymnospermic log.

Remarks: This species is characterized by golden brown to light brown fructifications; short celled hyphae devoid of clamps; thick-walled septate pseudocystidia; apically wide, 4-spored basidia and ellipsoid thin-walled basidiospores.

Ceriporia **Donk,**
Revis. Niederi. Homobasidio 2: 170, 1933.

Fructification annual, resupinate, soft, waxy separable from host; pore surface white, yellow cinnamon, green or purple; pores medium, round to angular; tubes in one layer, concolorous with the pore surface. Context thin, homogenous, non-xanthochroic. Hyphal monomitic; generative hyphae hyaline to subhyaline, thin to thick-walled, branched, simple septate acyanophilous. Cystidia absent. Basidia clavate, hyaline, thin-walled, 4-sterigmate. Basidiospores cylindrical to allantoids, hyaline, 4-spored, thin-walled, nonamyloid.

Twenty species, widespread

Lit.: Pieri & Rivoire (*Bull. Soc. Myco. Fr.* **113**: 193, 1984), Gilbertson & Hemms (*Mem. N. Y. bot. Gdn.* 89: 81, 2004; Hawai)

Type Species: *Polyporus viridens* Berk. & Br. 1861

Habitat: Dead Wood

Himalaya: One

Ceriporia viridians (Berk. & Br.) Donk, [as *Ceraporia*] Med. Bot. Mus. Univ. Utrecht 9: 171, 1933. Plate 6.25b, Fig. 6.65j–l

Fructification annual, resupinate, small in circular patches initially, confluing to become wide, easily separable from wood, soft waxy when fresh, brittle and shrivelled on drying. Pore surface yellow when fresh, yellowish brown with greenish tint on drying, rimose on drying; pores round to angular, splits on drying, 3–5 per mm; dissepiment equal; 65–135 µm thick, tubes in one layer, concolorous with the pore surface up to 2.5 mm deep. Context thin, white when fresh, creamish on drying, fibrous, homogenous, non-xanthochroic, up to 0.5 mm thick.

Hyphal system monomitic; generative hyphae hyaline to subhyaline, thin to thick-walled, without clamps, stiff and friable, dichotomously branched, run parallel and agglutinate, 3–6.4 µm in diameter. Cystidia absent. Basidia collapsed on drying. Basidiospores thin-walled, hyaline, smooth, cylindric, non-amyloid, 3.5–5 × 1.4–2.3 µm.

Distribution: Bhutan: Chimakothi, Dam side forests, Thimphu; H.P: Kullu, Kinnaur- Rohtang, Keylong, Shimla-Tara Devi.

Collection examined: SSV 21256, IBP 37219.

Substratum: On surface of under decaying angiospermic log, wood.

Remarks: This species is characterized by resupinate, soft, waxy, easily separable fructification; yellow with greenish tint pore surface; medium, 3–5 per mm pores; thin white homogenous context; monomitic hyphal system with thin to thick-walled, simple septate, dichotomously branched generative hyphae and hyaline, thin-walled, cylindric basidiospores.

Ceriporiopsis Domanski,
Acta Soc. Bot. Pol. 32 (4):731.1963

Fructification annual, soft, fleshy, resupinate, margin sometimes slightly raised, effused, sometimes form many small patches. Pore surface white, changes to ochraceous when touched, on drying brown to dark brown. Pore angular to round. Context thin to inconspicuous, ochraceous, homogenous, non-xanthochroic. Hyphal system monomitic; generative hyphae hyaline, thin to thick-walled, branched, septate, clamped. Cystidia absent. Basidia clavate, 4-spored, cyanophilous, clamps present at the base. Basidiospores cylindrical, slightly bent, smooth, hyaline, faintly cyanophilous, non-amyloid.

Twenty five, widespread

Lit.: Ryvarden & Iturriaga (*Mycol.* **95**: 1066, 2003; Venezuela)

Type Species: *Poria gilvescens* Bres. 1908

Habitat: Wood

Himalayas: One

Ceriporiopsis gilvescens (Bres.) Domański, Acta Soc. Bot. Pol. 32: 731 (1963) = *Tyromyces gilvescens* (Bres.) Ryv., Norw. Jl. Bot. 20: 10, 1973. Fig. 6.65m–p

Fructification annual, soft, fleshy, resupinate, margin slightly raised, effused, forming small patches. Margin white thin, soft when fresh, pore layer reaches the margin in older fructification, entire to irregular, curved on drying. Pore surface white, changes to ochraceous when touched, on drying brown to dark brown. Pores angular to round, entire, 4–6 per mm; Pore mouth smooth; pore tubes ochraceous; up to 4 mm deep dissepiments entire, 40–55 µm wide; tubes non-stratose. Context thin to inconspicuous, ochraceous, homogenous, non-xanthochroic.

Hyphal system monomitic; generative hyphae hyaline, thin to thick-walled, branched, septate, clamped agglutinate, 2–4.5 µm in diameter, hyphae covered with small crystals near the pore mouth. Cystidia absent. Basidia clavate, 4-spored, cyanophilous, clamps present at the base, up to 4.6 µm in diameter. Basidiospores cylindrical, slightly bent, smooth, hyaline, faintly cyanophilous, non-amyloid $3.5–4.8 \times 1.4–2$ µm.

Distribution: Bhutan-Thimphu; H.P.: Manali- reserve forests.

Collection examined: SSV 21771, IBP 37221.

Substratum: On decaying *Pinus* log.

Remarks: The species is characterized by annual, resupinate fructification which are soft fleshy pinkish white when fresh, changes brown to dark brown on drying; monomitic hyphal system with hyaline, thin to thick-walled generative hyphae; hyphae covered with crystals near the pore mouth; and hyaline, cylindrical basidiospores. The collection resembles with the description given by Ryvarden (1978) and Ryvarden & Johansen (1980).

Phanerochaete Karst.,
Bidr. Kann. Finl. Nat. Folk 48:426, 1889.
emend Donk, *Persoonia* 2: 223. 1962.

Fructifications resupinate, pelliculose-membranous when young becoming membranous-ceraceous at maturity, loosely adnate whitish to yellowish to orange brown; hymenial surface smooth to tuberculate or hydnoid; margin thinning, fibrillose to rhizomorphic. Hyphal system monomitic, subicular hyphae almost running parallel to the substrate, thick-walled, sparingly branched, subhymenial hyphae thin-walled, branched without clamps, subicular hyphae sometimes clamped, often 2 or more at the same septum, basal hyphae slightly thick-walled, branched at wide angles and somewhat loosely woven. Gloeocystidia absent. Cystidia present or absent. Basidia clavate to clavate-cylindrical, 4- spored without basal clamp. Basidiospores ellipsoid, subhyaline, thin-walled, smooth, nonamyloid.

Sixty five species, widespread.

Lit.: Donk (*Persoonia* **2:** 223, 1962), Martinez & Nakason (*Sydowia* **57:** 94, 2005; Uruguay).

Type species: *Thelephora alnea* Fr. 1821.

Habitat: Dead wood.

Himalayas: Seven

Key to species

1. Cystidia absent 2
1. Cystidia present 3
2. Hymenial surface white to cream, rhizomorphs poorly developed .... *P. tuberculata*
2. Hymenial surface pale ochraceous to orange yellow, rhizomorphs well developed .... *P. galactites*
3. Hymenial surface smooth to somewhat tuberculate 4
3. Hymenial surface greyish to greyish orange, Hymenial surface velutinous .... *P. velutina*
4. Cystidia numerous, heavily encrusted 5
4. Cystidia scattered, barely encrusted/not encrusted 6
5. Spores 4–5 × 2.5–3 μm; hymenial surface cream ochre .... *P. affinis*
5. Spores 4.5–6 × 2–3.5 μm; hymenial surface reddish in KOH .... *P. laevis*
6. Cystidia not encrusted .... *P. deflectans*
6. Cystidia encrusted in the apical part, spores 5.2–6.2 × 2.6–3.2 μm .... *P. sordida*

Phanerochaete affinis (Burt) Parm., Consp. Syst. Cort. 84: 1968. Fig. 6.66a–c

Fructification resupinate, membranous, adnate, often arising as small colonies which may later become confluent; hymenial surface deep cream to cream ochre, smooth farinose when observed under the lens, not creviced; margin thinning, adnate, paler concolorous. Subiculum yellow to orange in section, composed of a basal zone of repent hyphae and an upper zone of semi-erect hyphae, which are encrusted with crystalline matter, crystals soluble in 3 % KOH sol.

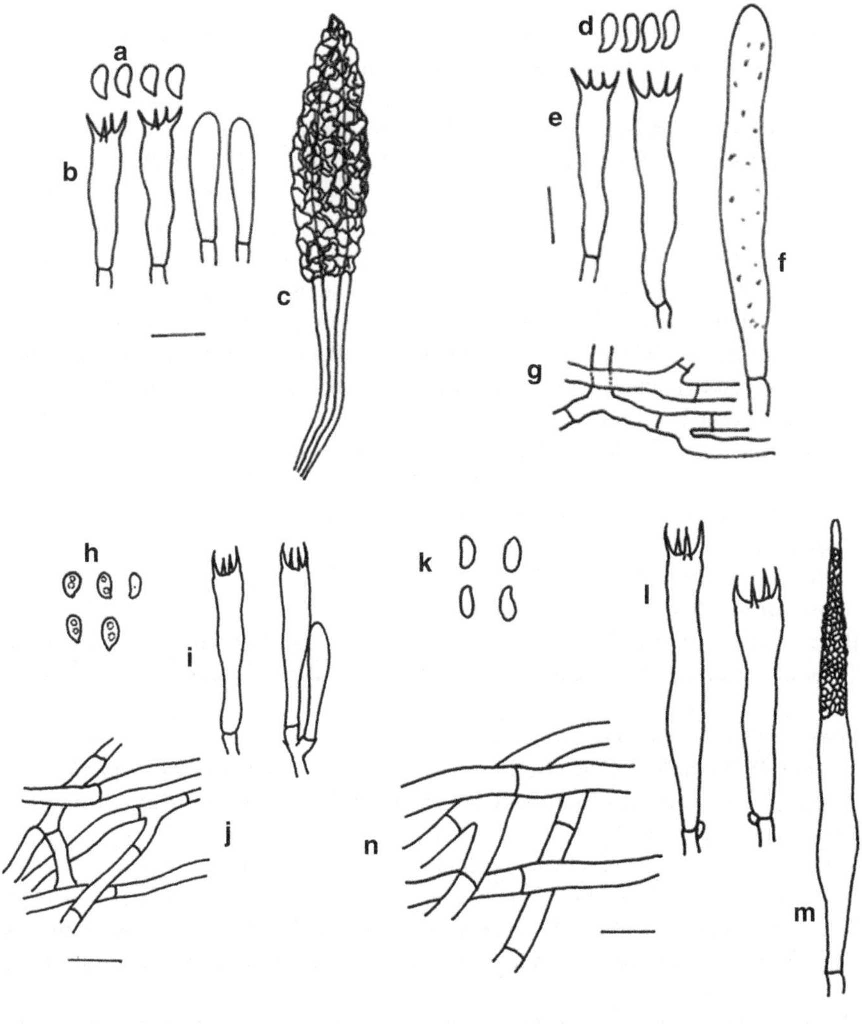

Fig. 6.66 (**a–c**) *Phanerochaete affinis* (**a**) Basidiospores, (**b**) Basidia, (**c**) Encrusted cystidia; (**d–g**) *Phanerochaete deflectens* (**d**) Basidiospores, (**e**) Basidia, (**f**) Cystidia, (**g**) Generative hyphae; (**h–j**) *Phanerochaete galactites* (**h**) Basidiospores, (**i**) Basidia, (**j**) Generative hyphae; (**k–m**) *Phanerochaete laevis* (**k**) Basidiospores, (**l**) Basidia, (**m**) Encrusted cystidia, (**n**) Generative hyphae

Hyphal system monomitic, hyphae 2.8–5.6 µm wide, branched, septate, clamps absent. The walls subhyaline, thin to slightly thick. Cystidia 35–85×6–10 µm, fusiform to subfusiform, arising from different parts of the subiculum, immersed or projecting out of the hymenium, the walls subhyaline, thick, impregnated with abundant, subhyaline crystals especially in the upper part. Basidia 20–28×4.5–5.2 µm, clavate, 4-spored. Basidiospores 4.5–5.8×2.6–3 µm, ellipsoid, shortly apiculate, thin-walled, subhyaline, non-amyloid.

Distribution: H.P.: Dalhousie- Bakrota, Manali, Shimla- Glen.

Collection examined: SSR: 5091, IBP 37223.

Substratum: stumps under mixed forest.

Remarks: The characteristic features of this species are membranous fructifications and shape and size of basidiospores and cystidia.

Phanerochaete deflectens (P. Karst.) Hjortstam, *Windahlia* 17: 58, 1987=*Grandinia deflectans* P. Karst., Bidr. Känn. Finl. Nat. Folk 37: 239, 1882.

Fig. 6.66d–g

Fructification resupinate, adnate, effused; hymenial surface smooth to tuberculate, pale orange to brownish orange when fresh, cracking white to pale orange on drying; margin thinning, paler concolorous to indeterminate.

Hyphal system monomitic; generative hyphae up to 3.3 µm wide, branched, septate, without clamps, thin-walled; basal hyphae parallel to the substrate, loosely interwoven; subhymenial hyphae vertically arranged and densely united. Cystidia 43.0–63.0×6.3–8.0 µm, cylindrical, thin-walled, with oily contents and without basal clamps. Basidia 17.0–25.0×4.3–5.0 µm, clavate, 4-sterigmate, without basal clamps. Basidiospores 5.1–5.8×2.7–4.1 µm, ellipsoid, apiculate, smooth, thin-walled, acyanophilous, inamyloid.

Distribution: H.P.-Shimla.

Collection examined: IBP 42963.

Substratum: On decaying stump of *Cedrus deodara.*

Remarks: This species is marked by presence of smooth to tuberculate fructification, cylindrical, thin-walled cystidia and ellipsoid basidiospores.

Phanerochaete galactites (Bourd. & Galz.) J. Eriksson & Ryv., in Eriksson, Hjortstam & Ryvarden, Cort. N. Europe 6: 1005, 1978=*Corticium rhodoleucum* Bourd. ssp. *galactites* Bourd.& Galz., Hym. de France p.189, 1928.

Plate 6.25c, Fig. 6.66h–j

Fructification resupinate, adnate, detectable from the substrate, effused, soft when fresh, brittle and cracking on drying, turning the fibrillose white subiculum clearly visible; hymenial surface smooth, whitish to yellowish white or pale-orchraceous; margin generally somewhat fibrillose and continuing in to rhizomorphs.

Hyphal system monomitic; generative hyphae normally without clamps, but rarely some clamps are present on the subicular hyphae, which may occur singly or in pairs; subicular hyphae up to 8.7 µm wide, sparingly branched, thin to thick-walled, hyphae next to substrate are more or less parallel to it; subhymenial hyphae 2.2–3.5 μm wide, thin-walled, ramified in to dense texture. Cystidia 52–65 × 7–8 μm, narrowly conical, with a tapering base, thin-walled initially to thick-walled, encrusted when mature especially in the upper half, immersed to somewhat projecting out of the hymenium. Basidia 26–45 × 4.8–5.8 μm, narrowly clavate, 4-sterigmate, without a basal clamp; sterigmata up to 4.8 μm long. Basidiospores 4–5.2 × 2.5–3.5 μm, ellipsoid to sub cylindrical, smooth, thin-walled, non-amyloid, acyanophilous, usually with one or more oil drops.

Distribution: West Bengal: Darjeeling, H.P.: Kinnaur, Sunder Nagar.

Collection examined: GSD 19104, IBP 37225, 37226.

Substratum: On decaying bark of a decaying angiospermic branch, on dead decaying stump.

Remarks: This species was first reported by Bourdot and Galzin (1928) from France as species of *Corticium*. Eriksson et al. (1978) shifted the species to *Phanerochaete*. This species is characterized by loosely attached, thin fructification, presence of rhizomorphs, encrusted cystidia, narrowly clavate, generally 4-spored basidia and ellipsoid basidiospores. It is a new record for N. W. Himalayas as it is recorded for the first time from Himachal Pradesh.

Phanerochaete laevis (Fr.) J. Eriksson & Ryv., in Eriksson, Hjortstam & Ryvarden, Cort. N. Europe 5: 1007 (1978) = *Thelephora laevis* Fr., Syst. mycol. 1: 451, 1821. Plate 6.25d, Fig. 6.66k–m

Fructification resupinate, adnate, widely effused, ceraceous when fresh, membranous after drying, generally with rhizomorphs, thin to somewhat thick-walled; hymenial surface smooth to somewhat tuberculate.

Hyphal system monomitic; generative hyphae branched, thin to somewhat thick-walled, 6.6 μm wide. Cystidia apically encrusted, narrowly clavate 42–76 × 4.2–5.8 μm Basidia 32–46.2 × 4.6–5.6 μm, clavate, 4-sterigmate, without basal clamp. Basidiospores 4.6–6.0 × 2.5–3.4 μm, ellipsoid, smooth, thin-walled, non-amyloid, acyanophilous.

Distribution: West Bengal: Darjeeling, Dhotrey.

Collection examined: GSD 19157, 19301.

Substratum: On decaying angiospermic branches and twigs.

Remarks: This species was first reported by Fries (1821) as *Thelephora laevis*. Burt (1902) transferred it to *Peniophora* and also erected a new species *Peniophora affinis* which was similar to *P. laevis* in all the characters except the encrustation on some of the basal hyphae. Eriksson et al. (1978) shifted *Peniophora laevis* to *Phaenerochaete laevis*.

Phanerochaete sordida (Karst.) J. Eriksson & Ryv., in Eriksson, Hjortstam and Ryvarden, Cort. N. Europe vol.5: 1023, 1978 = *Corticium sordidum* Karst., Medd. Soc. F. Fl. fenn 9: 65, 1882. Plate 6.25e, Fig. 6.67a–d

Fructification resupinate, adnate, widely effused, ceraceous when fresh, membranous after drying; hymenial surface smooth to somewhat tuberculate, at first creamish, with age greyish-yellow to greyish-orange; margin abrupt to thinning, somewhat fibrillose.

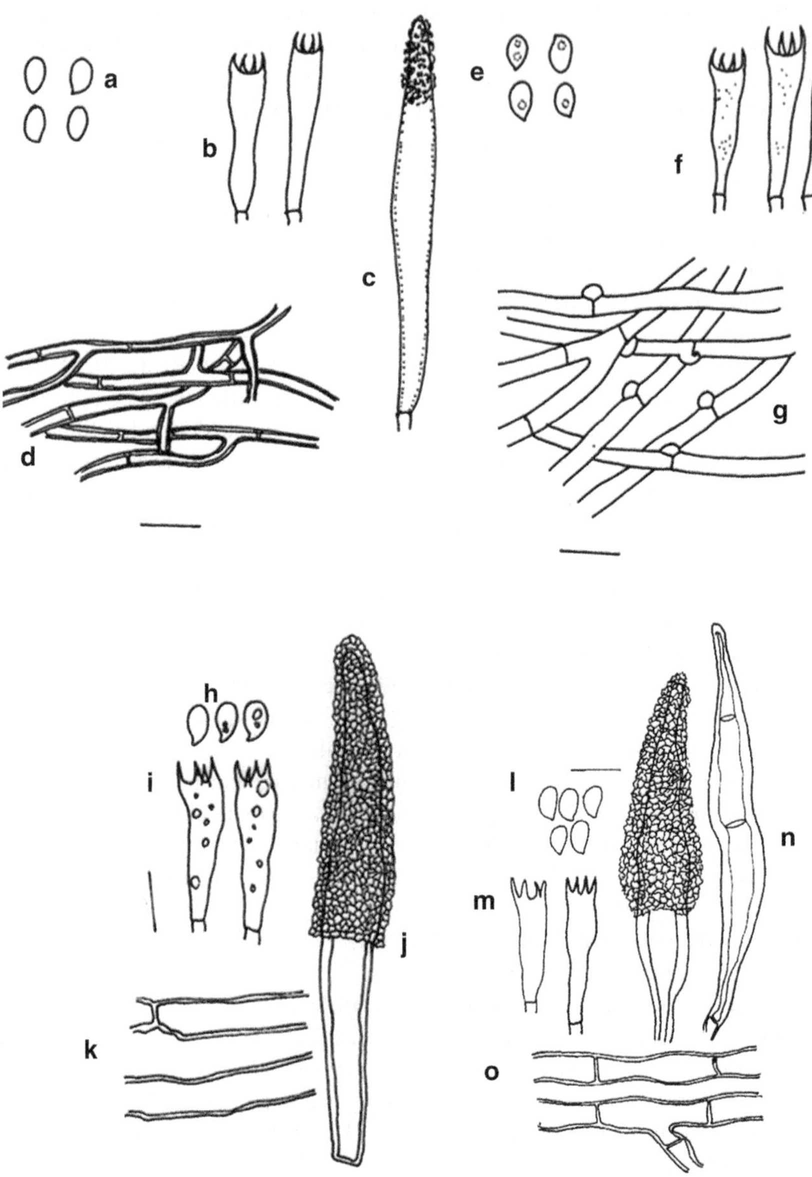

Fig. 6.67 (**a**–**d**) *Phanerochaete sordida* (**a**) Basidiospores, (**b**) Basidia, (**c**) Encrusted cystidia, (**d**) Generative hyphae; (**e**–**g**) *Phanerochaete tuberculata* (**e**) Basidiospores, (**f**) Basidia, (**g**) Generative hyphae; (**h**–**k**) *Phanerochaete velutina* (**h**) Basidiospores, (**i**) Basidia, (**j**) Basidioles, (**k**) Generative hyphae; (**l**–**o**) *Phlebiopsis darjeelingensis* (**l**) Basidiospores, (**m**) Basidia, (**n**) Cystidia, (**o**) Generative hyphae

Hyphal system monomitic; generative hyphae without clamps, subhymenial hyphae thin– walled, 2–3.2 µm, richly branched into a dense texture; subicular hyphae up to 6.2 µm wide, thick-walled, branched at wide angles, hyphae next to substrate parallel to it. Cystidia 64–78 × 7.8–10.4 µm, fusiform, apices obtuse, at first thin-walled, smooth, when mature thick-walled, encrusted in the apical part, immersed to somewhat projecting out of the hymenium. Basidia 21.8–28.4 × 3.5–4.6 µm, narrowly clavate, 4-sterigmate, without basal clamp. Basidiospores 5.2–6.2 × 2.6–3.2 µm, ellipsoid to subcylindrical, smooth, thin-walled, non-amyloid, acyanophilous.

Distribution: Arunachal Pradesh: West Kameng, Bomdila; Bhutan: Thimphu, Ha; H.P.: Manali, Kinnaur-Keylong.

Collection examined: GSD 19749, 19750, IBP 37227

Substratum: On decaying angiospermic branches and stumps.

Remarks: This was first reported from Finland as *Corticium sordium* by Karsten. Eriksson and Ryvarden transferred it to *Phanerochate.* It exhibits marked variation in colour of the fructification, size and shape of cystidia and size of basidiospores. But the nature of subicular hyphae, which are branched at wide angles, running in all directions, resulting in an interwoven hyphae but open in texture is a constant character. The above collection resembles the description given by Eriksson et al. (1978). It is being reported for the first time from N. W. Himalayas.

Phanerochaete tuberculata (Karst.) Parm., Consp. Syst. Cort. 83: 1968 = *Corticium tuberculatum* Karst., Hedwigia **35**: 45. 1895. Plate 6.26a, Fig. 6.67e–g

Fructification resupinate, adnate, widely effused, ceraceous when fresh and wet, membranous after drying; hymenial surface white to cream when fresh turning yellow on drying, smooth to somewhat tuberculate often cracking deeply on drying; margin thinning, adnate, white to paler concolorous. Subiculum subhyaline in section, composed of a basal zone of somewhat loosely woven hyphae and an upper zone of more compactly arranged semi-erect hyphae.

Hyphal system monomitic; generative hyphae branched, 2.5–5.5 µm wide, thin to somewhat thick-walled, septate without clamps. Cystidia absent. Basidia 22–30 × 5.0–6.8 µm, clavate, 4-sterigmate, without basal clamp. Basidiospores 5.0–6.6 × 3.0–4.2 µm, ellipsoid, smooth, thin-walled, non-amyloid, acyanophilous.

Distribution: H.P.: Kullu, Manali, Shimla; A.P.: West Kameng; Meghalaya: Shillong; W.B: Darjeeling.

Collection examined: GSR 5029, IBP 37228, 37230.

Substratum: On log under angiosperms, angiospermic twigs.

Remarks: This species has been earlier recorded from the N.W. Himalayas (Rattan 1977) and from the eastern Himalayas. This species is characterized by smooth to finely tuberculate hymenial surface, absence of cystidia and broadly ellipsoid to ovoid basidiospores.

Plate 6.26 (**a**) *Phanerochaete tuberculata*. (**b**) *Phlebiopsis darjeelingensis*. (**c**) *Phlebiopsis flavidoalba*. (**d**) *Phlebiopsis gigantea*. (**e**) *Phlebiopsis himalayensis*. (**f**) *Phlebiopsis ravenelii*

Phanerochaete velutina (DC.) Karst., Krit. Ofvers. Finl. Basidsv. Tillagg 3: 33, 1898 = *Thelephora velutina* DC in de Candole & Lamarck, Fl. france., Edn 3 5/6: 33(1815). Fig. 6.67h–k

Fructification resupinate, adnate, effused; smooth to somewhat tuberculate, creamish initially greyish to greyish-yellow or greyish orange when wet, becoming greyish-red on drying; margin fibrillose, paler concolorous, often with some rhizomorphs.

Hyphal system monomitic; generative hyphae normally devoid of clamps, but rarely some clamps are present in subicular hyphae; basal hyphae up to 8 μm wide, sparsely branched, thin to thick- walled, almost parallel to the substrate; subhymenial hyphae up to 4.8 μm wide thin- walled, much branched forming a dense texture. Cystidia 98–130 × 11–15 μm, subcylindrical- subfusiform, numerous, thick-walled, encrusted, especially in upper half, immersed or projecting out of the hymenium and under lens giving it a velutinous surface. Basidia 27.5–34.2 × 5–6.2 μm, subclavate, without a basal clamp, 4-sterigmate; sterigmata up to 4 μm long. Basidiospores 4.8–6.5 × 2.5–3.5 μm, mostly ellipsoid with oblique apiculus, rarely ovoid to subcylindrical, smooth, thin-walled, non-amyloid, acyanophilous, with oily contents.

Distribution: Bhutan: Thimphu, Dochula; H.P.: Mandi, Kinnaur.

Collection examined: GSD 19581, IBP 37231, 37232.

Substratum: On a decayed log.

Remarks: This species is characterized by different coloured fructification, thick-walled encrusted cystidia imparting velutinous nature to the hymenium, ellipsoid basidiospores with oblique apiculus. It is recorded for the first time from Indian Himalayas.

Phlebiopsis Jülich,
Persoonia 10(1):137, 1978.

Fructification resupinate, effused, adnate or loosening and rolling off from the subiculum on drying; margin indeterminate; hymenium smooth, tuberculate or with peg-like projections; subiculum variable, well-developed or very scarce. Hyphal system monomitic, hyphae hyaline, simple septate, compactly arranged, basal hyphae thick-walled, subiculum well developed. Subhymenial hyphae narrower, thin-walled, arranged to form a firm texture. Cystidia numerous, conical, richly encrusted in the upper portion, in the basal part naked, with thickened walls. Basidia clavate, apically dilated, normally with 4 sterigmata without basal clamp. Basidiospores narrowly ellipsoid, oblong, thin-walled, smooth, non-amyloid, acyanophilous.

Eleven species, world-wide

Lit.: Burdsall (*Mcol. Mem.* **10**, 1985), Hjortstam et al. (*Syn. Fung.* **20**: 42, 2005; Venezuela) = *Phanerochaete* fide.

Type species: *Thelephora gigantea* Fr. 1815

Habitat: Wood

Himalayas: Six

Key to species

1. Basidiospores ellipsoid to suballantoid.. *P. gigantea*
1. Basidiospores ellipsoid 2
2. Hymenial surface cream to pale yellow; smooth to somewhat tuberculate, no reaction with KOH 3
2. Hymenial surface greyish yellow; turns purplish with 3 % KOH.. 4
3. Generative hyphae with clamps 5
3. Generative hyphae without clamps, basidiospores 4.6–5.5 × 2–3.2 µm .... *P. ravenelii*
4. Basidiospores 5.2–7.1 × 3.8–4.6 µm, cystidia 75–100 × 12–14 µm.. *P. flavodoalba*
4. Basidiospores 5–6.3 × 3.5–4.8 µm, cystidia 40–75 × 13–18 µm.. *P. darjeelingensis*
5. Basidiospores 3.9–5.0 × 2.8–3 µm.. *P. himalayensis*
5. Basidiospores 5.1–7.2 × 2.5–3.9 µm.. *P. mussooriensis*

Phlebiopsis darjeelingensis Dhingra, Nova Hedwigia, 44: 222, 1987. Plate 6.26b, Fig. 6.67l–o

Fructification resupinate, adnate, effused up to 320 µm thick in section, ceraceous when fresh, hard on drying; hymenial surface yellowish-white to dull yellow when fresh, pale orange to light orange after drying, even to somewhat tuberculate; margin thinning to abrupt, adnate or often rolling off the substratum on drying.

Hyphal system monomitic; generative hyphae branched, septate, without clamps; basal zone thick-walled, compactly packed, 3–5.8 µm wide, running almost parallel to the substratum; subhymenial hyphae thinner, thin-walled, vertically densely packed to almost agglutinated. Cystidia 40.8–75 × 13.5–17.9 µm (with encrustation), 40–72 × 10.5–12.8 µm (without encrustation), subfusiform to conical, with subobtuse to obtuse apices, thick-walled, heavily encrusted in the upper half, immersed or projecting up to 40 µm out of the hymenium, rarely secondarily septate. Basidia 17–22.5 × 5.7–6.2 µm, clavate to subclavate, 4-sterigmate, without a basal clamp; sterigmata up to 4.5 µm long. Basidiospores 5–6.3 × 3.5–4.8 µm, broadly ellipsoid, smooth, thin-walled, nonamyloid, acyanophilous.

Distribution: West Bengal: Darjeeling, Siliguri.

Collection examined: GSD 19199.

Substratum: On decaying angiospermic branch.

Remarks: This species is characterized by thick-walled, compactly packed basal hyphae; subfusiform to conical cystidia with encrustation in the upper part, 4-sterigmaye basidia and broadly ellipsoid basidiospores. This collection resembles with *Phlebiopsis gigantea* and *Phlebiopsis roumeguerii* in many respects, but differs in having broadly ellipsoid basidiospores in contrast to oblong-narrowly ellipsoid-subcylindrical basidiospores and broader basidia

(17–22.5 × 5.75–6 μm clavate to subclavate in comparison to 16–22 × 4–5 μm in *P. gigantea* and 18–22(–30) × 4–5 μm in *P. roumeguerii* narrowly clavate, apically dilated). It also differs from *P. himalayensis* in having ceraceous when fresh, horny when dry fructification, hymenial surface not changing in 3 % KOH sol, massive cystidia, clavate to subclavate, broader basidia and larger size of the basidiospores.

Phlebiopsis flavidoalba (Cooke) Hjortstam, Windahlia 17: 58 (1987) = *Phanerochaete flavidoalba* (Cooke) Rattan, Biblthca mycol. 60: 262, 1977 = *Peniophora flavidoalba* Cooke, Grevillea 8: 21. Plate 6.26c, Fig. 6.68a–d

Fructification resupinate, adnate, widely effused, ceraceous when fresh, membranous after drying up to 240 μm thick in section; hymenial surface cream to pale yellow, smooth to somewhat tuberculate, often becoming areolately cracked at maturity; margin thinning to more or less abrupt, adnate, concolorous.

Hyphal system monomitic; generative hyphae branched, thin to somewhat thick-walled, with clamps, basal zone of repent hyphae and an upper zone of partly erect hyphae. Cystidia 75.2–100.2 × 12.0–14.3 μm, conical to subcylindrical, thick-walled, encrusted, immersed to somewhat projecting. Basidia 20.8–25.2 × 6.5–7.2 μm, clavate, 4-sterigmate. Basidiospores 5.2–7.1 × 3.8–4.6 μm, ellipsoid, smooth, thin-walled, non-amyloid, acyanophilous.

Distribution: Punjab: Pathankot; H.P.: Bilaspur; U. K.: Rudarprayag.

Collection examined: SSR 5197, IBP 42173.

Substratum: On bark of *Dalbergia sisso*and *Mangifera indica.*

Remarks: The species is of rare occurrence in the Himalayas. Earlier, Thind and Rattan (1973b) reported it from the Shiwaliks as *Peniophora flavidoalba.* In 1977, Rattan shifted this species to *Phaenerochaete.* It resembles well the description of *P. flavidoalba* as given by Rattan (1977). It is being recorded from Himalayas for the first time.

Phlebiopsis gigantea (Fr.) Jülich, Persoonia 10(1): 137, 1978 = *Thelephora gigantea* Fr., Observ = *Phanerochaete gigantea* (Fr.) S.S. Rattan, Biblthca Mycol. 60: 260(1977). Plate 6.26d, Fig. 6.68e–h

Fructification resupinate, adnate, effused, ceraceous to corneous on drying; hymenial surface smooth or tuberculate, greyish white to buff, sometimes cracked and rolled up from the subiculum when dry; margin indeterminate.

Hyphal system monomitic; generative hyphae branched, septate, without clamps, well developed subiculum. Cystidia massive, thick-walled, 52–78 × 10–14 μm encrusted in the apical half projecting beyond the hymenium. Basidia narrowly clavate, apically dilated, 26–30 × 4–5 μm, 4-sterigmate, without basal clamp. Basidiospores narrowly ellipsoid to suballantoid, smooth, thin-walled, 6.5–8 × 3–3.5 μm, non-amyloid, acyanophilous.

Distribution: H.P.: Mahasu- Narkanda; Kinnaur- Nichhar; Dalhousie- Lower Bakrota; Kalpa- Pangi nullah; Kullu- Gojra, Rohtang. Arunachal Pradesh: West Kameng, Bomdila, Shergaon, Bhutan: Thimphu, Chankaphug, Begana; U. K.: Mussoorie-Dehradun.

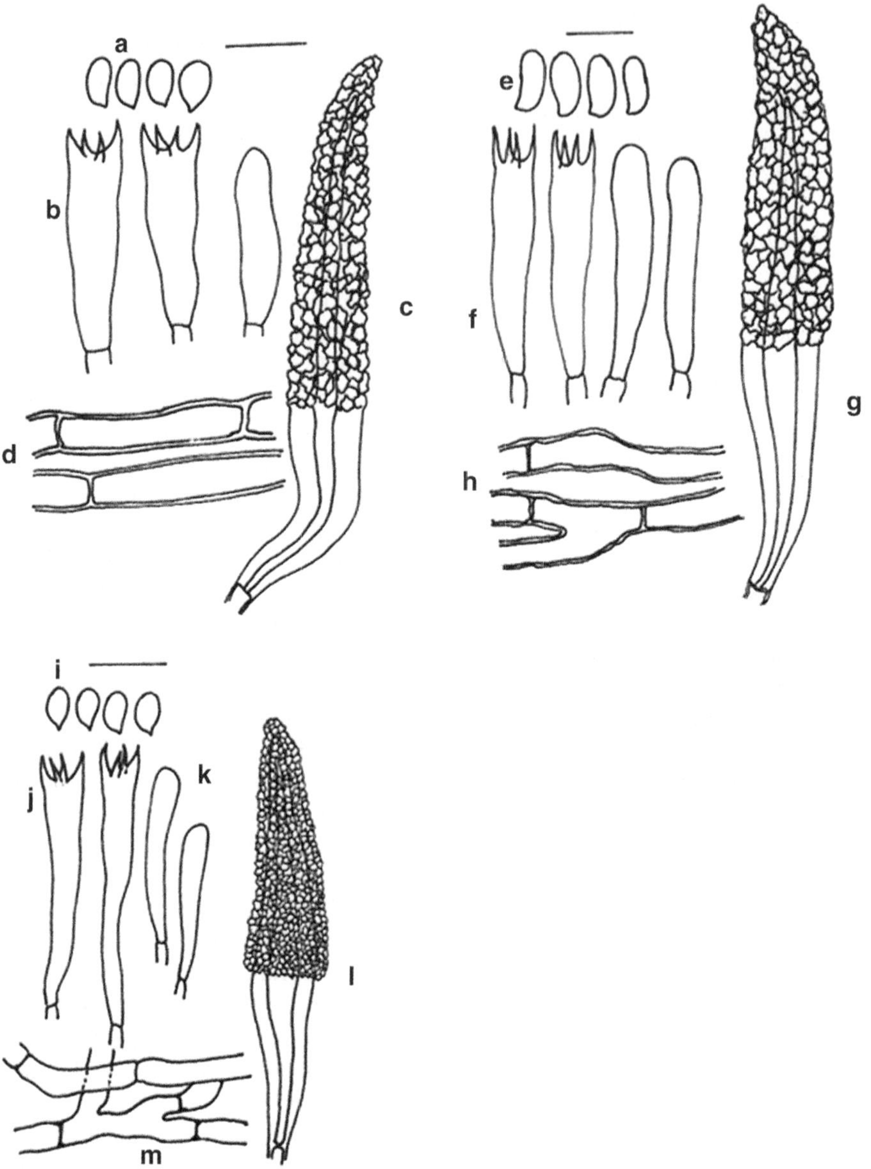

Fig. 6.68 (**a–d**) *Phlebiopsis flavidoalba* (**a**) Basidiospores, (**b**) Basidia, (**c**) Encrusted cystidia, (**d**) Generative hyphae; (**e–h**) *Phlebiopsis gigantea* (**e**) Basidiospores, (**f**) Basidia, (**g**) Encrusted cystidia, (**h**) Generative hyphae; (**i–m**) *Phlebiopsis himalayensis* (**i**) Basidiospores, (**j**) Basidia, (**k**) Basidioles, (**l**) Encrusted cystidia, (**m**) Generative hyphae

Collection examined: RW: 5021; SSR: 5090, 5329, GSD 19810, IBP 37233.

Substratum: stumps and logs of *Abies pindrow*, decorticated stump and bark of *Cedrus deodara*, log of *Pinus gerardiana*, logs under conifers, stumps under mixed forest.

Remarks: This is a commonly found species in the Himalayas as first reported from India by Bagchee and Bakshi (1954) as *Peniophora gigantea* (Fr.) Massee. Rattan (1977) shifted this species to *Phaenerochaete.* The collections are quite typical of the species described by Eriksson et al. (1981) and Rattan (1977). The species is new record for Uttarakhand.

Phlebiopsis himalayensis Dhingra, Nova Hedwigia, 44(1–2): 222, 1987. Plate 6.26e, Fig. 6.68i–m

Fructification resupinate, adnate, effused, ceraceous; hymenial surface smooth or tuberculate, greyish-white to reddish-grey or flesh coloured when fresh, greyish-orange to brownish-orange when dried, turn purplish on putting a drop of 3 % KOH sol.

Hyphal system monomitic; generative hyphae branched, septate, without clamps; basal zone composed of somewhat thick-walled, irregularly branched and loosely interwoven hyphae, followed by a zone of thin- walled, compactly packed horizontal hyphae, subhymenium composed of compactly packed to somewhat agglutinated, vertical hyphae. Cystidia 35.2–65 × 7.8–9 μm, numerous, fusiform, thick-walled, heavily encrusted in the upper half. Basidia 18–32 × 3.5–4.8 μm, narrowly clavate, apically somewhat dilated, 4-sterigmate, without basal clamp. Basidiospores 3.9–5.0 × 2.8–3.0 μm ellipsoid, smooth, thin-walled, non-amyloid, acyanophilous.

Distribution: Arunachal Pradesh: West Kameng, Bomdila, Shergaon, Basti side; West Bengal: Darjeeling, Siliguri.

Collection examined: GSD 19862, 19888, 19202.

Substratum: On decaying angiospermic branches.

Remarks: This species is characterized by ceraceous fructification, hymenial surface changing purplish in 3 % KOH sol, thick-walled, encrusted cystidia; clavate, apically dilated, 4-sterigmate basidia and ellipsoid, thin-walled basidiospores.

Phlebiopsis ravenelii (Cooke) Hjortstam, Windahlia 17: 58 (1987) = *Phlebiopsis roumeguerii* (Bres.) Jülich & Stalpers, Verh. Kon. Ned. Akad. Wet. Ser. 2. vol.74: 190, 1980 = *Corticium roumeguerii* Bres. Fungi trid. 2 p.36, 1892 = *Phlebia roumegueri* (Bers.) Donk Fungus 27: 18. 1957. Plate 6.26f, Fig. 6.69a–e

Fructification resupinate, membranous-ceraceous to ceraceous, adnate, widely effused; hymenial surface white to cream or whitish grey, smooth to minutely pilose or tuberculate, continuous, not creviced or rarely cracking in thicker parts on drying; margin thinning to more or less abrupt, sometimes fibrillose, adnate, concolorous. Subiculum subceraceous, composed of basal zone of compactly arranged repent hyphae and an upper zone of semi-erect hyphae with numerous cystidia embedded among them.

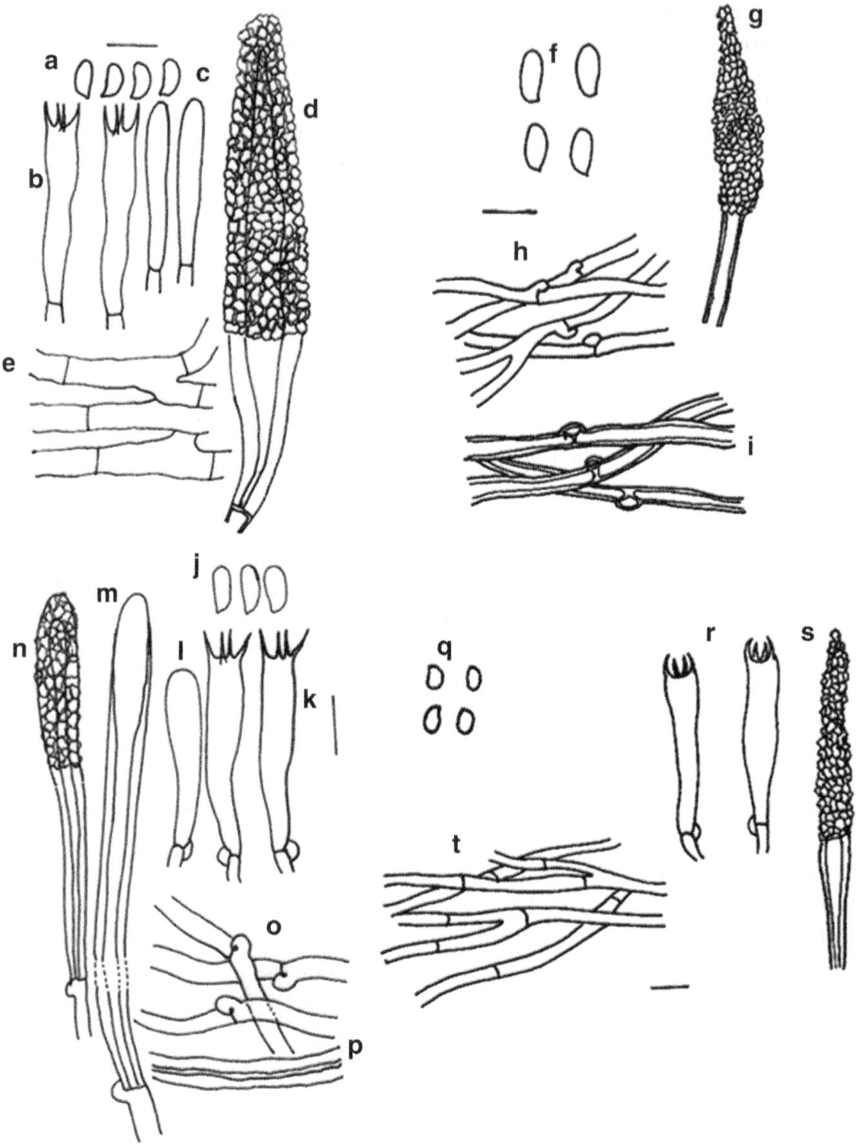

Fig. 6.69 (**a–e**) *Phlebiopsis ravenelii* (**a**) Basidiospores, (**b**) Basidia, (**c**) Basidioles, (**d**) Encrusted cystidia, (**e**) Generative hyphae; (**f–i**) *Porostereum crassum* (**f**) Basidiospores, (**g**) Encrusted cystidia, (**h**) Thin-walled generative hyphae, (**i**) Thick-walled generative hyphae (**j–p**) *Porostereum spadiceum* (**j**) Basidiospores, (**k**) Basidia, (**l**) Basidioles, (**m–n**) Cystidia hyphae, (**o**) Generative hyphae, (**p**) Skeletal hyphae; (**q–t**) *Rhizochaete filamentosa* (**q**) Basidiospores, (**r**) Basidia, (**s**) Encrusted cystidia, (**t**) Generative hyphae

Hyphal system monomitic; generative hyphae 2.7–5.2 µm wide, branched, septate, clamps absent, the walls thin to slightly thick, subhyaline. Cystidia (lamprocystidia) 50–82 × 8.6–15.2 µm, fusiform to subconical with subobtuse to acute apices, immersed or projecting out of the hymenium, the walls thick, subhyaline, impregnated with white crystalline matter especially in the upper half, abundant and always arranged in overlapping rows in the upper part of the context, hymenium forming a thick cystidial zone which is sharply delimited from the basal part, thickness of this cystidial zone increases with age. Basidia 20–25.2 × 3.6–4.6 µm, clavate, 4-spored. Basidiospores 4.6–5.5 × 2–3.2 µm, ellipsoid, minutely apiculate, the walls thin, subhyaline, smooth, non-amyloid.

Distribution: H.P.: Narkanda, Mahasu; U.K.: Hemkunt; J&K: Sanasar; Nepal: Kathmandu. Meghalaya: Cherrapunjee, Mawsmai, Shillong, Umsao, Mawphlang.

Collection examined: GSR 5004, 5006, IBP 37235, L 37236.

Substratum: On bark of *Abies pindrow*, stump of *Cedrus deodara.*

Remarks: This is a very common species inhabiting coniferous logs and stumps. It is easily identified by the membranous-ceraceous fructifications, abundant cystidia arranged in overlapping layers and small ellipsoid basidiospores.

Porostereum Pilát,
Bull. trimmest. Soc. Mycol.
France 52(3):330, (1937) [1936]

Fructification resupinate, effused-reflexed to pileate, loosely adnate, often arising as small circular or orbicular colonies which may coalesce later and become widely effused; upper surface sterile, tomentose, more or less zonate; hymenial surface smooth, tuberculate, pinkish brown to light brown or greyish brown; margin thick, often becoming reflexed, concolorous. Hyphal system dimitic or rarely monomitic; generative hyphae thin-walled, without clamps; skeletal hyphae brown, thick-walled; cystidia pale brown, thick-walled, apically encrusted; basidia narrowly clavate, 4-sterigmate; basidiospores oblong, smooth, thin-walled, non-amyloid, acyanophilous.

Fifteen species, widespread

Lit.: Ryvarden (*Syn. Fung.* **18:** 76, 2004.)

Type Species: *Porostereum phellodendri* Pilát, 1936

Habitat: Wood

Himalayas: Two

Key to species

1. Hyphal system dimitic or rarely monomitic; generative hyphae without clamps .... *P. crassum*
1. Hyphal system dimitic; generative hyphae with clamps .... *P. spadiceum*

Porostereum crassum (Lév.) Hjortstam & Ryvarden, Syn. Fung. (Oslo) 4: 29 (1990) = *Lopharia crassa* (Lev.) Boidin, Bull Soc. Mycol. France 74: 479.1958. = *Thelephora crassa* Lev., Ann. Sci. Nat. Bot. 2: 209. 1844 = *Stereum umbrinum* Berk. & Curt., Grevillea 1: 164. 1873. Plate 6.27a, Fig. 6.69f–i

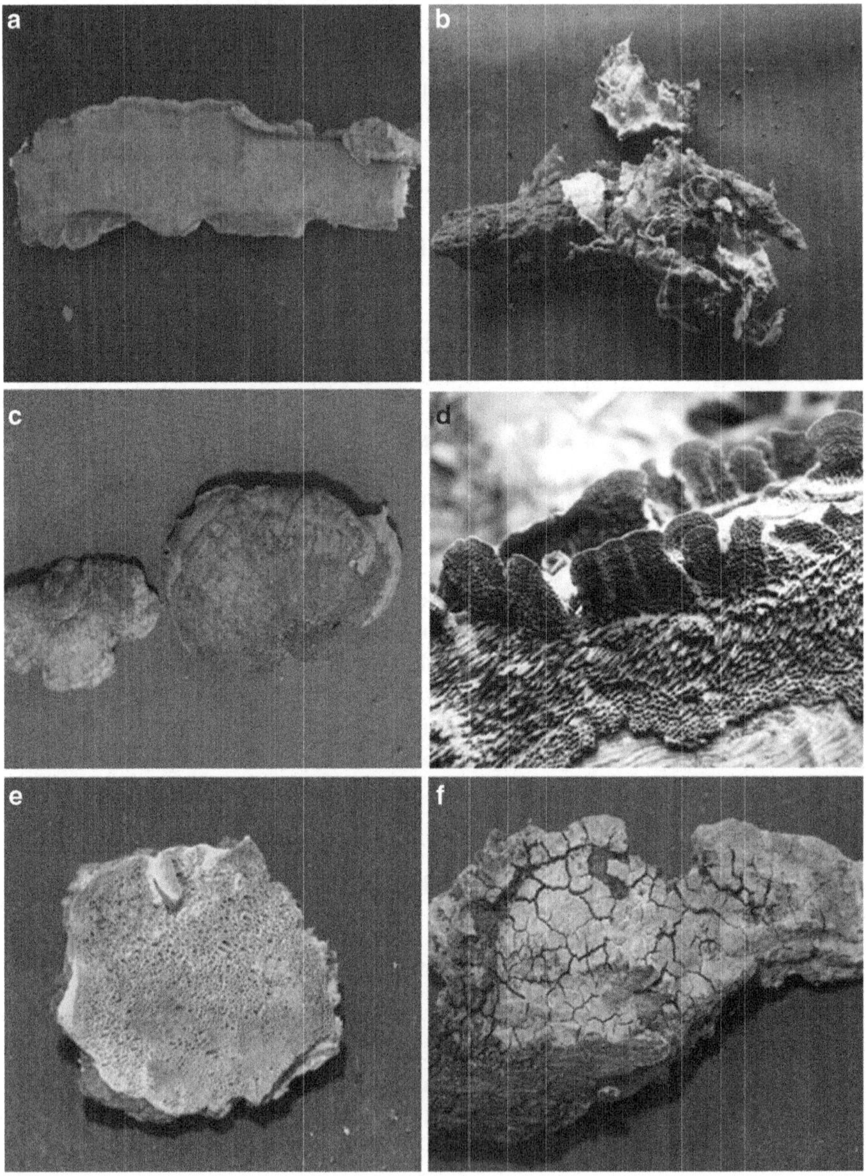

Plate 6.27 (**a**) *Porostereum crassum*. (**b**) *Rhizochaete filamentosa*. (**c**) *Coriolopsis caperata*. (**d**) *Datronia mollis*. (**e**) *Dichomitus leucoplacus*. (**f**) *Diplomitoporus crustulinus*

Fructification resupinate, effused-reflexed, submembranous to spongy-coriaceous, loosely adnate, often arising as small circular or orbicular colonies which may coalesce; hymenial surface pinkish brown to light brown or greyish brown, smooth, rarely cracking irregularly on drying; abhymenial surface finely tomentose, deep brown to greyish brown; margin thick, often becoming reflexed, concolorous.

Hyphal system monomitic, hyphae 3–4.5 μm wide, branched at wide angles, septate, clamps absent, the walls thin to moderately thick, subhyaline to ferruginous. Subicular hyphae are more or less of uniform diameter and tape-like. The subhymenial hyphae narrow, subhyaline becoming broader and pigmented towards the abhymenial side. Cystidia 80–120(150) × 7.5–12 μm, cylindrical, often arising from different parts of the subiculum, the walls thick, yellowish brown to light brown and covered with abundant subhyaline crystalline matter in the upper 2/3 part. Basidia 15–20 × 5–6 μm, clavate, 4-spored. Basidiospores 6.5–8.5 × 3–4 μm, ellipsoid-cylindrical to cylindrical, minutely apiculate, the walls thin, hyaline, smooth, non-amyloid.

Distribution: U.K.: Dehra Dun, H.P.: Rohtang.

Collection examined: IBP 37237, L 37238.

Substratum: On dead stump of *Quercus incana.*

Remarks: This species is characterized by the spongy membranous fructification, loosely interwoven subicular hyphae, generative hyphae without clamps, lack of cuticle, thick-walled cystidia, 4-sterigmate basidia, clavate, thin-walled, smooth, non-amyloid, acyanophilous basidiospores.

Porostereum spadiceum (Pers.) Hjortstam & Ryvarden, Syn. Fung. (Oslo) 4: 51 (1990) = *Lopharia fulva* (Lev.) Boidin, Bull. *Soc. Linn. Lyon* 28: 213. 1959.

Fig. 6.69j–p

Fructifications effused-reflexed to pileate but often occurring in resupinate forms as well, membranous, adnate, often arising as small colonies which may coalesce later and become effused and reflexed, up to 1 mm thick in section. Pileus flabelliform to umbonate; upper surface camel brown to light brown, tomentose, azonate to concentrically zonate, zones of erect and appressed tomentum; hymenial surface greyish brown to smoky brown, smooth to somewhat rough; margin thinning, paler concolorous to concolorous.

Hyphal system dimitic; generative hyphae 3–5.3 μm wide, branched, septate, clamped, the walls thin, subhyaline to tinted brown; skeletal hyphae 3–8.5 μm wide, unbranched, aseptate, the walls subhyaline to light brown, thick. Pseudocystidia 3.5–8 μm wide, of variable length, cylindrical, immersed or projecting out of the hymenium, the walls thick often leaving a narrow lumen, subhyaline when young but become light brown with age, unincrusted or slightly incrusted especially near the apices. The pseudocystidia are prolongations of skeletal hyphae which run horizontally in the subiculum and then curve in to the hymenium. Some of them eventually become embedded in the thickening hymenium and their place is taken by new ones which arise in the hymenium. These pseudocystidia are of limited growth often with a clamp at the base and are often mistaken for a true cystidium. Basidia 35.2–45×6–7.8 μm, clavate, 4-spored. Basidiospores 7–7.8×3.5–4.2 μm, ellipsoid, minutely apiculate, the walls hyaline, thin, smooth, non-amyloid.

Distribution: H.P.: Kinnaur- Nichhar; Kullu- Bandal; U.K.: Mussoorie, dhanulti, Hemkunt, Gobind dham.

Collection examined: RW: 5022, 5054, IBP 37240.

Substratum: fallen logs and rotten stumps.

Remarks: This species is marked by the absence of cuticle, non-amyloid basidiospores and characteristic pseudocystidia.

Rhizochaete Gresl.,
Nakasone & Rajchenb. Mycologia 96(2): 261, 2004.

Fructification resupinate, effused, pellicular to membranous, subceraceous fresh, coriaceous when dry hymenophore smooth to slightly tuberculate, velutinous, yellowish, orange or brownish coloured, tuning red to violet in KOH sol. Hyphal system monomitic; generative hyphae branched, branched at wide angles, with clamps or simple septa, numerous, walls thin to moderately thick, usually encrusted with dark yellow to yellowish brown granules that dissolve in KOH turning solution pale violet; hyphae in the basal part are broader, slightly thick-walled and more loosely interwoven. Cystidia cylindrical to subfusiform, thin to thick-walled, encrusted. Basidia narrowly clavate, 4-sterigmate. Basidiospores ellipsoid, smooth, thin-walled, non-amyloid, acyanophilous.

Six species, widespread

Lit.: Greslebil et al. *(Mycol.* **96:** 261, 2004)

Type Species: *Rhizochaete brunnea* Gresl., Nakasone & Rajchenb. 2004

Habitat: Wood

Himalayas: One

Rhizochaete filamentosa (Berk. & M.A. Curtis) Gresl., Nakasone & Rajchenb., Mycologia 96(2): 267 (2004) = *Phanerochaete filamentosa* (Berk.& Curt.) Parm. Consp. Syst. Cort. 83. 1968 = *Corticium filamentosum* Berk. & Curt., Grevillea 1: 178. 1873. Plate 6.27b, Fig. 6.69q–t

Fructification resupinate, loosely adnate, yellow ochraceous, widely effused, up to 800 μm thick in section; hymenial surface yellowish and purplish in 3 % KOH solution, smooth, not creviced or rarely creviced in thicker parts; margin fibrillose to rhizomorphic, separable, concolorous.

Hyphal system monomitic; generative hyphae branched, branched at wide angles, without clamps, numerous, walls thin to moderately thick, subhyaline. Cystidia encrusted. Basidia narrowly clavate, 4-sterigmate. Basidiospores 4–5×2.5–3 μm, ellipsoid, smooth, thin-walled, non-amyloid, acyanophilous.

Distribution: H.P.: Kullu- Khanag, Jalori; Dalhousie- Lover's walk; Chamba- Khijjiar; Shimla- Tara Devi, Chail; Mahasu- Narkanda, Bagi; J&K: Pehalgam- Batkote.

Collection examined: RW: 5035, 5051; SSR: 5114, IBP 37239.

Substratum: stumps of *Quercus incana* and *Rhododendron arboretum*, logs of *Abies pindrow*, fallen logs, stumps under angiosperms and mixed forest.

Remarks: This species is characterized by separable, yellow to orange fructification, presence of numerous rhizomorph and yellowish brown crystals in the hymenium, which are soluble in 3 % KOH and produce wine red exudates. It was first reported from India by Bagchee and Bakshi (1954) as *Peniophora filamentosa* (Berk. & Curt.) Burt. Later, Rattan (1977) collected it from different localities in N.W. Himalayas. This species is widely distributed in Himalayas. The above collection resembles closely the description of *Phanerochaete filamentosa* as given by Rattan (1977) and Eriksson et al. (1978).

Family- Polyporaceae (Table 6.6)
***Cinereomyces* Julich,**
Biblthca. Mycol. 85: 396, 1982

Fructifications annual, resupinate, membranous-coriaceous to corky when fresh but become hard and rigid on drying, adnate, widely effused; pore surface yellowish brown to pinkish brown when fresh but fades on drying, smooth to more or less uneven, not creviced; margin thinning, adnate, white to paler concolorous. Pores not stratose. Context composed of somewhat loosely woven hyphae. Hyphal system dimitic; skeletal hyphae unbranched, aseptate, thick-walled, subhyaline; generative hyphae branched, septate, clamped, thin-walled, subhyaline. Cystidia, gloeocystidia or hyphal pegs absent. Basidiospores allantoids, minutely apiculate, the walls thin, subhyaline, smooth, non-amyloid.

One species, widespread
Type Species: *Polyporus lindbladii* Berk. 1872.
Habitat: Dead logs
Himalayas: One

Cinereomyces lindbladii (Berk.) Julich, Biblthca Mycol. 85: 400 (1982)=*Diplomitoporus lindbladii* (Berk.) Gilbn. & Ryv., Mycotaxon 22: 364, 1985=*Polyporus lindbladii* Berk., Grevillea 1: 54, 1972=*Poria cinerascens* (Bres.) Sacc. Et Syd., Syll. Fung. 161, 1902=*Polyporus cinerascens* Bres., in Strass., Zool.-Bot. Ges. Vienna Verh. 50: 361. 1900. Fig. 6.70a–c

Fructifications annual, resupinate, membranous-coriaceous to corky when fresh but become hard and rigid on drying, adnate, widely effused; pore surface yellowish brown to pinkish brown when fresh but fades on drying, smooth to more or less uneven, not creviced; margin thinning, adnate, white to paler concolorous. Pores not stratose; pore mouths oval to angular but become irpicoid due to the unequal growth of dissepiments especially on slanting surfaces. Context composed of somewhat loosely woven hyphae.

Hyphal system dimitic; skeletal hyphae 2–6.2 μm wide, unbranched, aseptate, the walls thick, subhyaline; generative hyphae 1.7–4.6 μm wide, branched, septate, clamped, the walls thin, subhyaline. Cystidia, gloeocystidia or hyphal pegs absent. Basidia not observed. Basidiospores 3–5.3 × 1.8–2.3 μm, allantoids, minutely apiculate, the walls thin, subhyaline, smooth, non-amyloid.

Distribution: H.P.: Dalhousie.
Collection examined: IBP 30005.

Table 6.6 Synoptic comparison of genera of Polyporaceae

Name	Fructification	Pores	Hyphal system	Cystidia	Basidiospores
Cineromyces	Annual, resupinate, membranous, ceraceous to corky becoming hard and rigid	Oval to angular becomes irpicoid due to annual growth of dissepiments	Dimitic	Absent	Allantoids minutely apiculate, smooth, non-amyloid
Coriolopsis	Annual, pileate and sessile, rarely resupinate	Entire, round to entire, small to moderately large	Trimitic	Absent	Cylindrical to oblong ellipsoid, non-amyloid
Daedaleopsis	Annual, sessile, solitary to imbricate coriaceous, hard on drying	Daedaloid	Trimitic	Absent	Ellipsoid, smooth, non-amyloid
Datronia	Annual, sessile, resupinate to pileate, solitary, occasionally imbricate, coriaceous to hard	Irregular, circular, angular to irpiciform	Trimitic	Absent	Cylindric to fusoid, non-amyloid
Dichomitus	Annual to perennial, resupinate, coriaceous to woody, adnate	Round to angular	Dimitic	Absent	Ellipsoid, non-amyloid
Earliella	Annual to biennial, resupinate, effused-reflexed to pileate	Elongated to sinuous	Trimitic	Absent	Cylindrical to oblong ellipsoid, non-amyloid
Epithelopsis	Resupinate membranous to ceraceous becoming horny toothed, adnate	Teeth present	Monomitic	Gloeocystidia subfusiform to cylindrical	Cylindrical to subfusiform
Favolus	Annual, pileate stipitate to substipitate	Large hexagonal to elongated	Dimitic	Absent	Cylindrical, non-amyloid
Fomes	Perennial, sessile, effused-reflexed to pileate	Round to subangular	Dimitic	Absent	Cylindric ellipsoid, non-amyloid

Grammothele	Annual, resupinate, effused, adnate	Angular, irpicoid to sinuous	Dimitic	Absent	Ellipsoid to cylindrical, non-amyloid
Haplopilus	Annual, resupinate to pileate, broadly sessile to dimidiate	Round to angular	Monomitic	Absent	Broadly ellipsoid, non-amyloid
Hexagonia	Annual to perennial, sessile to pileate, corky to woody	Circular to somewhat angular,	Trimitic	Absent	Cylindrical, non-amyloid
Lenzites	Annual, resupinate to pileate, smooth to hirsute	Elongated to lamellate, angular to daedaleoid	Trimitic	Absent	Cylindrical, non-amyloid
Lignosus	Annual, centrally stipitate	Circular to angular to irregular	Trimitic	Absent	Ellipsoid, non-amyloid
Lopharia	Effused refelexed to pileate, coriaceous or membranous coriaceous,upper surface tomentose to strigose	Absent	Monomtic or dimitic	Skeletocystidia cylindricalto fusiform	Subglobose to ellipsoid amyloid or inamyloid
Loweporus	Annual to perennial, resupinate to pileate	Round to angular	Trimitic	Absent	Ellipsoid to truncate weakly dextrinoid
Macrohyphoria	Annual to perennial, resupinate, effused brittle to hard	Round to angular	Monomitic	Absent	Ellipsoid to ovoid nonamyloid
Microporellus	Annual, centrally or laterally stipitate, rarely sessile	Round to rarely angular	Dimitic	Absent	Globose to subglobose, non-amyloid
Microporus	Annual, stipitate, pileus circular	Round to oval	Trimitic	Absent	Allantoids to elliptical, non-amyloid
Navisporus	Annual, sessile with a broad base, resupinate or reflexed and woody	Circular	Dimitic	Present, extending beyondhymenial layer, dextrinoid	Navicular to fusiform, non-amyloid

(continued)

Table 6.6 (continued)

Name	Fructification	Pores	Hyphal system	Cystidia	Basidiospores
Nigroporus	Annual to perennial, pileate to resupinate, coriaceous when fresh	Round to angular	Dimitic	Absent	Allantoids to broadly ellipsoid, non-amyloid
Nigrofomes	Perennial, pileate, dimidiate, applanate	Very small, isodiametric to circular	Monomitic	Rare, ventricose, dark brown	Broadly ellipsoid, non-amyloid
Perenniporia	Annual to perennial, resupinate, sessile, applanate	subglobose	Dimitic/trimitic	absent	Weakly truncate or pip shaped, non-amyloid
Polyporus	Annual, stipitate, soft coriaceous	Round to angular	Dimitic	Absent	Cylindrical, non-amyloid
Poria	Annual, resupinate, soft fragile to cartilaginous or woody	Angular or oval becoming irpicoid	Mono-, di or trimitic	Present or absent	Ellipsoid to subglobose
Pycnoporus	Annual, sessile, pileus dimidiate to flabelliform	Circular	Trimitic	Absent	Short cylindric to ellipsoid, non-amyloid
Rhodonia	Annual, resupinate, adnate, coriaceous to hard and brittle	Round to angular	Monomitic	Absent	Ellipsoid to sub allantoids, minutely apiculate, nonamyloid
Skeletocutis	Annual, resupinate to effused-reflexed, reflexed portion forming a small pileus	Round to irregular	Dimitic	Absent	Cylindrical to allantoids, non-amyloid

Trametes	Annual, pileate, solitary or imbricate coriaceous, corky and tough	Round to angular	Trimitic	Absent	Cylindrical to ellipsoid, non-amyloid
Trametopsis	Annual, resupinate, soft to brittle	Angular to elongated, splits on drying	Dimitic	Absent	Ellipsoid, hyaline, smooth, apiculate, nonamyloid
Trichaptum	Annual, sessile, effused-reflexed	Round to irregular	Dimitic	Abundant, clavate, encrusted with irregular crystals	Cylindric ellipsoid, non-amyloid
Tyromyces	Annual, resupinate to pileate	Round to angular	Monomitic	Absent or present	Allantoid to ellipsoid, non-amyloid

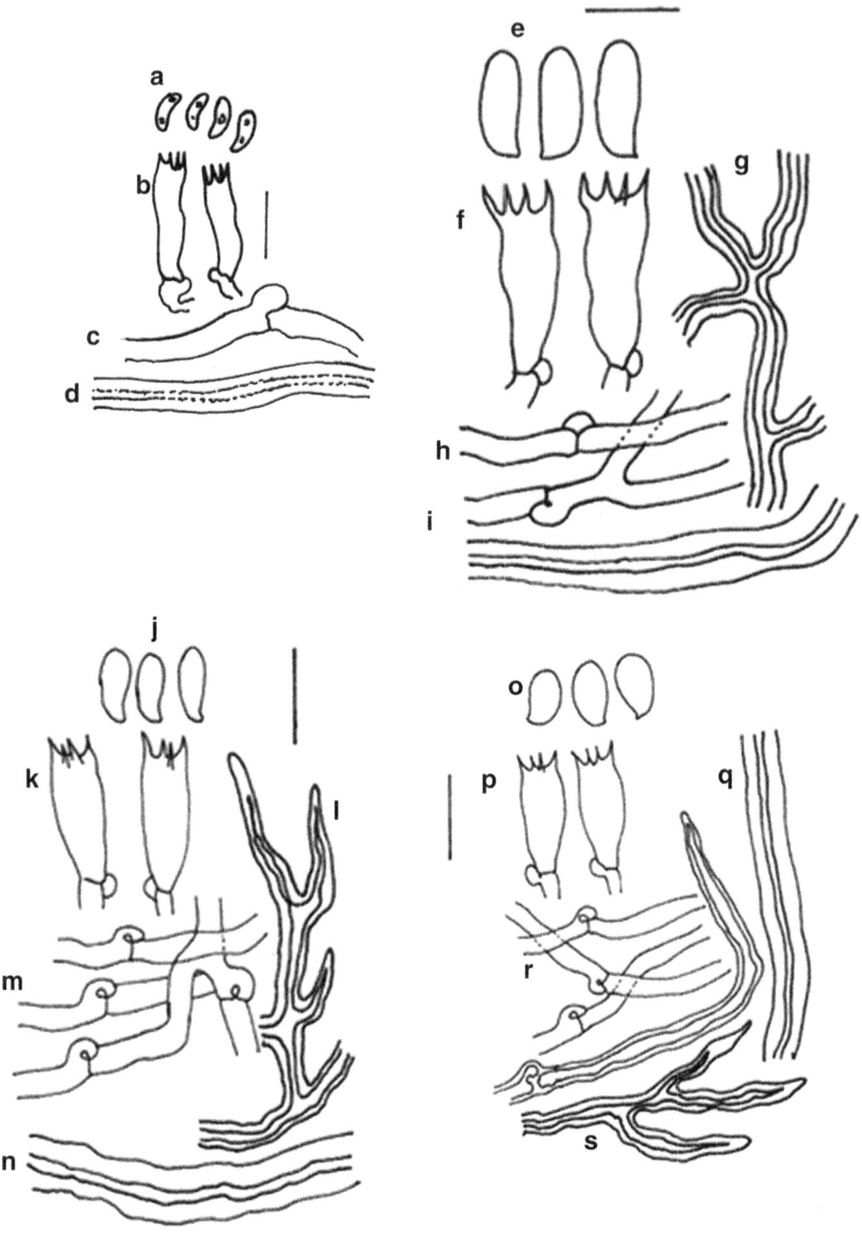

Fig. 6.70 (**a–c**) *Cinereomyces lindbladii* (**a**) Basidiospores, (**b**) Basidia, (**c**) Basidioles, (**d**) Skeletal hyphae; (**e–i**) *Coriolopsis caperata* (**e**) Basidiospores, (**f**) Basidia, (**g**) Binding hyphae, (**h**) Generative hyphae, (**i**) Skeletal hyphae; (**j–n**) *Coriolopsis occidentalis* (**j**) Basidiospores, (**k**) Basidia, (**l**) Binding hyphae, (**m**) Generative hyphae, (**n**) Skeletal hyphae (**o–s**) *Coriolopsis telfairii* (**o**) Basidiospores, (**p**) Basidia, (**q**) Skeletal hyphae, (**r**) Generative hyphae, (**s**) Binding hyphae

Substratum: On decaying coniferous log, on stump of *Pinus excelsa.*

Remarks: The species occurs on conifers in Western Himalayas and is marked by the yellowish brown or pinkish brown color of the pore surface, dimitic hyphal system and small allantoids basidiospores. It resembles being species of *Junghuhnia* but differs in lacking incrusted cystidia.

Coriolopsis **Murril,**

Bull. Torrey Bot. Club 32(7): 358, 1905.

Fructification annual, pileate and sessile, rarely resupinate, pileus velutinate to hirsute, more rarely glabrous, zonate to azonate, yellowish to umber brown, often greyish when hirsute; pore surface concolorous; pores entire, round to entire, small to moderately large. Context golden to umber brown. Hyphal system trimiti; generative hyphae hyaline, thin-walled, with clamps; skeletal hyphae thick-walled to solid, cyanophilous, hyaline, ochraceous to deep golden brown; binding hyphae thick-walled concolorous with skeletal hyphae. Cystidia absent. Basidiospores hyaline, cylindrical to oblong ellipsoid, smooth, thin-walled, negative in Melzer's reagent, 6–12 μm long. On hard wood causing a white rot.

Seventeen Species, widespread

Lit.: Peglar (*The polypores [Bull. trimes. Soc. Mycol. Fr. Suppl.],* 1973*),* Ryvarden & Johansen (*Prelim. Polyp. Fl. E. Afr.:* 315, 1980)

Type Species: *Polyporus occidentalis* Kl. Lin. 1833

Habitat: Dead Wood

Himalayas: Three

Key to species

1. Fructification mostly pileate, applanate;
   pores surface and tubes dark brown .... *C. caperata*
1. Fructification mostly resupinate, effused reflexed;
   pore surface yellowish to reddish brown 2
2. Pores regular, 0.5–6 per mm .... *C. occidentalis*
2. Pores angular to dentate to irpicoid ....*C. telfairii*

Coriolopsis caperata (Berk.) Murrill, North. Am. Flora 9(2): 77, 1908. Plate 6.27c, Fig. 6.70e–i

Fructification annual, pileate, coriaceous when fresh, tough and hard on drying, single or imbricate, attached by lateral base. Pileus sessile, applanate, dimidiate with contracted base, elongated, reflexed with decurrent pores, conchate to flabelliform up to 2.6–7 × 2.2–4.8 × 1.2 cm; upper surface light brown, dark brown to reddish brown, blackish on drying, broadly sulcate fine tomentose in few zones, with age becomes glabrous and shiny, harden due to agglutination of upper hyphae and form a thin dark cuticle; margin thin, sharp, smooth, brown, lighter than the upper surface, sterile below, up to 1 mm wide. Pore surface yellowish brown to cinnamon brown; pores round to angular, 3–5 per mm; dissepiment unequal, 68–156 μm thick. Context yellowish brown, upper context denser, brown, with a thin black crust above.

Hyphal system trimitic; generative hyphae thin-walled, hyaline, branched, septate, clamped, cyanophilous, 1.5–2 μm in diameter; skeletal hyphae golden brown, thick-walled, rarely branched, aseptate, acyanophilous, 2.6–4.6 μm in diameter; binding hyphae light brown, thick-walled to almost solid with narrow lumen, much branched, aseptate, 1.6–3.3 μm in diameter. Cystidia absent. Basidiospores thin-walled, hyaline, smooth, cylindrical, slightly curved, non-amyloid, 7–9 × 2–2.8 μm.

Distribution: A.P.: West Kameng, U.K.: Channua-Bageshwar.

Collection examined: SSV 21405, 21772, L 38025.

Substratum: On decaying angiospermic stem.

Remarks: These collections are quite typical of the species and closely resemble the description given by Ryvarden & Johansen (1980). *Coriolopsis helvola* (Fr.) Ryv. is close to *C. caperata* in having dimidiate to flabelliform fructifications and hyaline, thin-walled, cylindrical, 7.6–9 × 3–3.6 μm basidiospores. However, former differs in having bigger, angular, 2–3 pores per mm; thinner (1–4 mm), ochraceous to light brown context; yellow to pale yellow skeletal hyphae and binding hyphae; and rusty brown upper surface. This species is new record for Himalayas.

Coriolopsis occidentalis (Ki.) Murr., Bull. Torrey Bot. CI. 32(7): 358, 1905 = *Polyporus occidentalis* KI., Linnaea 8: 486, 1833. Fig. 6.70j–n

Fructification substipitate to sessile, reflexed, imbricate, hard and brittle on drying, dimidiate to aplanate 2.5–13 × 2.0–8.0 × 0.1–1 cm; upper surface yellowish brown, reddish brown to sometimes greyish brown, concentrically zonate, velvety tomentose to hirsute; margin even, sometimes incurved on drying; context yellowish brown, 0.5–5 mm; hymenial surface brown, pores round, regular or in some cases irregular, pore tubes 0.5–6 mm long; basidia and basidiospores not observed, hyphal pegs present in hymenium; hyphae (1) nearly hyaline to light brown, slightly thick to thick-walled, with lumen broad, narrow or obliterated, unbranched, aseptate, 3–7.5 μm broad, (2) thick-walled with lumen usually obliterated, much branched, aseptate, 1–3.6 μm broad and (3) thin to slightly thick-walled, often collapsing, branched, usually simple septate, sometimes with clamp connections, 1.6–4 μm broad.

Distribution: A.P.: West Kameng; U.K.: Dehra Dun, Nanital.

Collection examined: IBP 42201.

Substratum: On *Cocos nucifera, Sterculia alata*, dead decaying log.

Remarks: This species was reported by Bakshi (1971) as *Polyporus occidentalis* from Dehra Dun previously.

Coriolopsis telfairii (Kl.) Ryv. [as *telfairii*], Norw. Jl. Bot. 19(3–4): 230 (1972) = *Coriolopsis zeylanicus* (Berk.) Roy & De, Polyporaceae of India (Dehra Dun): 52(1996) = *Polyporus zeylanicus* Berk., Ann. Mag. Nat. Hist. 10: 374, 1843.

Fig. 6.70o–s

Fructification annual, resupinate to reflexed, mostly imbricate, sometimes laterally fused, flabelliform to reniform, flexible, up to 8×0.4 cm; pileus surface dirty brown to straw coloured, densely hirsute to strigose hirsute, hairs in some zones more dense towards the base and thinner towards the margin, mostly biforked or triforked, up to 5 mm long; margin lobed. Context yellowish; pore surface yellow to brownish, pores angular to dentate to irpicoid.

Hyphal system trimitic; generative hyphae hyaline, clamped, mostly thick-walled, few thin-walled, 1.5–4 μm wide; skeletal hyphae abundant, golden yellow, thick-walled to solid, flexous, sometimes apically biforked, with occasional pseudo-septa, 2.7–4 μm wide; binding hyphae equally abundant as skeletal hyphae, hyaline to yellowish, much branched, with long and short tortuous branches, thick-walled to solid, 1.3–3.0 μm wide. Basidia clavate, 11.8–15×5.3–6.6 μm, 4-sterigmate. Basidiospores hyaline, thin-walled, oblong-ellipsoid, 6.6–10.0×4–4.6 μm.

Distribution: U.K.: Kumaon, Nanital, Chakrata.

Collection examined: IBP 42187.

Substratum: On decaying angiospermic twigs.

Remarks: The species has earlier been recorded by Roy & De (1996) as occurring in Dhoni forest (Kerala), Maharashtra and Uttarakhand.

Daedaleopsis J. Schröt,
Krypt.-FI. Sch. 3. 1(25–32): 492 (1888) [1889]

Fructification annual, sessile, attached by broad lateral base, solitary to imbricate, coriaceous when fresh, drying hard and rigid; upper surface cream to deep creamish when dry, glabrous, uneven, concentrically sulcate; margin obtuse, entire, concolorous with upper surface. Pore surface creamish to creamish brown, sometimes daedaloid. Context creamish to creamish brown, homogenous, inconspicuously azonate, reaction with KOH sol. Hyphal system trimitic; generative hyphae hyaline, thin-walled, septate, branched, clamps not common, acyanophilous; skeletal hyphae hyaline to subhyaline, thick-walled to solid, long, unbranched, acyanophilous; binding hyphae hyaline, thick-walled, profusely branched, acyanophilous. Cystidia absent. Basidia clavate, 4-spored. Basidiospores, hyaline, thin-walled, smooth, non-amyloid, ellipsoid.

Six species widespread

Lit.: Bernichia et al. (*Mycol. Res.* 110: 14, 2006; Neolithic DNA)

Type species: *Boletus confragosus* Bolt. 1791

Habitat: Wood

Himalayas: One

Daedaleopsis confragosa (Bolton) J. Schrot, in Cohn, Krypt.-Fl. Schlesien (Breslan) 3.1 (25–32): 492 (1888). Fig. 6.71a–f

Fructification annual, sessile, attached by broad lateral base, solitary to imbricate, *coriaceous* when fresh, drying hard and rigid, light in weight. Pileus sessile, appalanate; upper surface cream to deep creamish when dry, glabrous, uneven, concentrically sulcate; margin obtuse, entire, concolorous with upper surface.

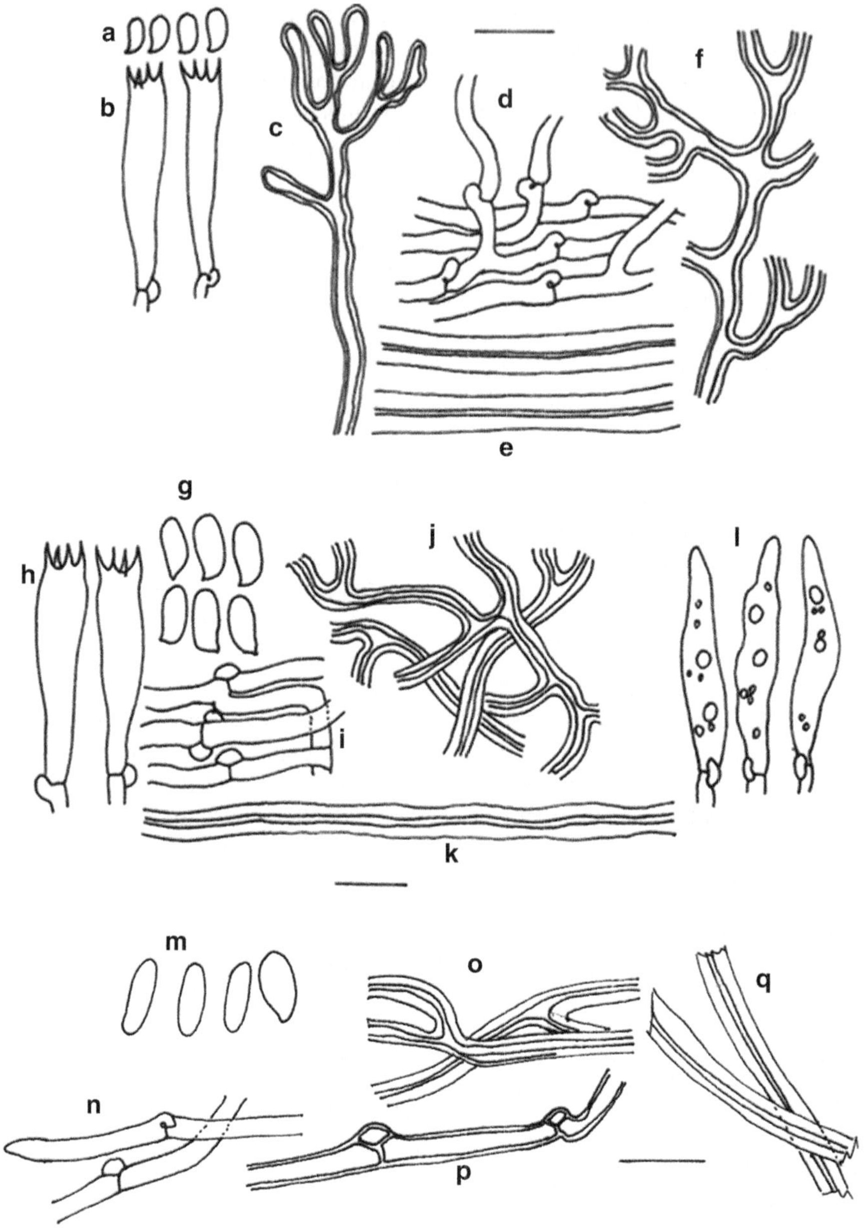

Fig. 6.71 (**a**–**f**) *Daedaleopsis confragosa* (**a**) Basidiospores, (**b**) Basidia, (**c**) Binding hyphae, (**d**) Generative hyphae, (**e**) Skeletal hyphae, (**f**) Binding hyphae; (**g**–**l**) *Datronia mollis* (**g**) Basidiospores, (**h**) Basidia, (**i**) Generative hyphae, (**j**) Binding hyphae, (**k**) Skeletal hyphae, (**l**) Cystidia; (**m**–**q**) *Datronia scutellata* (**m**) Basidiospores, (**n**) Thin walled generative hyphae, (**o**) Binding hyphae, (**p**) Thick walled generative hyphae, (**q**) Skeletal hyphae

Pore surface creamish to creamish brown, mostly lamellate, sometimes daedaloid. Context creamish to creamish brown, homogenous, inconspicuously azonate, reaction with KOH.

Hyphal system trimitic; generative hyphae hyaline, thin-walled, septate, branched, clamps not common, acyanophilous, 2.2–3.2 μm in diameter; skeletal hyphae hyaline to subhyaline, thick-walled to solid, long, unbranched, acyanophilous, 2.2–5.7 μm in diameter, binding hyphae hyaline, thick-walled, profusely branched, acyanophilous, 2.9–4.2 μm in diameter. Cystidia absent. Basidia not seen. Basidiospores, hyaline, thin-walled, smooth, non-amyloid, ellipsoid, 6.5–7.8 × 2.9–3.5 μm.

Distribution: U.K.: Rishikesh.

Collection examined: Sharma 6286.

Substratum: On logs and wood stall.

Remarks: This species is characterized by dimidiate, sessile, imbricate, cream coloured fructifications with creamish, glabrous, concentrically zonate upper surface, lamellate or daedaloid hymenophore; trimitic hyphal system and hyaline, non-amyloid, ellipsoid basidiospores.

Datronia Donk,

Persoonia 4(3): 337, 1966.

Fructification annual, sessile, resupinate to pileate, solitary, occasionally imbricate, coriaceous-flexible when fresh, firm and rigid on drying. Pileus when present, soft with different shades of brown or blackish, at first velutinate to tomentose, with age becoming nearly glabrous, sulcate; margin thick. Context light brown; hymenial surface greyish to brownish grey, pores irregular, circular, angular, daedaloid or irpiciform. Hyphal system trimitic; generative hyphae hyaline to pale brown, thin-walled to slightly thick-walled, clamped, branched; skeletal hyphae subhyaline to golden brown, thick-walled to solid, sometimes corrugated, frequently branched at the apex; binding hyphae subhyaline to brown, thick-walled to subsolid. Basidia 4-sterigmate. Basidiospores hyaline, thin-walled, cylindric to ellipsoid.

Five species, widespread

Lit.: Ryvarden & Gilbertson *(Europe. Polyp.* 1: 230, 1993).

Habitat: Dead Wood

Type Species: *Datronia mollis* (Sommerf.) Donk, 1966

Himalayas: Three

Key to species

1. Pores only circular, 4–6 per mm .... *D. scutellata*
1. Pores not circular only 2
2. Pores angular to circular, 2–4 per mm .... *D. stereoides var. microspora*
2. Pores irregular to circular to angular,
   daedaleoid or irpiciform, 0.5–3.0 per mm .... *D. mollis*

Datronia mollis (Sommerf.) Donk, Persoonia 4(3): 338, 1966=*Daedalea mollis* Sommerf., Suppl. Fl Lapp (Oslo): 271, 1826. Plate 6.27d, Fig. 6.71g–l

Fructification annual, sessile, resupinate or effused-reflexed with a narrow free margin, solitary, occasionally imbricate, coriaceous-flexible when fresh, firm and rigid on drying; pileus surface with different shades of brown or blackish, at first velutinate to tomentose, with age becoming nearly glabrous, sulcate; margin thick, sterile below up to 1 mm. Context light brown, usually separated from the overlying tomentum by a narrow black zone, tough and coriaceous; hymenial surface greyish to brownish grey, pores irregular, circular, angular, daedaloid or irpiciform, 0.5–3.0 pores mm, pore tubes vertical to oblique, 1–2 mm long or more when the tubes are oblique.

Hyphal system trimitic; generative hyphae hyaline to pale brown, thin-walled to slightly thick-walled, clamped, branched, 1.5–4.6 µm wide; skeletal hyphae subhyaline to golden brown, thick-walled to solid, sometimes corrugated, frequently branched at the apex, few branched in the stem portion, some unbranched and often showing pseudosepta towards the distal end, 2–4 µm wide. Binding hyphae subhyaline to brown, thick-walled to subsolid, shortly to distantly branched with tapering branches, 1.8–3.0 µm wide, infrequently found in the context and occurring particularly towards the base of the pileus. Basidia 4-sterigmate, 16.2–22.7 × 4.2–6.9 µm. Basidiospores hyaline, thin-walled, cylindric, 5.7–8.5 × 2.8–4.2 µm. Hyphal pegs occasionally present.

Distribution: H.P-Kullu; Meghalaya: Shillong; U.K: Chakrata; West Bengal: Darjeeling.

Collection examined: IBP 42207.

Substratum: On decaying angiospermic log.

Remarks: The genus is usually characterized as having a duplex context with black line separating an upper tomentum and lower denser context and having dimitic hyphal system. But black line does not appear in all the fructification of *D. mollis* and it is also reported to be lacking in *D. glabra* (Ryvarden 1987). Moreover critical observations on large collections of *Datronia mollis* show that the branched thick-walled hyphae are typical binding hyphae indicating thereby that this species is trimitic as is *Datronia glabra* (Ryvarden 1987). The genus *Datronia* is therefore trimitic.

Datronia scutellata (Schwein.) Gilb. & Ryvarden, Mycotaxon 22(2): 364 (1985)=*Hexagonia scutellata* (Schw.) Roy & De, J. Mycopathol. Res. 36(1): 36 (1998).

Fig. 6.71m–q

Fructification annual or perennial, sessile, effused reflexed to pileate, pileus broadly to narrowly attached, usually small, corky to woody, scutellate or ungulate, whitish or brownish and tomentose when young, becoming glabrous later, forming black crust, concentrically zonate and radially wrinkled; margin round, entire or lobed; context light brown, corky, about 2.5 mm wide; hymenophore white to grey to wood coloured, plane or concave, pores regular, circular, 4–6 per mm, tubes indistinctly stratified, up to 2 mm long in each layer.

Hyphal system trimitic, generative hyphae hyaline, clamped, thin-walled, 2.4–4.4 µm wide, some subhyaline, slightly thick-walled, shortly branched and with projections, all firmly interlocked, found in crusty part of the pileus; skeletal hyphae yellowish brown, thick-walled, 3.2–4.1 µm wide; binding hyphae hyaline to subhyaline, much branched with short flexous branches, thick-walled to solid, 2.6–3.6 µm wide. Basidia 6–9 µm broad, 4-sterigmate. Basidiospores hyaline, cylindrical, thin-walled, smooth, gutulate, 7.0–10.0 × 3.0–3.5 µm.

Distribution: H.P.: Chamba, Dalhousie.

Collection examined: IBP 42014, 42017.

Substratum: On decaying *Cedrus deodara* log.

Remarks: The species is characterized by having annual to perennial, effused to reflexed to pileate, corky fructification; margin round, entire; trimitic hyphal system; cylindrical, smooth basidiospores.

Datronia stereoides (Fr.) Ryv., Blyttia 25: 168 (1967) var. *microspora* Prasher and Lalita var. nov. Fig. 6.72a–f

Mycobank MB812331

Fructifications annual, resupinate, reflexed pileate, coriaceous, laterally accreted to each other; pileus when present reflexed and elongated up to 6.8 mm wide; upper surface brown to dark brown, becoming dark on drying, glabrous or minutely tomentose, concentrically zonate; margin smooth, light brown to greenish brown when fresh, dark on drying, thin up to 1 mm, sterile below. Pore surface brown to greyish brown, uneven; pores angular to round, 2–4 per mm; pore tubes not stratified, light brown, up to 2.4 mm deep; dissepiment even, light brown, concolorous with the pore tubes, 95–185 µm wide. Context brown to dark brown, duplex, light brown near the tubes and dark brown towards the upper surface, up to 0.8 mm thick.

Hyphal system dimitic; generative hyphae hyaline, thin-walled, branched, septate, clamped, cyanophilous, 2.3–3.5 µm in diameter; skeletal hyphae subhyaline, brown to dark brown, black in groups, aseptate, straight, thick-walled to almost solid with narrow lumen, unbranched, 2–4 µm in diameter. Cystidia absent. Basidiospores hyaline to subhyaline, thin-walled, cylindric-ellipsoid to ellipsoid, non-amyloid, 6–7.4 × 2.6–3 µm.

Etymology: Refers to the significantly smaller spores than the species.

Distribution: H.P. Manali, U. K. Badrinath.

Collection examined: L 37241, IBP 37690- PAN (Holotype).

Substratum: On decaying angiospermic log.

Remarks: The above collections are typical of species and closely resemble the description given by Ryvarden (1968). However, Norway collection has bigger, 9–11.5 × 3–4.5 µm basidiospores. The spores in Himalayan collection are significantly smaller (6.74 × 2.6–3 µm as compared to 9–11.5 × 3–4.5 µm in the type) than the type description given by Ryvarden (1968) Since it is similar in other macro and micro morphological features, it is proposed as a new variety of *D. stereoides* on the basis of markedly smaller spores.

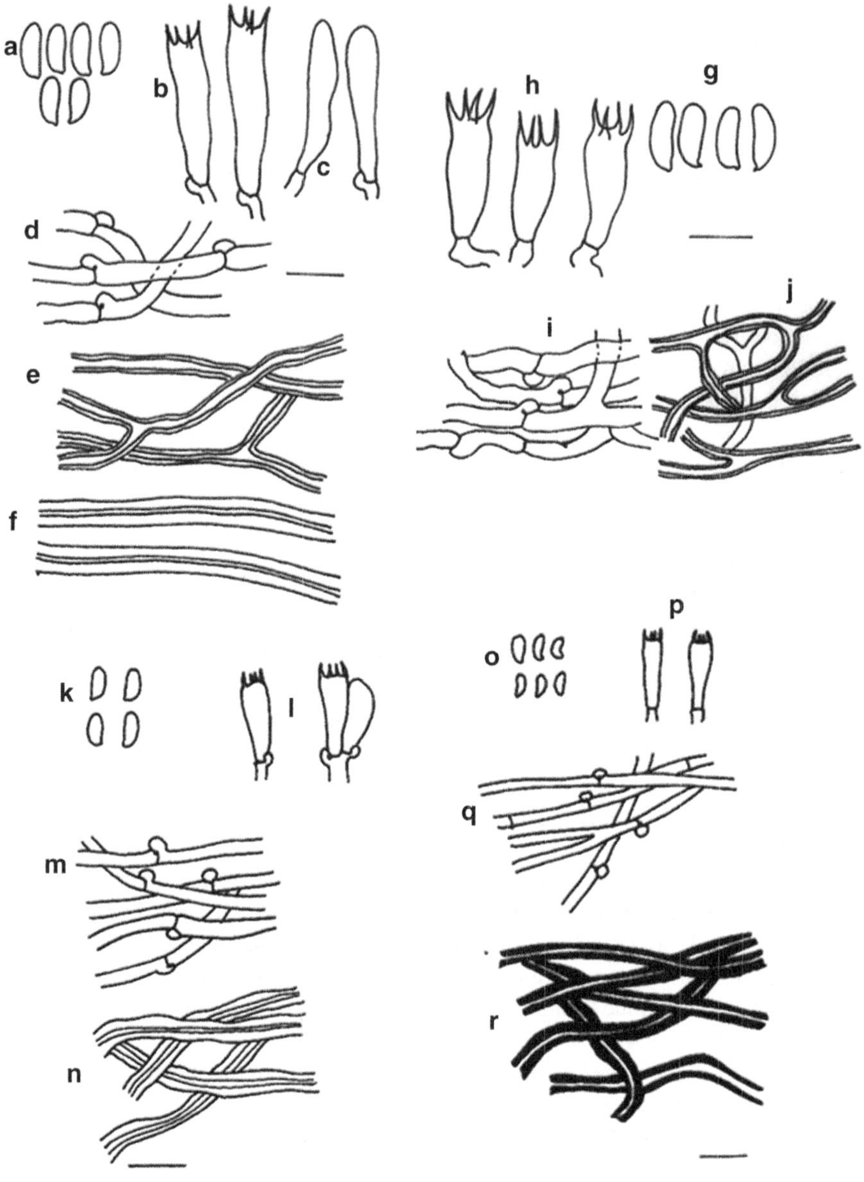

Fig. 6.72 (**a–f**) *Datronia stereoides* (**a**) Basidiospores, (**b**) Basidia, (**c**) Basidioles, (**d**) Generative hyphae, (**e**) Binding hyphae, (**f**) Skeletal hyphae; (**g–j**) *Dichomitus leucoplacus* (**g**) Basidiospores, (**h**) Basidia, (**i**) Generative hyphae, (**j**) Skeletal hyphae; (**k–n**) *Diplomitoporus crustulinus* (**k**) Basidiospores, (**l**) Basidia, (**m**) Generative hyphae, (**n**) Skeletal hyphae; (**o–r**) *Diplomitoporus rimosus* (**o**) Basidiospores, (**p**) Basidia, (**q**) Generative hyphae, (**r**) Skeletal hyphae

Dichomitus Reid,
Rev. Biol. 5: 149, 1965.

Fructification resupinate, annual to perennial, coriaceous when fresh turns woody on drying, adnate; upper surface in pileate fructification, white to cream, glabrous, azonate; margin thin or not present. Context thin, white, homogenous and non-xanthochroic. Hyphal system dimitic; generative hyphae hyaline, thin-walled, septate, branched, clamped; binding hyphae hyaline to subhyaline, aseptate, thick-walled to solid with narrow lumen. Cystidia and setae absent. Basidia clavate, 4-spored. Basidiospores cylindrical to ellipsoid, hyaline, thin-walled, smooth, non-amyloid.

Seven species, widespread

Lit.: Ipulet & Ryvarden (*Syn. Fung.* 20: 87; 2005 Uganda)

Habitat: Dead Wood

Type Species: *Trametes squalens,* Karst 1886

Himalaya: One

Dichomitus leucoplacus (Berk.) Ryvarden, Norw. Jl Bot. 24: 222 (1977) = *Poria leucoplaca* (Berk.) Sacc. Syll. fung. (Abellini) 6: 322 (1888). Plate 6.27e, Fig. 6.72g–j

Fructification annual, resupinate, adnate, soft coriaceous, becoming hard on drying in small patches of 3 × 4 cm rarely effused to form fructification 10 cm long, 3 cm wide and up to 2 mm thick; margin white, concolorous with pore surface, thin, fertile to sterile up to 2 mm wide. Pore surface white to cream, turns brown when touched, dark when dried; pores entire, round to subangular, elongated on oblique stratum, 4–5 per mm, 160–280 μm in diameter; dissepiments even, 45–100 μm thick; pore mouth veltinate; tubes in single layer, concolorous with the pore surface, up to 1 mm deep in section. Context thin, white, homogenous, non-xanthochroic up to 0.75 mm thick.

Hyphal system dimitic; generative hyphae hyaline, thin-walled, branched, septate, clamped, cyanophilous, 2.5–4 μm in diameter; binding hyphae hyaline, thick-walled to solid with narrow lumen, profusely branched, aseptate, cyanophilous, dominate in the dissepiments and context. 1–3.2 μm in diameter. Cystidia absent. Cystidioles cylindric, clavate, thin-walled, up to 3 μm wide. Basidia hyaline, clavate, 4-spored, cyanophilous, 17–25 × 6–8 μm. Basidiospores hyaline, thin-walled, smooth, ellipsoid, non-amyloid, 9.5–13.5 × 4–5 μm

Distribution: A.P.: West Kameng; H.P: Chamba; U.K.

Collection examined: SSV 21423, 21451, R 6140, IBP 37891.

Substratum: On decaying angiospermic twigs.

Remarks: This species occurs on angiospermic wood in West Kameng of Arunachal Pradesh and Uttarakhand. The species resembles with the description of the specimens as given by Ryvarden (1977) and Ryvarden and Johnsen (1980). It is the first record of the species from Himalayas.

Diplomitoporus Dom.,
Acta Soc. Bot. Pol. 39: 191, 1970.

Fructification annual, resupinate, fragile when fresh, hard on drying, adnate, broadly effused, deeply cracked after drying; margin white very narrow; pore surface pale yellow to sulphur yellow when fresh, yellowish brown on drying; pores angular, Context white, soft, homogenous. Hyphal system dimitic; generative hyphae hyaline, thin-walled, septate, branched, clamped, cyanophilous; skeletal hyphae hyaline, thick-walled, aseptate, unbranched. Cystidia absent, cystidioles present. Basidia clavate, 4-spored. Basidiospores hyaline, thin-walled, smooth, cylindric-ellipsoid, non-amyloid.

Eleven species, widespread

Lit.: Gilbertson & Ryvarden 9 (N. Amer. Polyp. 1: 240, 1986.)

Type Species: *Trametes flavescens* Bres. 1903

Habitat: Decayed wood

Himalayas: Two

Key to species

1. Pore surface pale yellow to sulphur yellow, yellowish brown on drying; pores angular .... *D. crustulinus*
1. Pore surface creamish brown when fresh, darkening on drying, uneven, dull; pores round to somewhat irregular .... *D. rimosus*

Diplomitoporus crustulinus (Bres.) Domański, Acta Soc. Bot. Pol. 39: 192 (1970) = *Antrodia crustulina* (Bres.) Ryvarden, Norw. Jl. Bot. 20: 8 (1973). Plate 6.27f, Fig. 6.72k–n

Fructification annual, resupinate, fragile when fresh, hard on drying, adnate, broadly effused, up to 4.8 cm long and up to 6.2 mm thick, smooth when fresh, deeply cracked after drying; margin white very narrow; pore surface pale yellow to sulphur yellow when fresh, yellowish brown on drying; pores angular. 3–5 per mm, 120–160 μm in diameter. Context white, soft, homogenous, non-xanthochroic, up to 2 mm thick; dissepiments concolorous, even, 85–150 μm thick; tubes in one layer, yellow up to 4.5 mm deep in section.

Hyphal system dimitic; generative hyphae hyaline, thin-walled, septate, branched, clamped, cyanophilous, 2.0–4.1 μm in diameter; skeletal hyphae hyaline, thick-walled, aseptate, unbranched, 2.4–5.9 μm in diameter. Cystidia absent, cystidioles present. Basidia clavate, 4-spored, cyanophilous, up to 4.7 μm broad, Basidiospores hyaline, thin-walled, smooth, cylindric-ellipsoid, non-amyloid, 5.8–6.8 × 2.6–3.2 μm.

Distribution: H.P.: Shimla- Narkanda, Manali-reserve forests, U.K.: Rudarprayag.

Collection examined: IBP 37242, 37243.

Substratum: On decaying gymnospermic log.

Remarks: This species is characterized by annual, resupinate, adnate, sulphur yellow, cracked fructification and cylindric-ellipsoid to ellipsoid basidiospores. It is a new record for the Himalayas/India.

Diplomitoporus rimosus (Murrill) Gilb. & Ryvarden, Mycotaxon 22(2): 364 (1985) = *Poria rimosa* Murrill, Mycologia 12(2): 91 (1920). Plate 6.28a, Fig. 6.72o–r

Fructification annual, resupinate, effused, 4–8 × 1–3.2 cm up to 2 mm thick, adnate, inseparable; margin thinning, creamish, narrow, fertile; pore surface creamish brown when fresh, darkening on drying, uneven, dull; pores round to somewhat irregular, 105–400 µm in diameter, 3–4 per mm; dissepiments 55–80 µm thick, equal of compactly arranged parallel hyphae, apices velutinate. Context creamish, homogenous of interwoven hyphae.

Hyphal system dimitic; generative hyphae hyaline, thin-walled, septate, branched, clamped, cyanophilous, 1.6–4.3 µm in diameter; skeletal hyphae subhyaline, thick-walled to solid, mostly unbranched, long, acyanophilous, 2.5–4.3 µm. Cystidia rare, subhyaline. Basidia hyphoid, hyaline, thin-walled, clavate, 2–4 spored, up to 4.2 µm in diameter. Basidiospores, hyaline, thin-walled, smooth, apiculate non-amyloid, broadly ellipsoid to ovoid, 3.2–4.2 × 2.1–2.9 µm.

Distribution: H.P.: Shimla- Shillaroo, Manali, Kullu.

Collection examined: Dhanda 6521, IBP 37246, 37247.

Substratum: Burnt logs in mixed forest.

Remarks: The species was first described from India by Dhanda (1977) as *Poria rimosa* Murr. The species is characterized by perennial, inseparable fructifications with cremish, often cracked, pore surface; 4–5 pores per mm, dimitic hyphal system; hyphoid cystidia and hyaline non-amyloid allantoids basidiospores.

Earliella Murrill,

Bullein Torrey Botanical Cl, 32(9): 478 (1905)

Fructification annual to biennial, resupinate, effused-reflexed to pileate, tough, pilear surface when present, glabrous, first white to cream, then with a reddish cuticle spreading from the base; pore surface white to cork-coloured, pores elongated and sinuous. Context white to wood-coloured. Hyphal system trimitic; generative hyphae with clamps; skeletal and binding hyphae hyaline. Cystidia absent. Basidiospores cylindrical to oblong ellipsoid, hyaline and negative in Melzer's reagent.

Monotypic

Lit.: Gilbertson & Ryvarden *(N. Amer. Polyp. 1:249, 1986)*

Type Species: *Earliella cubensis* Murrill, 1905

Habitat: Decay wood

Himalayas: One

Earliella scabrosa (Pers.) Gilb. & Ryvarden, Mycotaxon 22(2): 364 (1985) = *Trametes scabrosa* (Pers.) Cunn., Bull. N. Z. Dept. Sci. Industr. Res. 164: 162–163, 1985 = *Polyporus scabrosa* Pers., Gaudich, Voyage aut Monde. p. 172, 1827 = *Polyporus corrugatus* Pers. op. Cit. Earliella cubensis (Pers.) Murr., Bull. Torr. Bot. Cl. 32:478, 1905 = *Fomitopsis corrugata* (Pers.) Imaz., Bull. Tokyo Sci. Mus. 6: 92, 1943. Fig. 6.73a–f

Plate 6.28 (**a**) *Diplomitoporus rimosus*. (**b**) *Fomes fomentarius*. (**c**) *Fomes extensus*. (**d**) *Grammothele fuligo*. (**e**) *Hexagonia tenuis*. (**f**) *Lenzites betulina*

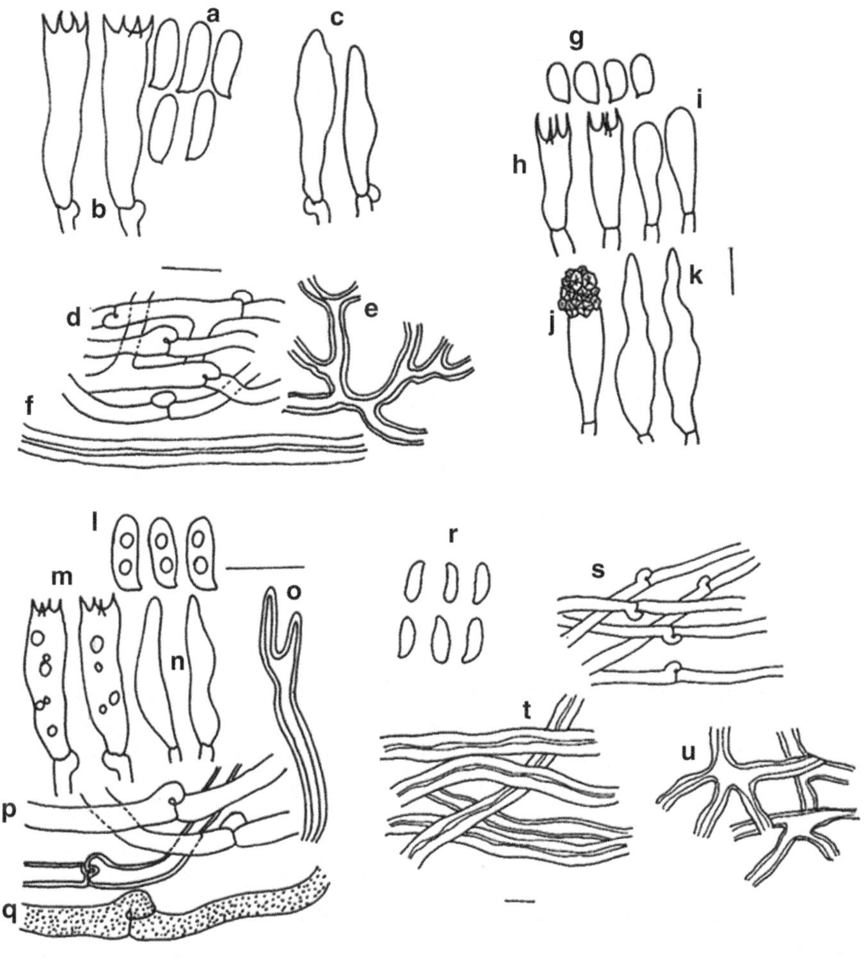

Fig. 6.73 (**a**–**f**) *Earliella scabrosa* (**a**) Basidiospores, (**b**) Basidia, (**c**) Basidioles, (**d**) Generative hyphae, (**e**) Binding hyphae, (**f**) Skeletal hyphae; (**g**–**k**) *Epithelopsis fulva* (**g**) Basidiospores, (**b**) Basidia, (**i**) Basidioles, (**j**) Encrusted cystidia, (**k**) Cystidia; (**l**–**q**) *Favolus tenuiculus* (**l**) Basidiospores, (**m**) Basidia, (**n**) Basidioles, (**o**) Binding hyphae, (**p**, **q**) Generative hyphae; (**r**–**u**) *Fomes fomentarius* (**r**) Basidiospores, (**s**) Generative hyphae, (**t**) Skeletal hyphae, (**u**) Binding hyphae

Fructification annual, sessile, effused, reflexed or resupinate, laterally extended, single, occasionally imbricate, leathery to corky, margin thin, entire to undulate, creamy white when fresh, drying to light buff; upper surface at first smooth, finely pubescent to glabrous, becoming cuticular later, concentrically zonate to subzonate in older parts, white when fresh, soon becoming creamy to light buff and thinly encrusted in "maroon" to "dark red" area in oldest part. Context white to biscuit colored; hymenial surface white when fresh, biscuit colored when dry, pores circular towards the margin, becoming irregular, tending to irpiciform or even daedaloid in the older part, 2–3 per mm, pore tubes up to 5 mm long.

Hyphal system trimitic; generative hyphae hyaline, thin-walled, clamped, 1.6–3.4 μm wide, few slightly brownish, thin-walled to slightly thick-walled, 2.1–3.3 μm wide showing irregular projections and forming cuticular cells of 8–10 μm width, the latter occurring in the maroon crustose crustose area of pileus surface. Skeletal hyphae hyaline, straight or tortuous, thick-walled to solid, 1.6–2.6 μm wide, branched, branches long, flexuous, with tapering ends, some short or coralloid, appearing in the pore field. Basidia long clavate, 16.1–30.1 × 6.0–7.0 μm. Basidiospores hyaline, cylindric, thin-walled, 5.2–9.2(−10.0) × 2.2–3.6 μm.

Distribution: A.P.: West kameng; U.K.: Dehra Dun, Nainital.

Collection examined: Chatrath 202, IBP 37244.

Substratum: On decaying angiospermic twigs.

Remarks: The diagnostic features of this species are resupinate to effused-reflexed corky fructification with maroon to dark red at upper surface, circular to daedloid to irpiciform hymenophore large rounded to irregular pores, trimitic hyphal system. It is a rare species in the Himalayas.

Epithelopsis Jülich,
Persoonia 8(4): 457, 1976.

Fructification resupinate, membranous membranous-ceraceous, adnate; hymenial surface toothed; teeth composed of fascicles. Context subhyaline. Hyphal system monomitic; hyphae septate, thin-walled, clamps present. Cystidia absent. Gloeocystidia subfusiform to cylindrical, flexous or beaded, apices acute to obtuse or capitates, thin-walled, subhyaline with granular contents stained deeply with phloxine, projecting out of hymenium. Basidioles cylindrical to subfusiform. Basidia clavate, 4-sterigmate. Basidiospores ellipsoid, cylindrical, shortly apiculate, thin-walled, smooth, non-amyloid.

Two species, widespread

Lit.: Lepp (*Australasian Mycologist* **23:** 53, 2004)

Type species: *Epithele fulva* G. Cunn. 1956

Habitat: Dead Wood

Himalayas: One

Epithelopsis fulva (G. Cunn.) Jülich, Persoonia 8(4): 457 (1976) = *Epithele fulva* G. Cunn., Trans. Roy. Soc. N.Z. 83: 631 (1956). Fig. 6.73g–k

Fructification resupinate, membranous to membranous ceraceous becoming somewhat horny on drying, adnate, arising as small circular colonies; hymenial surface cream becoming greyish cream with age, smooth to minutely but distinctly odontoid, continuous, often creviced irregularly on drying; margin thinning, adnate, concolorous. Teeth composed of compactly arranged hyphae which originate from the basal part of the context and project out of the hymenium, often covered with subhyaline crystals. Context composed of compactly arranged hyphae which may be agglutinated in part and impregnated with abundant crystals.

Hyphal system monomitic, hyphae 2.4–4.4 μm wide, branched, septate, clamped, the walls thin, subhyaline. Cystidia absent. Gloeocystidia 35.2–60.1 × 4.6–7.1 μm, subfusiform to cylindrical, flexuous or beaded, apices acute to obtuse or capitates, thin-walled, subhyaline with granular contents staining deeply with phloxine, immersed or projecting out of hymenium, naked but sometimes impregnated with few crystals at the top. Basidia clavate, 4-sterigmate, 24–28 × 6–8 μm. Basidiospores 8.6–9 × 4–5.3 μm, ellipsoid-cylindrical, shortly apiculate, thin-walled, subhyaline, smooth, nonamyloid.

Distribution: H.P.: Shimla.

Collection examined: SSR 4322.

Substratum: On bark of *Rhododendron arboretum.*

Remarks: The chief features of this species are white to pale coloured fructifications, presence of crystals on hyphal pegs, flexuous or beaded gloeocystidia and ellipsoid-cylindric basidiospores.

Favolus P. Beauv.,
Fl. Oware 1(8):74,1806.

Fructification annual, pileate, stipitate to substipitate, soft fleshy when fresh, brittle on drying, single to gregarious in a small cluster; upper surface white when young, later cream coloured, light brown on drying, glabrous, azonate; margin thin, smooth. Context white, soft, thin, homogenous. Hyphal system dimitic; generative hyphae hyaline, thin-walled, septate, branched, clamps absent, faintly cyanophilous; skeletal hyphae hyaline, thick-walled, aseptate, unbranched. Basidia clavate, 4-spored. Basidiospores hyaline, thin-walled, smooth, cylindrical, non-amyloid.

Two species, widespread

Lit.: Donk (*Persoonia* 1: 173, 1960)

Type Species: *Favolus hirtus* (Cooke) Imazeki 1943

Habitat: Wood

Himalayas: One

Favolus tenuiculus P. Beauv., Fl. Oware 1(8): 74 (1806) = *Polyporus tenuiculus* (Beauv.) Fr., Syst. Mycol. 1: 344, 1821 = *Favolus brasiliensis* (Fr.) Fr. Linnaea 5: 511, 1830. Fig. 6.73l–q

Fructification annual, pileate, soft fleshy when fresh, brittle on drying, single to gregarious in a small cluster; upper surface white when young, later cream coloured, light brown on drying, glabrous, azonate; margin thin, smooth, incurved on drying. Pore surface concolorous with the upper surface, variably large. Context white to cream, soft, homogenous.

Hyphal system dimitic; generative hyphae hyaline, thin-walled, septate, branched, clamps absent, faintly cyanophilous, 2.2–4.2 μm in diameter; skeletal hyphae hyaline, thick-walled, aseptate, unbranched, 2.4–7.0 μm in diameter. Basidia collapsed on drying. Basidiospores hyaline, thin-walled, smooth, cylindrical, 8.9–11.2 × 2.9–3.6 μm.

Distribution: U.K.: Dehra Dun.
Collection examined: L 42219.
Substratum: On dead log of *Shorea robusta.*
Remarks: This species is characterized by having annual, substipitate, soft, fleshy fructification, flabelliform to spathulate, white pileus; dimitic hyphal system with hyaline skeletal hyphae and hyaline smooth cylindric basidiospores. The species was first recorded from Eastern Himalaya. The above cited collection is quite similar to the description given by Bakshi (1971) and Ryvarden and Johnson (1980).

Fomes Fr.,
Summa Veg. Scand. 2: 319, 1849.

Fructification perennial, sessile effused-reflexed to pileate, hard and woody; upper surface covered with distinct crust. Pores small, tubes stratified. Context corky-woody, becomes dark in KOH sol, brown to maroon colour. Hyphal system trimitic; generative hyphae hyaline, thin-walled, septate, branched, clamps present; skeletal hyphae thick-walled, aseptate, sparsely branched; binding hyphae thick-walled, profusely branches. Setae and cystidia absent. Basidia clavate, 2–4 spored. Basidiospores cylindrical-ellipsoid, hyaline, thin-walled, non-amyloid.
Three Species, widespread
Lit.: Gilbertson & Ryvarden (*N. Amer. Polyp.* 1: 263, 1986)
Type Species: *Boletus fomentarius* L. 1753
Habitat: Decayed Wood
Himalayas: Three
Key to species

1. Upper surface brown to reddish brown 2
1. Upper surface grey, greyish brown to greyish black, smooth .... *F. fomentarius*
2. Pileus sessile, applanate to conchate;
basidiospores subglobose to slightly ellipsoid, yellowish brown .... *F. extensus*
2. Pileus sessile, applanate to conchate to convex;
basidiospores broadly ellipsoid to ovoid, brown.......................... *F. johnsonianus*

Fomes extensus (Lév.) Cooke, Grevillea 14(no. 69): 18 (1885) = *Phellinus extensus* (Lev´.) Pat., Essai Tax. P. 97, 1900. Plate 6.28c, Fig. 6.74a–d

Fructification perennial, sessile, coriaceous when fresh, hard and woody on drying, solitary, attached by narrow lateral base; pileus sessile, applanate to conchate; upper surface brown to reddish brown, blackish near the base, fine tomentose, becoming glabrous later distinct cuticle present, concentrically zonate cracked on drying; pore surface golden brown to brown, even, glancing; pores round to subangular, 6–8 per mm, 80–250 μm in diameter, dissepiments equal, 18–65 μm thick; pore mouth velutinate; tubes indistinctly stratified, golden brown in section, up to 3 mm deep in each layer. Context brown, fibrous, homogenous, xanthochroic, up to 6.5 mm thick, upper surface hard, black and white shiny crust.

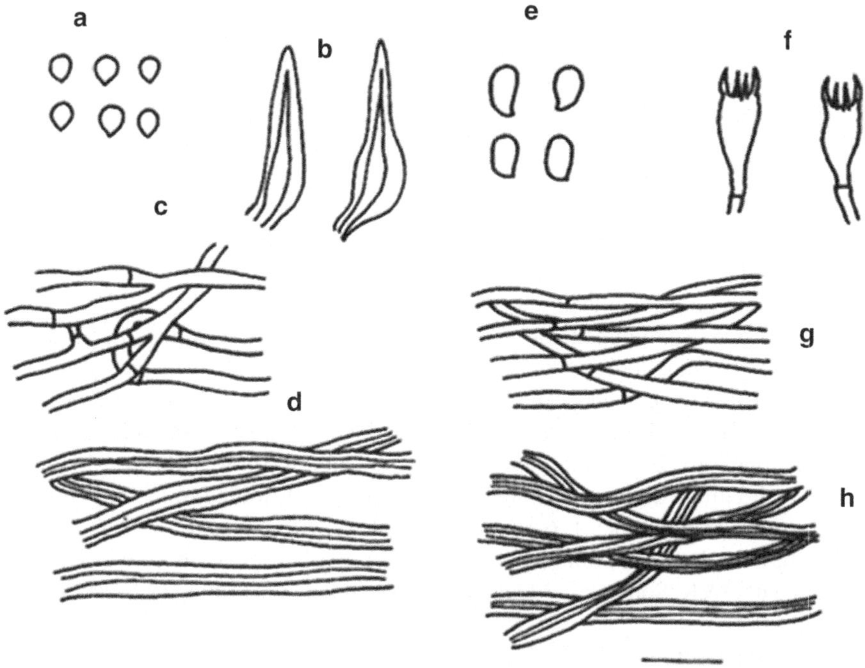

Fig. 6.74 (**a**–**d**) *Fomes extensus* (**a**) Basidiospores, (**b**) Setae, (**c**) Generative hyphae, (**d**) Skeletal hyphae; (**e**–**h**) *Grammothele fuligo* (**e**) Basidiospores, (**f**) Basidia, (**g**) Generative hyphae, (**h**) Skeletal hyphae

Hyphal system dimitic; generative hyphae hyaline, thin-walled, branched, septate, clamps absent, cyanophilous, 2–3 μm in diameter, skeletal hyphae light brown to dark brown, thick-walled, aseptate, unbranched, 2–4.2 μm wide, dissepiments 3–6 μm wide in context. Setae abundant, ventricose, thick-walled, brown 26–32 × 6.2–8.5 μm, apices acute. Basidia hyaline, thin-walled, clavate, 4-spored, 6.2–9.0 × 3.5–4 μm. Basidiospores yellowish brown, thin to slightly thick-walled, smooth, sub-globose to slightly ellipsoid, 3.6–4.5 × 2.5–3.5 μm.

Distribution: Meghalaya- Shillong, U.K.: Karanprayag- Chamoli.

Collection examined: R 6119, D 6654, K 7542, GSD 21076, IBP 37251.

Substratum: On decaying angiospermic logs.

Remarks: This species is characterized by perennial, broadly attached, woody hard fructifications with persistent, reddish brown tomentum; smaller pores; ventricose setae and yellowish brown, thick-walled basidiospores. *Phellinus nilgheriensis* (Mont.) Cunn. is closely related to *F. extensus* but differs in having thin-walled basidiospores; absence of setae; and more robust fructifications. It is a new record for Himalayas.

Fomes fomentarius (L.) Fr., Summa Veg. Scand. Section Post: 321, 1849. Plate 6.28b, Fig. 6.73r–u

Fructification perennial, sessile, effused-reflexed to pileate, attached by broad base, dimidiate to ungulate, woody hard; up to 12×18×15 cm in size; upper surface grey, greyish brown to greyish black, smooth, glabrous, azonate, sulcate, depicting different periods of growth, cuticle black 1–3 mm thick; margin obtuse, thick light brown to greyish brown, sterile below, 4 mm wide. Pore surface light brown to greyish brown; pores round, 2–4 per mm; pore mouth smooth; tubes distinctly stratified by thin brown zone, concolrous with pore surface, each layer up to 7 mm in section. Context light brown, tough, azonate, xanthochroic.

Hyphal system trimitic; generative hyphae hyaline, branched, septate, thin-walled, clamped, cyanophilous, 1.8–3.0 μm in diameter; skeletal hyphae pale yellow to yellowish brown, thick-walled, aseptate, unbranched, acyanophilous, 3.5–4.6 μm in diameter; binding hyphae yellowish brown, sinuous, aseptate, branched, acyanophilous, 2–4 μm in diameter. Cystidia absent. Basidia clavate, 4-spored, up to 4.8 μm in diameter. Basidiospores hyaline, thin-walled, smooth, cylindric-ellipsoid, 15.6–18.6×5.1–6.1 μm, non-amyloid.

Distribution: Bhutan: Thimphu; Nepal: Pokhra; H.P.: Solan, Shimla; A.P.: West Kameng; J&K: Sonamarg; Meghalaya: Khasi hills; U.K.: Kumaun, West Bengal: Darjeeling.

Collection examined: SSV 21321, 21613, IBP 37248, 37249, L 37250.

Substratum: On decaying angiospermic log.

Remarks: The species is very common in Himalayas and reported by Berkeley (1856) from Khasi hills. This species is characterized by perennial, hard and woody ungulate fructification, covered with hard and up to 2 mm thick crust, stratified tubes, yellowish brown context and trimitic hyphal system.

Fomes johnsonianus (Murr.) Lowe, Tech. Publ. N. Y. St. Univ. Coll. For. 80: 36 (1957)

Fructification perennial, effused-reflexed to pileate, imbricate, hard and woody, of medium weight; pileus conchate to applanate to sometimes convex; upper surface reddish-brown to almost black with age, densely pubescent but become almost glabrous with age; margin acute, entire, rarely lobed, paler concolorous, sterile below; pore dark brown, smooth and somewhat shiny. Context firm, fibrous, often separated by a black line from the tomentum. All parts turning brownish-black when touched with KOH sol.

Hyphal system dimitic; generative hyphae 2–3 μm wide, subhyaline, sparsely branched, septate, clamps absent, thin to slightly thick-walled; skeletal hyphae2.6–4.6 μm wide, brown, unbranched, aseptate, thick-walled, sometimes with capillary lumen. Pores indistinctly stratified, pores rounded, 60–150 μm in diameter, averaging 5–8 per mm; concolorous with the context, entire, finely velutinate. Setae 21–28×6.2–10.2 μm, ventricose with a swollen base and pointed apiced, brown, darkening in KOH sol, thick-walled with capillary lumen, often projecting out of the hymenium. Sporulating basidia not observed. Basidiospores 3.2–4×2.6–3 μm, broadly ellipsoid to ovoid, brown, smooth, non-amyloid.

Distribution: H.P.: Mahasu, Shimla; Punjab: Pathankot.
Collection examined: SSR 6155, RSD 6279.
Substratum: On stump under an angiospermic forest.
Remarks: The species is characterized by effused-reflexed to pileate fructification, yellowish-brown context darkening with KOH sol, presence of setae and brown coloured broadly ellipsoid to ovoid basidiospores.

Grammothele Berk. & M. A. Curtis,
J. Linn. Soc. Bot. 10:327 (1868) [1869]

Fructification annual, resupinate, effused, adnate. Pore surface cream, pinkish brown. Pores regular, angular, irpicoid, labryrinthine to sinuous; pore tubes in one layer. Context thin, brown, hyphae compactly arranged. Hyphal system diamitic; generative hyphae thin-walled, clamps present, branched; skeletal hyphae thick-walled to almost solid, pale brown, aseptate, unbranched. Dendrohyphidia present or absent. Cystidia absent. Basidia clavate, 4-spored, restricted to base of tubes, clavate, 4-spored. Basidiospores hyaline, smooth, thin to thick-walled, non-amyloid.
Eight Specie, widespread
Lit.: Ipulet & Ryvarden (*Syn. Fung.* **20:** 87, 2005; Uganda).
Type Species: *Grammothele lineata* Berk. & M.A. Curtis 1869.
Himalayas: One

Grammothele fuligo (Berk. & Broome) Ryvarden, Trans. Br. mycol. Soc. 73(1): 15 (1979)=*Porogramme revenlae* (Berk. & Br.) Pat. Essai Tax. Hym. P. 63. 1900=*Polyporus ravenalae* Berk. & Br., J. Linn. Soc. Lond. 14: 53. 1875. Plate 6.28d, Fig. 6.74e–h

Fructifications annual, resupinate, poroid, membranous-coriaceous, adnate, spreading up to 15×4.5 cm, often in small distinct patches up to 350 μm in section; pore surface whitish grey to bluish grey not creviced formed by the sterile ridges which arise from the basal part of the context, fertile at the base only(pseudopores); margin thinning, adnate, paler concolorous; pores angular, up to 210 μm deep, 5–8 per mm. Context light bluish grey, composed of compactly arranged hyphae.
Hyphal system dimitic; generative hyphae branched, clamps present, 1.7–2.1 μm wide, septate, thin-walled, subhyaline; skeletal hyphae 2.2–3 μm wide, sparsely branched, walls brownish grey to fuscus, thick present in the context as well as sterile tubes, aseptate, thick-walled. Dendrohyphidia present, prominent in sterile tubes, collapse on drying, arising from generative hyphae, strongly branched at the apices. Hymenium is restricted to the base of tubes. Basidia clavate, hyaline, thin-walled, cyanophilous, 12–15×4–5.5 μm. Cystidia absent. Basidia not seen. Basidiospores ellipsoid to cylindrical, thin-walled, 7.3–11.1×2.7–3.6 μm, subhyaline, non-amyloid, smooth.
Distribution: Haryana: Surajpur (near Chandigarh), Chandigarh P.U., A.P.: West Kameng, Bomdilla, Tipi, U.K.: Rajgiri-Uttarkashi.
Collection examined: IBP 37253, 37254.
Substratum: On the leaf bases of *Phoenix dactylifera.*

Remarks: The species is marked by light grey to bluish grey fructification occurring on date palm. It has long ellipsoid-cylindrical basidiospores. The skeletal hyphae are abundantly present. The species is recorded for the first time from Uttarakhand.

Hapalopilus (Pers.) P. Karst.,
Revue mycol., Toulouse 3(no. 9): 18(1881)

Fructification annual, resupinate to pileate and then broadly sessile to dimidiate, soft when fresh, brittle when dry, orange, salmon, reddish brown, purplish or cherry red in contact with KOH; pores round to angular, small to medium. Hyphal system monomitic; generative hyphae with clamps. Cystidia absent. Basidiospores ellipsoid to cylindrical, smooth, hyaline and thin-walled, negative in Melzer's reagent.

Ten species, widespread

Lit.: Corner (Beih. Nova Hedw. 96: 68, 1989), Ko et al. (Mycol 93: 270, 2011).

Type species: *Polyporus nidulans* Fr.1821

Habitat: On dead wood

Himalayas: One

Hapalopilus croceus (Pers.) Bondartsev & Singer, Annales Mycologici, 39(1): 52, 1941.

Fructification annual, pileate, broadly attached, up to 17 × 17 cm wide and 5.8 cm thick, soft and watery when fresh, shrinking to great extent and becoming hard and resinous after drying; pilear surface first bright orange and finely velutinous becoming brownish orange and smooth to scrupose with tufts of agglutinated hyphae, purplish when touched with KOH; pore surface bright reddish-orange when fresh, brownish when dry, pores angular, 2–3 per mm, tubes concolorous with the pore surface. Context bright orange, spongy and watery when fresh, darker orange to brownish resinous and hard when dry, but distinctly lighter than the tubes.

Hyphal system monomitic; generative hyphae hyaline, clamped, throughout the fructification covered with grains, crystals and granules, moderately branched, in the context partly as agglutinated strands of hyphae or more branched, up to 5 μm wide, in trama more irregularly organised and up to 6 μm wide. Basidia clavate, 16–24 × 3.8–6.9 μm, 4-spored. Basidiospores broadly ellipsoid, 4–6.8 × 3–4.4 μm.

Distribution: H.P.: Manali-reserve forests.

Collection examined: IBP 37268.

Substratum: On decaying angiospermic log, on bark of *Cedrus deodara* **.**

Remarks: The species is characterized by orange coloured fructification; ellipsoid basidiospores, monomitic hyphal system.

Hexagonia Fr.,
Fl. Scan. prodr.: 496 (1838)

Fructification annual to perennial, sessile to pileate, corky to woody; pileus smooth, hirsute with long dark hairs; pores angular. Context thin and dark brown, black in KOH sol. Hyphal system trimitic; generative hyphae hyaline, septate, clamps absent; skeletal and binding hyphae thick-walled to solid. Cystidia absent. Basidiospores hyaline, cylindrical, smooth, thin-walled, non-amyloid

Sixteen species, widespread

Lit.: Jülich (*Persoonia* 12: 107, 1984)

Type Species: *Favolus hirtus* P. Beauv, 1806

Habitat: Decayed wood

Himalayas: Two

Key to species

1. Basidiospores 8.5–9.5 × 3.5–4 μm .... *H. badia*
1. Basidiospores 10–15.2 × 4–6.3 μm .... *H. tenuis*

Hexagonia badia (Berk.) Imaz., Bull. Govt. For. Exp. Sta. 57: 112, 1952 = *Trametes badia* Berk., Hook. Lond. Journ. Bot. 1: 151, 1842. Fig. 6.75a–e

Fructification annual, solitary or imbricate, sessile with a thick base, coriaceous or corky, dimidiate, 4.8–8.2 × 3–5.4 × 0.5–1 cm; upper surface cinnamon brown to dark reddish brown, subzonate to zonate, glabrous, often thinly crusty, sometimes showing small raised areas with age; margin sterile, thin or thick; context brown, more or less fibrous and silky; hymenophore dark brown, pores circular to somewhat angular, 1–3 per mm, tubes up to 0.25 cm long.

Hyphal system trimitic; generative hyphae clamped, usually hyaline, thin-walled to thick-walled, some pale brown, often broken at clamps, showing lateral outgrowths of coralloid outlook, occurring in the lower context, 2.9–4.3 μm wide; skeletal hyphae subhyaline to yellowish brown, thick-walled, usually aseptate, occasionally showing 2–3 septa at the apex, and also few short apical branches, 4.3–7.2 μm wide; binding hyphae hyaline to yellowish brown, thick-walled to solid, much branched, 1.5–3.2 μm wide, some with coralloid branches and occurring particularly in the lower context and also in trama, a few with little longer flexuous branches and occur on the upper context. Hairs on the surface composed of longitudinally arranged skeletal hyphae, while the cutis below is a weakly differentiated plectenchyma composed of closely interwoven skeletal hyphae together with shortly branched generative and binding hyphae. Subhyaline to yellowish brown cystoid hyphae formed of terminal ends of skeletal hyphae abundant, penetrating through the hymenial layer. Basidia clavate, 20–28.4 × 5.6–6.8 μm. Basidiospores hyaline, cylindrical, smooth, 8.5–9.5 × 3.5–4 μm. Hyphal pegs abundant.

Distribution: Assam: Lokhra Hills.

Collection examined: GSR 9989.

Substratum: On decaying angiospermic log.

Remarks: This is a rare species in the Himalayas as has only been recorded from the Eastern Himalayas.

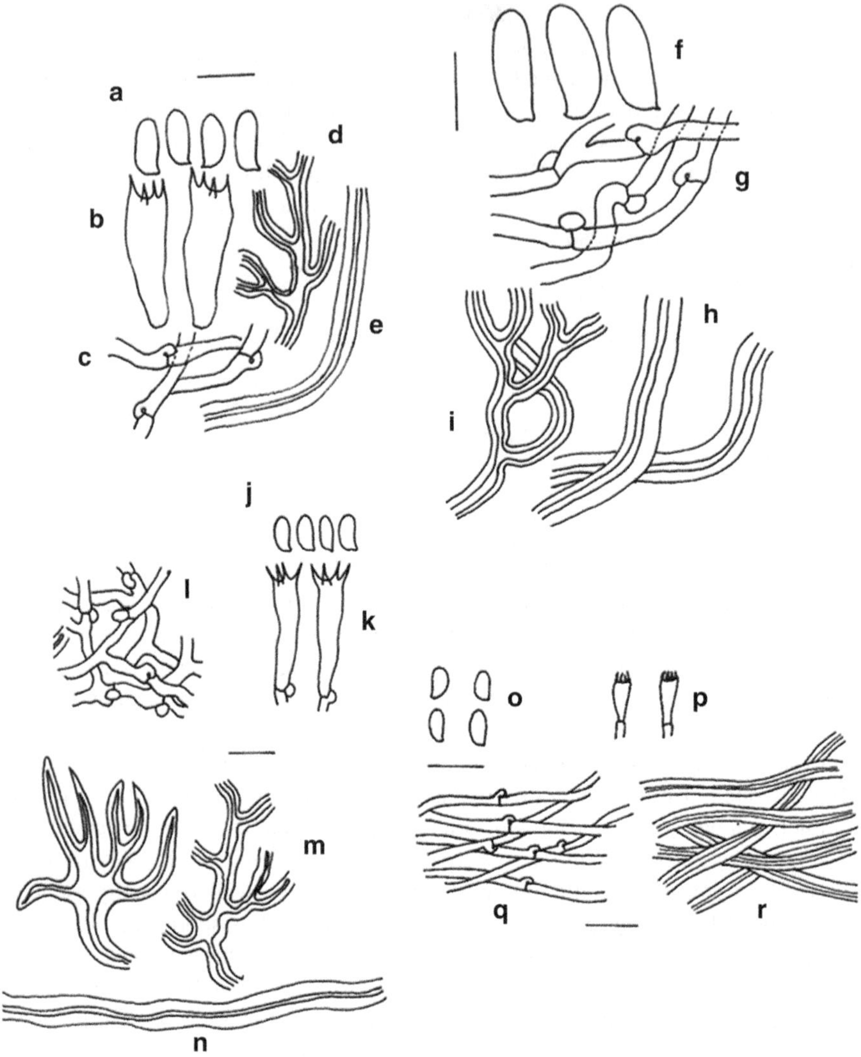

Fig. 6.75 (**a**–**e**) *Hexagonia badia* (**a**) Basidiospores, (**b**) Basidia, (**c**) Generative hyphae, (**d**) Binding hyphae, (**e**) Skeletal hyphae; (**f**–**i**) *Hexagonia tenuis* (**f**) Basidiospores, (**g**) Generative hyphae, (**h**) Skeletal hyphae, (**i**) Binding hyphae; (**j**–**n**) *Lenzites betulina* (**j**) Basidiospores, (**k**) Basidia, (**l**) Generative hyphae, (**m**) Binding hyphae, (**n**) Skeletal hyphae; (**o**–**r**) *Lenzites eximia* (**o**) Basidiospores, (**p**) Basidia, (**q**) Generative hyphae, (**r**) Skeletal hyphae

Hexagonia tenuis (Hook.) Fr., Epicr. syst. mycol. 498, 1838 (1836–1838) = *Boletus tenuis* Hook. in Kunth, Syn. pl. 1:10, 1822. Plate 6.28e, Fig. 6.75f–i

Fructification annual, solitary, sessile, effused-reflexed, occasionally resupinate, applanate, coriaceous; upper surface of pileus glabrous, smooth or rugose, light brown to chestnut brown, often with blackish or reddish brown crustose areas towards the base; margin thin, acute, entire, often undulate. Context brown, fibrous, tubes long.

Hyphal system trimitic; generative hyphae hyaline, thin-walled, branched, clamped, 1.6–3.2 µm wide, not very common; skeletal hyphae abundant, subhyaline to yellowish brown, straight or flexuous, thick-walled to solid, occasionally with septa towards the apex, 3–6.3 µm wide; binding hyphae abundant everywhere, hyaline to subhyaline, thick-walled to solid, much branched, branches mostly short and coralloid, a few freely branched but short and flexuous, 1.7–3.2 µm wide. reddish brown, thin-walled cuticular cells with irregular projections occur in the crustose area at the base on the pileus surface. Basidia narrow clavate, 4-sterigmate, 15–23×8–10.5 µm. Subhyaline to pale brown cystoid hyphae formed on apical ends of skeletal hyphae present, more commonly in sterile pore mouth. Basidiospores hyaline, thin-walled, cylindric, 10–15.2×4–6.3 µm.

Distribution: U.K.: Kumaun; W.B.: Darjeeling; H.P.: Shimla, Rampur.

Collection examined: IBP 37270, 37271, L 37275.

Substratum: On decaying angiospermic log.

Remarks: This species is characterized by fructification with trimitic hyphal system, abundant brown skeletal and binding hyphae, great many of the binding hyphae being of coralloid type, brown cuticular cells or plectenchymatic tissue in the crustose area on the pileus surface, usually large pores and basidiospores.

Lenzites Fr.,
Fl. Scan.: 339, 1836

Fructification annual, resupinate to pileate; pileus smooth to hisute, white to greyish with age. Pore surface white to yellowish. Hyphal system trimitic; generative hyphae hyaline, clamps present, thin-walled; skeletal hyphae hyaline, thick-walled to solid; binding hyphae strongly branched, hyaline. Basidiospores cylindrical, hyaline, thin-walled, smooth, non-amyloid.

Six species, widespread

Lit.: Gilbertson & Ryvarden (*N. Amer. Polyp.* 1: 424, 1987)

Type Species: *Agaricus betulina* L.1753

Habitat: Dead wood

Himalayas: Three

Key to species

1. Pores large, elongated to distinctly lamellate 2
1. Pores medium, angular, sinuous to daedaleoid 3
2. Upper surface whitish cream to greyish partly with concentrically sulcate zones, tomentose when young, later glabrous .... *L. betulina*
2. Upper surface creamish brown to light brown, reddish brown at the base, zonate .... *L. eximia*
3. Fructification long and wide, whitish to wood coloured with pinkish tint, hymenophore splitting into spines .... *L. stereoides*
3. Fructification very large, white to ochraceous with greyish tint; hymenophore poroid, daedaleoid .... *L. elegans*[a]

[a]Extra limital, not included in the text

Lenzites betulina (L.) Fr., Epicr. Syst. Mycol. (Upsaliae): 405(1838) [1836–1838] Plate 6.28f, Fig. 6.75j–n

Fructification annual, sessile, solitary and imbricate, coriaceous when fresh, firm and hard on drying, semicircular, attached by a broad base. Pileus sessile, applanate, dimidiate, 2–8 × 1–5.4 × 0.25–1.4 cm; upper surface whitish cream to greyish partly with concentrically sulcate zones, tomentose when young, later glabrous; margin acute, entire, fertile below, Pore surface cream to creamish brown, lamellate, lamellae dichotomously branched towards the margin, older lamellae undulating, entire 12–15 per cm, up to 12 mm deep in section, up to 0.3 mm thick, slightly thinning apices, even to dentate. Context white to cream, azonate, non-xanthochroic, up to 2 mm thick.

Hyphal system trimitic; generative hyphae hyaline, thin to thick-walled up to 2.5 μm wide, branched, clamps present, cyanophilous; skeletal hyphae subhyaline, thick-walled to solid, unbranched, 3.9–4.1; binding hyphae, hyaline, thick-walled, branched, 2.3–3.1 μm in diameter, acyanophilous. Cystidia absent Basidia clavate, 12.9–14.5 × 3.5–6 μm. Basidiospores, hyaline, ellipsoid, thin-walled, smooth, 5.1–5.5 × 2–2.6 μm.

Distribution: Bhutan: Paro; Nepal: Pokhra; A.P: West Kameng; H.P.: Daranghati; W.B.: Darjeeling, U.K.: Nainital, Mussoorie; H.P.: Solan, Kausali.

Collection examined: SSV 21372, IBP 37276, L 37277, 37278.

Substratum: On decaying angiospermic twigs.

Remarks: This is a widely distributed species in India/Himalayas. It is characterized by annual, broadly attached fructification; lamellate pore surface; densely hirsute and concentrically zonate upper surface; trimitic hyphal system.

Lenzites eximia Berk. & Curt., Hooker J. Bot. Kew Gard. Misc. 6: 134 (1854). Plate 6.29a, Fig. 6.75o–r

Fructification annual, broadly attached, laterally imbricate, coriaceous when fresh, hard on drying; pileus sessile, dimidiate, applanate; upper surface creamish brown to light brown, reddish brown at the base, zonate; margin acute, concolorous, even, incurved on drying. Pore surface cream to greyish brown, lamellate, 12–16 per cm, entire, dentate up to 1 cm deep in section, lamellae, creamish brown. Context light brown, azonate darkening in KOH sol up to 5 mm thick.

Hyphal system trimitic; generative hyphae thin-walled, septate, clamped, acyanophilous, 2.3–4.2 μm in diameter; skeletal hyphae subhyaline, thick-walled having narrow lumen, aseptate, sparsely branched, acyanophilous, 3.2–7.2 μm in diameter; binding hyphae hyaline, thick-walled, aseptate, branched, acyanophilous, 2.5–3.5 μm. Cystidia absent. Basidia collapsed on drying. Basidiospores hyaline, thin-walled, cylindrical to allantoids, 5.6–7.2 × 1.6–1.9, smooth, non-amyloid.

Distribution: A.P.: West Kameng, H.P.: Kasauli, Solan; U. K.: Devprayag-Tehri.

Collection examined: SSV 21499, IBP 37279.

Substratum: On decaying angiospermic log.

Remarks: The species is marked by greyish brown, lamellate pore surface; smooth, cylindrical to allantoids basidiospores. It is close to *L. tricolor* but the latter differs in having darken upper surface, blackish, brown, dichotomously branched

Plate 6.29 (**a**) *Lenzites eximia*. (**b**) *Lenzites stereoides*. (**c**) *Macrohyporia inflata*. (**d**) *Microporellus obovatus*. (**e**) *Microporus affinis*. (**f**) *Microporus xanthopus*

lamellae and is attached by a short lateral base with a hump at the base. The species is new record for North Western Himalayas.

Lenzites stereoides (Fr.) Ryv., Norw. J. Bot. 19(3–4): 232, 1972=*Daedalea stereoides* Fr. Nov. Symb. Mycol. 1: 99, 1851. Plate 6.29b, Fig. 6.76a–e

Fructification effused, effuse-reflexed or usually sessile with a narrow base, single or imbricate, corky; upper surface white to buff, sometimes tobacco brown, rarely blackish at base, glabrous, smooth or fine, closely wrinkled or with small raised areas, consisting of tufted tomentose hairs, zonate. Context pinkish buff, soft, corky; hymenial surface pinkish-brown, pores irregular to daedaloid or irpicoid, rarely subcircular, extending up to margin; pore tubes light pink, nearly straight, not forming a distinct layer.

Hyphal system dimitic; generative hyaline, thin-walled, or slightly thick-walled, branched, septate with clamps, 1.4–3 μm broad; skeletal hyphae hyaline to pale, flexuous, thick-walled, unbranched or sparsely branched, aseptate, 2–4.6 μm broad. Basidiospores hyaline, thin-walled, cylindric to cylindric-ellipsoid, 3.6–6 × 1.9–22.2 μm.

Distribution: U.K.: Dehra Dun, NDBR; West Bengal: Darjeeling.

Collection examined: IBP 42954, 42955.

Substratum: On stump of *Shorea robusta, Albizzia stipulata.*

Remarks: The species is characterized by having effused-reflexed corky fructification; hyphal system dimitic, generative hyphae clamped, thin-walled, skeletal hyphae thick-walled; basidiospores thin-walled, cylindric to cylindric-ellipsoid.

Lignosus Lioyd ex Torrend

Brotéria, sér. bot. 18: 121 (1920)

Fructification annual, centrally stipitate; pileus brown to white; pores small to large. Context white. Hyphal system trimitic; generative hyphae septate, branched, clamps present; skeletal hyphae and binding hyphae thick-walled. Cystidia absent. Basidiospores ellipsoid, smooth, hyaline, non-amyloid.

Five species, widespread

Lit.: Ryverden & Johansen (*Prelim. Polyp. Fl. E. Afr.:* 405, 1980), Douanla-Meli & Langer (*Mycotaxon* **86:** 389, 2003; Cameroon).

Type Species: *Polyporus sacer* Fr. 1838.

Habitat: Dead wood

Himalayas: One

Lignosus sacer (Fr.) Ryv., Norw. J. Bot. 19: 232, 1972=*Polyporus sacer* Fr., Epicr. Syst. Mycol. p. 436, 1838.

Fructification annual, solitary or in small group, centrally stipitate, arising from a sclerotium; stipe brown, finely tomentose, glabrous with age, laccate, up to 19 cm long and 1 cm in diameter; pileus circular, slightly umbonate, sometimes depressed in centre, woody brown to snuff brown, coriaceous, when young finely concentrically tomentose but soon glabrous with a thin crust and often sulcate, up to 10.0 cm in diameter and 0.3 cm thick; margin thin and sharp; context white

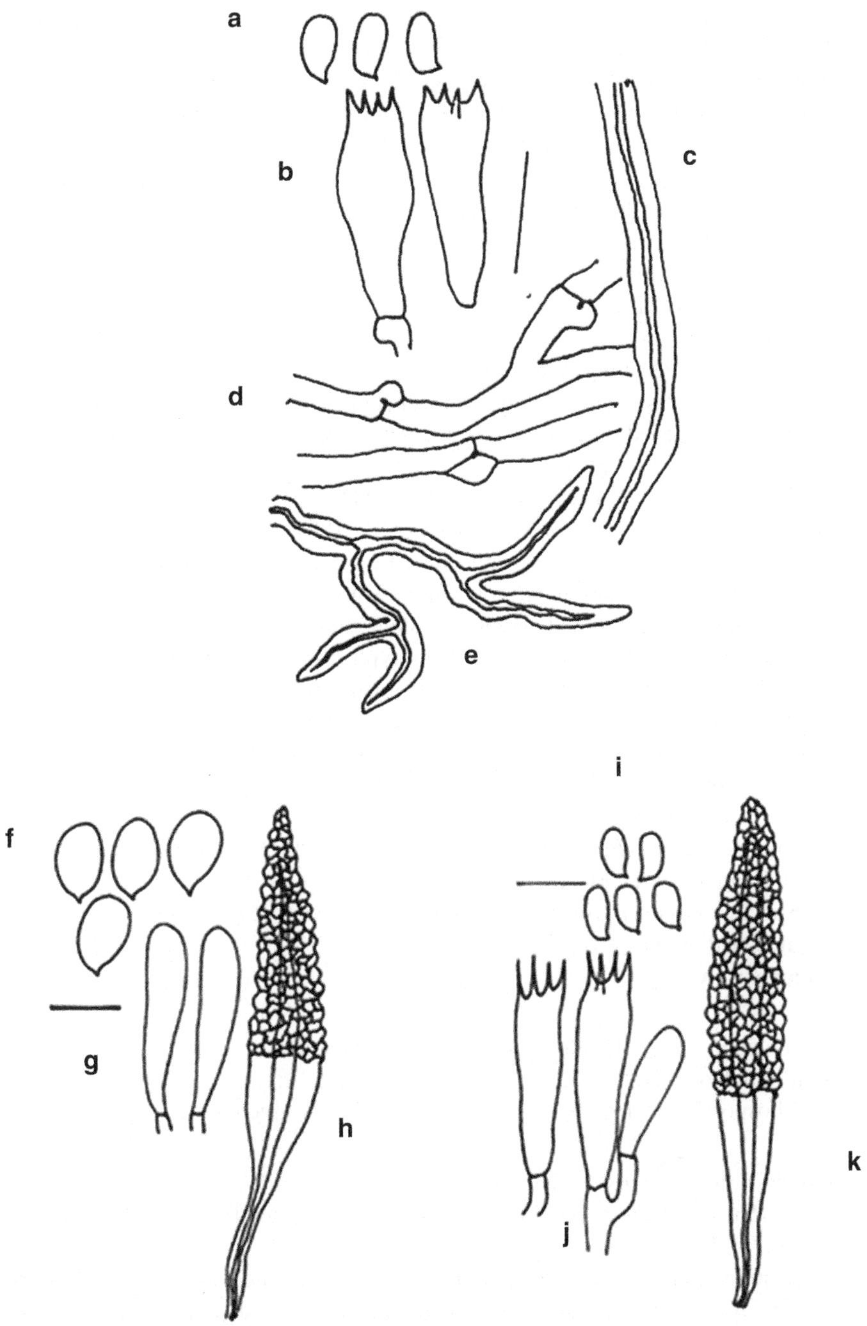

Fig. 6.76 (**a**–**e**) *Lenzites stereoides* (**a**) Basidiospores, (**b**) Basidia, (**c**) Skeletal hyphae, (**d**) Generative hyphae, (**e**) Binding hyphae; (**f**–**h**) *Lopharia cinerascens* (**f**) Basidiospores, (**g**) Basidioles, (**h**) Encrusted cystidia; (**i**–**k**) *Lopharia papyrina* (**i**) Basidiospores, (**j**) Basidia, (**k**) Encrusted cystidia

or yellowish, 1–2 mm thick; hymenial surface brownish, pores circular to angular to irregular, 4–5 per mm, pore tubes yellowish, up to 2 mm long.

Hyphal system trimitic. Generative hyphae hyaline, thin-walled, clamped, 2–3 μm wide in context and hymenium but wider up to 10 μm wide, thick-walled to subsolid and yellowish in the tomentum, and stipe, some forming brownish, irregular interlocking projections in upper surface of pileus. Skeletal hyphae straight to flexuous, hyaline, thick-walled to solid, 1.5–5.0 μm wide. Binding hyphae hyaline, much branched, thick-walled to solid, 2.5–4.5 μm wide. Basidia hyaline, clavate, 4-sterigmate, 15–20 × 5–7 μm. Basidiospores hyaline, thin-walled, smooth, broadly ellipsoid, 5.0–7.0 × 3.0–4.5 μm. Hyphal pegs numerous, white.

Distribution: Assam.

Substratum: On decaying angiospermic log.

Remarks: The author did not collect any specimen during the fungal forays. The description is based on literature as given by Roy and De (1996).

Lopharia Kalchbr. & MacOwan,
Grevillea 10(54): 58, 1881

Fructifications effused-reflexed to pileate, coriaceous or membranous-coriaceous; upper surface tomentose to strigose, rarely smooth; hymenial surface smooth to tuberculate, rarely toothed. Hyphal system monomitic or dimitic; generative hyphae with or without clamps; skeletal hyphae thick-walled, hyaline. Cystidia or skeletocystidia cystidia cylindrical to fusiform, thick-walled, subhyaline to light brown, scantly or heavily incrusted with subhyaline crystals. Gloeocystidia absent. Basidia clavate to clavate-cylindrical, 4-spored. Basidiospores subglobose to ellipsoid, the walls smooth, thin, subhyaline, amyloid or non-amyloid.

Thirteen Species, widespread

Lit.: Boidin & Gilles (*Bull. Soc. Mycol. Fr.* 118: 91, 2002), Hjorstatm & Ryvarden (Syn. Fung. 4:19, 1990; Key).

Type Species: *Lopharia lirellosa* Kalchbr. & Macowan, 1881

Habitat: Dead wood

Himalayas: Two

Key to species

1. Basidiospores 10.2–13 × 6.2–7 μm, broadly ellipsoid with rounded ends .... *L. cinerascens*
1. Basidiospores 5–8.4 × 3.0–4.3 μm, ellipsoid, shortly apiculate .... *L. papyrina*

Lopharia cinerascens (Schw.) Cunn., Trans. Roy. Soc. N.Z. 83: 622.1956 = *Thelephora cinerascens* Schw., Trans. Amer. Phil. Soc. 4: 167. 1832. Fig. 6.76f–h

Fructification resupinate to effused-reflexed, loosely adnate, widely effused; pileus absent, if present then an upturned margin to reduced; upper surface cream brown, tomentose, azonate; hymenial surface light brown to violaceous brown, smooth to rough or tuberculate, sometimes imperfectly to distinctly toothed; margin thinning, loosely adnate, paler concolorous. Subiculum with a thick brown cuticle bearing tomentum on the abhymenial side, composed mainly of horizontal hyphae in thin fructification but as the fructifications grow thicker they develop a strong zone of vertical hyphae with numerous cystidia embedded in it.

Hyphal system dimitic; generative hyphae 2–4.2 µm wide, branched, septate, clamps absent, the walls thin, subhyaline; skeletal hyphae 2–5.3 µm wide, unbranched, aseptate, the walls moderately thick to thick, light brown. Tomentose hyphae 2–4 µm wide, unbranched, aseptate, the walls tinted brown, moderately thick (up to 1 µm). Cystidia 80–200 × 15.5–28 µm, subconical to fusiform, with a long narrow pedicel, immersed or projecting out of the hymenium, the walls thick, tinted brown in the basal half and subhyaline in the apical half, heavily incrusted in the upper 2/3 part with subhyaline crystals. Basidia collapsing after spore discharge, 4-spored. Basidiospores 10.2–13 × 6.2–7 µm, broadly ellipsoid with rounded ends, the walls subhyaline, smooth, thin-walled, non-amyloid.

Distribution: H.P.: Mahasu; U.K.: Nainital; J&K: Bhadarwah; Nepal: Kathmandu.

Collection examined: SSR 5376, 5376, IBP 37283.

Substratum: On dead stump of *Quercus incana*.

Remarks: The species is characterized by dimitic hyphal system, simple-septate generative hyphae, large incrusted cystidia with tinted bases, broadly ellipsoid basidiospores.

Lopharia papyrina (Mont.) Boidin, Bull. Soc. Linn. Lyon 28: 210. 1959 = *Stereum papyrina* Mont., in Sagr. Hist. Cuba Pl. Cell. 374. 1842. Fig. 6.76i–k

Fructification resupinate to effused-reflexed, membranous, adnate; hymenial surface Sayal to snuff brown, smooth, not creviced; margin thick, often reflexed, concolorous; abhymenial surface tomentose, concentrically zonate, zones of erect and appressed tomentum, cinnamon buff. Subiculum composed of somewhat loosely woven hyphae.

Hyphal system monomitic, hyphae 2.6–4.5 µm wide, branched at wide angles, septate, clamps absent, the moderately thick, subhyaline to tinted yellow. Cystidia 49.5–68 × 7.8–20 µm, subfusiform, immersed or projecting out of the hymenium, the walls thick, light brown, covered with abundant subhyaline crystals especially in the apical half. Basidia not observed, collapsed. Basidiospores 5–8.4 × 3.0–4.3 µm, ellipsoid, shortly apiculate, the walls thin, subhyaline, smooth, nonamyloid.

Distribution: U.K.: Dehra Dun.

Collection examined: IBP 37284.

Substratum: On dead stump of *Quercus incana*.

Remarks: This species was earlier reported from Dehra Dun (U.K.) occurring on *Shorea robusta* by Bagchee and Bakshi (1954). It seems very close to *L. crassa* but differs in having shorter and broader cystidia.

Loweporus J. E. Wright, Mem. of the New York Botanical Garden, 28 (1): 225, 1975.

Fructification annual to perennial, resupinate to pileate, coriaceous to hard and woody. Pileus (when present) broadly attached, dimidiate with contracted base, applanate; upper surface glabrous to fine tomentose, concentrically sulcate, brown to greyish brown; pores small to medium, entire, round to angular; tubes concolorous with the pore surface. Hyphal system trimitic; generative hyphae hyaline, thin-walled, branched, septate, clamps present, cyanophilous; skeletal hyphae yellowish to brown, thick-walled to almost solid with narrow lumen, aseptate, unbranched, dextrinoid; binding hyphae hyaline to light yellow, thick-walled, branched, aseptate, dextrinoid. Cystidia absent. Basidia hyaline, thin-walled, clavate, 4-spored. Basidiospores hyaline to pale yellowish, thin to thick-walled, ellipsoid to truncate.

Type Species: *Polyporus lividus* Kalchbr. ex Cooke, 1882

Habitat: Decayed wood

Himalayas: Two

Key to species

1. Basidiospores oval to truncate, 4.4–6.1 × 3.3–4.6 μm .... *L. lividus*
1. Basidiospores ellipsoid to truncate, 4.7–5.9 × 3.1–4.3 μm .... *L. tephroporus*

Loweporus lividus (Kalchbr. Ex Cooke) J. E. Wright, Mem. N. Bot. Gdn 28(1): 225 (1976) = *Poria lividus* (Kalch.) Wright. (loc. Cit.) = *Polyporus lividus* Kalch., Grevillea 10: 103, 1882 = *Fomes lividus* (Kalch.) Sacc., Syll. Fung. 6: 206, 1888. Fig. 6.77a–d

Fructification usually widely effused, rarely pileate at margin, hard, brittle; upper surface black. Context brown to sepia, fibrous; hymenial surface whitish, greyish, pale brown to brown; margin darker, thin, fertile, usually cracking, pores usually angular to irpicoid, appearing in 5–7 rows or parallel layers along the length of the fructification, stratified, white.

Hyphal system monomitic; hyphae (1) nearly hyaline to dull brown, thick-walled, unbranched or distantly branched, sometimes swollen, incrusted, 3–5 μm broad, common, (2) similar but much branched, 2.6–3.3 μm broad, few, and (3) hyaline, thin-walled, branched, rarely septate with clamps, 2–3 μm broad, few. Basidia clavate, 5–6 μm broad. Basidiospores hyaline or nearly so, smooth, oval, sometimes truncate at one end, 4.4–6.1 × 3.3–4.6 μm.

Distribution: W.B.: Calcutta.

Substratum: On decaying angiospermic log.

Remarks: A specimen of *Fomes pallidus* Petch in the Bose Herbarium (2584) showed the same characters as *Loweporus lividus* and the former may be synonymous as also noted by Lowe (1957). The specimen has been reported by Roy & De (1996) but it has not been collected during fungal foray.

Loweporus tephroporus (Mont.) Ryvarden, in Ryvarden & Johansen, Prelim. Polyp. Fl. E. Afr. (Oslo): 416 (1980) = *Poria tephropora* (Mont.) Ryv., Fungiflora: 416, 1980.

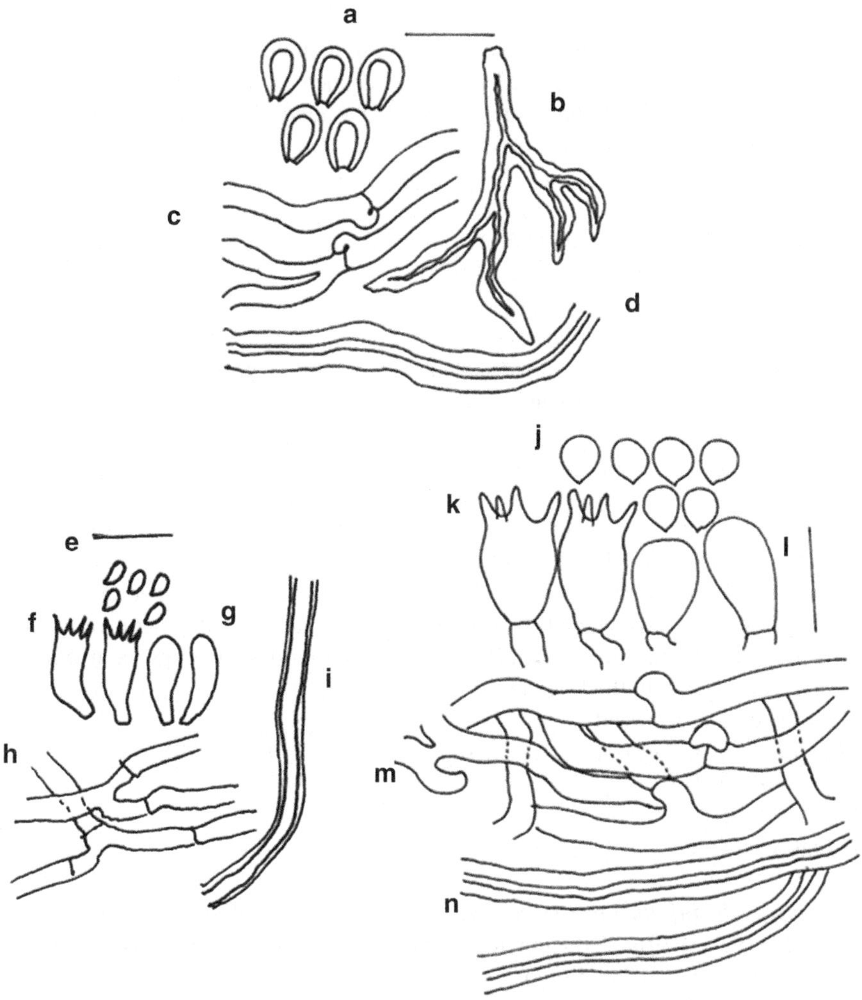

Fig. 6.77 (**a–d**) *Loweporus lividus* (**a**) Basidiospores, (**b**) Setae, (**c**) Generative hyphae, (**d**) skeletal hyphae; (**e–i**) *Macrohyporia inflata* (**e**) Basidiospores, (**f**) Basidia, (**g**) Basidioles, (**h**) Generative hyphae, (**i**) Skeletal hyphae; (**j–n**) *Microporellus obovatus* (**j**) Basidiospores, (**k**) Basidia, (**l**) Basidioles, (**m**) Generative hyphae, (**n**) Skeletal hyphae

Fructification perennial, resupinate, adnate, widely effused, up to 12 cm long, 3.8 cm broad and 5.8 mm thick, woody hard, usually on vertical surface; margin yellowish grey, thinning, narrow to wide, cracks on drying. Pore surface yellowish grey to brown, rimose on drying, even, dull; pores round, 5–6 per mm, 85–150 µm in diameter; dissepiment, equal 60–155 µm thick; tubes indistinctly stratified, light brown, up to 2 mm deep in each strata. Subiculum thin, greyish brown, fibrous, homogenous, xanthochroic up to 0.4 mm thick.

Hyphal system trimitic; generative hyphae hyaline, thin-walled, branched, septate with clamps, faintly cyanophilous, 2.2–4.5 μm in diameter; skeletal hyphae pale brown, thick-walled, straight, aseptate, unbranched, acyanophillous, 2.7–5.2 μm in diameter; binding hyphae subhyaline, thin-walled, flexuous, branched, aseptate, 1.6–3.7 μm in diameter. Cystidia and cystiodioles absent. Basidia collapsed. Basidiospores hyaline, thin to thick-walled, smooth, broadly ellipsoid, weakly dextrinoid, truncate, 4.7–5.9 × 3.1–4.3 μm.

Distribution: Meghalaya: about 60 km from Garampani, towards Haflong.

Collection examined: GSD 21067, 21070.

Substratum: On decaying angiospermic log.

Remarks: This species is characterized by perennial, yellowish grey to brown, rimose fructification; dextrinoid skeletal and binding hyphae; truncate dextrinoid spores which are ellipsoid and hyaline.

Macrohyporia I. Johans. & Ryvarden,
Trans. Brit. mycol. Soc. 72(2): 192, 1979

Fructification annual to perennial, resupinate in small patches to widely effused, brittle to hard when dry; Pore surface cream to ochraceous, dull; pores 1–5 mm or larger. Context thin. Hyphal system monomitic or dimitic; generative hyphae simple septate, thin-walled; binding hyphae or thick-walled generative hyphae dominant in trama and context. Basidiospores subglobose to ellipsoid, hyaline to pale yellow, thin to thick-walled, non-amyloid.

Three species widespread

Lit.: Johansen & Ryvarden (*Trans. Bri. Mycol. Soc.* 72: 192, 1979)

Type Species: *Polyporus dictyporus* Cooke, 1883

Habitat: Decaying wood

Himalayas: One

Macrohyporia inflata Overh. ex I. Johans. & Ryvarden, Trans. Br. mycol. Soc. 72(2): 192 (1979) = *Poria inflata* Overh., Bull. Pennsylvania Agaricult. st. 13: 123, 1939. Plate 6.29c, Fig. 6.77e–i

Fructification annual, resupinate, widely effused, adnate, inseparable, soft; margin white when fresh, creamish on drying, narrow to wide, fibrillose, thinning; pore surface creamish pink to brown, even, dull; pores round to angular about 3–4 mm; pore mouth becoming thin and fimbriate. Context creamish to pale yellow, dull, thin, fimbriate, firm; tubes not stratified, brittle, up to 11 mm deep, pale yellow in section.

Hyphal system monomitic; generative hyphae hyaline, thin-walled to thick-walled, septate, branched, clamps absent, cyanophilous, 2.8–8.4 μm. in diameter. Cystidia absent. Basidia collapsed, not observed. Basidiospores, hyaline, thin-walled, smooth, apiculate non-amyloid, broadly ellipsoid to ovoid, 3.0–4.1 × 2.2–2.98 μm.

Distribution: H.P.: Dalhousie, U.K.: Dehra Dun.

Collection examined: Dhanda 6880.

Substratum: On stump of *Shorea robusta*.

Remarks: The species is characterized by annual, resupinate, separable fructification, white when fresh, cream to sordid brown; pore surface on drying, non-xanthochroic context, monomitic hyphal system and hyaline non-amyloid basidiospores.

Microporellus Murr.,
Bull. Torrey Bot. Club 32: 483, 1905.

Fructification annual, solitary or aggregated in groups, centrally or laterally stipitate, rarely sessile, stipe gradually reduced to a long tapering base, soft, coriaceous when fresh, hard and brittle on drying. Pileus round, flabelliform to spathulate, upper surface finally tomentose when young, smooth due to agglutination of hyphae with age, faint to distinctly zonate, cream to light brown. Stipe solid, light to dark brown, finely velvety when young, becoming glabrous finally, expanded down into a mycellial disc at the base. Pore surface white to light golden brown; pores small; tubes in one layer. Context white to light brown, homogenous and non-xanthochroic. Hyphal system dimitic; generative hyphae septate, hyaline, thin to thick-walled, branched, clamps present or absent; skeletal hyphae present in hymenium but absent in the context, thick-walled, unbranched. Cystidia absent. Basidia clavate, 4-spored. Basidiospores globose to subglobose, hyaline, smooth, thin-walled, non-amyloid.

Nineteen species, widespread

Lit.: Reid (*Microscopy* **32**: 452, 1975), Decock & Ryvarden (*Czech. Mycol.* **54**: 19, 2002)

Habitat: Decaying Wood

Type Species: *Microporellus dealbatus* Berk. & Curt. 1853

Himalayas: Two

Key to species

1. Basidiospores thin-walled, globose to subglobose, 3.5–4.8 × 3.0–4.4 µm .... *M. obovatus*
1. Basidiospores thin to slightly thick walled, ovoid to broadly ellipsoid, 6–8.4 × 5–6.2 µm .... *M. violaceo-cinerascens*

Microporellus obovatus (Jungh.) Ryv., Norw. J. Bot. 19(3–4): 232, 1972.
Plate 6.29d, Fig. 6.77j–n

Fructifications annual, laterally stipitate, solitary or in small groups of two to four with fused pilei and free stipes, soft coriaceous when fresh, hard and brittle on drying. Pileus flabelliform to spathulate, up to 5 cm long, 2.8 cm wide and 2.2 mm thick; upper surface light brown to yellowish to greyish brown, darken on drying, tomentose when young, glabrous due to agglutination of hyphae with age, radially striate on drying; margin papery, thin, sterile below, brown, incurved on drying. Stipe up to 2.8 cm long and 1–2.5 mm thick, yellowish brown, fine velvety to glabrous, concolorous with pileus, smooth when fresh, radially wrinkled on drying, expanding gradually into a mycelial disc at the base, solid. Pore surface cream to light brown when fresh, entire; pores small, round to rarely angular, thin-walled, 6–7 per mm, 58–98 μm in diameter; dissepiment entire, 25–75 μm thick; tubes in one layer, brown, 1.5 mm deep in section. Context thin, brown, homogenous, non-xanthochroic.

Hyphal system dimitic; generative hyphae hyaline to subhyaline, thin to thick-walled, branched, septate, clamped, 1.4–4.8 μm in diameter; skeletal hyphae hyaline to subhyaline, thick-walled, aseptate, unbranched, 3.0–4.2 μm in diameter. Cystidia absent. Basidia clavate, hyaline, 4-spored, 9–11.3 × 6–7.8 μm. Basidiospores hyaline, thin-walled, smooth, globose to subglobose, non-amyloid, 3.5–4.8 × 3.0–4.4 μm.

Distribution: Meghalaya: Cherrapunji, Mawsmai falls; H.P.: Shimla-Tara Devi.

Collection examined: SSV 21865, 21864, IBP 37286.

Substratum: On dead angiospermic stump.

Remarks: The species is characterized by laterally stipitate, coriaceous fructifications; spathulate to flabelliform pilei; fine tomentose, brown to reddish brown upper surface; dimitic hyphal system and hyaline, thin-walled, globose to subglobose basidiospores. The description of the species as given above matches closely with that given by Ryvarden & Johansen (1980). It is a new record for Himalayas.

Microporellus violaceo-cinerascens (Petch) David & Rajchenberg, Mycotaxon 22: 303, 1985. Fig. 6.78a–e

Fructification annual, solitary or in small groups, stipitate, stipe central or eccentric, usually bulbous at the base, covered with violet hairs, cylindrical to somewhat flattened, ash grey, attached to a sclerotium at the base; pileus circular to dimidiate with depression at the centre, coriaceous when fresh, hard on drying; upper surface cream to vinaceous brown, minutely tomentose, irregularly wrinkled; margin acute, slightly wavy or lobed. Context brownish, tough; hymenial surface cream brown to greyish brown, pores circular to angular, 1–2 per mm, somewhat decurrent, pore tubes yellowish brown, up to 1 mm long.

Hyphal system dimitic; generative hyphae hyaline, thin-walled, clamped, 3–6.7 μm wide; skeletal hyphae subhyaline, thick-walled to subsolid, unbranched, dextrinoid, 4–8.3 μm wide. Basidia hyaline, clavate, 5–6.2 μm wide. Basidiospores hyaline, thin- to slightly thick-walled, smooth, weakly dextrinoid, ovoid to broadly ellipsoid, apiculate, one guttulate, 6–8.4 × 5–6.2 μm. Cystidia abundant, ventricose, subhyaline, thick-walled, some encrusted at the apex, 16–32.2 × 8–14.2 μm.

Distribution: U.K.: Dehra Dun, Nainital.

Collection examined: IBP 37287.

Substratum: On decaying angiospermic log.

Remarks: The species is characterized by annual, stipitate fructification; circular pileus; dimitic hyphal system; smooth, ovoid to broadly ellipsoid basidiospores.

Microporus P. Beauv.,
Fl. Oware.: 1:12, 1805.

Fructification annual, stipitate, solitary to gregarious. Pileus circular, flabelliform to spathulate, smooth to hirsute, zonate; upper surface reddish brown; pore surface white to cream; pores round and entire, 5–9 per mm. Context white, tough,

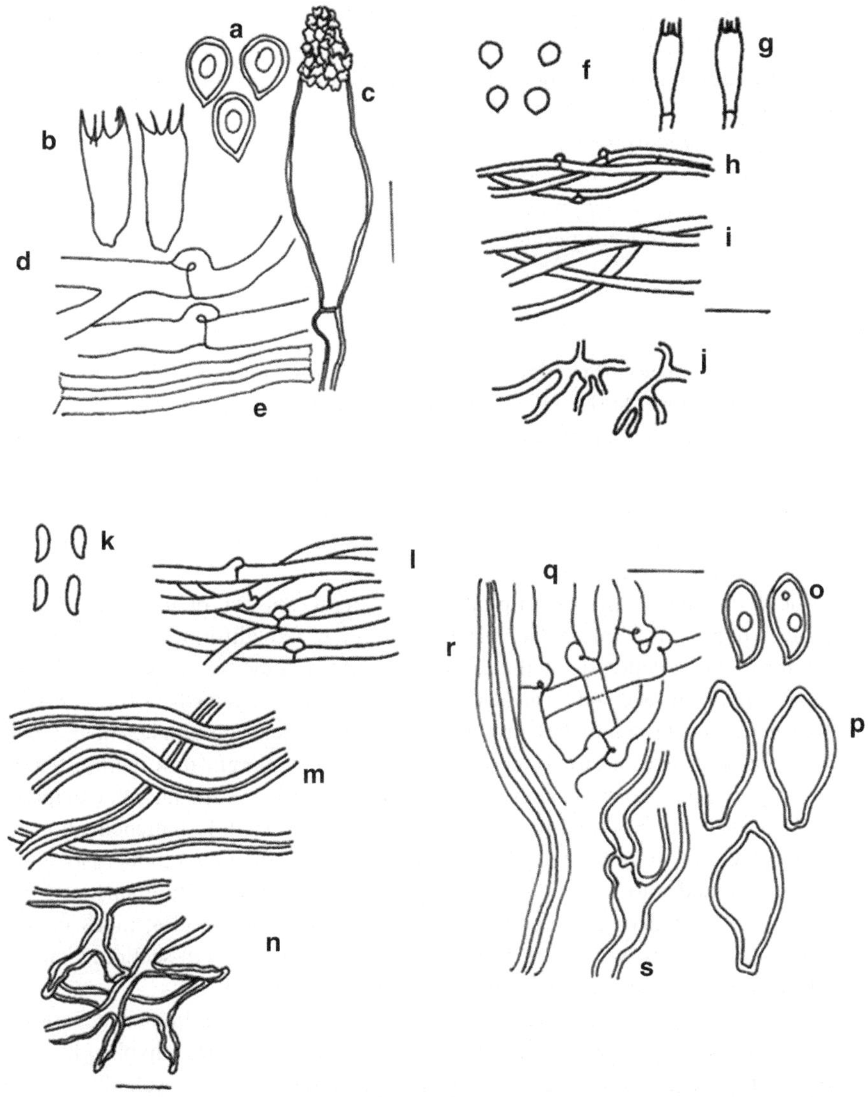

Fig. 6.78 (**a**–**e**) *Microporellus violaceo-cinerascens* (**a**) Basidiospores, (**b**) Basidia, (**c**) Cystidia, (**d**) Generative hyphae, (**e**) Skeletal hyphae; (**f**–**j**) *Microporus affinis* (**f**) Basidiospores, (**g**) Basidia, (**h**) Generative hyphae, (**i**) Skeletal hyphae, (**j**) Binding hyphae; (**k**–**n**) *Microporus xanthopus* (**k**) Basidiospores, (**l**) Generative hyphae, (**m**) Skeletal hyphae, (**n**) Binding hyphae; (**o**–**s**) *Navisporus floccosus* (**o**) Basidiospores, (**p**) Cystidia, (**q**) Generative hyphae, (**r**) Skeletal hyphae, (**s**) Thick-walled generative hyphae

homogenous, non-xanthochroic. Hyphal system trimitic; generative hyphae septate, branched, clamps present; skeletal and binding hyphae thick-walled. Cystidia absent. Basidiospores allantoids to elliptical, smooth, thin-walled, non-amyloid.

Eleven Species, widespread

Lit.: Peglar *The polypores* [*Bull. Brit. Mycol. Soc.* Suppl.], 1973, Ryvarden & Johansen (*Prelim. Polyp.* Fl. E. Afr.: 429, 1980)

Type Species: *Microporus perula* Beauv., 1804

Habitat: Wood

Himalayas: Two

Key to species

1. Stipe lateral; upper surface tomentose; basidiospores tomentose........... *M. affinis*
1. Stipe central; upper surface glabrous and polished; basidiospores smooth, cylindric to ellipsoid.................................. *M. xanthopus*

Microporus affinis (Blume & T. Nees) Kuntze, Revis. gen. pl. (Leipzig) 3(2): 494 (1898)=*Microporus flebelliformis* (Fr.) Pat., Revis. Gen. Pl. (Leipzig) 3(2): 496, 1898. Plate 6.29e, Fig. 6.78f–j

Fructifications annual, pileate, stipitate, stipe lateral, solitary or imbricate, coriaceous when fresh, hard on drying. Pileus stipitate, flabelliform, semi-circular to reniform, flat or depressed in the centre up to 6.8 cm long, 5.2 cm wide and 5 mm thick; upper surface with various shades of brown, yellowish brown, pinkish brown, reddish brown to dark brown, tomentose, hyphae agglutinate and become glabrous with age, narrowly concentrically zonate; margin thin, entire, wavy, slightly incised, incurved on drying, up to 1.6 mm wide. Stipe lateral, expanded both towards the base and the pore surface, round to slightly flattened, short to almost lacking, at first greyish brown, later almost black with a thin crust, at the base expanded into a circular disc, covered with a cuticle. Pore surface light cream, dark on drying; pores round, entire, small, 7–8 per mm; tubes in one layer, concolorous with the pore surface up to 1.8 mm deep in section. Context white, homogeneous, non-xanthochroic up to 2.8 mm thick.

Hyphal system trimitic; generative hyphae hyaline, thin-walled, branched, septate, clamps present, cyanophilous, 1.9–2.6 μm in diameter; skeletal hyphae hyaline in the dissepiment and context, yellowish in the tomentum, thin-walled to almost solid with a narrow lumen, straight, unbranched, aseptate, acyanophilous, 3–5.2 μm wide in dissepiment, and up to 7.8 μm wide in tomentum; binding hyphae hyaline, thick-walled, aseptate, branched, acyanophilous, 2–3 μm in diameter. Hyphal peg present up to 15 μm, project into pore cavity. Basidia hyaline, clavate, 4-spored, 4.6–5.2 μm in diameter. Basidiospores brown, slightly thick-walled, smooth, ovoid, non-amyloid, 2.4–3.5 μm in diameter.

Distribution: Bhutan- Thimphu; Nepal- Kathmandu; A.P.: Manipur; Meghalaya: Shillong; Mizoram, Assam: Lokhra hills; U.K.: Nainital; W.B.: Darjeeling; H.P.: Solan, Kasauli.

Collection examined: SSV 21410, 21618, IBP 37287.

Substratum: On decaying angiospermic twigs.

Remarks: This species is characterized by annual, stipitate, flabelliform to reniform fructifications; lateral, reddish brown to black stalk; pale yellow context; and brown, ovoid basidiospores. This species was first reported from India by Berkeley (1856) as *Polyporus flabelliformis* Koltzsch from Darjeeling (West Bengal) and Sikkim.

Microporus xanthopus (Fr.) Kuntze, Rev. Gen. Pl. 3: 494, 1898. Plate 6.29f, Fig. 6.78k–n

Fructifications annual, solitary or aggregated in small groups, coriaceous when fresh, hard on drying, centrally or laterally stipitate, sometimes two or more fructifications grow closely and from a united common imbricate pileus with several stipes. Pileus infundibuliform when young, turns to almost plane with age with umbilicate centre up to 13 cm in diameter and up to 2 mm thick; upper surface yellowish brown to reddish brown to maroon, glabrous and shining when fresh, concentrically zonate, zones of light and dark brown colour; margin thin, acute, plane or slightly curved on drying, paler concolorous with upper surface, entire or slightly wavy, sterile below. Pores white to cream coloured, smooth, even; entire and minute 5–8 per mm, round to oval 80–220 µm in diameter; dissepiments 45–135 µm thick; equal, Stipe up to 3.2 cm long and 8 mm in diameter, glabrous, covered with a thin, light yellowish to light brown cuticle, solid, cylindrical with broad mycelial disc at the base and up to 8 mm in diameter. Context white, thin, homogeneous, non-xanthochroic, up to 1.6 mm thick.

Hyphal system trimitic; generative hyphae hyaline, thin-walled, branched, septate, clamped, acyanophilous, 2–2.7 µm in diameter; skeletal hyphae hyaline to subhyaline, dominating in dissepiment and context, thick-walled to almost solid with narrow lumen, aseptate, acyanophilous, 3–5.6 µm in diameter; binding hyphae subhyaline, profusely branched, thick-walled, short, acyanophilous 1.4–3.2 µm in diameter. Cystidia and setae absent. Basidiospores hyaline, thin-walled, smooth, cylindric to ellipsoid, non-amyloid, 6.2–7 × 1.9–2.7 µm.

Distribution: Bhutan, Nepal, A.P., Meghalaya, W.B.: Drajeeling, H.P.: Shimla, Manali, Kullu; U.K.: Nainital.

Collection examined: SSV 21395, 21404, 21588, IBP 37288, 42957.

Substratum: On decaying angiospermic twigs.

Remarks: This species is characterized by centrally stipitate fructifications with yellowish stipe; concentrically zonate, reddish brown to maroon, glabrous, smooth upper surface; small, 5–8 per mm pores; white, non-xanthochroic context; trimitic hyphal system; and hyaline, narrowly ellipsoid basidiospores.

Navisporus Ryv. in Rev. & John,
Prelim. Polypore Fl. East Afr. 443, 1980.

Fructification annual, solitary, sessile with a broad base, resupinate or reflexed and woody, somewhat sapy when fresh, on drying very hard and light in weight; margin thick, rounded and sterile below; upper surface usually smooth, glabrous, buff brown to yellowish to reddish brown, azonate. Context light buff or pale cinnamon, fleshy when fresh, suberose on drying, distinctly zoned; hymenial surface white to cinnamon brown, pores mostly circular, pore tubes reddish brown. Hyphal system dimitic; generative hyphae hyaline, much branched, with clamp connections, thin-walled or slightly thick-walled; skeletal hyphae hyaline, thick-walled, more easily visible in fresh material, dextrinoid, flexous, usually with distinct lumina, often irregularly inflated, unbranched, occasionally with two or more much narrower tortuous apical branches. Basidia large clavate, 4-sterigmate, collapsed in dry fructification. Basidiospores hyaline, slightly thick-walled, smooth, navicular to fusiform, guttulate. Cystidia present, extending beyond the hymenial layer, hyaline, slightly thick-walled, swollen and fusoid, weakly dextrinoid.

Six species, widespread

Lit.: Ryvarden & Iturriaga (*Mycol.* 95: 1066, 2003; Venezuela)

Type Species: *Trametes floccosa* Bres. 1896.

Habitat: Dead wood

Himalayas: One

Navisporus floccosus (Bres.) Ryv., Prelim. Polypore Fl. East Afr., 443, 1980 = *Trametes floccosa* Bres., Ann. Roy. Inst. Bot. Roma 6: 179, 1896. Fig. 6.78o–s

Fructification annual, resupinate or reflexed, solitary, sessile with a broad base, large, usually hard and woody, somewhat sapy when fresh, on drying very hard and light in weight; margin thick, rounded; upper surface usually smooth, glabrous, buff brown to yellowish to reddish brown, azonate. Context light buff or pale cinnamon, fleshy when fresh, suberose on drying, distinctly zoned; hymenial surface white to cinnamon brown, pores mostly circular, at places elongated; pore tubes reddish brown.

Hyphal system dimitic; generative hyphae hyaline, much branched, with clamp connections, frequently inflated, thin-walled or slightly thick-walled, 2–4.2 μm wide, often empty, collapsed, partly gelatinized; skeletal hyphae hyaline, thick-walled, more easily visible in fresh material, dextrinoid, flexous, usually with distinct lumina, often irregularly inflated, unbranched, occasionally with two or more much narrower tortuous apical branches, 3.7–7.5 μm wide. Basidia large clavate, 25–32.0 × 5.8–9.8 μm, 4-sterigmate, collapsed in dry fructification. Basidiospores hyaline, slightly thick-walled, smooth, navicular to fusiform, guttulate, 11.4–12.8 × 4.4–5.6 μm. Cystidia present, extending beyond the hymenial layer, hyaline, slightly thick-walled, swollen and fusoid, weakly dextrinoid, 35.1–42 × 14.5–21.2 μm.

Distribution: W.B.: Calcutta, Darjeeling.

Collection examined: Sharma 8878.

Substratum: On decaying angiospermic log.

Remarks: The specimen has been collected only from Eastern Himalayas. It has been reported earlier by Roy & De (1980).

Nigroporus Murrill,
Bull. Torrey Bot. Club 32: 361, 1905.

Fructification annual to perennial, pileate to resupinate, coriaceous when fresh. Pileus if present, sessile, dimidiate, applanate; upper surface dark brown, reddish brown to violaceous brown, globose, azonate to zonate. Pore surface dark brown to brownish black; pores small, rounded to angular. Tubes not stratose. Context reddish brown to violaceous brown, homogenous, non-xanthochroic. Hyphal system dimitic; generative hyphae thin-walled, septate, clamps present; skeletal hyphae hyaline to light brown, thick-walled. Cystidia or setae absent. Basidiospores hyaline, thin-walled, allantoids to broadly ellipsoid, smooth, non-amyloid.

Three Species, widespread

Lit.: Peglar (*The polypores* [*Bull. Brit. Mycol. Soc. Suppl.*], 1973), Ryvarden & Iturriaga (*Mycol.* 95: 1066, 2003; Venezuela).

Type Species: *Polyporus vinosus* Berk. 1852

Habitat: Dead Wood

Himalaya: Two

Key to species

1. Basidia 8.2–14.2 × 3–4 µm; basidiospores thin-walled, broadly ellipsoid to subglobose, 3–5 × 2.7–4 µm .... *N. durus*
1. Basidia 6.2–9 × 3.6–4.3 µm; basidiospores allantoid, 3.1–4 × 1.3–1.6 µm .... *N. vinosus*

Nigroporus durus (Jungh.) Murr., Bull. Torr. Bot. Cl. 34:471, 1907 = *Polyporus durus* Praemiss. Flor. Crypt. Jav. Ins.: 62, 1838. Fig. 6.79a–e

Fructification perennial, solitary or imbricate, sessile, mostly dimidiate, sometimes applanate to ungulate, hard; margin thin and acute; upper surface finely tomentose, becoming glabrous with age, smooth or with minute pustules, not prominently zoned, at first pale brownish or dirty grey, becoming blackish with age. Context hard, dark violet to brown; hymenial surface brownish to grey when fresh, turning to chocolate to black when dry and old; pores oval to circular, vinaceous brown to blackish, often stratified.

Hyphal system dimitic; generative hyphae rare, hyaline, thin-walled, much branched, clamped, up to 4.2 µm wide; skeletal hyphae thick-walled to almost solid, brownish, unbranched, up to 7.2 µm wide. Basidia hyaline, thin-walled, clavate, 4-sterigmate, 8.2–14.2 × 2.7–4 µm. Basidiospores hyaline, thin-walled, smooth, broadly ellipsoid to subglobose, 3–5 × 2.7–4 µm.

Distribution: Assam: Khasi hills; H.P: Dalhousie.

Collection examined: 214, 215, IBP 37289.

Substratum: On decaying angiospermic log.

Remarks: The species is characterized by formation of clamped hyphae in the fructification and simple septate hyphae in culture.

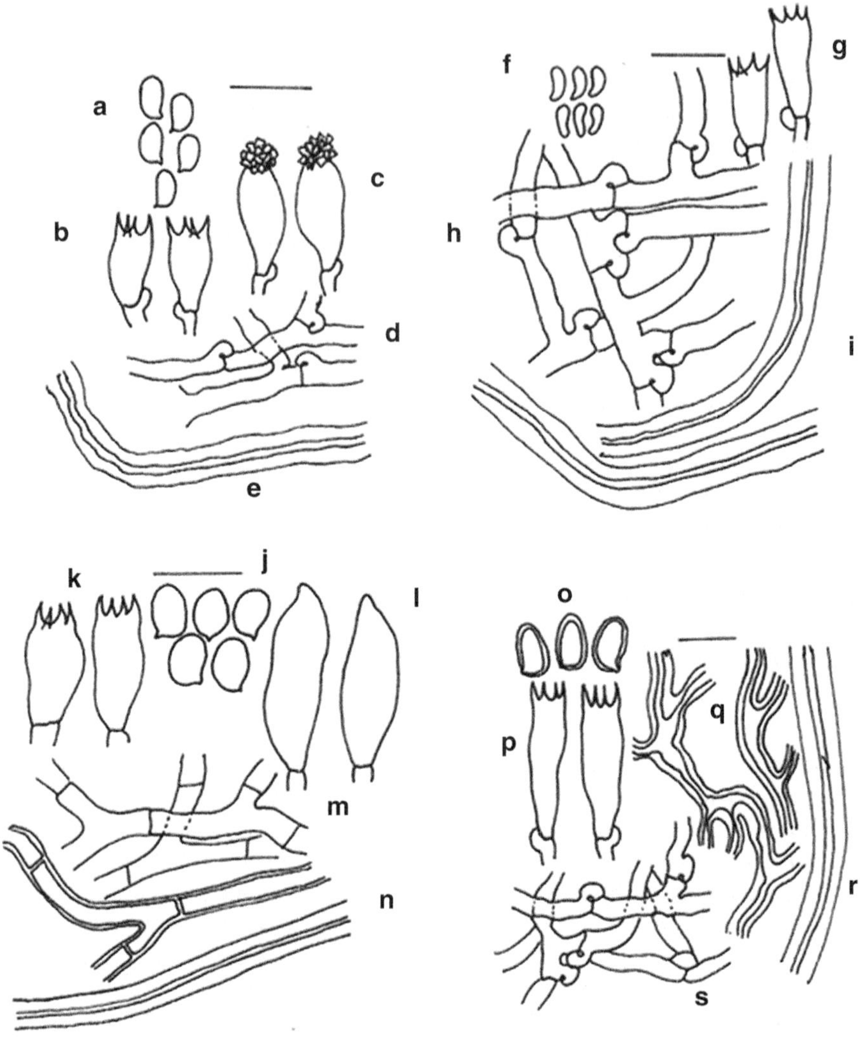

Fig. 6.79 (**a–e**) *Nigroporus durus* (**a**) Basidiospores, (**b**) Basidia, (**c**) Cystidia, (**d**) Generative hyphae, (**e**) Skeletal hyphae; (**f–i**) *Nigroporus vinosus* (**f**) Basidiospores, (**g**) Basidia, (**h**) Generative hyphae, (**i**) Skeletal hyphae; (**j–n**) *Nigrofomes melanoporus* (**j**) Basidiospores, (**k**) Basidia, (**l**) Cystidia, (**m**) Generative hyphae, (**n**) Skeletal hyphae; (**o–s**) *Perenniporia martia* (**o**) Basidiospores, (**p**) Basidia, (**q**) Binding hyphae, (**r**) Skeletal hyphae, (**s**) Generative hyphae

Nigroporus vinosus (Berk.) Murrill, Bull. Torrey Bot. Club 32: 361, 1905.
Plate 6.30a, Fig. 6.79f–i

Fructification annual, pileate, solitary, sessile to effused-reflexed, often with short lateral stalk, dimidiate to reniform, imbricate, rigid and brittle when dry. Pileus sessile; pores circular to angular or irregular, dissepiments thin, pore tubes up to

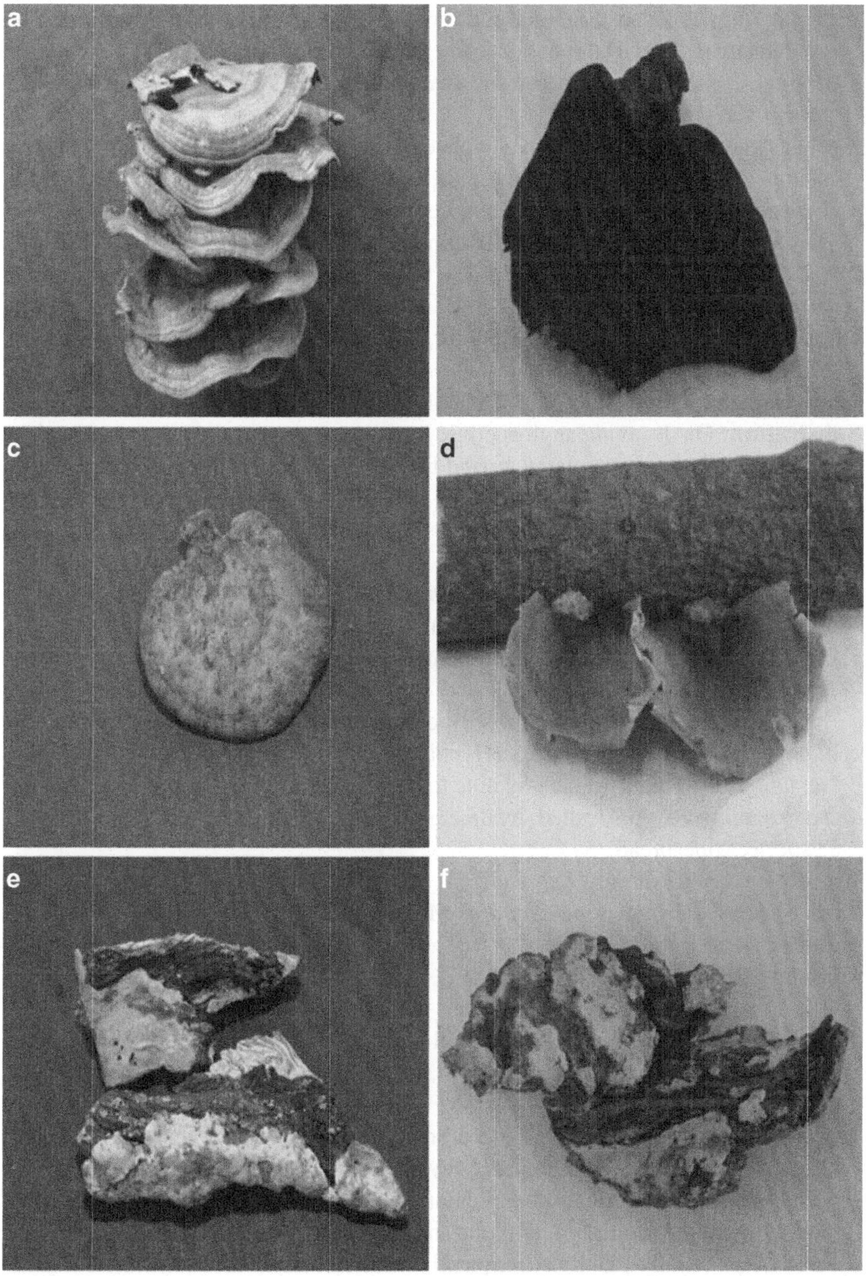

Plate 6.30 (**a**) *Nigroporus vinosus*. (**b**) *Polyporus grammocephlus*. (**c**) *Polyporus squamosus*. (**d**) *Polyporus varius*. (**e**) *Perenniporia fulviseda*. (**f**) *Perenniporia medulla-penis*

2 mm long; upper surface vinaceous brown to reddish brown, finely velutinate in younger part, then glabrous, concentrically sulcate; margin acute, incurved on drying. Context vinaceous brown; hymenial surface vinaceous brown to smoky brown.

Hyphal system dimitic; generative hyphae majority hyaline, thin-walled, branched, clamped and also showing simple septa, branched, 2.7–4.0 µm wide; skeletal hyphae thick-walled, straight or flexous, unbranched, pale bluish brown, 2.7–6.2 µm wide. Basidia hyaline, thin-walled, clavate, 4-sterigmate, 6.2–9.0×3.6–4.3 µm. Basidiospores hyaline, thin-walled, allantoids, smooth, 3.1–4.0×1.3–1.6 µm.

Distribution: Bhutan-Thimphu, Manipur, Assam, U.K.: Dehra Dun; H.P.: Kausauli, Shimla, Solan.

Collection examined: SSV 21279, 21281, IBP 37290.

Substratum: On decaying angiospermic log.

Remarks: This species is characterized by fructification having vinaceous- brown hymenophore, almost invisible pores and dimitic hyphal system with bluish-yellow skeletal hyphae.

Nigrofomes Murr.,
Bull. Torr. Bot. Club 31(8): 425, 1904.

Fructification perennial, pileate, dimidiate, applanate and hard on drying. Pileus glabrous, dark violaceous black and with distinct black cuticle. Pore surface black to dark violaceous purplish; pores very small, isodiametric to circular. Context dense, purplish black. Hyphal system monomitic; generative hyphae septate, thin to thick-walled, hyaline to dark brownish. Cystidia absent. Basidiospores ellipsoid, thin-walled, hyaline, non-amyloid.

One specie, widespread

Lit.: Roberts & Ryvarden *(Kew Bull.* 61: 55, 2006; Cameroon)

Type Species: *Polyporus melanoporus* Mont, 1892

Habitat: Dead wood

Himalayas: One

Nigrofomes melanoporus (Mont.) Murr. Bull. Torr. Bot. Club 31: 425, 1904=*Polyporus melanoporus* Mont., Ann. Sci. Nat. Ser. 2, 17: 127, 1842. Fig. 6.79j–n

Fructification perennial, sessile with a broad base, usually applanate and solitary, heavy and hard; hymenial surface greyish black, smooth, sometimes cracking at some places on drying, pores circular, 7–8 per mm, pore tubes blackish, distinctly stratified, up to 3.8 mm long in each layer; upper surface at first tomentose, soon usually glabrous, frequently with sulcate zones, sometimes slightly radiately cracking, purplish black, rough and tuberculate on ageing and with a thick black cuticle; margin thin and sharp. Context shiny, dark reddish brown, zones up to 3 cm thick;

Hyphal system monomitic; generative hyphae hyaline to subhyaline, simple septate, sparsely branched, thin-walled to thick-walled, some straight, unbranched, long, resembling skeletal hyphae, but showing very distinct septa, 2–5.2 µm wide. Cystidia very rare, ventricose, thick-walled, dark brown, 10.2–30.1 × 5.1–10 µm, Basidia clavate, 4-sterigmate, 12.2–15.2 × 4–5.2 µm. Basidiospores broadly ellipsoid, hyaline, 4.0–5.0 × 3.0–3.5 µm.

Distribution: Assam: Khasi hills; W.B.: Jalpaiguri.

Substratum: On decaying *Shorea robusta* log.

Remarks: The fungus is distinguished by its hard heavy and black fructification with zonate upper surface and reddish brown zonate context. The specimen has been reported by Roy & De (1996) but it has not been collected during fungal forays.

Pachykytospora Kotl. & Pouz.,
Česka Mykol. 17(1): 27–34, 1963.

Fructification annual to perennial, adnate, resupinate, effused, on substratum with rudimentary nodular pilei, tough when fresh. Pore surface white, cream to light brown. Pores round to angular; margin concolorous to lighter than pore surface narrow. Context light brown, coriaceous. Hyphal system trimitic; generative hyphae hyaline, thin-walled, septate, branched, clamped; skeletal hyphae subhyaline to yellowish thick-walled to almost solid with narrow lumen, unbranched to dichotomously branched; binding hyphae hyaline to yellowish, much branched, aseptate. Cystidia and setae absent. Basidia clavate, thin-walled, clamps present at the base. Basidiospores oblong-ellipsoid to broadly ellipsoid, hyaline, thick-walled, nonamyloid.

Three species, widespread

Lit.: Kotlabs & Pouzar (Česka Mykol 17:27–34, 1963, Key)

Type Species: *Polyporus tuberclosus* Fr. 1821

Habitat: Dead wood

Himalayas: One

Pachytospora papyracea (Cooke) Ryv., Norw. J. Bot. 19(3–4): 233, 1972.
Fig. 6.45a–f

Fructification annual, resupinate, effused, adnate, pulvinate, soft coriaceous when fresh, brittle on drying, 4 cm × 2.5 cm × 3 mm in size. margin white to cream, concolorous with pore surface, thinning, sterile, up to 2 mm wide, dissepiments up to 45 µm wide, tapering towards the pore mouth; tubes in one layer, whitish, up to 170 µm long. Pore surface white to cream when fresh, pale brown on drying; pores round to angular 3–4 per mm, pore mouth velutinate; tubes in one layer, whitish, dissepiments entire, thick. Context white to cream, soft thin, homogenous, non-xanthochroic, 0.7 mm thick.

Hyphal system trimitic; generative hyphae hyaline, thin to thick-walled, septate, clamped, branched, cyanophilous, 1.2–2.7 µm wide; skeletal hyphae hyaline to subhyaline, thick-walled to almost solid with narrow lumen, aseptate, cyanophilous, 2–3.8 µm in diameter; binding hyphae hyaline, branched with short branches, aseptate, 1.6–3.5 µm in diameter. Cystidia and setae absent. Basidia hyaline, thin-walled, clavate, 2–4 spored, up to 12 µm wide. Basidiospores hyaline, thin-walled, ellipsoid, ornamented with longitudinal striations, striae prominent and strongly cyanophilous, 9.7–14.7×4.5–7 μm.

Distribution: A.P.: West Kameng, Rupa, 14 km from Rupa towards Shergaon.

Collection examined: SSV 21756.

Substratum: On decaying angiospermic log, *Rhododendron* stems.

Remarks: The species is characterized by perennial, thick fructifications; pinkish pore surface and large pores, 1–2 pores per mm.

Perenniporia Murr.,
Mycologia 34: 595, 1942

Fructification annual to perennial, resupinate, sessile, applanate, solitary, mostly broadly attached, heavy, coriaceous when fresh hard on drying; margin thick; upper surface glabrous, usually concentrically sulcate, sometimes cracking with age, dirty brown to black with a crust. Context white when fresh, woody brown on drying; hymenial surface yellowish brown to blackish brown; pores subglobose, regular. Hyphal system dimitic or trimitic; generative hyphae hyaline, thin-walled, clamped; skeletal hyphae dominating, strongly dextrinoid, hyaline or pale brown and thick-walled to solid, unbranched; binding hyphae few, thick-walled to solid, dextrinoid. Cystidia absent. Basidia hyaline, clavate, 4-spored. Basidiospores weakly truncate or pip shaped, thin-walled, smooth, variously dextrinoid, truncate, acyanophilous.

Sixty Species, widespread

Lit.: Decock & Stalpers (Taxon 55: 227, 759, 2006)

Type Species: *Perenniporia medulla-panis* Jacquin.

Habitat: Wood

Himalayas: Five

Key to species

1. Basidiospores ovoid to truncate 2
1. Basidiospores subglobose to ellipsoid to truncate or pip shaped 3
2. Fructification hard and brittle on drying; pores round to angular, 4–5 per mm .... *P. fulviseda*
2. Fructification soft, corky on drying; pores 5–7 per mm .... *P. medulla-panis*
3. Fructification resupinate to effused-reflexed, easily separable, basidiospores 7 9.2×5–6.2 μm .... *P. mesoleuca*
3. Fructification perennial, sessile, applanate, solitary, mostly broadly attached, basidiospores 5.5–10×3.3–5.8 μm .... *P. martia*

Perenniporia fulviseda (Bres.) Dhanda [as *Perenniporis*] in Thind & Dhanda, Indian Phytopath, 33(3): 386 (1981)=*Poria fulviseda* Bres., Annls mycol. 18(1/3): 37, 1920. Plate 6.30e

Fructification annual to perennial, resupinate, membranous soft when fresh becoming brittle on drying; margin fibrillose to rhizomorphic, adnate, white to paler concolorous. Pore surface cream yellow to light yellowish brown when fresh but fades on drying, smooth; pores round to angular. Context subhyaline, cream brown, composed of somewhat compactly arranged hyphae.

Hyphal system dimitic; generative hyphae hyaline, thin-walled, septate, branched, clamped, acyanophilous, 2.0–4.0 µm. in diameter; skeletal hyphae sparsely branched, aseptate, walls thick, cyanophilous, subhyaline, 1.5–3.0 µm in diameter. Cystidia absent. Basidia clavate, 4-spored, 5–6 µm broad. Basidiospores, hyaline, thin-walled, minutely apiculate, usually truncate at the apex, non-amyloid, broadly ellipsoid to ovoid, 3.5–4.5 × 3.0–3.5 µm.

Distribution: H.P.: Dalhousie, Kullu.

Collection examined: SSR 6100, RSD 6215, IBP 37694.

Substratum: On stump of *Quercus incana.*

Remarks: The species is marked by the soft-membranous fructification, presence of rhizomorphs, dimitic hyphal system, dextrinoid, usually truncate basidiospores.

Perenniporia martia (Berk.) Ryvarden [as *'martius'*], Norw. J Bot. 19: 143 (1972) = *Poria martia* (Berk.) Ryv., Norw. J. Bot. 19:143, 1972. Fig. 6.79o–s

Fructification perennial, sessile, applanate, solitary, mostly broadly attached, hard, heavy; margin thick; upper surface glabrous, usually concentrically sulcate, sometimes cracking with age, dirty brown to black with a crust. Context white when fresh, woody brown on drying, often greyish black in old part; hymenial surface yellowish brown to blackish brown, pores subglobose, regular, 3–4 per mm.

Hyphal system trimitic; generative hyphae hyaline, thin-walled, clamped, 1.6–3.3 µm wide, mostly collapsed and scarcely found; skeletal hyphae dominating, strongly dextrinoid, hyaline or pale brown and thick-walled to solid, up to 3–5 µm wide; binding hyphae few, thick-walled to solid, dextrinoid, 2–4.4 µm wide. Cystidia thick-walled, capitate encrusted or clavate, arising from skeletal hyphae, 6–12 µm wide. Basidia not found. Basidiospores weakly truncate or pip shaped, thick-walled, smooth, variously dextrinoid, greatly varying in size, 5.5–10 × 3.3–5.8 µm.

Distribution: Meghalya: Shillong; H.P.: Kullu.

Collection examined: IBP 37293.

Substratum: On decaying angiospermic log.

Remarks: The species is characterized by having perennial, solitary, broadly attached fructification; yellowish brown to blackish brown hymenial surface; trimitic hyphal system; truncate, smooth, dextrinoid basidiospores.

Perenniporia medulla-panis (Jacq.) Donk, Persoonia 5(1): 76 (1967) = *Polyporus medullapanis* Jacq. ex Fr., Syst. Mycol. 1: 380. 1821. Plate 6.30f, Fig. 6.80a–e

Fructifications annual to perennial, resupinate, adnate to separable, often arising as small circular colonies which may coalesce later and become effused, corky when fresh, hard on drying; margin concolorous, thinning to more or less abrupt, entire; pores round to slightly angular, 5–7 per mm. Context white to cream becoming darker on drying.

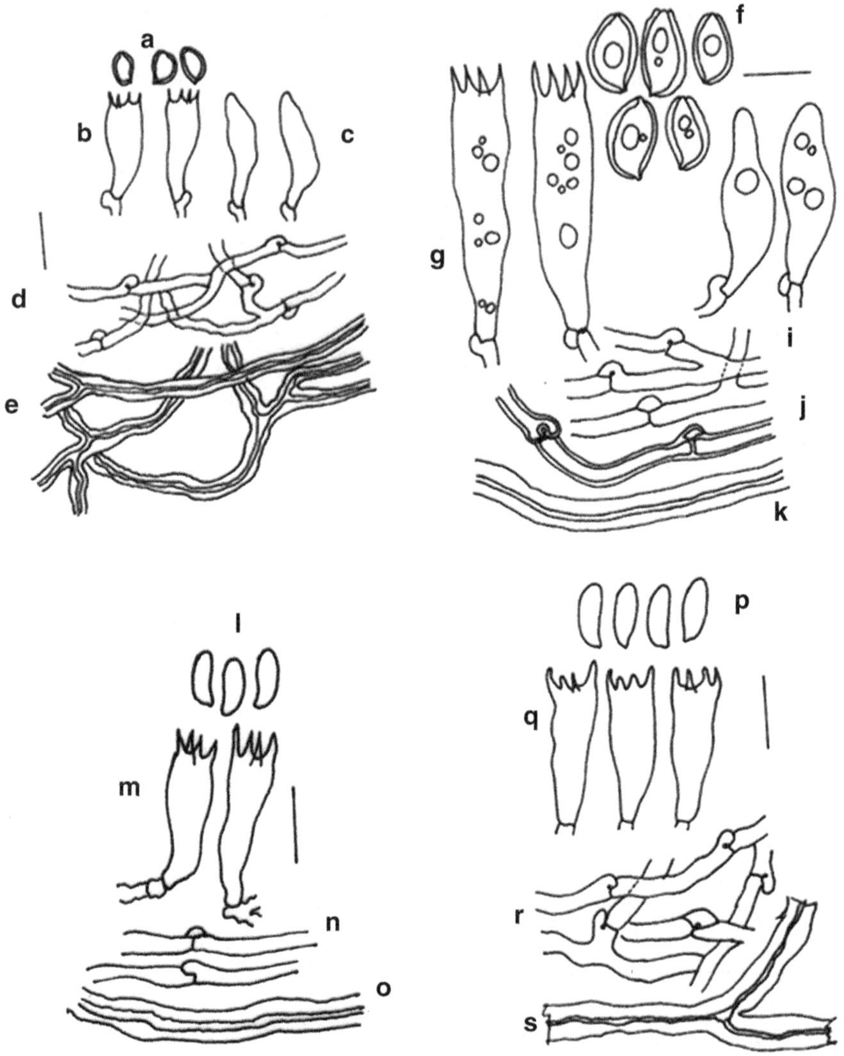

Fig. 6.80 (**a**–**e**) *Perenniporia medulla-panis* (**a**) Basidiospores, (**b**) Basidia, (**c**) Cystidia, (**d**) Generative hyphae, (**e**) Binding hyphae; (**f**–**k**) *Perenniporia ochroleuca* (**f**) Basidiospores, (**g**) Basidia, (**h**) Cystidia, (**i**) Thin-walled generative hyphae, (**j**) Thick-walled hyphae, (**k**) Skeletal hyphae; (**l**–**o**) *Polyporus arcularis* (**l**) Basidiospores, (**m**) Basidia, (**n**) Generative hyphae, (**o**) Skeletal hyphae; (**p**–**s**) *Polyporus blanchetianus* (**p**) Basidiospores, (**q**) Basidia, (**r**) Generative hyphae, (**s**) Skeletal hyphae

Hyphal system trimitic; generative hyphae 2.2–3 μm wide, hyaline, thin-walled, branched, septate, occasionally clamped also; skeletal hyphae 2–3.0 μm wide, thick-walled with capillary lumen, aseptate, unbranched to branched; binding hyphae subhyaline, thick-walled, aseptate, much branched. Pores stratose, each stratum up to 2.7 mm deep, rounded to oval; dissepiments 30–120 µm thick, equal, concolorous with the context. Hyphal pegs sometimes present. Sporulating basidia not observed. Basidiospores 4–4.56×3–3.6 µm, hyaline, broadly ellipsoid, smooth, firm-walled, dextrinoid.

Distribution: H.P.: Dalhousie; U.K.: Mussoorie, Nainital.

Collection examined: SSR 6107, MPS 6344, 6364.

Substratum: On stump of *Cedrus deodara.*

Remarks: This species appears to be common in the N. W. Himalayas. It can be recognized by the cream-colored fructification, stratose nature, trimitic hyphal system and dextrinoid basidiospores.

Perenniporia mesoleuca (Petch) Ryvarden, *Norw. Jl Bot.* **19**: 233 (1972)= Poria mesoleuca (Petch) Ginns, *Mycotaxon* 21: 331 (1984)

Fructification annual, mostly resupinate to effused-reflexed, coriaceous, widely effused forming patches, separable. Pileus effused-reflexed, small; upper surface creamish to creamish-yellow, sometimes greyish-cream, azonate, glabrous, smooth; margin concolorous with upper surface, obtuse, entire, in resupinate patches creamish, thinning to abrupt, not rhizomorphic. Pore surface creamish-yellow to light brown, uneven, rimose on drying; pores rounded to angular, 175–460 µm in diameter, about 3–4 per mm; dissepiments 27–68 µm thick, equal, entire, apices finely velutinate. Context creamish to creamish-yellow, homogenous, azonate, soft, fibrous, non-xanthochroic, up to 7 mm thick; tubes not stratose, creamish-brown in section, harder than context, up to 3 mm deep, deeper when obliquely positioned.

Hyphal system trimitic; generative hyphae hyaline, thin-walled, aseptate, clamped, clamps not very common, mostly collapsed in dry specimens, dextrinoid, 3–4.4 µm in diameter; skeletal hyphae thick-walled to solid, aseptate, cyanophilous, unbranched, 4.2–11.0 µm in diameter; binding hyphae subhyaline, thick-walled, branched, dextrinoid, 3.2–8.1 µm in diameter. Cystidioles not very prominent, broadly clavate, sometimes with narrowed apex, non-dextrinoid, 12.4–25×7.6–12.2 µm. Basidia not observed. Basidiospores hyaline, thin-walled, smooth guttulate, dextrinoid, cyanophilous, obovate to subglobose, somewhat narrowed and drawn out into apiculus, 7–9.2×5–6.2 µm.

Distribution: H.P.: Kausali.

Collection examined: Dhanda 6438, IBP 37667.

Substratum: At the base of an angiosperm.

Remarks: This collection is marked by the presence of dextrinoid basidiospores and trimitic hyphal system with dextrinoid skeletal hyphae.

Perenniporia ochroleuca (Berk.) Ryvarden, Norw. J. Bot.**19**: 233 (1972)=*Poria ochroleuca* (Berk.) Kotl. & Pouzar, Ceska Mykol. 13(1): 33 (1959). Fig. 6.80f–k

Fructification perennial, sessile, dimidiate, ungulate or applanate, hard and woody, pores decurrent; pileus surface whitish to pale brown to reddish brown, finely tomentose when young, with age becoming glabrous and blackening at the base, subzonate, sulcate; margin white, thin or thick. Context cream to pale brown, woody; hymenial surface light brown, pores circular, 3–5 per mm, pore tubes white, 1–7 mm long.

Hyphal system trimitic; generative hyphae hyaline, thin-walled, few thick-walled, clamped, 2.2–3.7 μm wide; skeletal hyphae hyaline, thick-walled to solid, unbranched, a few apically biforked, 2.6–3.7 μm wide; binding hyphae hyaline, thick-walled but showing lumen, branched with short branches, 2–3 μm wide, not abundant. Basidia 4-sterigmate, 25.2–30.2 × 8.1–11.1 μm. Basidiospores hyaline, ellipsoid-ovoid, truncate, thick-walled, one guttule, dextrinoid, 12.2–15 × 7–9.3 μm.

Distribution: A.P: West Kameng; H.P.: Kullu.

Collection examined: SSV 21819; IBP 37294.

Substratum: On decaying angiospermic log.

Remarks: The species is characterized by having perennial, sessile fructification; margin white; hymenial surface light brown; hyphal system trimitic; basidia 4-sterigmate; ellipsoid-ovoid, thick-walled basidiospores.

Polyporus Michel. ex. Adams
Familles des plantes 2: 10, 1763

Fructification annual, stipitate, pileus smooth to scaly, light to deep brown, soft coriaceous, hard when dry, pore surface white to cream; pores entire, round to angular. Pileus circular, dimidate, flabelliform; upper surface white, ochraceous, brown, greyish brown, tubes in one layer. Context white, light to deep brown, wrinkled. Hyphal system dimitic; generative hyphae hyaline, thin-walled, clamps present; binding hyphae arboriform to dendriform. Cystidia none. Basidia hyaline and clavate. Basidiospores cylindrical, thin-walled, hyaline, smooth, non-amyloid.

Twenty six species, widespread

Lit.: Stahl (*Biblthca Mycol.* 50, 1970.), Krüger & Gargas (*Fedds Repert.* 115: 530, 2004; phylogeny)

Type Species: *Boletus tuberaster* Jacq. ex. Pers., 1801

Habitat: Dead Wood, decayed twigs

Himalayas: Nine (Table 6.7)

Polyporus arcularis (Batsch) Fr., Syst. Mycol.1:342, 1821. Fig. 6.80l–o

Fructification annual, solitary to gregarious, soft when fresh, brittle on drying. Pileus round, weakly infundibuliform; upper surface white to cream when fresh, tomentose, azonate; margin white, incurved on drying. Stipe central, brown, glabrous to tomentose. Pore surface white to cream coloured when fresh, brown on drying; pores angular to elongated radially; dissepiments 45–162 μm thick; tubes in one layer, white. Context white, soft, homogenous.

Hyphal system dimitic; generative hyphae hyaline, 2.5–4 μm wide, thin-walled, septate, branched, cyanophilous; binding hyphae hyaline, thick-walled, branched, aseptate, 1.6–4 μm wide. Cystidia absent. Basidia clavate, cyanophilous, 4-spored, cyanophilous, 4.8–16 × 6.2–7.2 μm. Basidiospores hyaline, smooth, thin-walled, ellipsoid, 6.8–8.0 × 2–2.7 μm.

Table 6.7 Synopsis of *Polyporus* species

Name	Fructification	Stipe	Pores	Gleoeporus hyphae	Basidiospores
P. arcularis	Annual, growing singly or in tufts, fleshy to hard with age, pileus circular, white infundibuliform	Central minutely tomentose, black brown at the base	Hexagonal, 1–2 per mm	Present	Cylindric or ellipsoid, cylindric, hyaline, 5–9 × 2.2–3.8 μm
P. blanchetianus	Annual, solitary soft fleshy to friable, pileus flabelliform to spathulate, depressed in centre, brown	Lateral solid, orange white to concolorous with pileus attached to substrate with disc	Round to angular, 8–9 per mm	Absent	Cylindric, ellipsoid, subhyaline, 7–9 × 3–3.7 μm
P. badius	Annual, pileate soft coriaceous, brittle on drying. Upper surface reddish brown to black, margin cream coloured	Lateral initially fine tomentose	Round, 5–7 per mm	Absent	Cylindric ellipsoid, 5.3-7 × 2.3–3.2 μm
P. brumalis	Annual, solitary to common from a base, coriaceous. Pileus convex, upper surface yellowish brown or purplish brown, covered with scales	Central, bulbous at the base white to brownish minutely tomentose	Oval polygonal, 2–3 per mm	Present	Cylindric, 5.6–7.2 × 1–3.2 μm
P. melanopus	Annual, solitary coriaceous. Pileus convex, imblicate to infundibuliform yellow when fresh, brown on drying	Central or eccentric, chocolate brown to rusty brown, adpressed, velutinate	Round, 3–6 per mm	Absent	Ellipsoid, 6.2–8.4 × 2.6–3.7 μm
P. ostreiformis	Annual, sessile to effused reflexed, applanate to fan shaped, imbricate. Upper surface dirty white to pale pinkish	–	Round or angular, minute	Absent	Cylindric, 4.6–5.6 × 1.6–2.3 μm
P. squamosus	Annual, pileate. Pileus reniform to flabelliform, upper surface cream colored	Lateral smooth, covered with pores, near the pileus	Angular to irregular large, 0.5–1.5 mm in diameter	Absent	Oblong ellipsoid, 9.5–12.8 × 4–4.8 μm
P. varius	Annual, reflexed, spathulate, corky. Upper surface chocolate with deep brown venation closely zonate near margin	Substipitate	Subglobose to angular, 2–4 per mm	Absent	Cylindric, 7.4–12 × 3.2–4.2 μm

Distribution: A.P.: West Kameng; Meghalaya- Jowai.
Collection examined: SSV 21473, 21786, 21790.
Substratum: On decaying angiospermic twigs.
Remarks: The species is characterized by thicker and yellowish brown pileus; stipe stipitate; hyphal system dimitic; basidia clavate, 4-spored; ellipsoid, thin-walled basidiospores.

Polyporus badius Jungh., Syst. Verz. H.Zollinger auf Java gesamm. Pflanzen: 120 (1846) Fig. 6.81a–e

Fructification annual, pileate, laterally stipitate, single, soft, coriaceous when fresh, brittle on drying; pileus infundibuliform to flabelliform, depressed in the centre up to 13 cm wide, 20 cm long and 8 mm thick; upper surface reddish brown to black in the centre, glabrous, smooth, irregularly wrinkled on drying, azonate; margin cream coloured when fresh, reddish brown on drying, undulating, sharp, smooth, sterile below, incurved on drying up to 0.8 mm wide. Stipe lateral, black, initially fine tomentose, later glabrous, smooth expanded and flattened above with decurrent pores, round near the base. Pore surface cream coloured when fresh, light brown on drying, dull, even; pores small, round, 5–7 per mm, 90–145 μm in diameter; dissepiment equal, 30–95 μm thick; tubes in one layer, concolorous with the pore surface up to 1 mm deep. Context cream, fibrous, homogenous, non-xanthochroic, up to 6 mm thick.

Hyphal system dimitic; generative hyphae hyaline, thin-walled, branched, septate, clamps absent, cynophilous, 2.3–4.2 μm in diameter; binding hyphae hyaline, flexuous, thick-walled, aseptate, much branched, acyanophilous, 1.8–6.9 μm in diameter. Hyphal pegs present, projecting up to 25 μm into the pore cavity. Cystidia absent. Basidia thin-walled, hyaline, clavate, 4-spored, 15.6–18.2 × 6–7.2 μm. Basidiospores hyaline, thin-walled, smooth, cylindrical-ellipsoid, apiculate, non-amyloid, 5.3–7 × 2.3–3.2 μm.

Distribution: Bhutan- Thimphu, A.P-West Kameng; H.P.: Kullu.
Collection examined: SSV 21763, 21385, 21811, IBP 37296.
Substratum: On decaying angiospermic twigs.
Remarks: The species is characterized by annual, laterally stipitate fructifications; brown to reddish brown upper surface; white pore surface; medium and hyaline, smooth, ellipsoid, non-amyloid basidiospores. This species was previously reported from India by various workers (Currey 1874, Bose 1946). It is a common species in the N.W. Himalayas from where it was described in detail by Thind et al. (1957) as *Polyporus picipes*. This collection closely resembles to the description given by Thind et al. (1957) and Ryvarden (1978).

Polyporus blanchetianus Berk. & Mont., Ann. Sci. Nat. Ser. 3, Vol. 11: 238, 1849. Fig. 6.80p–s

Fructification annual, solitary, laterally stipitate, soft and fleshy when fresh, friable on drying. Pileus flabliform to spathulate, depressed in the centre; upper surface brown, greyish brown, glabrous, radially-striate; margin toothed, concolorous with upper surface, thin acute, fertile below, incurved on drying. Stipe lateral,

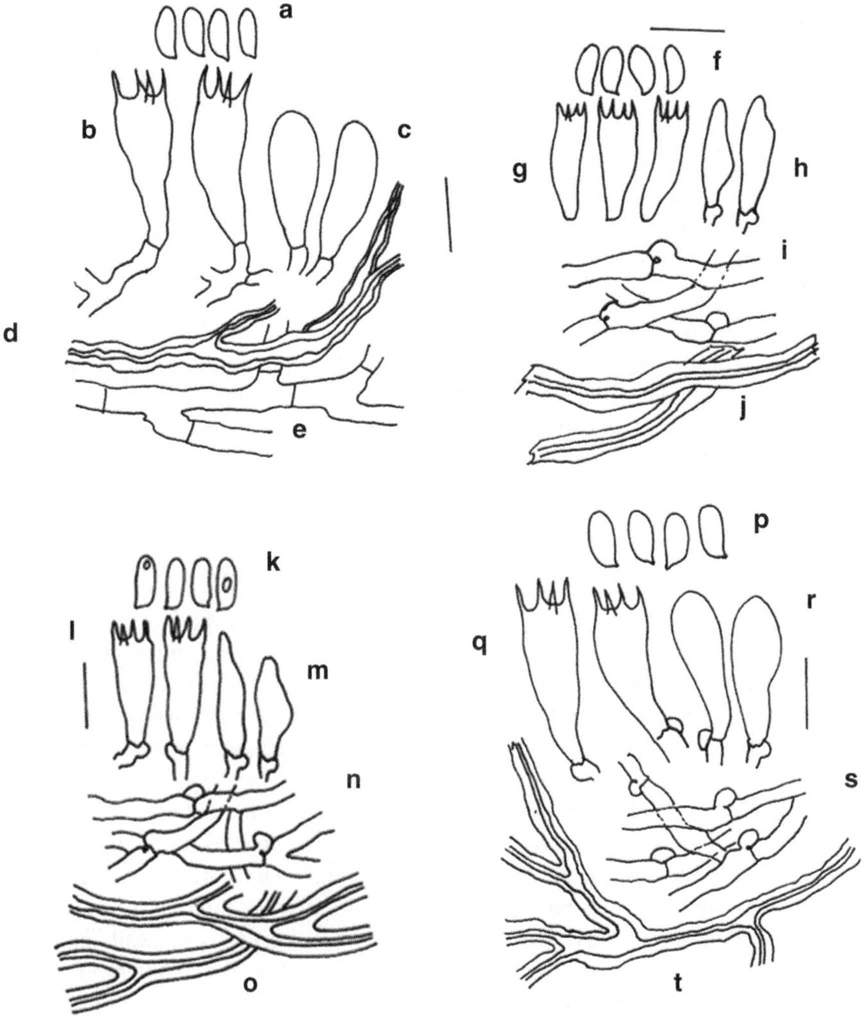

Fig. 6.81 (**a–e**) *Polyporus badius* (**a**) Basidiospores, (**b**) Basidia, (**c**) Basidioles, (**d**) Binding hyphae, (**e**) Generative hyphae; (**f–j**) *Polyporus brumalis* (**f**) Basidiospores, (**g**) Basidia, (**h**) Basidioles, (**i**) Generative hyphae (**j**) Skeletal hyphae; (**k–o**) *Polyporus grammocephalus* (**k**) Basidiospores, (**l**) Basidia, (**m**) Basidioles, (**n**) Generative hyphae, (**o**) Binding hyphae; (**p–t**) *Polyporus melanopus* (**p**) Basidiospores, (**q**) Basidia, (**r**) Basidioles, (**s**) Generative hypha, (**t**) Binding hyphae

solid, 1.6 cm long and 6.5 mm thick, orange white to concolorous with pileus, broad glabrous, attached to substrate with broad mycelia disc. Pore surface orange white, darker after drying, uneven; pores round to angular, 8–9 per mm, 115–175 µm in diameter; pore mouth velutinate, dissepiments 40–150 µm thick, equal; tubes not stratified, orange white, up to 1.5 mm deep in section. Context pale creamish, homogenous, non-xanthochroic, up to 1.4 mm thick.

Hyphal system dimitic; generative hyphae hyaline, thin-walled, branched, septate, clamped, cyanophilous, 2.4–3.2 μm in diameter; binding hyphae subhyaline to light coloured to yellowish brown at upper surface, thick-walled to almost solid with narrow lumen, branched, aseptate, acyanophilous, 2.5–4.6 μm in diameter. Hyphal pegs present, projecting up to 55 μm in to the pore cavity. Cystidia absent. Basidia hyaline, clavate, cyanophilous, 9.4–12.3 × 5.3–7.9 μm. Basidiospores subhyaline, thin-walled, non-amyloid, apiculate, cylindric-ellipsoid, 7–9.2 × 3.0–3.7 μm.

Distribution: A.P.: West Kameng; H.P.: Dalhousie, Kullu.

Collection examined: SSV 21677, 21867, 21667, IBP 37295.

Substratum: On decaying angiospermic twigs.

Remarks: The species is characterized by laterally stipitate fructifications; flabelliform to spathulate pilei; brown to greyish brown, radially striate upper surface; orange white stipe, attached to substratum by broad mycelia disc; orange white pore surface; small, 8–9 pores per mm; and hyaline, cylindric-ellipsoid basidiospores.

Polyporus brumalis (Pers.) Fr., Observ. Mycol. 2: 255 (1818). Fig. 6.81f–j

Fructification annual, usually solitary, sometimes several from a common base, soft coriaceous, centrally stipitate. Stipe frequently bulbous at the base and slightly flattened just below the pileus, white to brownish, may be minutely tomentose when fresh. Pileus convex, usually slightly depressed at the centre; upper surface yellowish brown or purplish brown, glabrous or shortly hispid and often covered with indistinct scales; margin thin, inturned with or without hairs. Context white, suberose; hymenial surface white, brownish on drying, pores circular, oval, polygonal or varying in shape, 2–3 per mm, pore tubes up to 2 mm long.

Hyphal system dimitic; generative hyphae clamped, hyaline, thin-walled, branched, usually 2.2–4.6 μm wide but many irregularly inflated up to 10 μm; on the pileus surface occur some generative hyphae 2.4–4.4 μm wide, thin-walled to slightly thick-walled, pale brown with black discontinuous deposits on walls and also small patches of cuticular-like cell formed of thin-walled closely interwined hyaline inflated generative hyphae. Gloeoporus hyphae present, up to 10 μm wide. Binding hyphae hyaline, thick-walled, often solid in old specimens, dendritic, sparingly branched, usually 3.5–6.0 μm wide producing whip-like branches, 1.5–2.0 μm wide from wider stems up to 15 μm wide. Basidia 2–4 sterigmate, clavate, 8.8–17.0 × 3.5–5.6 μm. Basidiospores hyaline, thin-walled, cylindric, with one or more guttulate, 5.6–7.2 × 2.1–3.2 μm. Cystidioles present, 8.5–17.0 × 3.2–4.3 μm.

Distribution: Meghalaya- Shillong; U.K.: Chakrata.

Collection examined: L 39998

Substratum: On rotten log.

Remarks: The species is characterized by annual, pileate fructification; upper surface yellowish brown; hyphal system dimitic; 2–4 sterigmate basidia.

Polyporus grammocephalus Berk., Lond. J. Bot. 1: 148, 1842. Plate 6.30b, Fig. 6.81k–o

Fructification annual, solitary, often few fructification grow nearer to form imbricate pilei, soft and fleshy when fresh, substipitate to short stipitate. Pileus dimidiate, flabelliform or spathulate up to 12×9×0.5 cm; upper surface glabrous, smooth, azonate, brown to reddish brown, radiate-striate, ridged or grooved; margin thin and incurved on drying, light brown, sterile, fertile below up to 1 mm wide, smooth and entire to wavy. Stipe lateral, short, up to 5 mm in diameter expanding into pileus above, smooth to minutely roughened, brown, fresh, reddish brown on drying, mycelia disc is present at the attachment with the substratum. Pore surface light brown, even; pores round to angular to split pores, decurrent, 3–5 per mm; dissepiment even, 95–180 µm thick; pore tubes in one layer up to 3.5 mm long in section. Context soft, cream to light brown, homogenous, up to 3.5 mm thick, non-xanthochroic.

Hyphal system dimitic; generative hyphae hyaline, thin-walled, branched, septate, clamped, slightly cyanophilous, 1.8–3.5 µm in diameter; binding hyphae subhyaline, thick-walled, branched, aseptate, acyanophilous, up to 2.5–4.4 µm in diameter. Cystidia absent. Basidiospores hyaline, thin-walled, smooth, apiculate, ellipsoid, nonamyloid, 5.6–8×2.6–3 µm.

Distribution: A.P.: West Kameng; U.K.: Nainital; H.P.: Kausali, Shilma-Tara Devi.

Collection examined: SSV 21433, 21520, 21804, IBP 37297.

Substratum: On decaying angiospermic twigs, on *Shorea robusta* log.

Remarks: The species is characterized by annual, stipitate to substipitate, flabelliform pilei; light brown, glabrous upper surface; reddish brown, solid stipe; 3–5 pore per mm; and hyaline, ellipsoid, non-amyloid basidiospores. Thind & Chatrath (1957) reported it for the first time from N.W. Himalayas. The collection fit well in the description given by Thind & Chatrath (1957). It is a new record for Eastern Himalayas.

Polyporus melanopus (Pers.) Fr., Syst. Mycol. 1: 347, 1821. Fig. 6.81p–t

Fructification annual, solitary, centrally or eccentrically stipitate, coriaceous, single; pileus convex, umblicate or infundibulloform up to 11 cm in diameter and up to 3.5 mm thick; upper surface yellow when fresh, yellowish brown on drying, azonate, glabrous, faintly radially striate, irregularly wrinkled on drying; margin undulating, strongly incurved on drying, lobed, sterile below, up to 1.2 mm wide. Stipe solid, up to 4.4 cm long and 1 cm in diameter, chocolate brown to rusty brown, adpressed, velutinate, glabrous later, long, longitudinally rugose on drying. Pore surface white to cream, brown on drying, decurrent; pores round, 3–6 per mm; dissepiments equal, 58–85 µm thick; tubes in one layer, concolorous with pore surface up to 2.5 mm deep in section. Context white, soft, homogenous, up to 1 mm thick, non-xanthochroic.

Hyphal system dimitic; generative hyphae, thin-walled, branched, septate, clamped, cyanophilous, 2.2–3.5 µm in diameter; binding hyaline to light yellow, thick-walled, much branched, tortuous, dominate the dissepiment and context, aseptate, acyanophilous, 2–6.4 µm in diameter. Cystidia absent. Basidia thin-walled, hyaline, clavate, 15.3–22.3 × 6.0–8.2 μm. Basidiospores hyaline, thin-walled, smooth, ellipsoid, nonamyloid, 6.2–8.4 × 2.6–3.7 μm.

Distribution: Meghalaya: Shillong, Mawphlang; W.B.: Darjeeling.

Collection examined: GSD 21042.

Substratum: On decaying angiospermic stump.

Remarks: The species is characterized by large, coriaceous, centrally stipitate fructification; yellow to yellowish brown upper surface; chocolate brown stipe, 3–6 per mm pores; dimitic hyphal system; and ellipsoid, 6–8.5 × 2.8–3.8 μm basidiospores.

Polyporus ostreiformis Berk., J. Linn. Soc. Bot. London, 16: 46, 1878 = *Daedalea ostreiformis* (Berk.) De., Can. J. Bot. 59: 1300, 1981. Fig. 6.82a–f

Fructification annual, sessile to effuse-reflexed, dimidiate, appalanate to sometimes fan shaped, imbricate, sometimes occurring singly, hard and rigid on drying, upper surface white to pinkish brown, occasionally greyish, smooth, glabrous, azonate; margin thick. Context white to pinkish brown; hymenial surface white or pale pinkish when fresh, black on drying; pores circular or angular.

Hyphal system trimitic; generative hyphae hyaline, branched, clamped, thin to thick-walled, 2.8–7.0 μm wide; skeletal hyphae hyaline, thick-walled to solid, 2.8–4 μm wide; binding hyphae solid, branches short with obtuse ends, 2–3 μm wide, present abundantly. Gloeoporus hyphae present, thin-walled, unbranched, aseptate or septate with clamps, 3–8 μm. Basidia clavate, 4-sterigmate, 11–17 × 4.3–6 μm. Basidiospores hyaline, thin-walled, smooth, cylindrical, apiculate, 4.3–5 × 1.5–2.8 μm.

Distribution: U.K.: Dehra Dun, Nainital.

Collection examined: L 39999

Substratum: On decaying angiospermic stem.

Remarks: The species is characterized by sessile to effused-reflexed, fan-shaped, imbricate fructification; hymenial surface white or pale pinkish; hyphal system trimitic; basidia clavate, 4-sterigmate; basidiospores cylindrical, smooth, apiculate.

Polyporus squamosus Huds.ex Fr., Syst. Mycol. 1: 343, 1821. Plate 6.30c, Fig. 6.82g–l

Fructification annual, pileate, soft fleshy, brittle on drying, laterally stipitate, single or growing in groups. Pileus reniform to flabelliform 3.8–7.2 cm in diameter and 0.5–2 cm thick; upper surface cream coloured when fresh, brown on drying, covered with concentric rows of adhering dark brown scales, irregularly wrinkled on drying; margin thin, undulating, dentate, incurved on drying. Stipe lateral, short, up to 2.4 cm long and 1.5 cm in diameter, smooth, covered with pores near the pileus; pore surface white when fresh, brown to rusty brown on drying; pores angular to irregular, large 0.5–1.5 mm in diameter; tubes in one layer, concolorous with the pore surface up to 6 mm deep. Context cream, soft spongy, homogenous, up to 1.6 cm thick, non-xanthochroic.

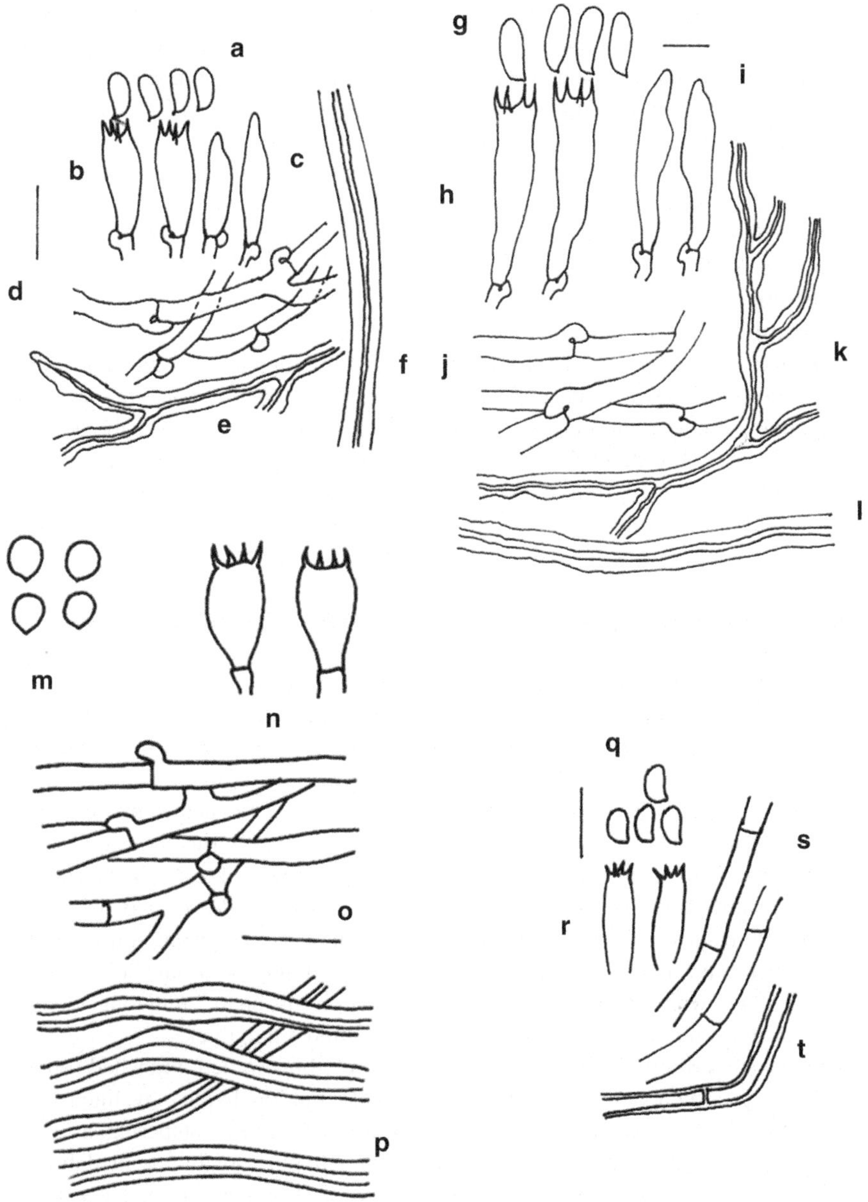

Fig. 6.82 (**a–f**) *Polyporus ostreiformis* (**a**) Basidiospores, (**b**) Basidia, (**c**) Basidioles, (**d**) Generative hyphae, (**e**) Binding hyphae, (**f**) Skeletal hyphae; (**g–l**) *Polyporus squamosus* (**g**) Basidiospores, (**h**) Basidia, (**i**) Basidioles, (**j**) Generative hyphae, (**k**) Binding hyphae, (**l**) Skeletal hyphae; (**m–p**) *Polyporus varius* (**m**) Basidiospores, (**n**) Basidia, (**o**) Generative hyphae, (**p**) Skeletal hyphae; (**q–t**) *Poria auricoma* (**q**) Basidiospores, (**r**) Basidia, (**s**) Thin-walled generative hyphae, (**t**) Thick-walled generative hyphae

Hyphal system dimitic; generative hyphae hyaline, thin-walled, branched, septate, clamps large, acyanophilous, 2–4.2 μm in diameter; binding hyphae hyaline, thick-walled, much branched in dissepiment, aseptate, 2–7.5 μm wide in dissepiments and 2–6 μm wide in the context. Hyphal pegs present, 34.7–45.8 × 20.7–30 μm, projecting into the cavity. Basidia hyaline, thin-walled, clavate 4-spored, 7–8.5 μm wide. Basidiospores thin-walled, hyaline, smooth, oblong-ellipsoid, apiculate, non-amyloid, 9.5–12.8 × 4.0–4.8 μm.

Distribution: W.B.: Darjeeling; U.K.: Dehradun, Nainital; H.P.: Kullu, Manali, Shimla, Solan.

Collection examined: IBP 37298.

Substratum: On decaying angiospermic log.

Remarks: The species is characterized by laterally stipitate fructification; upper surface covered with adhering, dark brown scales; large 1–2 pores per mm; and hyaline cylindrical basidiospores. The collection closely resembles the description given by Bakshi (1971) with slight variation in spore size. This species was recorded previously from India by Berkeley (1856) from Darjeeling (W. Bengal) and Sikkim; Henning (1900) from Penzi (H.P.) and Bakshi (1971) from Kullu (H.P.).

Polyporus varius (Pers.) Fr., Syst. Myco. 1: 352, 1821. Plate 6.30d, Fig. 6.82m–p

Fructification annual, pileate, stipitate, solitary or accreted into groups, soft coriaceous when fresh, hard on drying. Pileus flabelliform or infundibulum, up to 3.5 × 5.5 × 0.5 cm, flat or convex, depressed in the centre; upper surface glabrous, smooth, azonate, yellowish brown to brown, darken on drying, reddish brown in old specimens; margin reddish brown, even, lobed or toothed, thin, curved on drying, sterile below, up to 1.5 mm. Stipe lateral or eccentric, short up to 1 cm long, and 5 mm in diameter, cylindrical, distinctly separated from pileus, expanding upwards and with decurrent pores on the upper part, lower part black, distinct. Pore surface cream to light brown, darken on drying, even; pores angular, 4–5 per mm, 105–175 μm in diameter; pore mouth velutinate; dissepiment 52–140 μm wide, even, tubes in one layer, concolrous, up to 2 mm long. Context ochraceous, homogenous, non-xanthochroic, up to 3 mm thick.

Hyphal system dimitic; generative hyphae hyaline, thin- walled, branched septate, clamps present, cyanophilous 1.8–3.5 μm wide; binding hyphae hyaline to subhyaline, thick walled with narrow lumen, branched, branches tapering, aseptate, cyanophilous, 2.5–4.5 μm wide. In spite, binding hyphae are thick walled, brown, 3–5 μm wide. The black cuticle on the base of stipe consists of erect, thick-walled, dark brown hyphae strongly agglutinated. Basidia clavate, 4-spored, cyanophilous, 12.5–15 × 5–6.5 μm. Cystidia absent, cystidioles present. Basidiospores cylindric- ellipsoid, hyaline, thin-walled, smooth, faintly amyloid, 6.5–7.5 × 2.4–3 μm.

Distribution: Meghalaya, Jowai, Garampani; U. K.: Karanprayag-Chamoli.

Collection examined: GSD 21062, L 38035.

Substratum: On angiospermous decaying twig.

Remarks: The species is marked by annual, laterally stipitate fructifications; small, black stipe; cream pore surface; medium, 4–5 pores per mm; binding hyphae thick walled to almost solid with narrow lumen with tapering branches; and hyaline, cylindric-ellipsoid, faintly amyloid basidiospores. Bakshi (1971) reported this species from Lucknow (U.P.). This species is new report for India.

Poria Adans,
Fam. Des. Plantes 2:10.1763.

Fructifications resupinate, soft-fragile to cartilaginous or woody, white or variously coloured, pores well developed and fertile within, pore mouths sterile. Tissues not darkening or blackening in KOH sol. Hyphal system monomitic, dimitic or trimitic; generative hyphae with or without clamps. Pores angular or oval and becoming irpicicoid but not lamellate or deadaloid. Cystidia, Gloeocystidia or Hyphal pegs present or absent. Basidia clavate to cylindrical or utriform, 4-spored. Basidiospores subhyaline, smooth or finely echinulate, amyloid or non-amyloid.

Type Species: *Polyporus michelii* Fr. 1821

Habitat: Wood

Himalayas: Two

Key to species

1. Basidiospores hyaline to subhyaline, thin-walled, ellipsoid with one side somewhat flattened, minutely apiculate, 4.4–6.5 × 2.6–3.12 μm .... *P. auricoma*
1. Basidiospores hyaline, thin- to slightly thick-walled, smooth, broadly ellipsoid to subglobose, 4.6–6.1 × 4–4.6 μm .... *P. conferta*

Poria auricoma (Lév) Cooke, Grevilles 15: 26, 1886. Fig. 6.82q–t

Fructification annual, resupinate, widely effused, forming large patches, adnate, inseparable, with no particular taste; margin dirty white to creamish, thinning, myceloid; pore surface dirty white to creamish-brown, irregular, uneven; pores usually angular to irregular, pore mouths somewhat dentate, not typically irpiciform, about 2–3 per mm; dissepiments 68–135 μm thick, equal, entire, apices somewhat dentate. Context creamish homogenous, thin, non-xanthochroic; tubes not stratified, mostly obliquely placed on sloping surfaces, appearing irpiciform, creamish-brown in section.

Hyphal system monomitic; generative hyphae hyaline, thin to thick-walled, wall up to 1.5 μm thick, septate, clamps absent, branched, non-amyloid, acyanophilous, cyanophilous, 3.2–7.8 μm in diameter. Cystidia absent. Basidia hyaline, clavate, 4-spored, 4.3–5.7 μm in diameter. Basidiospores hyaline to subhyaline, thin-walled, smooth, non-amyloid, acyanophilous to weakly cyanophilous, ellipsoid with one side somewhat flattened, minutely apiculate, 4.4–6.5 × 2.6–3.12 μm.

Distribution: H.P.: Dalhousie; U.K.: Dehra Dun, Nainital.

Collection examined: Dhanda 6868.

Substratum: On stump of *Shorea robusta.*

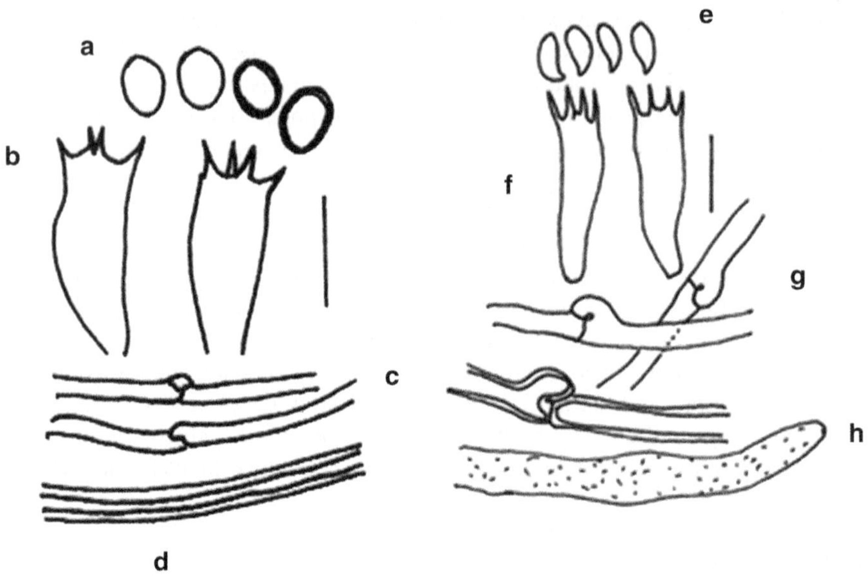

Fig. 6.83 (**a–d**) *Poria conferta* (**a**) Basidiospores, (**b**) Basidia, (**c**) Generative hyphae, (**d**) Skeletal hyphae; (**e–h**) *Rhodonia placenta* (**e**) Basidiospores, (**f**) Basidia, (**g**) Generative hyphae, (**h**) hyphae

Remarks: The species is characterized by annual, resupinate, widely effused fructification; margin dirty white to creamish, thinning, myceloid; pore surface dirty white to creamish-brown, irregular; hyphal system monomitic; basidia hyaline, 4-sterigmate; smooth thin-walled basidiospores.

Poria conferta Overh., Bull. Pa Agr. Exp. Sta., Tech. Bull. 418: 25, 1942. Fig. 6.83a–d

Fructification annual, resupinate, adnate, inseparable, effused; margin white, drying pale creamish, thinning to abrupt, reflexed at places, narrow. Pore surface white, drying pale creamish, uneven, cracking, dull; pores rounded to angular; dissepiments 41–109 μm thick, equal, apices velutinate. Context creamish, thin, non-xanthochroic, tubes not stratose, creamish in section, up to 1.5 mm deep.

Hyphal system dimitic; generative hyphae hyaline, thin-walled, branched, septate, clamped, cyanophilous, 1.5–2.6 μm in diameter; skeletal hyphae subhyaline, thick-walled, branched, aseptate, weakly cyanophilous, 1.2–4.0 μm in diameter. Cystidia absent. Basidia hyaline, clavate, 4-spored, 15.2–19.4 × 6.1–7.2 μm. Basidiospores hyaline, thin- to slightly thick-walled, smooth, broadly ellipsoid to subglobose, dextrinoid, often truncate at one end, 4.6–6.1 × 4–4.6 μm.

Distribution: H.P.: Shimla, Narkanda; U.K.: Uttarkashi

Collection examined: Dhanda 6453, IBP 37692.

Substratum: On log of *Juglens regia.*

Remarks: This species seems to be quite rare in the N. W. Himalayas. It is marked by annual, sometimes reviving, adnate fructifications; white pore surface, drying pale creamish, rounded to angular pores, 2–4 per mm; creamish context; dimitic hyphal system; and hyaline, smooth, broadly ellipsoid to subglobose, dextrinoid basidiospores. This species comes near *P. lenta* which, however, differs in having amyloid, smooth to asperulate basidiospores.

Pycnoporus Karst.,
Revue Mycol. 3(9): 18, 1981.

Fructification annual, sessile. Pileus dimidiate to flabelliform, cinnabar to orange; upper surface brightly coloured, initially soft and velvety, becoming glabrous and shining later, azonate to zonate, devoid of cortex. Pore surface concolrous with upper surface; pores circular, 4–6 per mm; tubes in one layer. Context corky, thin to thick with alternating bands of white and pale orange and darker orange or red, becoming black when touched with KOH sol. Hyphal system trimitic, generative hyphae thin-walled, branched, clamped, septate; skeletal hyphae thick-walled to solid, unbranched normally, aseptate; binding hyphae thick-walled to solid, with narrow lumen unbranched normally, aseptate, with narrow lumen, intensively tortuous and branched. Cystidia absent. Basidia clavate, 4-spored. Basidiospores hyaline, short cylindric to ellipsoid, non-amyloid.

Two species, widespread

Lit.: Nobles & Frew (*Can. J. Bot.* **40**: 987, 1962), Rebort & Ryvarden (*Kew. Bull*. **61:** 55, 2006; Cameroon)

Type Species: *Boletus cinnabarinus* Jacq. 1776

Habitat: Dead Wood

Himalaya: Three

Key to species

1. Fructification up to 3 mm thick, pores 5–6 per mm......................... *P. sanguineus*
1. Fructification up to 12 mm thick, pores 3–4 per mm..........................*P. coccineus*
1. Fructification up to 25 mm thick, pores 2–4 per mm.................... *P. cinnabarinus*

Pycnoporus cinnabarinus (Jacq.) P. Karst., Rev. Mycol. 3:18, 1881 = *Polyporus cinnabarinus* (Jacq.) Syst. Mycol. 1:371, 1821. Fig. 6.84a–f

Fructification annual, sessile to substipitate, solitary to imbricate. Pileus dimidiate to semicircular, leathery; upper surface of pileus reddish orange to apricot orange, darkening with age, glabrous, azonate; margin thick. Context red, fibrous to corky, zonate; hymenial surface coral red, pores circular to angular, 3–4 per mm, pore tubes 2–4 mm long.

Hyphal system trimitic; generative hyphae thin-walled, clamped, 3–4.8 μm wide; skeletal hyphae thick-walled, hyaline or pale orange, unbranched, frequently covered with orange granules, 5–8.2 μm wide; binding hyphae hyaline, thick-walled to solid, much branched, usually with short branches, 2.2–4 μm wide. Basidia 4-sterigmate, 4.4–5 μm in diameter. Basidiospores hyaline, short-cylindric, thin-walled, slightly curved, 5–7.2 × 2.4–3.4 μm. Hyphal peg present.

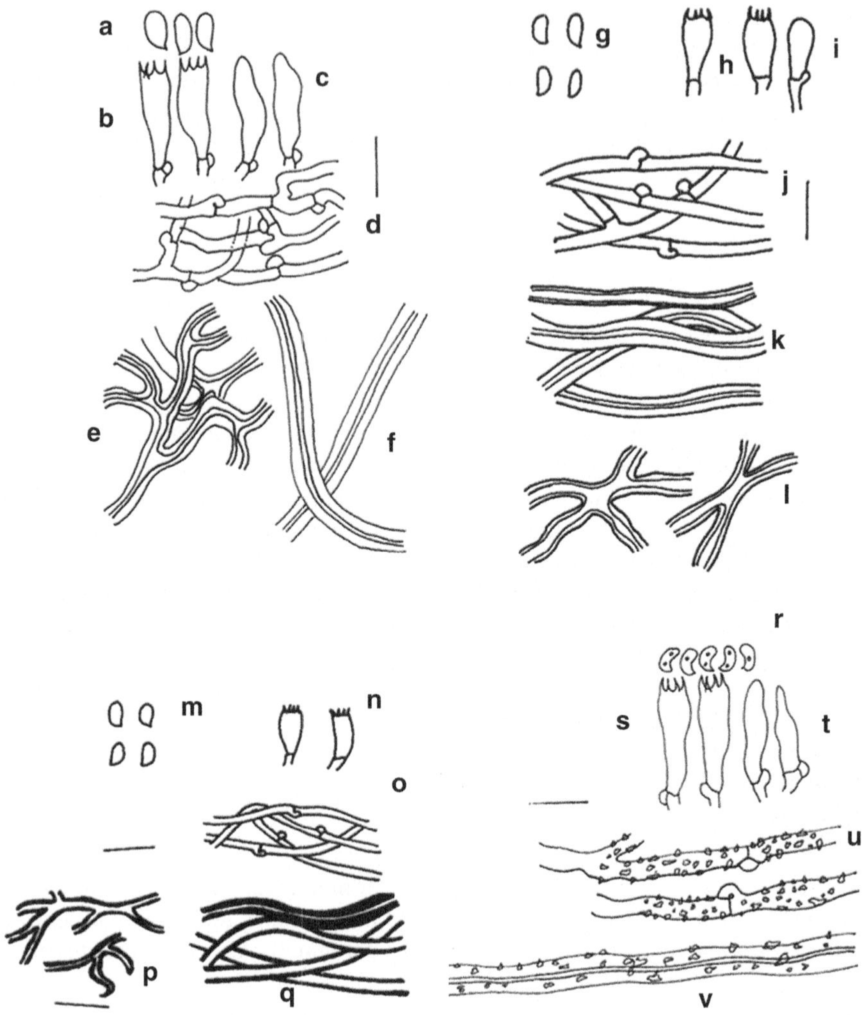

Fig. 6.84 (**a**–**f**) *Pycnoporus cinnabarinus* (**a**) Basidiospores, (**b**) Basidia, (**c**) Basidioles, (**d**) Generative hyphae, (**e**) Binding hyphae, (**f**) Skeletal hyphae; (**g**–**l**) *Pycnoporus coccineus* (**g**) Basidiospores, (**h**) Basidia, (**i**) Basidioles, (**j**) Generative hyphae, (**k**) Skeletal hyphae, (**l**) Binding hyphae; (**m**–**q**) *Pycnoporus sanguineus* (**m**) Basidiospores, (**n**) Basidia, (**o**) Generative hyphae, (**p**) Binding hyphae, (**q**) Skeletal hyphae; (**r**–**v**) *Skeletocutis amorpha* (**r**) Basidiospores, (**s**) Basidia, (**t**) Basidioles, (**u**) Generative hyphae, (**v**) Skeletal hyphae

Distribution: J&K: Sonamarg; H.P.: Mandi, Kullu.
Collection examined: IBP 37300.
Substratum: On decaying *Shorea robusta* log.
Remarks: The species is characterized by annual, sessile to substipitate fructification; hymenial surface coral red; trimitic hyphal system; basidia 4-sterigmate; short cylindric, hyaline basidiospores.

Pycnoporus coccineus (Fr.) Bond & Sing., Annals Mycol. 39(1): 59, 1941. Plate 6.31a, Fig. 6.84g–l

Fructification annual, sessile, pileate, solitary to imbricate, tough coriaceous when fresh, corky and flexible when dried. Pileus dimidiate, broadly attached or effused-reflexed, 5 × 12 × 1 cm; upper surface orange, reddish orange to brownish

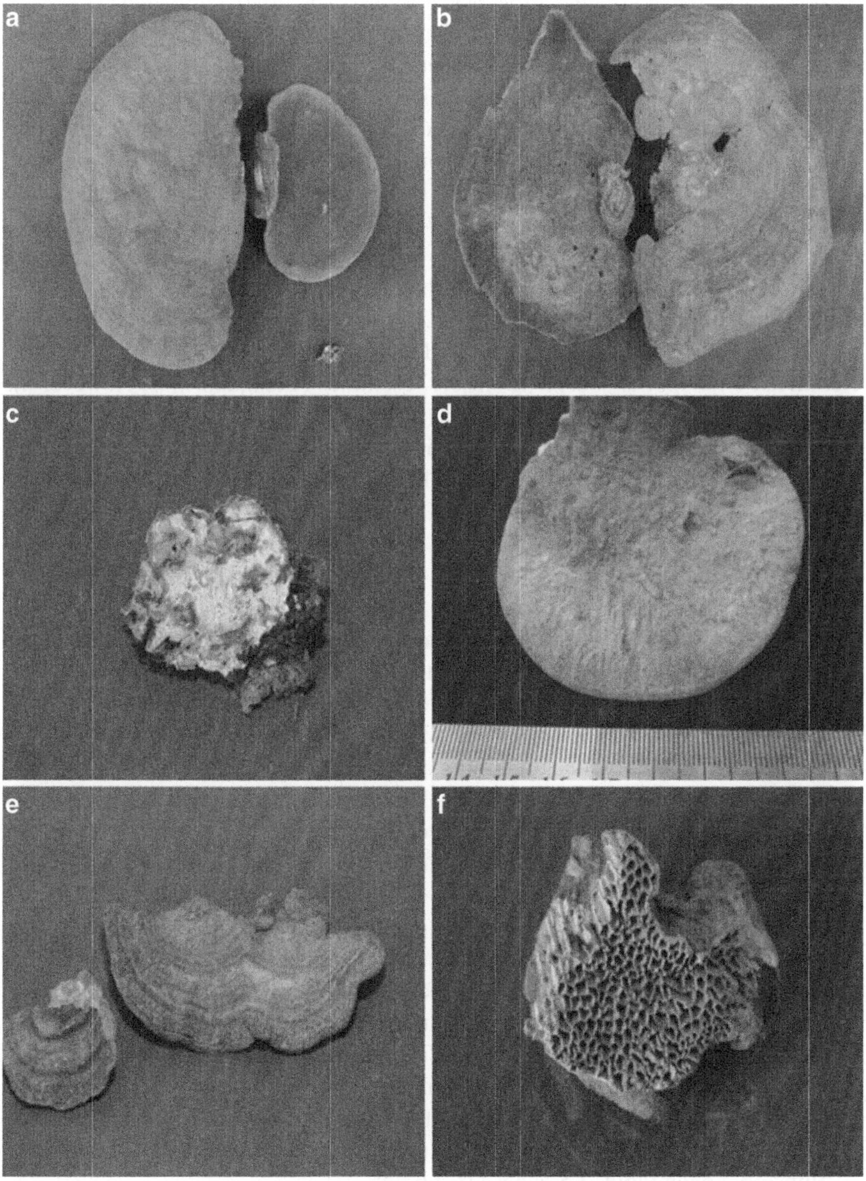

Plate 6.31 (**a**) *Pycnoporus coccineus*. (**b**) *Pycnoporus sanguineus*. (**c**) *Skeletocutis amorpha*. (**d**) *Trametes gibbosa*. (**e**) *Tramets hirsuta*. (**f**) *Trametes incerta*

orange, with white or grey trace in old specimens, azonate, weakly sulcate glabrous, irregularly wrinkled on drying; margin obtuse, orange or concolorous with the upper surface, entire, sterile below, up to 1.2 mm wide. Pore surface orange to reddish orange; pores angular 3–4 per mm, 135–240 μm in diameter; dissepiments 75–150 μm thick; pore mouth velutinate; tubes in one layer, orange, up to 5 mm deep in section. Context orange to reddish orange, tough, zonate, zones of white and orange, turns black in KOH sol, up to 5 mm thick.

Hyphal system trimitic; generative hyphae hyaline, thin-walled, clamped, 2.5–3.5 μm wide, faintly acyanophilous; skeletal hyphae hyaline to pale yellow, thick-walled, covered with orange granules, 3.5–6.5 μm wide, abundant; binding hyphae subhyaline, thick-walled to solid, much branched, mostly with short branches, 3.5–4.5 μm wide, abundant acyanophilous. Basidia hyaline thin-walled, clavate 4.5–5 μm in diameter. Basidiospores hyaline, thin-walled, short cylindric, smooth, slightly curved, non-amyloid, 4–5.2 × 2–2.2 μm. Hyphal pegs present.

Distribution: U.K.: Dehra Dun; H.P.: Solan.

Collection examined: IBP 37401, 37402.

Substratum: On decaying angiospermic log.

Remarks: *Pycnoporus coccineus* closely resembles *P. cinnabarinus*. However the latter differs in having layer pores (2–4 per mm) and larger basidiospores (5.3 × 3.5–3.9 μm). This species was first reported by Bakshi (1971) from Calcutta (W. Bengal) and Dehra Dun (U.K.). It is recorded for the first time from Himachal Pradesh.

Pycnoporus sanguineus (L.) Murr., Bull. Torrey bot. Club 31(8): 421, 1904. Plate 6.31b, Fig. 6.84m–q

Fructification annual, sessile to sub-stipitate, coriaceous when fresh, rigid and light weight on drying, solitary to imbricate. Pileus dimidiate to semicircular, narrowly attached to the substratum up to 5.8 cm long, 3 cm broad and 0.5 mm thick; upper surface fine tomentose when young and growing, smooth due to agglutinated hyphae with age, shining, glabrous, zonate, bright orange to reddish orange, light colored and dark colored zones alternate; margin concolorous or lighter than the pileus, entire or incised, acute, sterile below up to 2 mm wide. Pore surface red to reddish orange, even; pores circular, 4–6 per mm 110–190 μm in diameter; dissepiment thick, 45–150 μm thick; pore tubes in one layer up to 2 mm deep in section. Context tough and fibrous, white and orange places showing developmental stages, turn black with KOH sol.

Hyphal system trimitic; generative hyphae hyaline, thin-walled, branched, septate, clamped, cyanophilous, 1.3–3.6 μm in diameter; skeletal hyphae subhyaline, thick-walled, unbranched, 2.4–5.0 μm in diameter; binding hyphae thick-walled to almost solid with narrow lumen, much branched, branches short, aseptate, 2.0–3.5 μm in diameter. Cystidia absent. Basidia thin-walled, hyaline, clavate, up to 3.5 μm broad. Basidiospores hyaline, thin-walled, cylindrical to cylindric-ellipsoid, nonamyloid, 3.5–4.6 × 2.0–2.6 μm.

Distribution: Bhutan- Thimphu; Manipur- Ukhrul; Meghalaya: Shillong; U.K.: Dehra Dun, Mussoorie; H.P.: Shimla: Glen forests, Solan, Chandigarh.

Collection examined: SSV 21346, 212361, IBP 37403, L 37404.

Substratum: On decaying deciduous log.

Remarks: The species is characterized by thin and coriaceous light red to bright red, reddish orange fructifications; 4–6 pores per mm; context turning black with KOH; and hyaline, ellipsoid basidiospores. It is very close to *P. cinnabarinus* which has rugose, azonate and uneven upper surface; bigger, 2–3 per mm pores; and larger, 5.5–7 × 3.5–4 µm basidiospores. This is a common and widespread species in the Himalayas; Montagne (1842) from Darjeeling (West Bengal), Currey (1874) from Sikkim, Hennings (1900) from Bilaspur (H. P.), Theissen (1911) from Dehra Dun (U. K.), Bose (1919) and Benerjee (1947) from Calcutta (W. Bengal), and Thind et al. (1957) from N. W. Himalayas reported as *Polystictus sanguineus* (L.) Mey. Bakshi (1971) reported it as *Polyporus sanguineus* L.: Fr. who recorded its occurrence on dead wood of both hardwoods and conifers in tropical and subtemperate regions of India/Himalayas. This is an extremely variable species with regard to shape and color of the fructifications.

Rhodonia Niemelä,
Karstenia, 45(2): 79, 2005.

Fructifications annual, resupinate, adnate, coriaceous to widely effused; pore surface white to cream but turns woody brown on drying, not creviced; margin thinning, adnate, concolorous. Pores not stratose; pore mouths rounded to angular. Context composed of compactly arranged hyphae. Hyphal system monomitic, hyphae branched, septate, clamped, thin to moderately thick-walled. Cystidia, gloeocystidia or hyphal pegs absent. Basidia not observed. Basidiospores ellipsoid to suballantoid, minutely apiculate, the walls thin, subhyaline, smooth, non-amyloid.

Type Species: *Rhodonia placenta* (Fr.) Niemelä, 2005

Habitat: Wood

Himalayas: One

Rhodonia placenta (Fr.) Niemela, K.H. Larss. & Schigel, in Niemela, Kinnun, Schige & Larsson, Karstenia 45(2): 79 (2005) = *Poria monticola* Murr., Mycologia 12: 90. 1920. Fig. 6.83e–h

Fructifications annual, resupinate, adnate, coriaceous when fresh becoming hard and brittle on drying, widely effused; pore surface white to cream but turns woody brown on drying, not creviced; margin thinning, adnate, concolorous. Pores not stratose, up to 3 per mm deep; pore mouths rounded to angular; sometimes may collapse and pores may appear up to 1 mm wide. Context composed of compactly arranged hyphae.

Hyphal system monomitic, hyphae 3–5.4 µm wide, branched, septate, clamped, thin to moderately thick-walled. Cystidia, gloeocystidia or hyphal pegs absent. Basidia not observed. Basidiospores 5–6.1 × 3.2 µm, ellipsoid to suballantoid, minutely apiculate, the walls thin, subhyaline, smooth, non-amyloid.

Distribution: H.P.: Kullu.
Collection examined: SSR 5701.
Substratum: On stump of *Pinus excelsa.*
Remarks: This species is very close to *Poria placenta* but differs in having slightly curved or allantoid basidiospores and lacking pink colour on the pore surface. *Poria monticola* is North American in distribution. It has already been recorded by Bagchee and Bakshi (1951).

Skeletocutis Kotl. & Pouz.,
Česka Mykol. 12(2): 103, 1958.

Fructification annual, resupinate to effused-reflexed, the reflexed portion forming a small pileus, soft when fresh, brittle on drying; upper surface white, minutely tomentose. Pore surface white initially turning light orange on maturity; pores small round to irregular; tubes in one layer, concolorous with pore surface. Context white, duplex, upper part soft, lower gelatinous when fresh, hard on drying, non-xanthochroic. Hyphal system dimitic; generative hyphae hyaline, thin to thick-walled, septate, clamps present; skeletal hyphae hyaline, thick-walled, restricted to upper loose part of the context. Cystidia absent. Basidia clavate, 4-spored. Basidiospores hyaline, cylindrical to allantoids, smooth, thin-walled, non-amyloid.
Thirty species, widespread
Lit.: David (*Naturaliste* can. 109: 235, 1982), Roberts & Ryvarden (*Kew Bull.* 61: 55, 2006; Cameroon)
Type Species: *Polyporus amorphous* Fr. 1815
Habitat: Decayed wood
Himalaya: Two
Key to species

1. Hymenial surface pink to pinkish red; upper layer soft and cottony; pores angular, 5–7 per mm .... *S. amorpha*
1. Hymenial surface white to straw coloured; pores circular to angular, very small, 6–9 per mm .... *S. nivea*

Skeletocutis amorpha (Fr.) Kotl. & Pouz., Česka Mykol. 12(2): 103, 1958. Plate 6.31c, Fig. 6.84r–v

Fructification annual, resupinate, effused-reflexed with narrow pilei, solitary or imbricate with pilei up to 4.2 cm long, soft when fresh, separable and brittle when dried. Pileus up to 8.2 mm wide and 3 mm thick, soft; upper surface white, fine tomentose, azonate; margin entire, acute, concolorous with the upper surface, incurved when dried. Pore surface white when fresh, pale yellow on drying, uneven; pores angular, 70–220 μm in diameter, 5–7 per mm; pore mouth finely velutinate; dissepiments 75–125 μm thick; tubes in one layer, white, up to 2 mm deep in section, dark on drying. Context white, of two distinct layers: upper one whitish, very soft and cottony; lower one brown, waxy, dense gelatinous layer, becoming hard on drying, about 1.5 mm thick.

Hyphal system dimitic; generative hyphae thin-to thick-walled, septate, hyaline, clamped, 3–5.2 µm wide, strongly agglutinated in gelatinous layer and also in the trama, some typically encrusted at the ends of dissepiments; skeletal hyphae hyaline, thick-walled to subsolid, 3–5.1 µm wide, cyanophilous, occurring in uppermost layer of the pileus, and a few also in the gelatinous layer. Hyphal pegs present, projecting into pore cavity. Cystidia absent. Basidia 4-sterigmate, 9.8–14.2 × 3.1–4.1 µm. Basidiospores hyaline, thin-walled, allantoid, smooth, 4.2–5.2 × 1.5–1.8 µm. Cystidioles hyaline, fusiform or mucronate.

Distribution: Bhutan- Chimakothi, H.P.: Kullu, U.K.: Dehra Dun.

Collection examined: IBP 37404, 37405.

Substratum: On decaying logs, stumps of *Abies pindrow, Cedrus deodara, Picea smithiana, Pinus excelsa* etc.

Remarks: The fungus resembles *Gloeoporus dichrous* in that both have reddish and waxy hymenophore and tomentose upper surface; but *S. amorpha* is dimitic with encrustations on hyphae, while *G. dichrous* is monomitic lacking hyphal encrustations. It is of common occurrence on bark of stumps or logs of *Abies pindrow, Cedrus deodara, Pices smithiana, Pinus roxburghii, Pinus excelsa.*

Skeletocutis nivea (Jungh.) Keller, Persoonia 10(3): 353, 1979 = *Polyporus semipileatus* Peck., Rep. New York Sta. Mus. 34: 43, 1883 = *Polyporus niveus* (Jungh.) Cooke, Grevillea 14(71): 87, 1886. Fig. 6.85a–g

Fructification annual or occasionally biennial, usually resupinate to effused-reflexed, pileus often sessile, also may be imbricate, soft when fresh, hard on drying; upper surface of pileus whitish or ochraceous, finely tomentose or glabrous, azonate; margin thin. Context thin, whitish to straw coloured; hymenial surface white to straw yellow, pores circular to angular, very small, 6–9 per mm, pore tubes up to 2 mm long, in one or two layers.

Hyphal system dimitic; generative hyphae hyaline, thin- to thick-walled, clamped, smooth, many with characteristic encrustations which are very common in the pore mouths and dissepiment edges, frequently with short, thin- to thick-walled coralloid aseptate branches, more abundant in the trama, 1.5–4.1 µm wide; skeletal hyphae hyaline, thick-walled to solid, 2.5–5.5 µm wide. Basidia clavate, 7.5–11.3 × 3.8–4.5 µm. Basidiospores hyaline, thin-walled, smooth, very small, allantoid, 3–4 × 0.6–0.8 µm. Cystidioles abundant.

Distribution: U.K.: Chakrata.

Collection examined: IBP 42257.

Substratum: On decaying *Quercus dilata* log.

Remarks: This species can be recognised by very tiny pores and discoloured pore surface in older fructifications extremely tiny spores are diagnostic feature of the species.

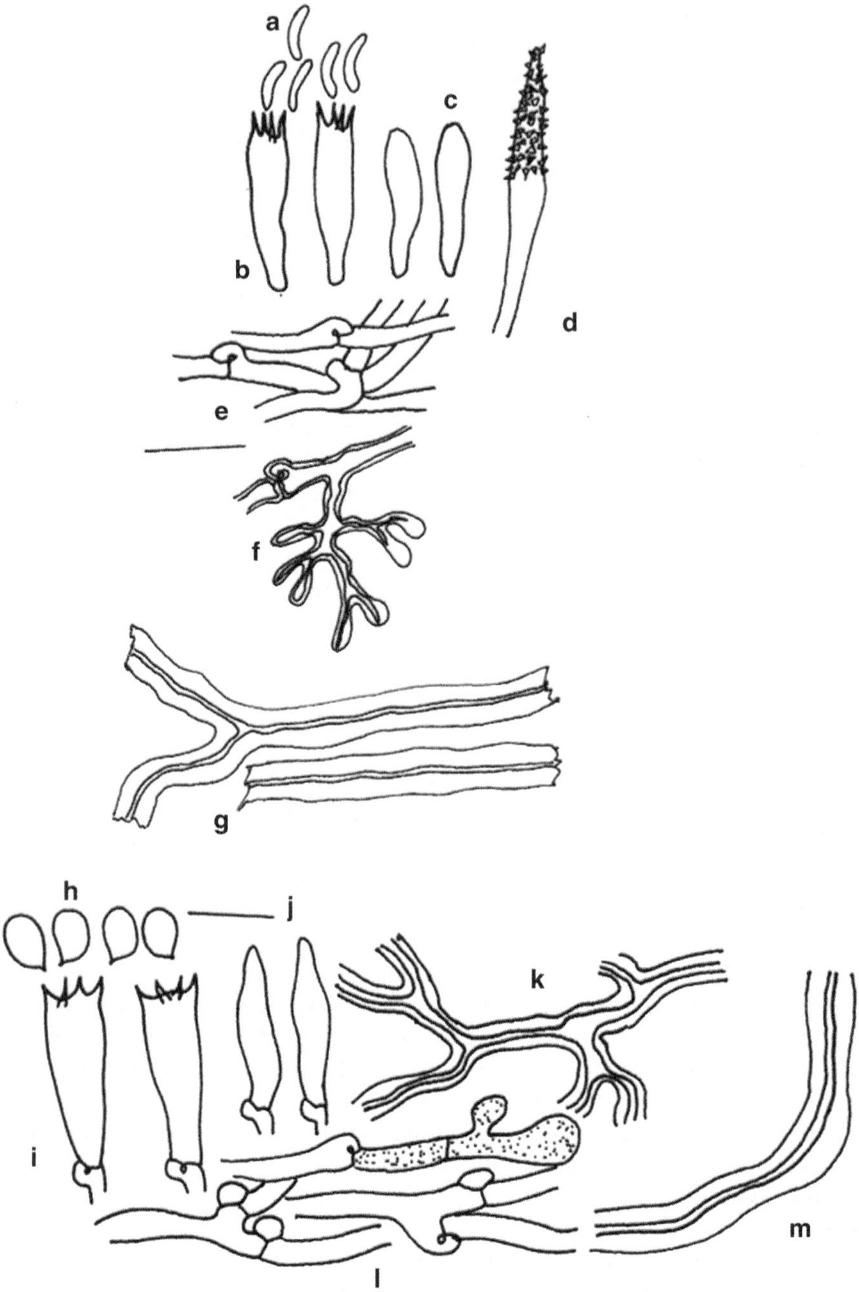

Fig. 6.85 (**a**–**g**) *Skeletocutis nivea* (**a**) Basidiospores, (**b**) Basidia, (**c**) Basidioles, (**d**) Encrusted cystidia, (**e**) Generative hyphae, (**f**) Hyphal pegs, (**g**) Skeletal hyphae; (**h**–**m**) *Trametes cingulata* (**h**) Basidiospores, (**i**) Basidia, (**j**) Basidioles, (**k**) Binding hyphae, (**l**) Generative hyphae, (**m**) Skeletal hyphae

Trametes Fr.,
Fl. Scan.: 339, 1836.

Fructification annual, pileate, solitary or imbricate, coriaceous, corky and tough when fresh hard on drying; upper surface smooth to hirsute, concentrically zonate or azonate, white, cream to brown, reddish brown, blue, greyish brown, violet grey. Pileus broadly attached to dimidiate. Pore surface white creamish, pale brown to dark brown; pores round to angular, sometimes radially elongated; tubes in one layer, cream or pale brown in section. Context white, cream to brown, tough, hard, non-xanthochroic. Hyphal system trimitic; generative hyphae hyaline, septate, clamped, branched; skeletal hyphae hyaline or light yellowish, thick-walled to solid with narrow lumen, aseptate, unbranched; binding hyphae thick-walled, branched, aseptate. Cystidia and setae absent. Basidia clavate, hyaline, 4-spored. Basidiospores cylindric to ellipsoid, hyaline, thin-walled, non-amyloid.

Fifty Species, widespread

Lit.: Haddow (*Trans. Brit. Mycol. Soc.* 22: 182, 1938), Zhang et al. (*Mcosystema* 25: 23, 2006)

Type Species: *Boletus suaveolens* L., 1753

Habitat: Dead Wood

Himalayas: Fourteen

Key to species

1. Pilear surface hirsute to tomentose 2
1. Pilear surface glabrous to velvety 12
2. Pilear surface tomentose 3
2. Pilear surface hirsute/hispid/villose 4
3. Fructification sessile or when short stipitate, attached Fto the substratum by a disc .... *T. incerta*
3. Fructification pileate/sessile, broadly attached to the substratum, basal absent .... *T. palisoti*
4. Pores 1–2 per mm 5
4. Pores 4–6 per mm 8
5. Fructification up to 25 mm thick, with an anise odour when fresh .... *T. suaveolens*[a]
5. Fructification up to 10 mm thick, without any odour 6
6. Context with black zone or black line below the upper loose tomentum; pores daedaleoid with age .... *T. maxima*[a]
6. Context without any black line; pores regular or semidaedaeloid 7
7. Pilear surface coarsely hispid and rough hair .... *T. trogii*[a]
7. Pilear surface hirsute to strigose 9
8. Pores becoming semidaedaleoid to dentate .... *T. cervina*[a]
8. Pores radially elongated; pore surface creamish to brownish .... *T. tephroleuca*[a]
9. Context with a black line below the tomentum 10
9. Context homogenous or duplex without dark line below the tomententum 11

10. Fructification thin, flexible, multizonate, tomentose to glabrous zones; pore surface white to cream .... *T. versicolor*
10. Fructification hirsute to tomentose, not multizonate; pore surface white to greyish .... *T. hirsute*
11. Fructification and pore surface greyish; pilear surface azonate; pores 5–6 per mm .... *T.radiato-rugosus*[a]
11. Fructification and pore surface cream to straw coloured, zonate, pores 3–4 per mm .... *T. pubescences*
12. Pores 1–2 per mm 13
12. Pores 4–8 per mm 15
13. Pores radially elongated, mostly on *Quercus* .... *T. gibbosa*
13. Pores not radially elongated, on various host 14
14. Pileus whitish grey with reddish black patches n ear the base, surface smooth .... *T. menziezii*
14. Pileus whitish to pale brown without any black patches, azonate, warts near the base .... *T. lactinea*
15. Pilear surface glabrous, smooth with greyish black areas; basidiospores broadly ellipsoid, 4–5.5 × 3–3.5 μm .... *T. cingulata*
15. Pilear surface without or with reddish cuticle spreading from the base; basidiospores cylindrical 16
16. Pilear surface azonate, white cream with red cuticle spreading from the base; pores 6–8 per mm .... *T. cubensis*
16. Pilear surface azonate to strongly multizonate, without any reddish cuticle; pores 3–5 per mm 17
17. Fructification very thin, papery 18
17. Fructification not pappery, thin 19
18. Basidiospores 7–10 μm long; pores 4–5 per mm .... *T. cotonea*
18. Basidiospores 4–6 μm long; pores 5–6 per mm .... *T. membranacea*
19. Pilear surface dull, multizonate; pores 3–4 per mm; hard on drying .... *T. ochracea*
19. Pilear surface semi glossy, azonate to zonate, rigid on drying .... *T. marianna*[a]

[a]Extra limital, not included in the text

Trametes cingulata Berk., Hooker's J. Bot. Kew Gard. Misc. 6: 164 (1854). Fig. 6.85h–m

Fructification sessile, reflexed or umbilicate, sometimes substipitate, corky; upper surface white to ochraceous, becoming blackish when old, finely matted tomentose, becoming glabrous later, usually smooth, sometimes rough or wrinkled, strongly concentrically zoned; hymenium surface white, becoming light brown when dry, with a silky sheen particularly when fresh, pores regular, circular to slightly angular, 3–5 per mm, pore wall thin; margin thin, pore tubes white, ochraceous when dry, slightly oblique, forming nearly an uniform layer, up to 3 mm long. Context white when fresh, ochraceous on drying, soft corky, indistinctly transversely zoned, up to 3 mm broad.

Hyphal system trimitic; generative hyaline, thin-walled, rarely branched, septate with occasional clamps, 1.4–2.4 μm broad; skeletal hyphae hyaline, thick-walled, with lumen small or absent, aseptate, unbranched, 3–5.2 μm broad; binding hyaline, thick-walled, branched, aseptate, 1.6–3.4 μm broad. Basidia clavate, 4–5 μm broad. Basidiospores hyaline, subglobose to short ellipsoid, apiculate, 3–5.2 × 3–3.4 μm.

Distribution: U.K.: Nainital.

Collection examined: IBP 42259.

Substratum: On *Cocos nucifera, Bambusa* sp., Angiospermic twigs.

Remarks: The fungus is characterized by its thin small substipitate fruit bodies, concentrically zoned surface, which becomes blackish at length.

Trametes cotonea (Pat. & Har.) Ryv., Norw. Jl. Bot. 19(3–4): 236, 1972 = *Polyporus cotoneus* Pat.& Har., Bull. Soc. mycol. Fr. 9:208, 1893. Fig. 6.86a–e

Fructification annual, mostly effused-reflexed with elongated pilei, sessile, often imbricate, flexible; upper surface white when fresh, pale ochraceous on drying, minutely pubescent to glabrous, soft; margin thin, wavy; hymenial surface cream to light vinaceous, sometimes with a greyish tint, pores circular to angular, about 3–4 per mm, pore tubes up to 2 mm long. Context white to pale ochraceous.

Hyphal system trimitic; generative hyphae hyaline, thin-walled, clamped, rarely slightly thick-walled, 1.3–3.4 μm wide, scarcely found. Basidiospores hyaline, thin-walled, cylindric to ellipsoid, sometimes shortly curved, 6.8–12.2 × 2.6–3.4 μm; hyphal pegs in hymenium and crystals in section; hyphae (1) hyaline, thick-walled, flexous, rarely branched, 2.1–4.4 μm diameter, common, (2) hyaline, slightly thick to thick-walled, profusely branched, 2.1–3.0 μm diameter and (3) thin to slightly thick-walled, rarely septate with clamp connections, less common, sometimes broken at nodes.

Distribution: U.K.: Chakrata, Nainital.

Collection examined: IBP 42703

Substratum: On decaying *Cedrus deodara* decaying log.

Remarks: The fructification is characterized by its thin, soft to touch, almost glabrous short pileus, hymenial surface white to vinaceous with sterile margin and long cylindrical-ellipsoid basidiospores.

Trametes cubensis (Mont.) Sacc., Syll. fung. (Abellini) 9: 198 (1891) = *Daedalea cubensis* (Mont.) Roy Can. J. Bot. 60: 1012, 1982 = *Polyporus cubensis* Mont., Ann. Sci. Nat., Bot. II, 8: 364, 1837 = *Fomitopsis cubensis* (Mont.) Wright & Deschamps, Rev. Invest. Agropec. Ser. 5, 12: 140, 1975. Fig. 6.86f–k

Fructification sessile, resupinate to reflexed, applanate, flexible when fresh, rigid on drying, single or imbricate; upper surface white, soon turning light brownish, becoming reddish to usually black at the base forming a distinct area, soft matted tomentose, azonate, nearly smooth, margin thin, entire. Context white to pale ochraceous, usually azonate, corky hymenium surface white when fresh, pale to light brown when dry, mostly even, sometimes with small raised areas at base, pores regular, round, 4–6 per mm, pore wall thick; margin thin, pore tubes ochraceous, slightly unequal, up to 5 mm long.

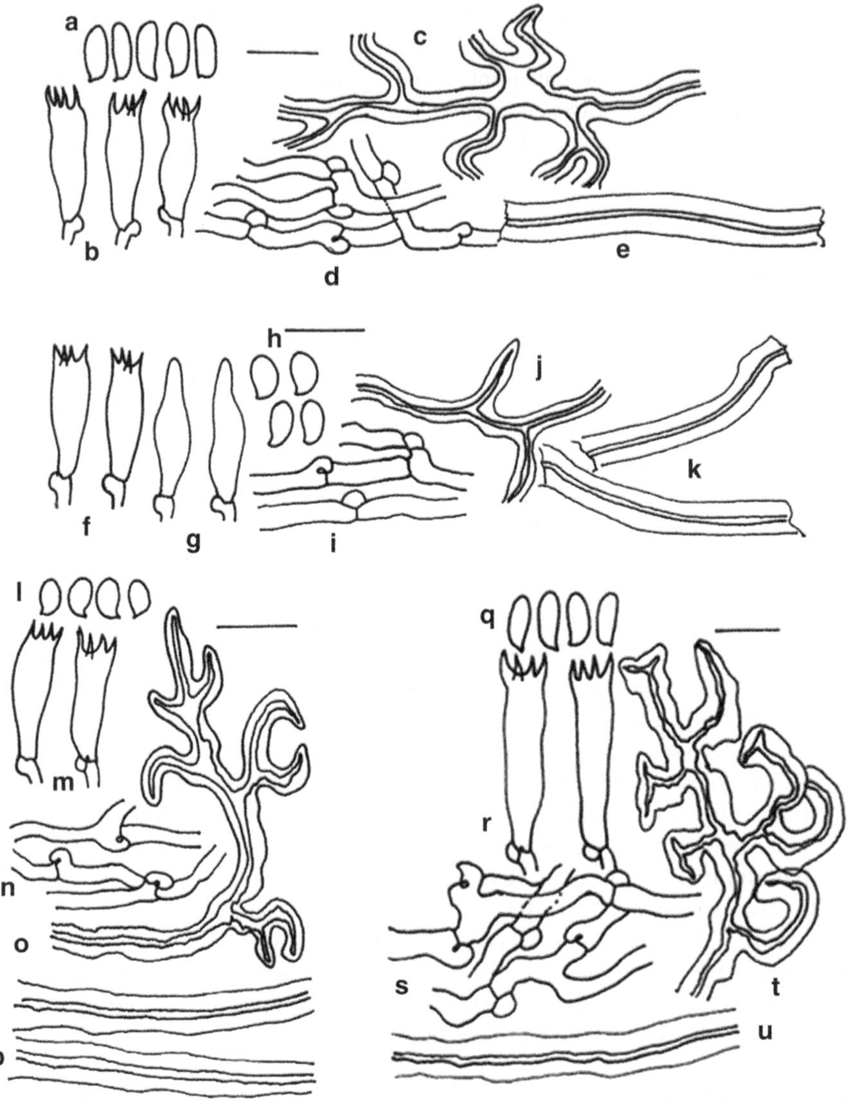

Fig. 6.86 (**a**–**e**) *Trametes cotonea* (**a**) Basidiospores, (**b**) Basidia, (**c**) Binding hyphae, (**d**) Generative hyphae, (**e**) Skeletal hyphae; (**f**–**k**) *Trametes cubensis* (**f**) Basidia, (**g**) Basidioles, (**h**) Basidiospores (**i**) Generative hyphae, (**j**) Binding hyphae, (**k**) Skeletal hyphae; (**l**–**p**) *Trametes gibbosa* (**l**) Basidiospores, (**m**) Basidia, (**n**) Generative hyphae, (**o**) Binding hyphae, (**p**) Skeletal hyphae; (**q**–**u**) *Trametes hirsuta* (**q**) Basidiospores, (**r**) Basidia, (**s**) Generative hyphae, (**t**) Binding hyphae, (**u**) Skeletal hyphae

Hyphal system dimitic; generative hyphae hyaline, thin-walled, branched, septate with clamps, 1.4–2.9 µm broad; skeletal hyphae hyaline, thick-walled, with lumen narrow or nearly absent, straight or flexous, aseptate, unbranched, 2.4–3.9 µm broad, common. Basidia clavate, 3–4.2 µm broad. Basidiospores hyaline, thin-walled, short cylindric to ellipsoid, 2.4–3.5 × 1.5–1.8 µm, sharp pointed structures, referred to as cystidia by Overholts (1953), 3–5 µm broad, present in hymenium.

Distribution: U.K.: Dehra Dun.

Collection examined: IBP 42268.

Substratum: On stump of *Shorea robusta, Eucalyptus maculata.*

Remarks: This species is characterized (particularly in old specimens) by black surface at the base, thin applanate fructifications and associated brown cuboidal decay.

Trametes gibbosa (Pers.) Fr., Epicr. Syst. Mycol. (Upsaliae): 492 (1838) [1836–1838]. Plate 6.31d, Fig. 6.86l–p

Fructification annual, sessile, applanate, imbricate; upper surface cream when fresh, tomentose, concentrically zonate, radially wrinkled on drying; margin sharp, even entire. Context white tough; hymenial surface creamish-white, pores angular, 2–3 per mm.

Hyphal system trimitic; generative hyphae hyaline, thin-walled, aseptate branched, clamped, acyanophilous, 2–2.5 µm in diameter; skeletal hyphae hyaline, thick-walled, unbranched, aseptate 3.7–4.1 µm in diameter; binding hyphae hyaline, thick-walled, aseptate branched 2.3–3.1 µm in diameter. Basidia, clavate, 11.1–12.7 × 6.5–4.6 µm. Basidiospores hyaline thin-walled, smooth, ellipsoid, slightly curved 3.7–5.1 × 2.2–3.6 µm.

Distribution: A.P.: West Kameng- Bhalukpong; U.K.: Chakrata, Chamoli; H.P.: Kullu, Solan, Sundernagar.

Collection examined: IBP 37409.

Substratum: On decaying angiospermic stem, wood.

Remarks: The species is characterized by annual, coriaceous, conchate fructification; white to cream upper surface; irregularly elongated pores which become sinuous towards the base; white context; trimitic hyphal system and hyaline, ellipsoid and slightly curved spores.

Trametes hirsuta (Wulfen) Lloyd, Mycol. Writ. 7: 1319 (1924) = *Coriolus hirsutus* (Wulfen) Pat. 1897 = *Coriolus velutinus* P. Karst., Trudy Troitsk. Otd. imp. russk. geogr. obsc. 8: 61 (1906). Plate 6.31e, Fig. 6.86q–u

Fructification annual, pileate, sessile, attached by broad lateral base, coriaceous when fresh, hard on drying, solitary to imbricate with resette like fructification with a contracted point of attachment from which pilei spreads in circular manner. Pileus sessile, applanate, dimidiate, 2–7.5 × 1.5–4.5 × 0.25–0.75 cm; upper surface white to light brown initially becoming, brown to greyish brown with age, densely hirsute, azonate to concentrically zonate, tomentose hyphae up to 3 mm long; margin lighter concolorous with upper surface, acute, incurved on drying entire to somewhat wavy, sterile below. Pore surface white to cream, darken on drying, even, dull; pores angular, large, 400–970 μm in diameter, 1–2 per mm; dissepiments thick, 80–320 μm thick, tubes in one layer, cream, up to 4 mm deep in section; pore mouth velutinate. Context white to cream, coriaceous, homogenous.

Hyphal system trimitic; generative hyphae hyaline, thin-walled, branched, clamps present, septate, cyanophilous, 2.2–3.4 μm in diameter; skeletal hyphae subhyaline, thick-walled to solid with narrow lumen, aseptate, unbranched, straight, long, acyanophilous, 3.2–6.4 μm in diameter; binding hyphae thick-walled, much branched, acyanophilous, 2.7–4.4 μm in diameter. Basidia hyaline, clavate, 4-spored, up to 5.5 μm in diameter. Basidiospores hyaline, thin-walled, smooth, cylindric-ellipsoid, curved on one side, non-amyloid, 6.0–7.2 × 1.9–3.0 μm.

Distribution: Bhutan- Chimakothi; Meghalaya; Tripura- Agartala; H.P.: Kullu, Pulga, Manali- Gulaba, Gojra, Jagatsukh; Chamba- Kilorgala- Dalhousie- Khajjiar, Banikhet, Lakkarmandi; Shimla- Baghi, the retreat, Narkanda, Khadrala, Rampur road; Dharamsala- Yol.

Collection examined: SSV 21253, RW 6055; R 6145, 6429, IBP 37410, 37411.

Substratum: On decaying angiospermic twigs.

Remarks: The species is characterized by brown to greyish brown, annual, coriaceous fructification with densely hirsute upper; white to cream pore surface, large pores (1–2 per mm); trimitic hyphal system and hyaline spores. It is very close to *T. velutinus* (Pers.: Fr.) Quel. but the latter differs in having smaller (3–4 per mm) pores, fine tomentose upper surface; smaller ellipsoid (4.5–5.6 × 1.5–2.1 μm) basidiospores. The species has been previously recorded from India as *Polyporus hirsutus* Wulf.: Fr. by Mitter and Tandon (1932) from Nainital (U.P.) and Bakshi from Dehra Dun (U.K.).

Trametes incerta (Curt.) Cke., Grevillea 15: 56, 1886 = *Coriolopsis sprucei* (Berk.) Roy & Mitra, Mycotaxon 26: 445, 1986 = *Daedalea sprucei* Berk., Hook. J. Bot., 8: 236, 1856. Plate 6.31f, Fig. 6.87a–f

Fructification perennial, sessile, pileate, effused reflexed, broadly attached, solitary, coriaceous when fresh, hard and woody on drying. Pileus flat to convex, applanate, semicircular to dimidiate, up to 8 cm long, 5 cm wide and 4 cm thick; upper surface light brown with pinkish tint, finely tomentose, becoming smoky brown in old specimens, concentrically zonate, uneven, radially wrinkled, irregularly cracking up on drying; margin thin, blunt, light concolorous with upper surface, acute, entire, fertile below. Pore surface pink to light orange, brown, uneven; pores angular to daedaleoid, large, radially elongated 5–9 per mm; tubes in one layer, up to 3.8 cm in section. Context light brown with pink tint, homogenous, xanthochroic.

Hyphal system trimitic; generative hyphae hyaline, thin-walled, septate, branched, clamped, cyanophilous, thin-walled 2.2–3.2 μm in diameter; skeletal hyphae subhyaline to light brown, thick-walled to solid, unbranched, aseptate, acyanophilous, 3.0–5.0 μm in diameter; binding hyphae subhyaline, thick-walled, extensively branched, acyanophilous, 2.2–2.6 μm in diameter. Basidiospores, hyaline, thin-walled, smooth, non-amyloid, ellipsoid to subglobose, 4.2–5.0 × 2.0–3.0 μm.

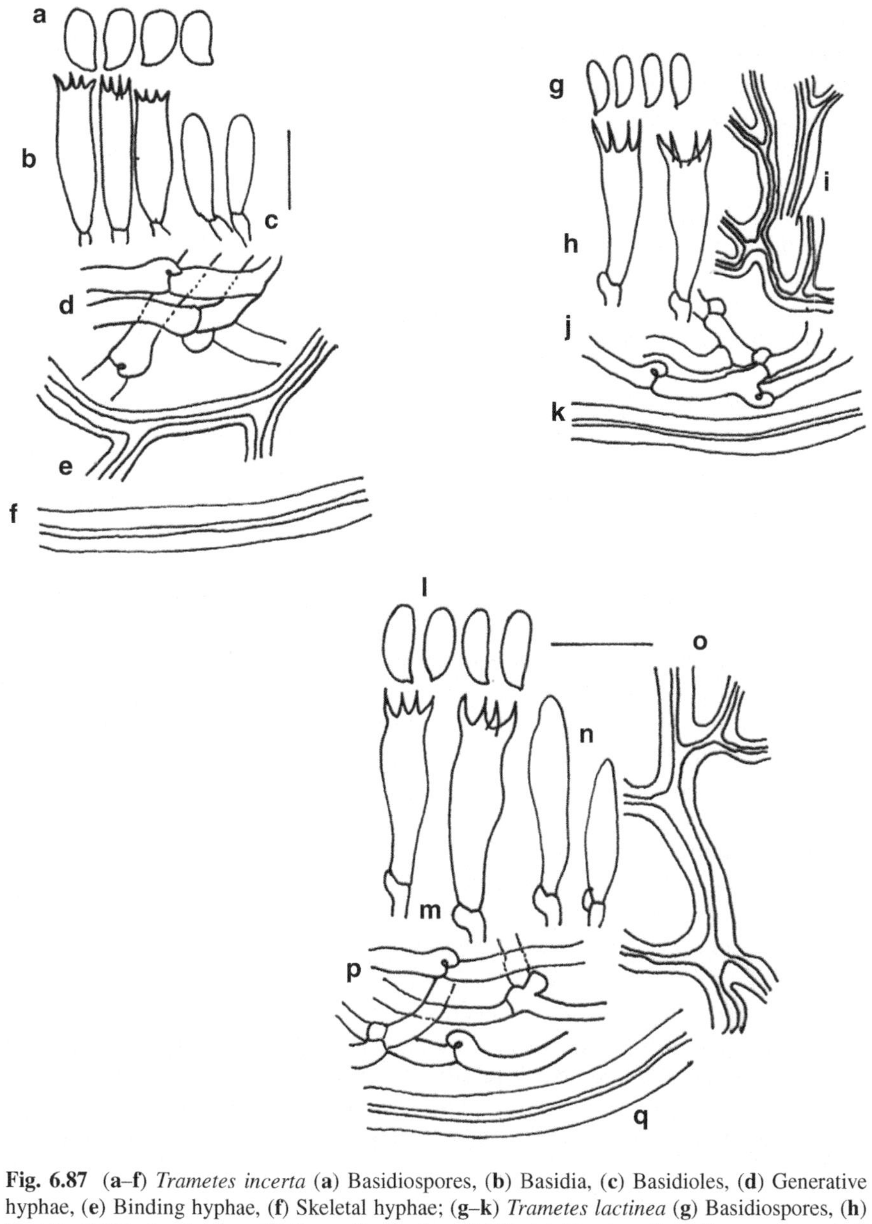

Fig. 6.87 (**a**–**f**) *Trametes incerta* (**a**) Basidiospores, (**b**) Basidia, (**c**) Basidioles, (**d**) Generative hyphae, (**e**) Binding hyphae, (**f**) Skeletal hyphae; (**g**–**k**) *Trametes lactinea* (**g**) Basidiospores, (**h**) Basidia, (**i**) Binding hyphae, (**j**) Generative hyphae, (**k**) Skeletal hyphae; (**l**–**q**) *Trametes menziesii* (**l**) Basidiospores, (**m**) Basidia, (**n**) Basidioles, (**o**) Binding hyphae, (**p**) Generative hyphae, (**q**) Skeletal hyphae

Distribution: U.K.: Nainital; Meghalaya: Mawphlang.
Collection examined: SSV 21405, 21772.
Substratum: On living trees of *Cassia fistula, Ficus hisipida, Lagerstromia parviflora, Shorea robusta, Terminalia tomentosa* and *Pinus roxburghii.*
Remarks: The species was described by Currey in (1874) as *Polyporus incerta* Currey based on a collection made by Sulpiz Kurz from Burma. From India, it was recorded from Nainital N.W. Himalaya. The above cited collection closely resembles with the description given by Bakshi (1971), Dhanda (1977) and Johansen (1980). It is a new record for Eastern Himalayas.

Trametes lactinea (Berk.) Sacc., Syll. fung. 6: 343 (1888). Plate 6.32a, Fig. 6.87g–k
Fructification annual, sessile, applanate, solitary to imbricate, mostly dimidiate to semicircular, broadly to narrowly attached, corky when fresh, hard on drying, 5.5–12.0 × 3.0–5.0 cm; margin thin or thick; upper surface white to cream, azonate, smooth, matted tomentose, velvety when fresh, orchraceous to tan on drying, typically uneven, nodulose when mature; context white, corky, 1.0–1.5 cm thick; hymenial surface white to cream, on drying woody brown, pores circular, 2–3 per mm, pore tubes up to 7 mm long.
Hyphal system trimitic; generative hyphae hyaline, thin to slightly thick-walled, clamped, occasionally collapsed, 1.4–3.3 μm wide; skeletal hyphae hyaline, mostly straight, thick-walled to solid, usually unbranched, rarely biforked apically, 2.4–4.4 μm wide; binding hyphae hyaline, slightly tortuous, branched, septate, branches long or short, mostly subsolid to solid, few thick-walled with narrow lumina, 1.5–3.2 μm wide. Basidia hyaline, 4 sterigmate, 12–17.9 × 6–7 μm. Basidiospores hyaline, cylindric-ellipsoid, 5.1–8.1 × 2.4–3.4 μm.
Distribution: U.K.: Kumaon; H.P.: Manali- Gulaba.
Collection examined: Dhanda 6617, IBP 37412, L 37413.
Substratum: On decaying *Shorea robusta, Pinus longifolia* log.
Remarks: The fungus is characterized by its thick, white fructifications which are azonate and velvety when young, becoming uneven and nodulose on drying.

Trametes menziesii (Berk.) Ryvarden, [as *'menzeisii'*], Norw. Jl Bot. 19(3–4): 236 (1972) = *Polyporus vittatus* Berk., *Lond. J. Bot.* 6: 505, 1847. Fig. 6.87l–q
Fructification annual to perennial, sessile, attached by broad lateral base, hard and corky when dry, light in weight. Pileus sessile, convex to subungulate; upper surface yellowish- brown to brown or brownish black, somewhat tomentose near the margin, later glabrous, rimose in the older parts; margin concolorous with upper surface, obtuse, entire. Pore surface yellowish brown to brown, even to uneven, dull; pore mostly rounded, occasionally irregular, with thick edges; occasionally breaking. Context yellowish brown, homogenous, darkening in KOH sol; tubes only indistinctly stratified when perennial, concolorous.
Hyphal system trimitic; generative hyphae hyaline to subhyaline, thin-walled, septate, clamped, branched, 2.6–3.4 μm in diameter; skeletal hyphae subhyaline to pale brown, thick-walled, aseptate, branched, 3–4.3 μm in diameter. Cystidia absent. Basidia and basidiospores not observed.

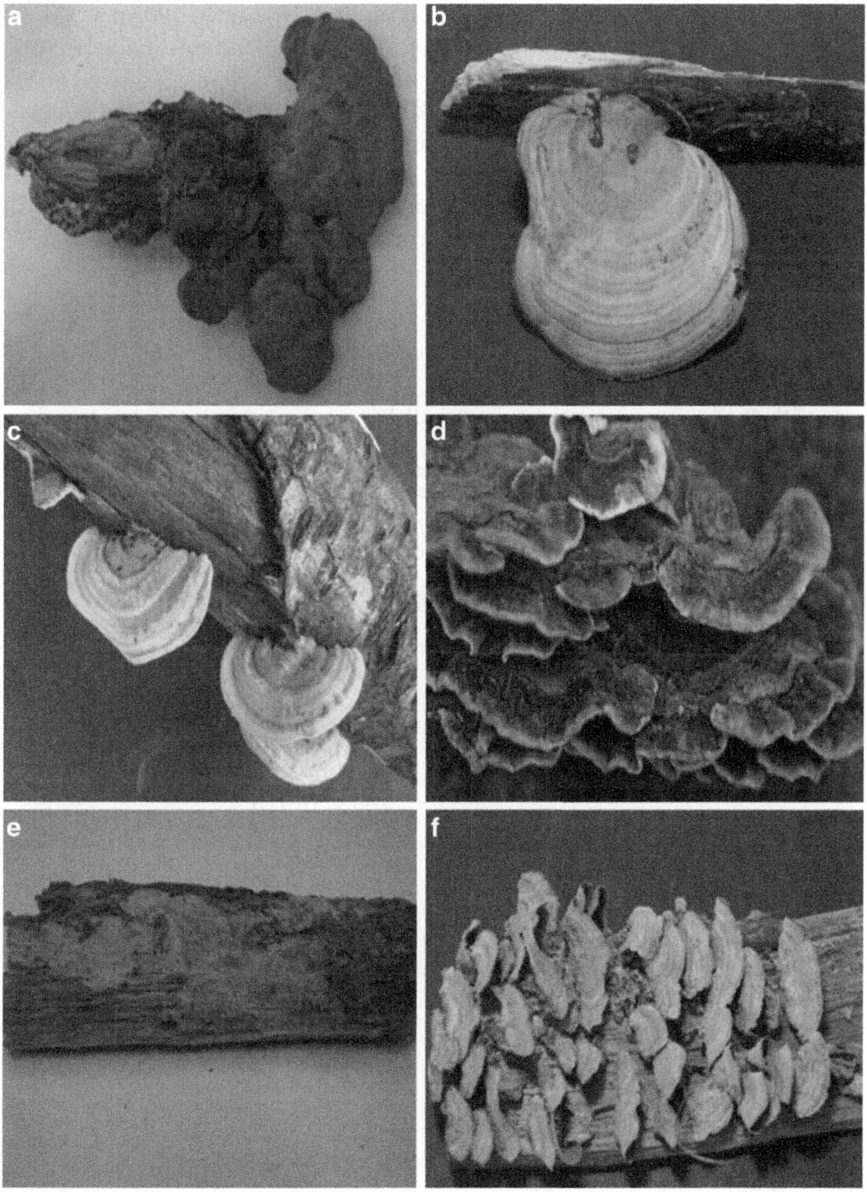

Plate 6.32 (**a**) *Trametes lactinea.* (**b**) *Trametes palisoti.* (**c**) *Trametes pubescens.* (**d**) T*rametes versicolor.* (**e**) *Trametopsis cervina.* (**f**) *Trichaptum abietnum*

Distribution: H.P.: Shimla- Khadrala.
Collection examined: Dhanda 6548, IBP 37413.
Substratum: stumps of *Quercus incana.*
Remarks: The species is characterized by annual to perennial fructification; trimitic hyphal system and absence of cystidia.

Trametes ochracea (Pers.) Gilb. & Ryvarden, N. Amer. Polyp., Vol. 2 Megasporoporia – Wrightoporia (Oslo): 752 (1987) = *Coriolus zonatus* (Nees) Quél., Enchir. fung. (Paris): 175 (1886). Fig. 6.88a–f

Fructification annual, pileate, sessile, broadly attached, sometimes spread in circular fashion to become rosette, coriaceous when fresh, hard on drying. Pileus applanate to flabelliform; upper surface yellowish brown, tomentose, concentrically zonate; margin entire, lobed, incurved on drying. Pore surface yellowish brown. Context white to cream, homogenous, azonate.

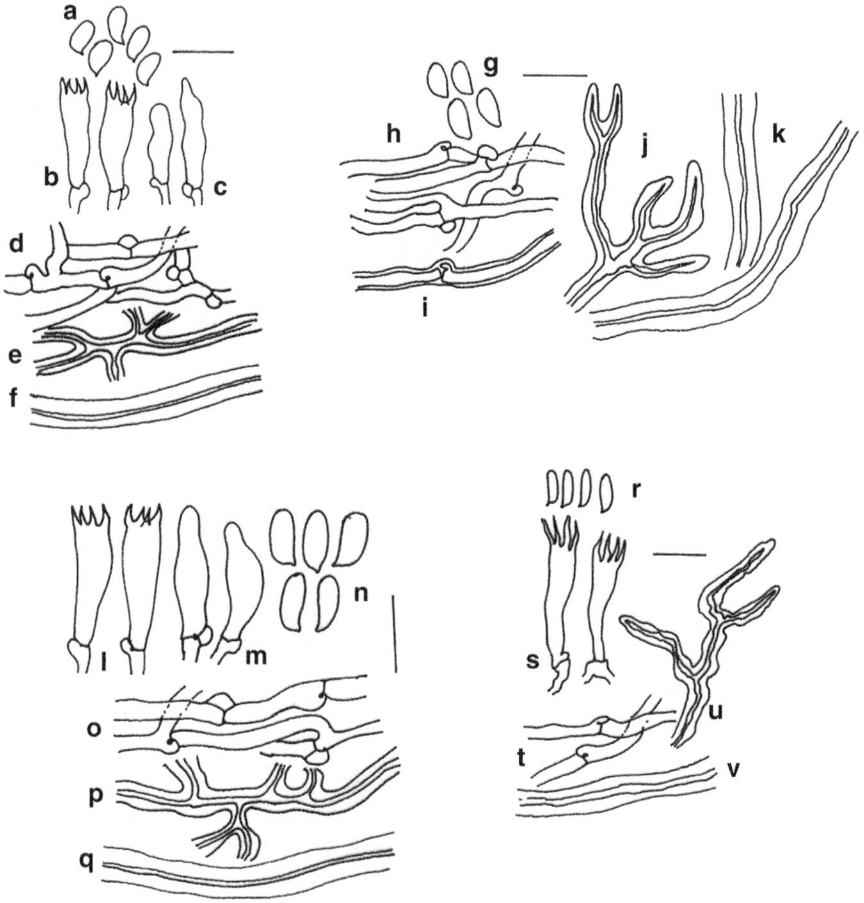

Fig. 6.88 (**a–f**) *Trametes ochracea* (**a**) Basidiospores, (**b**) Basidia, (**c**) Basidioles, (**d**) Generative hyphae, (**e**) Binding hyphae, (**f**) Skeletal hyphae; (**g–k**) *Trametes palisoti* (**g**) Basidiospores, (**h**) Thin-walled generative hyphae, (**i**) Thick-walled generative hyphae, (**j**) Binding hyphae, (**k**) Skeletal hyphae ; (**l–q**) *Trametes pubescens* (**l**) Basidia, (**m**) Basidioles, (**n**) Basidiospores, (**o**) Generative hyphae, (**p**) Binding hyphae, (**q**) Skeletal hyphae; (**r–v**) *Trametes versicolor* (**r**) Basidiospores, (**s**) Basidia, (**t**) Generative hyphae, (**u**) Binding hyphae, (**v**) Skeletal hyphae

Hyphal system trimitic; generative hyphae hyaline, thin-walled, septate, branched, clamped, acyanophilous, 2–4.2 µm in diameter; skeletal hyphae subhyaline, thick-walled, aseptate, long, acyanophilous, 3–4.2 µm in diameter. Cystidia absent. Basidia hyaline, clavate, 4-spored, 7–9.2 × 2.3–3.4 µm. Basidiospores hyaline, thin-walled, smooth, cylindrical, non-amyloid, 6.8–8.4 × 2.2–3 µm

Distribution: H.P.: Kinnaur- Kalpa, Tacklech, Kullu- Pulga Manali- Gojra, Jagatsukh, Gulaba; Chamba- sara; Shimla- Narkanda, Baghi; Dalhousie- Khajjiar.

Collection examined: RW 6022, 6056, 6070, 6071, 6081, 6089; SSR 6139; D 6410, 6477.

Substratum: logs, stumps, bark, rotten stumps, stumps under mixed forest, dead logs and stumps of *Quercus incana.*

Remarks: The species was firstly recorded from India as *Polystictus zonatus* (Nees.) Fr. by Theissen (1911) and Bose (1946) from Darjeeling and as *Polyporus zonatus* (Fr.) by Bakshi (1971). The above cited collections resembles with the descriptions as given by Dhanda (1977).

Trametes palisoti (Fr.) Imazeki, Bull. Gov. Forest Exp. Stn Tokyo 57: 120 (1952) = *Lenzites palisoti* (Fr.) Fr., Epicr. syst. mycol.: 404 (1838) [1836–1838] = *Lenzites palisoti* (Fr.) Fr., Epicr. syst. mycol.: 404 (1838) [1836–1838]. Plate 6.32b, Fig. 6.88g–k

Fructification annual, sessile or with a short stipe-like base, dimidiate, solitary, corky and flexible when fresh, more rigid when dry; margin thin, entire, acute; upper white to cream, finely tomentose when young, soon glabrous and smooth or with faint concentrically sulcate zones. Context white to cream, corky; stipe when present solid, attached to the substrate with a disc, white to cream; hymenial surface white to cream, variable, poroid to daedaloid to lamellate, pores 3 per mm.

Hyphal system trimitic; generative hyphae hyaline, thin-walled, branched, clamped, 1.6–2.3 µm wide, occasionally thick-walled to almost solid; skeletal hyphae hyaline, usually branched, few apically with arboriform type of branches, thick-walled to solid, 3.8 µm wide; binding hyphae hyaline, with short branches, branches mostly candelabra shaped, 2–4.5 µm wide. Basidia 4-sterigmate, 16.7–22.0 × 4.5–6.0 µm. Basidiospores hyaline, thin-walled, smooth, cylindrical to oblong-ellipsoid, 4.7–6.8 × 2.5–2.5 µm.

Distribution: Assam: Khasi Hills; U.K.: Dehra Dun, Mussoorie; W.B: Shantiniketan.

Collection examined: IBP 37281, 42962.

Substratum: On decaying angiospermic, *Shorea robusta* and *Quercus sp.* Log.

Remarks: The species is characterized by fructification having annual, sessile or with a short stipe-like base; hyphal system trimitic; basidia 4-sterigmate; smooth, cylindrical to oblong-ellipsoid basidiospores.

Trametes pubescens (Schmach.) Pilat, in Kavinia Pilat, Atlas Champ. I' Europe (Praha) 3: 268(1939) = *Trametes velutina* (Pers. ex Fr.) Cunn. N.Z.Dept. Sci. Industr. Res. Bull. 64: 173, 1965. Plate 6.32c, Fig. 6.88l–q

Fructification annual, sessile, pileate, solitary, soft and fleshy when fresh, light weight after drying; pileus applanate to dimidiate, 13 cm long, 8.5 cm broad and 5.25 mm thick; upper surface creamish white to tan coloured, even, azonate, finely pubescent when young, becomes glabrous with age; margin abrupt, even, concolorous with upper surface. Pore surface creamish white when fresh, darken on drying, uneven; pores angular to irregular, 3–4 per mm, 225–450 μm in diameter; pore tubes concolorous; dissepiments 90–280 μm wide. Context homogenous, non-xanthochroic, cream coloured, up to 10 mm thick.

Hyphal system trimitic; generative hyphae thin-walled, hyaline, branched, septate, clamped, cyanophilous, 2–3.4 μm wide; skeletal hyphae subhyaline, thick-walled to almost solid with narrow lumen, aseptate, straight to floccose, acyanophilous, 3.3–7.6 μm in diameter; binding hyphae hyaline, thick-walled to almost solid with narrow lumen, branched, aseptate, cyanophilous 2–3.3 μm in diameter. Cystidia absent. Basidiospores hyaline, thin-walled, smooth, cylindrical, non-amyloid, 5.0–7 × 1.7–2.6 μm.

Distribution: W.B.: Darjeeling, Meghalaya.

Collection examined: SSV 21036.

Substratum: On decaying angiospermic twigs.

Remarks: The species is characterized by annual, soft, fleshy, applanate to dimidiate, light weight fructifications; white and pubescent upper surface; white pore surface; large, 3–4 per mm, angular to irregular pores; context dense in the lower part; trimitic hyphal system with much branched binding hyphae; and hyaline smooth cylindrical, non-amyloid basidiospores.

Trametes versicolor (L.) Lloyd, Mycol. Notes (Cincinnati) 65: 1045 (1921) [1920] = *Coriolus versicolor* (L.) Quél., Enchir. fung. (Paris): 175 (1886). Plate 6.32d, Fig. 6.88r–v

Fructification annual, pileate, sessile, broadly attached by lateral base, coriaceous when fresh, hard on drying; Pileus dimidiate, applanate to flabelliform, adjacent pilei confluent forming larger pilei, solitary or imbricate, up to 5 cm long, 4.2 cm broad and 4 mm thick; upper surface tomentose, concentrically zonate, multicoloured, pale brown to dark brown, greyish blue, dark grey olive, glaborous zones alternating with tomentum; tomentum withering off in the form of regular concentric rings exposing shiny cortex below; margin acute, lighter concolrous with upper surface, wavy, sterile below, up to 1 mm wide. Pore surface white to cream when fresh, becoming dark on drying, even to uneven, dull; pores circular to angular, entire, 3–5 per mm, 130–330 μm in diameter; dissepiments equal, 45–150 μm thick; tubes in one layer, cream, up to 4 mm deep in section. Context white, homogenous, azonate, non-xanthochroic up to 2 mm thick, distinctly delimited on the upper surface by the dark zone formed due to thick-walled, parallel arranged skeletal hyphae.

Hyphal system trimitic; generative hyphae hyaline, thin-walled branched, clamped acyanophilous 2.1–3.8 μm in diameter; skeletal hyphae subhyaline thick-walled with narrow lumen, aseptate, branched, acyanophilous 3.8–4.7 μm in diameter; binding hyphae hyaline thick-walled, branched, acyanophilous 2.8–4.3 μm in diameter. Cystidia absent. Basidia clavate hyaline, 8.5–10 × 3.1–3.6 μm. Basidiospores thin-walled, smooth hyaline cylindrical, slightly curved on one side, 4.0–4.6 × 1.5–3.1 μm.

Distribution: Bhutan: Chimakothi; Nepal: Kathmandu; Manipur: Gnumba; Mizorum: Kolasib; West Bengal: Darjeeling; A.P; H.P.: Shimla- The Glen, Jakhu, Narkanda; U.K.: Nainital.

Collection examined: SSV 21423, SSV 21451, RW 6003, 6004, IBP 37414, 42967.

Substratum: On decaying angiospermic twigs.

Remarks: *Trametes versicolor* is highly variable species with respect to colour combinations on the upper surface. It is close to *T. azureus*, but the latter differs in the presence of grey zones and larger basidiospores. It is distributed throughout the Himalayas.

Trametopsis Tomsovský,
Czech Mycol. 60(1):7, 2008.

Fructification annual, resupinate, effused, soft when fresh, brittle on drying; margin uneven, white, thinning; upper surface light brown, glabrous, azonate; pore surface white to creamish, brown on drying; pores angular. Context white to cream, thin, soft, homogenous, non-xanthochroic. Hyphal system dimitic; generative hyphae thin to thick-walled, branched, septate, clamps present; skeletal hyphae hyaline, straight, unbranched to sometimes branched, thick-walled, aseptate, acyanophilous. Cystidia and Cystidioles absent. Basidia hyaline, thin-walled, clavate, 4-spored, cyanophilous. Basidiospores broadly ellipsoid, hyaline, thin-walled, smooth, apiculate, non-amyloid.

One species, widespread

Type Species: *Trametopsis cervina* (Schwein.) Tomsovský 2008

Habitat: Wood

Himalayas: One

Trametopsis cervina (Schwein.) Tomšovský, Czech Mycol. 60(1): 8 (2008) = *Coriolus cervinus* (Schwein.) Bondartsev 1953 = *Antrodia cervina* (Schwein.) Kotl. & Pouzar, Ceska Mykol. 37(1): 50 (1983). Plate 6.32e, Fig. 6.89a–e

Fructification annual, resupinate, effused, up to 25 cm long, soft when fresh, brittle on drying; margin uneven, white, thinning; upper surface light brown, glabrous, azonate. Pore surface white to creamish, brown on drying; pores angular to elongated splits on drying, 3–5 per mm. Context white to cream, thin, soft, homogenous, non-xanthochroic, up to 0.2 mm thick.

Hyphal system dimitic; generative hyphae thin to thick-walled, branched, 2.5–6.3 μm, septate, clamps present, cyanophilous; skeletal hyphae hyaline, straight, unbranched to sometimes branched, thick-walled, aseptate, acyanophilous, 3.3–5.2 μm in diameter; skeletal hyphae dominate the dissepiments and the context. Cystidia and Cystidioles absent. Basidia hyaline, thin-walled, clavate, 4-spored, cyanophilous, 11.8–13.9 × 4.7–5.2 μm. Basidiospores broadly ellipsoid, hyaline, thin-walled, smooth, apiculate, non-amyloid, 4–5.8 × 2.8–3.5 μm.

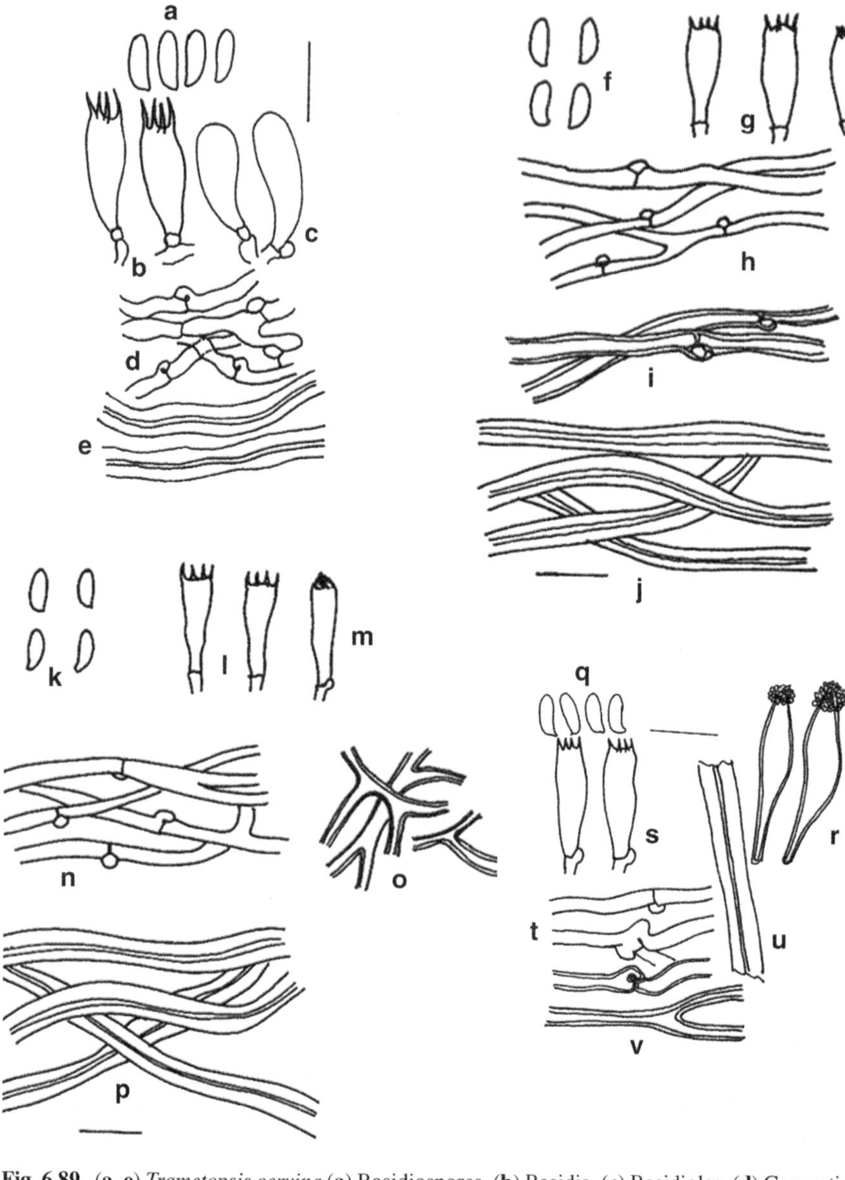

Fig. 6.89 (**a**–**e**) *Trametopsis cervina* (**a**) Basidiospores, (**b**) Basidia, (**c**) Basidioles, (**d**) Generative hyphae, (**e**) Skeletal hyphae; (**f**–**j**) *Trichaptum abietinum* (**f**) Basidiospores, (**g**) Basidia, (**h**) Thin-walled generative hyphae, (**i**) Thick-walled generative hyphae, (**j**) Skeletal hyphae; (**k**–**p**) *Trichaptum biforme* (**k**) Basidiospores, (**l**) Basidia, (**m**) Cystidia, (**n**) Generative hyphae, (**o**) Binding hyphae, (**p**) Skeletal hyphae; (**q**–**v**) *Trichaptum byssogenum* (**q**) Basidiospores, (**r**) Cystidia, (**s**) Basidia, (**t**) Generative hyphae, (**u**) Skeletal hyphae, (**v**) Binding hyphae

Distribution: H.P.: Shimla- Narkanda; Dalhousie- Lover's walk, Baloon, Dharamsala- Forsyth ganj.
Collection examined: IBP 37415, 42969, 42971.
Substratum: On decaying angiospermic log.
Remarks: This species is characterized by annual, resupinate, adnate fructification; white pore surface; white margin; angular pores; thick context; dimitic hyphal system; ellipsoid, smooth basidiospores.

Trichaptum Murrill,
Bull. Torrey bot. Club 31(11): 608, 1904

Fructification annual, sessile or effused-reflexed, thin; upper surface hispid to tomentose, dirty white, grey, greyish brown, concentrically zonate. Pore surface light violet, poroid, irpicoid or lamellate; pores medium sized, tubes in one layer. Context thin, homogenous, pale grey to light brown. Hyphal system dimitic or trimitic; generative hyphae hyaline, thin-walled, branched, septate, clamped; skeletal hyphae subhyaline, thick-walled with narrow lumen, straight; binding hyphae subhyaline, closely branched aseptate. Cystidia abundant, subhyaline, thin to thick-walled, subulate, apically encrusted. Basidia hyaline, clavate, 4-spored. Basidiospores hyaline, thin-walled, smooth, cylindric to cylindric-ellipsoid, non-amyloid.
Twenty species, widespread
Lit.: Macrae (*Can. J. Bot.* **45:** 1371, 1966), Hattori (*Mycoscience* **46:** 303, 2005)
Habitat: Dead wood
Type Species: *Polyporus trichomallus* Berk. & Mont. 1849
Himalaya: Four
Key to species

1. Hyphal system trimitic .... *T. biforme*
1. Hyphal system dimitic 2
2. Pores round to irregular 3
2. Pores irpicoid to poroid .... *T. byssogenum*
3. Cystidia 13–24 × 3.4–7 µm .... *T. abietinum*
3. Cystidia 21–35 × 5.2–8 µm .... *T. fusco-violaceum*

Trichaptum abietinum (Dicks) Ryv., Norw. J. Bot. 19: 237, 1972. Plate 6.32f, Fig. 6.89f–j

Fructification annual, resupinate to effused-reflexed, coriaceous when fresh, tough on drying attached by narrow lateral base. Pileus sessile, effused-reflexed, dimidiate to flabelliform; upper surface white to grey, tomentose to velutinate, azonate to faintly zonate; margin acute, concolorous with upper surface, entire, wide to narrow. Pore surface violet to violaceous brown when fresh, violet colour disappear on drying, uneven, dull; pores round to irregular near the margin. Context duplex, upper layer loose, white.

Hyphal system dimitic; generative hyphae hyaline, thin-walled, branched, clamps present, septate, acyanophilous, 2.1–3.4 µm in diameter; skeletal hyphae hyaline to subhyaline, thick-walled with narrow lumen, unbranched, aseptate, acyanophilous, 2.8–5.0 μm in diameter. Cystidia abundant, subhyaline, thin to thick-walled, clavate, encrusted with irregular crystals, 13.8–24.1 × 3.4–7.0 μm. Basidia hyaline, clavate, 4-spored, up to 5 μm in diameter. Basidiospores hyaline, thin-walled, smooth, cylindric-ellipsoid, 5.4–7.3 × 2.0–3.0 μm, non-amyloid.

Distribution: U.K.: Kumaun; Bhutan: Chukha Dam.

Collection examined: SSV 21270, IBP 37416, L 37417.

Substratum: On decaying angiospermic stump.

Remarks: The species is characterized by effused-reflexed to pileate fructifications, faintly concentrically zonate upper surface which is white to grey velutinate, violaceous brown pore surface; duplex context; clavate to subulate, apically encrusted cystidia and cylindric-ellipsoid basidiospores. It was first recorded from India by Bose (1934) as *Polystictus abietinus* (Dicks.) Fr. based on collection from Shillong (Meghalya). Dhanda (1977) recorded the species from several localities from N.W. Himalaya. The collection resembles with the description given by Thind and Dhanda (1977).

Trichaptum biforme (Fr.) Ryv., Norw. Jl. Bot. 19(3–4): 237, 1972 = *Trichaptum pargamenum* (Fr.) G.H.Cunn., Bull. N. Z. Dep. Scient. Ind. Res. 164:100, 1965. Plate 6.33a, Fig. 6.89k–p

Fructification annual, sessile attached by a narrow base, solitary or imbricate, dimidiate to flabelliform, petaloid, coriaceous, flexible; pileus surface finely tomentose to glabrous; pale bull to greyish, silky, finely zonate near the margin; margin acute, thin. Context pale bull, tough-fibrous; hymenial surface violet when fresh, later light brown; pores circular to angular, 4–5 per mm, pore tubes thin, edges often dentate or irpicoid.

Hyphal system trimitic; generative hyphae clamped, hyaline, thin-walled, branched, septate, 2.1–3.7 μm wide; skeletal hyphae subhyaline, thick-walled to solid, rarely branched, aseptate, acyanophilous, 2.7–5.3 μm wide; binding hyphae rare, subhyaline, thick-walled, much branched, 2–3.4 μm. Cystidia abundant, embedded or slightly projecting beyond the hymenium, hyaline to subhyaline, thin to thick-walled, capitates encrusted. Basidia clavate, 4-sterigmate, 11.9–22 × 3.8–5.2 μm. Basidiospores cylindric, hyaline, slightly curved, smooth, 5.1–6.7 × 2–2.4 μm.

Distribution: Bhutan: Chukha Dam; U.K.: Dehra Dun, Kumaun; H.P.: Shimla-Narkanda, Hattoo peak, The Glen, Dalhousie- Jandrighat spring, Manali- Gulaba.

Collection examined: SSV 21254, Dhanda 6421, 6515, SSR 6440, IBP 37418, 37419.

Substratum: On decaying angiospermic log.

Remarks: The species is characterized by violaceous hymenial surface, encrusted cystidia and cylindric basidiospores. It is close to *T. abietinum*, but *T. abietinum* grows on conifers, its pileus surface is less wide than the pileus of *T. biforme* which is often petaloid, has tomentose to smooth surface and grows on hardwoods.

Trichaptum byssogenum (Jungh.) Ryvarden, [as '*byssogenus*'], Norw. Jl Bot. 19(3–4): 237 (1972) = *Trichaptum venustum* (Berk.) G.H.Cunn., Bull. N. Z. Dep. Scient. Ind. Res. 164: 97, 1965. Fig. 6.89q–v

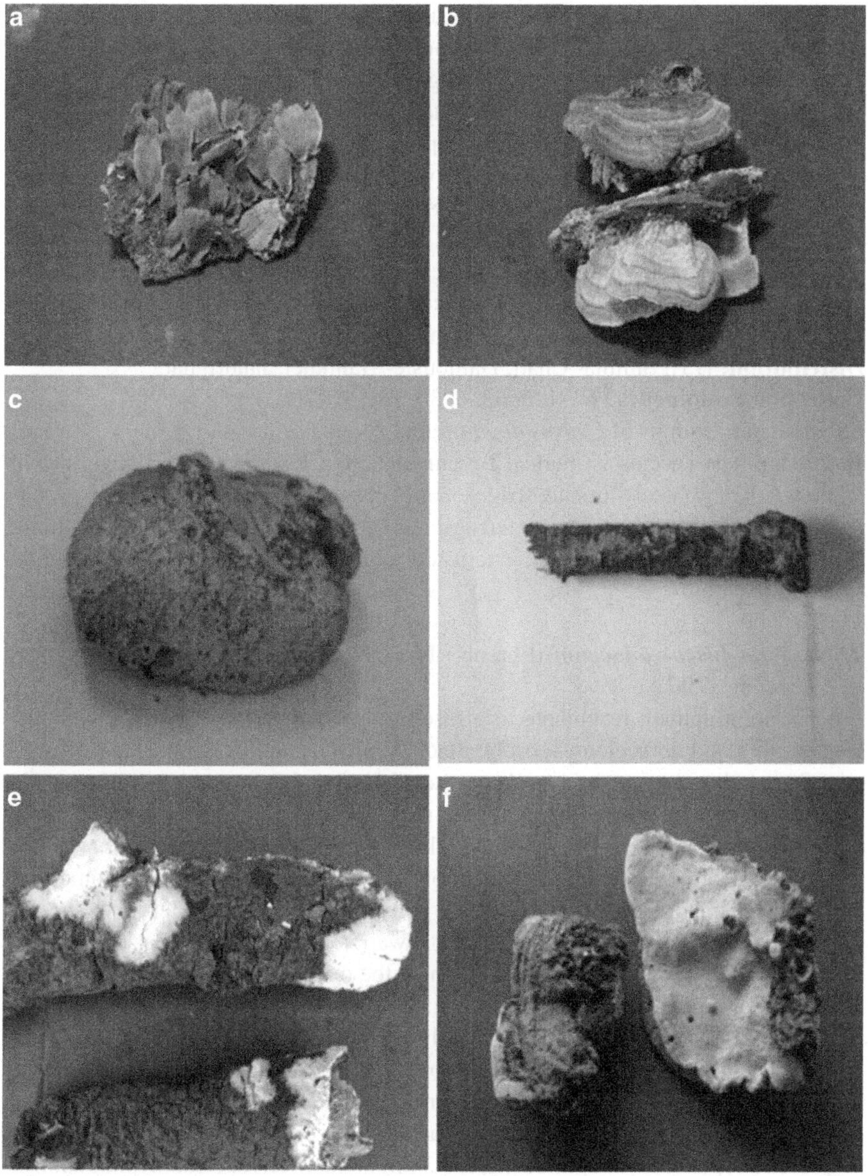

Plate 6.33 (**a**) *Trichaptum biforme.* (**b**) *Trichaptum fusco-violaceum.* (**c**) *Tyromyces lacteus.* (**d**) *Xenasma tulasnelloideu.* (**e**) *Xenasmatella vaga.* (**f**) *Heterobasidion annosum*

Fructification annual, mostly resupinate, sometimes resupinate, sometimes effused-reflexed, adnate; pileus effused-reflexed, drying hard and brittle; upper surface creamish to pale brown, matted tomentose, azonate; margin concolorous with upper surface, acute, entire, thinning in resupinate patches; pore surface creamish to greyish-orange, uneven, irpicoid, poroid near the margin; pores irpiciform, rounded to angular and shallow near margin, about 1–2 per mm. Context creamish, thin, non-xanthochroic, homogenous, up to 1 mm thick; tubes not stratose, irpiciform, tubes up to 5 mm long on sloping surface.

Hyphal system dimitic; generative hyphae hyaline, thin-walled, branched, septate, clamps absent, non-amyloid, acyanophilous, 3.6–3 µm in diameter; skeletal hyphae subhyaline, thick-walled to solid, branched, sometimes pseudoseptate, non-amyloid, acyanophilous, 3–5(−9.8) µm in diameter. Cystidia not abundant, subhyaline, thick-walled, incrusted all over, incrustation mild, dissolving in aq. KOH sol, cylindric-clavate, sometimes with pointed apices, 17–27.8 × 5.6–7.7 µm. Basidia hyaline, clavate, 9.8–14 × 4.2–4.9 µm. Basidiospores hyaline, thin-walled, smooth, non-amyloid, acyanophilous, ellipsoid, 4.2–5.6 × 2.8–3.1 µm.

Distribution: H.P.: Shimla- Chail, Dalhousie- Banikhet, Jandrighat.

Collection examined: D 6471, 6662, 6934, IBP 37420.

Substratum: Stumps of *Cedrus deodara* and *Quercus*.

Remarks: The species is marked by annual, often resupinate fructification with grey to bay, coarsely hirsute, strigose, only obscurely zonate upper surface; commonly irpicoid hymenophore, ferruginous or pallid purple pore surface; dimitic hyphal system; and mildly encrusted cystidia, hyaline, ellipsoid, smooth basidiospores.

Trichaptum fusco-violaceum (Ehrenb.) Ryv., Norw. J. Bot. 19: 237, 1972. Plate 6.33b, Fig. 6.90a–e

Fructification annual, resupinate to effused-reflexed, solitary to compactly imbricate, thin and coraceious, loosely attached with the substratum. Pileus effused-reflexed, dimidate, up to 4 cm long, 3 cm broad and 0.4 cm thick, adjacent pilei usually fused with each other; upper surface white to grey, tomentose, faintly zonate to azonate, convex, normally algae is present giving green tint and form light and dark zones; margin acute, concolorous with upper surface, sometimes brownish, incurved on drying, sterile below; pore surface violet to violaceous brown, decurrent, mostly irpiciform towards base, poroid near the margin, uneven, dull; pores round to irregular near the margin usually radially elongated, 2–3 per mm, hydnoid to irpicoid, dentate or toothed on maturity, light and dark zones. Context white to light creamish, non-xanthochroic, up to 1.5 mm thick, duplex; upper layer thin, soft and fibrous, lower layer thick and compact. Tubes in one layer, concolorous with pore surface 4 mm deep.

Hyphal system dimitic; generative hyphae hyaline, thin-walled, branched, clamps present, septate, acyanophilous, 2.4–4.3 µm in diameter; skeletal hyphae subhyaline thick-walled, unbranched, aseptate, acyanophilous, 3.2–5.2 µm in diameter. Cystidia scattered, subhyaline, thick-walled, cylindric-clavate, incrusted with irregular crystals, (Crystals dissolve in KOH sol) 21–35 × 5.2–8 µm. Basidia hyaline, clavate, 4-spored, up to 5.8 µm in diameter. Basidiospores hyaline, thin-walled, smooth, cylindric-ellipsoid, 5.0–7.2 × 2.4–3.0 µm, non-amyloid.

Distribution: H.P.: Dharamshala, Manali; Bhutan: Chukha Dam; U. K.: Raipur forests-Dehradun.

Collection examined: SSV 21273, IBP 37420.

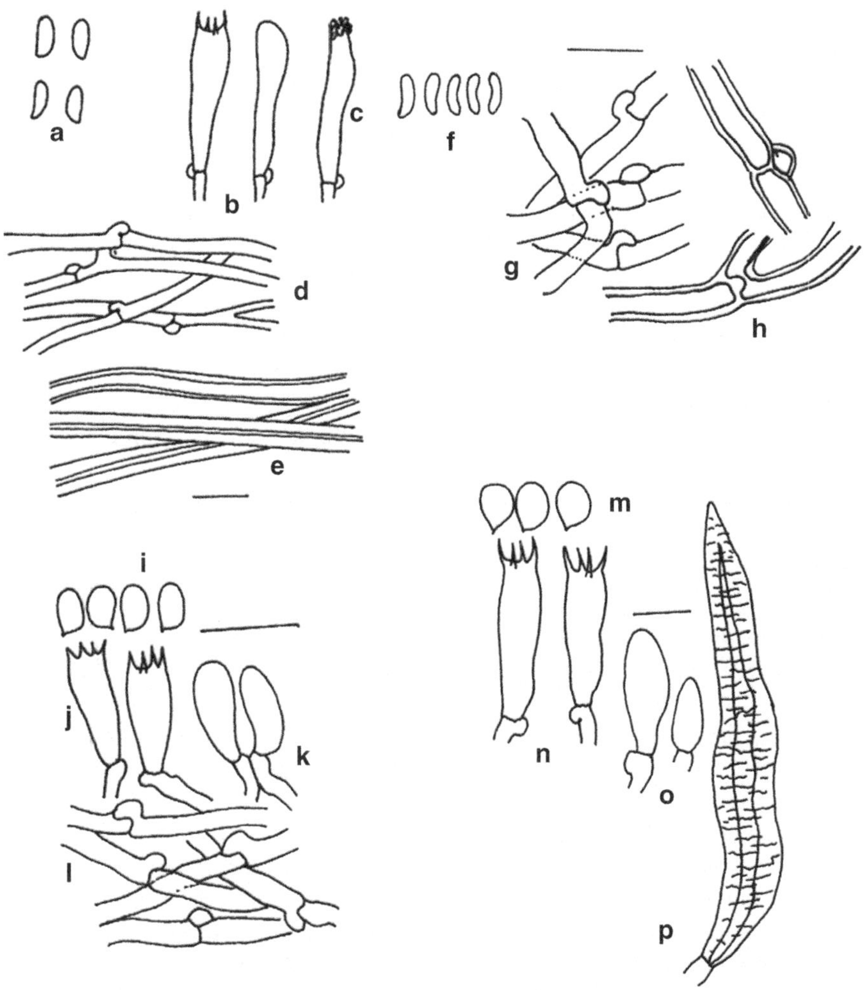

Fig. 6.90 (**a–e**) *Trichaptum fusco-violaceum* (**a**) Basidiospores, (**b**) Basidia, (**c**) Cystidia, (**d**) Generative hyphae, (**e**) Skeletal hyphae; (**f–h**) *Tyromyces lacteus* (**f**) Basidiospores, (**g**) Thin-walled generative hyphae, (**h**) Thick-walled generative hyphae; (**i–l**) *Tyromyces tephrus* (**i**) Basidiospores, (**j**) Basidia, (**k**) Basidioles, (**l**) Generative hyphae; (**m–p**) *Xenasma subclematidis* (**m**) Basidiospores, (**n**) Basidia, (**o**) Basidioles, (**p**) Cystidia

Substratum: On decaying *Pinus* log.

Remarks: The species was first recorded by Thind & Dhanda (1978) from Dharamshala (H.P.). The above collected sample resembles with the description given by Thind & Dhanda (1978). The species is characterized by annual, effused-reflexed to sessile, thin coriaceous fructifications with violaceous brown, irpiciform pore surface; dimitic hyphal system, thick-walled apically incrusted cystidia and hyaline cylindric-ellipsoid basidiospores. The species is new record for Uttarakhand.

Tyromyces P. Karst.
Revue mycol. **3(9)**: 17, 1881.

Fructification annual, resupinate to pileate, soft, fleshy when fresh, separable, white, reddish orange or brownish rarely bluish. Pileus sessile, effused, effused-reflexed, dimidiate, applanate or conchate, flabelliform; upper surface tomentose to glabrous, smooth. Pore surface white to cream, rarely reddish or greyish brown, even or uneven; pores small to large, round to angular; pore tubes in one layer. Context white, cream to brown, soft, succulent when fresh, fragile after drying, homogenous, non-xanthochroic. Hyphal system monomitic; generative hyphae hyaline, thin to thick-walled with narrow lumen, branched, septate, clamped. Cystidia absent or present. Basidia clavate, hyaline, 2–4 spored. Basidiospores hyaline, thin-walled, smooth, allantoids to ellipsoid, non-amyloid.

Thirty species, widespread.

Lit.: David (_Bull. Soc. Linn. Lyon_ **49**: 596, 1980; key 16 Afr. Spp.), Ipulet & Ryvarden (_Syn. Fung._ **20**: 79, 2005; tropical Afr.)

Habitat: Dead wood.

Type species: _Polyporus chioneus_ Fr. 1815

Himalayas: Two

Key to species

1. Basidiospores cylindrical-allantoid, 4.3–5.5 × 1–1.6 μm .... _T. lacteus_
1. Basidiospores ellipsoid, 3.3–5.0 × 2–3.7 μm .... _T. tephrus_

Tyromyces lacteus (Fr.) Murrill, N. Amer. Fl. (New York) 9(1): 36 (1907) = _Postia lactea_ (Fr.) Karst., Revue. Mycol., 3(9): 17(1818) = _Polyporus lacteus_ Fr., Syst. Mycol. 1: 359, 1821. Plate 6.33c, Fig. 6.90f–h

Fructification annual, effused-reflexed, solitary, sessile, watery-fleshy when fresh, hard and brittle on drying; margin acute, concolorous with upper surface, incurved on drying; upper surface glabrous, uneven, azonate, irregularly wrinkled, white to creamish slightly darken on drying. Pore surface pale creamish, uneven, darker than upper surface; pores angular, small to medium sized, 6–8 per mm; pore tubes non-stratified, dissepiments 34–75 μm. Context white, non-xanthochroic.

Hyphal system monomitic; generative hyphae hyaline to subhyaline, thick-walled, branched, septate, clamped, acyanophilous, 3.0–5.6 μm in diameter, Hyphal pegs present projecting up in to the cavity. Cystidia absent. Cystidioles fusoid to subulate present abundantly. Basidia hyaline, thin-walled, clavate, cyanophilous, 8.5–12.5 μm long. Basidiospores hyaline, thin-walled, 3.0–4.9 × 1.0–1.5 μm, cylindric-clavate to allantoids, amyloid.

Distribution: Bhutan-Chimakothi; U.K.: Mussoorie; H.P.: Narkanda, Shimla, Dalhousie- Kalatope.

Collection examined: Dhanda 6919, SSV 21252, IBP 37422, L 37424.

Substratum: On decaying _Pinus_ log.

Remarks: The above cited collection compare well with the description of _T. lacteus_ given by Lowe and Lombard (1973) except the hyphal system which is mentioned to be usually dimitic by these authors.

Tyromyces tephrus (Pat.) Ryvarden, Occ. Pap. Farlow Herb. Crypt. Bot. 18: 35 (1983) = *Poria tephra* Pat., Bull. Soc. mycol. Fr. 11(4): 208 (1895). Fig. 6.90i–l

Fructification annual resupinate, small to effused, adnate, soft coriaceous to tough; margin light brown, narrow, fimbriate, up to 2.8 mm wide; pore surface light brown when fresh darken on drying; pores angular split and elongated pores on oblique surface, 4–5 per mm; dissepiments thin, 25–65 μm; pore tubes thin, concolorous with pore surface up to 1.2 mm deep in section. Context light brown, coriaceous, thin, homogeneous, non-xanthochroic.

Hyphal system monomitic; generative hyphae hyaline to subhyaline, septate, branched, thin-walled, clamped, 2.2–4.5 μm in diameter, hyphae 2–2.7 μm in diameter, densely intricate in dissepiments. Cystidia absent. Basidia clavate, thin-walled, hyaline, cyanophilous, up to 6.2 μm in diameter. Basidiospores hyaline, ellipsoid, 3.3–5.0 × 2.0–3.7 μm, thin-walled, smooth, non-amyloid.

Distribution: Meghalaya: 14 km from Shillong towards Barapani; H.P.: Solan, Manali.

Collection examined: SSV 21134, IBP 37423.

Substratum: On decaying angiospermic stem.

Remarks: The species is characterized by annual, resupinate fructification; pore surface light brown but darken on drying; hyphal system monomitic; cystidia absent; basidia clavate, hyaline; ellipsoid, smooth basidiospores.

Family-Xenasmataceae

Key to genera

1. Fructifications usually occurring as a fine thin film of hyphae, basidiospores smooth or finely echinulate .... *Xenasma*
1. Fructification membranous, waxy gelatinous, basidiospores smooth or rarely warted .... *Xenasmatella*

Xenasma Donk,
Fungus Wageningen **27**: 25, 1957.

Fructifications resupinate, usually occurring as a fine thin film of hyphae which is difficult to discern with unaided eye, often waxy-gelatinous; hymenial surface smooth, lighter coloured; margin thinning to indistinct. Context subhyaline or pale coloured, wholly agglutinated or ceraceous. Hyphal system monomitic, hyphae often collapsed, agglutinated and difficult to discern, with or without clamps. Cystidia or Gloeocystidia present or absent. Basidia clavate-cylindrical to suburniform, arising directly from the repent hyphae and appearing bifurcate at the base (pleurobasidiate) but some are evidently with a septa at the base (pedicellate), 4-spored. Basidiospores globose to ellipsoid or suballantoid, smooth or finely echinulate, amyloid but mostly non-amyloid.

Nine species, widespread

Type Species: *Xenasma rimicola* (Karst.) Donk 1896

Lit.: Hjortstam et al. (*Cortic. N. Europ.* 8, 1988; key European spp.), Liberta (*Mycol.* 52: 884, 1962; key)

Habitat: Dead wood.

Himalayas: Two

Key to species

1. Cystidia present..*X. subclematidis*
1. Cystidia absent...*X. tulasnelloideum*

Xenasma subclematidis S.S. Rattan, Biblthca Mycol 60: 233, 1977. Fig. 6.90m–p

Fructification resupinate, membranous-ceraceous to ceraceous, closely adnate, widely effused; hymenial surface white to cream, smooth but appears somewhat farinose under the lens, continuous or occasionally cracking irregularly on drying; margin thinning to more or less indistinct, adnate, concolorous. Context composed of compactly agglutinated hyphae which are difficult to discern.

Hyphal system monomitic, hyphae 1.6–3.1 μm wide, sparsely branched, septate, clamped and thin-walled. Cystidia 59.8–75.2 × 10.1–13 μm, conical to subulate, arising from the different part of the context and often projecting out of the hymenium, the walls thick, subhyaline and usually impregnated with fine crystals especially in the apical half. Basidia 18.1–25.1 × 6.9–8 μm, utriform, often with a basal septum (pedicellate) but sometimes pleurobasidiate, 4-spored, sterigmata strong, somewhat curved and up to 6.5 μm long. Basidiospores 5.6–7.6 × 4.6–5.6 μm, globose to ovoid, minutely apiculate, thin-walled, smooth, non-amyloid or weakly amyloid.

Distribution: J&K.: Gulmarg; H.P.: Kullu, Dalhousie.

Collection examined: SSR 5764, IBP 37425.

Substratum: On log under conifers.

Remarks: This species is characterized by the type of cystidia and basidiospores. Thin and membranous-ceraceous to wholly agglutinated fructifications and occasional occurrence of pleurobasidia suggest it's close affinity with *Xenasma* wherein it comes near *X. clematidis* (Bourd. & Galz.) Liberta. In the latter species, however, the basidiospores wall is finely echinulate and nonamyloid. Moreover, the cystidia also have a rooting base.

Xenasma tulasnelloideum (Höhn. & Litsch.) Donk, Fungus, Wageningen 27: 26 (1957) = *Xenasmatella tulasnelloidea* (Hoehn. & Litsch.) Oberw., Syd. Ann. Mycol. 19 p.34, 1965 = *Corticium tulasnelloideum* Hoehn. & Litsch., Sber. Akad. Wiss., Wien. Math.-Nat. Klasse 117: 1118, 1908. Plate 6.33d, Fig. 6.91a–c

Fructification resupinate, adnate, effused, very thin, waxy gelatinous; hymenial surface smooth, light grey to grey; margin not distinctly marked.

Hyphal system monomitic; generative hyphae 1.2–2.1 μm wide, branched, septate, clamped, thin-walled, often gelatinized and difficult to observe. Cystidia absent. Basidia 11.8–19.9 × 5.6–7.6 μm, subclavate to cylindrical, pleurobasidiate, 4-sterigmate, with a basal clamp. Basidiospores 4.4–7 × 3.4–4.4 μm, broadly ellipsoid or ovoid or subglobose, thin-walled, prominently warted, non-amyloid, acyanophilous.

Distribution: Arunachal Pradesh: West Kameng, Bomdila; H.P.: Kullu, Dalhousie; U.K.: Karanprayag- Chamoli

Collection examined: GSD 19713, IBP 37426.

Substratum: On decaying angiospermic twigs.

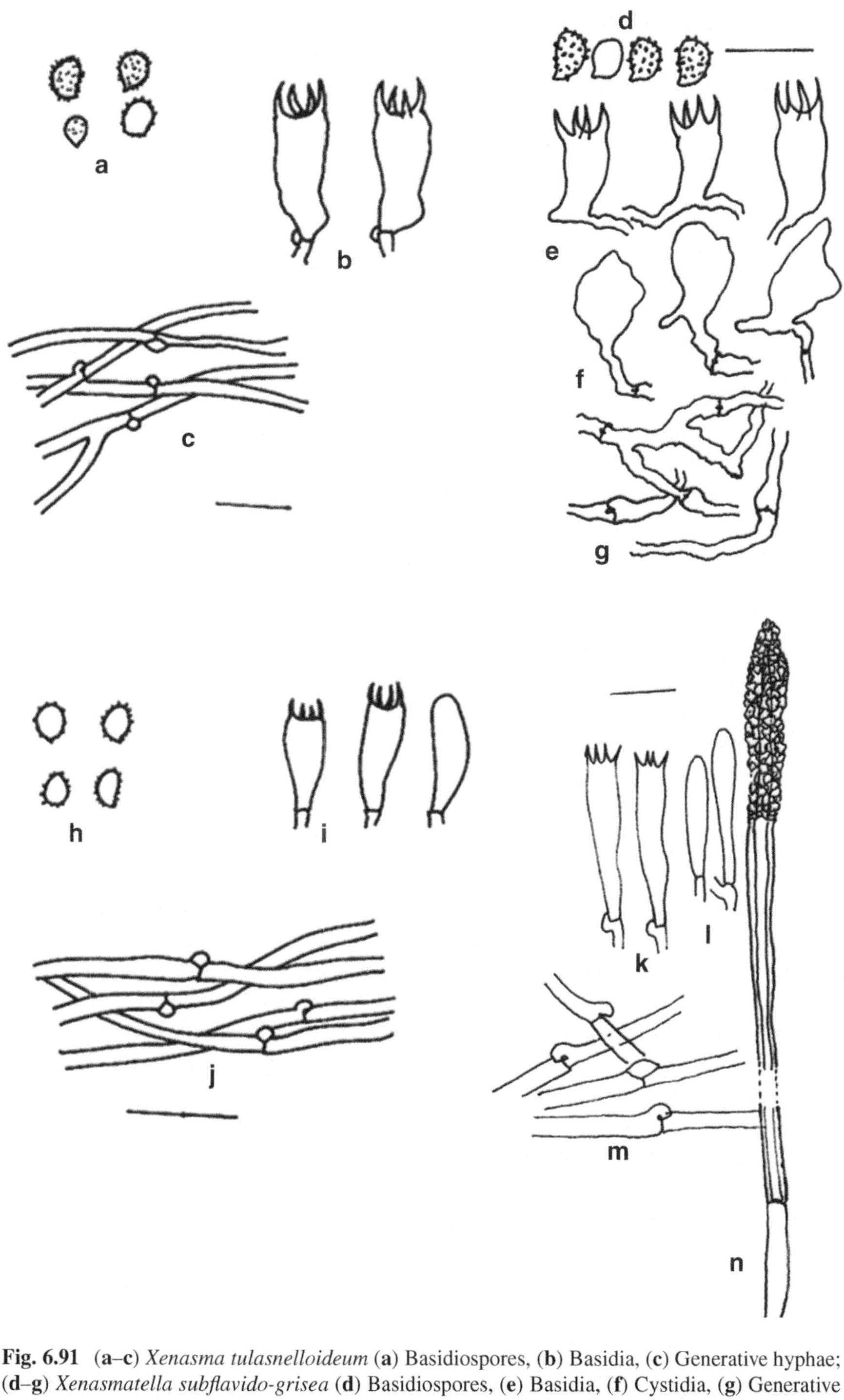

Fig. 6.91 (**a–c**) *Xenasma tulasnelloideum* (**a**) Basidiospores, (**b**) Basidia, (**c**) Generative hyphae; (**d–g**) *Xenasmatella subflavido-grisea* (**d**) Basidiospores, (**e**) Basidia, (**f**) Cystidia, (**g**) Generative hyphae; (**h–j**) *Xenasmatella vaga* (**h**) Basidiospores, (**i**) Basidia, (**j**) Generative hyphae; (**k–n**) *Amylostereum chailletii* (**k**) Basidia, (**l**) Basidioles, (**m**) Generative hyphae, (**n**) Encrusted cystidia

Remarks: This species is characterized by thin, waxy gelatinous fructification, clamped generative hyphae, subclavate to cylindrical, 4-spored, pleurobasidia and broadly ellipsoid to ovate or subglobose, prominently warted, non-amyloid basidiospores. The collection resembles the description of *Xenasmatella tulasnelloidea* as given by Oberwinkler (1965), Cunningham (1963) and Christiansen (1960) in almost all the characters. The species is new record for N.W. Himalayas.

Xenasmatella Oberw.,
Sydowia 19 (1–6):28, 1996.

Fructification resupinate, membranous, waxy gelatinous; hymenial surface smooth, generally light coloured; margin thinning. Hyphal system monomitic; generative hyphae branched, septate, clamped. Cystidia or gloeocystidia absent. Basidia clavate-cylindrical to suburniform, 4-sterigmate. Basidiospores globose, ellipsoid to suballantoid, smooth or rarely warted, thin-walled, amyloid, acyanophilous.

Fourteen species, widespread

Lit.: Hjortstam et al. (*Cortic. N. Euorp.* **8**, 1988; Sub *Phlebiella*)

Type Species: *Corticium subflavidogriseum* Litsch. 1941

Habitat: Wood

Himalayas: Two

Key to species

1. Fructification resupinate, effused, very-thin, membranous-ceraceous, hymenial surface greyish white to light grey .... *X. subflavido-grisea*
1. Fructification resupinate, floccose to pelliculose with numerous yellow rhizomorph over running the substratum; hymenial surface yellow to yellowish ochre .... *X. vaga*

Xenasmatella subflavidogrisea (Litsch.) Oberw. ex Jülich, Persoonia 10(3): 335 (1979) = *Corticium subflavidogriseum* Litsch., Ann. Myc. 39: 127, 1941. Fig. 6.91d–g

Fructification resupinate, effused, very thin, closely adnate, membranous-ceraceous; hymenial surface, greyish-white to light-grey; margin indistinct.

Hyphal system monomitic; generative hyphae branched, clamped, septate, thin-walled, 1.2–2.6 µm wide, often collapsed or agglutinated and difficult to observe. Basidia 8.1–16.1 × 4.4–6.2 µm, urniform to short-cylindrical, pleurobasidiate, 4-sterigmate. Basidiospores 4.1–6.1 × 3.1–3.4 µm, ellipsoid, thin-walled, warted, non-amyloid, acyanophilous.

Distribution: West Bengal: Darjeeling; H.P.: Kinnaur, Chamba-Khajjiar.

Collection examined: GSD 19245. IBP 37427.

Substratum: On a decaying angiospermic stump.

Remarks: The species is characterized by adnate fructification, clamped generative hyphae, 4-sterigmate pleurobasidia and warted, non-amyloid basidiospores. This collection resembles closely the type description of *X. subflavidogrisea*.

Xenasmatella vaga (Fr.) Stalpers, Stud. Mycol. 40: 37 (1996) = *Trechispora vaga* (Fr.) Liberta, Taxon 15: 319. 1966 = *Phlebia vaga* Fr., Syst. Mycol. 1: 428. 1821. Plate 6.33e, Fig. 6.91h–j

Fructification resupinate, floccose to pelliculose with numerous yellow rhizomorph over-running the substratum, loosely adnate, widely effused; hymenial surface yellow to yellow-ochre, discontinuous, not creviced; margin fibrillose to rhizomorphic, yellow, loosely adnate. KOH reaction tissues turning wine red in 3 % sol. Context composed of loosely woven hyphae and hyphal cordons.

Hyphal system monomitic, hyphae 1.4–4.4 µm wide, branched, septate, clamped, with ampulliform swellings near the septa, thin-walled, subhyaline to pale yellow. Rhizomorphic hyphae less frequently branched but are more coloured. Cystidia absent. Basidia 12.1–15.1 × 6.1–7.1 µm, clavate-cylindrical, frequently pleura basidiate, 4-spored. Basidiospores 5.1–5.4 × 3.1–3.4 µm, broadly ellipsoid to ovoid, shortly apiculate, thin-walled, subhyaline, finely echinulate, non-amyloid.

Distribution: H.P.: Khajjiar, Kullu, Dalhousie, Manali; U.K: Mussoorie, Lal tibba-Dehradun.

Collection examined: SSR 5819, IBP 37429, 37430

Substratum: On bark under mixed forest.

Remarks: This species is characterized by yellow, floccose to pelliculose fructification traversed with abundant yellow rhizomorph. The reaction of KOH sol is a characteristic feature of this species.

O- Russulales

Family- Amylostereaceae

Amylostereum Boidin,
Reveue Mycol. 23: 345, 1958.

Fructifications resupinate to effused-reflexed or occasionally pileate, somewhat coriaceous to cartilaginous; upper surface of the pileus smooth to strigose; hymenial surface smooth. Context with or without a cuticle on the abhymenial side, light brown in section, composed of compactly arranged hyphae. Hyphal system dimitic or monomitic; generative hyphae with clamps. Cystidia thick-walled, brown and often covered with subhyaline crystals. Gloeocystidia absent. Basidia clavate, 4-spored. Basidiospores ellipsoid, the walls thin, smooth, subhyaline, amyloid.

Four Species, widespread

Lit.: Boidin (1958), Legon and Pegler (Mycologist 16: 124, 2002), Larsson & Larsson (Mycol. 95, 1037, 2003).

Type Species: *Amylostereum chailletii* (Pers. ex Fr.) Boidin

Habitat: Dead Wood

Himalayas: One

Amylostereum chailletii (Pers.) Boidin, Reveu Mycol. Paris 23: 345. 1958 = *Thelephora chailletii* (Pers.) Mycol. Eur. 1: 125 (1822). Fig. 6.91k–n

Fructification effused-reflexed to pileate but mostly occurring in resupinate forms, membranous to membranous-coriaceous when fresh becoming hard and somewhat brittle on drying, loosely, adnate, often arising in small circular colonies which may remain discreet or coalesce later and become effused. Pileus when present small, flabelliform to conchate, simple or sometimes imbricate; upper surface deep brown to brownish black, concentrically zonate, zones of prominent ridges and grooves; hymenial surface whitish brown to light brown, smooth to finely tuberculate, rarely cracking irregularly on drying; margin acute in pileate forms often abrupt of cliff like in resupinate forms, concolorous with the hymenial surface. Context composed of a narrow basal zone of completely agglutinated hyphae and an upper zone of compactly arranged ascending hyphae.

Hyphal system dimitic; generative hyphae 2–4 μm, branched, septate, clamped at most septa, thin skeletal hyphae 2.4–4.4 μm wide, unbranched to rarely branched, aseptate, the walls subhyaline to light brown, thick, often leaving a narrow lumen;. Pseudocystidia 4–5.7 μm, of variable length being the prolongations of skeletal hyphae which curve into the hymenium, immersed or projecting up to 20 μm out of it, the walls light brown to brown, thick often leaving capillary lumen and finally incrusted with subhyaline crystals especially in the apical part. Some of the pseudocystidia eventually become buried (immersed) in the thickening hymenium and their place is taken by new one which arises in the hymenium. These are of limited growth, with a clamped at the base and are often mistaken for a true cystidium. Basidia 24.9–35.1 × 3.4–5 μm, clavate to clavate-cylindrical, 4-spored. Basidiospores 6.1–8.2 × 2.4–3 μm, ellipsoid to sub cylindrical, minutely apiculate, the walls subhyaline, thin, smooth, amyloid.

Distribution: H.P.: Narkanda, Kullu; U.K.: Mussoorie; J&K: Patnitop

Collection examined: GSR 5003, SSR 5406, IBP 37431.

Substratum: On dead stump of *Cedrus deodar*, *Abies pindrow*.

Remarks: This species is widely distributed in the coniferous forests of North Western Himalayas and is associated with white rot. The principal diagnostic features are thick, effused-reflexed fructification, dimitic hyphal system with clamped generative hyphae, finally incrusted light brown pseudocystidia and amyloid basidiospores.

Family-Bondarzewiaceae

Key to genera

1. Fructification annual, large, pileate, stipitate, soft fleshy.................. *Bondarzewia*
1. Fructification annual to perennial, resupinate to effused reflexed to pileate, sessile.. *Heterobasidion*

Bondarzewia Sing.,
Rev. Mycol.5:4, 1940.

Fructification annual, stipitate, imbricate or rosette agglomeration. Pileus dimidiate, flabelliform; upper surface azonate to zonate, glabrous cream to yellowish brown. Pore surface white to lemon yellow; pores large, uneven; tubes in one layer. Context fleshy, white to ochraceous, homogenous. Hyphal system dimitic; generative hyphae thin to thick-walled, branched, septate; skeletal hyphae thick-walled with narrow lumen, sinuous, unbranched, aseptate. Basidia thin-walled, clavate, 4-spored. Basidiospores globose to subglobose, hyaline, warted.

Three species, widespread

Type species: *Cerioporus montanus* Qué l 1888

Lit.: Corner (*Beih. Nova Hedwigia* 78: 205, 1984; key S. E. Asia spp.) Stalpers (Stud. Mycol. 40: 48, 1996)

Habitat: Dead wood

Himalayas: One

Bondarzewia mesenterica (Schaeff.) Kreisel, Feddes Repert. Spec. Nov. Regni Veg. 95(9–10): 699 (1984) = *Bondarzewia montana* (Quél.) Singer, Revue Mycol., Paris 5: 4 (1940). Fig. 6.92a–d

Fructifications annual, large, pileate, stipitate, soft fleshy, brittle on drying, several pilei fused to form a compound, imbricate to rosette agglomeration, up to 42 cm in diameter and 30 cm high. Pileus tongue-shaped, spathulate, flabelliform to infundibuliform, individual pilei 10–20 cm, wide and 10–19 cm long, pilei tapering downward gradually into a stipe, growing on a common base; upper surface finely tomentose to smooth, concentrically weakly zonate, radially rugose, light brown to yellowish brown, darker on drying; margin thin, concolorous with upper surface, fertile to sterile below, irregular, incurved on drying, up to 1 mm wide. Stipe irregular, short, lateral, covered with decurrent pores, arise from a common base up to 7 cm high and 5 cm thick. Pore surface white to cream, light brown on drying; pores large, uneven, angular to irregular, splits on drying, thin-walled 1–2 per mm; pore mouth incised to dentate; tubes in one layer, cream up to 1 cm deep in section. Context white to cream, soft fleshy, fibrous, homogenous, non-xanthochroic.

Hyphal system dimitic; generative hyphae hyaline, thin-walled to slightly thick-walled in context, branched, septate without clamps, straight to sinuous, cyanophilous, 2–5.8 μm in diameter; in dissepiments, generative hyphae thin-walled and intricately branching. In context these are filled with refractive contents; skeletal hyphae hyaline, thick-walled, aseptate, unbranched, sinuous, cyanophilous 2.5–4.2 μm in diameter. Basidia clavate, 4-spored, up to 6.9 μm in diameter. Basidiospores thin to slightly thick-walled, hyaline, ornamented, warted, cyanophilous, strongly amyloid, 5.3–8.1 × 5–7.2 μm, without ornamentation.

Distribution: A.P: West Kameng, Shergaon, 5 Km from Shergaon towards Rupa.

Collection examined: SSV 21501.

Substratum: On the roots of angiospermic tree.

Remarks: This species is characterized by annual, fleshy, large, compound fructifications with infundibulliform, tongue-shaped to spathulate pilei; dimitic hyphal system; and ornamented, very strongly amyloid basidiospores are the diagnostic features of *B. montana*.

Heterobasidion Brefeld,

Unters. Gesamtgeb. MyKol. 8: 154, 1888.

Fructification annual to perennial, resupinate to effused-reflexed to pileate, sessile. Pileus effused-reflexed to applanate; upper surface dark-brown, faintly zonate, finely tomentose; pores round to angular, small to medium sized. Tubes in many strata.

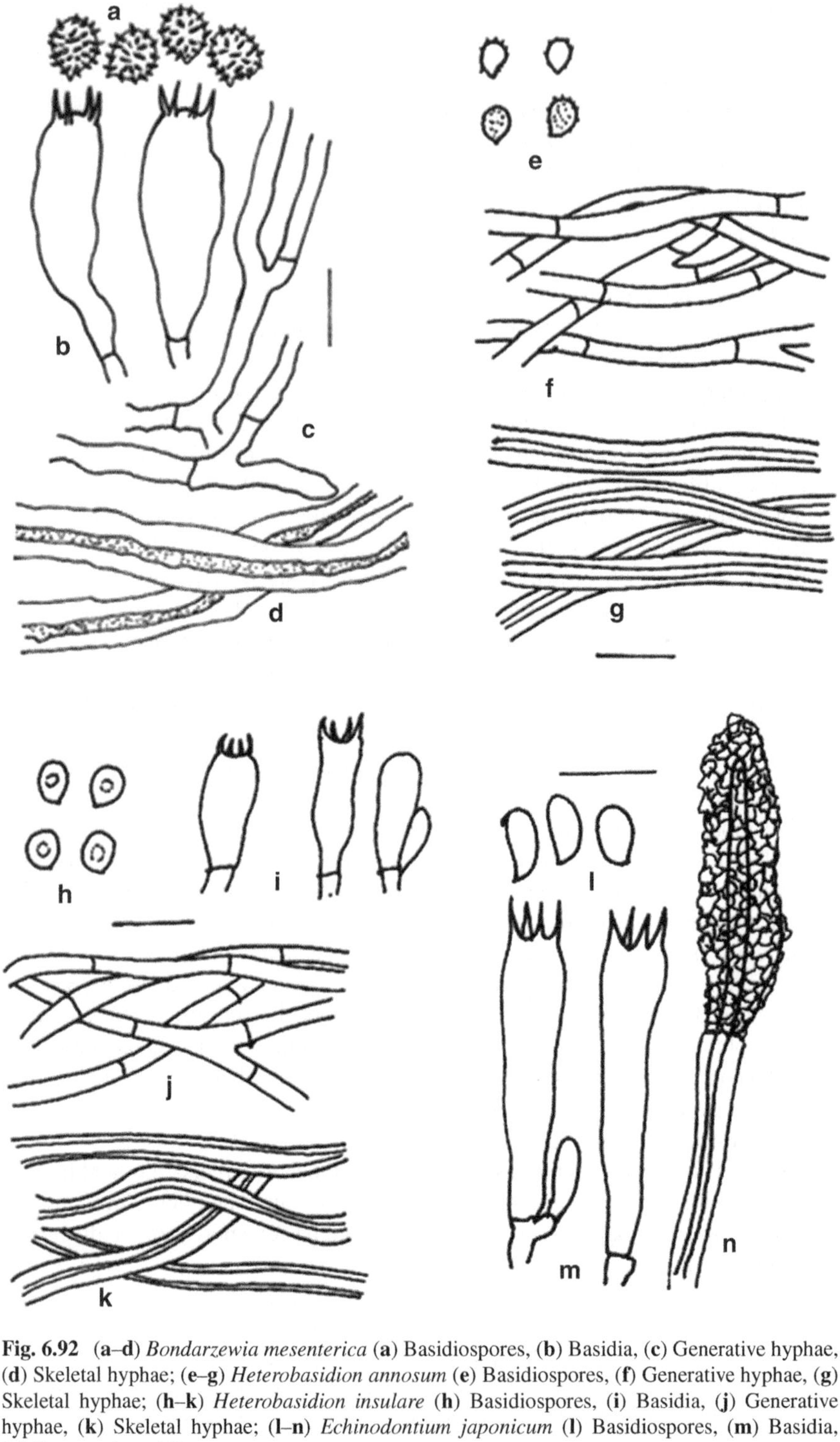

Fig. 6.92 (**a**–**d**) *Bondarzewia mesenterica* (**a**) Basidiospores, (**b**) Basidia, (**c**) Generative hyphae, (**d**) Skeletal hyphae; (**e**–**g**) *Heterobasidion annosum* (**e**) Basidiospores, (**f**) Generative hyphae, (**g**) Skeletal hyphae; (**h**–**k**) *Heterobasidion insulare* (**h**) Basidiospores, (**i**) Basidia, (**j**) Generative hyphae, (**k**) Skeletal hyphae; (**l**–**n**) *Echinodontium japonicum* (**l**) Basidiospores, (**m**) Basidia, (**n**) Encrusted cystidia

Context white to creamish, homogenous and non-xanthochroic. Hyphal system dimitic; generative hyphae hyaline, thin-walled, branched, septate, clamps absent, acyanophilous; skeletal hyphae dominate the context, hyaline, thick-walled with narrow lumen, rarely branched, aseptate, acyanophilous. Cystidia absent. Basidia clavate, 2–4 spored. Basidiospores hyaline, thin-walled, smooth, ovoid to subglobose to broadly ellipsoid.

Six Species, widespread

Lit.: Stalpers (*Stud. Mycol.* **40**: 79, 1996), Ota et al. (*Mycol.* **98**: 717, 2006)

Habitat: Decay wood

Type Species: *Polyporus annosus* Fr. 1821

Himalayas: Two

Keys to species

1. Fructification perennial; upper surface greyish brown.. *H. annosum* var. *indicum*
1. Fructification annual; upper surface reddish brown............................. *H. insulare*

Heterobasidion annosum var. *indicum* (Wakef.) Dhanda, in Thind & Dhanda, Indian Phytopath. 33(3): 386, 1980. Plate 6.33f, Fig. 6.92e–g

Fructification perennial, resupinate, effused-reflexed, to pileate, solitary or imbricate, mostly irregular, tough coriaceous when fresh, hard on drying, easily separable from the wood. Pileus when present, small, sessile, broadly attached; upper surface light brown to brown, reddish brown, greyish brown, black cuticle appear in the older portion, sulcate, concentrically zonate, finely tomentose, glabrous later on, radially wrinkled on drying; margin thin, wavy, acute to round, light brown, incurved on drying. Pore surface white to light cream when fresh, darken on drying, even, irregular; pores round to angular, elongated in oblique parts; tubes indistinctly stratified, concolorous with the pore surface. Context white to cream, fibrous and corky, homogenous, non-xanthochroic.

Hyphal system dimitic; generative hyphae hyaline, thin-walled, septate, without clamps, branched, faintly cyanophilous, 1.7–2.4 μm in diameter; skeletal hyphae hyaline to yellowish in upper surface of pileus, thick-walled to solid with narrow lumen, unbranched to branched, straight to tortuous, dextrinoid, 3–3.8 μm in diameter in dissepiments. Basidia hyaline, clavate, 4-sterigmate, up to 5.2 μm broad. Basidiospores hyaline, thin-walled, minutely asperulate, broadly ellipsoid to globose, nonamyloid, 4–4.7 × 3–3.8 μm.

Distribution: H.P.: Kullu, Dalhousie, Shimla, Kufri, Manali; Nepal: Pokhra; Meghalaya: Shillong; U.K.: Kumaun, Chakrata, Mussoorie, NDBR; J&K: Batote, Gulmarg.

Collection examined: IBP 37433.

Substratum: On decaying *Abies pindrow*, *Pinus excelsa* log.

Remarks: *Heterobasidion annosum* is reported to grow as root parasite in conifer forest in Northern hemisphere. In India it is very common as a saprophyte in the coniferous forest of the Himalayas. It is found to attack living *Cedrus deodara* only when the plantation is established on sites unsuitable for its growth (Bakshi 1955).

Heterobasidion insulare (Murrill) Ryvardan [as '*insularis*'], Norw. Jl. Bot. 19: 237 (1972)=*Trametes insularis* Murr., Bull. Torr. bot. Cl. 35:405, 1908. Plate 6.34a, Fig. 6.92h–k

Fructification annual, sessile, effused-reflexed, to pileate, imbricate with common base, soft coriaceous when fresh, hard and rigid on drying. Pileus effused-reflexed to sessile, dimidiate to elongated and irregular, sometimes conchate, up to 8 cm long, 6 cm wide and 2 cm thick; upper surface cream, brown to reddish brown, generally cream towards margin and brown to reddish brown near base, azonate to concentrically zonate, radially wrinkled, glabrous; margin acute to blunt, white to cream, entire, sterile below. Pore surface white to cream, even; pores subcircular, angular to irregularly elongated, 2–6 per mm dissipiment, white, 72–160 µm equal, entire, apices, velutinate; tubes in one layer. Context white to cream, azonate, homogenous, non-xanthochroic, delimited on the upper surface with a thin reddish brown crust.

Hyphal system dimitic; generative hyphae hyaline, thin-walled, septate, without clamps, branched, cyanophilous, 2.7–4.6 µm in diameter; skeletal hyphae sub-hyaline, thick-walled to solid with narrow lumen, aseptate, unbranched, acyanophilous, nonamyloid, 3.6–5 µm in diameter. Cystidia absent. Basidia hyaline, thin-walled, clavate to cylindric clavate, 2–4 spored, 14–18.4 × 5–6.4 µm. Basidiospores hyaline, smooth, thin-walled, subglobose to broadly ellipsoid, 5.5–8.6 × 4.2–6 µm.

Distribution: Bhutan: Thimphu; Nepal: Kathmandu; Manipur: Imphal; West Bengal: Darjeeling; U.K.: Kumaun, NDBR; H.P.: Shimla, Narkanda, Solan, Manali.

Collection examined: SSV 21299, IBP 37434.

Substratum: On decaying *Pinus* stump, gymnospermous log.

Remarks: This is a widely distributed species and earlier it was reported by Bakshi (1971) as *Trametes insularis* Murr. and mentioned it to be of common occurrence on logs and stumps of conifers in the temperate and sub temperate regions of Western Himalayas.

Family- Echinodontiaceae

Key to genera

1. Hyphal system monomitic .... *Echinodontium*
1. Hyphal system trimitic .... *Laurila*

Echinodontium Eill. & Ev., Bull. Torrey bot. Club 27; 49. 1900.

Fructifications effused-reflexed to dimidiate, conchate, perennial, corky to woody; hymenial surface toothed. Spines rigid, terrete, cream to cream brown. Context coloured but not reacting with KOH sol. (non-xanthochroic). Hyphal system monomitic, hyphae clamped, thin to thick-walled to almost solid. Cystidia (or Pseudocystidia) present. Gloeocystidia absent. Basidia clavate, 4-spored. Basidiospores ellipsoid, the walls slightly thick, subhyaline, smooth to minutely roughened, amyloid.

Five Species, widespread

Plate 6.34 (**a**) *Heterobasidion insulare*. (**b**) *Echinodontium japonicum*. (**c**) *Laurilia sulcata*. (**d**) *Asterostroma cervicolor*. (**e**) *Asterostroma musicola*. (**f**) *Dichostereum pallescens*

Lit.: Thomas (*Publ. Dep. Agric. Can. For. Bio. Div.*1041, 1958) Larsson & Larsson (*Mycol.* **95**: 1037, 2003)
Type Species: *Fomes tinctorum* Ell. & Ev. 1895
Habitat: Dead wood
Himalayas: One

Echinodontium japonicum Imazeki, J. Jap. Bot. 11: 520. 1935. Plate 6.34b, Fig. 6.921–n

Fructifications perennial, resupinate to effused-reflexed, often arising as small colonies which may coalesce latter and become widely effused, adnate, corky to woody. Pileus rudimentary, a number of times represented by an upturned margin; upper surface brownish-black to almost black, concentrically zonate, zones corresponding to yearly growth, hard and bark-like; hymenial surface toothed, cream to cream brown, not creviced; margin acute, adnate, cream brown to light brown. Teeth up to 12 mm long, subulate, terrete when young but become somewhat flattened, angular and confluent, brittle, apices acute, sterile. The new teeth arise from the previous years agglutinated teeth forming a brown tissue. Trama of the teeth surrounded by cream brown hymenium. Subiculum light brown, poorly developed to almost obliterated.

Hyphal system monomitic; generative hyphae 2–3.4 µm wide, branched, septate, clamped, the walls thin to slightly thick, subhyaline; sclerified generative hyphae 2.5–5 µm wide, sparsely branched, aseptate to occasionally septate, clamped, the walls thick to almost solid, pale yellow. The sclerified hyphae are common in the subiculum and trama and some of these may curve in to the hymenium as pseudocystidia. Pseudocystidia 5.3–7.5 µm broad, hyphoid to cylindrical, with obtuse apices, pale yellow with subhyaline crystalline matter, thick-walled to almost solid, immersed or projecting out of the hymenium. Gleoecystidia absent. Basidia 25–32 × 5.5–6.4 µm, clavate to clavate-cylindrical, 4-spored. Basidiospores 5.4–7 × 3.4–4.5 µm, ellipsoid, minutely apiculate, the walls subhyaline, slightly thick, smooth to finely rough, amyloid.

Distribution: H.P.: Shimla, Mahasu, Chamba; J&K: Bhadarwah.
Collection examination: HSK 4014, 4032, IBP 37435.
Substratum: On bark of coniferous log.
Remarks: This species is characterized by perennial, woody, pileate and hydnoid fructifications, encrusted skeletocystidia and slightly thick-walled amyloid spores.

Laurilia Pouz.,
Česká Mycol. 13(1): 14, 1959.

Fructification perennial, resupinate to, effused reflexed and with corky consistence, abhymenial surface of young specimens consists of a brown tinder-like layer, in old specimens almost black and hard; hymenial surface lighter, tuberculate; trama consisting of two layers, the upper tinder and the true trama, which corresponds to a subiculum, two layers being separated by a thin resinuous, hard layer, visible as a dark line in the section; hymenium light-coloured, more or less tuberculate, not hydnoid; texture dense and hard, light-coloured. Hyphal system trimitic with skeletal hyphae thick-walled sparsely branched, binding hyphae thick-walled and branched; generative hyphae thin-walled, branched, septate, clamped; metuloid cystidia numerous, thick-walled, encrusted; basidia clavate, with 4-sterigmata and basal clamp; spores globose, somewhat thick-walled, echinulate, amyloid.

Two Species, world-wide

Lit. : Gross (*Mycopath. Mycol. appl.* **24**: 1, 1964), Chamuris (*Mycol. Mem.* **14**, 1988; key).

Type species: *Stereum sulcatum* Burt. 1902

Habitat: Wood

Himalayas: One

Laurilia sulcata (Burt) Pouzar, [as '*sulatum*'] Česká Mycol. 13: 14, 1959 = *Stereum sulcatum* Burt, in Peck, Ann. Rep. Reg. N. Y. St. Mus. Ann. Rep. 54 p.154, 1901. Plate 6.34c, Fig. 6.93a–f

Fructification resupinate, perennial, adnate, effused, with corky consistence; upper side in young specimens covered with a brown, tinder-like tomentum, in old specimens dark brown to blackish; hymenial surface smooth to tuberculate, orange grey; margin thinning in young fructification, thickening with time, in old specimens consisting of parallel ridges due to the receding hymenium every year.

Hyphal system trimitic; generative hyphae 2–3.8 μm wide, branched, septate, clamped, thin-walled; skeletal hyphae 2.5–3.5 μm wide, thick-walled, aseptate, without clamps; binding hyphae 2–2.9 μm wide, profusely branched, thick-walled, without clamps; the upper tinder-layer composed mainly of brown skeletal hyphae, subiculum of pale yellowish, horizontal skeletal hyphae with numerous binding hyphae and a few generative hyphae between them; subhymenium of vertical skeletal hyphae, irregular binding hyphae, generative hyphae and cystidia. Cystidia 48–105 × 7.5–10.5 μm, apically conical, thick-walled, encrusted. Gloeocystidia 26–58 × 4.2–5.0 μm, tubular, tapering towards apex. Basidia 31–43 × 5.2–6.0 μm, clavate, 4-sterigmate, with a basal clamp. Basidiospores globose, thick-walled, echinulate, 4.3–6.0 × 4.2–5.0 μm, amyloid, non-cyanophilous.

Distribution: Bhutan: Thimphu, Paro, D'Dzong.

Collection examined: GSD 19642.

Substratum: On a decaying angiospermic log.

Remarks: The new genus *Laurilia* was erected by Pouzar (1959) on the basis of *Stereum sulcatum*. Eriksson and Ryvardan (1976) made a number of collections from N. Europe and remarked that *L. sulcatum* is one of the most characteristic stereoid fungi differing from *Echinodontium* in having dimorphic trama as compared to uniform trama, trimitic texture of subiculum as well as subhymenium in comparison to dimitic subicular trama and monomitic subhymenial trama and subhyaline to brown colour of the skeletal hyphae in comparison to brick red colour. The description of the above collected specimen resembles with that of Eriksson and Ryvardan (1976).

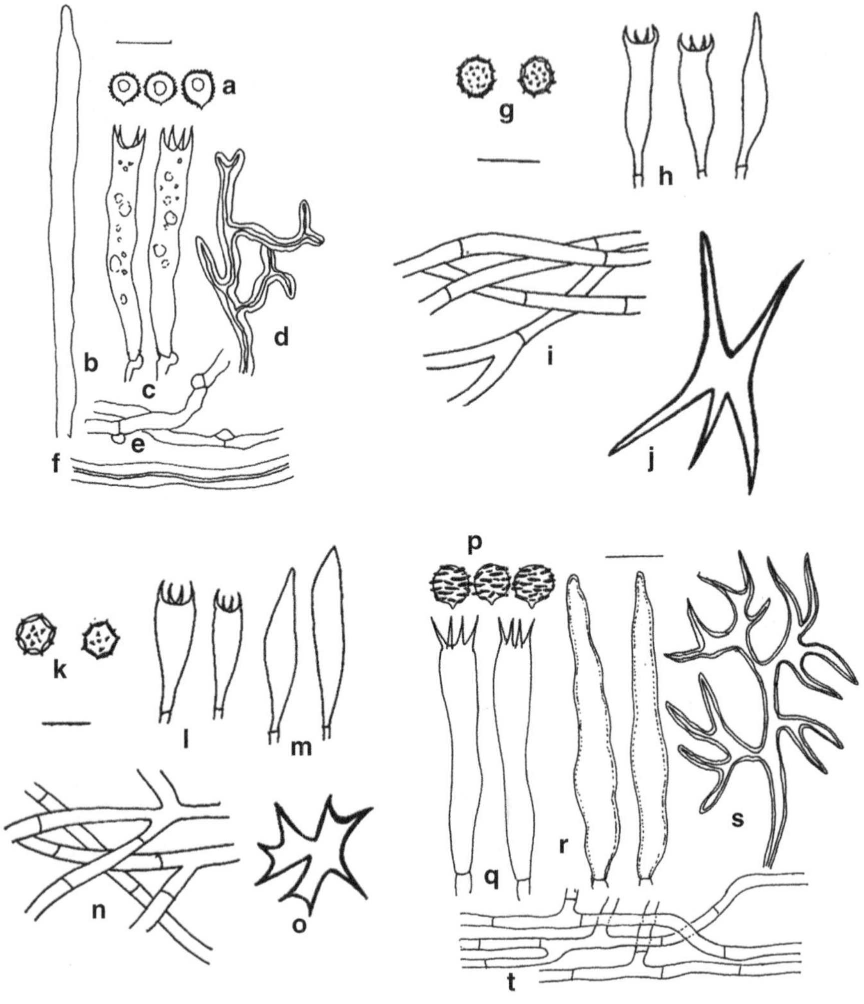

Fig. 6.93 (**a**–**f**) *Laurilia sulcata* (**a**) Basidiospores, (**b**) Cystidia, (**c**) Basidia, (**d**) Binding hyphae, (**e**) Generative hyphae, (**f**) Skeletal hyphae; (**g**–**j**) *Asterostroma cervicolor* (**g**) Basidiospores, (**h**) Basidia, (**i**) Generative hyphae, (**j**) Asterosetae; (**k**–**o**) *Asterostroma muscicola* (**k**) Basidiospores, (**l**) Basidia, (**m**) Cystidia, (**n**) Generative hyphae, (**o**) Asterosetae; (**p**–**t**) *Dichostereum effuscatum* (**p**) Basidiospores, (**q**) Basidia, (**r**) Cystidia, (**s**) Dichohyphidia, (**t**) Generative hyphae

Family- Lachnocladiaceae

Key to genera

1. Hymenial surface smooth, dichohyphidia present .... *Asterostroma*
1. Hymenial surface uneven or tuberculate, dichohyphidia present or absent......... 2
2. Hyphal system dimitic with clamps may or may not present......... *Scytinostroma*
2. Hyphal system monomitic with clamps may or may not present........................ 3
3. Cystidial elements present .... *Dichostereum*
3. Cystidial elements may or may not present .... *Vararia*

Asterostroma Massee,
J. Linn. Soc. Bot. 25(170): 154. 1889.

Fructifications resupinate, pelliculose to spongy and soft; hymenial surface smooth, brown. Context soft, full of asterosetae and hyphae. Hyphal system monomitic, hyphae branched, septate, clamps absent, the walls thin, subhyaline, nondextrinoid. Hymenial setae absent; asterosetae subhyaline to light brown, not or faintly darkening in KOH sol. Cystidia absent. Gloeocystidia present, cylindrical to subfusiform. Basidia utriform, 4-spored. Basidiospores globose to subglobose, the walls subhyaline, amyloid or nonamyloid, usually ornamented.

Fourteen species, widespread

Lit.: Boidin (*Bull. Soc. Mycol. Fr.* **113**: 269, 1997; key), Wagner (*Mycotaxon* 79: 235, 2001)

Type species: *Corticium apalum* Berk. & Br. 1875

Habitat: Wood

Himalayas: Two

Key to species

1. Asterosetae with more but long rays .... *A. cervicolor*
1. Asterosetae with more but shorter rays .... *A. muscicola*

Asterostroma cervicolor (Berk. & Curt.) Mass. J. Linn. Soc. Bot. 25(170): 155. 1889 = *Corticium cervicolor* Berk. & Curt., Grevillea 1: 179. 1873. Plate 6.34d, Fig. 6.93g–j

Fructification resupinate, annual loosely adnate, widely effused, soft, up to 45 mm long and 1 mm thick, pinkish yellow to golden brown to ochraceous, pulverulent; hymenial surface smooth, context soft, full of asterosetae and generative hyphae; margin thinning, fibrillose, loosely adnate, white to paler concolorous.

Hyphal system monomitic; generative hyphae branched, simple septate, clamps absent, thin-walled; asterosetae subhyaline to brown with 3–5 unbranched rays, rays subulate, 30–80 × 3–4.5 µm, rigid; gleocystidia present, cylindrical to subfusiform, with subhyaline granular contents staining deeply with Phloxin. Basidia generally utriform, 4-spored, 28–40 × 5–6 µm. Basidiospores globose to subglobose, 4.5–6 µm, warted, subhyaline, prominently apiculate, smooth or ornamented, amyloid.

Distribution: H.P.: Mahasu- Narkanda, Shimla, Manali, Chamba; U.K.: Ranikhet; A.P.: West kameng, Bomdila; Bhutan: Thimphu, Wangdi.

Collection examined: SSR 5115, 5379, IBP 37436, L 37437.

Substratum: On rotten stumps, fallen oak tree stump.

Remarks: This species was first reported from India by Thind and Adlakha (1956) from Mussoorie hills. Later Rattan (1977) collected this species from the various localities in the N.W. Himalayas. It is distinguished by loosely adnate fructification; large, generally unbranched asterosetae, cylindrical to subfusiform gleocystidia, utriform, 4-sterigmate basidia and globose to subglobose, warted, amyloid basidiospores.

Asterostroma muscicola (Berk. & M. A. Curtis) Massee, J. Linn. Soc., Bot. 25(no. 170): 155 (1889)=*Hymenochaete musicolum* Berk. & Curt., J. Linn. Soc. Bot. 10: 334.1868. Plate 6.34e, Fig. 6.93k–o

Fructification resupinate, pelliculose to spongy-crustose, adnate to somewhat separable, widely effused; hymenial surface ochre-yellow to ochre-brown, smooth or sometimes with minute tubercles, occasionally cracking on drying; margin fibrillose to rhizomorphic, loosely adnate, white to paler concolorous. Rhizomorphic strands common, branched and concolorous with the hymenial surface. Context composed of loosely arranged asterosetae and hyphae.

Hyphal system monomitic, hyphae 1.4–3.2 µm wide, branched, septate, clamps absent, walls thin, hyaline. Asterosetae 30–110 µm across, composed of 5–7 aculate rays, attached radially to a slightly inflated boss, rays subulate, often branched once or twice, the walls up to 2 µm thick, light brown. Asterosteae present in the hymenium (Asterophyses sensu Cunningham, 1963) are smaller, with shorter, narrower rays and usually much paler in colour to almost subhyaline. Gloeocystia 44.8–55 × 8–9.4 µm, subcylindrical to clavate, with subhyaline granular contents, thin-walled, subhyaline. Basidia 28–38 × 5.5–7.5 µm, utriform, 4-spored. Basidiospores 5–6.1 µm across, globose to subglobose, strongly apiculate, thick-walled, subhyaline and warted.

Distribution: H.P.: Mahasu- Narkanda, Dalhousie; U.K.: Hemkunt; J&K: Bhadarwah.

Collection examined: SSR: 5496, 5433, IBP 37438.

Substratum: On log of *Abies pindrow*, stump of *Cedrus deodara.*

Remarks: The characteristic features of the species are warted, amyloid basidiospores and asterostetae with branched rays. It can be differentiated from the allied *A. cerivicolor*, by the presence of small asterosetae in the hymenium. Such structures are either absent or poorly represented in the later.

Dichostereum Pilát,
Ann. Mycol., 24(3/4): 223, 1926.

Fructification resupinate, membranous to crustose, fleshy tough when fresh turning hard and brittle on drying, adnate; hymenial surface clay to pale yellow, smooth; margin indeterminate to almost abrupt, adnate, concolorous. Context somewhat darkening in KOH sol., Hyphal system monomitic, branched, septate, clamped. Dichohyphidia scanty in the hymenium but fairly common in the context, profusely branched, lower branches thick, ultimately branchlets tapering and spiny, the walls subhyaline to pale brown, thick, strongly dextrinoid. Gloeocystidia abundant. Basidia utriform, 4-spored. Basidiospores globose to rarely subglobose, prominently apiculate. Thin-walled, subhyaline to tinted brown, warted, warts discrete or tending to elongate forming an incomplete network, amyloid.

Eleven species, widespread

Lit.: Boidin & Lanquetin (Mycotaxon, 6:277, 1977), Boidin & Lanquetin (BSMF 96: 381, 1980; Key), Kotkova (*Mikol. Fitopatol.* **38**: 40, 2004; Russia)

Type Species: *Asterostromella dura* Bourdot & Galzin

Habitat: Wood

Himalayas: Four

Keys to species

1. Dendrohyphidia scanty ....*D. pallescens*
1. Dendrohyphidia abundant.. 2
2. Basidiospores forming network of warts .... *D. brevisporum*[a]
2. Basidiospores with irregular warts, not forming network.................................. 3
3. Basidia clavate *D. effuscatum*
3. Basidia clavate, subutriform to utriform 4
4. Hymenial surface pinkish white to pale red, gloeocystidia thin-to thick walled...*D. peniophoroides*[a]
4. Hymenial surface pale orange to greyish orange, gloeocystidia thin-walled 5
5. Gloeocystidia up to 90.0 μm long...*D. rhodosporum*
5. Gloeocystidia up to 77.0 μm long..*D. kenyense*

[a]Extra limital, not included in the text

Dichostereum effuscatum (Cooke & Ellis) Boidin & Lanq., Mycotaxon 6(2): 284 (1977) = *Vararia effuscata* (Cooke & Ellis) Rog. & Jacks. Farlowia 1: 290. 1943 = *Corticium effuscatum* Cooke & Ellis, Grevillea 9: 103. 1881. Fig. 6.93p–t

Fructification resupinate, membranous to crustose, fleshy tough when fresh turning hard and brittle on drying, adnate up to 3 mm thick in section; hymenial surface clay to pale yellow, smooth, continuous when fresh but becomes areolately cracked on drying; margin indeterminate to almost abrupt, adnate, concolorous. Context subhyaline in section, somewhat darkening in KOH sol., composed of compactly arranged more less agglutinated dichohyphidia and hyphae.

Hyphal system monomitic, hyphae up to 3.4 μm wide, branched, septate, clamped, thin-walled, subhyaline, non dextrinoid, often collapsing and difficult to discern. Dichohyphidia scanty in the hymenium but fairly common in the context, profusely branched, lower branches thick, up to 2 μm wide, ultimate branchlets tapering and spiny, the walls subhyaline to pale brown, thick, strongly dextrinoid. Gloeocystidia abundant in the context and hymenium, 40.5–60 × 5.7–7.5 μm, subventricose or cylindrical to somewhat flexuous with obtuse apices, with subhyaline granular contents, immersed or projecting out of the hymenium, thin-walled to moderately thick especially in the basal part, subhyaline. Basidia collapsing after spore discharge, 30 × 5.4–6 μm broad, utriform, 4-spored. Basidiospores 5–6.9 μm across, globose to rarely subglobose, prominently apiculate.

Distribution: U.K.: Dehra Dun, Nainital; H.P.: Kullu, Mandi- Sundernagar.

Collection examined: SSR 5849, IBP 37439.

Substratum: On stump of *Shorea robusta*, rotten hardwood stumps.

Remarks: The characteristic features of this species are scanty dichohyphidia, abundant gloeocystidia and amyloid, warted basidiospores.

Dichostereum kenyense Boidin & Lanq., Bull. trimmest. Soc. mycol. Fr. 96: 394, 1981. Fig. 6.94a–e

Fructification resupinate, adnate, effused up to 240 μm thick in section; hymenial surface smooth, whitish to yellow white; margin thinning, darker to indeterminate.

Hyphal system monomitic; generative hyphae up to 3.2 μm wide, septate, branched, thin-walled, clamped, basal hyphae parallel to the substrate. Dendrohyphidia present in hymenium and subhymenium, branched, thick-walled. Cystidia 46.2–63.0 × 5.4–6.5 μm, subcylindrical, to fusiform, thin-walled with oily contents, positive to sulfovanillin, with basal clamp. Basidia 32–41 × 5.4–7.2 μm 4-sterigmate, subclavate with basal clamp sterigmata up to 5.6 μm long. Basidiospores 7.3–9.5 × 6.2–7.2 μm, thin-walled, apiculate, subglobose to globose, warted, acyanophilous, inamyloid.

Distribution: H.P.: Kinnaur-Keylong.

Collection examined: IBP 37443, 37444.

Substrate: Stump of *Pinus wallichiana.*

Remarks: This species is characterized by thin-walled gloeocystidia having oil contents, positive to sulfovanillin subglobose to globose and warted basidiospores. Earlier it was reported by Boidin and Lanquetin (1980) from East Africa. It is a new record for Himalayas.

Dichostereum pallescens (Schwein.) Boidin & Lanq., Mycotaxon 6(2): 284 (1977) = *Vararia pallescens* (Schw.) Rog. & Jacks. Farlowia 1: 310. 1943 = *Thelephora pallescens* Schw., Trans. Am. Phil. Soc. n.s. 4:167. 1832. Plate 6.34f, Fig. 6.94f–i

Fructification resupinate, adnate, widely effused, membranous, annual or perennial up to 2 mm thick in section; hymenial surface light brown to dark brown, smooth to tuberculate, context subhyaline, light brown in section; margin determinate, full of dichohyphidia, dichohyphidia present in context and hymenium.

Hyphal system monomitic; generative hyphae branched, 1.6–2.4 μm wide, septate, clamps present, thin-walled; gleocystidia present. Basidia utriform, 4-spored; basidiospores globose, warted, prominently apiculate, amyloid, 6–7 μm in diameter.

Distribution: H.P.: Chamba; U.K.: Chakrata.

Collection examined: SSR: 5128, 5414, IBP 37440, L 37441.

Substratum: On stump of *Quercus incana.*

Remarks: The characteristic features of the species are brown coloured fructification, light brown to brown dichohyphidia and amyloid warted basidiospores. The closely related species *D. rhodosporum* differs in having light coloured fructification with subhyaline slender dichohyphidia and subhyaline basidiospores with simple warts which do not have the tendency to form reticulum.

Dichostereum rhodosporum (Wakef.) Boidin & Lanq., Mycotaxon 6(2): 284 (1977) = *Vararia rhodospora* (Wakef.) Cunn. Proc. Linn. Soc. NSW 77: 291. 1953 = *Asterostromella rhodospora* Wakef., Bull. Misc. Inf., Kew 8: 372. 1915.
Plate 6.35a, Fig. 6.94j–m

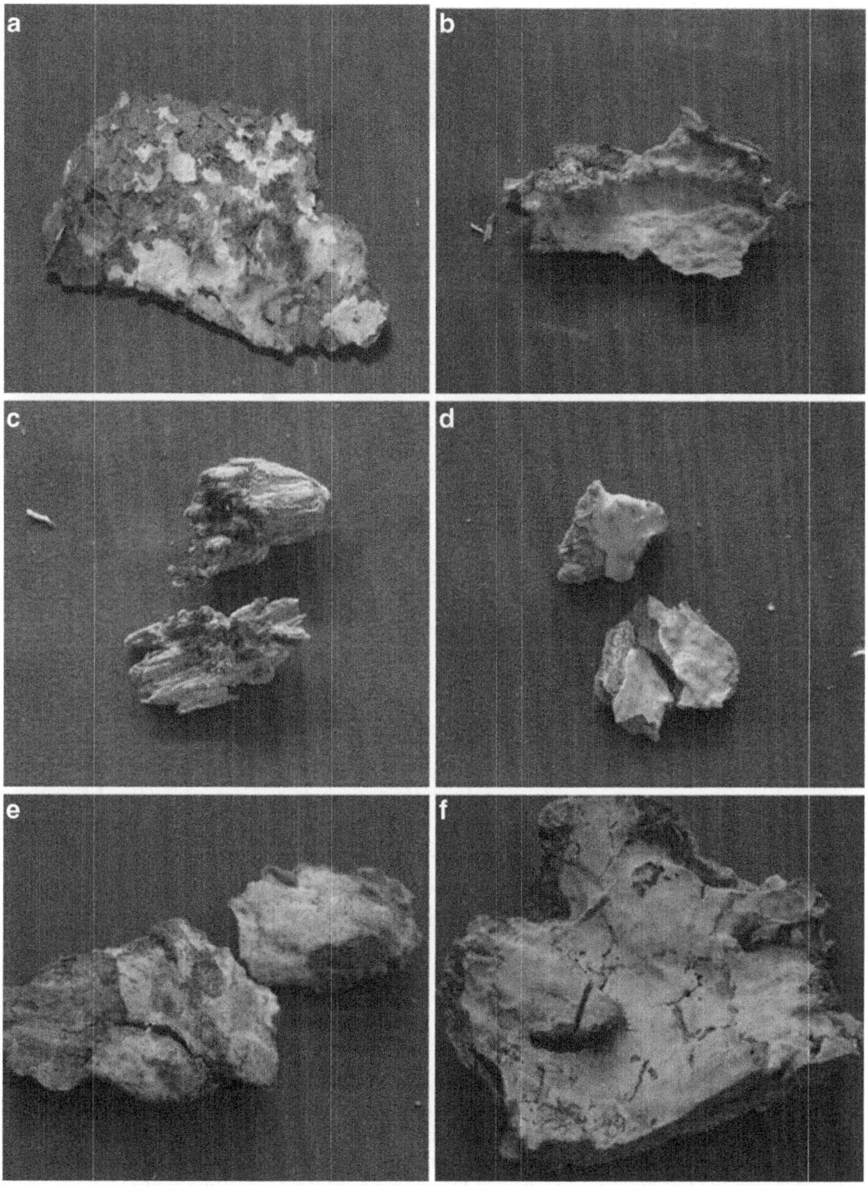

Plate 6.35 (**a**) *Dichostereum rhodosporum*. (**b**) *Scytinostroma crassum*. (**c**) *Scytinostroma cystidiatum*. (**d**) *Scytinostroma duriusculum*. (**e**) *Scytinostroma ochroleucum*. (**f**) *Scytinostroma portentosum*

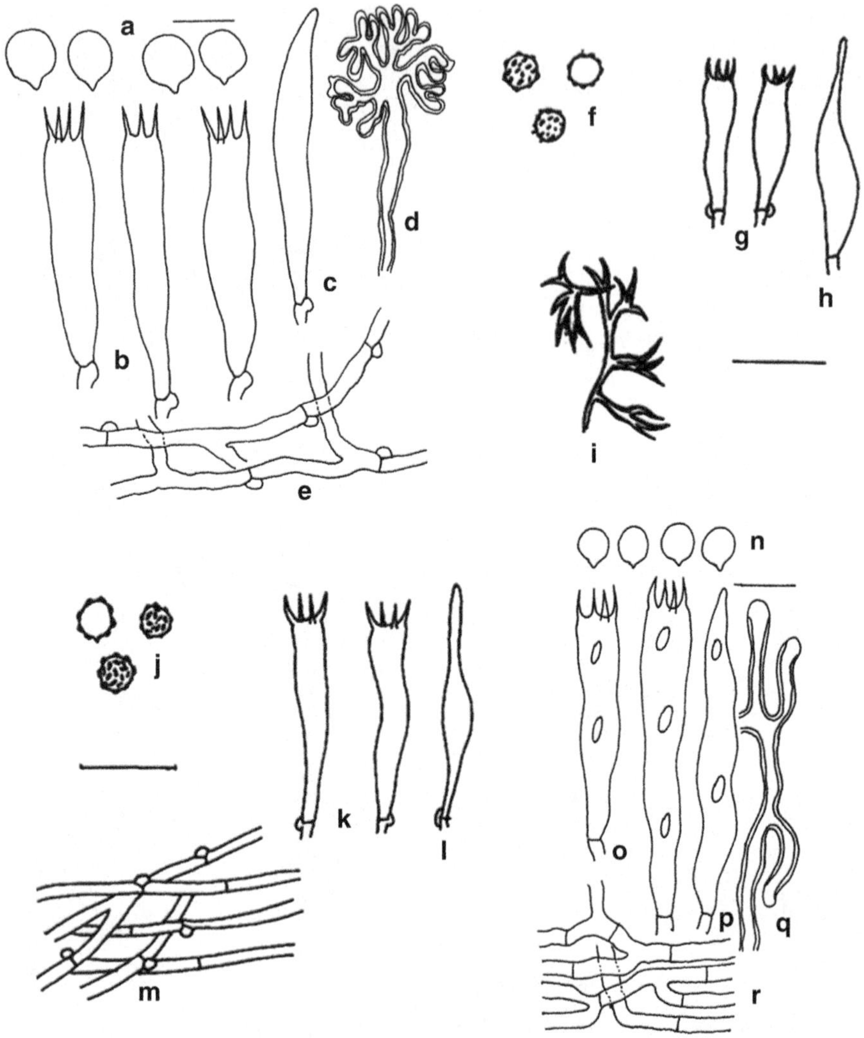

Fig. 6.94 (**a**–**e**) *Dichostereum kenyense* (**a**) Basidiospores, (**b**) Basidia, (**c**) Cystidia, (**d**) Dendrohyphidia, (**e**) Generative hyphae; (**f**–**i**) *Dichostereum pallescens* (**f**) Basidiospores, (**g**) Basidia, (**h**) Cystidia, (**i**) Dendrohyphidia; (**j**–**m**) *Dichostereum rhodosporum* (**j**) Basidiospores, (**k**) Basidia, (**l**) Cystidia, (**m**) Generative hyphae; (**n**–**r**) *Scytinostroma albocinctum* (**n**) Basidiospores, (**o**) Basidia, (**p**) Cystidia, (**q**) Dendrohyphidia, (**r**) Generative hyphae

Fructification resupinate, membranous, adnate, often forming irregular colonies when growing on bark but become effused on decorticated logs; hymenial surface deep cream to cream with pink tinge, smooth somewhat farinaceous under the lens, not creviced or rarely crack irregularly on drying; margin thinning, adnate, concolorous. Context subhyaline full of dichohyphidia and immersed basidiospores.

Hyphal system monomitic; generative hyphae clamped, branched, septate, thin-walled, subhyaline, full of dichohyphidia and immersed basidiospores, 1.4–2.6 μm wide, non dextrinoid. Dicohyphidia present in context and hymenium. These are abundant, branched, branches irregular, ultimate branchlets short ending in dichotomus spines, tips acute to subobtuse, walls thick, subhyaline, strongly dextrinoid, cyanophilous, dichohyphidia present in the context are with longer internodes, where as those in the hymenium are with shorter internodes, thick-walled, strongly dextrinoid, cyanophilous, gloeocystidia cylindrical to subfusiform with subhyaline granular contents, immersed or projecting out of the hymenium. Basidia utriform, 4-spored, 35–44.8 × 5–6 μm. Basidiospores 6–7.8 μm in diameter, globose, strongly apiculate, thin-walled to thick-walled, warted, warts irregular and prominent, amyloid.

Distribution: H.P.: Kullu, Manal, Shimla, Dalhousie, Chamba; U.K.: Nainital; J&K: Patnitop.

Collection examined: GSR 5023, IBP 37442.

Substratum: On stump of *Cedrus deodara & Abies pindrow.*

Remarks: The species is one of the commonly found resupinate fungi in coniferous forests of N.W. Himalayas. It is characterized by light coloured fructification, presence of subhyaline dichohyphidia and apiculate basidiospores with amyloid warts. The warts are discrete and do not form network.

Scytinostroma Donk,
Fungus 26(1–4): 19. 1956.

Fructifications resupinate, membranous-coriaceous to leathery, sometimes stratose; hymenial surface smooth to tuberculate or rarely toothed. Hyphal system dimitic; generative hyphae branched, septate, clamps present or absent, the walls subhyaline, thin, non-dextrinoid and acyanophilous; skeletal hyphae sparsely branched, aseptate, the walls subhyaline to faintly yellow or brown, thick, dextrinoid and cyanophilous. Gloeocystidia and Cystidia present or absent. Basidia tubular to utriform, 4-spored, broadly clamped or simple septate. Basidiospores globose to ellipsoid or cylindrical, the walls thin, subhyaline, smooth or finely echinulate, amyloid or nonamyloid, acyanophilous.

Thirty species, widespread

Lit.: Boidin & Lanquetna (Bibltheca Mycol. 114, 1987, Key), Stalpers (Stud. Mycol. 40: 103, 1996, Key), Larsson & Larsson (*Mycol.* 95: 1037, 2003)

Type Species: *Corticium portentosum* Berk. & Curt. 1873

Habitat: Deadwood

Himalaya: Nine (Table 6.8)

Scytinostroma albocinctum (Berk. & Broome) Boidin & Lanq., Kew Bull. 31: 621, 1976 = *Stereum albocinctum* Berk. & Broome, J. Linn. Soc. Bot. 14: 66, 1873.

Fig. 6.94n–r

Fructification resupinate, effused, adnate, up to 380 μm thick in section; hymenial surface smooth to tuberculate, orange grey to pale brown orange when fresh, pale orange to greyish orange on drying, darkening in 3 % KOH sol; margin thinning, whitish to paler concolorous.

Table 6.8 Synopsis of *Scytinostroma* species

Name	Fructification	Hymenial surface	Cystidia	Gloeocystidia	Basidia, basidiospores
S. albocinctum	Resupinate, effused, adnate	Hymenial surface smooth to tuberculate orange grey to pale brown	Non encrusted	Absent	Basidia clavate without clamp; Basidiospores globose to subglobose, smooth, 7.2–8 × 6.2 μm
S. alutum	Resupinate, effused, adnate	Hymenial surface smooth orange to pale orange	Non encrusted	Absent	Basidia clavate without clamp; Basidiospores globose to subglobose uniguttulate, smooth 4.6–6.2 × 5–5.8 μm
S. crassum	Resupinate, coriaceous, stratose, adnate, membranous, effused	Hymenial surface smooth to pruinose, cream brown	Absent	Present subfusiform to cylindrical	Basidia utriform; Basidiospores ellipsoid, smooth, 6.8–9 × 5–5.8 μm
S. cystidiatum	Resupinate, membranous, adnate, effused	Hymenial surface smooth to tuberculate, white to cream	Encrusted all over except base	Present subfusiform	Basidia utriform; Basidiospores globose, smooth, 4–4.5 μm
S. duriusculum	Resupinate, adnate, coriaceous, membranous	Hymenial surface smooth creamish to yellowish white	Absent	Present with granular contents	Basidia utriform; Basidiospores globose, smooth, 4–4.5 μm in dia.
S. ochroleucum	Resupinate, membranous, loosely adnate, effused	Hymenial surface smooth to tuberculate cream yellow	Absent	Present	Basidia utriform; Basidiospores ellipsoid, smooth, 8–12 × 6–8 μm
S. portentosum	Resupinate, membranous, effused	Hymenial surface smooth to rough, cream to pale yellow	Absent	Absent	Basidia utriform; Basidiospores globose, smooth, 4.5–6 μm in dia.
S. pulverulentum	Resupinate, membranous, effused	Hymenial surface smooth, creamish white to yellowish white	Absent	Present subfusiform to cylindrical, 28–42 × 4.5–6 μm	Basidia utriform with basal clamp; Basidiospores ellipsoid to ovoid smooth, 5–5.9 × 4.6–5.1 μm
S. rhizomorpharum	Resupinate, membranous, effused	Hymenial surface smooth, white to cream with rhizomorphs	Absent	Absent	Basidia utriform; Basidiospores cylindrical

Hyphal system dimitic; generative hyphae up to 4.2 µm wide, septate, thin-walled, without clamps; skeletal hyphae up to 5.2 µm wide, thick-walled, aseptate, irregularly to dichotomously, more profusely branched in subhymenium. Cystidia subfusiform 67–85.0×6.7–8.6 µm, thin-walled, without clamps, oily contents, positive to sulfovanillin. Basidia 27.0–68×7.2–8.8 µm, clavate, apiculate, 4-sterigmate, without basal clamp with oil globules. Basidiospores 7.2–8.0×6.2 µm, sterigmata up to 6.4 µm, long, globose to subglobose, thin-walled, amyloid, smooth, acyanophilous.

Distribution: H.P.: Kinnaur, Dalhousie.

Collection examined: IBP 37445.

Substrate: On angiosperm twigs.

Remarks: This species is characterized by presence of globose to subglobose basidiospores and presence of gloeocystidia.

Scytinostroma alutum Lanq., Bull. mens. Soc. linn. Lyon, Num. Spéc. 53: 187, 1984. Fig. 6.95a–e

Fructification resupinate, adnate effuse, up to 225 µm thick in section; hymenial surface smooth, orange grey to pale orange when fresh, brownish orange on drying; margin thinning, byssoid, concolorous to indeterminate.

Hyphal system dimitic; generative hyphae up to 3.2 µm wide, septate, thin-walled, without clamps, arising parallel to the substratum; skeletal hyphae up to 6.2 µm wide, thick-walled, dichotomously to irregularly branched, aseptate, abundantly present in context, cyanophilous. Cystidia 55–68×5–6 µm, subfusiform with tapering apex, thin-walled, with oily contents, positive to sulfovanillin, without clamps. Basidia 39–47×5.4–6.2 µm, 4-sterigmate, clavate without basal clamp, sterigmata up to 4.8 µm long. Basidiospores 4.6–6.2×4.0–5.2 µm, apiculate, globose to subglobose, thin-walled, amyloid, smooth, acyanophilous, uniguttulate.

Distribution: H.P.: Dalhousie.

Collection examined: IBP 37446.

Substrate: On fallen stick of *C. deodara*.

Remarks: This species is characterized by presence of gloeocystidia with oily content, positive to sulfovanillin, globose to subglobose basidiospores and generative hyphae without clamps. It is a new record for India/Himalayas.

Scytinostroma crassum (S. S. Rattan) Hjortstam, Mycotaxon 54: 192 (1995)=*Scytinostroma odoratum* forma *crassum* Rattan, Trans. Brit. Mycol. Soc. 63: 6. 1974. Plate 6.35b, Fig. 6.95f–j

Fructification resupinate, coriaceous, stratose, adnate, widely effused, membranous; hymenial surface cream brown to yellow brown, smooth or pruinose; margin thinning to determinate, adnate, paler concolorous to concolorous.

Hyphal system dimitic; generative hyphae branched, septate, clamps are absent, walls thin, acyanophilous, basal hyphae parallel to the substrate, less branched; subhyaline hyphae vertical, more branched, 1–2.4 µm wide; skeletal hyphae 1.5–2.5 µm wide, dichotomously branched with branches more profuse in the

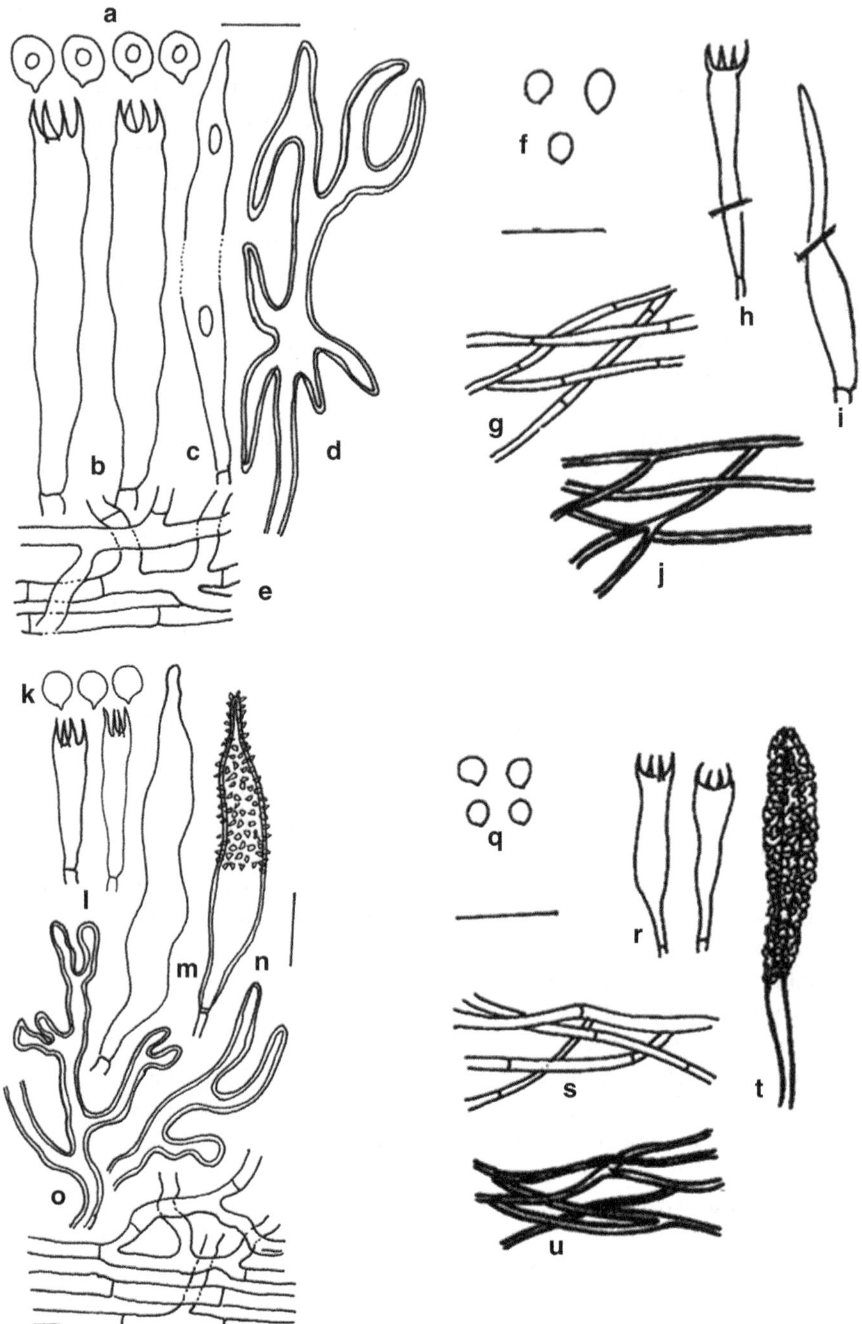

Fig. 6.95 (**a–e**) *Scytinostroma alutum* (**a**) Basidiospores, (**b**) Basidia, (**c**) Cystidia, (**d**) Dendrohyphidia, (**e**) Generative hyphae; (**f–j**) *Scytinostroma crassum* (**f**) Basidiospores, (**g**) Generative hyphae, (**h**) Basidia, (**i**) Cystidia, (**j**) Skeletal hyphae; (**k–p**) *Scytinostroma cystidiatum* (**k**) Basidiospores, (**l**) Basidia, (**m**) Cystidia, (**n**) Encrusted cystidia, (**o**) Dendrohyphidia, (**p**) Generative hyphae; (**q–u**) *Scytinostroma duriusculum* (**q**) Basidiospores, (**r**) Basidia, (**s**) Generative hyphae, (**t**) Encrusted Cystidia, (**u**) Skeletal hyphae

hymenium, aseptate, thick-walled, dextrinoid, cyanophilous. Gloeocystidia 45–92×6.5–9 μm, subfusiform to cylindrical, with subhyaline granular contents staining deeply with phloxine, becoming glassy in mature fructification, aseptate or rarely with retraction septa, immersed and just reaching the hymenial surface, thin-walled. Basidia utriform, 4-spored, 24.8–40×3–3.9 μm. Basidiospores broadly ellipsoid, shortly apiculate, smooth, 6.8–9×5–5.8 μm, non-amyloid.

Distribution: H.P: Dalhousie Shimla-Narkanda, Manali, Shimla-Narkanda, Manali; J&K: Patnitop; U.K.: Rudarprayag.

Collection examined: SSR 5156, IBP 37447.

Substratum: On wood of *Picea smithiana*, *Cedrus deodara*, Coniferous stump, logs of *Abies pindrow*.

Remarks: This species is confined to coniferous forests of North-western Himalayas. It is characterized by thick, coriaceous, widely effused fructification, presence of gloeocystidia and larger non-amyloid basidiospores. It is new report for Uttarakhand.

Scytinostroma cystidiatum Boidin, Bull. Jard. Bot. Etat. Brux. 30: 285. 1960.
Plate 6.35c, Fig. 6.95k–p

Fructification resupinate, membranous, adnate, widely effused, up to 430 μm thick in section; hymenial surface white to cream, smooth to somewhat tuberculate, not creviced; margin thinning, paler concolorous, adnate.

Hyphal system dimitic; generative hyphae 1–2.4 μm wide, branched, septate, clamps absent, the walls subhyaline, thin, nondextrinoid and acyanophilous, basal hyphae parallel to the substrate, less branched; subhymenial hyphae vertical, more branched; skeletal hyphae 1–2.4 (3) μm wide, dichotomously branched, branches more profuse in the hymenial zone, non septate, the walls subhyaline, thick, dextrinoid and cyanophilous. Cystidia 40–100×7–14 μm, cylindrical to subfusiform, often arising from different regions immersed or projecting out of the hymenium, heavily incrusted all over except at the base, incrustation subhyaline and soluble in 10 % KOH sol., the walls thick, subhyaline. Gloeocystidia 62–78×9–11 μm, subfusiform, present both in the context and hymenium, thin-walled with oil content, positive to sulfovanillin. Basidia 24.8–35×4.4–5 μm, utriform, 4-sterigmate up to 5 μm long. Basidiospores 4.4–5 μm in diameter, globose, shortly apiculate, the walls subhyaline, smooth, amyloid.

Distribution: H.P.: Pinjore, Solan; J&K: Jammu; Punjab: Pathankot.

Collection examined: SSR: 5758, IBP 37447.

Substratum: On root stock of *Prosopis sp.*

Remarks: This species is characterized by globose, amyloid basidiospores and presence of incrusted basidia which differentiates it from the allied *S. portentosum*, *S. durisuculum* and *S. hemidichophyticum*.

Scytinostroma duriusculum (Berk. & Br.) Donk, Fungus 26: 20. 1956=*Stereum duriusculum* Berk. & Br., J. Linn. Soc. Bot. 14: 66. 1873. Plate 6.35d, Fig. 6.95q–u

Fructification resupinate, adnate, membranous, coriaceous, widely effused, up to 470 μm in section; hymenial surface smooth, creamish-white to yellowish-white; margin thinning, adnate, white to paler concolorous. Context faintly stratose, strata being formed by the compactly arranged lateral branches of skeletal hyphae.

Hyphal system dimitic; generative hyphae branched, septate, clamps absent, thin-walled, up to 3.8 μm wide, non dextrinoid, acyanophilous; skeletal hyphae thick-walled,1–2 μm wide, aseptate, branched, branching lateral sometimes dichotomous, more frequently branched in the hymenium. Gloeocystidia thin-walled, 50–95 × 6–8 μm, clavate, cylindrical to subfusiform with subhyaline dense granular contents, aseptate, immersed or projecting out of the hymenium. Basidia thin-walled, with basal clamp, 4-sterigmate, 25–40 × 4–4.5 μm, utriform. Basidiospores globose, prominently apiculate, 4–4.5 μm in diameter, smooth, thin-walled, the walls subhyaline, amyloid.

Distribution: H.P.: Mahasu; U.K.: Nainital; J&K: Bhadarwah.

Collection examined: GSR 5009, SSR 5213, IBP 37447.

Substratum: On angiospermic twigs, bark of *Quercus incana*, twigs of *Berberis* sp..

Remarks: This species was first reported in India by Thind and Rattan (1968) from the N.W. Himalayas. This species is characterized by dimitic hyphal system, generative hyphae devoid of clamps; thin-walled, negative to sulfovanillin gloeocystidia, 4-sterigmate, utriform basidia and smooth, globose, amyloid basidiospores. Above collection resembles *S. duriusculum* as described by Thind and Rattan (1968). It is very similar to *S. portentosum* except for the absence of gloeocystidia in *S. portentosum*.

Scytinostroma ochroleucum (Bers. & Torrend.) Donk, Fungus 26: 20. 1956 = *Gloeocystidium ochroleucum* Bers. & Torrend., in Torrend., Broteria Ser. Bot. 11: 81. 1913. Plate 6.35e, Fig. 6.96a–d

Fructification resupinate, membranous, loosely adnate, widely effused, up to 280 μm thick in section, egg-yellow; hymenial surface smooth to tuberculate, cream yellow to egg-yellow, not creviced. Subuiculum composed of basal zone of compactly arranged, agglutinated, repent hyphae and an upper zone of loosely woven semi-erect hyphae.

Hyphal system dimitic; generative hyphae subhyaline, 2–3.5 μm wide, branched, septate, without clamps, thin-walled, non-dextrinoid, acyanophilous; skeletal hyphae 1.5–2.5 μm wide, branched, branches dichotomous to irregular, ultimate branchlets ending in hymenium are with tapering ends. Gloeocystidia thin-walled, negative to sulfovanillin. Basidia utriform, 4-sterigmate. Basidiospores 8.1–12.2 × 6.1–8.2 μm, ellipsoid to broadly ellipsoid, smooth, thin-walled, non-amyloid, acyanophilous.

Distribution: H.P.: Manali, Kullu.

Collection examined: GSR 5070, IBP 37448.

Substratum: On coniferous stump.

Remarks: This species is fairly common in the Himalayas. Earlier, Rattan (1977) has reported it from the N.W. Himalayas. This species is characterized by the loosely adnate, egg-yellow, thin fructification; thin-walled, generative hyphae without clamps, dextrinoid, cyanophilous, thick-walled skeletal hyphae, thin-walled, negative to sulfovanillin gleocystidia, 4-sterigmate, utriform basidia and large, non-amyloid basidiospores. Above collection is quite typical of *S. ochroleucum* as described by Rattan (1977).

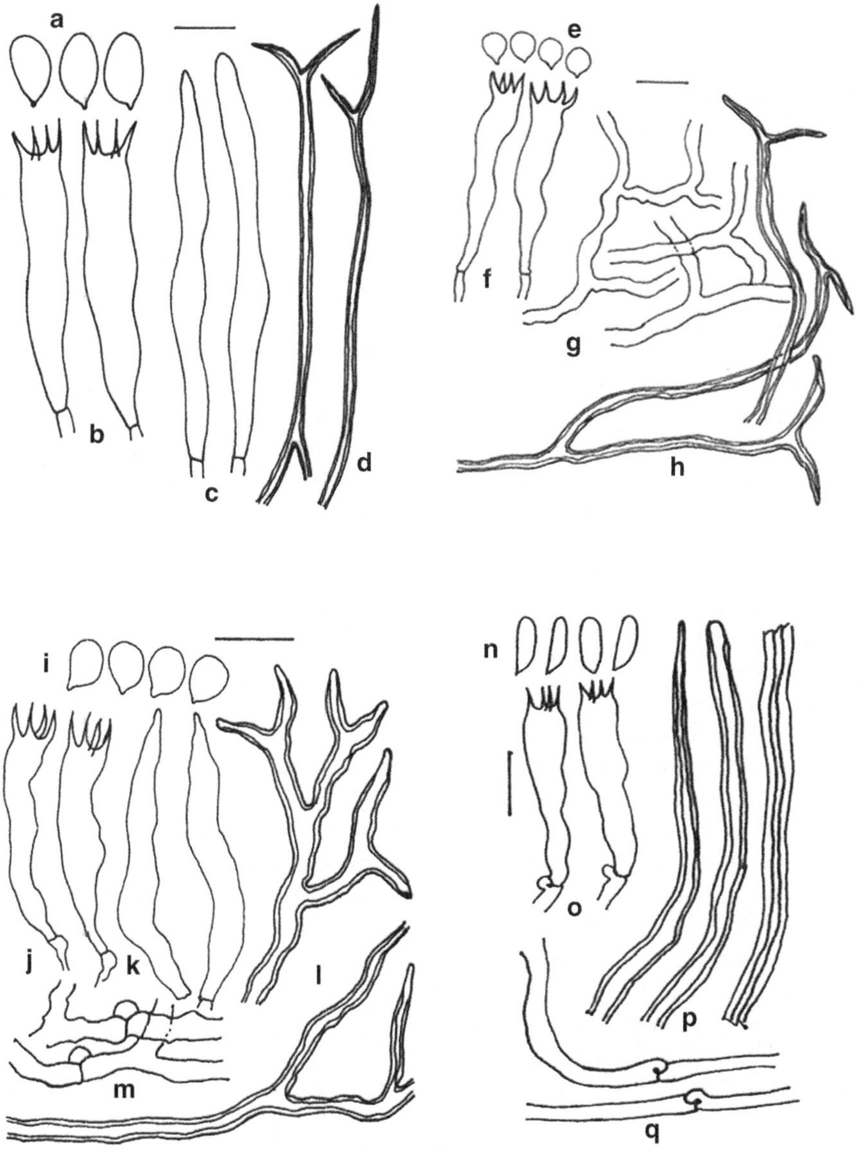

Fig. 6.96 (**a–d**) *Scytinostroma ochroleucum* (**a**) Basidiospores, (**b**) Basidia, (**c**) Cystidia, (**d**) Skeletal hyphae; (**e–h**) *Scytinostroma portentosum* (**e**) Basidiospores, (**f**) Basidia, (**g**) Generative hyphae, (**h**) Skeletal hyphae; (**i–m**) *Scytinostroma pulverulentum* (**i**) Basidiospores, (**j**) Basidia, (**k**) Cystidia, (**l**) Skeletal hyphae, (**m**) Generative hyphae; (**n–q**) *Scytinostroma rhizomorpharum* (**n**) Basidiospores, (**o**) Basidia, (**p**) Skeletal hyphae, (**q**) Generative hyphae

Scytinostroma portentosum (Berk. & Curt.) Donk, Fungus 26: 20. 1956 = *Corticium portentosum* Berk. & Curt., in Berk., Grevillean 21: 3. 1873. Plate 6.35f, Fig. 6.96e–h

Fructification resupinate, adnate, membranous, widely effused; hymenial surface cream to pale yellow, smooth to rough; margin thin when young, adnate, concolorous. Context subhyaline, stratose.

Hyphal system dimitic; generative hyphae 1.2–2 μm wide, septate, branched, clamps are absent, thin-walled acyanophilous; skeletal hyphae 1.5–2.5 μm wide, branched, aseptate, thick-walled, cyanophilous. Basidia utriform, 4-spored, 25–35 × 4.5–5.5 μm. Basidiospores globose, mainly apiculate, thin, smooth, subhyaline walls, 4.5–6 μm in diameter.

Distribution: H.P.: Kullu; U.K: Chakrata, Nanital; J&K: Bhadarwah.

Collection examined: GSR 5365, SSR 5413, L 37449.

Substratum: On stump under an angiospermic forests.

Remarks: The species is very common and reported (Rehill and Bakshi 1965) from the Mussoorie hills. It differs from *S. durisuculum* in thicker fructification broader skeletal hyphae and mainly, absence of gloeocystidia.

Scytinostroma pulverulentum Boidin & Dhingra, In Boidin & Lanquetin, Le Genre *Scytinostroma* Donk. (Basidiomycetes, Lachnocladiaceae), Bibliotheca Mycologia 114: 94, 1987. Fig. 6.96i–m

Fructification resupinate, membranous, coriaceous, effused; hymenial surface smooth, creamish-white to yellowish-white with orange tint; margin thinning, determinate to indeterminate.

Hyphal system dimitic; generative hyphae septate, branched, thin-walled, up to 4.2 μm wide, dextrinoid, acyanophilous; skeletal hyphae aseptate, thick-walled, cyanophilous, look like as dichohyphidia when branched with shorter internodes. Gloeocystidia subfusiform to subcylindrical, thin-walled, 28–42 × 4.5–6 μm. Basidia utriform, thin-walled, with basal clamp, 22.9–32.2 × 4.4–5.8 μm. Basidiospores ellipsoid to ovoid to subglobose, subhyaline, smooth, thin-walled, 5.0–5.9 × 4.6–5.1 μm, non-amyloid, acyanophilous.

Distribution: Bhutan: Thimphu, H.P.: Manali, Kullu.

Collection examined: GSD 19598, IBP 37450.

Substratum: On gymnospermic wood.

Remarks: This species is characterized by branched skeletal hyphae, presence of dichohyphidia, dimitic hyphal system and generative hyphae with clamps. The species differs from *Scytinostroma rhizomorpharum* in having dimitic hyphal system and clamped generative hyphae.

Scytinostroma rhizomorpharum Rattan, Trans. Brit. Mycol. Soc. 63: 8–9. 1974. Plate 6.36a, Fig. 6.96n–q

Fructification resupinate, membranous arising as small colonies later become effused; hymenial surface white to cream, smooth; margin thin concolorous. Context subhyaline in section composed of loosely interwoven hyphae. Hyphal system dimitic; generative hyphae branched, septate, clamps are present,

Plate 6.36 (**a**) *Scytinostroma rhizomorpharum*. (**b**) *Vararia brevispora*. (**c**) *Vararia sphericospora*. (**d**) *Metulodontia indica*. (**e**) *Gloiothele lactescense*. (**f**) *Metulodontia nivea*

thin-walled, acyanophilous, 2–3.5 µm wide; skeletal hyphae sparsely branched, aseptate, thick-walled, 1.5–3.0 µm wide. Rhizomorphs are abundant white to cream, branched. Basidia utriform, 4-spored, 30–35 × 4.5–5.5 µm. Basidiospores cylindrical, shortly apiculate, smooth, thin and amyloid.

Distribution: U.K.: Hemkunt, Badrinath, Mussoorie.

Collection examined: SSR 5431, L 37451.

Substratum: On bark and wood of *Betula utilis*.

Remarks: The species is characterized by presence of abundant rhizomorphs, thin fructification and cylindrical non-amyloid basidiospores. It differs from *S. praestans* in having abundant rhizomorphs and smaller basidiospores.

Vararia P.Karst.,
Kritisk öf versigt af Finlands Basidsvampar
Tillagg 3: 32, 1898.

Fructifications resupinate, adnate, effused, membranous, annual or perennial; hymenial surface smooth to tuberculate, whitish to cream to ochraceous. Hyphal system monomitic, hyphae branched, septate, with or without clamps, thin-walled, subhyaline. Dichohyphidia dichotomously to irregularly branched, thick-walled, dextrinoid, cyanophilous. Gloeocystidia present or absent. Basidia utriform, 4-spored. Basidiospores globose to cylindrical, the walls thin to slightly thick, smooth or warted, amyloid or non-amyloid.

Vararia is differentiated from *Dichostereum* by the presence of smooth-walled basidiospores. It is also similar in certain aspects to *Scytinostroma*.

Fifty four species, widespread.

Lit.: Boidin & Lanquetin (*Bull. Soc. Mycol. Fr.* 91: 457, 1975), Stalpers (*Stud. Mycol.* 40: 103, 1996; Key).

Type Species: *Vararia investiens* (Schw.) Karst, 1898

Habitat: Dead wood

Himalayas: Five

Key to species

1. Generative hyphae without clamps .... *V. minidichophysa*
1. Generative hyphae with clamps 2
2. Hymenial surface pale or cream yellow 3
2. Hymenial surface pale orange to brownish to greyish orange 4
3. Basidia 26–30 × 4–4.6 µm; basidiospores ellipsoid .... *V. vassilievae*
3. Basidia 40–45 × 4–5 µm; basidiospores globose to subglobose.......... *V. brevispora*
4. Hymenial surface orange grey to pale orange;
   basidiospores broadly ellipsoid.......... *V. rugosispora*
4. Hymenial surface greyish orange to red haired;
   basidiospores globose .... *V. sphericospora*

Vararia brevispora S. S. Rattan, Bibliotheca Mycologica 60: 142 1977 = *Dichostereum brevisporum* (S. S. Rattan) Boidin & Lanq., Mycotaxon 6(2): 284 (1977). Plate 6.36b, Fig. 6.97a–e

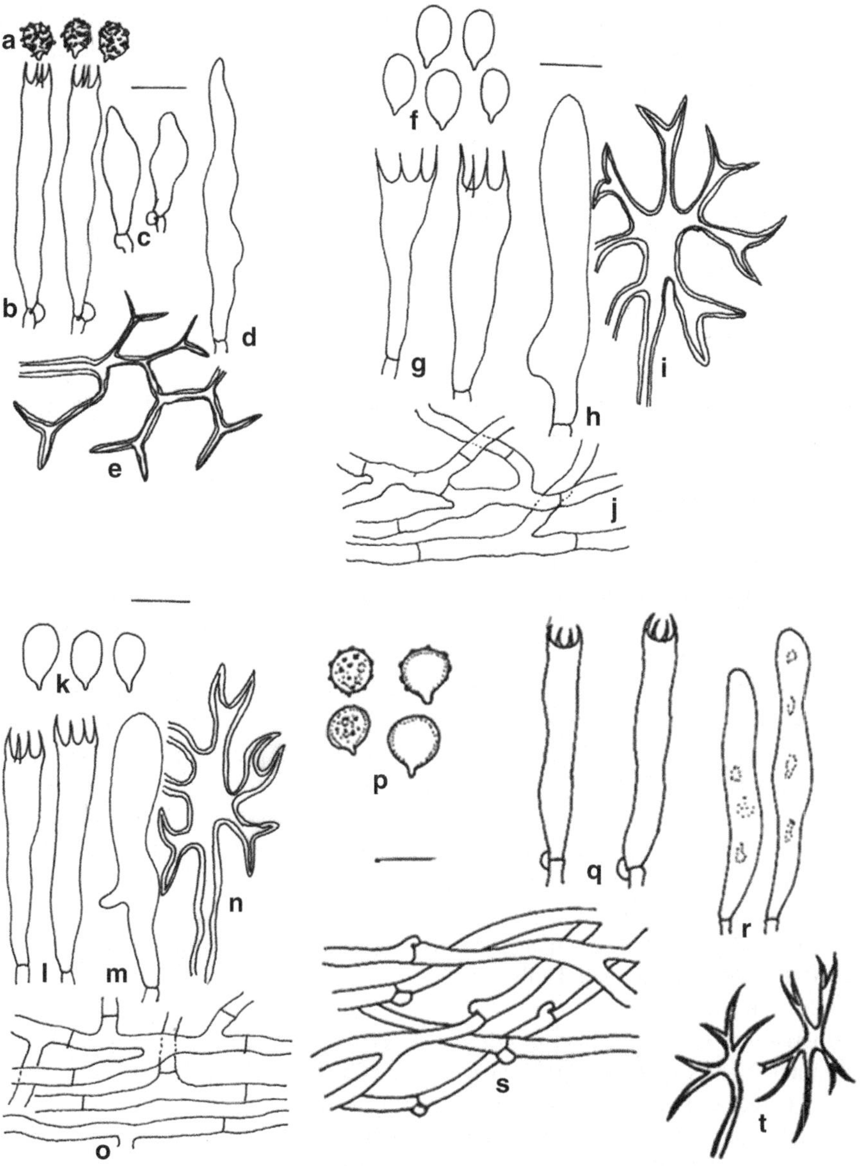

Fig. 6.97 (**a**–**e**) *Vararia brevispora* (**a**) Basidiospores, (**b**) Basidia, (**c**) Cystidia, (**d**) Gloeocystidia, (**e**) Dichohyphidia; (**f**–**j**) *Vararia minidichophysa* (**f**) Basidiospores, (**g**) Basidia, (**h**) Cystidia, (**i**) Dichohyphidia, (**j**) Generative hyphae; (**k**–**o**) *Vararia rugosispora* (**k**) Basidiospores, (**l**) Basidia, (**m**) Cystidia, (**n**) Dichohyphidia, (**o**) Generative hyphae; (**p**–**t**) *Vararia sphaericospora* (**p**) Basidiospores, (**q**) Basidia, (**r**) Cystidia, (**s**) Generative hyphae, (**t**) Dichohyphidia

Fructification resupinate, membranous, adnate, often arising as colonies and becomes effused; hymenial surface smooth to farinose, pale yellow; margin indeterminate, adnate, concolorous. Subiculum composed of compactly arranged dichohyphidia.

Hyphal system monomitic; generative hyphae branched, septate, thin-walled, clamps are present, subhyaline, 1.6–2.6 μm wide. Dichohyphidia are abundant. Gloeocystidia fusiform with apices, projecting out of the hymenium. Basidia utriform, 4-spored 40.1–45.2 × 4.1–5.1 μm. Basidiospores globose to subglobose, thin to thick-walled, warted, amyloid.

Distribution: H.P.: Kullu.

Collection examined: GSR 5060, IBP 37452.

Substratum: On rotten stump.

Remarks: The species resembles with *V. rhodospora* in having small basidiospores and presence of gloeocystidia.

Vararia minidichophysa Boid. & Lanq., Bull.trimest. Soc. mycol. France 91(4): 463, 1975. Fig. 6.97f–j

Fructification resupinate, adnate, effused up to 180 μm in section,; hymenial surface greyish to pale orange when fresh, cracked and pale orange to brownish orange on drying, smooth to tuberculate darkening in 3 % KOH sol; margin thinning, paler concolorous to indeterminate.

Hyphal systems monomitic; generative hyphae thin-walled, up to 3.5 μm wide, septate, without clamps; basal hyphae parallel to the substrate, less branched; subhymenial hyphae highly branched. Dendrohyphidia present dichotomously to irregularly branched, dark-brown, thick-walled with narrow trunk cyanophilous, present both in hymenium and subhymenium. Gloeocystidia 38–46 × 8.2–9.5 μm, thin-walled, smooth without basal clamp, oily contents negative to sulfovanillin. Basidia 32–45 × 65–8.2 μm, 4-sterigmate, clavate to subclavate, basal clamps absent, sterigmata up to 4.8 μm long. Basidiospores up to 13.2 × 7.8 μm, thin-walled, smooth, apiculate, broadly ellipsoid, inamyloid, acyanophilous.

Distribution: H.P.: Manali, Dalhousie.

Collection examined: IBP 37453.

Substratum: On fallen sticks of *Quercus leucotrichophora*.

Remarks: The species is marked by presence of broadly ellipsoid, inamyloid basidiospores, dichotomously to irregularly branched cyanophilous dendrohyphidia, subfusiform, negative to sulfovanillin, presence of gloeocystidia. It is a new record for Himalayas.

Vararia rugosispora Boidin, Lanq. & Gilles, Crytog. Mycol. 1(4): 328, 1980.
Fig. 6.97k–o

Fructification resupinate, adnate, effused, up to 280 μm thick in section; hymenial surface orange grey to pale orange when fresh, pale yellow to pale orange to greyish yellow and cracked on drying, smooth; margin thinning, paler concolorous to indeterminate.

Hyphal system monomitic; generative hyphae up to 4.5 μm wide, septate, without clamps, thin-walled; basal hyphae parallel to substrate, less branched; subhymenial hyphae vertical, highly branched. Dendrohyphidia dark brown in colour. thick-walled, dichotomously to irregularly branched, cyanophilous, dominant in hymenium and subhymenium. Gloeocystidia 52–64 × 8.5–13.2 μm, smooth, thin-walled, with oily content, negative to sulfovanillin, projects out from hymenium. Basidia 24–52 × 8.2–9.0 μm, clavate somewhat constricted without basal clamp, 4-sterigmate, sterigma up to 9.2 μm long. Basidiospores 10.3–13.0 × 6.0–7.2 μm, thin-walled, broadly ellipsoid, smooth, apiculate, inamyloid, acyanophilous.

Distribution: H.P.: Shimla, Manali.

Collection examined: IBP 37453.

Substrate: On decaying logs of *Quercus leucotrichophora.*

Remarks: The species is characterized by broadly ellipsoid basidiospores, presence of (Sulfovanillin -ve) gloeocystidia and presence of dichotomously to irregularly branched dendrohyphidia. It is a new record for Himalayas.

Vararia sphaericospora Gilbertson, Pap. Michigan Acad. Sciences Art Lett. 50: 176, 1965. Plate 6.36c, Fig. 6.97p–t

Fructification resupinate, stratose, adnate, effused, up to 430 μm thick in section, membranous; hymenial surface smooth, to farinose, cracks developing on drying, greyish-orange to red haired; margin thinning to indeterminate.

Hyphal system monomitic; generative hyphae branched, septate, clamped, subhyaline, thin-walled, up to 3.4 μm wide; dichohyphidia abundant, present both in context and hymenium, hyaline to light brown, thick-walled, dextrinoid, cyanophillous, branched, branches almost dichotomous; gloeocystidia 42–85 × 5.4–9.2 μm, subcylindrical to clavate, subventricose, thin-walled with oily contents, immersed, negative to sulfovanillin (S^-). Basidia 34–62 × 5.4–7.4 μm, subclavate, thin-walled, with a basal clamp, oily contents, 4-sterigmate, sterigma up to 4.8 μm long. Basidiospores globose, smooth, 6.4–9.2 μm in diameter, subhyaline, thin to thick-walled, non-amyloid, acyanophilous, uniguttulate.

Distribution: West Bengal: Darjeeling, Tiger hill; H.P.: Shimla-Glen forests, Manali; U.K.: Nainital.

Collection examined: IBP 37453.

Substratum: On stump of *Cryptomeria japonica*, dead twigs.

Remarks: This species is characterized by globose, non-amyloid basidiospores. It is a new record for Himachal Pradesh and a new record for N.W. Himalayas.

Vararia vassilieve Parmasto, Eesti NSV Tead. Akad. Toimet Biol. seer 14(2): 231. 1965. Fig. 6.98a–d

Fructification resupinate, membranous, adnate, widely effused up to 140 μm thick in section; hymenial surface cream-yellow, ochre-yellow, smooth to farinaceous, not creviced; margin thinning, adnate, white to paler concolorous. Subiculum composed of compactly arranged hyphae and dichohyphidia.

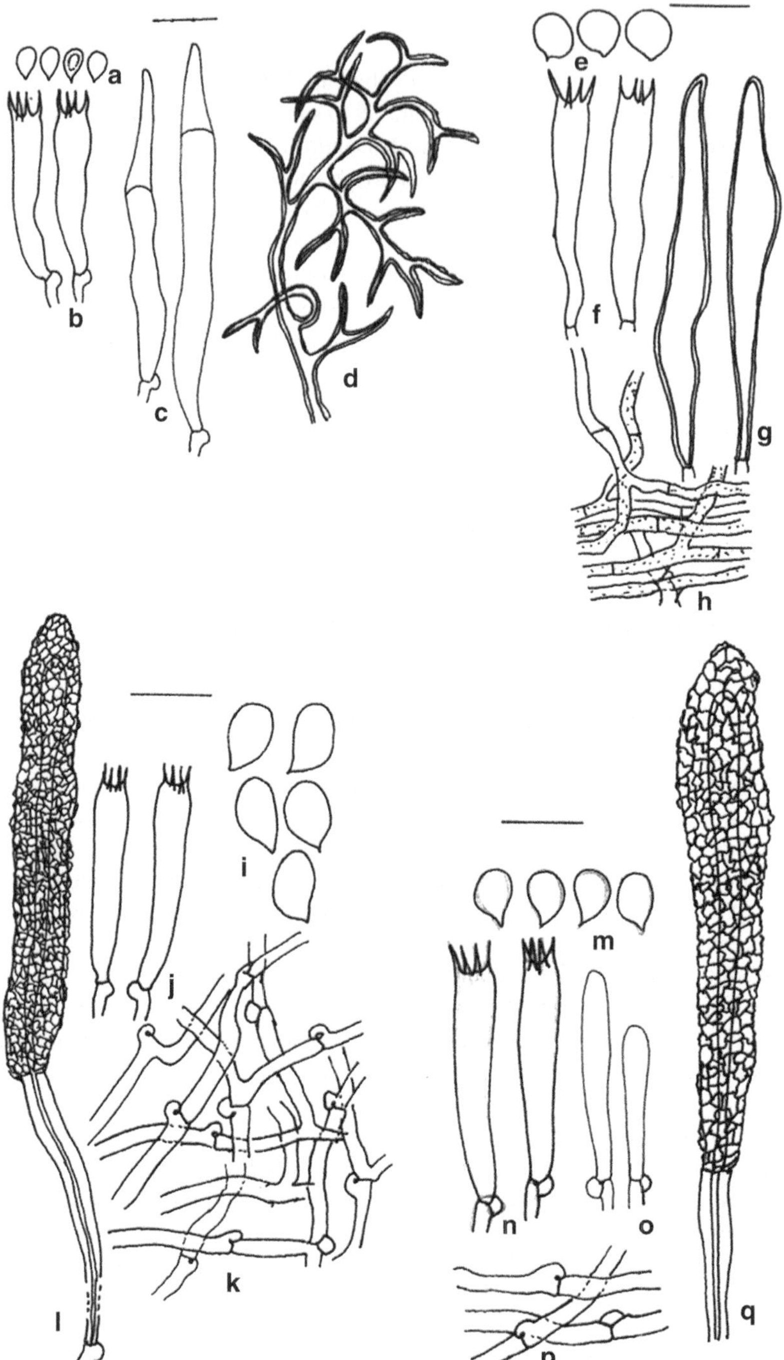

Fig. 6.98 (**a–d**) *Vararia vassilieve* (**a**) Basidiospores, (**b**) Basidia, (**c**) Cystidia, (**d**) Dichohyphidia; (**e–h**) *Gloiothele citrina* (**e**) Basidiospores, (**f**) Basidia, (**g**) Cystidia, (**h**) Generative hyphae; (**i–l**) *Metulodontia indica* (**i**) Basidiospores, (**j**) Basidia, (**k**) Generative hyphae, (**l**) Encrusted Cystidia; (**m–q**) *Metulodontia nivea* (**m**) Basidiospores, (**n**) Basidia, (**o**) Lamprocystidia, (**p**) Generative hyphae, (**q**) Encrusted Cystidia

Hyphal system monomitic; hyphae up to 3.2 µm, wide, branched, septate, clamped, branches subhyaline, thin-walled, non-dextrinoid. Dichohyphidia abundant, present both in the context and hymenium, branched with short internode, ultimate branchlet spiny, walls thin to moderately, subhyaline, dextrinoid, cyanophilous. Gloeocystidia 24.8–59.8 × 4–5.9 µm, fusiform to subcylindrical, with acute apices and subhyaline granular contents, immersed or projecting slightly out of the hymenium, thin-walled, subhyaline. Basidia 26–30 × 4–4.6 µm, utriform, 4-spored. Basidiospores 5–6.2 × 2.4–3 µm, ellipsoid, uniguttulate, guttule small, minutely apiculate, thin-walled, subhyaline, nonamyloid.

Distribution: H.P.: Shimla-Chadwick falls, Kullu, Mahasu; J&K: Gulmarg.

Collection examined: SSR5302, 5339, 5713, IBP 37454.

Substratum: On stump of *Abies pindrow*.

Remarks: The characteristic features of this species are nonamyloid, ellipsoid, smooth walled basidiospores and fusiform to subcylindrical gloeocystidia with acute apices. It shows close resemblance with *V. racemosa* (Burt.) Rog. & Jacks. but the basidiospores in that case are slender and appear more or less cylindrical. *Vararia ochroleuca* (Bourd. & Galz.) Donk also closely related species has larger basidiospores and hyphae without clamps.

Family-Peniophoraceae

Key to genera

1. Hyphae without clamps .... *Gloiothele*
1. Hyphae with clamps 2
2. Basidiospores amyloid .... *Vesiculomyces*[a]
2. Basidiospores inamyloid 3
3. Lamprocystidia brown over the entire length, basidiospores globose to subglobose .... *Duportella*[a]
3. Lamprocystidia hyaline or only brown at the basal part, basidiospores allantoid to suballantoid to broadly ellipsoid to ovoid to globose 4
4. Basidiospores broadly ellipsoid .... *Metulodontia*
4. Basidiospores allantoid to suballantoid to broadly ellipsoid to ovoid to globose .... *Peniophora*

[a]Extra limital, not included in this work

Gloiothele Bres.,

Annls mycol. 18(1–3): 44, 1920.

Fructification resupinate, adnate, effused; hymenial surface smooth to tuberculate; hyphal system monomitic; generative hyphae septate, branched, without clamps; gloeocystidia may or may not present, vesicular, with guttulate content. SA+. Basidia clavate to cylindrical, 4-sterigmate. Basidiospores globose to subglobose, smooth, apiculate, amyloid, thin-walled, acyanophilous.

Six species, wide spread

Lit.: Legon & Peglar (*Mycologist* 16: 177, 2002; *Gloiothele lactescence*), Larsson & Larsson (*Mycol.* 95: 1037, 2003; phylogeny)

Type species: *Poria lamellose* Henn. 1902

Habitat: Decayed wood

Himalayas: Two

Key to species

1. Basidiospores globose to subglobose..*G. citrina*
1. Basidiospores ellipsoid ....*G. lactescens*

Gloiothele citrina (Pers.) Ginns & G. W. Freeman, Biblthca. Mycol. 157: 55, 1994 = *Thelephora citrine* Pers., Mycol. Eur. 1: 136, 1822 = *Gloeocystidiellum citrinum* (Pers.) Donk, Fungus Wageningen 26: 9 (1956) = *Corticium citrinum* (Pers.) Pers., Fl.mycol.France (Paris): 6(1888). Fig. 6.98e–h

Fructification resupinate, adnate, effused, membranous, often arise as small colonies; hymenial surface creamish white to pale yellow to greyish yellow to greyish yellow to brownish orange when fresh, paler on drying, smooth to tuberculate; margin thinning, white, paler to concolorous.

Hyphal system monomitic; generative hyphae up to 4.0 µm wide, septate, thin-walled, branched, clamps absent; basal hyphae parallel to the substrate; subhymenial hyphae compactly packed. Cystidia absent. Gloeocystidia 51–80 × 8.5–11.8 µm, thin-walled, ends tapering, immersed, thin-walled. Basidia 30–44.5 × 6.2–6.9 µm, 4-sterigmate, clavate, without basal clamp having oily contents. Basidiospores 5.8–6.4 µm in diameter, apiculate, smooth, globose to subglobose, thin-walled, acyanophilous, amyloid.

Distribution: H.P.: Kullu.

Collection examined: GSR 5076, IBP 37446.

Substratum: On decaying stumps of *Quercus incana*, decaying stumps of *Cedrus deodara*.

Remarks: The species is characterized by the presence of vesicular gloeocystidia and globose, smooth, apiculate basidiospores. Rattan (1977) first described it and reported from Kullu as *Gloeocystidiellum citrinum*.

Gloiothele lactescens (Berk.) Hjortstam, Windahlia 17: 58 (1987). Plate 6.36e

Fructification resupinate, adnate, effused, up to 270 µm thick in section; hymenial surface smooth to tuberculate to cracked, orange white to greyish orange when fresh, greyish orange to brownish orange on drying; margins thinning, paler concolorous, to indeterminate.

Hyphal system monomitic; generative hyphae up to 3.5 µm wide, branched, septate, without clamps, thin-walled; basal hyphae parallel to the substrate, loosely interwoven; subhymenial hyphae vertical, denser. Basal zone with heavy encrustation which dissolves in 3 % KOH sol. Numerous projecting paraphysoid hyphae present. Gloeocystidia 74.0–86.0 × 7.0–8.2 µm, tubular, somewhat sinuous, thin-to somewhat thick-walled, with oily contents positive to sulfovanillin.

Basidia 24.0–39.0 × 4.1–5.2 µm, narrowly clavate, somewhat sinuous, thin-walled, 4-sterigmate, without basal clamp; stergmata up to 3.5 µm long. Basidiospores 5.2–6.4 × 2.8–3.1 µm wide, ellipsoid, smooth, thin-walled, amyloid, acyanophilous.

Distribution: U. K.: Haridwar.

Collection examined: L 42968.

Substratum: On angiospermous stump.

Remarks: This species is easily recognized by the cracked hymenial surface, generative hyphae without clamps, tubular cystidia and ellipsoid basidiospores. Thind and Rattan (1973a) were the first to describe it as *G. lactescens* following Boidin (1951) from district Dehradun (U.K.) and H. P. (Bhosle et al. 2005) and Ranadive et al. (2011) listed it from Western Ghats (Maharashtra), followed by Dhingra et al. (2011) from Bhutan in Eastern Himalaya.

Metulodontia Parm.,
Conp. Syst. Cort. p. 117. 1968.

Fructifications annual or perennial, resupinate, membranous to membranous-ceraceous; hymenial surface cream to light brown, smooth to hydnoid. Context subhyaline in section composed of somewhat compactly arranged hyphae. Hyphal system monomitic, hyphae branched, clamped, thin to slightly thick-walled, occasionally agglutinating and difficult to discern individually. Cystidia conical or fusiform, thick-walled, heavily incrusted (lamprocystidia). Basidia clavate-cylindrical, to utriform, 4-spored. Basidiospores ellipsoid to subcylindrical, the walls thin, subhyaline, smooth, nonamyloid, acyanophilous.

Two species, widespread

Lit.: Jülich (*Persoonia* **8**: 78, 1974).

Type Species: *Kneiffia nivea* Karst. 1896

Habitat: Dead wood

Himalayas: Two

Keys to species

1. Lamprocystidia 55–500 × 7–11 µm. Basidiospores 7–8 × 4–5 µm, broadly ellipsoid.. *M. indica*
1. Lamprocystidia 48–75 × 8.2–11 µm. Basidiospores 3.6–5.6 × 2.6–3 µm, broadly ellipsoid to ovoid.. *M. nivea*

Metulodontia indica (K. S. Thind & S. S. Rattan) Rattan, Biblthca Mycol. 60: 316. Basionym: *Peniophora indica* Thind & Rattan, Mycologia 65: 1250. 1973. Plate 6.36d, Fig. 6.98i–l

Fructifications resupinate, membranous, adnate, often arising in small colonies coalescing to form large patches; hymenial surface violet, smooth but appears pilose under the hand lens, continuous, somewhat shiny, not creviced; margin thinning, or occasionally abrupt, adnate, white to paler concolorous.

Hyphal system monomitic, hyphae 3.2–6.2 µm wide, branched, septate, clamped, the walls thin to slightly thick (up to 1 µm), subhyaline. Cystidia (lamprocystidia)

55–500×7–11 µm, cylindrical or lanceolate with a long narrow pedicle, usually arising from different parts of the context, immersed or projecting out of the hymenium, the walls subhyaline, thick (up to 4 µm), heavily impregnated with subhyaline crystals especially in the apical half. Basidia 35.2–40×5–7.6 µm, clavate, 4-spored, sterigma up to 5.2 µm long. Basidiospores 7–8×4–5 µm, broadly ellipsoid, minutely apiculate, thin-walled, subhyaline, smooth, non-amyloid, acyanophilous.

Distribution: Panjab: Chandigarh; H.P.: Baddi.

Collection examined: SSR 5218, IBP 42967.

Substratum: On wood of *Dalbergia sissoo*.

Remarks: This species is characterized by the membranous texture, distinct and clamped hyphae, presence of lamprocystidia and broadly ellipsoid basidiospores.

Metulodontia nivea (P. Karst.) Parm., Consp. Syst. Cort. p.118. 1968=*Kneiffia nivea* Karst., Hedwigia 35(4): 173. 1896. Plate 6.36f, Fig. 6.98m–q

Fructifications resupinate, membranous, adnate, widely effused; hymenial surface white to cream, smooth, continuous, not creviced; margin thinning, byssoid to fibrillose, adnate, white. Context composed of somewhat loosely woven hyphae.

Hyphal system monomitic, hyphae 2–4.2 µm wide, branched, septate, clamped, the walls thin to slightly thick or firm (up to 0.5 µm), subhyaline. Cystidia (lamprocystidia) 48–75×8.2–14 µm, clavate-cylindrical to cylindrical, immersed but usually projecting out of the hymenium, the walls subhyaline, thick, heavily impregnated with crystalline matter. Basidia 21–25.2×4–4.6 µm, clavate-cylindrical, 4-spored. Basidiospores 3.6–5.6×2.6–3 µm, broadly ellipsoid to ovoid, minutely apiculate, the walls thin, subhyaline, smooth, nonamyloid.

Distribution: H.P.: Khajjiar, Shimla, Chamba; U.K.: Mussoorie.

Collection examined: SSR 5134, 5420, 5556, IBP 37694.

Substratum: On stump under mixed forest, on stump of *Cedrus deodara*, *Quercus incana*.

Remarks: This species is characterized by the membranous fructifications, white to cream hymenial surface, presence of lamprocystidia and small basidiospores.

Peniophora Cooke, *Grevillea* **8**: 20, 1879

Fructification annual or perennial, lignicolous, resupinate, adnate or rolling off from the margins, effused or wart like, ceraceous; hymenial surface smooth or rarely tuberculate; brightly coloured in shades of yellow, red, grey violaceous brown; hyphal system monomitic; generative hyphae clamped or rarely without clamps; sulfocystidia or rarely dendrohyphidia present; basidia clavate to clavate-cylindrical, 4-sterigmate, with or without a basal clamp; basidiospores ellipsoid allantoid, smooth, thin-walled, non-amyloid, acyanophilous.

Sixty two species, widespread.

Lit.: Slysh (*Tech. Pub. Sta. Univ. Coll. Forestry, Syracuse* **83**, 1960; N.Y. State), Wu (*Mycotaxon* **85**: 187, 2003; Taiwan).

Type Species: *Peniophora quercina* (Pers. ex Fr.) Cooke.

Habitat: Dead wood
Himalayas: Nine
Keys to species

1. Basidiospores cylindrical .... *P. rhodocarpa*
1. Basidiospores allantoid to suballantoid to broadly ellipsoid to ovoid to subglobose 2
2. Basidia narrowly clavate .... *P. limitata*
2. Basidia clavate 3
3. Basidiospores broadly ellipsoid to ovoid to subglobose .... *P. ovalispora*[a]
3. Basidiospores not so 4
4. Hymenial surface violet to greyish violet .... *P. violaceolivida*
4. Hymenial surface greyish orange to reddish grey to greyish red to purplish grey to brownish grey 5
5. Gloeocystidia up to 200.0 μm long .... *P. incarnata*
5. Gloeocystidia up to 80.0 μm long 6
6. Basidiospores broadly ellipsoid, gloeocystidia rare .... *P. hallenbergii*[a]
6. Basidiospores allantoid to suballantoid, gloeocystidia common 7
7. Mostly on *Quercus* .... *P. quercina*
7. Mostly on other angiospermous hosts 8
8. Hyphae swell in 3 % KOH solution, with grainy crystals .... *P. rufa*
8. Hyphae not so, without grainy crystals 9
9. Basidiospores 7.8–9.0 × 2.8–3.5 μm .... *P. pithya*
9. Basidiospores 6.8–10 × 2.5–3.1 μm 10
10. Fructifications loosening from margins, basidia narrowly clavate .... *P. cinerea*
10. Fructifications not so, basidia clavate .... *P. suecica*[a]

[a]Extra limital, not included in the text

Peniophora cinerea (Fr.) Cooke, *Grevillea* **8**: 20. 1879. Fig. 6.99a–f

Fructification resupinate, membranous, adnate, often arising as small circular colonies which may coalesce later and becomes widely effused, up to 150 μm thick in section; hymenial surface cream with a violet tinge or pinkish buff, smooth to tuberculate, tubercles prominent but sparse, aerolately cracked in thicker parts; margin thinning, adnate, white to paler concolorous.

Hyphal system monomitic, hyphae 2.5–4 μm wide, branched, septate, clamped. Thin to thick-walled, subhyaline to light brown. Gloeocystidia absent. Cystidia 35–45 × 7–12 μm, subfusiform, arising from different layers of the subiculum, immersed or projecting out of the hymenium, thick-walled, subhyaline, heavily impregnated with subhyaline crystals. Basidia 25–29.8 × 5–6 μm, clavate-cylindrical, 4-spored, sterigmata up to 5 μm long. Basidiospores 7–9 × 2.3–3 μm, ellipsoid, minutely apiculate, thin-walled, subhyaline, smooth, non-amyloid.

Distribution: H.P.: Kullu- Jaggat such, Pulga; Chamba- Khijjiar; J&K: Pehalgam.

Collection examined: SSR: 5666, 5714, 5807, IBP 37454.

Substratum: Twigs under *Betula utilis*, fallen sticks under mixed forest, burnt angiospermic logs, cut stumps.

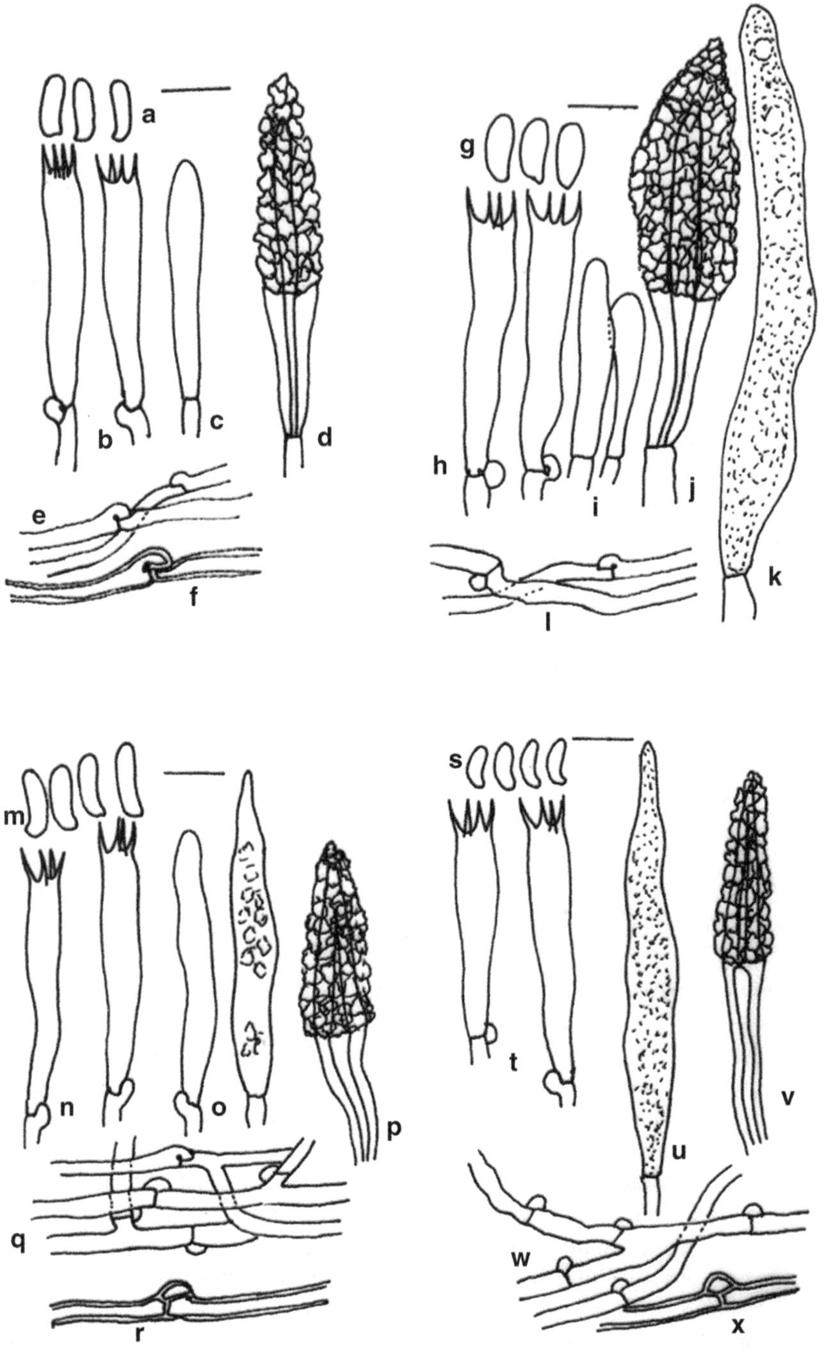

Fig. 6.99 (**a**–**f**) *Peniophora cinerea* (**a**) Basidiospores, (**b**) Basidia, (**c**) Cystidia, (**d**) Encrusted cystidia, (**e**) Thin-walled generative hyphae, (**f**) Thick-walled generative hyphae; (**g**–**l**) *Peniophora incarnata* (**g**) Basidiospores, (**h**) Basidia, (**i**) Basidioles, (**j**) Cystidia, (**k**) Gloeocystidia, (**l**) Generative hyphae; (**m**–**r**) *Peniophora limitata* (**m**) Basidiospores, (**n**) Basidia, (**o**) Gloeocystidia, (**p**) Cystidia (**q**) Generative hyphae, (**r**) Thick-walled generative hyphae; (**s**–**x**) *Peniophora pithya* (**s**) Basidiospores, (**t**) Basidia, (**u**) Gloeocystidia, (**v**) Cystidia, (**w**) Generative hyphae, (**x**) Thick-walled generatve hyphae

Remarks: This species is easily recognised by cream-violet to pinkish buff hymenial surface, absence of gloeocystidia and ellipsoid basidiospores.

Peniophora incarnata (Pers. ex. Fr.) Karst. Hedwigia 28: 27. 1889 = *Thelephora incarnata* Pers. ex Fr., Syst. Mycol. 1: 444. 1821. Fig. 6.99g–l

Fructification resupinate, membranous, adnate, arising, as small, colonies which may become confluent later and become widely effused, up to 200 µm thick in section; hymenial surface orange or bright red to reddish brown, smooth to finely tuberculate, continuous, not creviced; margin thinning, adnate, white to paler concolorous.

Hyphal system monomitic, hyphae 3.2–4.5 µm wide, branched, septate, clamped, thin to thick-walled, subhyaline to tinted brown. Cystidia 35–50 × 9–14 µm, conical to subfusiform, arising from different parts of the context, immersed or projecting out of the hymenium, thick-walled, subhyaline, heavily impregnated with abundant crystals. Gloeocystidia 35–90 × 7.5–11 µm, cylindrical to flexuous, immersed at different levels in the context, thin-walled, subhyaline. Basidia 30–40 × 5.5–6.5 µm, clavate, 4-spored. Basidiospores 8–9.8 × 2.5–3.5 µm, cylindrical, minutely apiculate, thin-walled, subhyaline, smooth, non-amyloid.

Distribution: N.W. Himalaya, H.P.: Kullu, Solan.

Collection examined: IBP 37455.

Substratum: On bark under a mixed forest.

Remarks: This species is rather rare in the N. W. Himalayas and is known earlier from a single specimen which was collected by W. Gollan in 1900 from the Mussoorie and later reported as *Corticium incarnatum* by Hennings (1901). The species is characterized by the colour of the hymenial surface, presence of gloeocystidia and cylindrical basidiospores.

Peniophora limitata (Chaillet ex Fr.) Cooke, Grevillea 8(no. 45): 21 (1879). Fig. 6.99m–r

Fruitbody resupinate, adnate, effused, up to 450 µm thick in section; hymenial surface in young specimens smooth, continuous, orange-grey, tuberculate and cracked in pieces, violaceous grey becoming irregularly on drying; margin dark, blackish, thinning.

Hyphal system monomitic; generative hyphae clamped, hyphae hyaline, thin-walled to thick-walled, pigmented brown; subhymenium of dense vertical hyphae together with enclosed cystidia stratified in thick specimen. Cystidia 27.8–45.0 × 6.4–9.0 µm, numerous, at first thin-walled, with oily contents, with time becoming thick-walled, encrusted in the apical half. Basidia 32.3–47.5 × 5.2–6.5 µm, narrowly clavate, 4-sterigmate, with a basal clamp. Basidiospores 6.0–10.2 × 2.8–3.5 µm, smooth, allantoid, thin-walled, non-amyloid, acyanophilous.

Distribution: A.P.: West Kameng, Wang Basti: Sikkim: Tashi view point, Bhutan: Thimphu, Chankahg: H.P.: Manali, Dalhousie.

Collection examined: GSD 19273, IBP 42970.

Substratum: On angiospermous wood.

Remarks: This species is characterized by orange-grey to brownish, cracked fructifications with dark to blackish margin; hyaline to pigmented brown, thin to thick walled, encrusted cystidia; narrowly clavate, 4-spored basidia and allantoids, thin-walled basidiospores.

Peniophora pini (Schleich.) Boidin, Revue de Mycologie, 21: 123, 1956.
Plate 6.37a, Fig. 6.100q–v

Fructification resupinate, ceraceous, effused; hymenophore smooth to tuberculate, reddish when young, concolorous with hymenial surface.

Hyphal system monomitic; generative hyphae thin-walled, clamped, 2.8–6 μm wide, hyaline yellowish to light brown, denser and vertically arranged in subhymenium. Dendrohyphidia absent. Cystidia of two types: (a) lamprocystidia present in subhymenium, naked when young (b) gloeocystidia thin-walled, variable in shape. Basidia subclavate, 22–40 × 5–6 μm, 4-sterigmate with basal clamp. Basidiospores smooth, thin-walled, hyaline, 6–8 × 2.5–3 μm, acyanophilous.

Distribution: H.P.: Kullu.

Collection examined: IBP 37671.

Substrate: On branches of gymnospermic tree.

Remarks: The species is characterized by presence of crustose spots, reddish-grey colour of fruiting body, presence of lamprocystidia, gloeocystidia and gelatinized walls of hyphae.

Peniophora pithya (Pers.) John Erikss., Symb. Bot. Upsal. 10: 45, 1950 = *Thelephora pithya* Pers., Myc. Eur. 1: 146, 1822. Plate 6.37b, Fig. 6.99s–x

Fruitbody resupinate, closely adnate, effused, up to 200 μm thick in section; hymenial surface smooth, continuous, light grey to medium grey when fresh, brownish grey in herbarium; margin darker brownish.

Hyphal system monomitic; generative hyphae up to 4.3 μm wide, branched, septate, clamped, young hyphae hyaline, thin-walled, old hyphae thick-walled, pigmented brown. Sub-hyphocystidia 56–113 × 8.0–13.0 μm, subcylindrical, thick-walled in the basal half gradually thinning above. Cystidia 31–50.0 × 6.8–9.8 μm, conical. Basidia 25.0–39.8 × 4.8–6.0 μm, clavate to subclavate, 4-sterigmate, with a basal clamp. Basidiospores 5.8–7.4 × 2.5–3.0 μm, smooth, allantoid, thin-walled, non-amyloid.

Distribution: West Bengal: Darjeeling; H.P.-Shimla, Manali.

Collection examined: GSD 19273, IBP 37458.

Substratum: On angiospermous wood.

Remarks: These collection resembles the description of *P. pithya*, as given by Eriksson et al. (1978), in all the macro and microscopic characters, but occurs on angiospermic host, as compared to the coniferous host.

Peniophora quercina (Pers. ex Fr.) Cooke, *Grev*. **8**: 20. 1879. Plate 6.37c, Fig. 6.100a–e

Fructifications resupinate, membranous to membranous-ceraceous, widely effused, adnate, up to 450 μm thick in section; hymenial surface reddish brown to pinkish

Plate 6.37 (**a**) *Peniophora pini*. (**b**) *Peniophora pithya*. (**c**) *Peniophora quercina*. (**d**) *Peniophora rufomarginata*. (**e**) *Aleurodiscus oakesii*. (**f**) *Aleurodiscus taxicola*

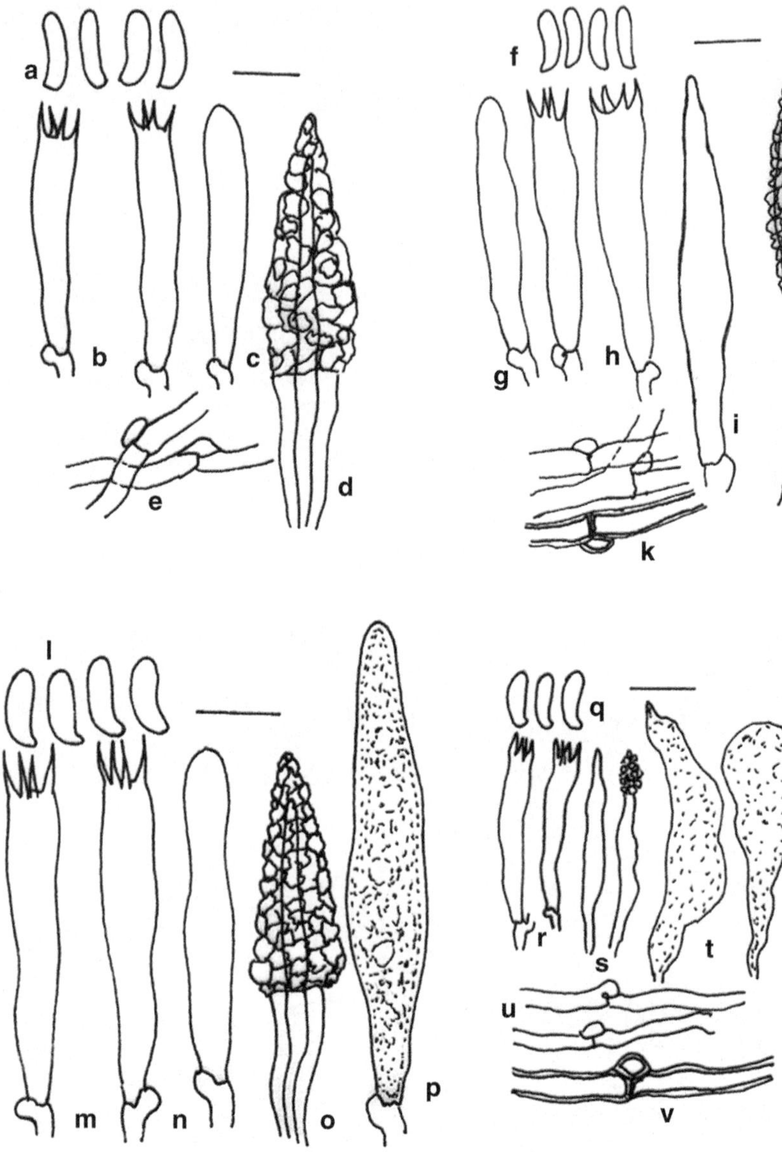

Fig. 6.100 (**a**–**e**) *Peniophora quercina* (**a**) Basidiospores, (**b**) Basidia, (**c**–**d**) Cystidia, (**e**) Generative hyphae; (**f**–**k**) *Peniophora rufomarginata* (**f**) Basidiospores, (**g**) Basidioles, (**h**) Basidia, (**i**–**j**) Cystidia, (**k**) Generative hyphae; (**l**–**p**) *Peniophora violaceolivida* (**l**) Basidiospores, (**m**) Basidia, (**n**) Basidioles, (**o**) Cystidia, (**p**) Gloeocystidia; (**q**–**v**) *Peniophora pini* (**q**) Basidiospores, (**r**) Basidia, (**s**) Lamprocystidia, (**t**) Gloeocystidia, (**u**) Thin-walled generative hyphae, (**v**) Thick-walled generative hyphae

brown when fresh but fades on drying, smooth to finely tuberculate but becomes rimoseor sulcate on drying, continuous, cracked on drying exposing the white context; margin thinning, byssoid, adnate, white to paler concolorous. Subiculum composed of basal zone (up to 150 thick) of compactly arranged, agglutinated repent hyphae and an upper zone (up to 300 μm thick) of somewhat agglutinated erect branches.

Hyphal system monomitic, hyphae up to 2.2–5.3 μm wide, branched, septate, clamped, thin-walled, subhyaline, naked or impregnated with fine crystals, often collapsing and agglutinating and difficult to discern; hyphae in the basal zone are more compactly arranged and tinted yellow. Cystidia 52–75 × 11.8–15 μm, fusiform to subcylindrical with acute to subobtuse apices, immersed at different levels in the subiculum or emerging out of the hymenium, thick-walled, subhyaline, heavily impregnated with subhyaline crystals. Gloeocystidia absent. Basidia 35.3–40 × 6.6–7 μm, clavate-cylindrical, 4-spored, sterigmata up to 5.5 μm long. Basidiospores 10–11.2 × 3–3.4 μm, cylindrical to slightly curved, minutely apiculate, the walls thin, subhyaline, smooth, non-amyloid.

Distribution: H.P.: Kullu- Pulga, Manali, Dalhousie.

Collection examined: SSR 5698, IBP 37459.

Substratum: Bark under a mixed forest.

Remarks: The species is characterized by subhyaline subiculum, reddish brown hymenial surface, absence of gloeocystidia and large cylindrical to curved basidiospores.

Peniophora rhodocarpa Rehill & Bakshi, Ind. For. Bull. (Dehra Dun) 242: 6. 1965 = *Lopharia rhodocarpa* (Rehill & B.K. Bakshi) S.S. Rattan Bibliotheca Mycologica 60:316.1977. Fig. 6.101a-e

Fructifications resupinate, loosely adnate, membranous to membranous-ceracerous, widely effused; hymenial surface pink to pinkish brown when fresh but usually fades on drying, smooth to finely tuberculate, continuous, not creviced; margin thinning, adnate, concolorous. Subiculum composed of a basal zone of repent hyphae and an upper zone of semi erect hyphae.

Hyphal system monomitic, hyphae 2.7–4 μm wide, branched, septate, clamped, thin to thick-walled, subhyaline to pigmented or light brown; hyphae are compactly arranged and more or less agglutinated especially in the cuticle. Tomentose hyphae 3–4 μm wide, sparsely branched, septate, clamped, moderately thick-walled, tinted brown, occurring singly but more often forming loose strands. Leptocystidia 51–90 × 6.5–9 μm, cylindrical to clavate-cylindrical, thin-walled, walls subhyaline but sometimes tinted brown in the basal part, usually smooth or rarely with a fine coating of subhyaline crystals, arising from the subhymenium and usually just emerging to projecting (up to 10 μm). Lamprocystidia 60–100 × 12–18 μm, fusiform to subcylindrical, thick-walled, subhyaline, heavily impregnated with subhyaline crystals, usually immersed in the upper part of subhymenium or rarely reaching up to the hymenial surface. Basidia 20.5–25 × 4–5 μm, clavate-cylindrical, 4-spored. Basidiospores 5–8.4 × 1.8–2.2 μm, suballantoid, minutely apiculate, the walls thin, subhyaline, smooth, non-amyloid.

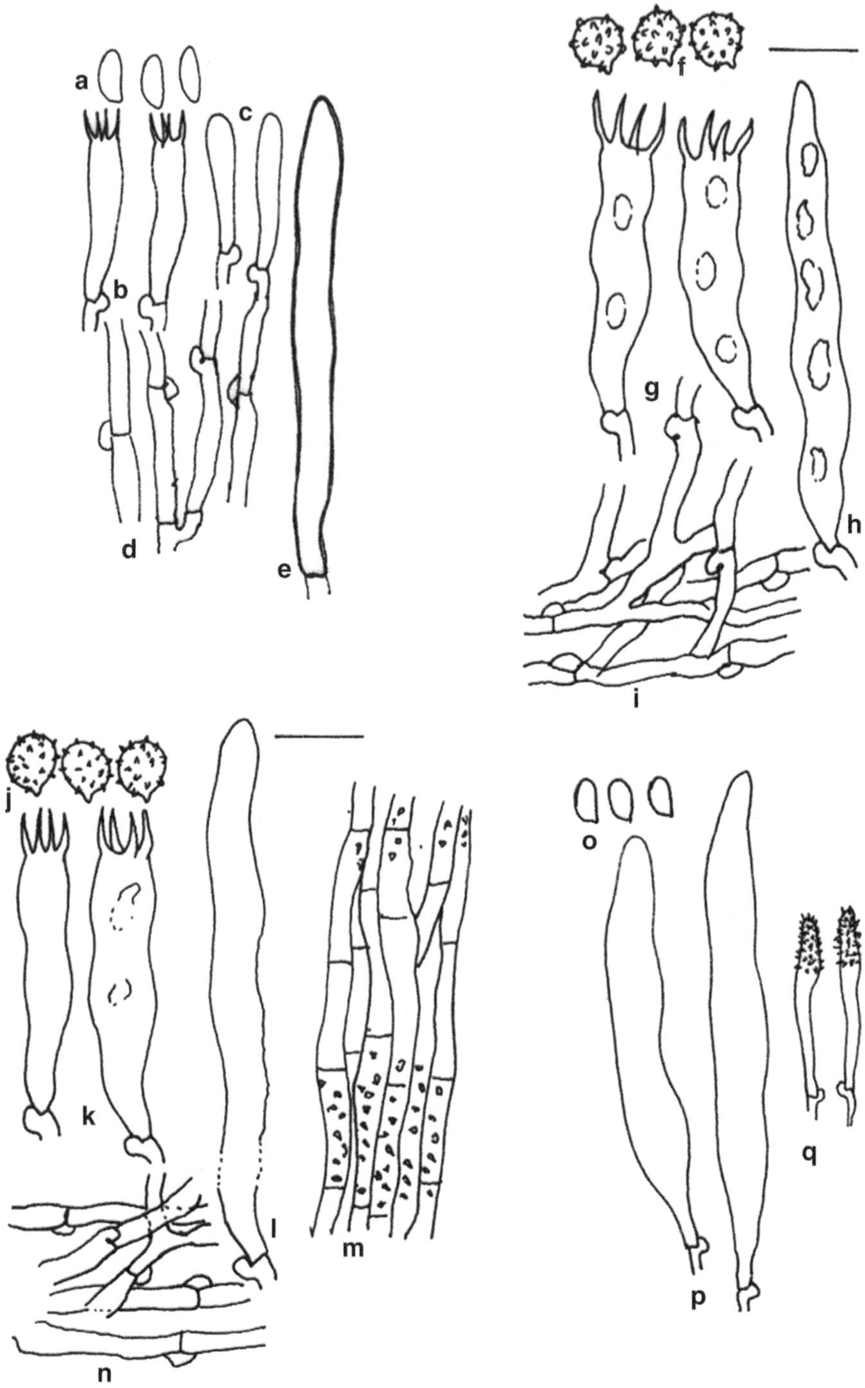

Fig. 6.101 (**a**–**e**) *Peniophora rhodocarpa* (**a**) Basidiospores, (**b**) Basidia, (**c**) Basidioles, (**d**) Generative hyphae, (**e**) Lamprocystidia; (**f**–**i**) *Boidinia furfuraceum* (**f**) Basidiospores, (**g**) Basidia, (**h**) Cystidia, (**i**) Generative hyphae; (**j**–**n**) *Boidinia lacticolor* (**j**) Basidiospores, (**k**) Basidia, (**l**) Gloeocystidia, (**m**) Basal hyphae, (**n**) Generative hyphae; (**o**–**q**) *Acanthofungus ahmadii* (**o**) Basidiospores, (**p**) Cystidia, (**q**) Basidia

Distribution: U.K.: Dehra Dun, Nainital.
Collection examined: L 37464, Rehill 6569.
Substratum: On dead twigs of *Castanea sativa*.
Remarks: This species is characterized by two types of cystidia, presence of cuticle and tomentum on the abhymenial side, monomitic hyphal system with clamped generative hyphae.

Peniophora rufomarginata (Pers.) Litsch., in Keissl., *Kryptog. Exs. Wien 2613, 1923*; *Ann. Nath. Mus. Wien* 36p. 76, 1923 = *Thelephora rufomarginata Pers., Mycol. eur.* Ip. 124, 1822. Plate 6.37d, Fig. 6.100f–k

Fructification resupinate, adnate, effused, up to 300 µm thick in section; hymenial surface smooth to more or less tuberculate, rose-grey to greyish- brown, fading on drying; margin in young state finally fibrillose, whitish, followed by a pale brown zone, the whitish fibrillose zone disappear with maturity.

Hyphal system monomitic; generative hyphae branched, septate, clamped, hyphae in margin and basidial bases thin-walled, other hyphae thick-walled, more or less pigmented brown. Cystidia 37.0–76.0 × 5.8–11.0 µm, developing to strongly encrusted metuloids. Basidia 25.4–35.0 × 5.2–6.0 µm, clavate to subclavate, 4-sterigmate and a basal clamp present. Basidiospores 4.8–8.0 × 2.3–3.0 µm, allantoid, smooth, thin-walled, non-amyloid.
Distribution: Bhutan: Thimphu, Ha, Kana.
Collection examined: GSD 19452.
Substratum: On decaying angiospermic log.
Remarks: This species is characterized by rose-grey to greyish-brown fruitbodies with light coloured margin, clamped generative hyphae, thick walled, encrusted cystidia, clavate to sub clavate, constricted, 4-spored basidia and allantoids, thin-walled basidiospores.

Peniophora violaceolivida (Sommf.) Mass. J. Linn. Soc. Bot. 25: 147. 1889 = *Thelephora violaceolivida* Sommf., F1. Lapp. Suppl. p. 283. 1826. Fig. 6.100l–p

Fructifications resupinate, membranous to compactly membranous, adnate, widely effused, up to 150 µm thick in section; hymenial surface smooth to finely tuberculate, violet to greyish violet, continuous, rarely cracking irregularly in its thicker parts; margin thinning, adnate, white to paler concolorous.

Hyphal system monomitic, hyphae 1.6–3 µm wide, branched, septate, clamped, the walls thin to slightly thick, subhyaline to tinted brown, often collapsing and agglutinating and difficult to discern. Cystidia 21–30 × 6–8.8 µm, conical to subfusiform, arising from different parts of context, immersed or projecting out of the hymenium, thick-walled, subhyaline, heavily impregnated. Gloeocystidia 5.2–8 µm broad, of variable length, cylindrical to flexous, immersed at different levels in the subiculum, thin-walled, subhyaline with granular contents staining deeply with phloxine. Basidia 20–24 × 4–5 µm, clavate-cylindrical, 4-spored, sterigmata up to 4.5 µm long. Basidiospores 8–10 × 2.4–3.3 µm, narrowly ellipsoid to allantoids, minutely apiculate, the walls thin, subhyaline, smooth, non-amyloid.
Distribution: U.K.: Pithoragarh, Bageshwar.
Collection examined: L 37461.

Substratum: Dead wood.

Remarks: The species is characterized by colouration of the hymenial surface and presence of gloeocystidia.

Family- Russulaceae

Boidinia Stalpers & Hjortstam,
Mycotaxon 14: 76, 1982.

Fructification resupinate, adnate, effused; hymenial surface smooth; hyphal system monomitic; generative hyphae septate,, clamps may or may not be present, hyphae branched; cystidia cylindrical to subfusiform, with crystalline encrustation which dissolve in KOH sol, thin-walled positive to sulfovanillin; basidia clavate to subclavate, 4-streigmate, with oily contents, with or without clamps; basidiospores globose to subglobose, apiculate, amyloid.

Ten species, wide spread

Lit: Wu & Buchanan (*Mycotaxon* 67: 123, 1998).

Type species: *Hyphochnus furfuraceus* Bresadola 1900

Habitat: Decayed wood

Himalayas: Two

Key to species

1. Hyphae with clamps.. *B. furfuraceum*
1. Hyphae without clamps.. *B. lacticolor*

Boidinia furfuraceum (Bres.) Donk, Fungus 26: 9, 1956 = *Hyphochnus furfuraceus* Bres., Fungi trid. 2: 97, 1900. Fig. 6.101f–i

Fructification resupinate, adnate, effused up to 290 μm thick in section; hymenial surface yellowish white to brownish orange when fresh, becoming pale yellow to greyish yellow on drying; margin thinning, paler concolorous.

Hyphal system monomitic; generative hyphae up to 3.6 μm wide, thin-walled, clamped, septate; basal hyphae parallel to the substrate, less branched; subhymenial hyphae highly branched, vertically oriented. Gloeocystidia 54–108 × 11–12.6 μm, thin-walled, with basal clamp, positive to sulfovanillin, projecting out of the hymenium. Basidia 23–35 × 11.0–12.4 μm, 4-sterigmate, obovate to clavate with basal clamp, with oily contents, sterigmata up to 5.5 μm long. Basidiospores 6.9 × 9.4 μm in diameter, thick-walled, apiculate, echinulate, acyanophilous, amyloid.

Distribution: H.P.: Mandi, Sundernagar, Bilaspur.

Collection examined: IBP 37465.

Substrate: On decaying angiospermous wood.

Remarks: This species is characterized by the presence of gloeocystidia and echinulate basidiospores.

Boidinia lacticolor (Bres.) Hjortstam & Ryvarden, in Hjortstam, Mycotaxon 28(1): 19, 1987 = *Gloeocystidiellum lacticolor* Bres., Hedwigia 56(4,5): 303, 1915. Fig. 6.101j–n

Fructification resupinate, adnate, effused; hymenial surface smooth and cracked on drying, yellowish white to pale orange; margin thinning, byssoid, paler concolorous. Hyphal cordons visible near the margins.

Hyphal system monomitic; generative hyphae up to 4.2 μm wide, thin-walled, without clamps, septate; basal hyphae parallel to the substrate, less branched; subhymenial hyphae highly branched. Hyphal cordons up to 45 μm wide, individual hyphae up to 3.8 μm wide with encrustation. Cystidia 65–98 × 7.0–7.9 μm, cylindrical to subfusiform thin-walled, positive to sulfovanillin with crystalline encrustations. Basidia 25.8–32.2 × 4.5–6.2 μm, clavate to subclavate, with oily content, 4-sterigmate, without basal clamp. Basidiospores 5 × 5.6 μm in diameter, thick-walled, globose to subglobose, verrucose, apiculate, acyanophilous, amyloid, uniguttulate.

Distribution: H.P.: Manali-reserve forests, Hadimba temple.

Collection examined: IBP 37466.

Substrate: Bark and cut stumps of *C. deodara*.

Remarks: The species is characterized by subglobose to globose, verrucose, amyloid basidiospores and thin-walled, cylindrical to subfusiform, cystidia, positive to sulfovanillin. It is a new record for Himalayas.

Family-Stereaceae

Key to genera

1. Hymenophore resupinate to effused to reflexed to pileate 2
1. Hymenophore resupinate 4
2. Fructification woody, acanthophyses present, encrusted acanthophysoid encrusted cystidia present/absent 3
2. Fructification woody, encrusted acanthophysoid cystidia always absent . *Stereum*
3. Acanthophysoid encrusted cystidia present .... *Xylobolus*
3. Acanthophysoid encrusted cystidia absent .... *Acanthofungus*
4. Rhizomorphic hyphae present.. *Amylosporomes*
4. Rhizomorphic hyphae absent 5
5. Pseudocystidia present .... *Scytinostromella*
5. Pseudocystidia absent 6
6. Gloeocystidia absent 7
6. Gloeocystidia present 8
7. Basidiospores amyloid .... *Conferticium*
7. Basidiospores nonamyloid .... *Chaetoderma*
8. Hyphal system monomitic 9
8. Hyphal system dimitic .... *Scytinostromella*
9. Basidiospore amyloid .... *Aleurodiscus*
9. Basidiospores non-amyloid .... *Scotoderma*

Acanthofungus Sheng H. Wu, Boidin & C.Y. Chien, Mycotaxon, 76: 154, 2000.

Fructification perennial, woody, stratose, often arising as small circular colonies which become effused and reflexed with age, often attached by a narrow base; upper surface dark brown, azonate to concentrically zonate; hymenial surface chalky white to grey, smooth to somewhat uneven occasionally cracked deeply;

margin pale brown, abrupt, soft, turning black when touched with KOH sol. Context composed of compactly arranged hyphae, light brown in section. Hyphal system dimitic; skeletal hyphae unbranched, aseptate or rarely with few retraction septa, thick-walled, subhyaline to pale brown; generative hyphae branched, septate, clamped, thin to thick-walled. Gloeocystidia cylindrical, with the clamp at the base, with subhyaline granular contents staining deeply with phloxine, the walls moderately thick. Acanthophyses clavate to cylindrical. Basidiospores ellipsoid, minutely apiculate, the walls thin, smooth, subhyaline, amyloid.

Three species, widespread

Lit.: Wu *et al.* (*Mycotaxon* **76**: 154, 2000)

Type Species: *Acanthofungus rimosus* Sheng H. Wu, Boidin & C.Y. Chien. 2000

Habitat: Wood

Himalayas: One

Acanthofungus ahmadii (Boidin) Sheng H. Wu, Boidin & C.Y. Chien, *Mycotaxon* 76: 159 (2000) = *Xylobolus ahmadii* (Boidin) Boidin, Revue Mycol., Paris 23: 341 (1958). Fig. 6.101o–q

Fructification perennial, woody, stratose, initially arising as small circular colonies later becoming effused and reflexed with age, attached by a narrow base. Pileus up to 120 mm long and 80 cm broad, thickness up to 7 mm at maturity; upper surface dark brown, azonate to concentrically zonate; hymenial surface chalky white when fertile but turns grey after sporulation, smooth to somewhat uneven occasionally cracked deeply but rarely forms frustules; margin pale brown, abrupt, soft, turning black when touched with KOH sol.

Hyphal system dimitic; generative hyphae 2–4.4 μm wide, branched, septate, clamped, thin to moderately thick-walled, subhyaline; skeletal hyphae 3–5.9 μm wide, unbranched, aseptate or rarely with few retraction septa, the walls thick (up to 3 μm), subhyaline to pale brown. Gloeocystidia 34–200 × 4.8–10.9 μm, cylindrical, immersed but continue to grow from the old hymenial strata to the developing one, often with the clamp at the base, with subhyaline granular contents staining deeply with phloxine, the walls moderately thick (up to 1 μm), subhyaline. Acanthophyses 19.8–30.1 × 2.7–3.8 μm, clavate to cylindrical beset with spine-like processes up to 1 μm long, spines few or many and often covering the upper part of the acanthophyses up to 12 μm, thin to thick-walled, subhyaline. Basidiospores 4.4–5.3 × 2.9–3.4 μm, ellipsoid, minutely apiculate, the walls thin, smooth, subhyaline, amyloid.

Distribution: H.P.: Narkanda, Mahasu, Rohtang, Kullu, Dalhousie; U.K.: Nainital, Mussoorie.

Collection examined: SSR: 5326, 5371, 5631, IBP 37467.

Substratum: On log of *Cedrus deodara*, *Quercus incana*, hard wood stumps.

Remarks: This is a widely distributed species in the North-Western Himalayas and is characterized by the woody texture of fructifications and presence of gloeocystidia and acanthocystidia. It occurs on both angiosperm and gymnospermous wood.

Aleurodiscus Rabenh. ex Schroet.,
In Cohn, Krypt.-Fl. Schlesien (Breslau) 3(1): 429. 1888 [1889].

Fructification resupinate to disciform or substereoid; hymenial surface smooth, white or variously coloured, continuous, usually not creviced. Hyphal system monomitic or dimitic, hyphae clamped or with simple septa; hymenium includes dendrohyphidia, acanthophyses, gloeocystidia, simple or branched basidioles and basidia. Cystidia absent. Basidia clavate, 4-spored, medium to large in size simple septate or with a clamp. Basidiospores generally large in size, smooth to finely echinulate, amyloid.

Fifty three species, widespread

Lit.: Larsson & Larsson (*Mycol.* **95**: 1037, 2003; phylogeny)

Type Species: *Peziza amorpha* Pers. 1801

Habitat: Dead wood

Himalayas: Five

Key to species

1. Hyphae without clamps.. 2
1. Hyphae with clamps.. 3
2. Basidiocarps greyish orange to dark brown.................................... *A. amorphous*
2. Basidiocarps reddish white to pale red to orange grey .... *A. oakesii*
3. Basidiocarps brownish grey when fresh,
   bluish grey on drying .... *A. lividocoeruleus*[a]
3. Basidiocarps orange grey 4
4. Basidiospores echinulate.. *A. taxicola*
4. Basidiospores smooth .... *A. lapponicus*

[a]Extra limital, not included in this work

Aleurodiscus amorphus (Pers.) J. Schröt., in Cohn, Krypt fl. Schlesien (Breslan) 3 (1): 429, 1888 = *Peziza amorpha* Pers., Syn. Meth. Fung 2: 657, 1801.

Fig. 6.102a–e

Fructification resupinate, effused, loosely adnate reflexed, up to 380 μm thick in section; hymenial surface, greyish to brownish orange, smooth; margins thinning, paler concolorous to indeterminate.

Hyphal system monomitic; generative hyphae up to 3.2 μm wide, without clamps, septate, thin-walled, encrusted; basal hyphae parallel to the substrate; subhymenial hyphae vertical highly branched. Cystidia 45–65 × 5.6–7.0 μm, thin-walled, sinuous to moniliform, basal clamp absent. Dendrohyphidia abundant with encrustation. Basidia 87.8.0–121.5 × 19.0–24.0 μm, 4-sterigmate, basal clamps absent. Basidiospores 24.0–29.8 × 17.2–21.9 μm, apiculate, thin-walled, broadly ellipsoid to ovoid, acyanophilous, acyanophilous.

Distribution: H.P.: Shimla, Narkanda, Shimla, Manali, Kinnaur.

Collection examined: IBP 37468.

Substrate: On decaying wood of *C. deodara*.

Remarks: The species is characterized by presence of dendrohyphidia, broadly ellipsoid to ovoid, amyloid basidiospores. The species was first described by Persoon (1801) as *Peziza amorpha*, later Schröter (1888) shifted it to genus *Aleurodiscus*.

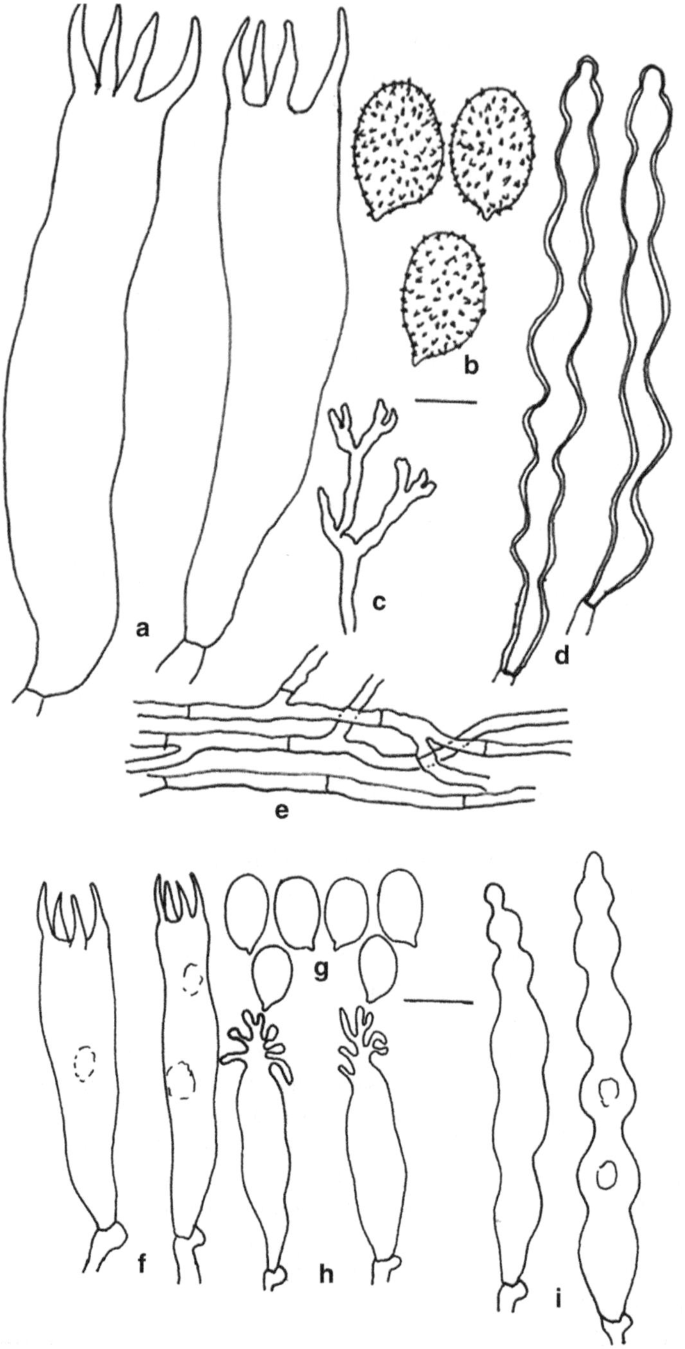

Fig. 6.102 (**a–e**) *Aleurodiscus amorphus* (**a**) Basidia, (**b**) Basidiospores, (**c**) Dendrohyphidia, (**d**) Cystidia, (**e**) Generative hyphae; (**f–i**) *Aleurodiscus lapponicus* (**f**) Basidia, (**g**) Basidiospores, (**h**) Acanthophyses, (**i**) Cystidia

Aleurodiscus lapponicus Litsch., Annls mycol. 42: 11, 1944. Fig. 6.102f–i

Basidiospores resupinate, adnate, effused, up to 870 µm thick in section; hymenial surface greyish orange, smooth to tuberculate, cracked; margin thinning, paler concolorous to abrupt.

Hyphal system monomitic; generative hyphae up to 4.2 µm wide, thin-walled, clamped, septate; basal hyphae parallel to the substrate; subhymenial hyphae highly branched vertical. Cystidia 58–68 × 7.4 µm, thin-walled, smooth, moniliform at the apex, with oily contents, negative to sulfovanillin. Acanthophyses present abundantly with apical finger like protuberances with basal clamp, smooth, basally. Basidia 29–45 × 6.5–9.3 µm, clavate, 4-sterigmate, with basal clamp. Basidiospores 8.6–10.0 × 4.8–6.0 µm, ellipsoid, thin-walled, apiculate, smooth, acyanophilous, amyloid.

Distribution: H.P.: Kinnaur, Sangla.

Collection examined: IBP 37469, D 37470.

Substrate: On decaying wood of *C. deodara*.

Remarks: The species is characterized by fusiform cystidia which are sinuous to somewhat moniliform at the apex, acanthophyses and broadly ellipsoid, smooth basidiospores. It is a new record for Himalayas.

Aleurodiscus oakesii (Berk. & Curt.) Hoehn. & Litsch., Sitz. Kais. Akad. Wiss. Wien. Math.-Nat. Klasse 116: 802. 1907. Plate 6.37e, Fig. 6.103a–c

Fructification disciform to substereoid, adnate, membranous to membranous-ceraceous, up to 650 µm thick in section; hymenial surface smooth to farinaceous, light-orange to greyish-orange or greyish-red; margin thinning, somewhat raised, paler concolorous to indeterminate.

Hyphal system monomitic; generative hyphae 1.4–3.8 µm wide, thin-walled, branched, septate, with or without clamps; basal zone composed of compactly packed repent hyphae parallel to the substrate, less branched and subhymenial zone of vertical much branched hyphae; acanthohyphidia with protuberance at the apical portion with a basal clamp up to 7 µm wide; gloeocystidia 39–75 × 4.2–11.3 µm subfusiform to fusiform with moniliform apices smooth thin-walled without basal clamp, negative to sulfovanillin. Basidia up to 69–138 × 21–25 µm, very large, subclavate, 4-sterigmate; sterigmata very large, with basal clamp. Basidiospores 24.8–30.8 × 18.1–21.9 µm, ellipsoid to subglobose, thin to somewhat thick-walled, verrucose, amyloid, acyanophilous, guttulate.

Distribution: U.K.: Nainital Uttarkashi; H.P.: Shimla; Bhutan.

Collection examined: SSR 5377, IBP 37470, 37471.

Substratum: *Quercus incana* twigs.

Remarks: The species is characterized by disciform fructification, tomentose hymenial surface; verrucose, amyloid, guttulate basidiospores. It is particularly restricted to oak trees.

Aleurodiscus taxicola Thind & Rattan, Mycolgia 65(6): 1255 (1974) [1973]. Plate 6.37f, Fig. 6.103d–g

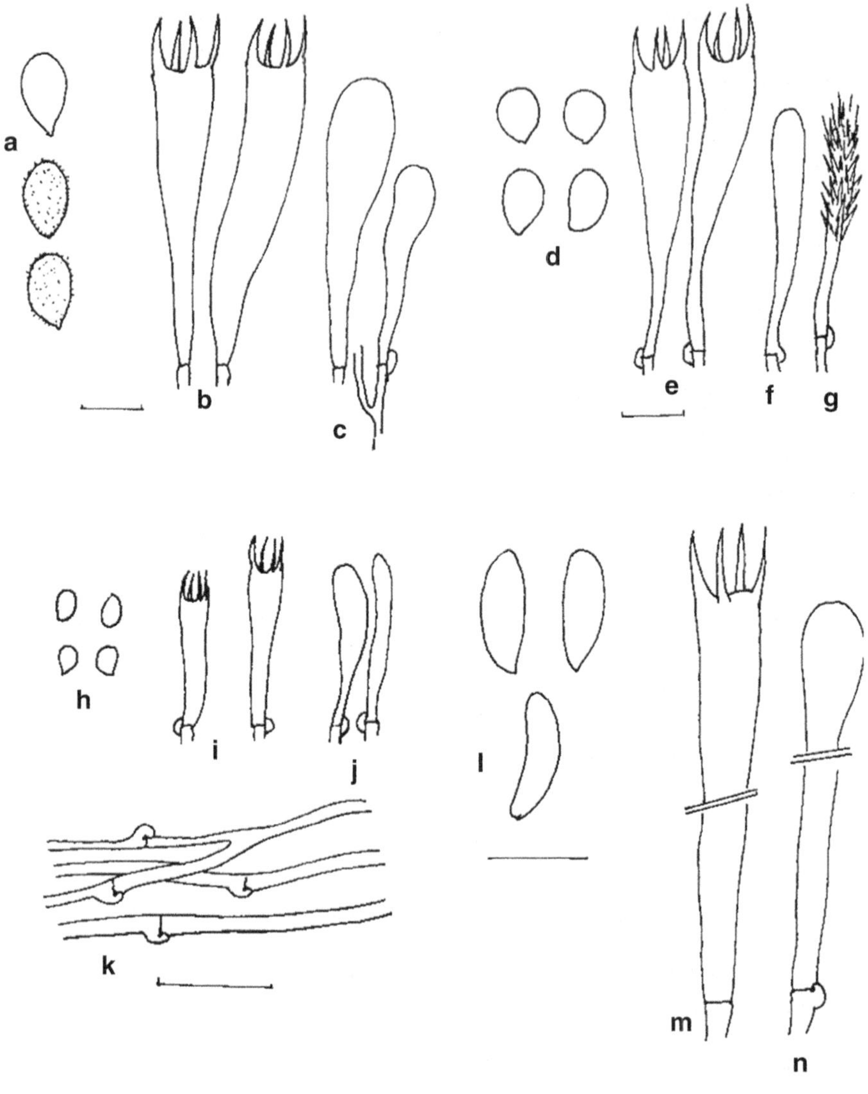

Fig. 6.103 (**a**–**c**) *Aleurodiscus oakesii* (**a**) Basidiospores, (**b**) Basidia, (**c**) Basidioles; (**d**–**g**) *Aleurodiscus taxicola* (**d**) Basidiospores, (**e**) Basidia, (**f**) Gloeocystidia, (**g**) Acanthophyses; (**h**–**k**) *Amylosporomyces echinosporus* (**h**) Basidiospores, (**i**) Basidia, (**j**) Basidioles, (**k**) Generative hyphae; (**l**–**n**) *Chaetoderma luna* (**l**) Basidiospores, (**m**) Basidia, (**c**) Cystidia

Fructifications resupinate, membranous, adnate, often arising as small pin-head colonies which may become effused, up to 2.8 × 1.4 cm; up to 480 μm thick in section; hymenial surface white to cream, smooth to somewhat farinose, not creviced or creviced in thicker parts; margin thin to more or less abrupt, adnate, concolorous.

Hyphal system monomitic, hyphae 2–3.5 μm wide, branched, septate, clamped, the walls thin, subhyaline. Basal zone comprising of repent hyphae and an upper zone of semi-erect hyphae densely impregnated with crystalline matter which often obscure the hyphal details. The hyphae become usually collapsed in mature fructifications. Gloeocystidia of variable length, 8–15.8 μm broad, fusiform, subhyaline with granular contents, thin-walled, immersed or just reaching the hymenial surface. Some gloeocystidia ending in hymenium are often beset with aculeate protuberances (acanthophysoid). Acanthophyses up to 7 μm broad, clavate, the walls thin, subhyaline, dextrinoid, echinulate. Hymenium thickening resulting in the immersion of abundant basidiospores in it. Basidia 64–84 × 11–17 μm, clavate with long tapering base, retraction septa common, 2–4 spored. Basidiospores 14.8–18.2 × 11.4–14 μm, broadly ellipsoid to ovoid, prominently apiculate, thin-walled, subhyaline, amyloid, finely echinulate.

Distribution: H.P.: Mahasu- Narkanda, Dalhousie.

Collection examined: SSR 5352, IBP 37472.

Substratum: On bark of *Taxus baccata*.

Remarks: The characteristic features of the species are white, adnate and resupinate fructifications; presence of gloeocystidia; thin-walled, clamped hyphae and broadly ellipsoid to ovoid, echinulate, amyloid basidiospores.

Amylosporomyces S.S. Rattan,
Biblthca Mycol. 60: 244 (1977)

Fructifications resupinate, membranous, thin, adnate; hymenial surface smooth, white to cream or pale coloured, continuous not creviced; margin thinning, white to paler concolorous. Reaction with KOH sol negative. Hyphal system monomitic, hyphae branched, septate, clamped, the walls thin to firm, distinct but occasionally collapsing in mature parts. Rhizomorphs present or absent. Rhizomorphic hyphae occasionally with obscure ampulliform swellings near septa. Gloeocystidia or Cystidia absent. Basidia clavate-cylindrical, 4-spored. Basidiospores ellipsoid to ovoid, minutely apiculate, the walls thin, subhyaline, finely echinulate, amyloid.

Two species, widespread

Lit.: Rattan (*Biblothca Mycol.* **60**: 244, 1977)

Type Species: *Amylosporomyces echinosporus* Rattan 1977

Habitat: Wood

Himalayas: One

Amylosporomyces echinosporus S.S. Rattan, [as '*echinosporum*'] Biblthca Mycol. 60: 245 (1977). Fig. 6.103h–k

Fructifications resupinate, submembranous to membranous, adnate, often arising in small patches restricted in growth; hymenial surface white to cream, smooth, continuous, not creviced; margin thinning to indeterminate, adnate, white to paler concolorous. Reaction with 3 % KOH sol. Subiculum composed of somewhat compactly arranged hyphae impregnated with abundant crystals.

Hyphal system monomitic, hyphae 2.2–3.2 μm wide, branched, septate, clamped, the walls subhyaline, thin to firm, distinct but often collapsing in mature parts.

Rhizomorphs white, branched, present mostly in the substratum transversing the crevices; rhizomorhic hyphae 1.6–4(5) µm wide, branched, septate, clamped, occasionally with obscure ampulliform swellings near the septa, the walls thin, subhyaline. Gloeocystidia or cystidia absent. Basidia 14.8–25 × 3.4–4.1 µm long. Basidiospores 3.4–4.4 × 2.6–3.1 µm, broadly ellipsoid to ovoid, shortly apiculate, the walls thin, subhyaline, finely echinulate, amyloid, acyanophilous.

Distribution: H.P.: Dalhousie, Manali.

Collection examined: SSR 5157, IBP 37474.

Substratum: On stump under conifers.

Remarks: This species is characterized by white to cream and submembranous to membranous fructifications; presence of rhizomorphs and shape and size of basidiospores. *Amylosporomyces humifaciens* is very similar but differs in the size of basidiospores.

Chaetoderma **Parm.,**
Consp. Syst. Cort. p. 86. 1968.

Fructifications resupinate, perennial, membranous-crustose, often arising in small colonies which may coalesce later; hymenial surface smooth to rough or uneven. Subiculum subhyaline in section, stratose. Hyphal system monometic, hyphae thin to slightly thick-walled, clamped at all septa. Cystidia long, cylindrical, thick-walled with narrow lumen. Gloeocystidia absent. Basidia clavate-cylindrical, the walls thin, smooth, subhyaline, non-amyloid, acyanophilous.

Two species, widespread

Lit.: Parmasto (*Consp. System. Corticiac.*: 86, 1968)

Type Species: *Peniophora luna* Romell ex D. P. Rogers & H. S. Jacks 1943

Habitat: Dead wood

Himalayas: One

Chaetoderma luna Romell ex D. P. Rogers & H. S. Jacks, Farlowia, 1(2): 320, 1943 = *Peniophora luna* Rom. ex Rog. & Jack., Farlowia 1: 320. 1943. Plate 6.38a, Fig. 6.103l–n

Fructification perennial, resupinate, membranous to membranous-crustose, stratose, often arising as small circular colonies which may become effused; hymenial surface white to cream, uneven, slightly wrinkled or sulcate but sometimes with faint concentric zonation in fresh specimens which tend to disappear on drying, finely powdery or hairy, continuous, not creviced but some thicker specimens may crack areolately on drying; margin thick, more or less abrupt and rounded or rarely acute, adnate, concolorous. Subiculum zonate or stratose, each stratum up to 340 µm thick composed of somewhat compactly arranged hyphae.

Hyphal system monomitic, hyphae 2–3.5 µm wide, branched, septate, clamped at all septa, the walls subhyaline, thin to slightly thick (up to 0.5 µm), naked but more often impregnated with crystalline matter. Cystidia 98–250 × 5–5.8 µm, cylindrical to subcylindrical, arising from different levels of the context and continue to grow with the hymenium, immersed or projecting out of the hymenium, the walls subhyaline, thick, lumen capillary throughout except near the apices

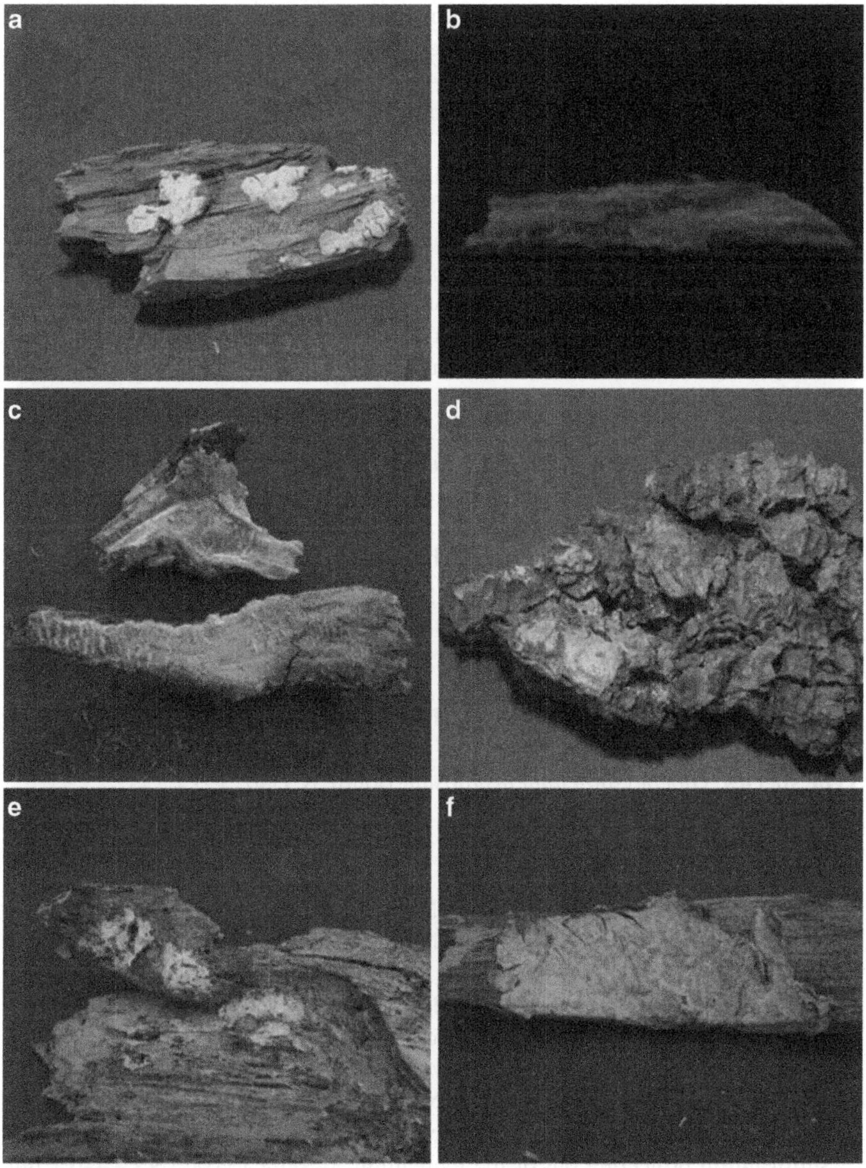

Plate 6.38 (**a**) *Chaetoderma luna.* (**b**) *Conferticium ochraceum.* (**c**) *Gloeocystidiellum sulcatum.* (**d**) *Scotoderma viridae.* (**e**) *Scytinostromella heterogenea* (**f**) *Stereum acanthophysatum*

where it widens gradually, naked but more often covered with a fine coating of subhyaline crystals which usually disappear in KOH sol. Gloeocystidia absent. Basidia 40–52 × 6.4–6.9 μm, clavate, 4-spored. Basidiospores 8.4–9.8 × 3.0–3.9 μm, ellipsoid to subcylindrical, shortly apiculate, thin-walled, subhyaline, smooth, non-amyloid, acyanophilous.

Distribution: H.P.: Kullu; J&K: Patnitop.

Collection examined: GSR 5039, SSR 5824, IBP 37475.

Substratum: On stump under conifers, on stump of *Abies pindrow*.

Remarks: This species is characterized by thick, perennial, stratose fructifications; peculiar type of cystidia which are long, cylindrical to sub cylindrical, usually finely incrusted and often arising from different levels of the context. The lumen is capillary throughout except near the apices where it widens gradually. This species is found on freshly cut stumps of conifers and it causes brown-rot.

Conferticium Hallenb.,
Mycotaxon 11(2): 447, 1980.

Fructification resupinate, effused, adnate; hymenial surface smooth to tuberculate; hyphal system monomitic; generative hyphae septate, branched, without clamps; cystidia cylindrical, thin-walled having oily contents; basidia 4-sterigmate, clavate, without basal clamp; basidiospores thin-walled, ellipsoid, smooth, amyloid, acyanophilous.

Four species, wide spread

Lit.: Ginns & Freeman (*Biblthca Mycol.* **157**, 1994), Larsson & Larsson (*Mycol.* 95: 1037, 2003; phylogeny)

Type species: *Gloeocystidiellum insidiosum* Bourd. & Galz. 1913

Habitat: Decayed wood

Himalayas: One

Conferticium ochraceum (Fr.) Hallenb., Mycotaxon 11(2): 448, 1980 = *Thelephora ochracea* Fr., Observ. Mycol. 1: 151, 1815. Plate 6.38b, Fig. 6.104a–e

Fructification resupinate, effused, adnate; hymenial surface pale orange to brownish orange when fresh, smooth to tuberculate, pale orange to greyish orange on drying; margin thinning, whitish to paler concolorous to indeterminate.

Hyphal system monomitic; generative hyphae up to 3.0 μm wide, without clamps, septate, thin-walled; basal hyphae parallel to the substrate. Cystidia 46.8–67 × 4.8–6.4 μm, thin-walled, cylindrical, with oily contents, positive reaction to sulfovanillin projecting out of the hymenium. Basidia 13.6–22.9 × 4.4–6.1 μm, clavate, repetitive, without basal clamp, 4-sterigmate, sterigmata up to 5.2 μm long. Basidiospores 5.5–7.0 × 3.1–3.8 μm, thin-walled, smooth, ellipsoid, apiculate, acyanophilous, amyloid.

Distribution: H.P.: Solan, Shimla.

Collection examined: IBP 37476.

Substrate: On decaying gymnospermous wood.

Remarks: The species is characterized in having gloeocystidia positive to sulfovanillin, repetitive basidia and ellipsoid, amyloid basidiospores.

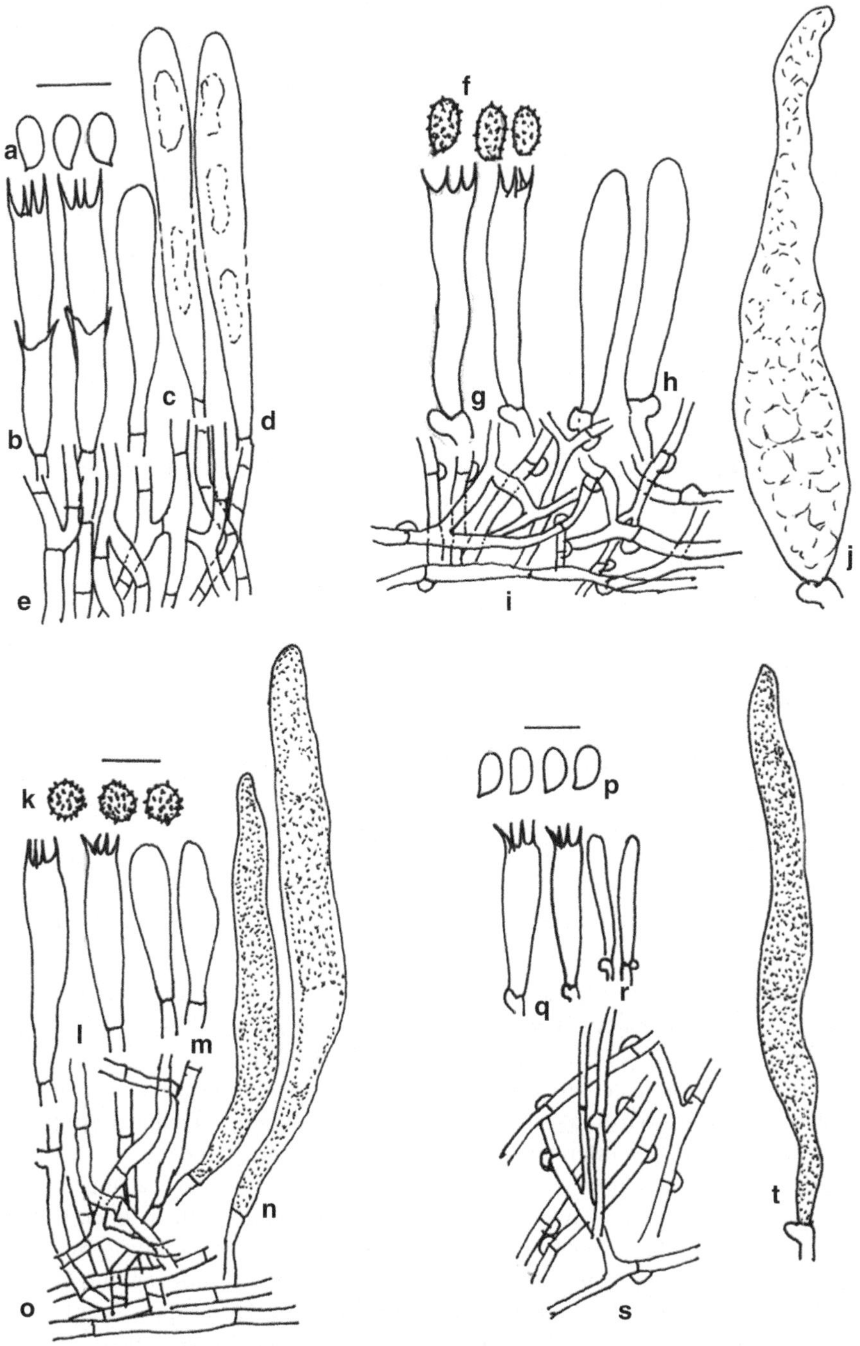

Fig. 6.104 (**a–e**) *Conferticium ochraceum* (**a**) Basidiospores, (**b**) Basidia, (**c**) Basidioles, (**d**) Cystidia, (**e**) Generative hyphae; (**f–j**) *Gloeocystidellum clavuligerum* (**f**) Basidiospores, (**g**) Basidia, (**h**) Basidioles, (**i**) Generative hyphae, (**j**) Cystidia; (**k–o**) *Gloeocystidiellum donkii* (**k**) Basidiospores, (**l**) Basidia, (**m**) Basidioles, (**n**) Cystidia, (**o**) Generative hyphae; (**p–t**) *Gloeocystidiellum luteocystidium* (**p**) Basidiospores, (**q**) Basidia, (**r**) Basidioles, (**s**) Generative hyphae, (**t**) Cystidia

Gloeocystidiellum Donk,
Meded, Ned. Mycol. Ver. 18–20: 156, 1931
emend Donk, *Fungus* 26: 8. 1956.

Fructifications resupinate, adnate, membranous to ceraceous; hymenial surface smooth to rarely odontoid, variously coloured, continuous or sometimes areolately cracked on drying; margin thinning to abrupt, concolorous. Subiculum subhyaline to light brown in section, composed of compactly arranged hyphae and gloeocystidia. Hyphal system monomitic, hyphae with or without clamps. Gloeocystidia abundant, immersed or rarely projecting out of the hymenium. Cystidia absent or rarely present. Basidia clavate, 4-spored. Basidiospores globose to ellipsoid, the walls smooth to finely echinulate, amyloid.

Seven species, widespread.

Lit.: Donk (*Fungus Wageningen* **26**: 8, 1956), Stalpers (Stud. Mycol. **40**: 59, 1996; Key)

Type: *Corticium porosum* Berk. & Curt. 1879

Habitat: Dead wood

Himalayas: Four

Key to species

1. Hyphal clamps absent 2
1. Hyphal clamps present 3
2. Hymenial surface pale cream turning dark cream on drying. Gloeocystidia 65–160 × 8–10 μm .... *G. donkii*
2. Hymenial surface cream turns greyish brown with drying. Gloeocystidia 35–85 × 4–8.5 μm .... *G. sulcatum*
3. Hymenial surface greyish orange turning brownish orange on drying. Gloeocystidia 46–70 × 9–13 μm .... *G. clavuligerum*
3. Hymenial surface deep orange turns pale yellow later on. Gloeocystidia 29–140 × 5–13 μm .... *G. leuteocystidium*

Gloeocystidellum clavuligerum (Höhn. & Litsch.) Nakasone, Mycotaxon 14(1): 320, 1982 = *Gloeocystidium clavigerum* Höhn. & Litsch., Sber. Akad. Wiss. Wien, Math.-naturw. KI., Abt. 1 115: 1906. Fig. 6.104f–j

Fructification resupinate, effused, loosely adnate; hymenial surface greyish orange when fresh, greyish orange to brownish orange on drying, smooth to tuberculate; margin thinning, indifferentiated, whitish to paler concolorous.

Hyphal system monomitic; generative hyphae 2.0–3.0 μm wide, clamped, thin-walled, septate; basal hyphae parallel to the substrate, thin to thick-walled. Gloeocystidia 48–72 × 8.0–12.5 μm, smooth, fusiform to subfusiform, thin-walled with basal clamp, with oily contents positive to sulfovanillin. Basidia 20–27.2 × 3.8–5.2 μm, cylindrical, with basal clamp. Basidiospores ellipsoid to subglobose, 4.8–5.2 × 3.8–4.3 μm, verrucose, apiculate, thin-walled, amyloid, acyanophilous.

Distribution: H.P.: Chamba- Dalhousie.

Collection examined: IBP 37477.

Substratum: On fallen stick of *Cedrus deodara.*

Remarks: The species is characterized by presence of gloeocystidia and ellipsoid basidiospores.

Gloeocystidiellum donkii S.S.Rattan, *Biblthca Mycol.* 60:111, 1977. Fig. 6.104k–o

Fructification resupinate, pelliculose-membranous to membranous, adnate, widely effused; hymenial surface white to pale cream when fresh but turn deep cream on drying, smooth, continuous but sometimes cracking deeply on drying often exposing the substratum below, somewhat slimy when fresh; margin thinning to more or less determinate, byssoid, paler concolorous.

Hyphal system monomitic; hyphae 2.4–4.6 µm wide, branched, septate, clamps absent, thin-walled, subhyaline, often collapsing in the upper part of the context. Cystidia absent. Gloeocystidia 65–160 × 6.8–10 µm, subfusiform to cylindrical, empty but more often with subhyaline granular contents staining deeply with Phloxine, immersed or projecting up to 15 µm out of the hymenium, thin-walled, subhyaline. Hymenium thickening resulting in the immersion of abundant basidiospores. Basidia 24.8–34.8 × 5.4–6.4 µm, clavate-cylindrical, 4-spored. Basidiospores 5–6.1 × 4.4–5 µm, subglobose, minutely apiculate, thin-walled, subhyaline, warted, wart short and blunt, amyloid.

Distribution: Nepal: Kathmandu, U.K.: Dehradun.

Collection examined: SSR 5488, IBP 37478.

Substratum: On rotten bamboo sticks.

Remarks: The species is characterized by absence of clamps, presence of abundant gloeocystidia; clavate basidia; cylindrical, subglobose to warted basidiospores.

Gloeocystidiellum luteocystidium var. *brevisporum* Rattan, *Biblthca Mycol.* 60:103, 1977. Fig. 6.104p–t

Fructification annual or perennial, resupinate, stratose, membranous-ceraceous, adnate but curl away from the substratum on drying, often arising as small circular colonies which may coalesce later and becomes effused; hymenial surface deep orange when young fades to pale yellow in older parts, smooth to finely granulose; margin thinning to almost abrupt, adnate to separable, concolorous.

Hyphal system monomitic, hyphae 2–4.6 µm wide, branched, septate, clamped, thin-walled to thick-walled, subhyaline. Cystidia absent. Gloeocystidia 29.8–140.1 × 5.1–13.2 µm, clavate-cylindrical to cylindrical, with orange coloured contents but these may disappear with age, slightly thick-walled, immersed or rarely projecting slightly out of the hymenium. Basidia 5.6 µm broad, clavate, 4-spored. Basidiospores 4.4–6.4(8.4) × 4–4.4 µm, narrowly ellipsoid and slightly flattened on the adaxial side, minutely apiculate, thin-walled, subhyaline, smooth, amyloid.

Distribution: H.P.: Dalhousie- Dunera, Narkanda; U.K.: Badrinath.

Collection examined: SSR 5145, IBP 37479.

Substratum: Fallen twigs under angiospermic forest.

Remarks: This species is characterized by the thick, stratose, orange coloured fructification and narrowly ellipsoid basidiospores.

Gloeocystidiellum sulcatum (Rehill & Bakshi) Boidin, Cah. Maboke **4**: 10. 1966. Plate 6.38c, Fig. 6.105a–e

Fructification resupinate, annual to perennial, adnate, membranous, widely effused; hymenial surface cream when young becomes brown to finally greyish brown

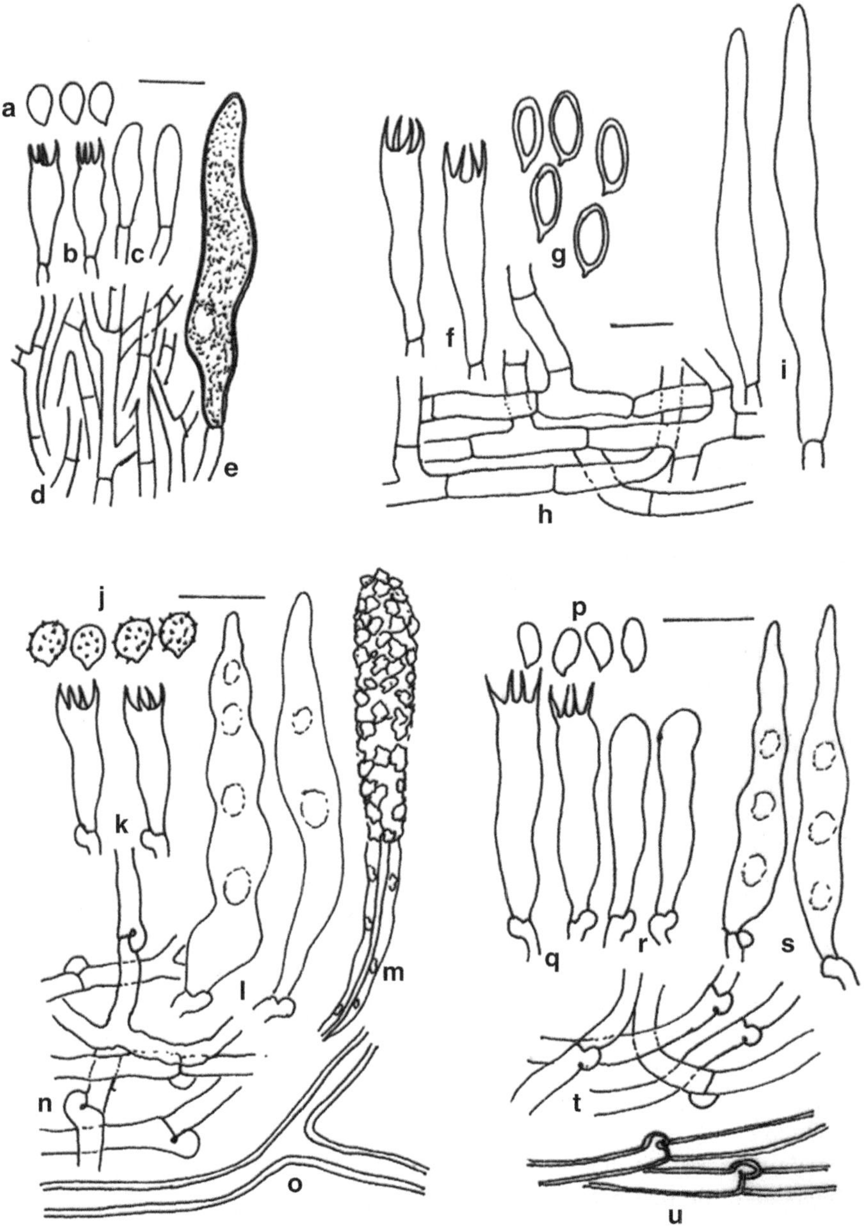

Fig. 6.105 (**a–e**) *Gloeocystidiellum sulcatum* (**a**) Basidiospores, (**b**) Basidia, (**c**) Basidioles, (**d**) Generative hyphae, (**e**) Cystidia; (**f–i**) *Scotoderma viride* (**f**) Basidia, (**g**) Basidiospores, (**h**) Generative hyphae, (**i**) Cystidia; (**j–o**) *Scytinostromella heterogenea* (**j**) Basidiospores, (**k**) Basidia, (**l**) Pseudocystidia, (**m**) Cystidia, (**n**) Generative hyphae, (**o**) Skeletal hyphae; (**p–u**) *Scytinostromella olivaceoalba* (**p**) Basidiospores, (**q**) Basidia, (**r**) Basidioles, (**s**) Cystidia (**t**) Thin-walled Generative hyphae, (**u**) Thick-walled generative hyphae

with age, tuberculate; margin thin, adnate. Subiculum subhyaline, full of crystalline matter.

Hyphal system monomitic, generative hyphae 2–3.2(−4.6) μm wide, septae, clamps are absent, thin-walled, branched. Cystidia absent. Gloeocystidia present. 34.8–84.8 × 4–8.5 μm, abundant, clavate-cylindrical to cylindrical with obtuse apices, thin to thick-walled, projecting out of the hymenium. Basidia clavate, 4-spored, 16–25 × 3.6–4.6 μm. Basidiospores broadly ellipsoid, minutely apiculate, the walls are thin, subhyaline, smooth, amyloid.

Distribution: H.P.: Dalhousie- Bakrota, Jandrighat, Banikhet, Kalatope, Chamba-Khijjiar; Shimla- Kufri, Summer hill, Mashobra; Mahasu- Narkanda, Bagi; Kullu- Pulga, Naggar, Jaggat sukh, Rohtang, Gojra.

Collection examined: SSR 5088, 5089, 5104, IBP 37480.

Substratum: Stumps and logs of *Cedrus deodara*, fallen logs of *Abies pindrow*, logs and stumps under a mixed and coniferous forest.

Remarks: The species is characterized by annual to perennial fructification; presence of gloeocystidia; clavate basidia; ellipsoid, apiculate smooth basidiospores.

Scotoderma Jülich,
Proc. Kon. Ned. Akad. Wetensch.
Ser. C. Biol. Med. Sci. 77:149, 1974.

Fructification resupinate, adnate, effused, membranous; hymenial surface smooth to somewhat velutinate, light brown to chestnut to yellowish brown; margin thinning, indeterminate. Hyphal system monomitic; generative hyphae branched, septate, without clamps, thin to somewhat thick-walled. Gloeocystidia cylindrical, strangulated near the apex, thin-walled, negative to Sulfovanillin. Basidia subclavate, 4-sterigmate, without a basal clamp. Basidiospores ovoid, with rounded apices, smooth, thick-walled, inamyloid, cyanophilous.

Monotypic, widespread.

Lit.: Jülich (*Proc. K. Ned. Akad. Wet.* Ser. C, Biol. Med. Sci. **77**: 149, 1974).

Type Species: *Corticium viride* Berk. 1855

Habitat: Wood

Himalayas: One

Scotoderma viride (Sacc.) Jülich, Proc. K. Ned. Akad. Wet., Ser. C, Biol. Med. Sci. 77(2): 151 (1974) = *Coniophora viridis* Sacc., Sylloge Fungorum 6: 649, 1888. Plate 6.38d, Fig. 6.105f–i

Fructification resupinate, adnate, effused, membranous; hymenial surface smooth to somewhat velutinate, light brown to chestnut to yellowish brown; margin thinning, indeterminate.

Hyphal system monomitic; generative hyphae up to 3.8 μm wide, branched, septate, without clamps, thin to somewhat thick-walled, somewhat celled hyphae up to 9 μm wide also present. Gloeocystidia 40.2–59.0 × 5.5–9.4 μm, cylindrical, strangulated near the apex, thin-walled, negative to sulfovanillin. Basidia 21.8–34.5 × 7.9–10.4 μm, subclavate, 4-sterigmate, without a basal clamp. Basidiospores 10.0–14.9 × 7.0–9.5 μm, ovoid, with rounded apices, smooth, thick-walled, inamyloid, cyanophilous.

Distribution: H.P.: Shimla- Chaupal.
Collection examined: IBP 37481.
Substratum: On log of *Cedrus deodara.*
Remarks: This species is characterized in having generative hyphae without clamps and presence of gloeocystidia. It was first described by Berkeley (1855) from Australia as *Thelephora viridis.* Saccardo (1888) reported it from New Zealand and shifted it to genus *Coniophora.* Cunningham (1963) also described it as a species of *Coniophora.*

Scytinostromella Parm.,
Consp. Syst. Cort.p.171, 1968.

Fructifications resupinate, pelliculose to membranous; hymenial surface smooth, white to cream or ochraceous; margin fibrillose to rhizomorphic. Context subhyaline in section. Hyphal system dimitic; generative hyphae subhyaline, thin-walled, clamped; skeletal hyphae subhyaline. Pseudocystidia are the prolongations of skeletal hyphae which is curve in to the hymenium, immersed or projecting out, usually incrusted. Gloeocystidia present. Basidia subutriform, 4-spored. Basidiospores ellipsoid to ovoid, finely echinulate, amyloid.
Five Species, world-wide
Lit.: Freeman & Petersen (Mycol, **71**: 85, 1979; Key), Spirin (*Novosti Sistematiki Nizshikh Nov. sist. Niz. Rast.* **36**: 66, 2002; Russia).
Type species: *Peniophora heterogena* Bourd. & Galz. 1912
Habitat: Wood
Himalayas: Two
Key to species

1. Hymenial surface yellowish white turning light ochraceous on drying; Hyphal system dimitic.. *S. heterogenea*
1. Hymenial surface deep cream to pale ochre; Hyphal system monomitic .... *S. olivaceoalba*

Scytinostromella heterogenea (Bourd. & Galz.) Parm., Consp. syst. Cort. P.171, 1968 = *Peniophora heterogena* Bourd. & Galz., Bull. Soc. Mycol. France 28: 393, 1913 [1912]. Plate 6.38e, Fig. 6.105j–o

Fructification resupinate, loosely adnate, effused; hymenial surface smooth when fresh, cracking on drying, yellowish-white when fresh, yellowish-white to light ochraceous in herbarium; margin thin.

Hyphal system dimitic; generative hyphae subhyaline, branched, septate, clamped, thin-walled, 2–3.8 µm wide; skeletal hyphae up to 4.2 µm wide, unbranched, aseptate, subhyaline, thick-walled, lumen narrow or indistinct, encrusted in the apical portion. Pseudocystidia are the prolongation of the skeletal hyphae which curve into hymenium. Gleocystidia 50.2–100.2 × 6.9 µm, fusiform to subcylindrical, thin-walled, immersed, with a basal clamp. Basidia 18.2–22.2 × 5.2–6.0 µm, subclavate, 4-sterigmata, sterigmata up to 5.5 µm long. Basidiospores 2.4–4.5 × 2.4–3.4 µm, broadly ellipsoid to subglobose, subhyaline, thin to somewhat thick-walled, warted, amyloid.

Distribution: Bhutan: Thimphu, Paro, Chailella; H.P.: Kullu, Chamba-Dalhousie.
Collection examined: GSD 19442, IBP 37484.
Substratum: On a decaying log.
Remarks: It is characterized by loosely adnate fructification, fibrillose margin, dimitic hyphal system, generative hyphae with clamps, presence of pseudocystidia, gleocystidia, positive to sulfovanillin and broadly ellipsoid to subglobose, warted, amyloid basidiospores. This species is of rare occurrence in the Himalayas. Rattan (1977) has reported it from the N.W. Himalayas.

Scytinostromella olivaceoalba (Bourdot & Galzin) Ginns & M.N.L. Lefebvre, Mycol. Mem. 19: 141 (1993) = *Confertobasidium olivaceoalbum* (Bourdot & Galzin) Jülich, Willdenowia, Beih. 7: 167 (1972). Fig. 6.105p–u

Fructification resupinate, loosely adnate, pelliculose to atheloid, when young, submembranous to crustose with age, arising in small colonies; hymenial surface deep cream to pale ochre, smooth, continuous but occasionally cracking irregularly on drying; margin thinning, loosely adnate, paler concolorous. Rhizomorphs abundant in the substratum, pinkish brown to light brown, branched, some rhizomorphs near the margin and basal part of the fructification seem to be the prolongation of hyphal cordons. Context composed of loosely woven hyphae and hyphal cordons.

Hyphal system monomitic, hyphae 1.6–3.6 µm wide, branched at wide angles, septate, clamped, thin-walled to slightly thick, subhyaline. Basal hyphae light brown, somewhat compactly arranged and often organised in to hyphal cordons, slightly thick-walled and less frequently branched while the subhymenial hyphae are subhyaline, loosely woven, thin-walled and more frequently branched. Basidia 10.2–14.2 × 3–4.4 µm, clavate, 4-spored. Basidiospores 3.4–4.4 × 2.3–2.6 µm, broadly ellipsoid, minutely apiculate, thin-walled, subhyaline, smooth, nonamyloid

Distribution: H.P.: Kullu, Chamba- Dalhousie.
Collection examined: SSR 5683, IBP 37485.
Substratum: On dead stump of *Cedrus deodara, Abies pindrow*.
Remarks: The species is characterized by atheloid fructifications, presence of light brown basal hyphae, rhizomorphs and shape and size of basidiospores.

Stereum Hill ex. Pers.,
Neues Mag. Bot. 1: 10, 1794.

Fructifications annual or perennial, effused-reflexed to pileate but often occurring in resupinate form as well, coriaceous; hymenial surface smooth, cream to yellowish brown, often discolouring of bleeding on bruising. Context composed of compactly arranged parallel hyphae, with a thick brown cuticle carrying tomentum on the abhymenial side. Hyphal system dimitic; generative hyphae branched, septate, clamps absent, the walls thin to moderately thick, subhyaline; skeletal hyphae unbranched to rarely branched, aseptate, or rarely with few retraction septa, thick-walled, subhyaline. Conducting hyphae or cystidial hyphae present, often arising from the upper part of the context and curving in to the hymenium, immersed or projecting slightly out of it. Gloeocystidia absent. Basidia clavate to clavate-cylindrical, 4-spored. Basidiospores ellipsoid to subhyaline, thin-walled, subhyaline, smooth, amyloid, acyanophilous.

Twenty species, widespread
Lit.: Eicker & Louw (*South African Journal of Botany* 64: 30, 1998; S. Afr.)
Type Species: *Thelephora hirsute* Willd. 1787
Habitat: Dead wood
Himalaya: Eight
Key to species

1. Hymenium stratose, composed of 2 or more layers............................ *S. rugosum*
1. Hymenium not stratose 2
2. Acanthophyses present.. 3
2. Acanthophyses absent 4
3. Orange coloured conducting hyphae present......................... *S. acanthophysatum*
3. Orange coloured conducting hyphae absent .... *S. peculiare*
4. Acutocystidia present.. *S. subtomentosum*
4. Acutocystidia absent 5
5. Hymenial surface bleed on bruising 6
5. Hymenial surface does not bleed on bruising 7
6. Mostly on angiosperms .... *S. gausapatum*
6. Mostly on conifers .... *S. sanguinolentum*
7. Fructification effused, reflexed to pileate... *S. hirsutum*
7. Fructification mostly pileate .... *S. ostrea*

Stereum acanthophysatum Rehill & Bakshi, For. Bull. Dehra Dun, n.s 250: 6. 1966. Plate 6.38f, Fig. 6.106a–f

Fructification annual or sometimes reviving, resupinate to effused–reflexed or pileate, loosely adnate, arising as small colonies; hymenial surface smooth, cream to orange to yellowish-brown, smooth to finely tuberculate.

Hyphal system dimitic; generative hyphae branched, septate, 2.2–4.1 μm wide, without clamps, thin to thick-walled; skeletal hyphae unbranched, aseptate, 3.4–4.4 μm wide, thick-walled; conducting hyphae present are of orange coloured content, acanthophyses present. Basidia clavate, to clavate-cylindrical, 4-spored, 20.2–25.1 × 3.9–4.9 μm. Basidiospores ellipsoid to suballantoid, 6.2–8.9 × 3.4–4.6 μm, thin-walled, smooth, amyloid, acyanophilous.

Distribution: U.K.: Chakrata.

Collection examined: SSR 7171, 7170, IBP 37486.

Substratum: On dead stump of *Cedrus deodara*.

Remarks: This species is characterized by soft texture of fruiting body, presence of acanthohyphidia, 4-spored basidia; ellipsoid, amyloid basidiospores. The species was earlier recorded from N.W. Himalaya by Rehill and Bakshi (1966).

Stereum gauspatum (Fr.) Fr. Hym. eur.: 638. 1874 = *Thelephora gauspata* Fr., Elench fung. 1: 171. 1828. Plate 6.39a, Fig. 6.106g–k

Fructifications resupinate to effused-reflexed to pileate often arising as small circular colonies which may coalesce, loosely adnate. Pileus narrowly reflexed to flabelliform, strongly imbricate; upper surface light brown to greyish brown, tomentose to hirsute, faintly concentrically zonate, zones of erect and appressed

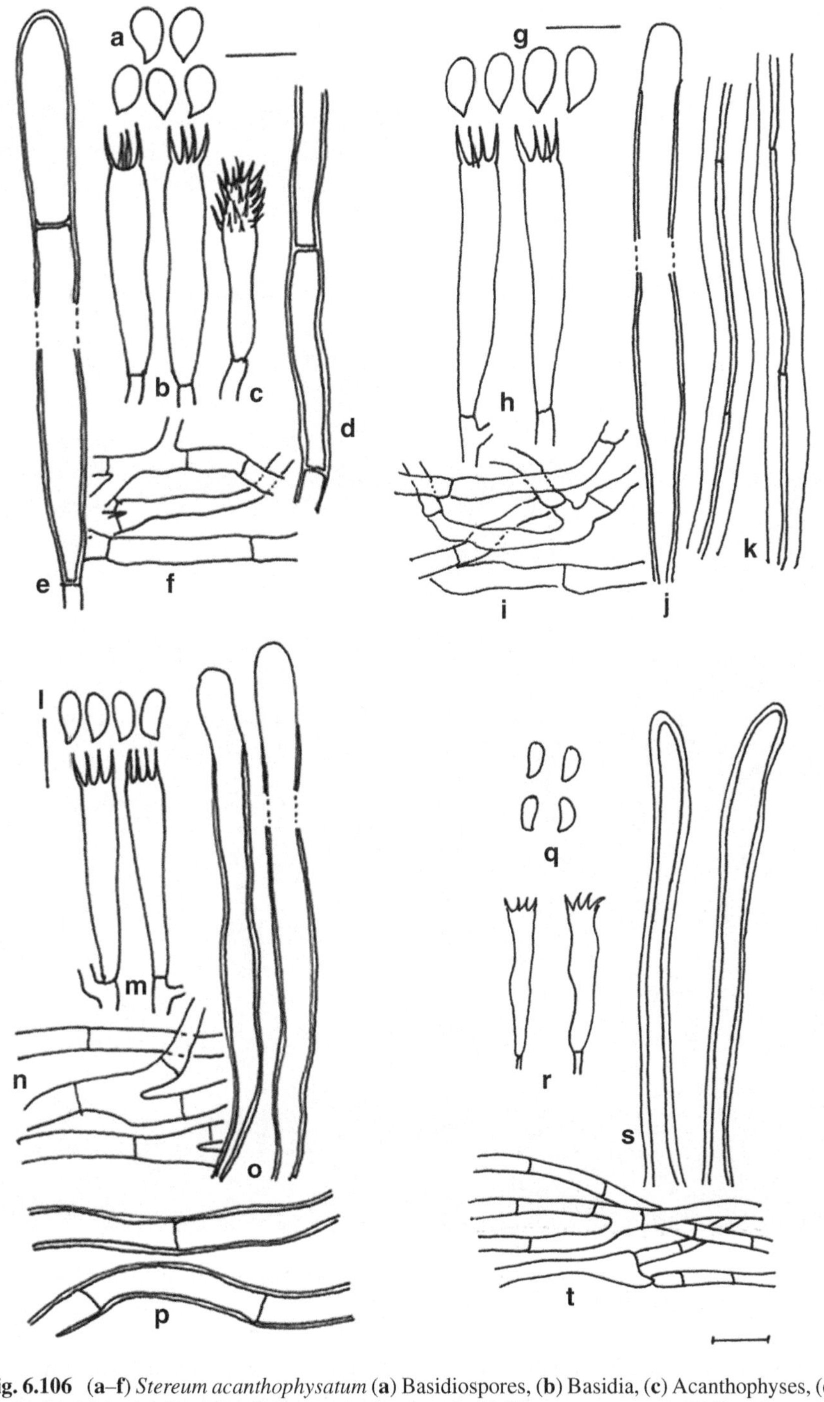

Fig. 6.106 (**a**–**f**) *Stereum acanthophysatum* (**a**) Basidiospores, (**b**) Basidia, (**c**) Acanthophyses, (**d**) Skeletal hyphae, (**e**) Cystidia, (**f**) Generative hyphae; (**g**–**k**) *Stereum gauspatum* (**g**) Basidiospores, (**h**) Basidia, (**i**) Generative hyphae, (**j**) Cystidia, (**k**) Skeletal hyphae; (**l**–**p**) *Stereum hirsutum* (**l**) Basidiospores, (**m**) Basidia, (**n**) Generative hyphae, (**o**) Cystidia, (**p**) Skeletal hyphae; (**q**–**t**) *Stereum ostrea* (**q**) Basidiospores, (**r**) Basidia, (**s**) Cystidia, (**t**) Generative hyphae

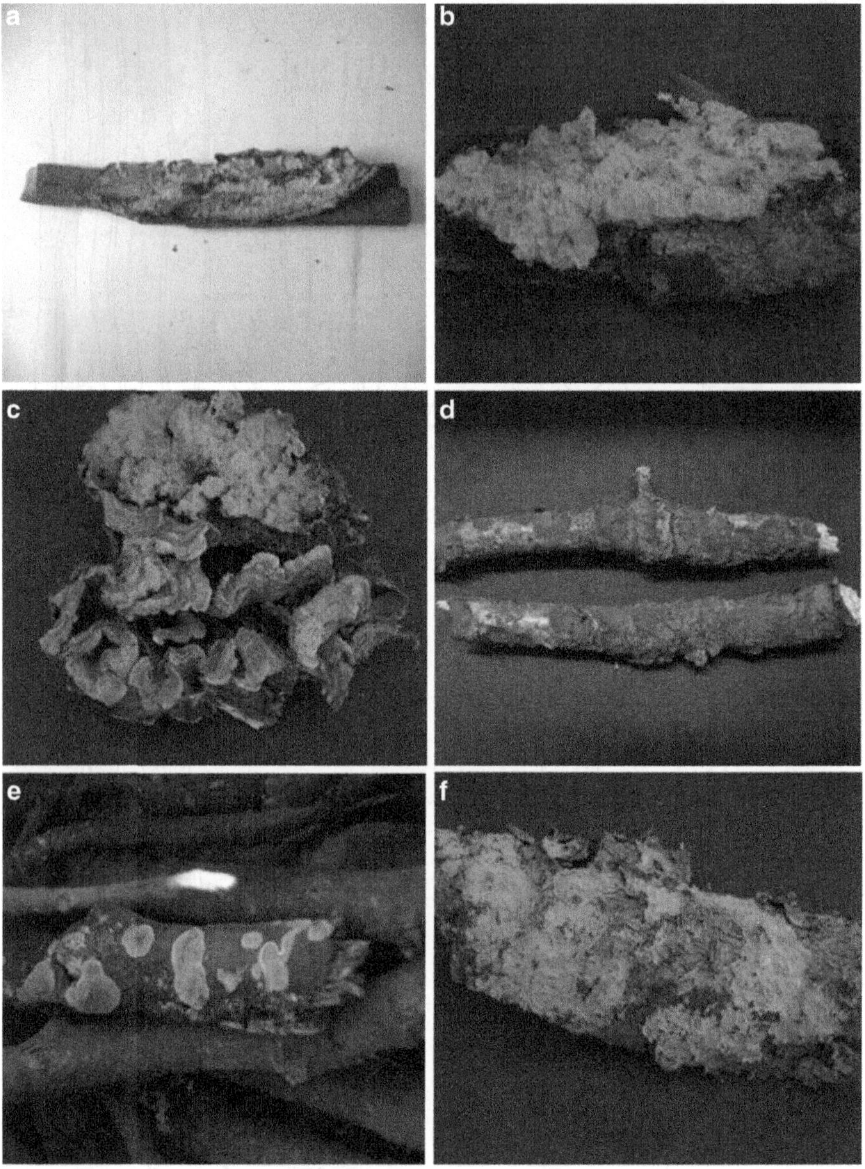

Plate 6.39 (**a**) *Stereum gauspatum.* (**b**) *Stereum hirsutum.* (**c**) *Stereum ostrea.* (**d**) *Stereum peculiare.* (**e**) *Stereum rugosum.* (**f**) *Stereum sanguinolentum*

tomentum, strongly radially sulcate; hymenial surface deep cream with a red tinge when fresh but turns greyish brown on drying, smooth, often reddening or bleeding when fresh specimens are cut or bruised; margin thinning or acute, concolorous. Context composed of compactly arranged hyphae, with a thick brown cuticle on the abhymenial side.

Hyphal system dimitic; generative hyphae 2–3(4.4) μm wide, branched, septate, clamps absent, the walls thin, subhyaline; skeletal hyphae 2.9–7.1 μm wide, sparsely branched, aseptate or rarely with few retraction septa, clamps absent, unevenly thick, often leaving little or no lumen;. Conducting hyphae 8–11 μm wide, cylindrical to flexuous, with orange colored contents, often arising from the upper layers of context and curved in to the hymenium, immersed or projecting slightly out of it, the walls thin, subhyaline. Basidia clavate, usually with one or rarely 2–3 spine-like projections at the top. Basidia 31.1–35×4.6–5.1 μm clavate, 4-spored. Basidiospores 4.9–7.5×2.6–3 μm, ellipsoid to suballantoid, shortly apiculate, the walls thin, subhyaline, smooth, amyloid.

Distribution: H.P.: Shimla; U.K.: Mussoorie; J&K: Bhadarwah, Nepal: Kathmandu.

Collection examined: GSR 5001, SSR 5395, IBP 37487.

Substratum: On dead stump of *Quercus incana.*

Remarks: This species is characterized by strongly imbricate fructifications, presence of conducting hyphae and aculeate tipped basidioles. The hymenium of fresh specimens when cut or bruised not only discolors but also bleeds profusely.

Stereum hirsutum (Willd.) Pers., Observ. Mycol. 2: 90 (1800) [1799] Arrangm. Brit. P1.1: 653. 1821 = *Thelephora hirsuta* Willd. Fl. berol. Prodr.: 397 (1787). Plate 6.39b, Fig. 6.106l–p

Fructification resupinate to effused–reflexed, loosely adnate; pileus narrowly reflexed, upper surface strongly hirsute, concentrically zonate, light brown; hymenial surface pale brown, smooth; margin thinning.

Hyphal system dimitic, generative hyphae branched, 2.6–4.6 μm wide, septate, without clamps, thin to thick-walled; skeletal hyphae unbranched, 5.2–7.6 μm wide, aseptate, thick-walled; absence of conducting hyphae; cystoids' hyphae cylindrical to hyphoid. Basidia clavate, 24.8–29.8×4.2–4.8 μm, 4-spored. Basidiospores ellipsoid, 4.6–5.4×2.3–2.9 μm, minutely apiculate, thin-walled, smooth, amyloid.

Distribution: H.P.: Mahasu; U.K.: Nainital.

Collection examined: GSR 5017, SSR 5373, IBP 37488, L 37489.

Substratum: On dead stump of *Quercus incana.*

Remarks: This species is characterized by having strongly hirsute fructification and presence of cystidoid hyphae, not radially sulcate upper surface and absence of conducting hyphae. The description of the above collected sample resembles with the description given by Rattan (1977). The species is commonly found in Himalayas. Various workers recorded this species from Himalayas. Rattan (1977) recorded it from N.W. Himalayas.

Stereum ostrea (Blume & Nees) Fr. Epicr. syst. mycol.: 547 (1838) = *Thelephora ostrea* Blume & Nees Nova Acta Acad. Caes. Leop- Carol. 13: 13(1826) = *Stereum fasciatum* (Schw.) Fr. Epicr. syst. mycol. 546 (1838) = *Stereum lobatum* (Kuntz.) Fr., Epicr. 547. 1838. Plate 6.39c, Fig. 6.106q–t

Fructification resupinate, effused-reflexed to dimidiate or laterally sessile, coriaceous to leathery. Pileus flabelliform to petaliform, imbricate rarely; upper surface strongly tomentose, concentrically zonate, multi-coloured surface of pileus; hymenial surface cream yellow to yellow-ochre, smooth; margin acute, paler concolorous. Context subhyaline in section, composed of compactly arranged hyphae, with a thick brown cuticle on the abhymenial side.

Hyphal system dimitic; generative hyphae 2.5–4.5 μm wide, branched, septate, without clamp, thin to moderately thick-walled; skeletal hyphae thick-walled, rarely branched, 5.3–8.2 μm wide, aseptate, generally subhyaline. Basidia clavate, 5.2–6.2 × 2.2–3.3 μm, 4-spored, collapsing after spore discharge. Basidiospores ellipsoid, minutely apiculate, thin-walled, smooth, subhyaline, amyloid.

Distribution: H.P.: Dalhousie; U.K.: Nainital; J&K: Bhadarwah; Nepal: Kathmandu.

Collection examined: SSR 5376, SSR 5376, IBP 37490.

Substratum: On dead stump of *Quercus incana.*

Remarks: This species is characterized by dimidiate to laterally sessile fructification, concentrically zonate, multicoloured surface of the pileus, cylindrical to hyphoid cystidiate hyphae, clavate basidia and ellipsoid, amyloid basidiospores.

Stereum peculiare Parmasto, Boidin & Dhingra, Persoonia 10(3): 311, 1979. Plate 6.39d

Fructification resupinate, uneven to tuberculate with scattered, irregular finger-like outgrowth; pseudocystidia present.

Hyphal system dimitic; generative hyphae branched, septate, without clamp, thin to moderately thick-walled; skeletal hyphae thick-walled, rarely branched, aseptate, generally subhyaline. Basidia clavate, 4-spored; basidiospores ellipsoid to suballantoid, thin-walled, smooth, subhyaline, amyloid, 9.2–10.2 × 2.6–3.5 μm.

Distribution: A.P.: West Kameng, Dirang, Bomdilla, Shergaon.

Collection examined: GSD 19787, 19789.

Substratum: On decaying angiospermic branches.

Remarks: This species is closely related to *Stereum acanthophysatum* Rehill. & Bakshi, but differs in having smooth to tuberculate or with finger-like outgrowths fructification in comparison to smooth to finely tuberculate fructification, presence of gleocystidia, absence of conducting hyphae with orange-coloured contents, larger acanthophyses and ellipsoid to suballantoid basidiospores in contrast to ellipsoid basidiospores.

Stereum rugosum Pers., Neus Mag. Bot. 1: 110. 1794. Plate 6.39e, Fig. 6.107a–e

Fructifications effused-reflexed, adnate, membranous to coriaceous, arising in small resupinate colonies which grow later and may become effused-reflexed; hymenial surface smooth, pinkish brown to yellowish brown or greyish brown, often bleeding on bruising in fresh specimens; margin rather abrupt, adnate, concolorous. Abhymenial surface light yellowish brown, tomentose when young but tomentum often disappears with age. Context composed of compactly arranged hyphae, with a well developed cuticle on the abhymenial side.

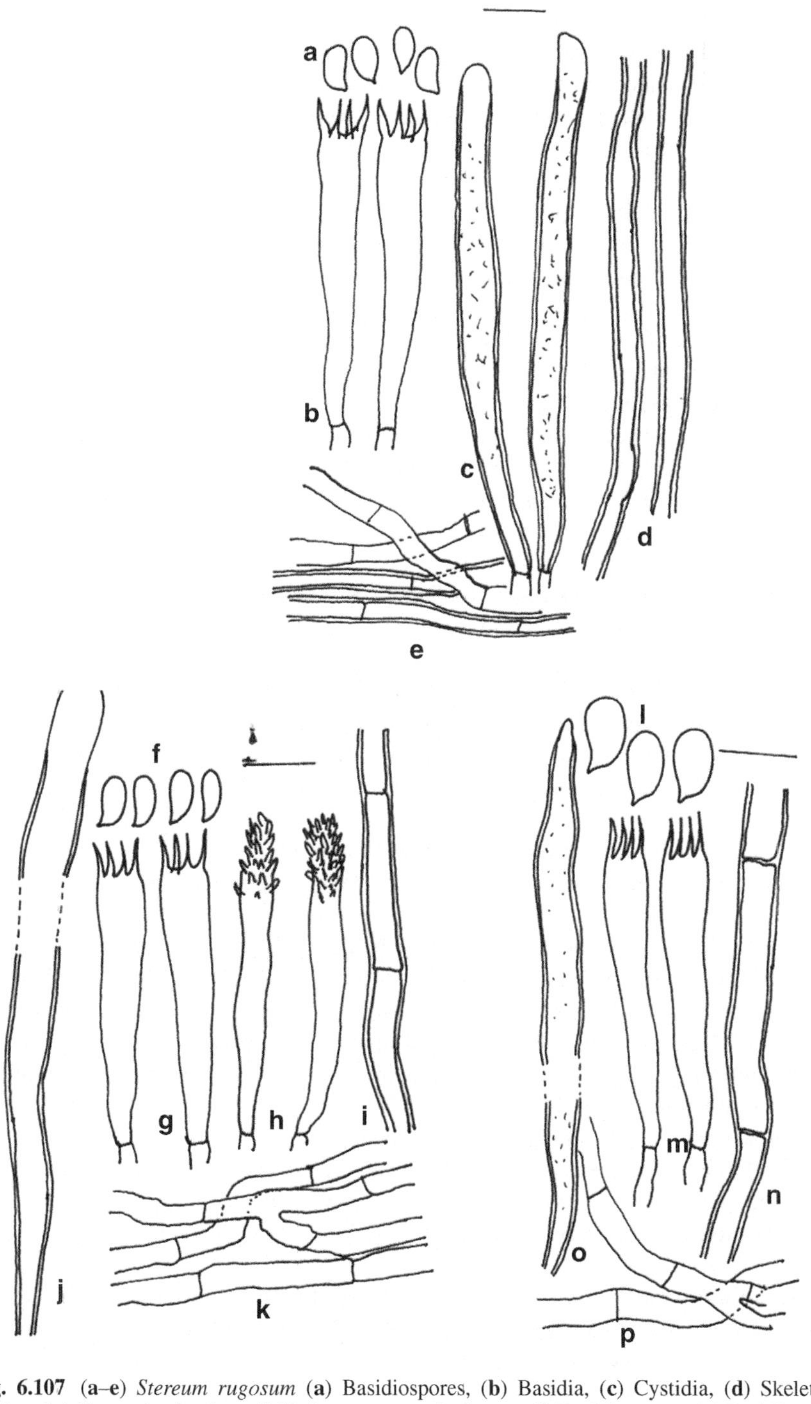

Fig. 6.107 (**a–e**) *Stereum rugosum* (**a**) Basidiospores, (**b**) Basidia, (**c**) Cystidia, (**d**) Skeletal hyphae, (**e**) Generative hyphae; (**f–k**) *Stereum sanguinolentum* (**f**) Basidiospores, (**g**) Basidia, (**h**) Acanthophyses, (**i**) Skeletal hyphae, (**j**) Cystidia, (**k**) Generative hyphae; (**l–p**) *Stereum subtomentosum* (**l**) Basidiospores, (**m**) Basidia, (**n**, **p**) Generative hyphae, (**o**) Cystidia

Hyphal system dimitic; generative hyphae 1.6–3.1 μm wide, branched, septate, clamps absent, the walls thin to moderately thick, subhyaline; skeletal hyphae 3.6–7 μm wide, sparsely branched, aseptate, the walls subhyaline, thick. Hymenium thick and stratose. Conducting hyphae of variable length, 5–8.2 μm broad, cylindrical to flexuous, arising from the upper layers of the context and traversing through different hymenial strata, immersed or projecting slightly out of it, often containing yellow-orange contents, the walls thin, subhyaline. Basidia 5.2–6.2 μm broad, clavate, 4-spored. Basidiospores 5.9–7.4 × 2.9–3.4 μm, ellipsoid to suballantoid, minutely apiculate, the walls thin, subhyaline, smooth, nonamyloid.

Distribution: H.P.: Narkanda, Shimla; U.K.: Chakrata.

Collection examined: IBP 37493.

Substratum: On stump of Oak.

Remarks: This species was reported by Bagchee and Bakshi (1954) from the oak forests of Narkanda (H.P.) and Chakrata (U.K.) hills. It is quite close to *S. sanguinolentum* and *S. gausapatum* but differ in having thicker fructifications and stratose hymenium.

Stereum sanguinolentum (Alb. & Schw.) Fr., Epicr. syst. mycol.: 549 (1838) [1836–1838] = *Thelephora sanguinolenta* Alb. & Schw., Consp. fung.: 274–75. 1805.

Plate 6.39f, Fig. 6.107f–k

Fructification resupinate to effused reflexed to pileate, loosely adnate, arise as small colonies which later on coalesce and become widely effused; pileus narrowly reflexed to flabelliform, faintly concentrically azonate, hirsute towards the base, un-even, confined to coniferous wood; hymenial surface deep cream with red tinge when fresh becomes greyish brown on drying.

Hyphal system dimitic, generative hyphae branched, septate, 2.2–3.6 μm wide, without clamps, thin to thick-walled; skeletal hyphae unbranched, aseptate, thick-walled, 3.1–8.2 μm wide. Conducting hyphae arising from the upper layer of subiculum of different length. Basidia clavate, 30.2–35.2 × 4.1–5.1 μm cylindrical, 4-spored; aculeate tipped basidioles. Basidiospores ellipsoid to suballantoid, 5.6–7.1 × 2.9–3.3 μm, thin-walled, minutely apiculate, hyaline, smooth, amyloid.

Distribution: H.P.: Mahasu; U.K.: Mussoorie; J&K: Batote; Nepal: Gosainkund.

Collection examined: GSR 5012, SSR 5404.

Substratum: On dead stump of *Cedrus deodara*.

Remarks: This species is characterized by aculeate basidioles and its presence confined to conifers only. The features of the above collected sample resembles with the description given by Rattan (1977).

Stereum subtomentosum Pouzar, Česká Mykol. 18: 147, 1964. Fig. 6.107l–p

Fructification resupinate to pileate, reflexed, effused; pilei fan shaped to spathulate with tapering base, lobed; hymenial surface orange to brownish orange; margin thinning to indeterminate.

Hyphal system monomitic; two types of hyphae; hymenial hyphae up to 3.2 μm, thin-walled, highly branched. Cystidia is of two types: (i) Pseudocystidia thick-walled, arise from the cortex, (ii) Acutocystidia thin-walled, acute apex, present in the hymenium. Basidia 24.2–33.0 × 3.6–5.2 μm, 4-sterigmate, clavate without basal clamp. Basidiospores 8.7–12.2 × 4.2–5.4 μm, smooth, thin-walled, apiculate, cylindrical, acyanophilous, amyloid.

Distribution: H.P.: Shimla, Narkanda, Manali.

Collection examined: IBP 37494.

Substrate: On wood of *C. deodara.*

Remarks: This species is characterized by spathulate fructification, clavate basidia, cylindrical, amyloid basidiospores.

Xylobolus P. Karst.,
Meddel. Soc. Fauna Fl. Fenn. 6: 11, 1881.

Fructifications resupinate to effused-reflexed or pileate, annual to perennial, stratose; upper surface of the pileus concentrically zonate, hard and bark-like; hymenial surface smooth to finely tuberculate, rarely cracked. Subiculum generally obliterates in thicker fructifications due to the increasing pressure of developing hymenial strata. Hyphal system dimitic; skeletal hyphae subhyaline to light brown; generative hyphae with or without clamps. Stratose with acanthophyses, incrusted acanthophyses, acanthophysoid incrusted cystidia, pseudocystidia or gloeocystidia. Basidia clavate, 4-spored. Basidiospores ellipsoid, the walls thin, smooth, subhyaline and amyloid.

Three species, widespread

Lit.: Solheim & Hofton (*Sopp og Nyttevekster* 1: 54, 2005)

Type Species: *Thelephora frustulata* Pers. 1801

Habitat: Dead Wood

Himalaya: Two

Key to species

1. Fructification resupinate, breaks into frustules; incrusted cystidia absent .... *X. frustulatus*
1. Fructification pileate, not break into frustules; incrusted cystidia present .... *X. subpileatus*

Xylobolus frustulatus (Pers.) P. Karst, Boidin, Revue Mycol. 23: 341. 1958 = *Thelephora frustulata* Pers. Syn. meth. fung. 2: 577 (1801). Plate 6.40a, Fig. 6.108a–e

Fructification resupinate to effused-reflexed, perennial, woody, stratose, adnate, up to 5 mm thick in section; hymenial surface cream to pale yellowish cream when young but becomes yellow-ochre with age, smooth to finely tuberculate; margin concolorous with the hymenial surface but becomes brownish black on the abhymenial side.

Hyphal system dimitic; generative hyphae branched, septate, with or without clamp; skeletal hyphae subhyaline to light brown, thick-walled. Cystidia or gleocystidia absent. Basidia clavate-cylindrical, 5.3–6.2 μm wide, 4-spored. Basidiospores ellipsoid, subhyaline, thin-walled, 4–4.2 × 2.2–3.1 μm, smooth, amyloid.

Distribution: H.P.: Dalhousie; Nepal: Kathmandu; J&K: Patnitop.

Collection examined: SSR: 5169, SSR 5205, IBP 37496.

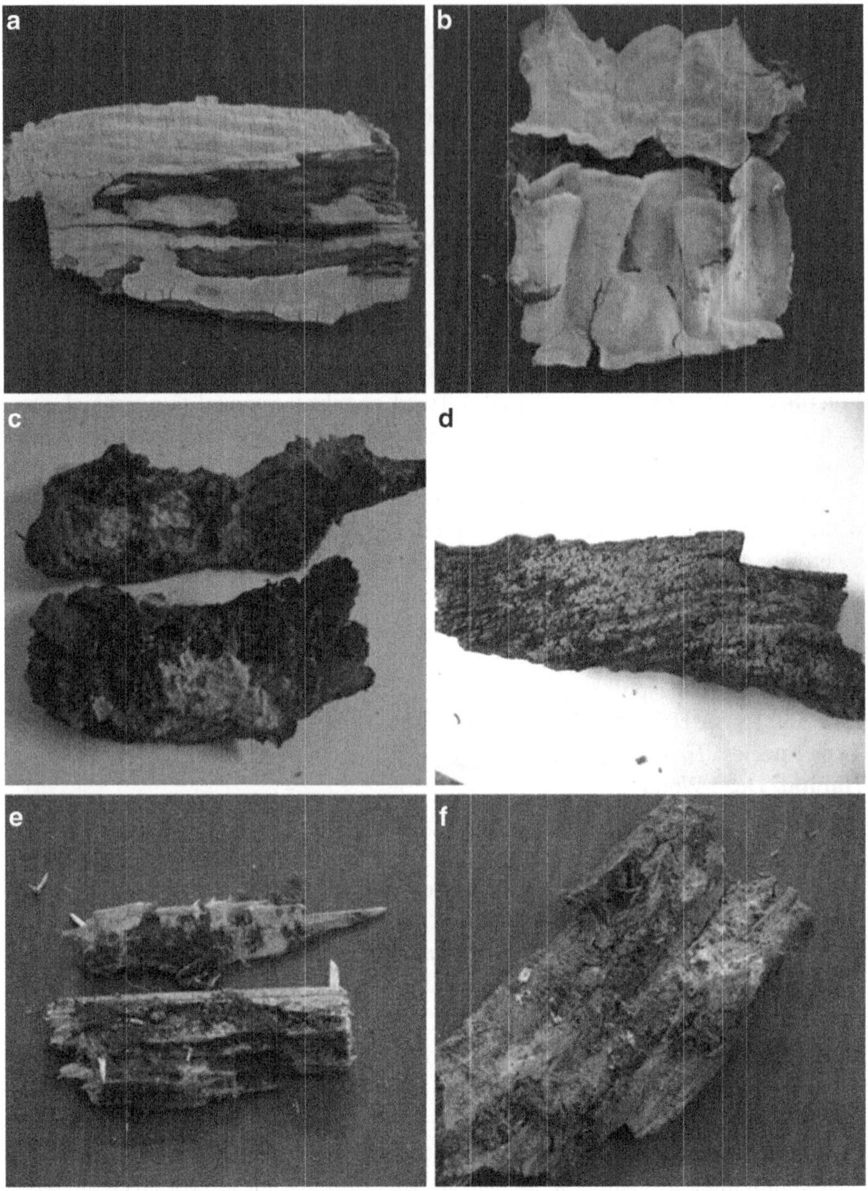

Plate 6.40 (**a**) *Xylobolus frustulatus*. (**b**) *Xylobolus subpileatus*. (**c**) *Wrightoporia lenta*. (**d**) *Amaurodon viridis*. (**e**) *Tomentella botryoides* (**f**) *Tomentella calcicola*

Substratum: On stump of *Cedrus deodara*.

Remarks: This species is characterized by its thick, resupinate fructification which break up into frustules. In India, it was first reported by Bagchee and Bakshi (1954) from N.W. Himalayas and Madras. Later Rattan (1977) reported it from the N.W. Himalayas on the basis of many collections.

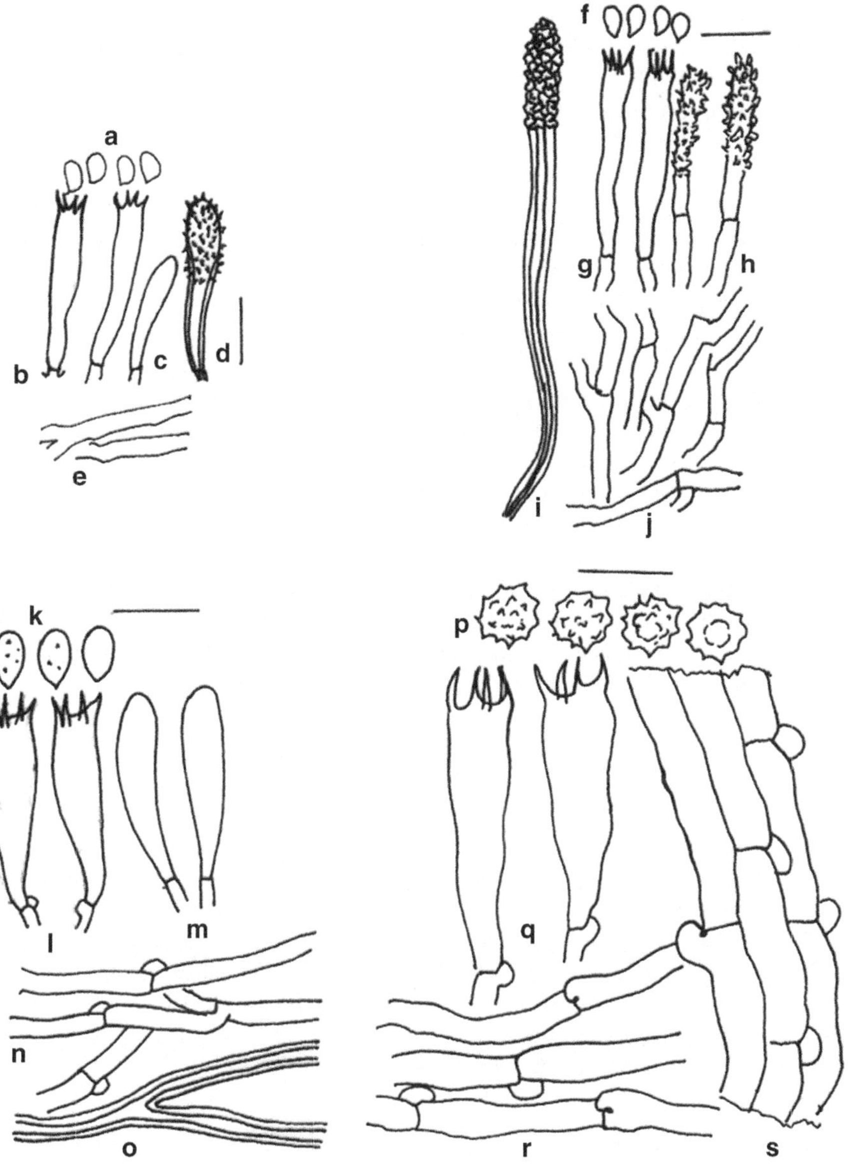

Fig. 6.108 (**a–e**) *Xylobolus frustulatus* (**a**) Basidiospores, (**b**) Basidia, (**c**) Basidioles, (**d**) Acanthophyses, (**e**) Generative hyphae; (**f–j**) *Xylobolus subpileatus* (**f**) Basidiospores, (**g**) Basidia, (**h**) Acanthophyses, (**i**) Cystidia, (**j**) Generative hyphae; (**k–o**) *Wrightoporia lenta* (**k**) Basidiospores, (**l**) Basidia, (**m**) Basidioles, (**n**) Generative hyphae, (**o**) Skeletal hyphae; (**p–s**) *Amaurodon viridis* (**p**) Basidiospores, (**q**) Basidia, (**r**) Generative hyphae, (**s**) Basal hyphae

Xylobolus subpileatus (Berk. & Curt.) Boidin, Revue Mycol. 23: 341. 1958 = *Stereum subpileatum* Berk & Curt., Hook. J. Bot. And Kew Gard. Misc. 1: 238. 1829. Plate 6.40b, Fig. 6.108f–j

Fructification resupinate, effused-reflexed to pileate, perennial, woody, form large patches; hymenial surface stratose, deep cream to yellowish brown, smooth to unevenly tuberculate, rarely creviced; upper surface yellowish brown, brown to finely blackish brown, concentrically zonate; margin acute, paler concolorous.

Hyphal system dimitic; generative hyphae branched, 2–3.5 μm wide, septate, without clamp, subhyaline; skeletal hyphae subhyaline to light brown, 3–4.8 μm wide, thick-walled, unbranched, aseptate, often curving in to the hymenium as pseudocystidia. Cystidia present, cylindrical, thick, encrusted, arising from upper part of the context from the generative hyphae heavily with subhyaline crystalline matter specially in the upper part of the context. Acanthophyses 28–43 × 3–6 μm, clavate to clavate-cylindrical, covered with fine crystals. Gleocystidia absent. Basidia clavate-cylindrical, 4-spored. Basidiospores ellipsoid, shortly apiculate, subhyaline, thin-walled, smooth, amyloid.

Distribution: H.P.: Kullu; Nepal: Gosainkund; U.K.: Mussoorie.

Collection examined: SSR: 5169, 5205, IBP 37497.

Substratum: On fallen log, on stump of *Quercus incana*.

Remarks: The species is characterized by large, effused-reflexed to pileate, woody fructification, presence of numerous acanthohyphidia, clavate basidia and ellipsoid, smooth, thin-walled, amyloid basidiospores.

Family- Wrightoporiaceae

Wrightoporia Pouzar Z.

Ceská Mykol. 20(3): 1731966.

Fructification annual, resupinate, widely effused, soft to tough; pore surface cream, ochraceous to fulvous, pores round to angular or irregular, 2-6 mm, dissipiment thin to rather thick, margin myceloid. Context thin. Hyphal system dimitic; generative hyphae clamped, hyaline and thin-walled; skeletal hyphae dominating, thick-walled and dextrinoid. Cystidia none. Basidiospores broadly elliptical, small, smooth or with small warts, distinctly amyloid, thin to rather thick-walled.

Twenty three species, widespread

Lit.: Stalpers (Stud. Mycol. 40:129, 1996; key)

Type species: *Poria lenta* Overh. 1946

Himalayas: One

Wrightoporia lenta (Overh. & J. Lowe) Pouzar, Česká Mykol. 20: 173 (1966) = *Poria lenta* Overh. & Lowe, Mycologia 38(2): 210, 1946. Plate 6.40c,
Fig. 6.108k–o

Fructification annual, resupinate, effused, adnate to separable, without any distinct taste; margin white to creamish, thinning, myceloid, narrow to wide. Pore surface white to creamish when fresh, drying creamish-yellow to pale brown, uneven, dull; pores rounded angular, equal composed of compact parallel hyphae, torn at apices, finely velutinate. Context white, thin, soft, of interwoven hyphae not darkening in KOH; tubes not stratose, creamish to pale brown in section, soft and brittle when dry.

Hyphal system dimitic; generative hyphae hyaline, thin-walled, branched, septate, clamped, cyanophilous, 2.2–6.4 μm in diameter; skeletal hyphae subhyaline, thick-walled to highly thick-walled, branched, aseptate, acyanophilous, 2.2–5.7 μm in diameter. Hymenium composed of basidia and basidioles. Cystidia absent. Basidia hyaline, thin-walled, clavate, 2–4 spored, 12.4–14.1 × 5.5 μm; sterigmata short to long. Basidiospores hyaline to subhyaline, thin-walled smooth to asperculate, broadly ellipsoid to obovoid, some uniguttulate, apiculate, apiculus small, non-amyloid, 4.2–5.5(−7.6) × 2.2–4.5 μm, large pores 2–3 per mm.

Distribution: J&K: Patnitop, H.P.: Narkanda, Shimla.

Collection examined: Dhanda 6522, 6587, IBP 37498.

Substratum: On log of *Cedrus deodara.*

Remarks: The species is characterized by annual, widely effused, white to creamish fructification, creamish context; dimitic hyphal system; hyaline, smooth to asperculate and amyloid broadly ovoid to subglobose basidiospores.

O- Thelephorales
Family- Thelephoraceae

Key to genera

1. Fructification bluish when fresh, deep to yellowish green when dry; spores turning violet or bluish violet in 3 % KOH sol........................ *Amaurodon*
1. Fructification not coloured as above; spores never turning violet or bluish violet in 3 % KOH sol.. 2
2. Fructification strictly effused 3
2. Fructification not strictly effused 4
3. Hymenial surface smooth; hyphae septate; spores hyaline in 3 % KOH; hyphal cords present ....*Pseudotomentella*
3. Hymenial surface smooth; granulose to hydnoid; hyphae clamped, rarely septate; spores mostly brownish in 3 % KOH sol; hyphal cords present or absent.................................*Tomentella*
4. Hymenial surface tubular.. *Boletopsis*[a]
4. Hymenial surface smooth ....*Thelephora*

[a]Extra limital, not included in this work

Amaurodon J. Schröt.,
Krypt. –Fl. Von. Schlesien 3–1(4):461, 1888

Fructifications resupinate, loosely adnate, often arising as small circular colonies which may coalesce later and become widely effused; hymenial surface yellowish green to deep green, smooth to tuberculate or distinctly toothed, more or less continuous; margin thinning to determinate, loosely adnate, paler concolorous. Context light green in section, composed of loosely woven hyphae. Hyphal system monomitic, branched at wide angles, septate, clamped, thin-walled. Hyphal cordons absent or very obscurely developed. KOH reaction tissues turning black when mounted in 3 % sol, in sectional preparations tissues appear blue to bluish green and produce an ink-blue exudates. Cystidia absent. Basidia clavate to clavate-cylindrical, rarely with a median retraction septum, 4-spored. Basidiospores globose to subglobose, the walls thin, finely echinulate, spines small, blunt and more or less hemispherical, nonamyloid.

Nine species, widespread

Lit.: Agerer & Bougher (*Aust. Syst. Bot.* **14**: 599, 2001)

Type Species: *Sistotrema viridae* Alb. & Schwein. 1805

Habitat: Wood

Himalayas: One

Amaurodon viridis (Alb. & Schwein.) J. Schröt., in Cohn, Krypt.-Fl. Schlesien (Breslau) 3.1(25–32): 461 (1888) [1889] = *Tomentella chlorina* (Mass.) Cunn. Proc. Linn. Soc. NSW 77: 279. 1953. Plate 6.40d, Fig. 6.108p–s

Fructifications resupinate, floccose to submembranous, loosely adnate, often arising as small circular colonies coalesce later and become widely effused, up to 500 μm thick in section; hymenial surface yellowish green to deep green, smooth to tuberculate or distinctly toothed, more or less continuous; margin thinning to determinate, loosely adnate, paler concolorous. Teeth ranging from small protuberances to 0.5(0.75) mm long, discrete when fresh but tend to coalesce on drying, yellowish green to deep green, subulate to subcylindrical, apices acute to subobtuse, fertile all over. Subiculum light green in section, composed of loosely woven hyphae.

Hyphal system monomitic, hyphae 2.4–4.5 μm wide, branched at wide angles, septate, clamped, thin-walled, tinted yellow to pale citrine when mounted in water or 3 % NH_4OH but turns blue to bluish green in KOH sol. The basal hyphae are slightly broader, darker in colour, sparsely branched and distantly septate than the subhymenial ones. Cystidia absent. Basidia 24.8–34.8 × 5–7 μm, clavate to clavate-cylindrical, rarely with a median retraction septum, 4-spored, sterigmata up to 4 μm long. Basidiospores 4.2–5.6 μm in diameter, globose to subglobose, thin-walled, appear pale yellow to pale citrine when seen in water mounts but turn blue in KOH sol mounts, finely echinulate, spines small, blunt and more or less hemispherical, non-amyloid.

Distribution: H.P.: Mahasu- Narkanda; Kullu- Soja, Gojra, Jaggat sukh, Rohtang, Chamba- Khijjiar, Dalhousie, Kalatope.

Collection examined: K: 4031, 4041, 4064, IBP 37499.

Substratum: Wood, stumps and logs of conifers, logs of *Picea smithiana* and *Abies pindrow*, bark and slash of *Abies pindrow*.

Remarks: The species is characterized by yellowish green to deep green fructifications, presence of teeth and small, globose basidiospores with blunt, hemispherical spines or warts. The reaction of KOH sol on tissues is also an important diagnostic feature.

Pseudotomentella Svrček,
Ceská Mykol. 12(2):67 1958

Fructifications resupinate, floccose to soft-spongy, loosely adnate to separable, widely effused; hymenial surface pinkish brown to light brown, smooth to obscurely colliculose, almost continuous; margin thinning to sometimes abrupt, loosely adnate. Subiculum light brown in section. Hyphal system dimitic; skeletal hyphae unbranched, aseptate or occasionally developing some retraction septa also, the walls thin to moderately thick, tinted brown, present only among hyphal cordons; generative hyphae wide, moderately branched, septate, clamps absent, the walls thin, subhyaline. Hyphal cordons abundant. Basidia utriform or subcylindrical, 2–4 spored. Basidiospores globose to subglobose, thin-walled, hyaline in 3 % KOH sol, non-amyloid.

Twelve Species, widespread

Lit.: Larsen (*Mycol.* 66167, 1974; key), Martini & Hentic (*BSMF* 119: 19, 2003; France).

Type Species: *Hypochnus mucidulus* Karst. Svrcek 1882

Habitat: Decay Wood

Himalayas: Two

Key to species

1. Hyphal system monomitic, hymenial surface light orange to brownish .... *P. tristis*
1. Hyphal system dimitic, hymenial surface pinkish grey to light brown .... *P. mucidula*

Pseudotomentella mucidula (Karst.) Svrček, Česka Mykol. 12: 68. 1958 = *Hypochnus mucidulus* Karst., Bidr. Kann. Finl. Nat. Folk 37:163. 1882. Fig. 6.109a–e

Fructifications resupinate, floccose to soft-spongy, loosely adnate to separable, widely effused; hymenial surface pinkish brown to light brown, smooth to obscurely colliculose, almost continuous; margin thinning to sometimes abrupt, loosely adnate, white to paler concolorous. Subiculum composed of loosely woven hyphae and hyphal cordons.

Hyphal system dimitic; generative hyphae 1.4–4.0 μm wide, moderately branched, septate, clamps absent, thin-walled, subhyaline skeletal hyphae 1.6–2.6 μm wide, unbranched, aseptate or occasionally developing some retraction septa also, thin to thick-walled, tinted brown, present only among the hyphal cordons. Hyphal cordons abundant, unbranched or occasionally branched and are of two types: (a) subhyaline to pale brown, being composed of generative hyphae only and (b) dark brown, being composed of skeletal hyphae only. Tissues darkening appreciably in 3 % KOH sol. Basidia 36–58 × 7–8 μm, sphaeropedunculate when young but become utriform or subcylindrical at maturity, often with a median retraction septa, 2–4 spored. Basidoles 2.6–3 μm wide, subclavate to filiform. Basidiospores 7–8.3 μm across, globose to subglobose, more or less regular in outline, thin-walled, subhyaline, warted, non-amyloid.

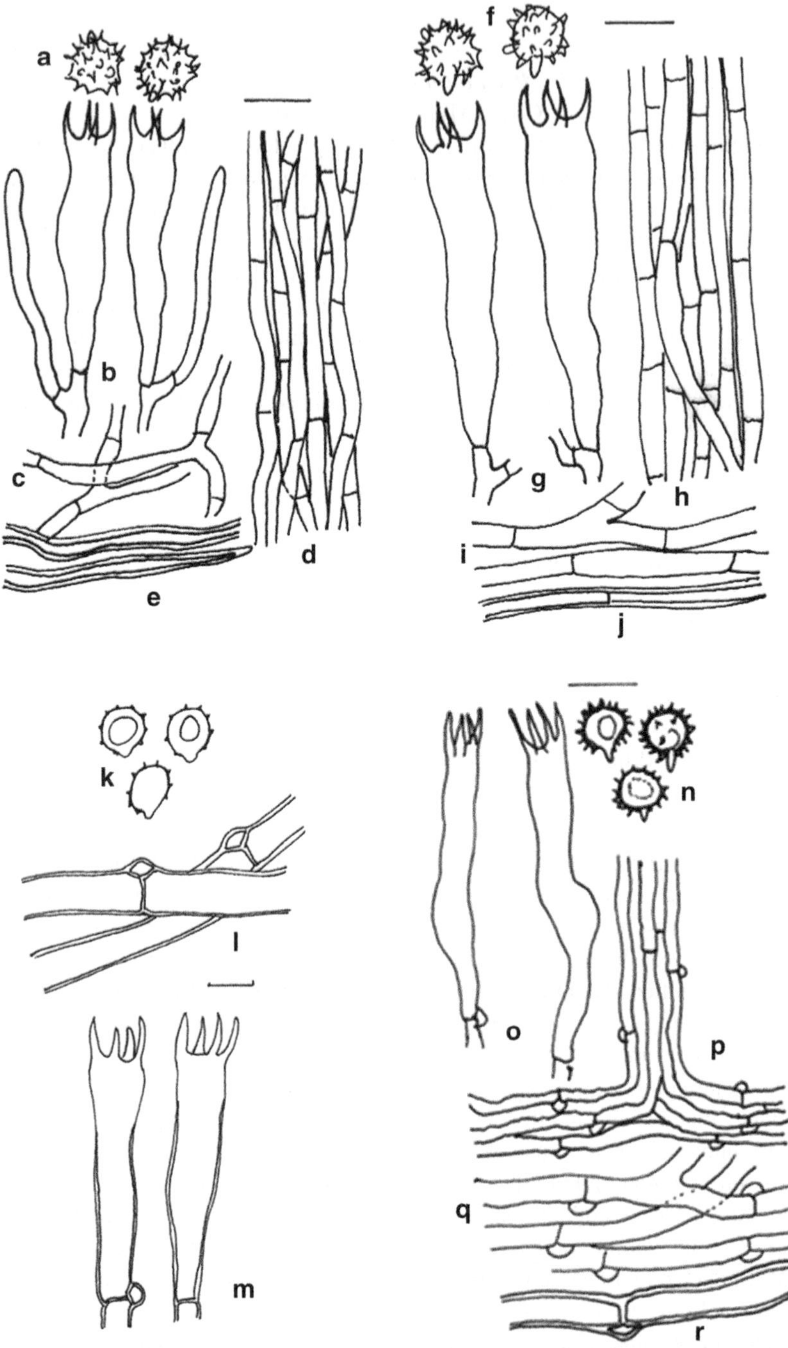

Fig. 6.109 (**a**–**e**) *Pseudotomentella mucidula* (**a**) Basidiospores, (**b**) Basidia, (**c**) Generative hyphae, (**d**) Hyphal cordons, (**e**) Skeletal hyphae; (**f**–**j**) *Pseudotomentella tristis* (**f**) Basidiospores, (**g**) Basidia, (**h**) Hyphal cordons, (**i**) Generative hyphae, (**j**) Skeletal hyphae; (**k**–**m**) *Thelephora atra* (**k**) Basidiospores, (**l**) Generative hyphae, (**m**) Basidia; (**n**–**r**) *Tomentella albomarginate* (**n**) Basidiospores, (**o**) Basidia, (**p**) Hyphal cordons, (**q**) Thin-walled generative hyphae, (**r**) Thick-walled generative hyphae

Distribution: H.P.: Khadrala, Manali.
Collection examined: SSR 5643, IBP 37500.
Substratum: On coniferous slash.
Remarks: The species is characterized by dimitic hyphal system with simple-septate hyphae, subhyaline and large basidiospores with dichotomously tipped warts and sphaeropedunculate to subcylindrical basidia.

Pseudotomentella tristis (P. Karst.) M. J. Larsen, Nova Hedwigia 22(1–2): 613, 1971 = *Hypochnus subfuscus* subsp. Tristris P. Karst., Meddn. Soc. Fauna Flora fenn. 9: 71, 1882. Fig. 6.109f–j

Fructification resupinate, loosely adnate, smooth to grandinoid; hymenial surface light orange to brownish when fresh, brown on drying; margin thinning, paler concolorous to indeterminate.

Hyphal system monomitic; generative hyphae septate without clamps; basal hyphae parallel to the substrate, septate, loosely interwoven, thick-walled; subhymenial hyphae up to 3.6 μm wide, compactly arranged, thin-walled, septate, highly branched. Cystidia absent. Basidia 45–81 × 8.2–11 μm, 4-sterigmate, clavate to subclavate, without clamp. Basidiospores 7.5–11.3 μm across, apiculate, globose to subglobose, warted.

Distribution: H.P.: Dalhousie, Chamba, Shimla-Narkanda.
Collection examined: IBP 37501.
Substrate: On decaying gymnospermous wood.
Remarks: The species is characterized by smooth, brownish orange fructification; clavate basidia and globose to subglobose, warted basidiospores.

Thelephora Ehrh. ex Willd.,
Fl. Berol. Prodr.: 396, 1787

Fructifications resupinate, loosely adnate, effused to membranous; hymenial surface dark brown, smooth, black in KOH sol; margin thinning, concolorous with hymenial surface or indeterminate. Hyphal system monomitic; basal hyphae loosely arranged, parallel to the substrate, sparsely branched, distantly septate, clamped, thick-walled; subhymenial hyphae up to 8 μm wide, compactly packed, vertically arranged, profusely branched, comparatively closely septate, clamped, thin-walled. Hyphal cordons absent. Cystidia absent. Basidia subclavate, 4-spored, with a basal clamp. Basidiospores globose to subglobose, often elongated along one axis, thin to moderately thick-walled, aculeolate to echinulate, pale brown, guttulate.

Fifty species, widespread
Lit.: Zecchin (*Rivista Di Mycologia* **48**: 243, 2005)
Type Species: *Thelephora terrestris* Ehrh. 1787
Habitat: Wood
Himalayas: One

Thelephora atra Weinm., Hym. à Gast. Imp. Ross. Obs. (Petropoli): 636 (1836) = *Tomentella atramentaria* Rostr., Bot. Tidsskr. 19: 41, 1894. Fig. 6.109k–m

Fructifications resupinate, loosely adnate, effused, mucedinoid to submembranous to membranous; hymenial surface dark brown, smooth, black in KOH; subiculum greyish brown to dark brown; margin thinning, concolorous with hymenial surface or indeterminate.

Hyphal system monomitic; basal hyphae up to 9.8 μm wide, loosely arranged, parallel to the substrate, sparsely branched, distantly septate, clamped, olive brown, thick-walled; subhymenial hyphae up to 8 μm wide, compactly packed, vertically arranged, profusely branched, comparatively closely septate, clamped, yellowish grey, thin-walled. Hyphal cordons absent. Cystidia absent. Basidia 46.9–70 × 9–12.9 μm, clavate to subclavate, 4-spored, with a basal clamp, mostly with oily contents, subhyaline to yellowish grey; sterigmata up to 12.5 μm long. Basidiospores 7.2–11.5 μm across, globose to subglobose, becoming irregular in outline, often elongated along one axis, thin to moderately thick-walled, aculeolate to echinulate, pale brown, guttulate.

Distribution: H.P.: Chamba- Kalatope.

Collection examined: IBP 37601.

Substratum: Decaying gymnospermic stump.

Remarks: The species is characterized by bark brown hymenial surface and 7.2–11.5 μm across, aculeolate to echinulate basidiospores.

Tomentella Pers. ex.Pat.,
Hym. Eur. P. 154. 1887.

Fructifications resupinate, arachnoid to floccose, tomentose, pelliculose or submembranous, loosely constructed and easily separable; hymenial surface yellow, brown, green, red or rarely black, discontinuous or occasionally continuous, smooth to tuberculate or distinctly toothed. Hyphal system monomitic or dimitic; generative hyphae mostly distinct, rarely collapsing, often with tinted walls, simple-septate to nodose septate, with or without clamp. Hyphal cordons present or absent. KOH reaction tissues often darkening and changing colour in 3 % sol. Gloeocystidia absent. Cystidia present or absent. Basidia clavate to subcylindrical, 2–4 spored, often developing one or rarely more retraction septa. Basidiospores globose to ellipsoid, regular but more commonly irregular in outline to distinctly lobed, the walls thin to moderately thick, subhyaline or tinted some shade of yellow-brown or fuscus, nonamyloid, echinulate, spines mostly simple or rarely dichotomously, branched or forked.

Eighty species, widespread, ectomycorrhizal.

Lit.: Wakefield (*Trans. Bri. Mycol. Soc.* 53: 168 1993; key)

Type species: *Tomentella ferruginea* Pers. 1801

Habitat: Dead wood

Himalayas: Thirty six

Key to species

1. Hyphal system dimitic 2
1. Hyphal system monomitic 5
2. Hymenophore toothed.. *T. calcicola*
2. Hymenophore tuberculate to toothed.. 3
3. Hymenophore tuberculate to toothed, basidiospores7.8–9 μm............. *T. crinalis*
3. Hymenophore smooth to finely granulose, basidiospores 6.8–8 μm.................. 4
4. Hymenial surface rusty brown to yellowish brown .... *T. umbrinospora*
4. Hymenial surface yellowish brown to ochraceous......................... *T. ferruginea*
5. Hyphae with clamps.. 7
5. Hyphae with or without clamps 6
6. Hyphae simple septate .... *T. badia*
6. Hyphae without clamps, basal hyphae with clamps .... *T. radiosa*
7. Cystidia present.. 8
7. Cystidia absent.. 13
8. Hyphal cordons absent.. 9
8. Hyphal cordons present.. 12
9. Cystidia subfusiform to acuminate .... *T. galzinii*
9. Cystidia hyphoid, fusiform or acuminate... 10
10. Fructification orange brown to brown.. *T. subclavigera*
10. Fructification greyish brown 11
11. Cystidia fusiform .... *T. pyrole*
11. Cystidia obclavate .... *T. clavigera*
12. Cystidia arise from subhymenium hyphae.. *T. muricata*
12. Cystidia arising from basal hyphae.. *T. pilosa*
13. Basidiospores regular in outline.. 14
13. Basidiospores irregular to lobed 18
14. Basidiospores aculeolate or echinulate 15
14. Basidiospores aculeate... 16
15. Hyphal walls swell up in KOH solution .... *T. stuposa*
15. Hyphal walls do not swell up in KOH solution .... *T. lapida*
16. Fructification up to 750 μm thick in section .... *T. cladii*[a]
16. Fructification up to 350 μm thick in section 17
17. Hymenial surface brownish orange to greyish orange,
basidiospores subglobose... *T. kalatopii*
17. Hymenial surface light brown to dark brown,
basidiospores globose to subglobose .... *T. atramentaria*[a]
18. Basidiospores irregular in outline... 19
18. Basidiospores irregular to lobed 31
19. Hymenial surface smooth to farinaceous, yellow ochre
to yellow brown, basidiospores irregular....................................... *T. bryophila*
19. Hymenial surface is not with above combination... 20
20. Basidiospores up to 7 μm in diameter.. 21
20. Basidiospores more than 8 μm.. 24

21. Hymenial surface pink or purple in colour 22
21. Hymenial surface orange brown to greyish brown to dark brown 23
22. Hymenial surface pinkish brown .... *T. cinerescens*
22. Hymenial surface purplish .... *T. micospora*[a]
23. Basidia 18–25 µm in length .... *T. griseoumbrina*
23. Basidia 35–40 µm in length .... *T. elisii*
24. Hyphal cordons present 25
24. Hyphal cordons absent 26
25. Basidia 40–55 µm, clavate .... *T. testaceogilva*
25. Basidia 37–79 µm, narrowly clavate to subcylindrical .... *T. botryoides*
26. Fructification colour changes on putting a drop of 3 % KOH solution 27
26. Fructification colour not changes with KOH solution 29
27. Basidiospores up to 7 µm in diameter .... *T. coerulea*
27. Basidiospores 7.5–10 µm in diameter 28
28. Basidia 45–60 µm, clavate to cylindrical to utriform .... *T. indica*
28. Basidia 50–70 µm, clavate to cylindrical .... *T. subcorticiodes*
29. Fructification hypochnoid to arachnoid .... *T. griseovilacea*
29. Fructification mucedinoid 30
30. Basidiospores up to 11 µm across .... *T. brevispina*
30. Basidiospores up to 14 µm across .... *T. terrestris*
31. Hymenial surface smooth to tuberculate, orange to brownish orange .... *T. albomarginata*
31. Hymenial surface smooth to granular, yellowish brown to greyish brown to olive brown 32
32. Hyphae less than 5 µm wide 33
32. Hyphae more than 5 µm wide 34
33. Hymenial surface light brown, subiculum light brown to dark brown .... *T. puberula*
33. Hymenial surface light orange to orange to brown, subiculum paler concolorous .... *T. bicolor*
34. Hyphal cordons absent 35
35. Basidiospores warted .... *T. unicusa*
35. Basidiospores echinulate 36
36. Basidiospores with granular material red in water mount .... *T. lateritia*
36. Basidiospores without granular material, subhyaline to hyaline in water mount .... *T. scobinella*
37. Hyphal cordons up to 90 µm, basidiospores 7.5–10 µm .... *T. himalayana*
37. Hyphal cordons up to 40 µm, basidiospores 5–9 µm 38
38. Hymenial surface brown to soot brown, basidia 30–35 µm .... *T. punicea*
38. Hymenial surface olive brown, basidia 35–55 µm .... *T. olivascens*

[a]Extra limital, not included in this work

Tomentella albomarginate (Bourdot & Galzin) M. P. Christ., Mycologia 62: 134, 1960 = *Tomentella spongia* f. *albomarginate* Bourdot & Galzin, Bull. trimmest. Soc. Mycol. Fr. 40: 154, 1924. Fig. 6.109n–r

Fructification resupinate, loosely adnate; hymenial surface light orange to brownish orange to greyish brown when fresh, pale brown to dark brown on drying, smooth to tuberculate; margin thinning, paler concolorous to indeterminate.

Hyphal system monomitic; generative hyphae clamped, septate; basal hyphae parallel to the substrate, septate, thick-walled, less branched; subhymenial hyphae compactly packed, thin-walled, septate, highly branched. Cystidia absent. Basidia 38.9–47.2 × 4.5–8.5 μm, 4-sterigmate, clavate to subclavate, with basal clamp. Basidiospores 6.7–8.4 × 6.4–8.5 μm, apiculate, echinulate, irregular.

Distribution: H.P.: Shimla, Kullu.

Collection examined: IBP 37602.

Substrate: On decaying gymnospermous wood.

Remarks: The species is characterized in having smooth to tuberculate hymenial surface; monomitic hyphal system; presence of hyphal cordons; clavate to subclavate basidia; irregular, echinulate basidiospores.

Tomentella badia (Link) Stalpers, Revue Mycol., Paris 39(2): 98 (1975) = *Tomentella fimbriata* Christ., Dansk Bot. Arkv. 19: 258. 1960. Fig. 6.110a–d

Fructification resupinate, floccose to submembranous, loosely adnate, widely effused; hymenial surface light brown to chocolate brown, smooth to finely granulose, discontinuous, somewhat velvety in appearance; margins indeterminate to more or less abrupt, loosely adnate, concolorous.

Hyphal system monomitic; basal hyphae 4–7 (8.5) μm wide, moderately branched, branches at wide angles, septate, septa at long intervals, clamps absent, uninflatted or rarely inflated between the septa, thick-walled (0.8 μm), tinted brown; subhymenial hyphae 3–6.8 μm wide, profusely branched, branches at wide angles, septate, septa at relatively short intervals, clamps absent, often inflated between the septa, thin-walled, subhyaline or faintly tinted brown, becomes black when mounted in 3 % KOH sol, hymenium and subhymenium appears faintly green in section. Cystidia absent. Basidia 60–91 (111) × 9–12.5 (14) μm, clavate-cylindrical, 4-spored, sterigmata up to 8.5 μm long. Basidiospores 8.6–11.1 × 7.1–8.4 μm, subglobose but often elongated along one axis, irregular to lobed, uniguttulate, thin to moderately thick (0.8 μm), tinted brown, non-amyloid, echinulate.

Distribution: H.P.: Narkanda.

Collection examined: SSR 5672, 8530.

Substratum: On slash of *Pinus excelsa*.

Remarks: This species is characterized by simple-septate hyphae, characteristic reaction of KOH sol and relatively large basidia and basidiospores, fibrous subiculum and smooth to finely granulose hymenial surface with velvety appearance.

Tomentella bicolor (G.F. Atk. & Burt) Bourdot & Galz., Bull. trimmest. Soc. Mycol. France 40(1): 132, 1924 = *Hypochnus bicolour* Atk. & Burt., Ann. Missouri, Bot. Gard. 3: 229, 1916. Fig. 6.110e–h

Fructification resupinate, loosely adnate, arachnoid to submembranous; hymenial surface grey to olive grey, smooth to colliculose, greenish grey in KOH sol; subiculum arachnoid, paler concolorous; margins thinning, arachnoid, concolorous with subiculum, or indeterminate.

Hyphal system monomitic; basal hyphae 3.5 µm wide, loosely intermingled, branched, septate, clamped, with encrusting material, greyish green, thin-walled; subhymenial hyphae up to 3.5 µm wide, compactly packed, branched, septate, clamped, hyaline, thin-walled. Hyphal cordons wide, branched; individual hyphae up to 3.8 µm wide, thin-walled, hyaline. Cystidia absent. Basidia 50.0–67 × 6.0–11 µm, clavate, 4-spored, with a basal clamp, with oily contents, sometimes pale green in KOH sol. Basidiospores 6.2–9.5 µm across, irregular in outline to lobed, thin to somewhat thick-walled, echinulate, subhyaline to yellowish grey in KOH sol, uniguttulate.

Distribution: H.P.: Chamba- Dalhousie.

Collection examined: IBP 37603.

Substratum: decayed log of gymnosperms.

Remarks: The species is characterized by loosely adnate, arachnoid fructification; light orange to greyish orange, smooth to granulose hymenial surface; monomitic hyphal system; presence of hyphal cordons; basidia clavate; irregular to lobed basidiospores. The description of the species resembles with Larsen (1974) in most of the characters, but for somewhat smaller length of basidia (i.e. 23.4–35.4 × 5.7–7.7 µm as compared to 35.0–45.0 × 5.0–7.0 µm).

Tomentella botryoides (Schw.) Bourd. & Galz. Bull. Soc. Mycol.Fr. 40: 159. 1924. Plate 6.40e, Fig. 6.110i–k

Fructification resupinate, loosely adnate, arachnoid to floccose, effused; hymenial surface snuff-brown to smoky brown, smooth to finely granulose; margin indeterminate, loosely adnate and paler concolorous.

Hyphal system monomitic; basal hyphae 3–5.8 µm wide, sparsely branched, distantly septate, clamps present, walls thin, in KOH sol the section turns blue; subhymenial hyphae 2.8–4.8 µm wide, closely branched, clamps present, septate. Hyphal cordons present, light brown to golden brown. Cystidia absent. Basidia 39.8–65 × 7–8.4 µm, 4-spored, clavate, sterigmate slender. Basidiospores 7–8.8 µm, irregular, thin to thick-walled, echinulate, nonamyloid.

Distribution: H.P.: Mahasu- Narkanda, Bagi, Manali; U.K.: Mussoorie-Dehradun.

Collection examined: SSR 5364, 5869, IBP 37605.

Substratum: logs of *Abies pindrow* and bark under conifers.

Remarks: The species is characterized by resupinate, effused fructification; hymenial surface snuff-brown; hyphal system monomitic; cystidia absent; basidia clavate, 4-sterigmate; irregular thin-walled basidiospores.

Tomentella brevispina (Bourd. & Galz.) Larsen, Mycologia 62: 136, 1970 = *Tomentella spongiosa f. brevispina* (Bourd. & Galz.) Svrcek Ceska Mykol. 12(2): 75, 1958. Fig. 6.110l–o

Fructifications, resupinate, loosely adnate, effused, granulose, hypochnoid to mucedinoid, up to 620 μm thick in section; hymenial surface pale yellow to yellowish brown when fresh, light brown to brown on drying, subiculum fibrous; margins thinning, pale concolorous.

Hyphal system monomitic; generative hyphae septate, clamped, basal hyphae thick-walled, branched at right angles and of two types on the basis of color (i) dark brown hyphae close to substratum, up to 13 μm wide, and (ii) yellowish brown hyphae between subhymenial hyphae and dark brown hyphae, up to 8.4 μm wide; subhymenial hyphae up to 7.2 μm wide, compact, branched, clamped, thin to moderately thick-walled, subhyaline. Hyphal cordons absent. Basidia 40.8–53.9 × 10.2–12.2 μm, clavate to subcylindrical, clamped at the base, 4-sterigmate. Basidiospores 9.4–10.9 μm across, regular in outline, usually globose, sometimes subglobose, apiculate, pale to medium brown, aculei up to 2.5 μm long.

Distribution: H.P.: Solan, Kullu.

Collection examined: IBP 37607.

Substratum: On decayed angiospermic twigs.

Remarks: This species is characteristic in having two types of subicular hyphae, on the basis of which it has been differentiated from closely related *T. bresadolae*. It was first reported by Larsen (1974) from France.

Tomentella bryophila (Pers.) Larsen Mycologia Mem. 4: 51. 1974 = *Tomentella pallidofulvus* Peck., Litsch. Ost. Bot. Zeitschr. 88: 131. 1939. Fig. 6.111a–e

Fructification resupinate, subpelliculose to membranous, somewhat spongy, loosely adnate, widely effused; hymenial surface yellow ochre to yellow brown, smooth to somewhat farinaceous, continuous; margins indeterminate, loosely adnate, paler concolorous.

Hyphal system monomitic; basal hyphae up to 3.5–5 μm wide, sparsely branched, distantly septate, clamped, clamps prominent and almost at all septa, thin to thick –walled (0.5–0.8 μm) slightly tinted yellow; subhymenial hyphae 2.5–5 (6) μm wide, branched, closely septate, clamped, often developing few retraction septa also, the walls subhyaline to tinted yellow, thin to moderately thick-walled, becoming progressively thin-walled and lighter colored towards the hymenium. Hyphal cordons absent. Tissues darkening or becoming blackish when mounted in 3 % KOH sol. Cystidia absent. Basidia 45.2–55.2 × 8.4–10.4 μm, clavate-cylindrical, often with a median retraction septum, 4-spored, contents subhyaline but sometimes turn yellow in KOH sol. Basidiospores 7.8–10 μm across, subglobose to irregular in outline, moderately thick-walled (0.9 μm), tinted yellow, non-amyloid, echinulate.

Distribution: H.P.: Dalhousie- Banikhet, Mahasu- Narkanda.

Collection examined: SSR: 5831, 5877, IBP 37609.

Substratum: On leaves of *Quercus incana* and logs under confers.

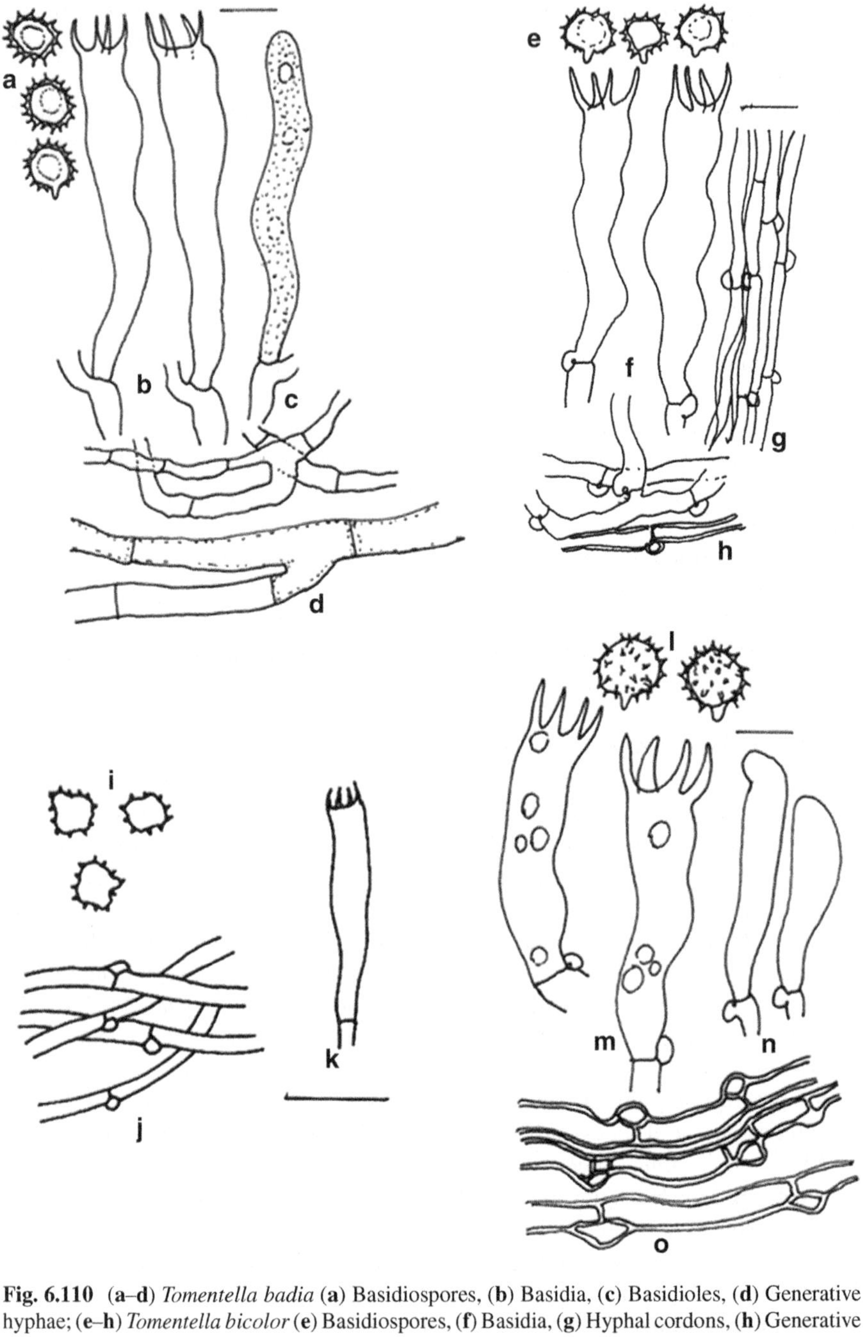

Fig. 6.110 (**a**–**d**) *Tomentella badia* (**a**) Basidiospores, (**b**) Basidia, (**c**) Basidioles, (**d**) Generative hyphae; (**e**–**h**) *Tomentella bicolor* (**e**) Basidiospores, (**f**) Basidia, (**g**) Hyphal cordons, (**h**) Generative hyphae; (**i**–**k**) *Tomentella botryoides* (**i**) Basidiospores, (**j**) Generative hyphae, (**k**) Basidia; (**l**–**o**) *Tomentella brevispina* (**l**) Basidiospores, (**m**) Basidia, (**n**) Basidioles, (**o**) Generative hyphae

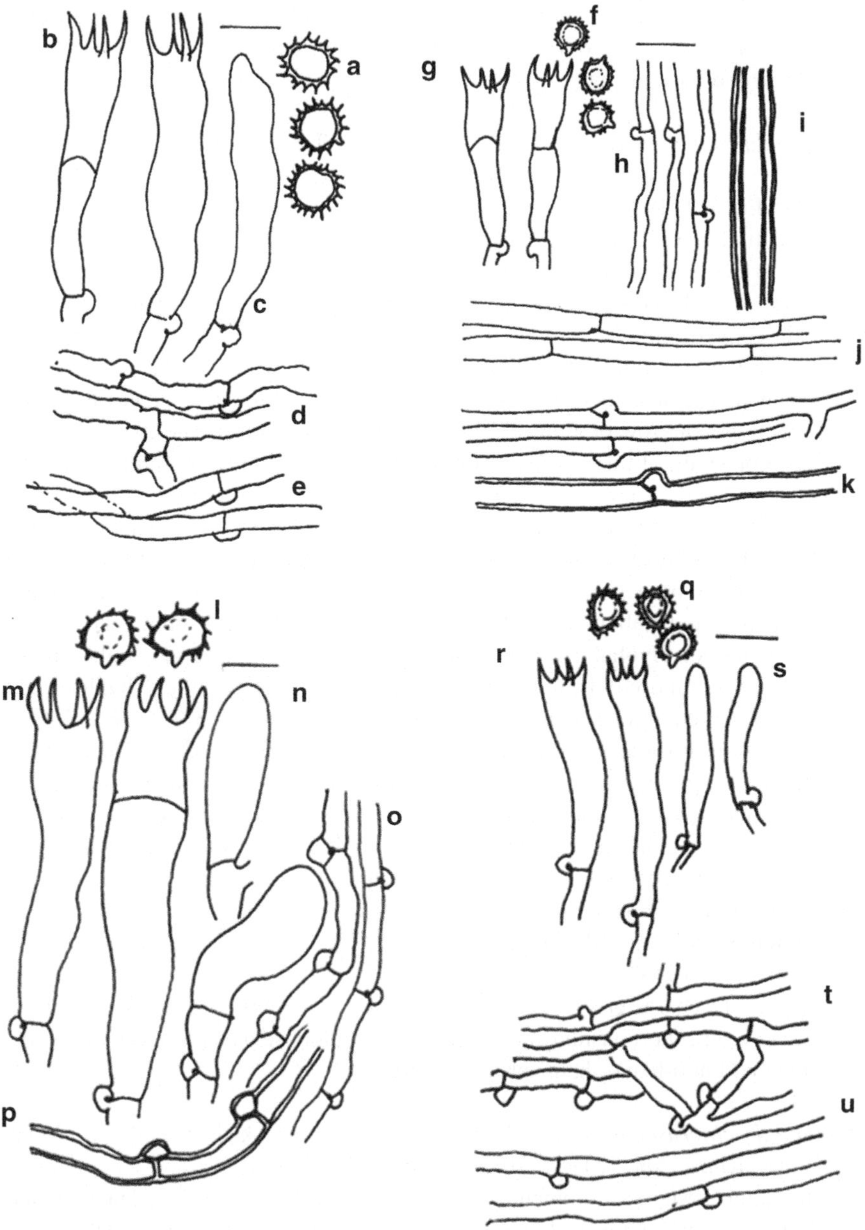

Fig. 6.111 (**a**–**e**) *Tomentella bryophilla* (**a**) Basidiospores, (**b**) Basidia (**c**) Basidioles, (**d**) Subhymenial hyphae, (**e**) Basal hyphae; (**f**–**k**) *Tomentella calcicola* (**f**) Basidiospores, (**g**) Basidia, (**h**) Subhymenial hyphae, (**i**) skeletal hyphae, (**j**) Simple septate generative hyphae, (**k**) Generative hyphae with clamps; (**l**–**p**) *Tomentella clavigera* (**l**) Basidiospores, (**m**) Basidia, (**n**) Cystidia, (**o**–**p**) Generative hyphae; (**q**–**u**) *Tomentella coerulea* (**q**) Basidiospores, (**r**) Basidia, (**s**) Basidioles (**t**) Subhymenial hyphae, (**u**) Basal hyphae

Remarks: This species is characterized by subpelliculose to membranous fructification, yellow ochre to yellow brown hymenial surface and yellow color of hyphae and basidiospores.

Tomentella calcicola (Bourd. & Galz.) M. J. Lars., Taxon 16:511, 1967. = *Caldesiella ferruginosa* var. *Calcicola* Bourd. & Galz., Hym. France p. 471.1928. Plate 6.40f, Fig. 6.111f–k

Fructification resupinate, submembranous, loosely adnate, widely effused; hymenial surface yellow ochre to light brown, distinctly toothed; margin thinning to abrupt, loosely adnate, concolorous. Teeth gregarious terrete when fresh but coalesce on drying.

Hyphal system dimitic; skeletal hyphae 1.5–2.5(3) µm wide, unbranched to rarely branched, aseptate to occasionally producing few retraction septa, the walls thick, lumen narrow or almost obliterated, bright yellow or pale citrine; generative hyphae are of three types: (a) basal hyphae 4–5 µm wide, rarely branched, simple-septate, or rarely clamped, the walls thin to slightly thick, subhyaline to tinted brown; (b) subhymenial hyphae 1.5–4 µm wide, branched, septate, clamped, clamps prominent and almost at all septa, the walls thin to slightly thick, tinted brown; (c) tramal hyphae 1.5–3 µm wide, branched, septate, clamped, clamps abundant and at all septa, the walls thin, subhyaline or very faintly tinted. Hyphal cordons abundant, up to 40 µm broad, light brown, branched, composed of exclusively of either skeletal or generative hyphae only. Tissues darkening or becoming blackish when mounted in 3 % KOH sol. Cystidia absent. Basidia 26–34 × 6.2–7 µm, clavate-cylindrical, often with a median retraction septum, 4-spored, sterigmata up to 5.4 µm long, contents subhyaline but sometimes turn yellow in KOH sol. Basidiospores 5.8–6.9 µm across, subglobose to irregular in outline, moderately thick-walled (0.8 µm), tinted yellow, non-amyloid, echinulate, spines acute and up to 1.5 µm long.

Distribution: H.P.: Bagi, Kullu.

Collection examined: SSR 5643, IBP 37110.

Substratum: On coniferous slash.

Remarks: The characteristic feature of this species are dimitic hyphal system, presence of teeth, short basidia and globose to sub-globose basidiospores with short, blunt and knob-like or hemispherical warts.

Tomentella cinerascens (Karst.) Höhn. & Litsch., Sber Akad. Wiss. Wein., Mathnaturw. Klarse Abt. 1, 15: 1570. 1906. Fig. 6.112d–g

Fructification resupinate, tomentose to submembranous, loosely adnate, widely effused, up to 200 µm thick in section; hymenial surface pinkish brown, smooth to finely granulose, discontinuous; margin thinning, white to paler concolorous, adnate. Subiculum subhyaline in section, composed of loosely to somewhat compactly woven hyphae and hyphal cordons.

Hyphal system monomitic, hyphae 1.5–5.5 µm wide, branched, branches sparse and at wide angles, distinctly septate, clamps at most septa but some septa evidently with or without clamps, thin to moderately thick-walled (0.8 µm), subhyaline to pale yellowish brown. Tissues darkening slightly when mounted in 3 % KOH sol. Cystidia absent. Basidia 40–48.9×7–7.4 μm, clavate- cylindrical, rarely with a median retraction septum, 4-spored, sterigmata up to 4.5 μm in length. Basidiospores 5–6 μm across irregular to somewhat lobed, thin-walled, tinted brown, non-amyloid, echinulate.

Distribution: H.P.: Mahasu- Khadrala.

Collection examined: SSR: 5639, IBP 37114.

Substratum: Stump of *Abies pindrow*.

Remarks: This species is characterized by the pinkish brown colour of the hymenial surface, subhyaline context, subhyaline to lighter coloured cordons and small, somewhat lobed basidiospores.

Tomentella clavigera Litsch., In Svrce Sydowia 14: 192. 1960. Fig. 6.111l–p

Fructifications resupinate, mucedinous adherent; hymenial surface greyish brown to dark brown, smooth to granulose, subiculum fibrous; margins apparently fimbricate, concolorous with subiculum.

Hyphal system monomitic; generative hyphae branched, septate, clamped; basal hyphae up to 4.8 μm wide, horizontally woven, dark brown, thin-walled, distantly septate; subhymenial hyphae up to 7 μm wide, loosely and vertically arranged, subhyaline to yellowish, thin-walled. Hyphal cordons absent. Cystidia 33.2–50.2×4.7–14.5 μm, arising from subhymenial hyphae, obclavate, may be septate, septa with or without clamps, subhyaline to yellowish. Basidia 55–67×10.2–12.1 μm, clavate, clamped at base, 4-sterigmate, retraction septa present; sterigmata up to 8 μm long. Basidiospores 6.5–10.5 μm across, irregular to lobed, frequently elongate along one axis, echinulate, apiculate, yellowish brown.

Distribution: H.P.: Solan, Kullu, Manali.

Collection examined: IBP 37611.

Substratum: On decayed angiospermic twigs.

Remarks: The species is characterized by the presence of cystidia and irregular to lobed, frequently elongate along one axis basidiospores. It is a new record for Himalayas.

Tomentella coerulea (Bres.) Höhn. & Litsch., Wiesner Festschrift (Wein.): 77, 1908. Fig. 6.111q–u

Fructifications resupinate, floccose to submembranous, loosely adnate, widely effused; hymenial surface ochre-brown to light brown, discontinuous, uneven to tuberculate, tubercles prominent and often crowded, concolorous; margin indeterminate, loosely adnate. Subiculum subhyaline to light brown in section, composed of somewhat loosely arranged hyphae in basal part but rather compact to more or less agglutinated in the subhymenial part and the trama of tubercles.

Hyphal system monomitic; basal hyphae 2–3.4 μm wide, rather scanty, sparsely branched, branches at the wide angles and often arising from the clamps, septate, clamped, clamps almost at all septa, the walls thin, subhyaline. Hyphal cordons absent. Tissue becomes black when mounted in 3 % KOH sol. Cystidia absent.

Basidia 29.8–40.1 × 5.6–7.1 μm, clavate to clavate-cylindrical, rarely with a median retraction septum, 4-spored. Basidospores 6.2–7.2 × 4.4–5 μm, subglobose to irregular in outline, uniguttulate, thin to thick-walled, tinted brown, non-amyloid, echinulate.

Distribution: H.P.: Kullu- Pulga, Kullu.

Collection examined: SSR 5712, IBP 37612.

Substratum: Bark of *Cedrus deodara.*

Remarks: The species is characterized by having colliculose to tuberculate hymenial surface, narrow subicular hyphae, characteristic reaction of KOH sol and shape and size of basidiospores. The collection is quite typical of the species but differs in having slightly shorter basidia and smaller basidiospores.

Tomentella crinalis (Fr.) Larsen, Taxon 16: 511. 1967. Fig. 6.112a–c

Fructifications resupinate, floccose to subpellicose, loosely adnate, widely effused; hymenial surface light brown to ochre-brown, tuberculate to distinctly toothed, discontinuous; margin thinning, byssoid to fibrillose, loosely adnate, paler concolorous. Teeth gregarious, terrete when fresh but tend to coalesce on drying. Subiculum light brown in section, composed of loosely woven hyphae and hyphal cordons.

Hyphal system dimitic; generative hyphae are of three types: (a) basal hyphae 4.6–5 μm wide, sparsely branched, septate, clamps absent; (b) tramal hyphae 2.4–4.4 μm wide, branched, septate, clamps present, the walls thin and subhyaline; (c) 2.4–5.5 μm wide, branched, septate, clamped, clamps prominent and almost at all septa, thin-walled, subhyaline or faintly tinted brown, present mostly in the upper part of the subiculum; skeletal hyphae 1.4–3 μm wide, sparsely branched, aseptate but often developing few retraction septa, clamps absent, the walls tinted yellow or pale citrine, thick, lumen narrow or almost obliterated; thin to thick-walled (1 μm), tinted yellow or brown. Hyphal cordons abundant, branched, composed exclusively of either the skeletal hyphae or generative hyphae only. Tissues darkening slightly when mounted in 3 % KOH sol. Cystidia absent. Basidia 40.2–50.2 × 9–11 μm, clavate, cylindrical, often with a median retraction septum, 4-spored. Basidiospores 8–9.2 μm across, irregular in outline to somewhat lobed, the walls thin to moderately thick-walled, tinted brown, non-amyloid, echinulate, the spines subacute and usually dichotomously branched.

Distribution: H.P.: Kullu- Pulga, Gojra, Mahasu- Narkanda.

Collection examined: K 4056, 4063, SSR 4364, IBP 37613.

Substratum: Angiospermic wood, decaying angiospermic stumps and stumps under mixed forest.

Remarks: The species is characterized by the dimitic hyphal system and nature of basidiospores which are large, irregular in outline to somewhat lobed and possess dichotomously tipped spines.

Tomentella ellisii (Sacc.) Jülich & Stalpers, Verh. K. ned. Akad. Wet., 2 Sectie 74: 236 (1980) = *Tomentella ochracea* (Sacc.) Larsen Mycologia 66(1): 37(1974). Fig. 6.112h–j

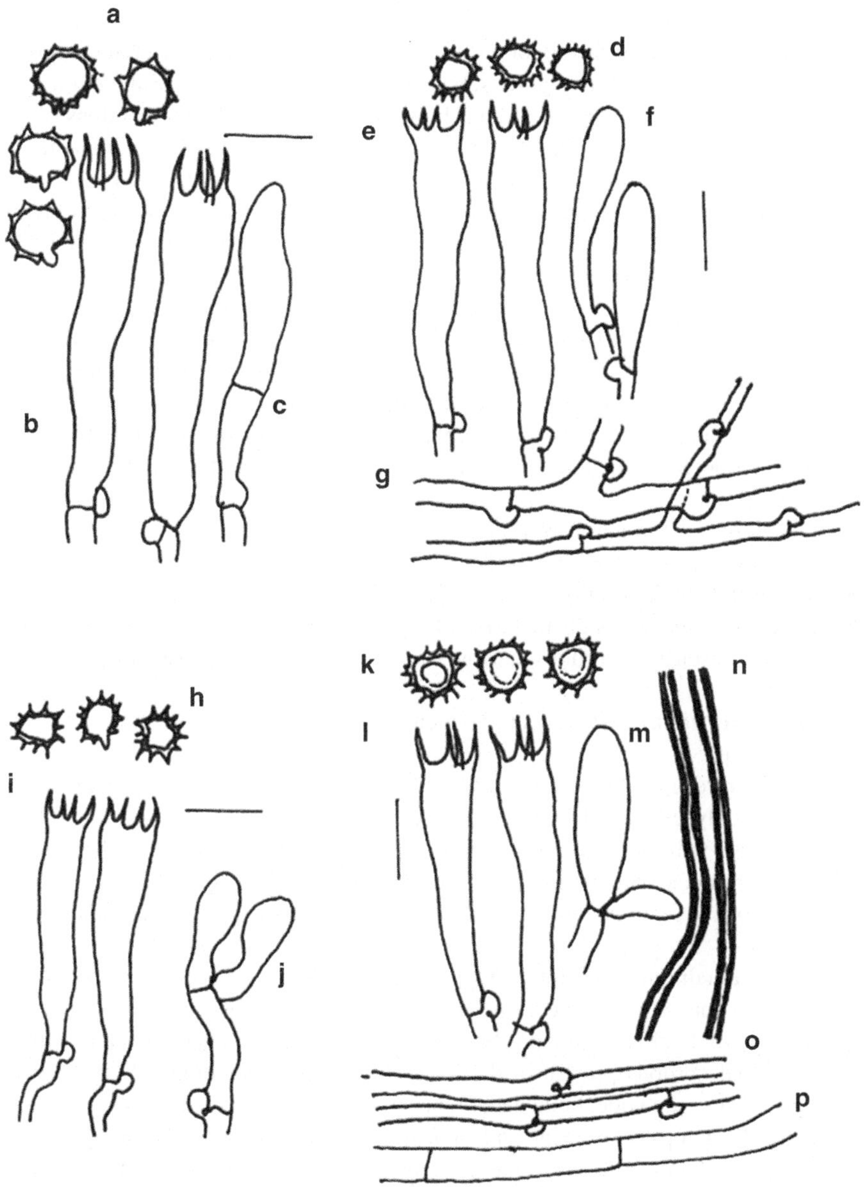

Fig. 6.112 (**a**–**c**) *Tomentella crinalis* (**a**) Basidiospores, (**b**) Basidia (**c**) Basidioles; (**d**–**g**) *Tomentella cinerascens* (**d**) Basidiospores, (**e**) Basidia, (**f**) Basidioles, (**g**) Generative Hyphae; (**h**–**j**) *Tomentella ellisii* (**h**) Basidiospores, (**i**) Basidia, (**j**) Basidioles; (**k**–**p**) *Tomentella ferruginea* (**k**) Basidiospores, (**l**) Basidia, (**m**) Basidioles, (**n**) Skeletal hyphae, (**o**–**p**) Generative hyphae

Fructifications resupinate, arachnoid to floccose, loosely adnate to separable, widely effused; hymenial surface light brown to brown, smooth to finely granulose, discontinuous; margin thinning, loosely adnate, paler concolorous. Subiculum pale brown in section, composed of loosely woven hyphae and hyphal cordons.

Hyphal system monomitic; basal hyphae 2–2.4(5) μm wide, branched, branches at wide angles, septate, clamped, simple septate, the walls thin (0.6 μm), tinted brown; subhymenial hyphae 2–3.8 μm wide, profusely branched, septate, clamps abundant, thin-walled and subhyaline. Hyphal cordons rather uncommon, up to 25 μm broad, light brown, unbranched to sparsely branched, composed of compactly arranged and somewhat agglutinated basal hyphae. Tissues somewhat darkening when mounted in 3 % KOH sol. Cystidia absent. Basidia 32–39.8 × 5.5–7 μm, clavate-cylindrical, rarely with a median retraction septum, 4-spored, sterigmata up to 5 μm in length. Basidiospores 5.3–6.4 × 4.4–5.2 μm, often elongated along one axis, irregular in outline and distinctly lobed, the walls thin to moderately thick, tinted brown, non-amyloid, echinulate, spines acute and up to 1 μm long.

Distribution: H.P.: Mahasu- Narkanda.

Collection examined: SSR 5363, IBP 37615.

Substratum: Coniferous log.

Remarks: The species is characterized by having light brown to brown hymenial surface, presence of cordons, narrow basidia and small lobed basidiospores.

Tomentella ferruginea (Pers.) Pat., Hyménomyc. Eur. (Paris): 154, 1887.

Fig. 6.112k–p

Fructification resupinate, loosely adnate, to separable in small pieces, submembranous, widely effused; hymenial surface yellowish brown to ochre, smooth to colliculose with hemispherical warts or papillae, margin thinning, fibrillose to rhizomorphic, adnate, yellow to ochraceous. Subiculum yellowish brown to deep ochraceous in section, composed of somewhat loosely woven hyphae and hyphal cordons.

Hyphal system dimitic; generative hyphae are of three types: (a) basal hyphae 3–6.3 μm wide, sparsely branched, distantly septate, clamps absent or rarely present, the walls subhyaline to tinted yellow thin to moderately thick, confined mostly to the hyphal cordons present in the lower part of the subiculum; (b) generative hyphae 2.4–3 μm wide, branched, branches common and at acute angles, septate, clamped, clamps prominent and almost at all septa, the walls subhyaline to faintly tinted brown; (c) subhymenial hyphae 2–3.2 μm wide, profusely branched, closely septate, clamped, clamps prominent and almost at all septa, thin-walled, subhyaline; skeletal hyphae 1.4–2.4 μm wide, unbranched to rarely branched, aseptate but occasionally developing few retraction septa, clamps absent, the walls tinted yellow or pale citrine, thin to moderately thick, confined mostly to the hyphal cordons; hyphal cordons abundant, ochre to light brown, branched. Tissues turning black when mounted in 3 % KOH sol, in sectional preparations hymenium and upper part of subhymenium appear green. Cystidia absent. Basidia 34–40 × 5.5–7 μm, clavate to clavate-cylindrical, subhyaline but contents turn green to bluish green in KOH sol, 4-spored. Basidiospores 6.3–7.3 μm across, irregular in outline to distantly lobed, thin to thick-walled, tinted brown, nonamyloid, echinulate.

Distribution: H.P.: Mahasu- Narkanda, Chamba- Khijjiar; U.K.: Mussoorie.

Collection examined: SSR 5613, 5823, IBP 37616.

Substratum: Bark of *Cedrus deodara*, bark under mixed forest and coniferous trees.

Remarks: The species is characterized by smooth to colliculose hymenial surface, characteristic reaction of KOH sol and irregular to lobed basidiospores.

Tomentella galzinii Bourd., in Bourd. & Galzin, Bull. trimmest. Soc. mycol. France 40(2): 143, 1924 Fig. 6.113a–e

Fructification resupinate, adnate, arachnoid to submembranous; hymenial surface greyish brown, smooth, dark brown to black in KOH sol; subiculum light yellow to yellowish grey; margin thinning, arachinoid, concolorous with subiculum, or indeterminate.

Hyphal system monomitic; generative hyphae clamped, septate, branched; basal hyphae up to 6.8 μm wide, loosely arranged, sparsely branched, septate, clamped, subhyaline to greyish yellow, thin to moderately thick-walled; subhymenial hyphae up to 7 μm wide, compactly packed, highly branched, septate, clamped, hyaline, thin-walled. Hyphal cordons absent. Cystidia up to 36.7–58.0 × 4.3–7.5 μm, acuminate, with a clamp at the base, hyaline, thin-walled, projecting up beyond the basidia. Basidia 34.5–54.9 × 6.8–7.5 μm, clavate-cylindrical, 4-spored with a basal clamp, with only contents, often with transverse septa, hyaline. Basidiospores 7.5–10.5 μm across, irregularly globose to lobed, thick-walled, echinulate to rarely aculeate, yellowish grey in water, pale brown in KOH sol.

Distribution: H.P.: Chamba- Khajjiar.

Collection examined: GSD 1736, IBP 37617.

Substratum: Dead wood.

Remarks: The species is characterized by having short, narrow and small, globose to subglobose basidiospores, having greyish brown fructifications and acuminate cystidia.

Tomentella griseoumbrina Litsch., Fungi Exs. Suec.: 24 (1936). Fig. 6.113f–j

Fructification resupinate, adnate, widely effused; hymenial surface whitish brown to light brown, smooth to finely granular; margin indeterminate, adnate, concolorous. Context brown, composed of loosely woven hyphae.

Hyphal system monomitic; basal hyphae 2–3.4 μm wide, branched, septate, clamped, profusely branched at wide angles; subhymenial hyphae scanty, 2.5–3.5 μm, wide, branched, septate, thin to slightly thick-walled. Tissue gets darken in KOH sol. Cystidia absent. Basidia clavate to clavate-cylindrical, 18–24.8 × 4.0–5.1 μm long, 4-spored. Basidiospores 3.9–4.5 μm, globose to subglobose, thin-walled, non-amyloid, finely echinulate.

Distribution: H.P.: Mahasu- Narkanda, Kullu.

Collection examined: SSR 5540, IBP 37618.

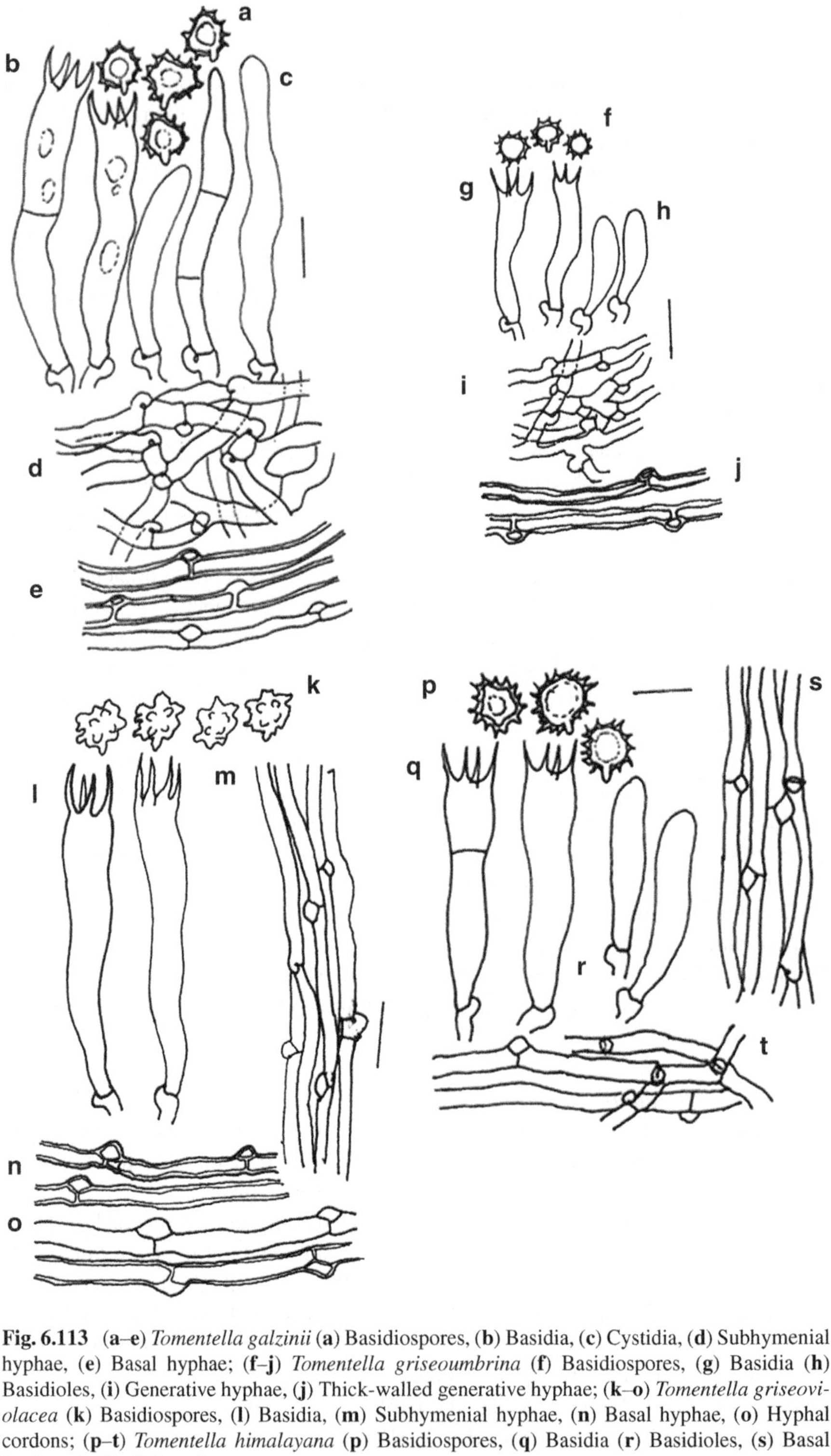

Fig. 6.113 (**a**–**e**) *Tomentella galzinii* (**a**) Basidiospores, (**b**) Basidia, (**c**) Cystidia, (**d**) Subhymenial hyphae, (**e**) Basal hyphae; (**f**–**j**) *Tomentella griseoumbrina* (**f**) Basidiospores, (**g**) Basidia (**h**) Basidioles, (**i**) Generative hyphae, (**j**) Thick-walled generative hyphae; (**k**–**o**) *Tomentella griseoviolacea* (**k**) Basidiospores, (**l**) Basidia, (**m**) Subhymenial hyphae, (**n**) Basal hyphae, (**o**) Hyphal cordons; (**p**–**t**) *Tomentella himalayana* (**p**) Basidiospores, (**q**) Basidia (**r**) Basidioles, (**s**) Basal hyphae, (**t**) Hyphal cordons

Substratum: Wood of *Cedrus deodara*.

Remarks: The species is characterized by short, narrow basidia and small, globose to subglobose basidiospores.

Tomentella griseoviolacea Litsch., Annls mycol. 39: 375, 1941. Fig. 6.113k–o

Fructification resupinate, effused, hypochnoid to arachnoid, loosely adnate; hymenial surface brownish grey, smooth; margin thinning; paler concolorous to indeterminate.

Hyphal system monomitic; generative clamped, septate; basal hyphae parallel to the substrate, 5.5 µm thick-walled, septate, distinctly septate; subhymenial hyphae up to 3.5 µm wide, septate, thick-walled, clamped, yellowish brown. Hyphal cordons present, branched, septate, clamped, thick-walled, yellowish brown. Cystidia absent. Basidia 41.2–54.1 × 6.8–8.2 µm 4-sterigmate with basal clamp. Basidiospores 6.1–8.8 µm across, globose to subglobose, echinulate, apiculate.

Distribution: H.P.: Mandi, Sundernagar.

Collection examined: IBP 37619.

Substrate: Twigs and fallen branches.

Remarks: The species is characterized by arachnoid fructification, clavate basidia and flobose to subglobose, echinulate basidiospores.

Tomentella himalayana S. S. Rattan, Biblth. Mycol. 60: 54, 1977. Fig. 6.113p–t

Fructification resupinate, subpelliculose to pelliculose, loosely adnate, widely effused and hymenial surface light brown to snuff-brown, smooth to finely farinose, mostly continuous; margin thinning, loosely adnate, paler concolorous. Subiculum light brown in section, composed of loosely woven hyphae and hyphal cordons supporting a compact, more or less subceraceous hymenium.

Hyphal system monomitic; basal hyphae 3.4–6.6 µm wide, branched, branches less and often arising below the septum, distantly septate, clamped, uninflated, the walls thin to thick-walled, slightly tinted brown; subhymenial hyphae 2.4–4.2 µm wide, profusely branched, closely septate, clamped, somewhat inflated between the septa, the walls thin and subhyaline. Hyphal cordons abundant, up to 90 µm broad, light brown, branched, composed of basal hyphae only. Tissues turn black when mounted in 3 % KOH sol. Cystidia absent. Basidia 39.8–50(70) × 8.4–11 µm, clavate to clavate-cylindrical, mostly with the median retraction septum and often collapsing after spore discharge especially above the septum, 4-spored, sterigmata up to 8 µm in length. Basidiospores 7.6–10.2 µm across, irregular in outline and distantly lobed, uniguttulate, thin-walled, subhyaline to tinted brown, nonamyloid, echinulate.

Distribution: H.P.: Shimla-Mahasu, Khadrala, Narkanda.

Collection examined: SSR 5346, 5620.

Substratum: slash under conifers and slash of *Abies pindrow*.

Remarks: The species is characterized by subpelliculose to pelliculose texture, light brown to snuff-brown hymenial surface, presence of hyphal cordons and shape and size of basidia and basidiospores.

Tomentella indica S. S. Rattan, Biblth. Mycol. 60: 62, 1977. Fig. 6.114a–e

Fructification resupinate, floccose to submembranous, loosely adnate, widely effused; hymenial surface light brown to snuff-brown, smooth to farinaceous, discontinuous; margin thinning, loosely adnate, paler concolorous to concolorous. Subiculum light brown in section, composed of loosely woven hyphae.

Hyphal system monomitic; basal hyphae 4.4–8(10) μm wide, branched, distantly septate, clamped, thin-walled to thick-walled (1.5 μm), tinted yellow or ochre; subhymenial hyphae 4–7.2 μm wide, profusely branched, closely septate, clamped, thin to thick-walled. The basal hyphae are long-celled with thicker and darker walls but these become progressively short-celled, thin-walled and lighter coloured towards the hymenium. Hyphal cordons absent. Tissues darkening or almost turning black when mounted in 3 % KOH sol, basidial contents turn blue. Basidia 44.8–60 × 8–11.4 μm, clavate-cylindrical to utriform, often with a median retraction septum, 4-spored. Basidia quite variable in shape, sometimes hyphoid with septa bearing clamps and looking more cystidioles. Basidiospores 7.2–10.1 μm across, subglobose to irregular in outline, thin to thick-walled (0.8 μm), tinted brown, non-amyloid, echinulate.

Distribution: H.P.: Kullu- Naggar, Jaggat sukh; Mahasu- Bagi, Dalhousie- Banikhet, Kalatope; U.K: Hemkunt.

Collection examined: RW5071; R 5350, 5727, IBP 37620.

Substratum: Bark of *Abies pindrow* and *Picea smithiana*, leaves of *Quercus incana*, logs under a mixed forest, stumps of *Cedrus deodara*.

Remarks: The species is characterized by floccose to membranous fructification; light brown to snuff brown hymenial surface; monomitic hyphal system; subglobose to irregular shaped basidiospores.

Tomentella kalatopii Dhingra & Malka 1994. in (eds.) T. A. Sarma, S. S. Saini, M. L. Trivedi & M. Sharma North-West Himalayan Thelephoraceae (Basidiomycetes) Genus *Tomentella* from Dalhousie Hills *Current Researches in Plant Sciences* 43–56. Fig. 6.114f–h

Fructification resupinate, adnate, effused, sub membranous to membranous; hymenial surface brownish orange to greyish brown, smooth, greyish brown in KOH sol; subiculum light yellow to almost white; margin thinning, fibrillose, concolorous with subiculum or indeterminate.

Hyphal system monomitic; basal hyphae up to 7.5 μm wide, loosely arranged, almost parallel to the substrate, branched, distantly septate, clamped, subhyaline, thick-walled; subhymenial hyphae up to 6.3 μm wide, compactly packed, vertically arranged, branched, closely septate, clamped, subhyaline to hyaline, thin-walled. Hyphal cordons absent. Cystidia absent. Basidia 34.9–58.2 × 6.4–12.2 μm, clavate to subclavate, 4-spored, with or without oily contents, usually with transverse septa, hyaline. Basidiospores 6.2–8.5 μm across, subglobose to rarely globose, regular in outline, usually elongated along one axis, thin to somewhat thick-walled, aculeolate.

Distribution: H.P.: Chamba- Kalatope, Shimla.

Collection examined: IBP 37621, 37623.

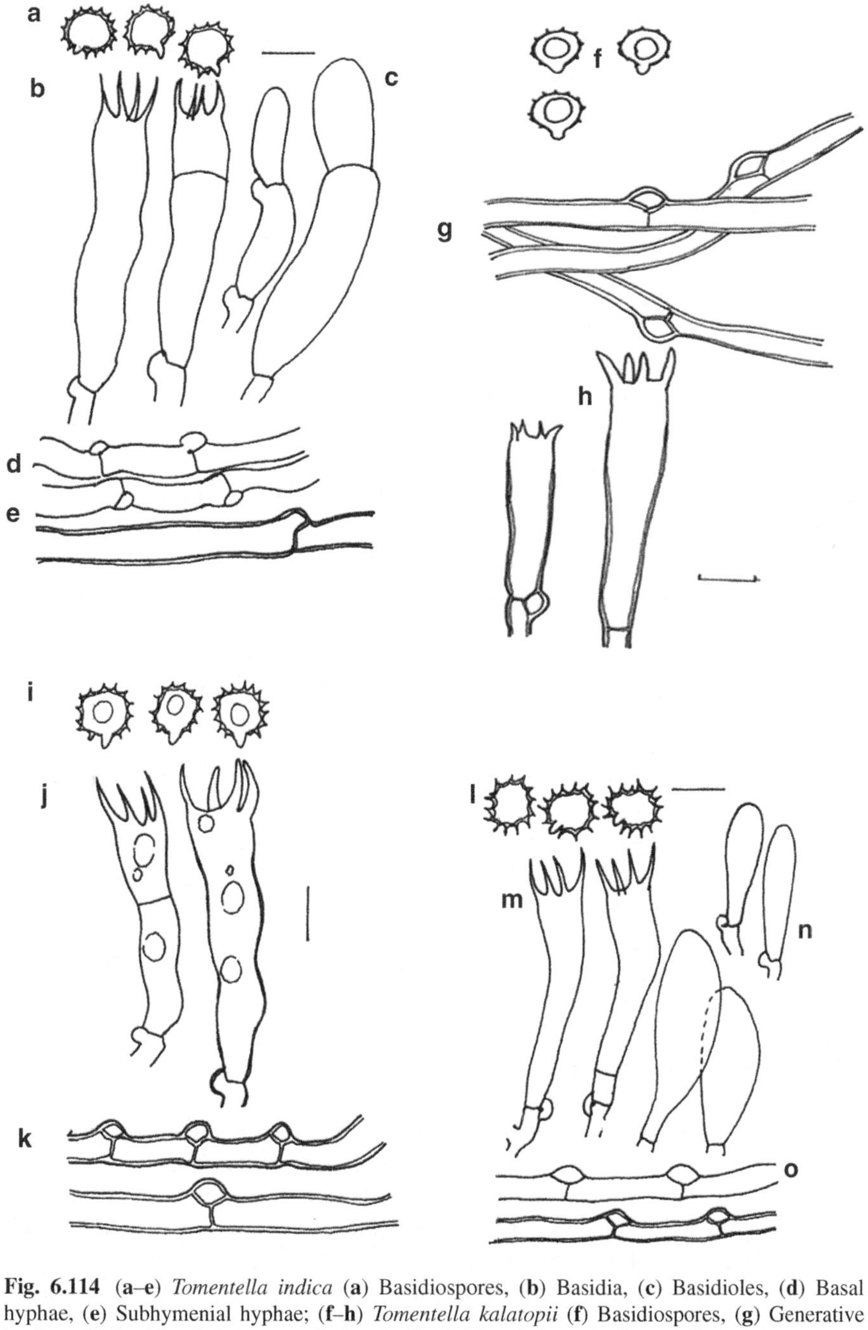

Fig. 6.114 (**a–e**) *Tomentella indica* (**a**) Basidiospores, (**b**) Basidia, (**c**) Basidioles, (**d**) Basal hyphae, (**e**) Subhymenial hyphae; (**f–h**) *Tomentella kalatopii* (**f**) Basidiospores, (**g**) Generative hyphae, (**h**) Basidia; (**i–k**) *Tomentella lapida* (**i**) Basidia, (**j**) Basidia, (**k**) Generative hyphae; (**l–o**) *Tomentella lateritia* (**l**) Basidiospores, (**m**) Basidia, (**n**) Basidioles, (**o**) Generative hyphae

Substratum: Decaying gymnospermic stump/twig.

Remarks: The species is characterized in having light yellow to almost white subiculum and margins, subglobose to rarely globose, regular, usually elongated along one axis, aculeolate, light green basidiospores and in lacking the hyphal cordons.

Tomentella lapida (Pers.) Stalpers, [as '*lapidum*'], Stud. Mycol. 24: 65 (1984) = *Tomentella ramosissima* (Berk. & Curt.) Wakef., Mycologia 52(6): 927, 1960 = *Zygodesmus ramosissimus* Berk. & Curt., Grevillea 3: 145, 1875 = *Tomentella fuliginea* (Burt) Bourd. & Galz. *Bull.trimest. Soc. mycol. Fr. 40*: 153.1924.
Fig. 6.114i–k

Fructification resupinate, loosely adnate, effused, mucedinoid to submembranous; hymenial surface greyish brown to dark brown, granulose, brownish grey to black in KOH sol; subiculum fibrillose dark brown; margins thinning, concolorous with subiculum or indeterminate.

Hyphal system monomitic; basal hyphae up to 7.2 μm wide, loosely arranged, parallel to substrate, branched, distantly septate, clamped, golden yellow, bluish green in KOH sol, thin to thick-walled; subhymenial hyphae up to 6.0 μm wide, loosely arranged, vertically arranged, profusely branched, closely septate, clamped, thin to moderately thick-walled. Hyphal cordons absent. Cystidia absent. Basidia 32.3–53 × 7.8–11.2 μm, clavate to subclavate, 4-spored, with a basal clamp; crystalline material present on walls, bluish green in KOH sol. Basidiospores 6.4–9.6 μm across, globose to subglobose, regular in outline, thin to moderately thick-walled, aculeolate.

Distribution: H.P.: Kullu- Naggar, Jaggat sukh; Mahasu- Bagi; Dalhousie- Banikhet. Chamba- Kalatope.

Collection examined: RW 5071; SSR 5350, 5727, 5832, IBP 37622.

Substratum: On bark and logs under mixed forest, on bark of *Abies pindrow* and *Picea smithiana*, Onleaves of *Quercus incana*.

Remarks: This species is characterized by the presence of crystalline material on the walls of hyphae, basidia and basidiospores, which is responsible for the change in color of these structures to bluish green in KOH sol.

Tomentella lateritia Pat., Cat. Rais. Pl. Cellul. Tunisie (Paris): 63 (1897).
Fig. 6.114l–p

Fructification resupinate, floccose to submembranous, adnate, often arising as small circular colonies which may coalesce and grow later and become effused, up to 125 μm thick in section; hymenial surface reddish brown or pinkish brown, granulose to minutely colliculose, rarely smooth, discontinuous; margins indeterminate, adnate, concolorous. Subiculum subhyaline in section, composed of loosely woven hyphae which are mostly collapsed and difficult to discern in mature fructifications.

Hyphal system monomitic; hyphae 3–6.8 μm wide, branched, septate, clamped, clamps prominent and almost at all septa, thin to moderately thick-walled, subhyaline to tinted brown. Hyphal cordons absent. Tissues darkening slightly when mounted in 3 % KOH sol, subhymenial hyphae, basidia and basidiospores contain a reddish granular material visible in water mounts but disappear when sections are mounted in KOH sol. Cystidia absent. Basidia 34.8–45 × 8–8.4 µm, clavate to clavate-cylindrical, rarely developing a median retraction septum, 4-spored, sterigmata slender and up to 8 µm long. Basidiospores 7.4–9.4 µm across, irregular in outline to distinctly lobed, thick-walled, tinted brown, non-amyloid.

Distribution: H.P.: Mahasu- Narkanda.

Collection examined: SSR 5863.

Substratum: On bark under coniferous forest.

Remarks: The species is characterized by reddish brown to pinkish brown hymenial surface and presence of vesicles in the subhymenial zone; presence of reddish granular contents among the subhymenial hyphae. These granular contents are best seen in water mounts but disappear when sections are mounted in KOH sol.

Tomentella muricata (Ell. & Ev.) Wakef., Mycologia 52(6): 924, 1962 [1960] = *Zygodesmus muricatus* Ell. & Ev., Bull. Torrey bot. Club 11: 17, 1884. Fig. 6.115a–f

Fructification resupinate, loosely adnate, effused, arachnoid to byssoid; hymenial surface brown, granulose, black in KOH sol; subiculum arachnoid, concolorous with hymenium to somewhat dark colored; margins thinning, arachnoid, concolorous with subiculum.

Hyphal system monomitic; basal hyphae up to 6.7 µm wide, loosely arranged, branched at wide angles, septate, clamped, minutely warted to spinulose, yellowish brown, thin to moderately thick-walled; subhymenial hyphae up to 5.1 µm wide, compactly arranged, branched at wide angles, septate, clamped, without encrustations, hyaline, thin-walled. Hyphal cordons up to 70 µm wide, branched; individual hyphae up to 5.5 µm wide, thin to slightly thick-walled, subhyaline. Cystidia 72.3–104 × 10.0–13.1 µm, often slender and hyphoid with somewhat expanded apex, with a basal clamp, hyaline, arising from subhymenial zone. Basidia 27.8–38.4 × 7.8–10.3 µm, clavate, 4-spored, with a basal clamp, occasionally with a transverse septum, subhyaline; sterigmata up to 7 µm long. Basidiospores 5.4–8.2 µm across, irregularly globose to subglobose, usually elongated along the axis, thick-walled, echinulate to rarely aculeate.

Distribution: H.P.: Chamba- Khajjiar.

Collection examined: IBP 37624.

Substratum: Decaying angiospermic log.

Remarks: The species is characterized in having arachnoid to byssoid fructification, conspicuous cordons and cystidia arising from the subhymenial hyphae.

Tomentella olivascens (Berk. & M. A. Curtis) Bourdot & Galzin, Hyménomyc. de France (Sceaux): 477 (1928) = *Zygodesmus olivascens* Berk. & Curt., Grevillea 3(28): 145 (1875). Fig. 6.115g–j

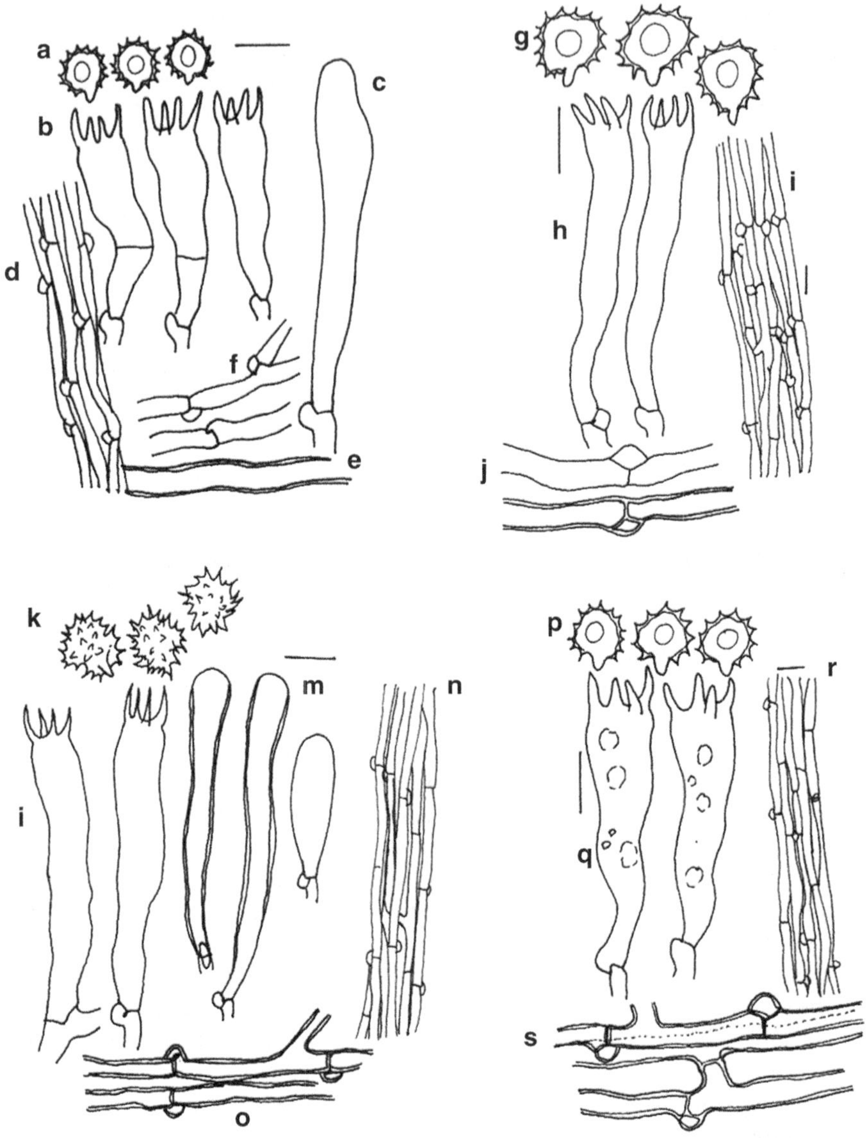

Fig. 6.115 (**a**–**f**) *Tomentella muricata* (**a**) Basidiospores, (**b**) Basidia, (**c**) Cystidia, (**d**) Hyphal cordons, (**e**) Skeletal hyphae, (**f**) Generative hyphae; (**g**–**j**) *Tomentella olivascens* (**g**) Basidiospores, (**h**) Basidia, (**i**) Hyphal cordons, (**j**) Generative hyphae; (**k**–**o**) *Tomentella pilosa* (**k**) Basidiospores, (**l**) Basidia, (**m**) Cystidia, (**n**) Hyphal cordons, (**o**) Generative hyphae; (**p**–**s**) *Tomentella puberula* (**p**) Basidiospores, (**q**) Basidia, (**r**) Hyphal cordons, (**s**) Generative hyphae

Fructification resupinate, loosely adnate, mucedinoid to submembranous; hymenial surface olive brown, smooth, dark green in KOH sol; subiculum arachnoid to byssoid, pale yellow to greyish yellow to brownish orange to light brown; margins thinning, fibrillose, concolorous with subiculum or indeterminate.

Hyphal system monomitic; basal hyphae up to 5.6 µm wide, loosely interwoven, branched, septate, clamped, subhyaline to pale yellow, thin to somewhat thick-walled; subhymenial hyphae up to 4.3 µm wide, compactly packed, vertically arranged, branched, septate, clamped, hyaline, sometimes with a pale yellow tinge, thin-walled. Hyphal cordons abundant, up to 21 µm wide, branched; individual hyphae up to 3.5 µm wide, hyaline to pale yellow, thin-walled. Cystidia absent. Basidia 34.8–54.0×5.8–9.6 µm, clavate, 4-spored, with a basal clamp, with or without oily contents, subhyaline. Basidiospores 5.5–9.1 µm across, irregular in outline to lobed, thin to somewhat thick-walled, echinulate, greyish yellow to greyish green in water, yellowish grey in KOH sol, uniguttulate.

Distribution: H.P.: Chamba- Dalhousie.

Collection examined: IBP 37625, 37627.

Substratum: Decaying gymnospermic stump.

Remarks: The species is marked by olive brown hymenial surface, presence of hyphal cordons and irregular to lobed, echinulate basidiospores.

Tomentella pilosa (Burt) Bourd. & Galz. Bull. trimest. Soc. Mycol. Fr. **40**: (2): 160 1924. Fig. 6.115k–o

Fructification resupinate, floccose to subpelliculose, loosely adnate, arachnoid, widely effused; hymenial surface light brown to greyish brown, smooth to finely granulose, discontinuously; margin thinning, adnate, paler concolorous. Subiculum arachnoid to byssoid, pale yellow to grayish yellow to brownish orange to light brown; margins thinning, fibrillose, concolorous with subiculum or indeterminate.

Hyphal system monomitic; hyphae 2.4–4(5.54) µm wide, branched, branches at wide angles, septate, clamps almost at all the septa, thin-walled, subhyaline to brown. Hyphal cordons abundant, 21–51 µm broad, light brown, branched. Tissues darkening in 3 % KOH sol. Cystidia 34.8–90(15)×2.8–4.1 µm, cylindrical with obtuse or capitates apex, often arising from the basal hyphae with a clamp at the base and may project out of the hymenium, thin to thick-walled, subhyaline, smooth or finely crystals incrusted. Basidia 40.1–64.8×7–10(12) µm, clavate to clavate-cylindrical, occasionally with median retraction septum, 4-spored. Basidiospores 8.2–10.4 µm across, irregular in outline to indistinctly lobed, thin to thick-walled (0.8 µm), tinted brown, nonamyloid.

Distribution: H.P.: Mahasu- Narkanda, Khadrala; Dalhousie- Kalatope.

Collection examined: SSR 5366, 5641, 5791, IBP 37268.

Substratum: On bark under coniferous forest and bark of *Abies pindrow*.

Remarks: The species is characterized by presence of hyphal cordons, long capitates, cystidia and large basidiospores.

Tomentella puberula Bourd. & Galz., Bull. Soc. mycol. France 40(2): 150, 1924. Fig. 6.115p–s

Fructification resupinate, loosely adnate; hymenial surface light brown, granulose to colliculose, greyish brown in KOH sol; subiculum arachnoid, light brown to dark brown; margins indeterminate.

Hyphal system monomitic; basal hyphae up to 4.6 µm wide, loosely arranged, almost parallel to the substrate, branched, septate, clamped, yellowish grey, thin to slightly thick-walled; subhymenial hyphae up to 3.2 µm wide, compact, vertically arranged, branched, septate, clamped, hyaline to subhyaline, thin-walled. Hyphal cordons few, unbranched, up to 20 µm wide; individual hyphae up to 4.8 µm wide, dull yellow, thin-walled. Cystidia absent. Basidia 25.8–47.9 × 7.8–11.0 µm, clavate, 4-spored, with a basal clamp, with or without oily contents, subhyaline; sterigmata up to 5.0 µm long. Basidiospores 5.6–8.4 µm across, irregular to irregularly globose, sometimes lobed slightly thick-walled, aculeolate to echinulate.

Distribution: H.P.: Chamba- Dalhousie, Kullu, Manali.

Collection examined: IBP 37629, 37630.

Substratum: Decaying gymnospermic stump.

Remarks: The species is characterized in having subicular hyphae less than 5 µm wide, occasional presence of hyphal cordons and lack of color of hymenial surface reaction when exposed to KOH sol.

Tomentella punicea (Alb. & Schwein.) J. Schröt., in Cohn, Krypt.-Fl. Schlesien (Breslau) 3.1(25–32): 420 (1888). Plate 6.41a, Fig. 6.116a–e

Fructifications resupinate, arachnoid to tomentose, loosely adnate, somewhat brittle on drying, widely effused; hymenial surface deep ochre to ochre-brown, smooth to finely granulose, discontinuous; margins indeterminate, loosely adnate, paler concolorous to concolorous. Subiculum light yellow in section, composed of loosely woven hyphae and hyphal cordons.

Hyphal system monomitic; basal hyphae up to 2–4.9 µm wide, sparsely branched, branches at wide angles, distantly septate, clamped, thin to thick-walled (0.5 µm), subhyaline to tinted brown; subhymenial hyphae 2–4.2 µm wide, profusely branched, closely septate, clamped, the walls thin, subhyaline. Hyphal cordons abundant, up to 40 µm broad, branched, composed of compactly arranged more or less agglutinated basal hyphae. Tissues somewhat darkening when mounted in 3 % KOH sol. Cystidia absent. Basidia 30.1–45.1 × 7.1 µm, clavate-cylindrical, often with a median retraction septum, 4-spored. Basidiospores 7.2 µm across, thin to moderately thick-walled (0.8 µm), tinted brown, nonamyloid, echinulate.

Distribution: H.P.: Dalhousie- Jandrighat.

Collection examined: SSR 5165, IBP 37631.

Substratum: stumps under mixed forest.

Remarks: The species is characterized by arachnoid to tomentose and somewhat brittle fructifications, presence of hyphal cordons, narrow basidia and small lobed basidiospores.

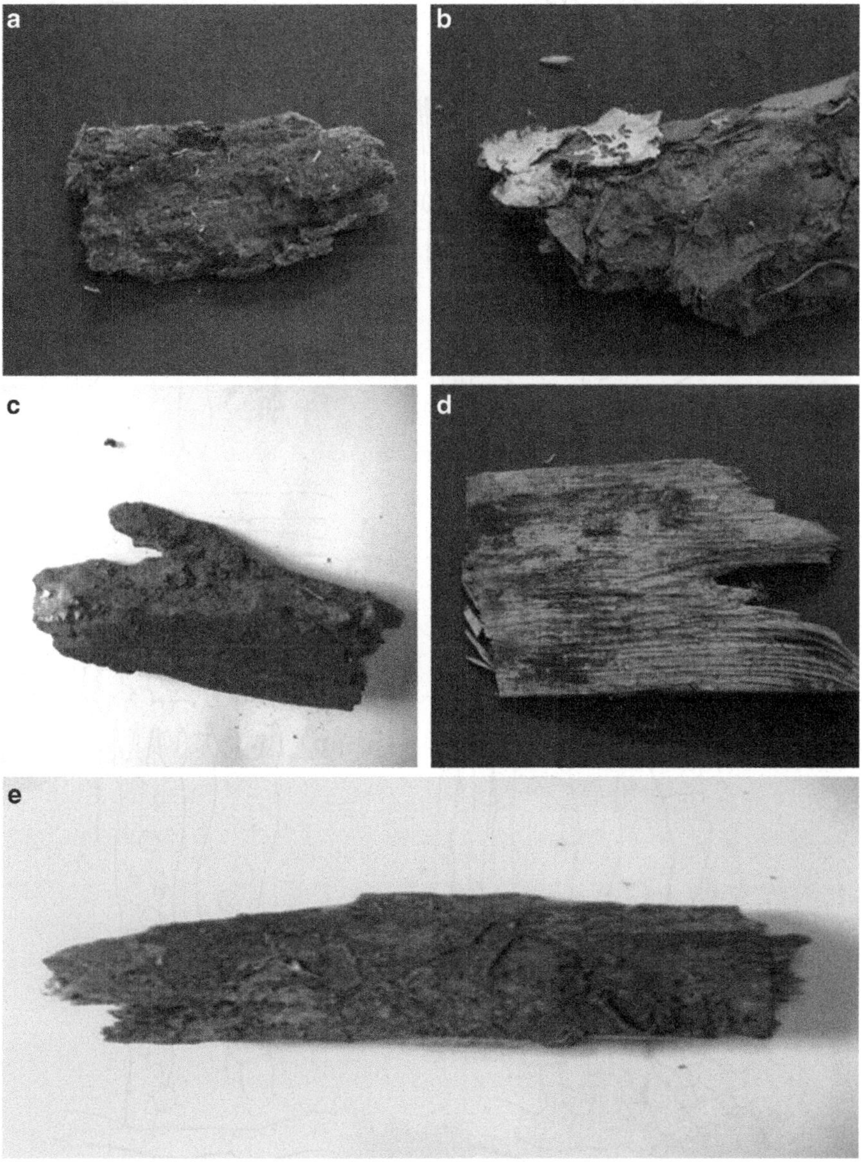

Plate 6.41 (**a**) *Tomentella punicea.* (**b**) *Tomentella radiosa.* (**c**) *Tomentella scobinella.* (**d**) *Tomentella stuposa.* (**e**) *Tomentella terrestris*

Tomentella pyrolae (Ellis & Hlst.) M. J. Larsen, Tech. Publ. N. Y. St. Univ. Coll. For. 93: 105, 1968 = *Zygodesmus pyrolae* Ellis & Halst., J. Mycol. 6(1): 34, 1890.
Fig. 6.116f–i

Fructification resupinate, adnate, loosely adnate; hymenial surface grey to greyish brown when fresh, dark brown on drying; margin thinning white, paler concolorous to indeterminate.

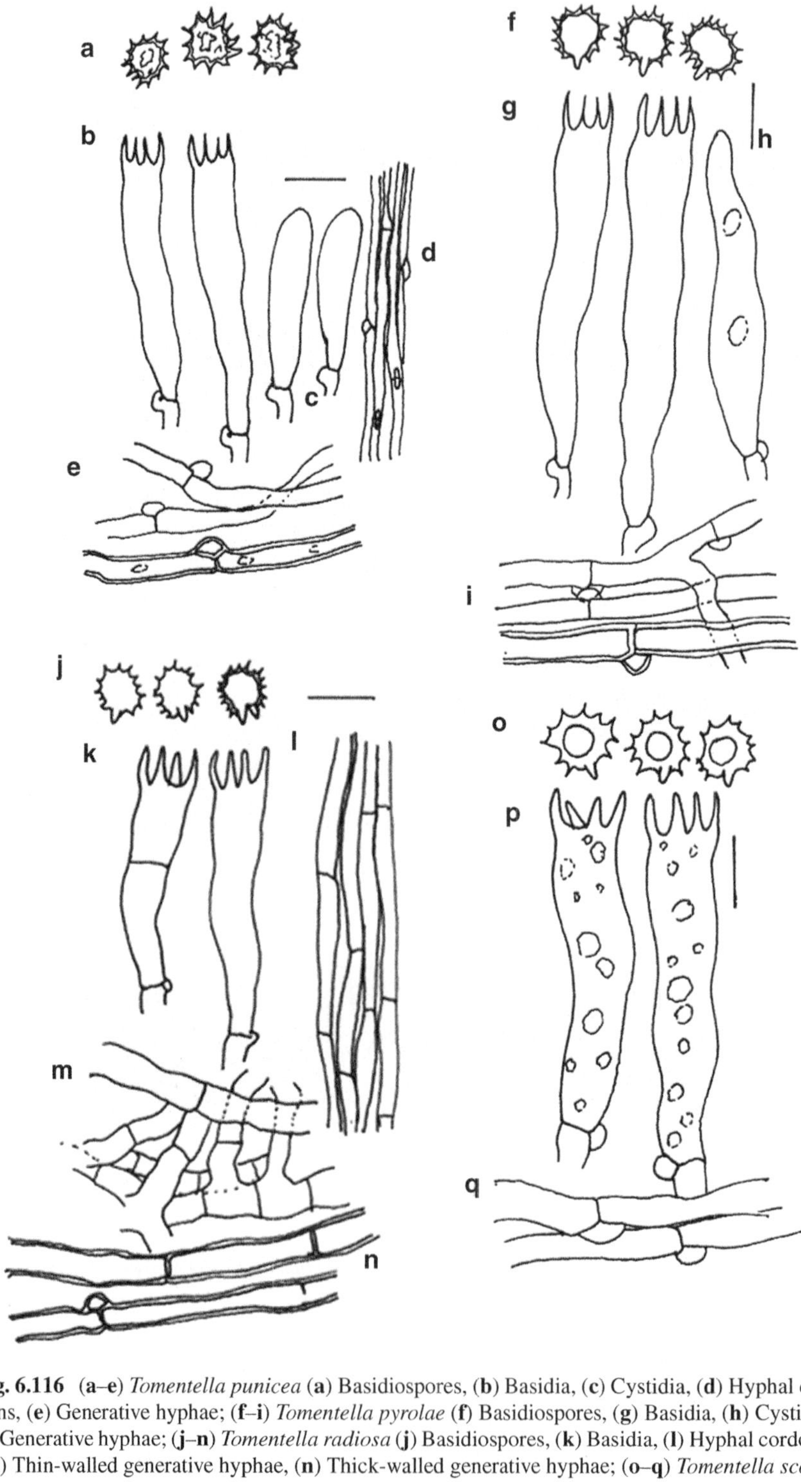

Fig. 6.116 (**a**–**e**) *Tomentella punicea* (**a**) Basidiospores, (**b**) Basidia, (**c**) Cystidia, (**d**) Hyphal cordons, (**e**) Generative hyphae; (**f**–**i**) *Tomentella pyrolae* (**f**) Basidiospores, (**g**) Basidia, (**h**) Cystidia, (**i**) Generative hyphae; (**j**–**n**) *Tomentella radiosa* (**j**) Basidiospores, (**k**) Basidia, (**l**) Hyphal cordons, (**m**) Thin-walled generative hyphae, (**n**) Thick-walled generative hyphae; (**o**–**q**) *Tomentella scobinella* (**o**) Basidiospores, (**p**) Basidia, (**q**) Generative hyphae

Hyphal system monomitic; generative hyphae septate, clamps present; basal hyphae parallel to the substrate, septate, thick-walled, less branched; subhymenial hyphae compactly arranged, thin-walled, 4.9 µm wide, profusely branched. Cystidia 45–60 × 7–11 µm, thin-walled, fusiform with basal clamp, smooth. Basidia 48.2–74.0 × 6.8–11.2 µm, 4-sterigmate, clavate to subclavate with basal clamp. Basidiospores 7–10 µm across, globose to subglobose, apiculate, echinulate, pale brown.

Distribution: H.P.: Chamba, Manali.

Collection examined: IBP 37631.

Substrate: On decaying gymnospermous wood.

Remarks: The species is marked by having echinulate and globose to subglobose basidiospores.

Tomentella radiosa (P. Karst.) Rick, Brotéria, sér. 3 2: 79 (1934) = *Tomentella cladii* Wakef., Trans. Brit. mycol. Soc. 53(2): 179, 1969. Plate 6.41b, Fig. 6.116j–n

Fructifications resupinate, loosely adnate, effused, mucedinoid to submembranous; hymenial surface dark brown, smooth, black in KOH sol. Subiculum almost concolorous with the hymenial surface; margins indeterminate.

Hyphal system monomitic; basal hyphae up to 4.9 µm wide, loosely arranged, parallel to the substrate, branched, distantly septate, clamped, greyish brown, thin to thick-walled; subhymenial hyphae up to 5 µm wide, compactly packed, vertically arranged, highly branched, closely septate, clamped, subhyaline to yellowish grey, thin-walled. Hyphal cordons present, dull brown, septate, clamps absent. Cystidia absent. Basidia 26.8–54.8 × 9.1–11.2 µm, clavate-cylindrical, 4-spored, with a basal clamp, with or without oily contents, subhyaline; sterigmata up to 8.4 µm in length. Basidiospores 7–8.4 µm across, normally subglobose, sometimes elongated along one axis, thin to moderately thick-walled.

Distribution: H.P.: Chamba- Dalhousie, Shimla-Tara Devi.

Collection examined: IBP 37632, 37633.

Substratum: Decaying gymnospermic stump.

Remarks: This species is characterized by dark brown hymenial surface clamped basal hyphae, subglobose, sometimes elongated basidiospores.

Tomentella scobinella Cunn., Trans. Roy. Soc. New. Zeal. 84(3): 485, 1957. Plate 6.41c, Fig. 6.116o–q

Fructification resupinate, arachnoid, separable, effused, up to 270 µm thick in section; hymenial surface greyish-brown, discontinuous, farinose under lens; margin whitish to paler concolorous, arachnoid, thinning.

Hyphal system monomitic; generative hyphae branched, septate, clamped; subicular hyphae somewhat thick-walled, sparsely branched, tinted brown, up to 6 µm wide; subhyaline hyphae subhyaline to tinted pale brown, up to 5 µm in diameter, thin to slightly thick-walled. Basidia 43–57 × 9–11 µm, subclavate, 4-sterigmate, with oily contents and a basal clamp; sterigmata up to 6 µm long. Basidiospores 6–8 µm across, irregular in outline to somewhat lobed, pale brown to brown, thick-walled, echinulate, acyanophilous; spines small, generally 1 µm long.

Distribution: A.P.: West Kameng, Bomdilla, Shergaon; U.K: Nainital.
Collection examined: GSD 19783, IBP 37635.
Substratum: On angiospermic wood.
Remarks: The species is characterized by having loosely attached, arachnoid fructification, greyish-brown colour of the hymenial surface, thick-walled, tinted brown basal hyphae, subclavate basidia and thick-walled, echinulate, acyanophilous, irregular in outline to somewhat lobed basidiospores. The species is new record for Uttarakhand.

Tomentella stuposa (Link) Stalpers, Stud. Mycol. 24: 86 (1984) = *Tomentella ruttneri* Ltsch. *Bull. Soc. mycol. Fr.* 49: 67. 1933 = *Tomentella bresadolae* (Brinkm.)1924 = *Hypochnus bresadola* Brinkm. In Bresadola, Ann. Mycol. 1(2): 10, 1903. Plate 6.41d, Fig. 6.117a–c
Fructification resupinate, floccose to subpellicular, widely effused; hymenial surface brown to snuff-brown; margin indeterminate to abrupt, loosely adnate, paler concolorous. Subiculum light brown, composed of loosely interwoven hyphae.
Hyphal system monomitic, hyphae 3.8–7.3 µm wide, branched at wide angles, clamped, septate, thin to thick-walled. Tissues get darken in KOH sol. Basal hyphae are slightly broader, darker in colour. Hyphal cordons absent. Cystidia absent. Basidia 51–65 × 9–12.5 µm, clavate-cylindrical, 4-spored. Basidiospores 9.3–11.4 µm across, globose to sub globose, regular, echinulate, non-amyloid.
Distribution: H.P.: Solan, Mahasu- Narkanda.
Collection examined: IBP 37636.
Substratum: On a decayed angiospermic twigs, On Bark of *Abies pindrow*, Onbark and slash of conifers.
Remarks: The species is characterized by globose to subglobose basidiospores. The spores having reaction with KOH sol on hyphal wall. The above collected sample resembles with the description given by Rattan (1977).

Tomentella subclavigera Litsch., Bull. trimmest. Soc. Mycol. France **49**: 57, 1933. Fig. 6.117d–h
Fructification resupinate, adnate, tomentose to arachinoid, adherent, submembranous to membranous; hymenial surface light yellow to deep yellow when fresh, brown on drying, minutely granulose; margins thinning.
Hyphal system monomitic; generative hyphae up to 7.5 µm wide, septate, clamped; basal hyphae, parallel to the substratum, branched, septate, dark brown to golden brown; subhymenial hyphae vertically arranged, profusely branched, closely septate, clamped, thin-walled, subhyaline to pale yellow. Hyphal cordons absent. Cystidia 57–59.8 × 3.3–7.1 µm, hyphoid, expanded at the apex, septate, clamped, subhyaline, with basal clamp. Basidia clavate, 4-sterigmate, sterigmata up to 7.5 µm long. Basidiospores 10.4–12.8 µm across, globose to subglobose, regular, thick-walled, aculeate, apiculate, pale brown to brown.
Distribution: H.P.: Solan, Sirmour, Manali.
Collection examined: IBP 37637, 37638.
Substratum: On decayed angiospermic twigs.

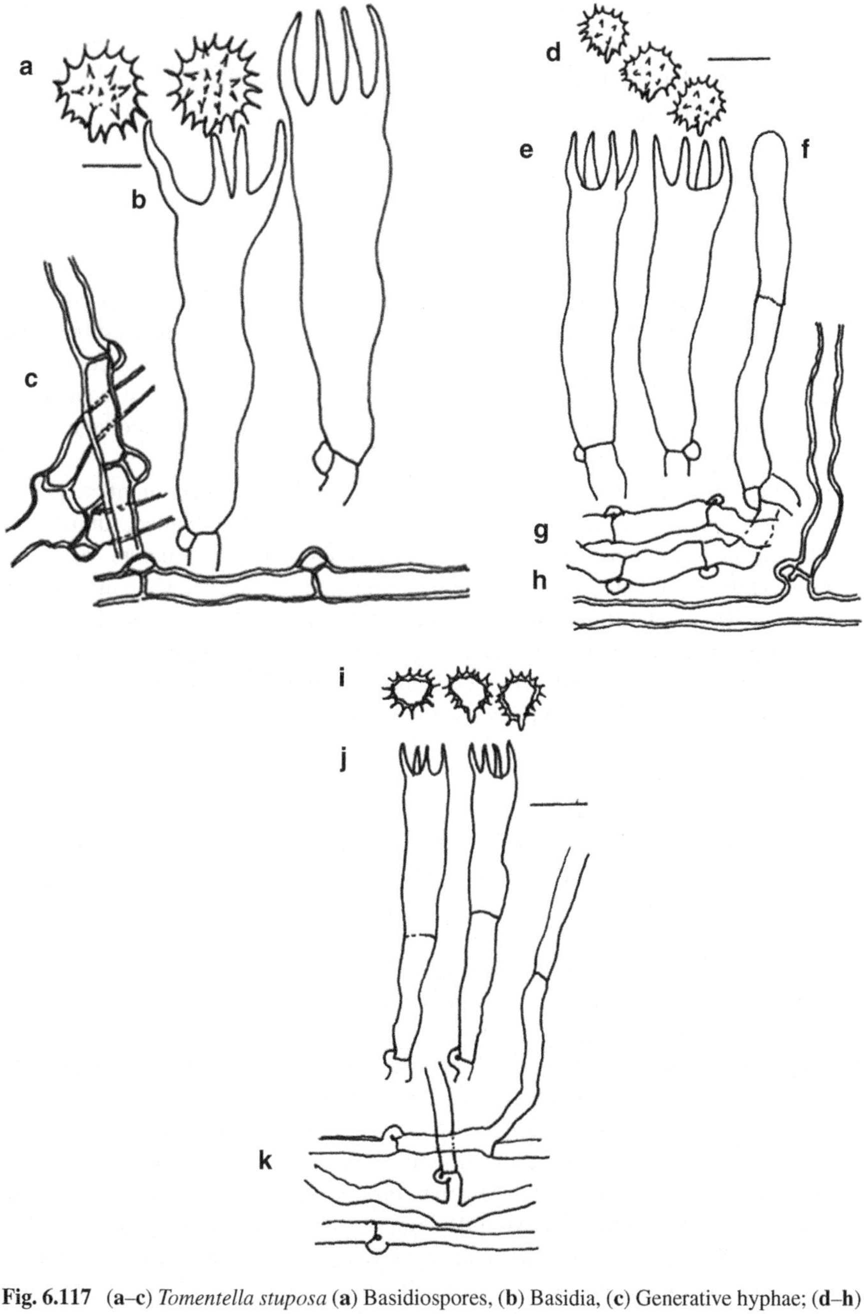

Fig. 6.117 (**a–c**) *Tomentella stuposa* (**a**) Basidiospores, (**b**) Basidia, (**c**) Generative hyphae; (**d–h**) *Tomentella subclavigera* (**d**) Basidiospores, (**e**) Basidia, (**f**) Cystidia, (**g**) Thin-walled generative hyphae, (**h**) Thick-walled generative hyphae; (**i–k**) *Tomentella subcorticoides* (**i**) Basidiospores, (**j**) Basidia, (**k**) Generative hyphae

Remarks: The species is characterized by having tomentose to arachnoid, submembranous to membranous fructification; monomitic hyphal system; presence of cystidia; basidia clavate, 4-sterigmate with basal clamp; globose to subglobose, apiculate basidiospores.

Tomentella subcorticioides S.S. Rattan, Biblthca Mycol. 60: 53 (1977). Fig. 6.117i–k

Fructifications resupinate, submembranous to membranous, loosely adnate, widely effused, up to 300 μm thick in section; hymenial surface pinkish brown to light brown, continuous, smooth, often cracking irregularly on drying; margin thinning to indeterminate, white to paler concolorous, adnate. Subiculum subhyaline to paler brown in section, fibrous, composed of a basal layer of compactly arranged repent hyphae and an upper layer of somewhat loosely woven semi-erect hyphae. The basal layer often becomes agglutinated in part producing few obscure hyphal cordons.

Hyphal system monomitic, hyphae 2.8–4.5(6) μm wide, branched, uninflated, septate, clamped, clamps prominent and almost at all septa, thin-walled, subhyaline. The hyphae are distinct in younger parts but tend to collapse in mature fructifications. Hyphal cordons absent. Tissues become dark to blackish when mounted in 3 % KOH sol. Cystidia absent. Basidia 52–55(70) × 7.4–8.5 μm, clavate to clavate-cylindrical, 4-spored, sterigmata up to 8.5 μm in length. Basidiospores 5.8–7.8 μm across, irregular, the walls moderately thick, slightly tinted brown, nonamyloid, echinulate, spines acute and up to 1 μm long.

Distribution: H.P.: Mahasu- Narkanda; U.K.: Mussoorie.

Collection examined: SSR 5528.

Substratum: On slash of *Abies pindrow*.

Remarks: The species is characterized by corticoid fructifications submembranous to membranous texture, subhyaline and more or less compactly arranged hyphae and continuous hymenium and small basidiospores.

Tomentella terrestris (Berk. & Broome) M.J. Larsen, Mycol. Mem. 4: 105 (1974) = *Zygodesmus terrestris* Berk. & Br., Ann. Mag. nat. Hist. Ser. 5 7: 130(1881). Plate 6.41e, Fig. 6.118a–d

Fructification resupinate, effused, adnate, arachnoid; hymenial surface continuous, smooth, dark brown; margins thinning, paler concolorous.

Hyphal system monomitic; generative hyphae up to 5.0 μm wide, septate, clamped; subicular hyphae almost parallel to the substrate, yellowish-brown, thick-walled, sparsely branched; subhymenial hyphae mostly vertical, pale yellowish-brown, much branched, thin to slightly thick-walled. Cystidia none. Basidia 43–68 × 15–18 μm, subutriform, narrowing sharply near the basal septum, thin to somewhat thick-walled in the basal half, tinted pale yellowish-brown, with a basal clamp, 4-sterigmate; sterigmata up to 8 μm long. Basidiospores 6.8–8.2 μm across, globose to subglobose, irregular to lobed, thick-walled, echinulate, nonamyloid, mostly with one guttule.

Distribution: West Bengal: Darjeeling.

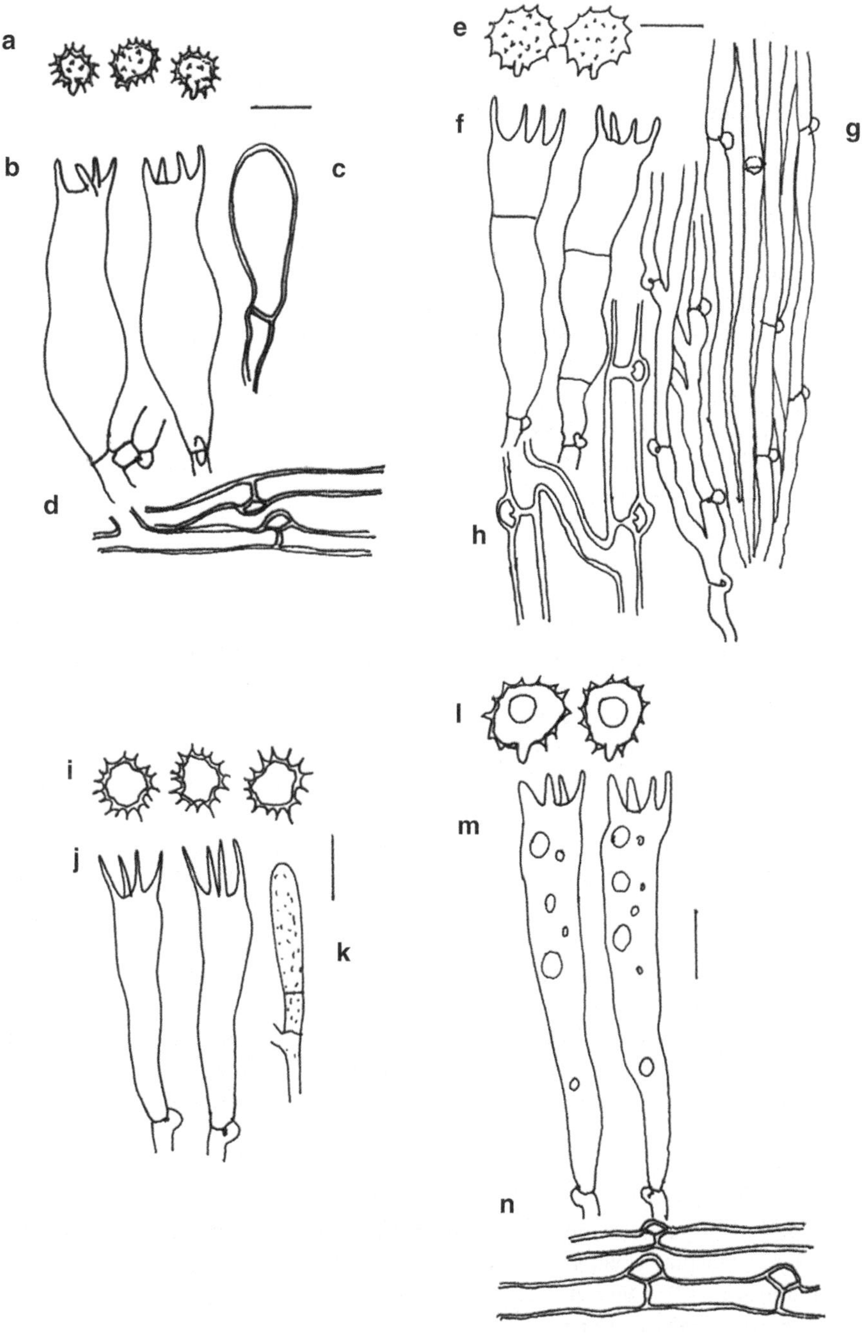

Fig. 6.118 (**a**–**d**) *Tomentella terrestris* (**a**) Basidiospores, (**b**) Basidia, (**c**) Cystidia, (**d**) Generative hyphae; (**e**–**h**) *Tomentella testaceogilva* (**e**) Basidiospores, (**f**) Basidia, (**g**) Hyphal cordons, (**h**) Generative hyphae; (**i**–**k**) *Tomentella umbrinospora* (**i**) Basidiospores, (**j**) Basidia, (**k**) Basidioles; (**l**–**n**) *Tomentella unicusa* (**l**) Basidiospores, (**m**) Basidia, (**n**) Generative hyphae

Collection examined: GSD 19317.

Substratum: On decaying wood.

Remarks: The species is characterized by arachnoid fructification; subutriform basidia and especially tomentellaoid basidiospores.

Tomentella testaceogilva Bourdot & Galzin, Bull. trimmest. Soc. mycol. Fr. 40(2): 149, 1924. Fig. 6.118e–h

Fructification resupinate, adherent, effused, mucedinoid; hymenial surface brownish orange to brown when fresh, start shrinks on drying; margin thinning, paler concolorous to indeterminate.

Hyphal system monomitic; generative hyphae clamped, septate; basal hyphae parallel to the substrate, thick-walled, distantly septate, dark brown; subhymenial hyphae 5.9 µm wide, thin-walled, subhyaline, more branched. Cystidia absent. Basidia 41–55.5 × 8.0–10.0 µm, clavate, 4-sterigmate, with or without oily contents, with basal clamp. Basidiospores 8.9–10.2 µm across, echinulate, apiculate, globose to subglobose.

Distribution: H.P.: Solan.

Collection examined: IBP 37642.

Substrate: On bark of *R. arboretum*.

Remarks: The species is characterized by having irregular globose to subglobose basidiospores. It was first described by Bourdot and Galzin (1924).

Tomentella umbrinospora M.J. Larsen, State Univ. College of forestry at Syracause Univ. Tech. Publ., 93: 61 (1968). Fig. 6.118i–k

Fructification resupinate, floccose to subpelliculose, loosely adnate, widely effused; hymenial surface brown to yellow-brown, smooth to finely granulose, discontinuous; margin indeterminate or rarely thinning, loosely adnate, paler concolorous. Subiculum light brown in section, composed of loosely woven hyphae and hyphal cordons.

Hyphal system dimitic; generative hyphae 1.5–3(5) µm wide, branched, branches at wide angles, septate, clamped, clamps prominent and almost at all septa, thin-walled, subhyaline or rarely tinted yellow; skeletal hyphae 1.5–2(2.5) µm wide, the walls thin to slightly thick (0.8 µm), tinted yellow or citrine;. Hyphal cordons abundant, up to 25 µm broad, light brown, unbranched or occasionally branched, composed exclusively of either skeletal hyphae or generative hyphae only. KOH reaction tissues darkening appreciably when mounted in 3 % sol. Cystidia absent. Basidia 36–54 × 7–8 µm, clavate to clavate-cylindrical, often with a median retraction septum, 4-spored. Basidiospores 7–7.6 µm across, distinctly lobed, thin-walled to moderately thick (1 µm), light brown to umbrinose, non-amyloid, echinulate, spine slender.

Distribution: H.P.: Mahasu- Narkanda.

Collection examined: SSR: 5365, IBP 37643.

Substratum: logs under a coniferous forest.

Remarks: The species is characterized by smooth, rusty brown to yellow-brown hymenial surface, dimitic hyphal system and light brown to umbrinose basidiospores.

Tomentella unicusa Dhingra & Malka 1994 in Sarma et al. Current Researches in Plant Sci. 43–56, 1994. Fig. 6.118l–n

Fructification resupinate, effused, loosely adnate, mucedinoid to submembranous; hymenial surface brownish orange, smooth, black in KOH sol; subiculum concolorous with fertile area; margins thinning, arachnoid, concolorous, or indeterminate.

Hyphal system monomitic; basal hyphae up to 7.2 µm wide, loosely arranged, parallel to substrate, branched, septate, usually clamped, greyish yellow to dull yellow, thin to thick-walled; subhymenial hyphae up to 5.0 µm wide, compactly packed, vertically arranged, branched, septate, clamped at few septa, subhyaline, thin-walled. Hyphal cordons absent. Basidia 46.0–56.0 × 9.0–13.0 µm, clavate to subclavate, 4-spored, with or without oily contents, subhyaline. Basidiospores 7.0–9.0 µm across, irregular in outline to lobed, thick-walled, subhyaline to yellowish grey with brown wall, uniguttulate, warted.

Distribution: H.P.: Chamba.

Collection examined: IBP 37644.

Substratum: On gymnospermic stump.

Remarks: The species is characterized by mucedinoid to submembranous fructification, brownish orange hymenial surface usually clamped subicular hyphae, rarely clamped subhymenial hyphae, absence of hyphal cordons, clavate to subclavate basidia and irregular to lobed, warted basidiospores with warts usually becoming bifurcate.

O-Trechisporales
Family-Hydnodontaceae

Key to genera

1. Cystidial elements present 2
1. Cystidial elements absent 3
2. Cystidia subulate, with discrete crystals, basidiospores narrowly cylindrical or narrowly subfusiform .... *Subulicystidium*
2. Cystidia subconical to subcylindrical, thick-walled, covered with dendroid hyphae, basidiospores sigmoid to vermicular .... *Tubulicium*[a]
3. Basidiospores cyanophilous .... *Fibriciellum*
3. Basidiospores acyanophilous 4
4. Basidiospores verrucose .... *Trechispora*
4. Basidiospores smooth 5
5. Basidia 4-sterigmate 6
5. Basidia 4–6 sterigmate, basidiospores ellipsoid to subcylindrical to suballantoid .... *Sistotremastrum*
6. Basidiospores subglobose .... *Brevicellicium*
6. Basidiospores ellipsoid .... *Fibrodontia*

[a]Extra limital, not included in this work.

Brevicellicium K-H.Larss.& Hjortst.
Mycotaxon 7: 117, 1978.

Fructification resupinate, adnate, thin to thick-walled, membranous, ceraceous, when dried somewhat brittle, smooth to granulose. Hyphal system monomitic, cystidia absent, but sometimes with smooth, subglobose cells (sphaerocysts), arising from the basal hyphae. Basidia clavate to subclavate, 4-sterigmate, with a basal clamp. Basidiospores subglobose to ellipsoid, often lacrimiform, with a distinct apiculus, acyanophilous.

Thirteen species, world-wide

Lit.: Larsson & Hjortstam (*Mycotaxon* **7**: 117, 1978), Hjortstam (*Mycotaxon* **79**: 181, 2001; tropical and subtropical species).

Type species: *Corticium exile* Jacks.1950

Habitat: Wood

Himalayas: One

Brevicellicium olivascens (Bres.) K-H.Larss. & Hjortst., Mycotaxon 7: 119, 1978 = *Odontia olivascens* Bres., Fungi Trid. 2, p.36, 1892. Fig. 6.119a–d

Fructification resupinate, effused, adnate; hymenial surface smooth to tuberculate often cracked, greyish white to greyish orange to brownish orange when fresh, light brown on drying; margin thinning, paler concolorous to indeterminate.

Hyphal system monomitic; generative hyphae septate, clamped; basal hyphae up to 3.8 μm wide, parallel to substrate, long-celled, thick-walled, less branched; subhymenial hyphae up to 5.7 μm, vertical, short-celled, thin-walled, richly branched. Cystidia none. Basidia 11.0–13.5 × 5.3–6.5 μm, clavate to subclavate, 4-sterigmate with basal clamp. Basidiospores 5.3 × 4.1 μm in diameter, acyanophilous, inamyloid.

Distribution: A.P.: West Kameng, Bomdila.

Collection examined: GSD 19759.

Substratum: On a decaying angiospermic log.

Remarks: This species is characterized by thin fructifications, clamped generative hyphae, long-celled basal hyphae, short and broad celled subhymenial hyphae, small (10–15 × 6–7.5 μm) subcylindrical, somewhat constricted, 4-sterigmate, basidia with basal clamps and subglobose to globose (4.5–5.25 × 4–4.75 μm), smooth, thin-walled, non-amyloid, acyanophilous basidiospores.

Fibriciellum J. Erikss. & Ryvarden
The Cort. of N. Eu. 3: 373, 1975.

Fructification resupinate, effused, thin; margin mostly fibrillose, often with rhizomorphs in the periphery and in the substrate; hymenium glabrous, rather smooth, ochraceous as mature; hyphal system dimitic, generative hyphae thin-walled, short celled, with clamps and anastomoses, densely united, skeletal hyphae thick-walled, narrow, fiber-like, straight and with rare clamps and ramifications, all hyphae stained in cotton-blue, especially the skeletal, less evident for the generative hyphae, owing to their thin walls; basidia clavate, rather short, with 4 sterigmata; spores small, ellipsoid, with thin or slightly thickened walls, somewhat stained in cotton blue.

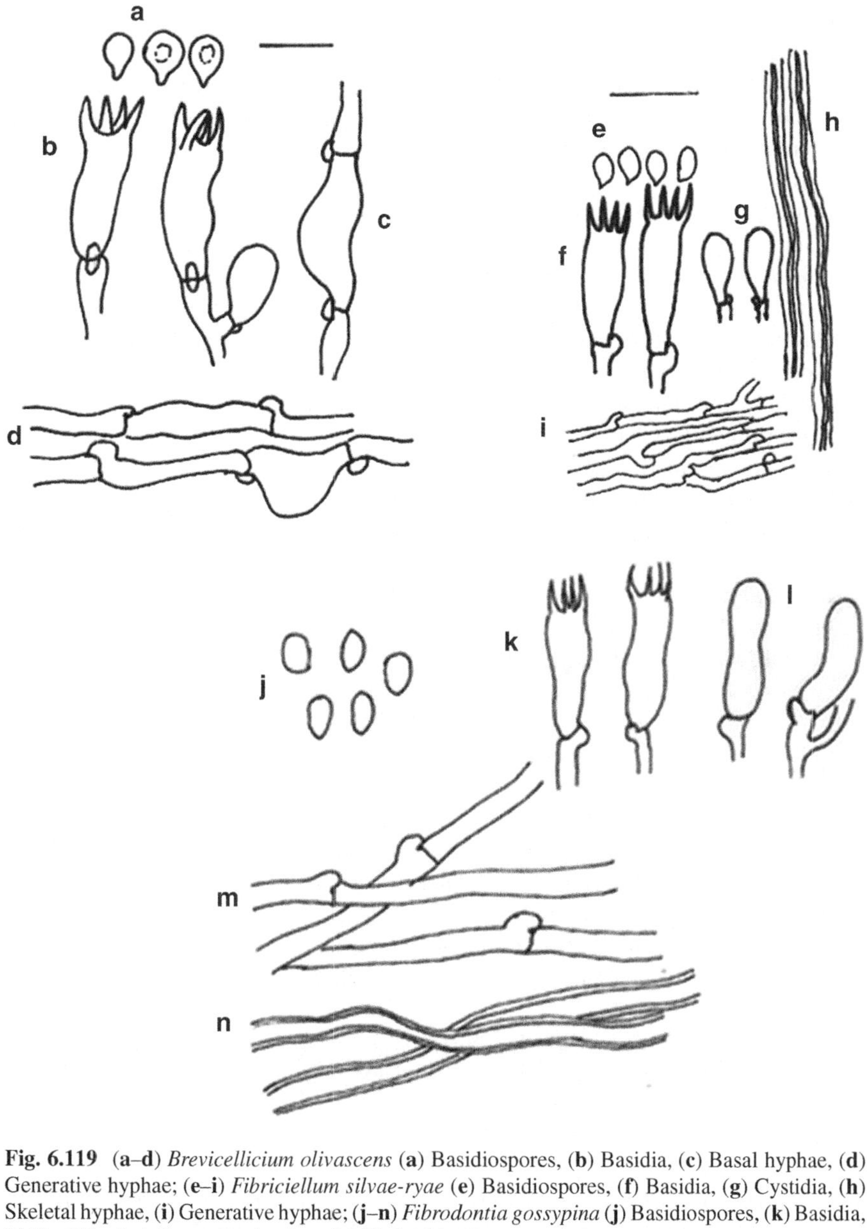

Fig. 6.119 (**a–d**) *Brevicellicium olivascens* (**a**) Basidiospores, (**b**) Basidia, (**c**) Basal hyphae, (**d**) Generative hyphae; (**e–i**) *Fibriciellum silvae-ryae* (**e**) Basidiospores, (**f**) Basidia, (**g**) Cystidia, (**h**) Skeletal hyphae, (**i**) Generative hyphae; (**j–n**) *Fibrodontia gossypina* (**j**) Basidiospores, (**k**) Basidia, (**l**) Cystidia, (**m**) Generative hyphae, (**n**) Skeletal hyphae

Monotypic, Europe.
Lit.: Eriksson & Ryvarden (*Cortic. N. Europ*. **3**: 373, 1975)
Habitat: Dead wood.
Type species: *Fibriciellum silvae-ryae* Erikss. & Ryv. 1975
Himalayas: One.

Fibriciellum silvae-ryae Erikss. & Ryv., Cort. N. Europ III: 375, 1975. Fig. 6.119e–i

Fructification resupinate, effused, loosely adnate, membranous; hymenial surface glabrous, creamish to pale yellow to light orange, smooth, fresh, but cracked on drying, cracks superficial to deep; margins variable, indeterminate to finely fibrillose with conspicuous radiating rhizomorphs.

Hyphal system dimitic, generative hyphae 2–3.3 μm wide, thin-walled, septate, clamped, usually straight in subicular layer and much branched and densely united in the subhymenium; skeletal hyphae 2.5 μm wide, straight and very sparsely ramified. Cystidia none. Basidia 9.0–11.5 × 3.5–4.6 μm clavate to subclavate, thin-walled, 4-sterigmate, with a basal clamp; sterigmata up to 2.5 μm long. Basidiospores 3.2–4.3 × 2.5–4.8 μm ellipsoid, thin-walled, smooth, minutely apiculate, non-amyloid, weakly cyanophilous.

Distribution: H.P.: Chamba- Khajjiar, Kullu.

Collection examined: IBP 37646.

Substratum: Mixed forest, *Cedrus* needles and gymnospermic twigs.

Remarks: The species is characterized by fibrillose margins in young fructifications which become indeterminate in mature fructifications. Hyphal system monomitic; basal hyphae sparsely branched; subhymenial hyphae richly branched, densely united, short-celled. Ellipsoid to cylindrical, thin-walled, smooth, non-amyloid, acyanophilous basidiospores.

Fibrodontia Parm., Consp. Syst. Cort. p.174, 1968.

Fructification resupinate, rather loosely adnate, easily detached in pieces, white to pale ochraceous-isabellinous, ceraceous when alive, in the dry state soft but fibrous; hymenial surface odontoid, with short and dense, usually cylindrical aculei, projecting hyphae; cystidia absent. Hyphal system dimitic, all hyphae with clamps, generative hyphae thin-walled; skeletal hyphae thick-walled. Basidia suburniform, with 4-sterigmata and basal clamps. Basidiospores spores ellipsoid, thin-walled, inamyloid, noncyanophilous.

Five Species, world-wide

Lit.: Langer (*Biblthca Mycol.***154**, 1994).

Type species: *Fibrodontia gossypina* Parm. 1968

Habitat: Wood

Himalayas: One

Fibrodontia gossypina Parm., Consp. Syst. Cort. p.207, 1968 = *Hyphodontia gassypina* (Parm.) Hjort. Mycotaxon 39: 416 (1990). Fig. 6.119j–n

Fructification resupinate, adanate, effused, ceraceous when fresh, when dried soft but fibrous; hymenial surface greyish-white when fresh, pale ochraceous on drying, odontoid; margin not well differentiated. Subiculum composed of loosely interwoven generative hyphae and skeletal hyphae.

Hyphal system dimitic; generative hyphae 1.8–2.8 μm wide, distinct, richly branched, septate, clamped; skeletal hyphae 2.8–3.8 μm wide, unbranched, aseptate, without clamps. Skeletal hyphae in the aculei arise from the generative hyphae in the subiculum. Cystidia none. Basidia 12–17 × 4–5 µm, suburniform, with a basal clamp, 4-sterigmate; sterigmata up to 4.5 µm long. Basidiospores 4–5.8 × 3–4.8 µm, ellipsoid to broadly ellipsoid, smooth, thin-walled, non-amyloid, acyanophilous.

Distribution: Arunachal Pradesh: West Kameng, Bomdila.

Collection examined: GSD 19669.

Substratum: On a decaying angiospermic stump.

Remarks: This species is characterized by odontoid hymenial surface, dimitic hyphal system, clamped generative hyphae, suburniform basidia and ellipsoid to broadly ellipsoid basidiospores.

Sistotremastrum J. Erikss.,
Symb. bot. Upsal. 16 (no. 1): 62 (1958)

Fructifications resupinate, loosely adnate, membranous, effused; hymenial surface smooth to tuberculate, white to faintly coloured. Subiculum subhyaline to pale coloured in section. Hyphal system monomitic, hyphae distinct, clamped, usually with union-like swellings near the septa. Cystidia or gloeocystidia absent. Basidia clavate-cylindrical or tubular, 4–8 spored. Basidiospores ellipsoid to cylindrical, thin walled, subhyaline, smooth, non-amyloid.

Three species, widespread

Lit.: Eriksson (Symb. bot. Upsal. 16 (1):62, 1958.)

Type Species: *Sistotremastrum suecicum* Litsch. ex J. Erikss. (1958)

Habitat: Decay wood

Himalayas: One

Sistotremastrum niveocremeum (Höhn. & Litsch.) J. Erikss., Symb. bot. Upsal. 16(1): 62 (1958). Fig. 6.120a–d

Fructification resupinate, loosely adnate, effused; hymenial surface smooth to tuberculate, greyish white when fresh, yellow to greyish yellow on drying; margin thinning, paler concolorous to indeterminate.

Hyphal system monomitic; generative hyphae up to 3.8 µm wide, septate, clamped, thin-walled; basal hyphae parallel to the substrate, sparsely branched with oily content; subhymenial hyphae vertical, branched. Cystidia none. Basidia 10–19 × 3.9–4.5 µm, obconical to tubular, constricted. Basidiospores smooth, thin-walled, non-amyloid, acyanophilous, 5–7.5 × 2.5–3.75 µm.

Distribution: Bhutan: Thimphu, Bunakha; H.P.: Kullu; U.K.: Karanprayag-Chamoli.

Collection examined: GSD 19542, IBP 37648.

Substratum: On a decaying angiospermic stump.

Remarks: The species is characterized by thin fructification; clamped generative hyphae, broadly clavate 6–8 sterigmate basidia with a basal clamp, ellipsoid to subcylindrical, smooth, thin-walled, non-amyloid, acyanophilous basidiospores. It is a new record for Uttarakhand.

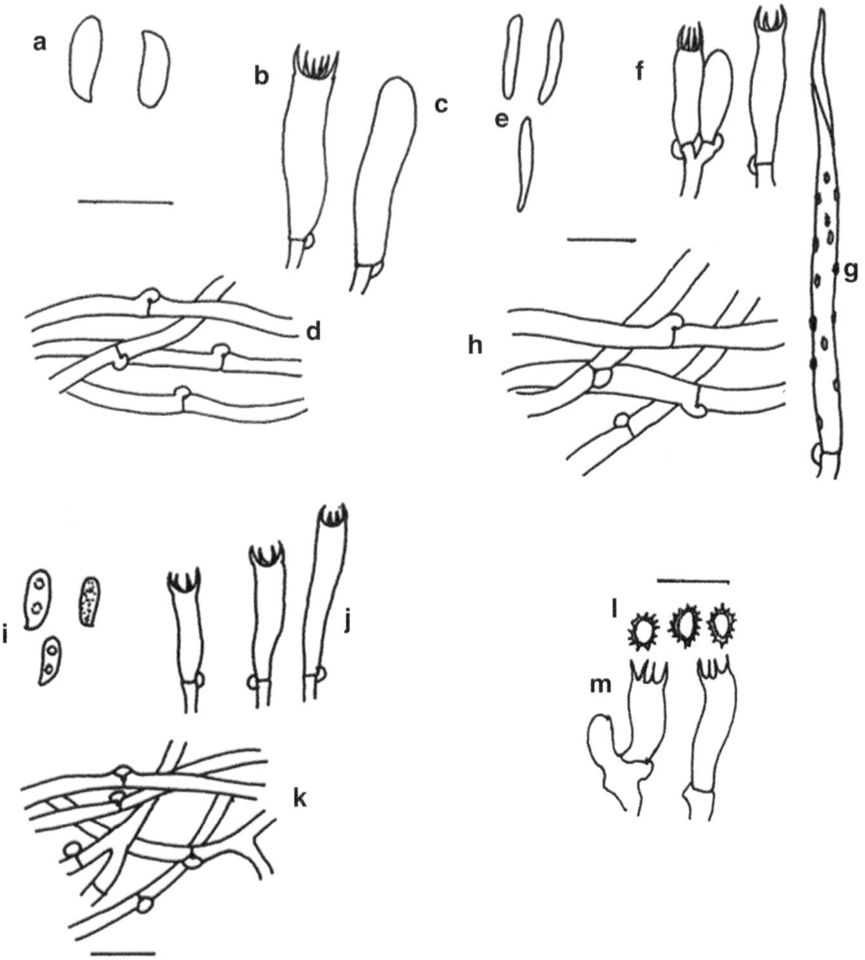

Fig. 6.120 (**a–d**) *Sistotremastrum niveocremeum* (**a**) Basidiospores, (**b**) Basidia, (**c**) Cystidia, (**d**) Generative hyphae; (**e–h**) *Subulicystidium longisporum* (**e**) Basidiospores, (**f**) Basidia, (**g**) Cystidia, (**h**) Generative hyphae; (**i–k**) *Subulicystidium meridens* (**i**) Basidiospores, (**j**) Basidia, (**k**) Generative hyphae; (**l**, **m**) *Trechispora allnicola* (**l**) Basidiospores, (**m**) Basidia

Subulicystidium Parmasto,
Consp. Syst. Cort. p. 120.1968.

Fructifications resupinate, floccose to pelliculose, loosely adnate, usually thin; hymenial surface white to cream, smooth. Subiculum composed of loosely woven hyphae, somewhat fibrous. Hyphal system monomitic, hyphae firm-walled, subhyaline, branched at wide angles and interwoven, clamped at all septa. Cystidia subulate, thin but firm walled, ornamented with regularly spaced flat crystals. Gloeocystidia absent. Basidia 2–4 spored, pleurobasidiate, short cylindrical. Basidiospores cylindric to fusiform, subhyaline, smooth, non-amyloid, acyanophilous.

Seven species, widespread
Lit.: Duhem & Michel (*Cryptog. Mycol.* **22**: 163, 2001)
Type Species: *Hypochnus longisporus* Pat., 1894
Habitat: Decayed Wood
Himalayas: Two
Key to species

1. Basidiospore subfusiform to curved cylindrical .... *S. longisporum*
1. Basidiospore ellipsoid to subcylindrical to suballantoid .... *S. meridens*

Subulicystidium longisporum (Pat.) Parm., Consp. Syst. Cort. p. 121.1968 =*Hypochnus longisporus* Pat., J. Bot. Paris 8: 221. 1894. Fig. 6.120e–h

Fructification resupinate, floccose to subpelliculose, loosely adnate, arising in small patches which may coalesce and grow later; hymenial surface white to cream, smooth but appears farinose under a lens, discontinuous, not cerviced. Subiculum subhyaline in section, composed of loosely woven hyphae.

Hyphal system monomitic, hyphae 1.8–2.9 µm wide, branched at wide angles, septate, clamped, clamps prominent and almost at all septa, thin to slightly firm, subhyaline. Cystidia 42–72 × 2.5–3.5 µm, subulate with acute apices, walls thick in the basal part but become progressively thin-walled towards the apex, subhyaline. Basidia 10–14 × 4.8–7.5 µm, short cylindrical, pleurobasidiate, 4-spored, sterigmata up to 4.5 µm in length. Basidiospores 13.0–15.5 × 2.8–3.5 µm, subfusiform to curved cylindrical, minutely apiculate, the walls thin, smooth, subhyaline, nonamyloid, acyanophilous.

Distribution: H.P.: Kullu; J&K: Pehalgam; U.K.: Dehra Dun.
Collection examined: SSR 5652, IBP 37649.
Substratum: On stump of *Cedrus deodara* and *Shorea robusta.*
Remarks: This species is characterized by the texture of the fructification, characteristic cystidia, presence of pleurobasidia and very long, narrow basidiospores.

Subulicystidium meridense Oberw., Bibliotheca Mycologica 61: 343, 1977. Fig. 6.120i–k

Fructification resupinate, effused, loosely adnate, floccose to subpelliculose; hymenial surface white to creamish, smooth to tuberculate; margin not differentiated.

Hyphal system monomitic; generative hyphae septate, clamped, branched, 2–3 µm wide, thin to somewhat thick-walled. Cystidia 64–78 × 4–5 µm, subulate with acute to subobtuse apices, thick-walled, characteristically impregnated with discrete, regularly arranged, flat crystals except in the apical portion where the wall is smooth and thin. Basidia 10.8–18 × 4–5 µm, sub-clavate, plurobasidia, generally 4-sterigmate; sterigmata up to 4.5 µm long. Basidiospores 5.8–8.0 × 2.5–3.2 µm, narrowly ellipsoid to subcylindrical or suballantoid, smooth, thin-walled, non-amyloid, acyanophilous, with oily content.

Distribution: Arunachal Pradesh: West Kameng, Bomdila, Shergaon, Basti side, West Bengal: Darjeeling, Siliguri; U.K.: Lal tibba-Mussoorie.
Collection examined: GSD 19857, 19201; L 38088.
Substratum: On decaying angiospermic branches.

Remarks: The species is characterized by resupinate, floccose to subpelliculose fructification; monomitic hyphal system; sub-clavate, pleurobasidiate basidia; narrowly ellipsoid to subcylindrical, smooth basidiospores. It is a new record for Uttarakhand.

Trechispora Karst.,
Hedwigia 29: 147. 1890.

Fructifications resupinate or pileate, sessile or stipitate; hymenial surface smooth to tuberculate, hydnoid or poroid. Subiculum composed of loosely or compactly arranged hyphae, hyphae distinct and not agglutinated or ceraceous. Hyphal system monomitic or dimitic; generative hyphae septae, clamped, usually with ampulliform swellings near the septa. Cystidia present or absent. Gloeocystidia absent. Basidia clavate to cylindrical, 4-spored. Basidiospores subhyaline to pale yellow, smooth to echinulate or tuberculate, nonamyloid, acyanophilous.

Forty six species, widespread

Lit.: Larsson The Genus *Trechispora* (Corticiaceae, Basidiomycetes) Dissertation Universität Göteborg, 1992

Type species: *Trechispora onusta* Karst. 1890

Habitat: Decay wood

Himalayas: Eight

Key to species

1. Hymenial surface poroid .... *T. mollusca*
1. Hymenial surface smooth to hydnoid 2
2. Basidiospores echinate/verrucose 3
2. Basidiospores smooth .... *T. mutabilis*
3. Hymenial surface smooth 4
3. Hymenial surface smooth to grandinoid to hydnoid 5
4. Basidiospores subglobose to lacrimoid, verrucose except for in apicular region .... *T. micospora*
4. Basidiospores ovoid to subglobose, echinulate .... *T. candidissima*
5. Hyphae with acicular crystals .... *T. praefocata*
5. Hyphae without acicular crystals 6
6. Hymenial surface pseudoporoid .... *T. fastidiosa*
6. Hymenial surface not so 7
7. Hymenial surface smooth to grandinioid .... T. farinacea
7. Hymenial surface grandinioid to hydnoid .... *T. alnicola*

Trechispora alnicola (Bourd. & Galz.) Liberta, Taxon 15. 318.1966 = *Grandinia alnicola* Bourd. & Galz., Bull. Soc. Mycol. Fr. 30: 254.1914. Fig. 6.120l,m

Fructification resupinate, membranous to subceraceous when fresh becoming somewhat horny and brittle on drying; loosely adnate, widely effused, up to 250 µm thick in section; hymenial surface cream to cream brown, smooth to finely colliculose, irregularly cracking on drying; margin thinning to almost abrupt, adnate when fresh but often curls away from the substratum on drying. Subiculum subhyaline in section, composed of compactly arranged somewhat agglutinated hyphae, impregnated with abundant crystalline matter and discharged basidiospores.

Hyphal system monomitic, hyphae 1.3–2.8 μm wide, branched, septate, clamped, ampulliform swelling near the septa not distinct, thin-walled, subhyaline. Cystidia absent. Basidia 12–14×3.5–4.3 μm, clavate-cylindrical, frequently pleurobasidiate, 4-spored, sterigmata up to 4 μm in length. Basidiospores 3.2–3.8×2–2.3 μm, broadly ellipsoid to ovoid, shortly apiculate, thin-walled, subhyaline, finely echinulate, non-amyloid.

Distribution: H.P.: Mahasu, Narkanda.

Collection examined: SSR 5628, IBP 37651.

Substratum: On stump of *Abies pindrow*.

Remarks: This species is characterized by membranous to ceraceous, horny or brittle on drying, widely effused fructification; pleurobasidiate 4-spored basidia, broad ellipsoid to ovoid basidiospores.

Trechispora candidissima (Schwein.) Bondartsev & Singer, Annls mycol. 39(1): 48 (1941) = *Poria candidissima* (Schwein.) Cooke, Grevillea 14(72): 110, 1886.

Fig. 6.121a–c

Fructification annual, resupinate, becoming effused up to 7 cm wide, up to 1.5 mm thick, separable, without any distinct taste; margin white to creamish, membranous to arachnoid, wide; pore surface white to creamish, dull, uneven; pores rounded to angular, averaging 2–4 per mm; dissepiments torn at their apices, somewhat irpicoid, finely velutinate. Subiculum white, soft, thin, poorly developed, not darkening in KOH; tubes poorly developed, fragile, creamish in section.

Hyphal system monomitic; generative hyphae hyaline, thin-walled, branched, often encrusted, septate, clamped, cyanophilous, 2–5.6 μm in diameter. Cystidia absent. Basidia hyaline, thin-walled, clavate, 2-4spored, 9–12×4–5.2 μm; steerigmate up to 2.8 μm in length. Basidiospores subhyaline, thin-walled, echinulate, ovoid to subglobose, non-amyloid, 2.5–4.1×2–3.1 μm.

Distribution: H.P.: Shimla- Shillaroo, Kullu- Pulga.

Collection examined: Dhanda 6522, 6587, IBP 37652.

Substratum: Burnt logs of *Pinus excelsa* and gymnospermic stumps and twigs.

Remarks: The species is characterized in having monomitic hyphal system with encrusted, clamped generative hyphae and echinulate spores.

Trechispora farinacea (Pers.) Liberta, Taxon 15: 318. 1966 = *Hydnum farinaceum* Pers. Syn. Meth. Fung. 2: 562 (1801). Fig. 6.121d–f

Fructifications resupinate, floccose-pelliculose when young becoming submembranous at maturity, loosely adnate, widely effused, up to 250 μm thick in section; hymenial surface cream to cream-yellow, hydnoid, often cracking irregularly forming discreet patches or frustules; margin cottony to sometimes fibrillose, adnate, white. Teeth gregarious, terrete, up to 1.5 mm long, subulate to subcylindrical, rarely flattened, apices acute to subobtuse, white to cream yellow. Subiculum subhyaline in section, composed of loosely woven hyphae impregnated with abundant subhyaline crystalline matter.

Hyphal system monomitic, hyphae 2–3.5 μm wide, branched, septate, clamped, frequently forming ampulliform swellings near the septa, thin-walled, subhyaline.

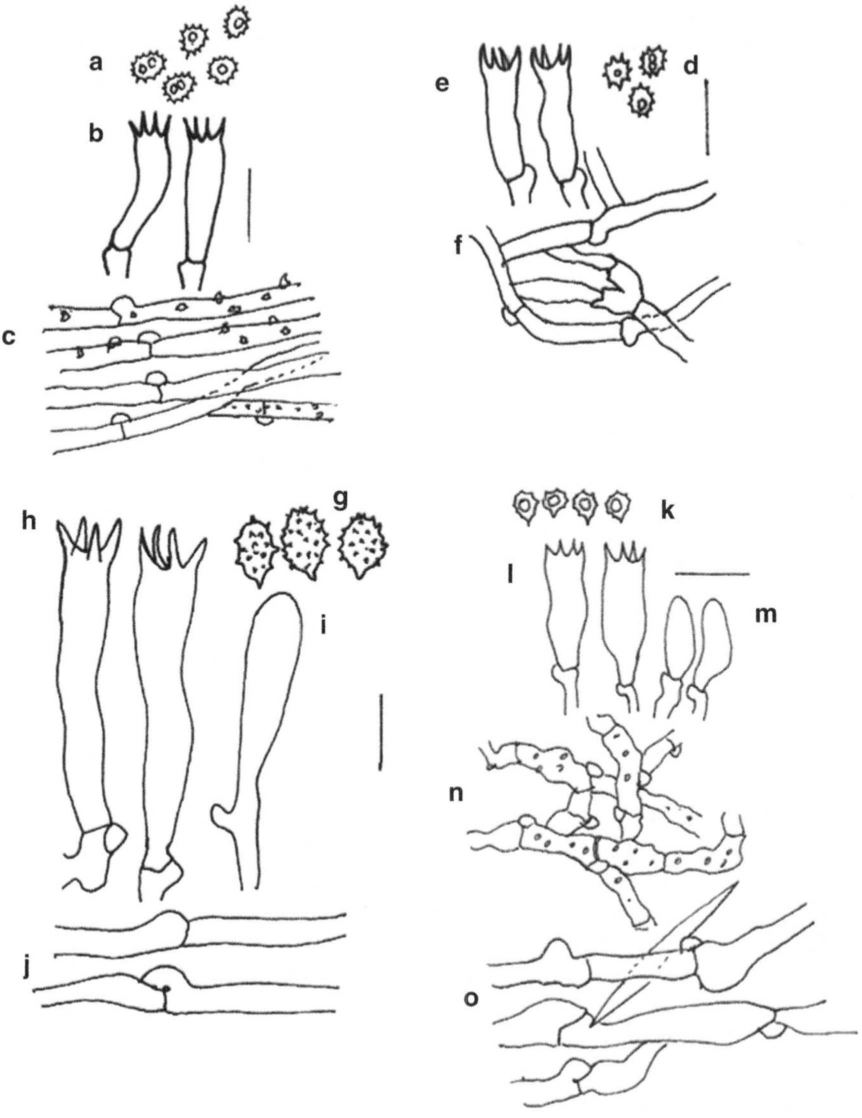

Fig. 6.121 (**a**–**c**) *Trechispora candidissima* (**a**) Basidiospores, (**b**) Basidia, (**c**) Generative hyphae; (**d**–**f**) *Trechispora farinacea* (**d**) Basidiospores, (**e**) Basidia, (**f**) Generative hyphae; (**g**–**j**) *Trechispora fastidiosa* (**g**) Basidiospores, (**h**) Basidia, (**i**) Cystidia, (**j**) Generative hyphae; (**k**–**o**) *Trechispora microspora* (**k**) Basidiospores, (**l**) Basidia, (**m**) Basidioles, (**n**–**o**) Subicular hyphae

Cystidia absent. Basidia 11.8–19.8 × 3–4 μm, clavate-cylindrical to subutriform, rarely pleurobasidiate, 4-spored, sterigmata up to 4.5 μm in length. Basidiospores 3.2–4.3 × 2.4–2.8 μm, broadly ellipsoid to ovoid, shortly apiculate, the walls thin, subhyaline, echinulate, non-amyloid.

Distribution: H.P.: Khajjiar, Chamba, Manali; U.K.: Mussoorie.

Collection examined: HSK 4104, 4242, IBP 37653.

Substratum: On coniferous stump.

Remarks: The species is characterized by soft, loosely woven to floccose-pelliculose fructifications, hydnoid hymenial surface and broadly ellipsoid to ovoid and echinulate basidiospores.

Trechispora fastidiosa (Pers.) Liberta, Taxon 15 (8): 318, 1966 = *Thelephora fastidiosa* Pers. ex Fr., Syst. Mycol. 1: 435, 1821. Fig. 6.121g–j

Fructifications resupinate, loosely adnate, effused, up to 250 μm thick in section, submembranous; hymenial surface yellowish-white, discontinuous and appears to be pseudoporoid under lens; margin thinning, byssoid, whitish in young fructification, abrupt in mature ones. Subiculum composed of compactly packed hyphae running almost parallel to the substrate.

Hyphal system monomitic; generative hyphae 2–3.2 μm wide, branched at wide angles, septate, clamped, thin-walled, often with ampulliform swellings near the septa, uninflated or inflated up to 5 μm. Cystidia none. Basidia 25–37.5 × 6–7 μm, narrowly clavate to subclavate, somewhat sinuous, 4-sterigmate; sterigmata up to 6.2 μm in length. Basidiospores 4.5–6 × 3.2–5 μm, broadly ellipsoid to ovoid, thin-walled, finely echinulate, non-amyloid, acyanophilous, generally with one oil drop.

Distribution: Meghalaya: Shillong, Mawphlong.

Collection examined: GSD 19013.

Substratum: On a decaying angiospermic stump.

Remarks: This species is characterized by submembranous fructification with byssoid margin, narrowly clavate to subclavate basidia and broadly ellipsoid to ovoid, finely echinulate basidiospores.

Trechispora microspora (Karst.) Liberta, Taxon 15(8): 319, 1966. Fig. 6.121k–o

Fructifications resupinate, effused, thin, loosely adnate; hymenial surface smooth, poroid under lens, white; margins thinning, indeterminate, fibrillose; cordons present in the subiculum and extending beyond the margins.

Hyphal system monomitic; generative hyphae up to 4 μm wide, branched, septate, clamped, thin-walled, subicular hyphae and hyphae of cordons with frequent ampulliform swellings near the septa, provided with acerose crystals; subhymenial hyphae short celled, much branched, of even width or slightly inflated. Cystidia none. Basidia 8.5–14.2 × 4.5–5.1 μm, sub-cylindrical to cylindrical, thin-walled, 4-sterigmate, clamped at the base; sterigmata up to 2.8 μm long. Basidiospores 2.5–3.4 × 2.3–2.8 μm, subglobose to more or less lacrimoid, prominently apiculate, thin-walled, verrucose except for in the apicular region, non-amyloid, acyanophilous, usually with single guttule.

Distribution: H.P.: Chamba- Khijjiar, Manali.

Collection examined: GSD: 1472, IBP 37655.

Substratum: On decaying angiospermic log.

Remarks: This species is characterized by having thin membranous, effused fructification; presence of cordons; monomitic hyphal system; absence of cystidia; subcylindrical to cylindrical basidia, thin-walled; subglobose to prominently apiculate basidiospores.

Trechispora mutabilis (Pers.) Liberta, Taxon 15: 319. 1966=*Hydnum granulosum* var. *mutabile* Pers., Mycol. Eur. 2: 184. 1825. Fig. 6.122a, b

Fructifications resupinate, membranous to subceraceous, adnate, widely effused, up to 75 μm thick in section; hymenial surface deep cream to cream brown, smooth to faintly colliculose or hydnoid, not creviced; margin cottony, adnate, white to paler concolorous. Subiculum subhyaline in section, composed of partly collapsed, interwoven hyphae impregnated with abundant crystalline matter.

Hyphal system monomitic, hyphae 2.8–3.8 μm wide, branched, septate, clamped, frequently with ampulliform swellings near the septa, uninflated to unevenly

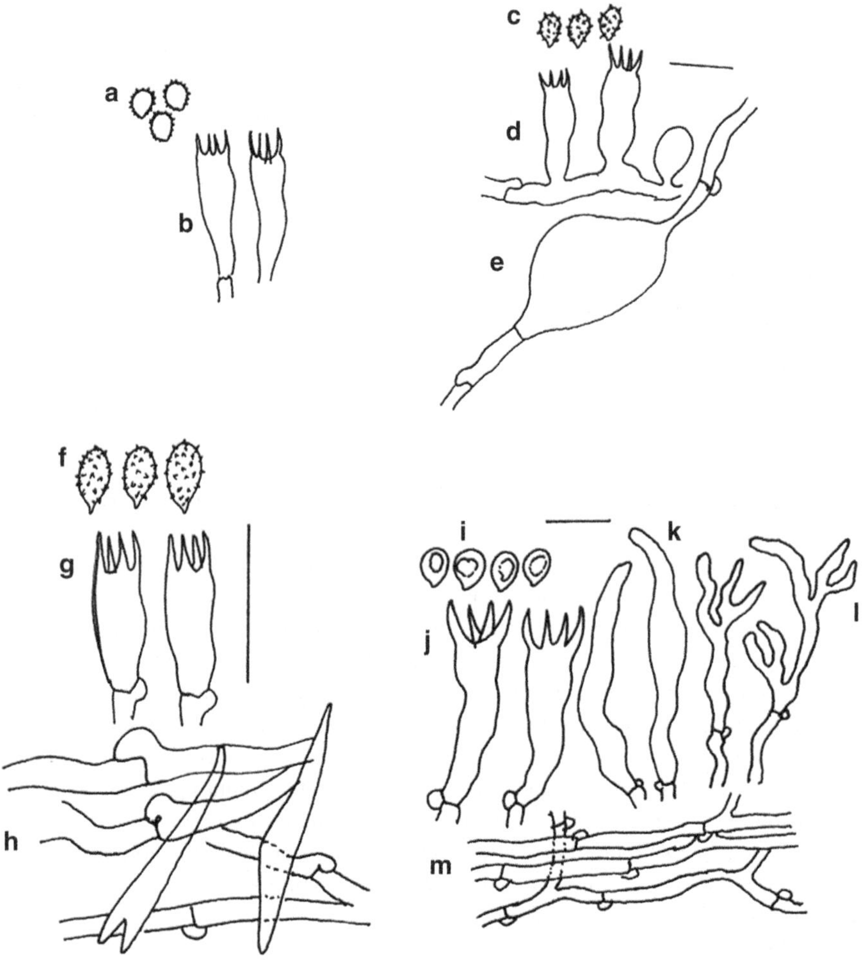

Fig. 6.122 (**a**–**b**) *Trechispora mutabilis* (**a**) Basidiospores, (**b**) Basidia,; (**c**–**e**) *Trechispora mollusca* (**c**) Basidiospores, (**d**) Basidia, (**e**) Generative hyphae with ampulliform swelling; (**f**–**h**) *Trechispora praefocata* (**f**) Basidiospores, (**g**) Basidia, (**h**) Generative hyphae; (**i**–**m**) *Dendrophlebia crassispora* (**i**) Basidiospores, (**j**) Basidia, (**k**) Cystidia, (**l**) Dendrohyphidia, (**m**) Generative hyphae

inflated, thin-walled, subhyaline. Cystidia absent. Basidia 10–14.8×5.5–5.8 μm, clavate-cylindrical, 4-spored, sterigmata up to 4.5 μm in length. Basidiospores 4–4.2×3–3.8 μm, globose to subglobose, shortly apiculate, the walls thin, subhyaline, smooth, non-amyloid.

Distribution: U.K.: Mussoorie; H.P.: Chamba- Khajjiar.

Collection examined: HSK 4244, IBP 37656.

Substratum: On decaying angiospermic stump.

Remarks: The species is characterized by membranous to subceraceous fructifications and subglobose, smooth basidiospores. It can be separated from *T. confinis* by the texture of the fructification, shape and size of basidia and basidiospores.

Trechispora mollusca (Pers.) Liberta, Can. J. Bot. 51: 1878. 1973 = *Polyporus mollusca* Pers.: Fr., Observ. Mycol. 1: 129 (1815). Fig. 6.122c–e

Fructifications resupinate, soft-floccose, loosely adnate to separable, widely effused; hymenial surface poroid, cream or white, not creviced; margin fibrillose, white, loosely adnate. Pores up to 1 mm deep, rounded to angular or sinulate, 2–4 per mm, dissepiments thin, fimbriate, often splitting and becoming lacerate or irpiciform. Subiculum subhyaline in section, composed of loosely woven hyphae.

Hyphal system monomitic, hyphae 2.8–4.3 μm wide, branched, septate, clamped, clamps not at all septa, frequently with ampulliform swellings near the septa, thin-walled, subhyaline. Tramal hyphae are similar to the subiculum hyphae but are less frequently branched and more uniform in diameter. Cystidia absent. Basidia 10.8–16×4.4–5.4 μm, clavate-cylindrical to subutriform, 4-spored, sterigmata up to 3.5 μm long. Basidiospores 3.4–4.1×2.4–3 μm, ellipsoid to ovoid, shortly apiculate, the walls thin, subhyaline, finely echinulate, non-amyloid.

Distribution: H.P.: Kullu, Manali, Chamba-Khajjiar.

Collection examined: SSR 6587, IBP 37657.

Substratum: On angiospermic stump.

Remarks: The species is characterized by soft-floccose fructifications, poroid hymenial surface and small ellipsoid to ovoid, finely echinulate basidiospores.

Trechispora praefocata (Bourdot & Galzin) Liberta, Taxon 15(8): 319, 1966 = *Corticium sphaerosporum* subsp. Praefocatum Bourdot & Galzin, Hymenomyc. De. France: 233, 1928. Fig. 6.122f–h

Fructification resupinate, effused, adnate; hymenial surface, greyish white to yellowish white when fresh turns pale yellow to greyish yellow on drying, arachnoid to byssoid; margin thinning, paler concolorous to indeterminate.

Hyphal system monomitic; generative hyphae up to 2.0 μm, clamped, septate, thin-walled; basal hyphae parallel to the substrate, with acicular crystals, less branched; subhymenial hyphae branched with ampullate septa. Basidia 6.2–10.2×4.3–5.8 μm, 4-sterigmate, subcylindrical to subclavate with clamp. Basidiospores 3.8–6.0×2.5–3.0 μm, apiculate, thin-walled, echinulate, inamyloid, acyanophilous.

Distribution: H.P.: Kangra, Chamba- Khajjiar.

Collection examined: IBP 37658, 37659.

Substrate: On decaying bark of angiospermous tree.

Remarks: The species is characterized by acicular crystals on hyphae and echinulate basidiospores.

Agaricomycetes
Incertae sedis
Dendrophlebia Dhingra G.S & Priyanka
Mycotaxon 116, 157–160.

Fructification resupinate, adnate, effused, ceraceous; hymenial surface smooth, continuous, yellowish to brownish; margin not differentiated. Hyphal system monomitic; generative hyphae septate, branched, clamped. Dendrophlebia present. Cystidia thin-walled, hyphoid. Basidia clavate to subclavate, 4-sterigmate with basal clamp. Basidiospores ellipsoid, smooth, thick-walled, inamyloid, acyanophilous.

One species, Himalayas

Type species: *Dendrophlebia crassispora* Dhingra & Priyanka, *Mycotaxon* **116**: 157–160, 2011.

Habitat: Wood

Himalayas: One

Dendrophlebia crassispora Dhingra & Priyanka, *Mycotaxon* **116**: 157–160, 2011. Fig. 6.122i–m

Fructification resupinate, adnate, effused; hymenial surface smooth, continuous yellow to corn-yellow, tissue becomes dark in contact with 3 % KOH sol; margin indifferentiated.

Hyphal system monomitic; generative hyphae up to 2.8 μm wide, branched, septate, clamped, agglutinated hyphae and dendrohyphidia are compactly arranged, covered by yellowish-brown crystalline matter. Dendrohyphidia irregularly branched, thin-walled present in context and hymenium. Cystidia hyphoid, thin-walled with basal clamp. Basidia 21–32 × 5–6.2 μm clavate to subclavate, often constricted or sinuous, 4-sterigmate with basal clamp. Basidiospores 4.8–5.2 × 3.2–4 μm ellipsoid to broadly ellipsoid or ovate, smooth, thick-walled, inamyloid, acyanophilous, with oil drops.

Distribution: A.P.: West kameng, Bombila; West Bengal: Darjeeling, Dhotrey.

Collection examined: GSD 19726 (Type).

Substratum: Decayed angiospermic twigs.

Remarks: The species is characterized by compact texture; clavate, 4-sterigmate basidia; hyphal system monomitic, clamped generative hyphae; presence of dendrohyphidia; ellipsoid to ovate basidiospores with oil drops.

Hallenbergia Dhingra G.S. & Priyanka,
Mycotaxon 118, 289–292, 2011.

Fructification resupinate, adnate, effused, membranous, ceraceous; hymenial surface smooth, continuous, cracks on drying; margin not differentiated. Hyphal system monomitic; generative hyphae thin-walled, septate, clamped; basal hyphae irregularly branched and interwoven into dense texture; subhymenial hyphae short-celled, compactly arranged. Cystidia absent. Basidia subclavate to suburniform, 4-sterigmate. Basidiospores ellipsoid to ovoid or subglobose, smooth, thick-walled, cyanophilous, inamyloid.

One species, Himalayas

Type species: *Hallenbergia singularis* Dhingra & Priyanka.

Habitat: Wood

Himalayas: One

Hallenbergia singularis Dhingra & Priyanka, Mycotaxon **118**: 289–292, 2011. Fig. 6.123a–c

Fructification resupinate, adnate, effused, membranous, ceraceous; hymenial surface smooth to farinosed, yellowish grey to orange grey, continuous when fresh, cracks on drying; margin not differentiated.

Hyphal system monimitic; generative hyphae up to 4.2 μm wide, thin-walled, septate, clamped; basal hyphae irregularly branched, covered by crystalline matter in the upper half; subhymenial hyphae short-celled, compactly arranged. Cystidia absent. Basidia 16–32 × 7.5–10 μm, ellipsoid, subclavate to suburniform, 4-sterigmate, oily contents present. Basidiospores 6–8 × 4.2–7.2 μm, broadly ellipsoid to ovate, smooth, somewhat thick-walled, cyanophilous, inamyloid.

Distribution: Bhutan: Thimphu.

Collection examined: GSD 19548 (Type).

Substratum: Decayed angiospermic twigs.

Remarks: The species is characterized by resupinate, adnate, ceraceous fructification; hyphal system monomitic; basidia ellipsoid subclavate to suburniform, 4-sterigmate; ellipsoid to ovate or subglobose, smooth basidiospores with oil drops.

Radulomycetopsis Dhingra, Priyanka & Kaur J., Mycotaxon 119, 133–136, 2011.

Fructification resupinate, adnate, effused, membranous-ceraceous; hymenial surface smooth to tuberculate, orange to red; margin fibrillose, paler concolorous. Hyphal system monomitic; generative hyphae without clamps, branched at wide angles. Cystidia thin to thick-walled. Basidia clavate to subclavate, 4-spored, without basal clamp. Basidiospores broadly ellipsoid to subglobose, smooth to thick-walled, non-amyloid, acyanophilous.

One species, widespread

Type species: *Radulomycetopsis cystidiata* Dhingra, Priyanka & Kaur J. 2011

Habitat: Wood

Himalayas: One

Radulomycetopsis cystidiata Dhingra, Priyanka & J. Kaur, Mycotaxon 119, 133–136, 2011. Fig. 6.123d–h

Fructification resupinate, adnate, effused, membranous-ceraceous; hymenial surface smooth to tuberculate, brownish orange in young stage but changes into red colour at maturity; margin thinning.

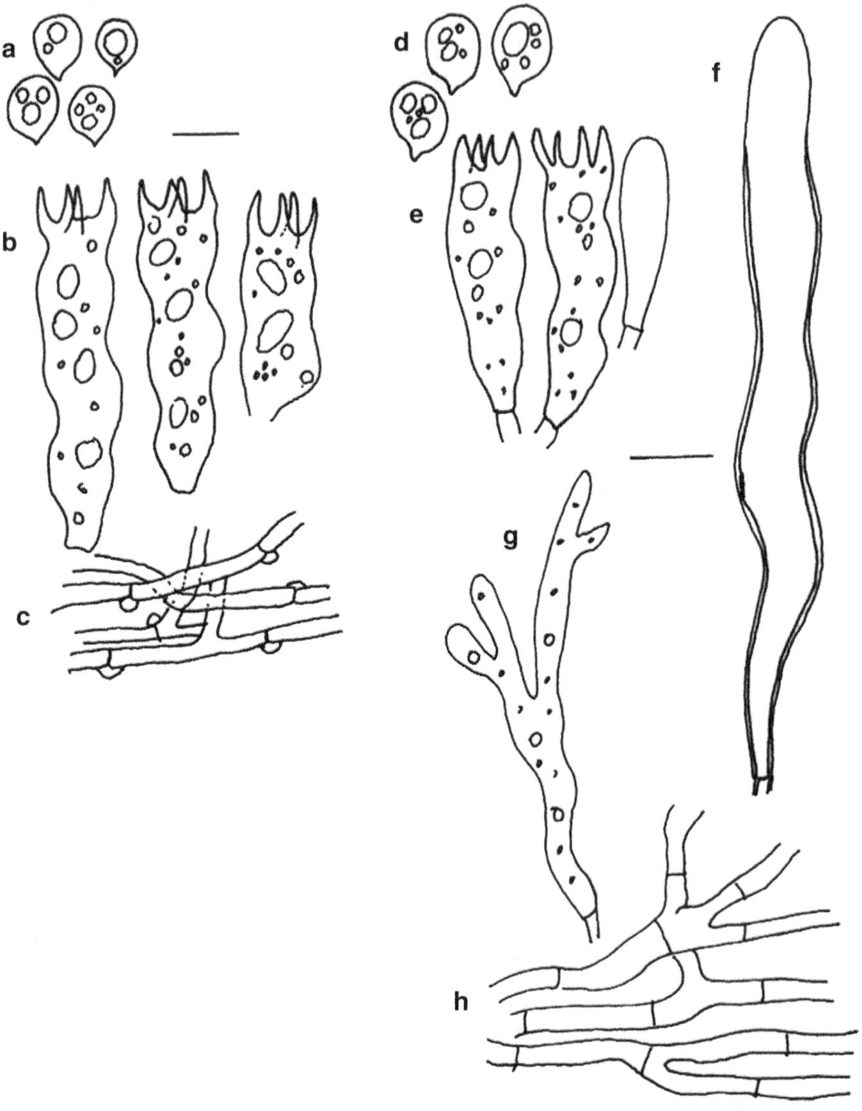

Fig. 6.123 (**a–c**) *Hallenbergia singularis* (**a**) Basidiospores, (**b**) Basidia, (**c**) Generative hyphae; (**d–h**) *Radulomycetopsis cystidiata* (**d**) Basidiospores, (**e**) Basidia, (**f**) Cystidia, (**g**) vertical arranged hyphae, (**h**) Generative hyphae

Hyphal system monomitic; generative hyphae up to 3 μm wide, thin-walled, septate, clamps absent; basal zone consist of branched hyphae; subhymenial zone composed of vertically arranged and branched hyphae. Cystidia 72–82 × 9.5–10.8 μm, cylindrical to subcylindrical, thin to thick-walled. Basidia 26–35 × 7.2–8.3 μm, clavate to subclavate, 4-sterigmate with oily content. Basidiospores 6–8.7 × 5.4–7.2 μm broadly ellipsoid to subglobose, thin to thick-walled, non-amyloid, acyanophilous.

Distribution: A.P.: West Kameng.
Collection examined: GSD 19239 (Type).
Substratum: On decayed angiospermous branches.
Remarks: The species is characterized by having adnate, effused, membranous-ceraceous fructification; hymenial surface smooth to tuberculate; hyphal system monomitic, generative hyphae without clamps; basidia clavate to subclavate; broadly ellipsoid to subglobose basidiospores.

References

Annonymous (1950) List of common names of Indian plant diseases. Indian J Agric Sci 20:107–142

Bagchee KD (1950) Progress of forest pathology in India during the quinquennium 1944–49. Indian Forester 76:216–220

Bagchee KD (1961) The fungal diseases of Sal (*Shorea robusta* Gaertn.) – IV. *Fomes caryophylli* (Rec.) Bres. A destructive heart rot of 'Sal'. Indian For Rec 2(3):25–58

Bagchee KD, Bakshi BK (1950) Some fungi as wound parasites of Indian trees. Indian Forester 76(6):244–253

Bagchee KD, Bakshi BK (1951) *Poria monticola* Murr on Chir (*Pinus longifolia* Roxb) in India. Nature 167:824

Bagchee KD, Bakshi BK (1954) Studies on Indian Thelephoraceae I. Some species of *Stereum*, *Peniophora* and *Corticium*. Indian For Bull 166:1–11

Bagchee KD, Singh U (1960) List of common names of fungi attacking Indian Forest trees, timber and the herbaceous and shrubby undergrowths and list of cultures of forest fungi. Indian For Rec 1(10):199–348

Bakshi BK (1955) Diseases and decays of conifers in the Himalayas. Indian Forester 81:779–797

Bakshi BK (1958) New records of Hymenomycetes in India. Indian Phytopathol 11:88

Bakshi BK (1971) Indian Polyporaceae. ICAR, New Delhi

Bakshi BK, Gibson S, Singh B (1958) Occurrence of *Tremetes sepium* in India. Can J Bot 36:603–606

Banerjee SN (1935a) A historical account of the classification of Thelephoraceae. Sci Cultiv 1:314–317

Banerjee SN (1935b) Thelephoraceae of Bengal-I. J Indian Bot Soc 14:13–48

Banerjee SN (1947) Fungus flora of Calcutta and suburbs-I. Bull Bot Soc Beng 1:37–54

Berkeley MJ (1854a) Decades XLI, XLIII. Indian Fungi. Hooker's J Bot Kew Gard Misc 6:129–143

Berkeley MJ (1854b) Decades XLIV. XLVI. Indian Fungi. Hooker's J Bot Kew Gard Misc 6:161–174

Berkeley MJ (1855) The botany of the Antarctic voyageII. In: Hooker JD (ed) Flora novae-zelandiae Part II, Flowerless plants. Lovell Reeve, London

Berkeley MJ (1856) Decades of fungi. Decas 1–62, nos. 1–620. Hooker's London J Bot 3–8:1844–1856

Berkeley MJ (1873) Notices of North American fungi. Grevillea 2(13):3–7

Bernicchia A, Gorjon SP (2010) Corticiaceae s.l. Fungi Europaei 12:1–1008

Bhosle SR, Lamrood PY, Vaidya JG (2005) Diversity of Aphyllophoraceous fungi from the Western Ghats of Maharashtra. In: Atri NS, Dhingra GS, Dargan JS (eds) The fungi- Diversity and conservation in India. Bishen Singh Mahendra Pal Singh, Dehra Dun, pp 103–113

Boidin J (1951) Disposition hemi-chiastobasidiee chez quelques Thelephoracees. CR Acad Sci 233:1667–1669

Boidin J (1970) Basidiomycètes de la République Centralafricaine II. – Les genres Botryobasidium Donk et Candelabrochaete nov gen. Cahiers de La Maboké 8:17–25
Boidin J, Lanquetin P (1980) Contribution A Letude du Genre *Dichostereum* Pilat (Basidiomycetes: Lachnocladiaceae). Bull Soc Mycol 96:381–406
Bose SR (1919) Description of fungi in Bengal-I. Proc Indian Ass Cultiv Sci 4:109–114
Bose SR (1921) Polyporaceae of Bengal- IV. Bull Carmichael Med Coll 2:1–5
Bose SR (1928) Polyporaceae of Bengal-VIII. J Dep Sci Calcutta Univ 9:27–34
Bose SR (1934) Polyporaceae of Bengal-X. J Dep Sci Calcutta Univ 11:1–18
Bose SR (1946) Polyporaceae of Bengal-XI. J Dep Sci Calcutta Univ 2:53–87
Bourdot H (1922) Add aux Cort Rev Sci Bourd Centr Fr 35(1):17
Bourdot H, Galzin A (1924) Hymenomycetes de France X. Phylacteries. Bull Soc Mycol Fr 40:105–162
Bourdot H, Galzin A (1928) Hymenomycetes de France. Sceaux, Paris
Bresadola G (1903) Fungi Polonici a cl. Viro B. Eichler lecti. Ann Mycol 1:65–121
Burt EA (1902) Annual report on the New York State Museum of Natural History. New York State Museum Albany, Regents of the University of the State of New York 54:954
Burt EA (1917) The Thelephoraceae of North America VIII. *Coniophora*. Ann Mo Bot Gard 4:237–269
Burt EA (1926) The Thelephoraceae of North America XV. *Corticium*. Ann Mo Bot Gard 13(3):173–354
Christiansen MP (1960) Danish Resupinate Fungi. II. Homobasidiomycetes. Dansk Bot Arkiv 19:57–383
Cooke MC (1880) Fungi of India. Grevillea 8:93–96
Cooke MC, Ellis JB (1879) New Jersey fungi. Grevillea 8(45):11–16
Cunningham GH (1963) The Thelephoraceae of Australia and New Zealand, vol 145, DSIR New Zealand Bull. R.E. Owen, Govt. printer, Wellington
Cunningham GH (1965) Polyporaceae of New Zealand, vol 164, New Zealand Department of Scientific and Industrial Research bulletin. R.E. Owen, Govt. printer, Wellington
Currey F (1874) On a collection of fungi made by Mr. Sulpiz Kurz, curator of the Botanical Garden Calcutta. Trans Linn Soc Lond II Ser Bot 1:119–131
Dhanda RS (1977) Studies on Polyporaceae of North-Western Himalayas. PhD thesis Panjab University Chandigarh India
Dhingra GS (1992) East Himalayan Coniophoraceae (Basidiomycetes) I. Some new reports. Geobios new reports, University of Jodhpur 11:118–122
Dhingra GS, Malka R (1994) North West Himalayan Thelephoraceae (Basidiomycetes) genus *Tomentella* from Dalhousie hills. In: Trivedi ML, Sharma M, Sarma TA, Saini SS (eds) Current researches in plant sciences. Bishan Singh Mahendra Pal Singh, Dehra Dun, pp 43–56
Dhingra GS, Priyanka, Singh AP (2009) Three new records of genus *Sistotrema* from India. J Indian Bot Soc 88:76–79
Dhingra GS, Priyanka, Kaur J (2011) A checklist of resupinate, non-poroid agaricomycetous fungi from North-East India and Bhutan. Synopsis Fungorum 29:22–70
Donk MA (1956) Notes on resupinate Hymenomycetes-III. Fungus 26:3–24
Eriksson J (1958) Studies in Corticiaceae (*Botryohypochnus* Donk, *Botryobasidium* Donk and *Gloeocystidiellum* Donk). Svensk Bot Tidskr 52:1–17
Eriksson J, Hjortstam K (1976) In: Eriksson J, Ryvarden L (eds) The Corticiaceae of North Europe, vol 4, p 665
Eriksson J, Ryvarden L (1973) The Corticiaceae of North Europe-II. Fungiflora, Oslo, Norway, pp 59–286
Eriksson J, Ryvarden L (1975) The Corticiaceae of North Europe-III. Fungiflora, Oslo, Norway, pp 287–546
Eriksson J, Ryvarden L (1976) The Corticiaceae of North Europe-IV. Fungiflora, Oslo, Norway, pp 549–886
Eriksson J, Strid Å (1975) Coronicium-Hyphoderma. In: Eriksson J, Ryvarden L (eds) Cortic N Eur, vol 3. Oslo, p 541

Eriksson J, Hjortstam K, Ryvarden L (1978) The Corticiaceae of North Europe-V. Fungiflora, Oslo, Norway, pp 890–1046
Eriksson J, Hjortstam K, Ryvarden L (1981) The Corticiaceae of North Europe-VI. Fungiflora, O slo, Norway, pp 1051–1276
Eriksson J, Hjortstam K, Ryvarden L (1984) The Corticiaceae of North Europe-VII. Fungiflora, Oslo, Norway, pp 1281–1449
Fidalgo O, Fidalgo MEPK (1968) Polyporaceae from Venezuela-I. Mem N Y Bot Gdn 17(2):1–34
Fries EM (1818) Observationes mycologicae, vol 2, pp 1–372
Fries EM (1821) Systema mycologicum I. Lundae, pp 1–520
Gilbertson RL, Ryvarden L (1986) North American polypores, vol 1. Fungiflora, Oslo, pp 1–436
Hennings P (1900) Fungi Indiae orientalis I. Hedwigia 39:150–153
Hennings P (1901) Fungi Indiae orientalis II. Hedwigia 40:323–342
Hjortstam K, Ryvarden L (1979) Notes on Corticiaceae (Basidiomycetes) V. Mycotaxon 10(1):201–209
Höehnel F (1912) Fragmente zur Mykologie XIV Mitteilung (Nr. 719 bis 792). Sitzungsberichte der Kaiserlichen Akademie der Wissenschaften Math.-naturw. Klasse Abt I 121:339–424
Jülich W (1971) Einige neue order unbekannte Corticiaceae (Basidiomycetes). Willdenowia 6:215–224
Jülich W (1972) Monographie der Athelieae (Corticiaceae: Basidiomycetes). Willdenowia 7:1–283
Karsten PA (1889) Kritisk öfversigt af Finlands Basidsvampar (*Basidiomycetes*, *Gastero*–and *Hymenomycetes*). Bidrag till Kännedom of Finlands Natur Folk 48:1–470
Larsen MJ (1974) A contribution to the taxonomy of the genus *Tomentella*. Mycologia Memoir 4:1–145
Lemke PA (1964) The genus *Aleurodiscus* (sensu lato) in North America. Can J Bot 42(6):723–768
Lowe JL (1942) The Polyporaceae of New York State (Except *Poria*), vol 60. Technical Publication of the State University College of Forestry at Syracuse University, Syracuse, New York, pp 1–128
Lowe JL (1957) Polyporaceae of North America: the genus *Fomes*, vol 80. Technical Publication of the State University College of Forestry at Syracuse University, Syracuse, New York, pp 1–96
Lowe JL (1966) Polyporaceae of North America. The genus *Poria*, vol 90. Technical Publication of the State University College of Forestry at Syracuse University, Syracuse, New York, pp 1–183
Lowe JL, Lombard FF (1973) On the identity of *Polyporus lacteus*. Mycologia 65(4):725–732
Massee G (1889) A monograph of the Thelephoraceae-I. J Linn Soc Bot 25:107–155
Mitter JH, Tandon RN (1932) Fungus flora of Nainital-I. J Indian Bot Soc 11:178–180
Montagne JFC (1842) Cryptogammae nilgherenses. Ann Sci Nat II Ser 18:12–23
Oberwinkler F (1965) Primitive basidiomyceten. Sydowia 19:1–72
Overholts LO (1953) Polyporaceae of the United States, Alaska and Canada. University of Michigan Press, Ann Arbor
Parmasto E (1968) Conspectus systematis corticiacearum. Institutum zoologicum et botanicum Academiae scientiarum R.P.S.S. Estonicae, Tartu
Persoon CH (1801) Synopsis methodica fungorum. Johnson Reprint Corp., Gottingen
Pouzar Z (1958) Nova genera macromycetum II. Ceská Mykologie 12(1):31–36
Pouzar Z (1959) New genera of higher fungi III. Ceska Mykol 13:10–19
Ranadive KR, Vaidya JG, Jite PK, Ranande VD, Bhosale SR, Rabba AS, Hakimi M, Deshpande GS, Rathod MM, Forutan A, Kaur M, Naik- Vaidya CD, Bapat GS, Lamrood P (2011) Checklist of Aphyllophorales from the Western Ghats of Maharashtra State, India. Mycosphere 2:91–114
Rattan SS (1977) The resupinate Aphyllophorales of the North Western Himalayas, vol 60, Bibliotheca Mycologica. J Cramer, Germany

Rehill PS, Bakshi BK (1965) Studies on Indian Thelephoraceae II. Indian species of *Peniophora* and *Corticium*. Indian For Bull Dehra Dun 242:1–31
Rehill PS, Bakshi BK (1966) Studies on Indian Thelephoraceae III. The genus *Stereum*. Indian For Bull Dehra Dun 250:1–20
Reid DA, Thind KS, Chatrath MS (1959) The polyporaceae of the Mussoorie Hills (India). IV. Trans Br Mycol Soc 42(1):40–44
Romell L (1911) Hymenomycetes of Lappland. Arkiv før Botanik 11(3):1–35
Roy A, De AB (1980) Studies on Indian Polypores-III. Morphological and cultural characters of *Trametes floccose*. Norw J Bot 27:297–300
Roy A, De AB (1996) Polyporaceae of India. IBD, Dehra Dun
Ryvarden L (1968) Flora over kjuker. Universitetsforlaget, Oslo, pp 1–96
Ryvarden L (1977) Type studies in the Polyporaceae 9. Species described by EM Wakefield. Mycotaxon 5(1):331–336
Ryvarden L (1978) The Polyporaceae of North Europe Vol II. *Innotus-Tyromyces*. Fungiflora, Oslo
Ryvarden L (1987) New and noteworthy polypores from tropical America. Mycotaxon 28(2):525–541
Ryvarden L, Johansen I (1980) A preliminary Polypore flora of East Africa. Fungiflora, Oslo
Saccardo PA (1888) Sylloge Fungorum omnium Hususque. Comniterum VI. Patavii
Schröter J (1888) Kryptogamen- Flora von Schlescien. Cramer, Lehre
Sharma JR (1985) Studies on Polyporaceae of Himachal Pradesh. J Econ Tax Bot 7:95–101
Sharma JR (1995) Hymenochaetaceae of India. Botanical Survey of India, Calcutta
Sharma JR (2000) Genera of Indian Polypores. Botanical Survey of India, Ministry of Environment and Forests, Calcutta
Sharma JR (2012) Aphyllophorales of Himalayas. Botanical Survey of India, Dehra Dun
Talbot PHB (1973) Aphyllophorales I: General characteristics; Thelephroid and Cupuloid Families. In: Ainsworth GC, Sparrow FK, Susman AS (eds) The fungi IVB. Academic Press, New York/ London, pp 327–349
Theissen F (1911) Fungi aliquot Bombayenses a Rev. Ed. Blatter Collecti. Ann Mycol 9:153–159
Thind KS, Adlakha KL (1956) Thelephoraceae of Mussoorie Hills- I. Indian J Mycol Res 2:57–64
Thind KS, Chatrath MS (1957) Polyporaceae of Mussoorie Hills-II. Res Bull Panjab Univ 125:431–442
Thind KS, Dhanda RS (1978) The Polyporaceae of India XI. Indian Phytopathol 31(4):462–472
Thind KS, Dhanda RS (1980) The Polyporaceae of India XIII. Indian Phytopathol 33(3):380–387
Thind KS, Rattan SS (1968) The Thelephoraceae of North Western Himalayas. Indian Phytopathol Soc Bull 4:15–24
Thind KS, Rattan SS (1970) The Thelephoraceae of India- III. The genus *Tubulicrinis* and *Hyphoderma*. Proc Indian Acad Sci 71:118–131
Thind KS, Rattan SS (1973a) The Thelephoraceae of India VI, VII. Indian Phytopathol 26(485–494):528–536
Thind KS, Rattan SS (1973b) The Thelephoraceae of India X. Mycologia 65:1250–1258
Thind KS, Bindra PS, Chatarth MS (1957) The polyporaceae of Mussoorie hills- III. Res Bull Panjab Univ 129:471–483
Thind KS, Rattan SS, Dhanda RS (1970) The polyporaceae of India-VI. Res Bull Panjab Univ 21(1–2):109–117

Chapter 7
Ecology

Abstract A brief introduction concerning the distribution of different genera and species in different regions of Himalayas is discussed. Neumerical data concerning the endemic species, common species in Eastern Himalayas and North-Western Himalayas is also provided.

7.1 Mycogeographical Distribution and Ecological Relationships

A total of 488 wood rotting fungi have been recorded from the Himalayas from different phytogeographic zones. Out of these 488 taxa, 425 have been reported from North Western Himalayas whereas 186 occur in Eastern Himalayas including the royal kingdom of Bhutan. An analysis of the inventory of these fungi (Tables 7.1 and 7.2) reveals that 63 taxa are endemic to Eastern Himalayas whereas 302 taxa have been recorded only from N.W. Himalayas where as 123 taxa are distributed throughout the Himalayas.

In the preceeding six decades much of the floristic studies on Indian/Himalayan fungi particularly with respect to higher fungi have been primarily focussed with respect to N.W. Himalayas. The main reason for this has been close proximity of N.W. Himalayas to major research centres of the country involved with the systematics of these fungi. The other major factor for this has been the lack of infrastructural support in the Eastern part of the country for such studies and the non approachable areas in the states of N.E. India. Inspite of massive virgin forests in the N.E. states in the Himalayan region and Assam Hills, there have been insuregencis in the recent part which also hindered and restricted to a larger extent the floristic studies in these areas. The floristic studies in the eastern Himalayas has been restricted to the forests in and around the main centres of urban areas where adequate protection has been provided by the Government to the mycologists. Due to calm and peaceful atmosphere now prevailing in these areas for quite some time and the infrastructural development carried out in these areas, the floristic studies on these fungi in this part of the Himalayas have been undertaken by some workers and it is hoped that with in a decade considerable compilations will start appearing from these areas.

I.B. Prasher, *Wood-rotting non-gilled Agaricomycetes of Himalayas*,
Fungal Diversity Research Series, DOI 10.1007/978-94-017-9858-7_7

Table 7.1 Mycogeographic interrelationship of species of wood rotting Agaricomycetes (with their respective families) of N.W. Himalayas

Name of the species	H.P	UK	J&K	N
Amylocorticiaceae				
Amylocorticium indicum	√		√	
Amyloathelia crassiucula		√		
Amyloxenasma grisellum	√	√		
Ceraceomyces bizonatus	√			
Ceraceomyces borealis	√			
Ceraceomyces cystidiatus		√		
Ceraceomyces fibuliger	√			
Ceraceomyces reidii	√			
Ceraceomyces sublaevis	√			
Ceraceomyces tessulatus	√			
Cyphellaceae				
Chondrostereum purpureum	√			
Physalacriaceae				
Cylindrobasidium evolvens	√			
Pterulaceae				
Aphanobasidium subnitens	√			
Radulomyces confluens var. macrobasidiata		√		
Radulomyces molaris	√	√		
Stephanosporaceae				
Cristinia helvetica	√	√		
Cristinia mucida		√	√	
Atheliaceae				
Amphinema byssoides	√		√	
Athelia decipiens	√	√		
Athelia teutoburgensis	√	√		
Athelopsis parvispora	√			
Atheleopsis subinconspicua	√	√		
Fibulomyces mutabilis	√			
Hypochniciellum ovoideum	√			
Leptosporomyces adnatus	√	√		
Leptosporomyces galzinii	√			
Leptosporomyces globosum	√			
Leptosporomyces thindii sp. nov.		√		
Leptosporomyces roseus var. *macrosporus*		√		
Coniophoraceae				
Coniophora arida	√	√	√	
Coniophora cordensis	√			
Coniophora fusispora	√	√		
Coniophora olivacea	√			
Coniophora puteana	√			

(continued)

Table 7.1 (continued)

Name of the species	H.P	UK	J&K	N
Hygrophoropsidaceae				
Leucogyrophana mollusca	√	√		
Leucogyrophana olivascens	√			
Serpulaceae				
Serpula himantoides	√			
Serpula lachrymans	√	√	√	
Botrybasidiaceae				
Botryobasidium candicans	√		√	
Botryobasidium subbotryosum	√			
Botryobasidium subcoronatum	√		√	√
Botryohypochnus isabellinus	√		√	
Ceratobasidiaceae				
Scotomyces subviolaceus	√	√		
Hydnaceae				
Paullicorticium delicatissimum	√			
Sistotrema binucleosporum	√			
Sistotrema brinkmannii	√			
Sistotrema lachrymisporum	√			
Sistotrema porulosum	√			
Sistotrema sernanderi	√			
Sistotrema subtrigonospermum	√			
Corticiaceae				
Corticium confine	√			
Dendrothele alliacea	√			
Dendrothele incrustans	√			
Dendrothele seriata	√			
Erythricium laetum	√			
Galzinia ellipsospora	√			
Galzinia incrustans	√			
Laeticorticium expallens	√			
Gloeophyllaceae				
Gloeophyllum carbonarium	√			
Gloeophyllum sepiarium		√		
Gloeophyllum subferrugineum		√		
Hericiaceae				
Dentipellis leptodon	√			
Laxitextum bicolour	√	√		
Lentariaceae				
Kavinia alboviridis	√		√	
Hymenochaetaceae				
Aurificaria indica		√		
Fomitiporia robusta		√		

(continued)

Table 7.1 (continued)

Name of the species	H.P	UK	J&K	N
Fuscoporia contigua	√			
Fuscoporia ferruginosa	√			
Fuscoporia senex	√			
Fuscoporia torulosa	√	√	√	
Hymenochaete fuscobadia		√		√
Hymenochaete leonina	√			
Hymenochaete luteobadia	√			
Hymenochaete mougeotii	√	√	√	√
Hymenochaete rubiginosa		√		
Hymenochaete				
semistuposa	√			
Inonotus cuticularis		√		
Inonotus dryadeus	√			
Inonotus tabacinus		√		
Onnia circinata	√		√	
Phellinus allardii		√		
Phellinus caryophylli	√			
Phellinus fastuosus		√		
Phellinus gilvus	√			
Phellinus grenadensis	√			
Phellinus johnsonianus		√		
Phellinus linteus		√		
Phellinus merrillii		√		
Phellinus nilgheriensis	√			
Phellinus sanfordii	√	√		
Phellinus xeranticus	√			
Phylloporia pectinata	√	√		
Phylloporia ribis	√	√		
Phylloporia spathulata	√			
Porodaedalea pini	√			
Tubulicrinis chaetophorus	√		√	
Tubulicrinis glebulosus	√	√		
Tubulicrinis hamatus	√	√		
Tubulicrinis strangulates	√			
Tubulicrinis subulatus	√		√	
Ricknellaceae				
Peniophorella pallida	√			
Peniophorella praetermissa	√	√	√	
Peniophorella pubera	√	√	√	
Peniophorella rude	√	√		
Peniophorella microtsugae sp. nov.		√		
Resinicium bicolor	√	√	√	√

(continued)

Table 7.1 (continued)

Name of the species	H.P	UK	J&K	N
Resnicium friabile	√			
Sidera lenis var. *minutisporus*		√		
Sidera lunata	√			
Schizoporaceae				
Basidioradulum tuberculatum	√	√		
Hyphodontia abieticola	√			
Hyphodontia altaica	√			
Hyphodontia alienata	√			
Hyphodontia alutaria	√		√	
Hyphodontia arguta	√	√	√	
Hyphodontia aspera		√		
Hyphodontia barbajovis	√			
Hyphodontia caulicystidiata	√			
Hyphodontia crustosa		√		
Hyphodontia efibulata	√			
Hyphodontia hastata	√			
Hyphodontia juniper	√			
Hyphodontia nespori		√		
Hyphodontia pallidula	√	√	√	
Hyphodontia papilosa			√	√
Hyphodontia propinqua	√			
Hyphodontia spathulata	√	√	√	√
Hyphodontia stipata		√		
Oxyporus cervinogilvus		√		
Oxyporus corticola		√		
Oxyporus populinus		√		
Oxyporus ravidus	√	√		
Schizopora paradoxa	√	√		
Xylodon pruni			√	
Xylodon rimosissimus	√			
Cystostereaceae				
Cystostereum murrayi	√			
Fomitopsidaceae				
Antrodia albida		√		
Antrodia gossypium	√			
Antrodia serialis	√		√	
Antrodia xantha	√	√		
Daedalea dickinsii		√		
Daedalea dochmia		√		
Daedalea flavida	√			
Daedalea gollanii		√		
Daedalea quercina		√		

(continued)

Table 7.1 (continued)

Name of the species	H.P	UK	J&K	N
Daedalea sulcata		√		
Dacryobolus costratus		√		
Dacryobolus karstenii	√	√	√	
Dacryobolus sudans	√			
Fomitopsis palustris		√		
Fomitopsis pinicola		√		
Fomitopsis rosea	√	√	√	
*Fomitopsis rubidu*a		√		
Fomitopsis rufolaccata	√			
Laetiporus sulphureus	√	√	√	
Parmastomyces corticola			√	
Phaeolus schweinitzii	√			
Postia caesia		√		
Postia ceriflua		√		
Postia guttulata	√	√		
Postia leucomallella	√	√		
Postia mappa		√		
Postia sericeomollis	√			
Ganodermataceae				
Ganoderma applanatum		√		
Ganoderma lucidum		√		
Ganoderma multiplicatum	√			
Ganoderma sessiliforme		√		
Meripileaceae				
Physisporinus rivulosus	√			
Rigidoporus crocatus		√		
Rigidoporus lineatus		√		
Rigidoporus microporus		√		
Rigidoporus ulmarius		√		
Rigidiporus vinctus		√		
Meruliaceae				
Abortiporus biennis	√	√		
Bjerkendra adusta	√	√		
Bjerkendra fumosa		√		
Cabalodontia queletii	√	√		
Cabalodontia subcretacea	√			
Crustoderma dryinum		√		
Flavodon flavus		√		
Gloeoporus dichorus	√			
Gyrophanopsis polonensis	√			
Hyphoderma argillaceum	√		√	
Hyphoderma macedonicum	√			

(continued)

Table 7.1 (continued)

Name of the species	H.P	UK	J&K	N
Hyphoderma occidentale			√	
Hyphoderma parvisporum	√			
Hyphoderma setigerum	√	√	√	
Hyphoderma sikkimium	√			
Hyphoderma singularibasidium	√			
Hypochnicium caucasicum		√		
Hypochnicium cystidiatum	√			
Hypochnicium longicystidiosum	√			
Hypochnicium lundellii	√			
Hypochnicium punctulatum	√	√	√	
Hypochnicium sphaerosporum	√			
Irpex consors	√	√		
Irpex lacteus	√	√		
Irpex vellereus	√	√		
Junghuhnia collabens	√	√		
Junghuhnia luteoalba		√		
Jughuhnia nitida	√	√		
Mycoacia fuscoatra		√		
Mycoacia stenodon		√		
Phlebia crassisubiculata	√			
Phlebia interjacenoides		√		
Phlebia livida	√			
Phlebia radiate	√			
Phlebia rufa	√			
Phlebia segregata	√	√		
Phlebia subserialis	√			
Phlebia subochracea	√			
Radulodon americanus	√		√	
Radulodon erikssonii	√			
Scopuloides hydnoides	√			
Sarcodontia delectans	√	√		
Sarcodontia spumea	√			
Sarcodontia pachyodon	√		√	
Steccherinum ciliolatum	√			
Steccharinum fimbriatum	√		√	
Steccherinum laeticolor	√			
Steccherinum ochraceum	√	√		
Phanerochaetaceae				
Antrodiella semisupina		√		
Antrodiella zonata	√			
Byssomerulius corium	√			
Ceriporia viridians	√			

(continued)

Table 7.1 (continued)

Name of the species	H.P	UK	J&K	N
Ceriporiopsis gilvescens	√			
Phanerochaete affinis	√			
Phanerochaete deflectans	√			
Phanerochaete galactites	√			
Phanerochaete sordida	√			
Phanerochaete tuberculata	√			
Phanerochaete velutina	√			
Phlebiopsis flavidoalba	√	√		
Phlebiopsis gigantea	√	√		
Phlebiopsis ravenelii	√	√	√	
Porostereum crassum	√	√		
Porostereum spadiceum	√	√		
Rhizochaete filamentosa	√		√	
Polyporaceae				
Cinereomyces lindbladii	√			
Coriolopsis caperata		√		
Coriolopsis occidentalis		√		
Coriolopsis telfairii		√		
Daedaleopsis confragosa		√		
Datronia mollis	√	√		
Datronia scutellata	√			
Datronia stereoides var. *microspora*		√		
Dichomitus leucoplacus	√	√		
Diplomitoporus crustulinus	√	√		
Diplomitoporus rimosus	√			
Earliella scabrosa		√		
Epithelopsis fulva	√			
Favolus tenuiculus		√		
Fomes extensus		√		
Fomes fomentarius	√	√	√	√
Fomes johnsonianus	√			
Grammothele fuligo		√		
Hapalopilus croceus	√			
Hexagonia tenuis	√	√		
Lenzites betulina	√	√		√
Lenzites eximia	√	√		
Lenzites stereoides		√		
Lopharia cinerascens	√	√	√	√
Lopharia papyrina		√		
Macrohyporia inflata	√	√		
Microporellus obovatus	√			
Microporellus violaceo-cinerascens		√		

(continued)

Table 7.1 (continued)

Name of the species	H.P	UK	J&K	N
Microporus affinis	√	√		√
Microporus xanthopus	√	√		√
Nigroporus durus	√			
Nigroporus vinosus	√	√		
Perenniporia fulviseda	√			
Perenniporia martia	√			
Perenniporia medulla-panis	√	√		
Perenniporia ochroleuca	√			
Perenniporia mesoleuca	√			
Polyporus badius	√			
Polyporus blanchetianus	√			
Polyporus brumalis		√		
Polyporus grammocephalus	√	√		
Polyporus ostreiformis		√		
Polyporus squamosus	√	√		
Polyporus varius	√	√		
Poria auricoma	√	√		
Poria conferata	√	√		
Rhodonia placenta	√			
Pycnoporus cinnabarinus	√		√	
Pycnoporus coccineus	√	√		
Pycnoporus sanguineus	√	√		
Skeletocutis amorpha	√	√		
Skeletocutis nivea		√		
Trametes cingulate		√		
Trametes cotonea		√		
Trametes cubensis		√		
Trametes gibbosa	√	√		
Trametes hirsute	√			
Trametes incerta		√		
Trametes lactinea	√	√		
Trametes menziesii	√			
Trametes ochracea	√			
Trametes palisotii		√		
Trametes versicolor	√	√		
Trametopsis cervina	√			
Trichaptum abietinum		√		
Trichaptum biforme	√	√		
Trichaptum byssogenum	√			
Trichaptum fusco-violaceum	√	√		
Tyromyces lacteus	√	√		
Tyromyces tephrus	√			

(continued)

Table 7.1 (continued)

Name of the species	H.P	UK	J&K	N
Xenasmataceae				
Xenasma subclematidis	√		√	
Xenasma tulasnelloideum	√	√		
Xenasmatella subflavido-grisea	√			
Xenasmatella vaga	√	√		
Amylostereaceae				
Amylostereum chailletii	√	√	√	
Bondzarwiaceae				
Heterobasidion annosum var. *indicum*	√	√	√	√
Heterobasidion insulare	√	√		√
Echinodontiaceae				
Echinodontium japonicum	√		√	
Lachnocladiaceae				
Asterostroma cervicolor	√	√		
Asterostroma muscicola	√	√	√	
Dichostereum effuscatum	√	√		
Dichostereum kenyense	√			
Dichostereum pallescens	√	√		
Dichostereum rhodosporum	√	√	√	
Scytinostroma albocinctum	√			
Scytinostroma alutum	√			
Scytinostroma crassum	√	√	√	
Scytinostroma cystidiatum	√		√	
Scytinostroma duriusculum	√	√	√	
Scytinostroma ochroleucum	√			
Scytinostroma portentosum	√	√	√	
Scytinostroma pulverulentum	√			
Scytinostroma rhizomorpharum		√		
Scytinostromella heterogenea	√			
Scytinostromella olivaceoalba	√			
Vararia brevispora	√			
Vararia minidichophysa	√			
Vararia rugosispora	√			
Vararia sphaericospora	√	√		
Vararia vassilievae	√		√	
Peniophoraceae				
Gloiothele citrina	√			
Gloiothele lactescens		√		
Metulodontia indica	√			
Metulodontia nivea	√	√		
Peniophora cinerea	√		√	
Peniophora incarnata	√			

(continued)

Table 7.1 (continued)

Name of the species	H.P	UK	J&K	N
Peniophora limitata	√			
Peniophora pithya	√			
Peniophora quercina	√			
Peniophora violaceolivida		√		
Peniophora pini	√			
Peniophora rhodocarpa		√		
Russulaceae				
Boidinia furfuraceum	√			
Boidinia lacticolor	√			
Stereaceae				
Acanthofungus ahmadii	√	√		
Aleurodiscus amorphous	√			
Aleurodiscus lapponicus	√			
Aleurodiscus oakesii	√	√		
Aleurodiscus taxicola	√			
Amylosporomyces echinosporus	√			
Chaetoderma luna	√		√	
Conferticium ochraceum	√			
Gloeocystidellum clavuligerum	√			
Gloeocystidiellum donkii		√		
Gloeocystidiellum luteocystidium var. *brevisporum*	√	√		
Gloeocystidiellum sulcatum	√			
Scotoderma viride	√			
Stereum acanthophysatum		√		
Stereum gauspatum	√	√	√	√
Stereum hirsutum	√	√		
Stereum ostrea	√	√	√	√
Stereum rugosum	√	√		
Stereum sanguinolentum	√	√	√	√
Stereum subtomentosum	√			
Xylobolus frustulatus	√		√	√
Xylobolus subpileatus	√	√		√
Wrightoporiaceae				
Wrightoporia lenta	√		√	
Thelephoraceae				
Amaurodon viridis	√			
Pseudotomentella mucidula	√			
Pseudotomentella tristis	√			
Thelephora atra	√			
Tomentella albomarginate	√			
Tomentella badia	√			
Tomentella bicolor	√			

(continued)

Table 7.1 (continued)

Name of the species	H.P	UK	J&K	N
Tomentella botryoides	√	√		
Tomentella brevispina	√			
Tomentella bryophila	√			
Tomentella calcicola	√			
Tomentella clavigera	√			
Tomentella coerulea	√			
Tomentella crinalis	√			
Tomentella cinerascens	√			
Tomentella ellisii	√			
Tomentella ferruginea	√	√		
Tomentella galzinii	√			
Tomentella griseoumbrina	√			
Tomentella griseoviolacea	√			
Tomentella himalayana	√			
Tomentella indica	√	√		
Tomentella kalatopii	√			
Tomentella lapida	√			
Tomentella lateritia	√			
Tomentella muricata	√			
Tomentella olivascens	√			
Tomentella pilosa	√			
Tomentella puberula	√			
Tomentella punicea	√			
Tomentella pyrolae	√			
Tomentella radiosa	√			
Tomentella scobinella		√		
Tomentella stuposa	√			
Tomentella subclavigera	√			
Tomentella subcorticioides	√	√		
Tomentella testaceogilva	√			
Tomentella umbrinospora	√			
Tomentella unicusa	√			
Hydnodontaceae				
Fibriciellum silvae-ryae	√			
Sistotremastrum niveocremeum	√	√		
Subulicystidium longisporum	√	√	√	
Subulicystidium meridense		√		
Trechispora alnicola	√			
Trechispora candidissima	√			
Trechispora farinacea	√	√		
Trechispora microspora	√			
Trechispora mutabilis	√	√		

(continued)

Table 7.1 (continued)

Name of the species	H.P	UK	J&K	N
Trechispora mollusca	√			
Trechispora praefocata	√			

HP Himachal Pradesh, *UK* Uttarakhand, *J&K* Jammu & Kashmir, *N* Nepal

The N.W. Himalayas is spread over three major states of the country which are Himachal Pradesh, Uttarakhand and Jammu and Kashmir. Some of the extralimital collections of the species which had been deposited in the PAN by the previous workers have also been included in the table 7.1. The forest types of these states supports both tropical and temperate vegetation; resembling Asian elements on one hand and European elements on the other hand; it is predictable that extensive floristic exercise will reveal much more fungal species than what is known today. This will add to knowledge regarding the non-wood forest products.

The distributional pattern of these fungi in the N.W. Himalayas reveal that out of the total diversity of these fungi, 78.58 % taxa have been reported from Himachal Pradesh whereas 46.28 and 13.88 % taxa have been recorded from Uttarakhand and Jammu & Kashmir. One-hundred and ninety-seven taxa of this particular group of fungi are endemic to H.P. whereas 83 & 3 taxa are restricted to the states of Uttarakhand and J & K respectively.

In comparison to the N.W. Himalayas which touches three states of the Indian Union, the Eastern Himalayas and Assam Hills spread through seven Indian states of Meghalaya, Mizoram, West Bengal, Arunachal Pradesh, Tripura, Nagaland and Assam, beside Bhutan which is outside the jurisdiction of Indian states. The diversity of the fungi is largest in Bhutan (45.6 % taxa) followed by Arunachal Pradesh (39.9 % taxa), W.B. (29.03 % taxa), Meghalaya (18.27 % taxa) and Assam (6.45 % taxa). A total of 2.6 % of the total diversity of the Eastern Himalayas has been reported from the Himalayan regions of Tripura and Mizoram.

The elements in this diversity of fungi clearly resemble the Chinese diversity due close geographical proximity in addition to the similarity in climatic conditions and Forest types.

Table 7.2 Mycogeographic interrelationship of species of wood rotting Agaricomycetes (with their respective families) of Eastern Himalayas

Name of the species	B	M	Mi	WB	AP	As	T	Me
Amylocorticiaceae								
Amyloxenasma allantosporum	√							
Amyloxenasma grisellum	√			√				
Ceraceomyces borealis	√			√	√			
Physalacriaceae								
Cylindrobasidium evolvens	√				√			
Atheliaceae								
Amphinema byssoides	√							
Fibulomyces cystoideus	√							
Coniophoraceae								
Coniophora arida	√							
Coniophora cordensis	√							
Coniophora fusispora	√							
Coniophora olivacea	√							
Hygrophoropsidaceae								
Leucogyrophana olivascens	√							
Leucogyrophana thimphina	√							
Serpulaceae								
Serpula himantoides	√			√	√			
Botrybasidiaceae								
Botryobasidium asperulum				√				
Botryobasidium danicum				√	√			
Botryobasidium subbotryosum	√							
Botryobasidium vagum	√			√				
Hydnaceae								
Paullicorticium indicum				√				
Sistotrema angustispora				√				
Corticiaceae								
Licrostroma subgiganteum	√				√			
Gloeophyllaceae								
Gloeophyllum abietinum	√				√			
Gloeophyllum sepiarium					√			
Gloeophyllum subferrugineum	√				√			
Gloeophyllum striatum						√		
Hymenochaetaceae								
Fomitiporia robusta					√			√
Fuscoporia contigua			√					
Fuscoporia ferrea					√			
Fuscoporia senex				√	√		√	√
Fuscoporia torulosa					√			
Inonotus albertinii					√			

(continued)

Table 7.2 (continued)

Name of the species	B	M	Mi	WB	AP	As	T	Me
Inonotus tenuicarinus								√
Onnia tomentosa					√			
Phellinus adamantinus	√							
Phellinus allardii	√				√			
Phellinus caryophylli	√				√			
Phellinus gilvus								√
Phellinus setulosus					√			
Phellinus xeranticus	√			√				√
Porodaedalea pini	√				√			√
Tubulicrinis glebulosus				√	√			
Tubulicrinis subulatus	√							
Rickenellaceae								
Peniophorella praetermissa	√				√			
Peniophorella pubera	√			√	√			
Peniophorella rude					√			
Schizoporaceae								
Alutaceodontia alutacea	√							
Hyphodontia nespori					√			
Hyphodontia propinqua				√				
Hyphodontia sambuci	√				√			
Oxyporus cervinogilvus						√		
Oxyporus corticola					√			
Schizopora flavipora					√			
Fomitopsidaceae								
Antrodia albida	√							
Antrodia gossypium					√			
Antrodia sinuosa					√			
Antrodia xantha	√				√			
Anomoporia bombycina	√							
Daedalea dickinsii						√		
Daedalea flavida				√	√	√		
Daedalea imponens								√
Daedalea quercina	√				√			
Daedalea sulcata				√				
Fomitopsis pinicola				√				
Fomitopsis rubida				√				
Laetiporus sulphureus						√		√
Postia caesia							√	
Postia ceriflua	√							
Postia guttulata					√			
Postia mappa					√			
Postia undosa					√			

(continued)

Table 7.2 (continued)

Name of the species	B	M	Mi	WB	AP	As	T	Me
Ganodermataceae								
Ganoderma applanatum	√				√			
Ganoderma lucidum	√	√	√	√	√		√	√
Ganoderma resinaceum	√							
Hyphodermataceae								
Intextomyces contiguus	√							
Meripileaceae								
Rigidoporus lineatus				√				
Rigidoporus microporus				√				
Rigidiporus vinctus								√
Meruliaceae								
Bjerkendra adusta	√	√	√		√		√	√
Bjerkendra fumosa	√							
Conohypha grandispora				√				
Crustoderma dryinum					√			
Flaviporus hydrophilus	√							
Flavodon flavus				√				
Gloeoporus thelephoroides				√				
Hyphoderma argillaceum	√			√	√			
Hyphoderma clarusproprietas				√				
Hyphoderma setigerum	√				√			
Hyphoderma sibiricum	√							
Hyphoderma sporulosum	√			√				
Hypochnicium caucasicum								√
Hypochnicium geogenium	√			√				
Irpex lacteus	√							
Junghuhnia luteoalba	√	√						
Junghuhnia nitida	√							
Junghuhnia zonata	√							
Mycoacia stenodon	√	√						
Phlebia interjacenoides	√							
Phebia microspora	√							
Phlebia rufa								√
Phlebia singularisa	√							
Phlebia thindii				√				
Sarcodontia delectans	√							
Phanerochaetaceae								
Byssomerulius corium	√	√			√			√
Candelabrochaete himalayana	√							
Ceriporia viridans	√							
Ceriporiopsis gilvescens	√							
Phanerochaete galactites				√				

(continued)

Table 7.2 (continued)

Name of the species	B	M	Mi	WB	AP	As	T	Me
Phanerochaete laevis				√				
Phanerochaete sordida	√				√			
Phanerochaete tuberculata				√	√			√
Phanerochaete velutina	√							
Phlebiopsis darjeelingensis				√				
Phlebiopsis gigantea	√				√			
Phlebiopsis himalayensis				√	√			
Phlebiopsis ravenelii								√
Polyporaceae								
Coriolopsis caperata					√			
Coriolopsis occidentalis					√			
Datronia mollis				√				√
Dichomitus leucoplacus					√			
Earliella scabrosa					√			
Fomes fomentarius	√			√	√			√
Fomes extensus								√
Grammothele fuligo					√			
Hexagonia badia						√		
Hexagonia tenuis				√				
Lenzites betulina	√			√	√			
Lenzites eximia					√			
Lenzites stereoides				√				
Lignosus sacer						√		
Loweporus lividus				√				
Loweporus tephroporus								√
Microporellus obovatus								√
Microporus affinis	√	√	√		√	√		√
Microporus xanthopus	√			√	√			√
Navisporus floccosus				√				
Pachytospora papyracea					√			
Nigroporus durus						√		
Nigroporus vinosus	√	√				√		
Nigrofomes melanoporus				√		√		
Perenniporia martia								√
Perenniporia ochroleuca					√			
Polyporus arcularis					√			√
Polyporus badius	√				√			
Polyporus blanchetianus					√			
Polyporus brumalis								√
Polyporus grammocephalus					√			
Polyporus melanopus				√				√
Polyporus squamosus				√				

(continued)

Table 7.2 (continued)

Name of the species	B	M	Mi	WB	AP	As	T	Me
Polyporus varius								√
Pycnoporus sanguineus	√	√						√
Skeletocutis amorpha	√							
Trametes gibbosa					√			
Trametes hirsuta	√						√	√
Trametes incerta								√
Trametes palisoti				√		√		
Trametes pubescens				√				
Trametes versicolor	√	√	√	√	√			
Trichaptum abietinum	√							
Trichaptum biforme	√							
Trichaptum fusco-violaceum	√							
Tyromyces lacteus	√							
Tyromyces tephrus								√
Xenasmataceae								
Xenasma tulasnelloideum					√			
Xenasmatella subflavido-grisea				√				
Bondarzewiaceae								
Bondarzewia mesenterica					√			
Heterobasidion annosum var. *microspora*								√
Heterobasidion insulare	√	√		√				
Echinodontiaceae								
Laurilia sulcata	√							
Lachnocladiaceae								
Asterostroma cervicolor	√				√			
Scytinostroma pulverulentum	√							
Scytinostromella heterogenea	√							
Vararia sphaericospora				√				
Peniophoraceae								
Peniophora limitata	√				√			
Peniophora pithya				√				
Peniophora rufomarginata	√							
Stereaceae								
Aleurodiscus oakesii	√							
Stereum peculiare					√			
Thelephoraceae								
Tomentella scobinella					√			
Tomentella terrestris				√				
Hydnodontaceae								
Brevicellicium olivascens					√			
Fibrodontia gossypina					√			
Sistotremastrum niveocremeum	√							

(continued)

Table 7.2 (continued)

Name of the species	B	M	Mi	WB	AP	As	T	Me
Subulicystidium meridense				√	√			
Trechispora fastidiosa								√
Agaricomycetes Incertae sedis								
Dendrophlebia crassispora				√	√			
Hallenbergia singularis	√							
Radulomycetopsis cystidiata					√			

B Bhutan, *M* Manipur, *Mi* Mizoram, *WB* West Bengal, *AP* Arunachal Pradesh, *As* Assam, *T* Tripura, *Me* Meghalaya

The manufacturer's authorised representative in the EU is Springer Nature Customer Service Centre GmbH, Europaplatz 3, 69115 Heidelberg, Germany. If you have any concerns regarding our products, please contact ProductSafety@springernature.com

Printed and bound by CPI Group (UK) Ltd, Croydon, CR0 4YY
30/09/2026
02225447-0018